T0203761

HANDBOOK OF
LAYERED MATERIALS

HANDBOOK OF
LAYERED MATERIALS

EDITED BY

SCOTT M. AUERBACH
University of Massachusetts Amherst
Amherst, Massachusetts, U.S.A.

KATHLEEN A. CARRADO
Argonne National Laboratory
Argonne, Illinois, U.S.A.

PRABIR K. DUTTA
The Ohio State University
Columbus, Ohio, U.S.A.

CRC Press
Taylor & Francis Group
Boca Raton London New York

CRC Press is an imprint of the
Taylor & Francis Group, an **informa** business

CRC Press
Taylor & Francis Group
6000 Broken Sound Parkway NW, Suite 300
Boca Raton, FL 33487-2742

First issued in paperback 2019

© 2004 by Taylor & Francis Group, LLC
CRC Press is an imprint of Taylor & Francis Group, an Informa business

No claim to original U.S. Government works

ISBN-13: 978-0-8247-5349-8 (hbk)
ISBN-13: 978-0-367-39444-8 (pbk)

This book contains information obtained from authentic and highly regarded sources. Reasonable efforts have been made to publish reliable data and information, but the author and publisher cannot assume responsibility for the validity of all materials or the consequences of their use. The authors and publishers have attempted to trace the copyright holders of all material reproduced in this publication and apologize to copyright holders if permission to publish in this form has not been obtained. If any copyright material has not been acknowledged please write and let us know so we may rectify in any future reprint.

Except as permitted under U.S. Copyright Law, no part of this book may be reprinted, reproduced, transmitted, or utilized in any form by any electronic, mechanical, or other means, now known or hereafter invented, including photocopying, microfilming, and recording, or in any information storage or retrieval system, without written permission from the publishers.

For permission to photocopy or use material electronically from this work, please access www.copyright.com (http:// www.copyright.com/) or contact the Copyright Clearance Center, Inc. (CCC), 222 Rosewood Drive, Danvers, MA 01923, 978-750-8400. CCC is a not-for-profit organization that provides licenses and registration for a variety of users. For organizations that have been granted a photocopy license by the CCC, a separate system of payment has been arranged.

Trademark Notice: Product or corporate names may be trademarks or registered trademarks, and are used only for identification and explanation without intent to infringe.

Library of Congress Cataloging-in-Publication Data
A catalog record for this book is available from the Library of Congress.

Visit the Taylor & Francis Web site at
http://www.taylorandfrancis.com

and the CRC Press Web site at
http://www.crcpress.com

HANDBOOK OF
LAYERED MATERIALS

EDITED BY

SCOTT M. AUERBACH
University of Massachusetts Amherst
Amherst, Massachusetts, U.S.A.

KATHLEEN A. CARRADO
Argonne National Laboratory
Argonne, Illinois, U.S.A.

PRABIR K. DUTTA
The Ohio State University
Columbus, Ohio, U.S.A.

CRC Press
Taylor & Francis Group
Boca Raton London New York

CRC Press is an imprint of the
Taylor & Francis Group, an **informa** business

CRC Press
Taylor & Francis Group
6000 Broken Sound Parkway NW, Suite 300
Boca Raton, FL 33487-2742

First issued in paperback 2019

© 2004 by Taylor & Francis Group, LLC
CRC Press is an imprint of Taylor & Francis Group, an Informa business

No claim to original U.S. Government works

ISBN-13: 978-0-8247-5349-8 (hbk)
ISBN-13: 978-0-367-39444-8 (pbk)

This book contains information obtained from authentic and highly regarded sources. Reasonable efforts have been made to publish reliable data and information, but the author and publisher cannot assume responsibility for the validity of all materials or the consequences of their use. The authors and publishers have attempted to trace the copyright holders of all material reproduced in this publication and apologize to copyright holders if permission to publish in this form has not been obtained. If any copyright material has not been acknowledged please write and let us know so we may rectify in any future reprint.

Except as permitted under U.S. Copyright Law, no part of this book may be reprinted, reproduced, transmitted, or utilized in any form by any electronic, mechanical, or other means, now known or hereafter invented, including photocopying, microfilming, and recording, or in any information storage or retrieval system, without written permission from the publishers.

For permission to photocopy or use material electronically from this work, please access www.copyright.com (http:// www.copyright.com/) or contact the Copyright Clearance Center, Inc. (CCC), 222 Rosewood Drive, Danvers, MA 01923, 978-750-8400. CCC is a not-for-profit organization that provides licenses and registration for a variety of users. For organizations that have been granted a photocopy license by the CCC, a separate system of payment has been arranged.

Trademark Notice: Product or corporate names may be trademarks or registered trademarks, and are used only for identification and explanation without intent to infringe.

Library of Congress Cataloging-in-Publication Data
A catalog record for this book is available from the Library of Congress.

Visit the Taylor & Francis Web site at
http://www.taylorandfrancis.com

and the CRC Press Web site at
http://www.crcpress.com

To my beautiful wife Sarah, the light of my life. —SMA

To my husband Joe for his unwavering support. —KAC

To T and N, for extending my limits. —PKD

Foreword

Although there are numerous families of lamellar solids, only a handful of them exhibit the kind of versatile intercalation chemistry that forms the basis of this book. In arriving at the content of this volume, the editors have accurately identified six classes of versatile layered compounds that are at the forefront of materials intercalation chemistry, namely, smectite clays, zirconium phosphates and phosphonates, layered double hydroxides (known informally as "hydrotalcites" or "anionic clays"), layered manganese oxides, layered metal chalcogenides, and lamellar alkali silicates and silicic acids. Graphite and carbon nanotubes have not been included, in part because this specialty area of intercalation chemistry is limited to one or two molecular layers of comparatively small guest species that are capable of undergoing electron transfer reactions with the host structure.

Six of the eleven chapters are devoted primarily to the intercalation chemistry of smectite clays, the most versatile among all lamellar compounds. Two of these chapters are devoted to the experimental and theoretical aspects of the clay structures and surface chemistry, including chemical catalysis. Organo clays and polymer-clay nanocomposites, the adsorption of nitroaromatic compounds of environmental significance onto clay surfaces, photochemical processes, and pillared clays and porous clay heterostructures are the subjects of the remaining four chapters. These six chapters provide detailed discussions of the factors that influence access to the intragallery surfaces of the clay host and the materials properties of the resulting intercalates.

The remaining five chapters provide succinct overviews of the intracrystal guest–host chemistry of the next most prominent families of lamellar structures. Originally recognized for their importance in the production of precipitated metal oxide catalysts, the layered double hydroxides are finding increasing importance in many areas of materials chemistry. Expanded materials applications are also being realized for the layered zirconium phosphates, which initially attracted attention only on the basis of their cation exchange properties, as well as for the layered metal chalcogenides. Having remarkably regular porosity and redox properties, the layered manganese oxides represent the newest family of lamellar compositions of contemporary research interest. Finally, the alkali silicates and their proton-exchanged derivatives form a robust family of layered materials that can be readily synthesized from silica gel and surface modified for use in many areas.

Overall, this is a timely, well-written book that deserves to be in every chemistry and materials science library, as well as on the personal bookshelves of new and established researchers engaged in the study of layered materials. It provides an excellent up-to-date reference source on the state of the art of important lamellar structures and their contemporary materials applications.

Thomas J. Pinnavaia
University Distinguished Professor of Chemistry
Michigan State University
East Lansing, Michigan, U.S.A.

Preface

This book is a comprehensive and dedicated source of information on clays and related layered materials. Related materials are defined for our purposes as those that share the ability to "pillar," i.e., materials in which permanent intracrystalline porosity can be created within the layers. While myriad layered materials exist, they do not all share this pillaring ability (graphite, for example), and therefore are not included in this book. The layered materials that form the core of this book certainly stand on their own merits in terms of their own particular chemistries, yet their applied technologies tend to be similar, including catalytic applications.

This handbook serves as a companion volume to Zeolite Science and Technology, which was published in 2003. The scope and philosophy of the two books is much the same because they share the same publisher and coeditors. Because both volumes are strong on the basics and fundamentals, they are handbooks rather than simple monographs on the most recent applications, which tend to become outdated quickly. It has been our intent to focus keenly on the fundamental properties of the materials. Advances made in recent years concerning synthesis, characterization, host–guest chemistry, and modern applications are included. The permanent intracrystalline porosity feature ties these layered materials together for adsorptive and catalytic applications, among others.

The subject matter of this book includes the following materials: clays, pillared clays and pillared clay heterostructures, together with layered analogs of zirconium phosphates and phosphonates, double hydroxides, manganese oxides, metal chalcogenides, and polysilicates (including alkali silicates and crystalline silicic acids). Discussion of their synthesis and/or natural occurrence, characterization, host–guest chemistry including pillaring, and their adsorptive and catalytic applications are all included.

The content of this Handbook reflects the fact that the bulk of knowledge in this field concerns clays. In Part 1, which concentrates on clays, there is considerable focus on clay–organic interactions because of their relevance to catalysis. A chapter on nitroaromatic compound sorption is included because it provides an excellent example of clay-organic interactions. Other pertinent topics include molecular modeling of surface chemistry and photochemical processes, including photocatalysis. Pillared clays and porous clay heterostructures are the subject of an entire chapter. Chapters in Part 2 cover synthesis, characterization, host–guest pillaring, sorption, and catalysis for each class of layered material.

The chapters evolve from basic introductory concepts appropriate for the nonspecialist to detailed presentations of state-of-the-art research. Referencing is extensive, and substantial effort was made to include references in tabular form for rapid access to the extensive literature available. Thus, the *Handbook of Layered Materials* is appropriate for beginning research students as well as more experienced practitioners. It is our sincere hope that this book inspires the next generation of researchers to tackle exciting and important issues regarding these remarkable materials.

Scott M. Auerbach
Kathleen A. Carrado
Prabir K. Dutta

Contents

Contributors

Pilar Aranda Consejo Superior de Investigaciones Cientificas (CSIC), Cantoblanco, Madrid, Spain

Akhilesh Bhambhani University of Connecticut, Storrs, Connecticut, U.S.A.

Stephen A. Boyd Michigan State University, East Lansing, Michigan, U.S.A.

Paul S. Braterman University of North Texas, Denton, Texas, U.S.A.

Kathleen A. Carrado Argonne National Laboratory, Argonne, Illinois, U.S.A.

Pegie Cool University of Antwerp, Antwerp, Belgium

Jason P. Durand University of Connecticut, Storrs, Connecticut, U.S.A.

Laura Espinal University of Connecticut, Storrs, Connecticut, U.S.A.

Luis-Javier Garces University of Connecticut, Storrs, Connecticut, U.S.A.

Sinue Gomez University of Connecticut, Storrs, Connecticut, U.S.A.

Nathan Hnatiuk University of Connecticut, Storrs, Connecticut, U.S.A.

Cliff T. Johnston   Purdue University, West Lafayette, Indiana, U.S.A.

Challa V. Kumar   University of Connecticut, Storrs, Connecticut, U.S.A.

Gerhard Lagaly   Universität Kiel, Kiel, Germany

Michael M. Lerner   Oregon State University, Corvallis, Oregon, U.S.A.

Jia Liu   University of Connecticut, Storrs, Connecticut, U.S.A.

Makoto Ogawa   Waseda University, Tokyo, Japan

Christopher O. Oriakhi   Hewlett-Packard Corporation, Corvallis, Oregon, U.S.A.

Sung-Ho Park   Lawrence Berkeley National Laboratory, Berkeley, California, U.S.A.

Eduardo Ruiz-Hitzky   Consejo Superior de Investigaciones Cientificas (CSIC), Cantoblanco, Madrid, Spain

Wilhelm Schwieger   Universität Erlangen-Nürnberg, Erlangen, Germany

José María Serratosa   Consejo Superior de Investigaciones Cientificas (CSIC), Cantoblanco, Madrid, Spain

Guangyao Sheng   University of Arkansas, Fayetteville, Arkansas, U.S.A.

Young-Chan Son   University of Connecticut, Storrs, Connecticut, U.S.A.

Garrison Sposito   Lawrence Berkeley National Laboratory, Berkeley, California, U.S.A.

Steven L. Suib   University of Connecticut, Storrs, Connecticut, U.S.A.

Brian J. Teppen   Michigan State University, East Lansing, Michigan, U.S.A.

Etienne F. Vansant   University of Antwerp, Antwerp, Belgium

Josanlet Villegas   University of Connecticut, Storrs, Connecticut, U.S.A.

Zhi Ping Xu   University of North Texas, Denton, Texas, U.S.A.

Faith Yarberry   University of North Texas, Denton, Texas, U.S.A.

1

Introduction: Clay Structure, Surface Acidity, and Catalysis

Kathleen A. Carrado

Argonne National Laboratory
Argonne, Illinois, U.S.A.

I. INTRODUCTION

Geologists, mineralogists, chemists, and soil scientists all approach the terms *clay* and *clay mineral* somewhat differently. Historically the term *clay* has referred to the small inorganic particles in the <2-μm portion of a soil fraction without regard to composition or crystallinity, and *clay minerals* has referred to the specific phyllosilicates (the term for sheet silicate structures) that are the layered, hydrous, magnesium or aluminum silicates in such a fraction. In 1995 an AIPEA report (1) defined clay as a "material composed primarily of fine-grained minerals which is generally plastic at appropriate water contents and will harden when dried or fired" and clay minerals as "phyllosilicate minerals and minerals which impart plasticity to clay and which harden upon drying or firing." For the purposes of this chapter the two terms will be used interchangeably, but the intent is always to refer to (either) *clay mineral* definition.

This chapter will provide only a brief description of clay mineral structures and properties, concentrating on those relevant to catalytic applications. There

This chapter has been created by the University of Chicago, an Operator of Argonne National Laboratory under contract no. W-31-109-ENG-38 with the U.S. Department of Energy. The U.S. Government retains for itself, and others acting on its behalf, a paid-up, nonexclusive, irrevocable worldwide license in said article to reproduce, prepare derivative works, distribute copies to the public, and perform publicly and display publicly, by or on behalf of the Government.

are entire books dedicated to these topics, however, some more recent than others, which can be accessed by the reader interested in more depth. Excellent resources include (in chronological order) books by Grim (2), Weaver and Pollard (3), Barrer (4), Brindley and Brown (5), Newman (6), Bailey (7), Velde (8), Moore and Reynolds (9), and Yariv and Cross (10).

II. STRUCTURES

Every clay mineral contains two types of sheets, tetrahedral (T) and octahedral (O); Figure 1 shows schematic representations of each. In the T sheets, each silicon atom is surrounded by four oxygen atoms, and the tetrahedrally coordinated silicon cations are linked to one another via covalent bonding through shared oxygens. These shared oxygens form a basal plane, and the remaining apical oxygens are shared with another layer of cations. The tetrahedral units arrange as a hexagonal network (Figure 1a, b) along this basal plane. The O sheets have cations (usually aluminum or magnesium) that are coordinated with six oxygens or hydroxyls, and these units are covalently linked into a sheet structure as well. In clays, T and O sheets are also covalently linked through the apical tetrahedral oxygens. Natural minerals made up of only O sheets include gibbsite $Al(OH)_3$ and brucite $Mg(OH)_2$. In the former, just two-thirds of the octahedra are filled with Al atoms. Clay derivatives of gibbsite are therefore referred to as dioctahedral. Brucite and its clay derivatives, on the other hand, are referred to as trioctahedral because all of the octahedra are filled with Mg atoms.

Al(III) can replace Si(IV) in the tetrahedral sheet and therefore cause a negative charge. In most clays this substitution is quite small, although in micas it can occur at levels up to 25%. Occasionally Fe(III) ions are present as well. More variation of cations is found in the octahedral layers. In dioctahedral minerals the most common substitution is Mg(II) for Al(III), and Li(I) most often replaces Mg(II) in trioctahedral minerals. Isomorphic substitutions by cations with higher charge (the most common situation) leaves a net negative charge on the octahedral sheets. Many other transition metal cations have been found in the octahedral layers, at trace levels or higher, including Fe(II), Fe(III), Ni(II), Zn(II), Mn(II), chromium, and titanium.

The types of polyhedral anions occur as: (1) oxygens in the tetrahedra, (2) oxygens that link cations between tetrahedra and octahedra, (3) hydroxyl and

FIGURE 1   A clay $[Si_4O_{10}]^{4-}$ tetrahedral (T) sheet in (a) top view and (b) side view, and a clay octahedral (O) sheet in (c) top view and (d) side view. The $[Al_4O_{12}]^{12-}$ dioctahedral top view is shown in (c); a $[Mg_6O_{12}]^{12-}$ trioctahedral top view would show a continuous sheet of octahedral units. (Adapted from Ref. 11.)

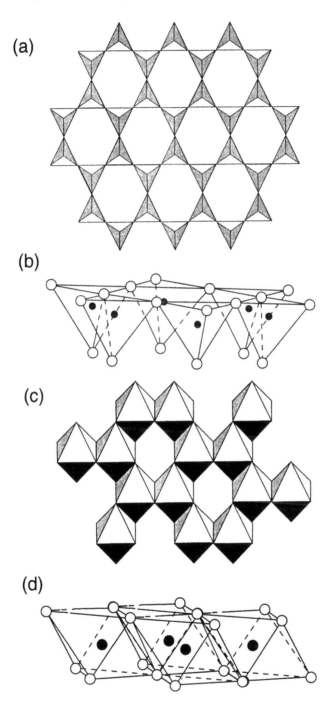

oxygen for linking octahedra to each other, and (4) hydroxyl when there is no linkage (apical octahedral positions not coordinated with tetrahedra). In some clay structures a certain percentage of the hydroxyls are replaced by fluoride anions.

Because of the net negative charge on many clay lattices due to isomorphous substitutions, there are charge-compensating cations between the basal oxygen planes. They are typically mono- or divalent and surrounded by water molecules. Since micas have a large amount of surface charge occurring within the tetrahedral plane (charge per unit cell = 1), the exchangeable cations tend to be tightly fixed in place and are almost exclusively potassium cations. Other types of clay minerals (e.g., smectites) with lower surface charges contain cations that are easily exchanged, and in natural clays they are usually calcium(II), sodium(I), or magnesium(II).

III. GROUPS AND SUBGROUPS

Table 1 shows the major clay mineral groups according to layer type and layer charge, which fall into the predominate layer types of 1:1 and 2:1 (T:O), along with ideal structural chemical compositions. The wide variety of clays found in nature occurs due to the myriad substitutions that can occur in T or O sites. Certain substitutions give rise to particular minerals with their own physical properties and stability (thermal and chemical) parameters. Illite is added as its own group, which has also been done by others (8,11). However, many state that this non-swelling mineral is a "clay mica" (12,13), with a layer charge that is less than that for a true mica (0.25–0.75 rather than 1), without a strict definition by the AIPEA as yet. Figure 2 provides a [010] crystallographic view (which, though still side-on in the a-b plane, differs from the view projected in Figures 1b and 1d) of some of the major clay mineral groups. The $c \cdot \sin \beta$ distance indicated is the basal (001) c-axis d-spacing (or d_{001}-spacing). Table 2 gives the chemical compositions analyzed for some representative, natural clay minerals for simple and rapid comparison.

A. Kaolin–Serpentine Group

One mineral layer of the 1:1 T:O serpentine–kaolin group is composed of one tetrahedral sheet that is condensed to a single octahedral sheet, and the minerals for the most part have a d-spacing of 7.1–7.3 Å. In kaolins the ideal structural formula is $[Al_4Si_4O_{10}](OH)_8$, with an octahedral sheet that is of the gibbsite type; in serpentines the ideal structural formula is $[Mg_6Si_4O_{10}](OH)_8$, with an octahedral sheet of the brucite type. The kaolin subgroup consists of the three polytypes (which means different stacking arrangements) kaolinite, dickite, and nacrite, as well as halloysite. The last is unique in that it contains, under certain

TABLE 1 Classification Scheme of Phyllosilicate Clay Minerals (T = Tetrahedral, O = Octahedral)

Structure type	Charge per unit cell	Group	Mineral examples	Ideal composition	Notes
1:1 (TO)	0	Kaolin–serpentine	Kaolinite, dickite, nacrite	$Al_4Si_4O_{10}(OH)_8$	Kaolin subgroup, dioctahedral, nonswelling
			Halloysite	$Al_4Si_4O_{10}(OH)_8 \cdot 4H_2O$	Kaolin subgroup, dioctahedral, swelling
			Chrysotile, antigorite, lizardite	$Mg_6Si_4O_{10}(OH)_8$	Serpentine subgroup, trioctahedral, nonswelling
2:1 (TOT)	0	Pyrophyllite–talc	Pyrophyllite	$Al_4Si_8O_{20}(OH)_4$	Dioctahedral, nonswelling
			Talc	$Mg_6Si_8O_{20}(OH)_4$	Trioctahedral, nonswelling
	0.5–1.2	Smectite	Beidellite	$[(Al_4)(Si_{7.5-6.8}Al_{0.5-1.2})O_{20}(OH)_4]Ex_{0.5-1.2}$	Dioctahedral, swelling
			Montmorillonite	$[(Al_{3.5-2.8}Mg_{0.5-1.2})(Si_8)O_{20}(OH)_4]Ex_{0.5-1.2}$	
			Nontronite	$[(Fe_{4.0})(Si_{7.5-6.8}Al_{0.5-1.2})O_{20}(OH)_4]Ex_{0.5-1.2}$	
			Saponite	$[(Mg_6)(Si_{7.5-6.8}Al_{0.5-1.2})O_{20}(OH)_4]Ex_{0.5-1.2}$	Trioctahedral, swelling
			Hectorite	$[(Mg_{5.5-4.8}Li_{0.5-1.2})(Si_8)O_{20}(OH)_4]Ex_{0.5-1.2}$	
	1.2–1.8	Vermiculite	Vermiculite	$[(Al_4)(Si_{6.8-6.2}Al_{1.2-1.8})O_{20}(OH)_4]Ex_{1.2-1.8}$	Dioctahedral, swelling
			Vermiculite	$[(Mg_6)(Si_{6.8-6.2}Al_{1.2-1.8})O_{20}(OH)_4]Ex_{1.2-1.8}$	Trioctahedral, swelling

(Continued)

TABLE 1 *Continued*

Structure type	Charge per unit cell	Group	Mineral examples	Ideal composition	Notes
		Illite	Illite	$[(Al_4)(Si_{7.5-6.5}Al_{0.5-1.5}) O_{20}(OH)_4]K_{0.5-1.5}$	Dioctahedral, nonswelling
			Glauconite	Illite rich in Fe	
	2	Mica	Muscovite	$[(Al_4)(Si_6Al_2) O_{20}(OH,F)_4]K_2$	Dioctahedral, nonswelling
			Celadonite	$[(Fe_2Mg_2)(Si_8) O_{20}(OH,F)_4]K_2$	
			Phlogopite	$[(Mg_6)(Si_6Al_2) O_{20}(OH,F)_4]K_2$	Trioctahedral, nonswelling
			Taenolite	$[(Li_2Mg_4)(Si_8) O_{20}(OH,F)_4]K_2$	Trioctahedral, lithium mica
	4	Brittle mica	Margarite	$[(Al_4)(Si_4Al_4) O_{20}(OH,F)_4]Ca_2$	Dioctahedral, nonswelling
2:1 channels or inverted ribbons	Variable	Palygorskite-sepiolite	Palygorskite	$[(Mg,Al)_4(Si_{7.5-7.75}Al_{0.5-0.25}) O_{20}(OH)_2(OH_2)_4]Ex_{var}$	Dioctahedral, nonswelling
			Sepiolite	$[(Mg,M)_8(Si,M')_{12} O_{30}(OH)_4(OH_2)_4]Ex_{var}$	Trioctahedral ($M = Al$, $Fe(III)$; $M' = Fe(II)$, $Fe(III)$, $Mn(II)$
2:1:1	Variable	Chlorite	Clinochlore		[TOT]O[TOT] structure

Source: Adapted from Ref. 11.

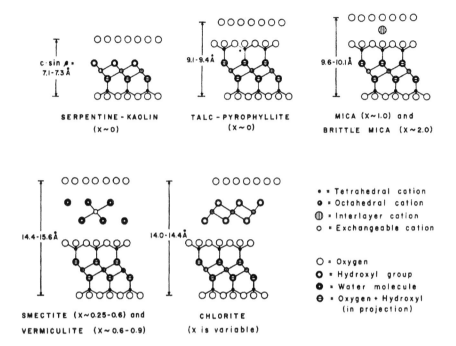

FIGURE 2 [010] view of structures of major clay mineral groups. 1:1 and 2:1 T:O layers are represented by the primary units in serpentine–kaolin and talc–pyrophyllite, respectively. The c·sin β distance is the basal (001) c-axis d-spacing. Layer charge per formula unit = X. (Reproduced with the permission of the Mineralogical Society of Great Britain and Ireland from Ref. 13.)

conditions, an interlayer of water molecules with a d-spacing of 10 Å. Also, halloysites have a tubular structure derived from rolled sheets rather than the typical planar structure. While in theory electrically neutral, in reality these 1:1 frameworks tend to carry a slight negative charge due to very low levels of isomorphous substitutions (see Table 2), which in turn accounts for their small cation exchange capacities (10–80 μeq/g for kaolinites, 90–95 μeq/g for halloysites) (12). Kaolinites are widely found in commercial products due partly to their low level of substitutions, which yields relatively chemically pure materials. Figures 3a and 3c provide scanning electron micrographs (SEM) of typical kaolinite (platy) and halloysite (tubular) (14) textural forms.

The serpentine subgroup is interesting in that these 1:1 layers can be of three types: planar, continuously curved to produce rolled or tubular structures, and planar structures comprising layers curved about the mean plane. The various

TABLE 2   Chemical Compositions of Various Representative Clay Minerals

Mineral	Source location	Chemical composition
Kaolinite	Keokuk, Iowa	$[(Al_{3.97}Fe(III)_{0.006}Mg_{0.019})(Si_{3.977}$ $Al_{0.023})O_{10}OH_8]Ca_{0.021}Na_{0.015}$ $K_{0.001}$
Pyrophyllite	Moore County, North Carolina	$[(Al_{3.94}Fe(III)_{0.06}Mg_{0.02})(Si_{7.94}Al_{0.06})$ $O_{20}OH_4]K_{0.01}$
Talc	Harford County, Maryland	$[(Mg_{5.77}Fe(III)_{0.03}Fe(II)_{0.09})(Si_{8.04})$ $O_{20}OH_4]K_{0.01}$
Beidellite	Black Jack Mine, Beidell, Colorado[1]	$[(Al_{3.96}Fe(III)_{0.04}Mg_{0.02})(Si_{6.96}Al_{1.04})$ $O_{20}OH_4]Na_{1.00}$
Montmorillonite	Upton, Wyoming[1]	$[(Al_{3.07}Ti_{0.01}Fe(III)_{0.40}Mg_{0.49})$ $(Si_{7.79}Al_{0.12})O_{20}OH_4]Na_{0.75}$
Nontronite	Clausthal, Zellerfeld, Germany[1]	$[(Fe(III)_{4.01}Mg_{0.07})(Si_{6.81}Al_{0.13}$ $Fe(III)_{1.06})O_{20}OH_4]Na_{1.02}$
Saponite	Milford, Utah[1]	$[(Al_{0.19}Fe(III)_{0.02}Mg_{5.72})(Si_{7.50}Al_{0.05})$ $O_{20}OH_4]Na_{0.42}$
Hectorite	Hector, California[1]	$[(Mg_{5.33}Li_{0.60})(Si_{7.98}Al_{0.02})O_{20}OH_4]$ $Na_{0.76}$
Vermiculite	Llano County, Texas[1]	$[(Al_{0.03}Fe(III)_{1.02}Fe(II)_{0.09}Mn_{0.02}$ $Mg_{4.63})(Si_{5.87}Al_{2.13})O_{20}OH_4]Na_{1.50}$
Illite	Jefferson Canyon, Montana[2]	$[(Al_{2.71}Ti_{0.06}Fe(III)_{0.52}Fe(II)_{0.15})$ $(Si_{7.28}Al_{0.72})O_{20}OH_4]K_{1.36}Na_{0.02}$
Muscovite	Georgia (from pegmatite)	$[(Al_{3.79}Ti_{0.03}Fe(III)_{0.04}Fe(II)_{0.10}$ $Mg_{0.11})(Si_{6.03}Al_{1.97})O_{20}OH_4]K_{1.72}$ $Na_{0.20}Ca_{0.01}$
Sepiolite	Akatani Mine, Niigata, Japan	$[(Mg_{6.98}Mn(II)_{0.55}Fe(III)_{0.25}Fe(II)_{0.16}$ $Al_{0.02})(Si_{11.77}Al_{0.23})O_{30}OH_4$ $(OH_2)_4]K_{0.01}$
Clinochlore[3]	Appennine region, Italy	$[(Al_{1.01}Fe(III)_{0.13}Fe(II)_{0.27}Mg_{10.66})$ $(Si_{6.69}Al_{1.31})O_{20}OH_{16}]$

[1] Na-saturated samples.
[2] Silver Hill, c.e.c. 150 μeq/g, <10% expandable layers.
[3] A chlorite mineral.
Source: Adapted from Refs. 11 and 12.

situations arise as a result of misfits between T and O layers due to isomorphous substitutions and the subsequent strains from size differences. Large amounts of Al(III) and/or Fe(III) lead to minerals of planar structure (e.g., lizardite), while correspondingly low levels leads to rolls from curved layers (needle-shaped), as occurs for chrysotile. A detailed discussion of this phenomenon and all of their resulting minerals can be found in Ref. 12.

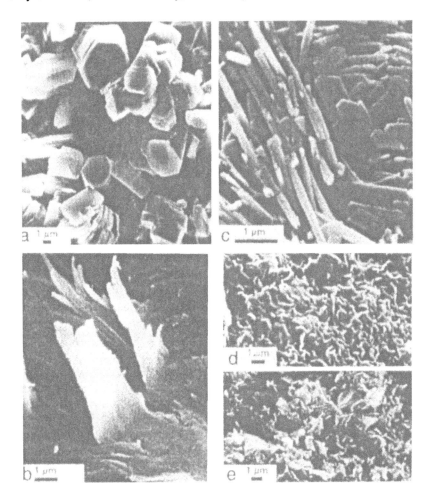

FIGURE 3 Scanning electron micrographs of (a) well-crystallized kaolinite in quartz geodes from Keokuk, Iowa; (b) flakes of illite curling upward from a shale cleavage surface; the flakes appear to form via coalescence of ribbon structures back from the growing ends (from Silver Hill, Montana); (c) a mixture of platy kaolinite and elongate halloysite from Kaoling, China; (d) Ca-montmorillonite from Chambers, Arizona; thin flexible plates join together and curl at the edges to yield a "cornflakes" texture; (e) Na-montmorillonite from Clay Spur, Wyoming. (Reprinted with the permission of Cambridge University Press from Ref. 14.)

B. Pyrophyllite–Talc Group

All of the 2:1 (TOT) minerals contain one central octahedral sheet condensed to two parallel tetrahedral sheets, via T silica oxygen apices to octahedral OH hydroxyl planes. The ideal structural formulas of pyrophyllite and talc are $[Al_4Si_8O_{20}](OH)_4$ and $[Mg_6Si_8O_{20}](OH)_4$, respectively, with a d-spacing of 9.1–9.4 Å, as indicated in Figure 2. The layers are attracted by van der Waals forces, and the exposed oxygen planes are perforated and highly hydrophobic. These 2:1 phyllosilicates vary the least from their ideal structure, and any charge built up by tetrahedral substitution is mostly neutralized by octahedral replacement.

C. Smectite Group

Smectite minerals also contain 2:1 TOT layers, but they differ from pyrophyllite–talc in that a more significant amount of isomorphous substitutions take place, enough to raise the charge per unit cell to 0.5–1.2. The net negative charge that arises from this is compensated for by interlayer, hydrated cations (primarily K, Na, Ca, and Mg). The charge interactions are somewhat diffuse because most of the charge arises from isomorphous substitutions in the octahedral layer. Because of this, the hydrated cations are loosely held and easily exchanged. In addition, large amounts of water can be accommodated between the layers, which gives rise to the swelling component that gives smectites their name. In addition, polar organic molecules are attracted by the exchangeable cations and can intercalate, causing swelling. Of all the clay minerals, smectites then become among the most chemically interesting for modification and application. Typical basal spacings vary from 10 to 20 Å (dehydrated state to containing multiple layers of water molecules; see Figure 2). In dilute aqueous suspensions, Li- and Na-smectites are able to dissociate (delaminate) to such a degree, up to 40 Å, that the negatively charged silicate layers and small positively charged cations produce polyelectrolyte properties. Other exchangeable cations yield smectite tactoids in dilute aqueous suspensions, which are several TOT parallel layers held together by electrostatic forces.

The two end members of this group with mainly tetrahedral substitutions are beidellite and saponite, which are di- and trioctahedral smectites, respectively. The corresponding end members with mainly octahedral substitutions are montmorillonite and hectorite. Another common smectite, nontronite, is an iron-rich mineral. Chemical compositions of various smectite samples are provided in Table 2. Montmorillonite is the most common mineral of this group; it is named for its location in Montmorillon, France. Figures 3d and 3e provide SEM views of the textural morphology of two montmorillonites. A common industrial mineral is bentonite, which is actually a montmorillonite of volcanic ash origin that contains a significant amount of impurities, such as cristobalite (α-quartz), that is intimately mixed with the clay.

D. Vermiculite Group

These minerals are very similar to 2:1 TOT smectites, except their layer charge densities are higher (1.2–1.8) due to greater levels of isomorphous substitution, which arises primarily from tetrahedral replacements. While water and polar organic molecules will intercalate, swelling is more limited than it is for smectites. In natural minerals the exchangeable cation is Mg, with small amounts of Ca and Na. Vermiculites are trioctahedral and occur frequently as large crystals with platy morphology, similar to micas, although they are softer and contain interlayer water. Relative to the parent micas from which they are derived, the negative layer charge of vermiculites is lower and the iron is oxidized. They have a positive octahedral charge (0.3–1.2) and a large negative tetrahedral charge (>2), which distinguishes them from high-charge saponites (12).

E. Illite Group

Illites are also 2:1 TOT clay minerals, dioctahedral for the most part, but they are nonexpanding. This is because the charge per unit cell at about 1.8 arises primarily from tetrahedral substitutions, and the exchangeable cations are mainly K ions. Interlayer K ions happen to fit very well into the ditrigonal holes of the basal oxygen planes. The planes then pack parallel, with K ion embedded into their surfaces, and the electrostatic forces are strong. Illites contain less K and more water than a true mica. See Figure 3b for a SEM image of an illite. Debate continues as to classification of this mineral (12), from a 1984 Clay Minerals Society report based on several criteria (15) to a recommendation in 1989 that the term be used based solely on chemical rather than physical parameters (16).

F. Mica and Brittle Mica Groups

Also 2:1 TOT minerals, the variety of isomorphous substitutions yields dozens of specific micas and brittle micas divided into di- and trioctahedral subgroups. True micas normally have K or Na exchangeable cations, while in brittle micas the main interlayer cation is Ca. The high charge densities of micas and brittle micas (2 and 4, respectively) leads to electrostatic forces that are so strong that interlayer polar molecules are not present (see Figure 2), and they are all consequently nonswelling. Reference 12 contains information about micas, current until 1987. Since that time, however, a vast amount of new data has been amassed regarding micas. A 1998 International Mineralogical Asoociation committee on micas found that the vague nomenclature of >300 mica species discovered over the past 200 years has been reduced to just 37 main species names in six series, and pointed out large gaps that require new chemical and structural data (17). In 2002 a comprehensive review was published containing crystal-chemical, petrological, and historical aspects of micas, and the reader is referred here for detailed specific information regarding these minerals (18).

G. Palygorskite–Sepiolite Group

These 2:1 TOT fibrous clay minerals are unique in that they possess a channel structure resulting from repeated inversions of the silicate layer. They therefore lack continuous octahedral sheets. Figure 4 gives two views of the mineral palygorskite, one showing the tunnel structure. This structure has been explained as "ribbons of 2:1 phyllosilicate units" linked at SiO_4 tetrahedral inversions about Si–O–Si bonds (12). This corner linking of ribbons yields a relatively open framework of channels running parallel to ribbon edges and along the fiber axes. Ribbon widths contain five and eight octahedral sites per structural unit for palygorskite and sepiolite, respectively. Sepiolite therefore has a slightly wider channel. Sepiolites are primarily trioctahedral, with Mg in these sites, whereas the ratio of Mg to trivalent cations can vary from 3:1 to 1:3 (along with vacancies) in palygorskites, making them more dioctahedral. Two perpendicular surfaces of the tunnels contain broken-bond surfaces, while the other two are made up of perforated oxygen planes. Bound water, zeolitic water, and a small amount of exchangeable cation is present within the tunnels.

H. Chlorite Group

The structural unit of a chlorite mineral consists of a 2:1 layer with the negative charge balanced by a positively charged octahedral hydroxide sheet in the interlayer. Two different types of octahedral layers are therefore present, one within the 2:1 TOT layers and the other between them. In di- and trioctahedral chlorites, both types of octahedral sheets are di- or trioctahedral, respectively. Di,trioctahedral chlorites have 2:1 dioctahedral layers and trioctahedral interlayers (the reverse mineral is not known). A detailed discussion of specific chlorite minerals can be found in Ref. 12. Refer to Figure 2 for the [010] crystallographic view of a chlorite and to Table 2 for the composition of clinochlore, one chlorite mineral.

IV. COMMON CHARACTERIZATION TOOLS

A. X-Ray Powder Diffraction (XRD)

The advent of the uniaxial powder diffractometer along with using oriented clay samples has allowed for routine and reliable analysis of clay minerals. Prior to these developments the small coherence domains in clay crystals (compounded by their sheet structure) and their small particle size (most are up to 10–20 times smaller than 2 μm) made them difficult to study. Diffraction effects are enhanced for the small crystallites by orienting them, or by having them all lie on the same plane by exploiting their sheet structure. As a result the basal spacing (001) and higher orders of reflection (00l) are often enhanced relative to other diffraction

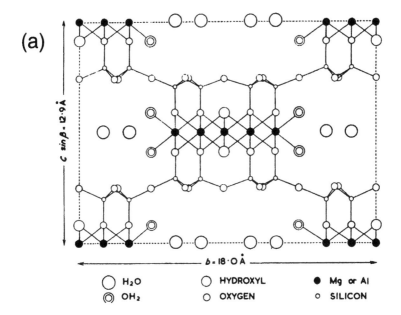

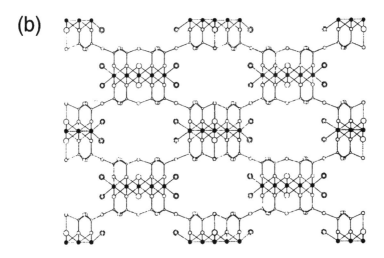

FIGURE 4   (a) [100] projection of the palygorskite structure and (b) an extended arrangement showing the tunnel structure. Sepiolite is very similar, except there are 8 Mg in the octahedral positions of each main block, rather than the 5 Mg, Al of palygorskite. (4a is reproduced with the permission of the Mineralogical Society of Great Britain and Ireland from Ref 13.)

planes (*hkl*, *hk*0), especially those without an *l* component. There are two standard polar molecules that are intercalated to determine the swelling properties of clays: Water yields a 5-Å increase in basal spacing, and ethylene glycol expands layers by 7 Å. In order to enhance the (*hk*) planes, nonoriented clay specimens are employed, where the least physical pressure possible is used to ensure that the powders are not packed. The two most useful XRD regimes for clay characterization are the basal spacing region (2–10° 2θ, Cu radiation) and the (060) reflection (near 60° 2θ, Cu radiation), which is diagnostic for several structural and compositional types. See Ref. 5 for a detailed presentation of clay crystal structures via XRD.

B. Thermal Analysis

There are several different types of water and hydroxyl groups associated with a clay matrix that is liberated with increasing temperature. Thermogravimetric analysis (TGA) is often used to qualitatively determine the binding energies of these different types of water, which can then be used to help characterize a mineral. In TGA, the mass of a sample is monitored vs. temperature while it is heated according to a program in a specific atmosphere (oxidizing, reducing, inert). At lowest temperatures, adsorbed water is lost from clay surfaces in defect sites or at broken bond sites; typical values for this are 1 wt% at 80–90°C (8). Zeolitic water is observed only for sepiolite–palygorskites, at 100–150°C. Interlayer water, which is associated with interlayer cations in smectites as monolayers (2.5 Å) or bilayers (5 Å), is liberated between 100 and 200°C. Dehydroxylation of the inner 2:1 or 1:1 lattice hydroxyl groups, also lost as water, is observed at 500–700°C. Derivative thermogravimetry (DTG), where the first derivative of the TG curve is plotted against temperature, is often more useful.

In differential scanning calorimetry (DSC), the heat flow rate (power) to a sample is monitored against temperature during programmed heating. Differential thermal analysis (DTA) allows for the observation of endothermic and exothermic events. In DTA, the difference in temperature between a sample and a reference material is monitored against temperature while the temperature of the sample in a specific atmosphere is programmed. Water losses are normally endothermic, and recrystallization (phase changes, etc.) are exothermic. Organic transformations from clay–organic complexes, under certain atmospheres (oxidizing or reducing), will also yield diagnostic endo- or exothermic events. Combining evolved gas analysis (EGA), mass spectrometry (MS), or IR with a thermal method can be a very powerful method of characterization.

Recently the thermal behavior of the organic components in organo-clay complexes has been thoroughly reviewed (19), which includes nearly all of the clays mentioned in Table 1 (except micas). The differences in thermal behavior of the separate organic and clay components within organo-clay complexes with those of the individual, pure compounds is used to draw conclusions regarding the character of organo-clay interactions. This also provides a means of differentiating simple clay/organic mixtures from actual clay-organic complexes.

C. Electron and Scanning Probe Microscopy

In scanning electron microscopy (SEM), the flux of secondary and backscattered electrons from a clay sample bombarded by an electron beam is used to form an intensity image. The resulting three-dimensional maps are very useful for identifying textures and morphologies, with resolution on the order of 0.01 μm. SEM shows clay crystal morphologies very clearly, as evidenced by the micrographs in Figure 3. Transmission electron microscopy (TEM) is used to image crystal shapes and sizes in essentially a two-dimensional plane. Clays are normally deposited on carbon-coated copper grids (8). In TEM the electron beam is absorbed to a greater extent by the clay than the carbon, creating a shadow image of the clay particle. Resolution is greater than for SEM. Further, the composition of individual crystallites can be estimated using energy-dispersive techniques. Figure 5a shows the TEM of illite crystal hexagons and illite/smectite mixed clay laths. Many clay structures have hexagonal symmetry and yield forms that are either hexagons or those derived from hexagonal prisms (8). High-resolution TEM (or HRTEM) has a resolution of <2 Å, allowing the imaging of atomic planes in clays. Also visible is unit layer stacking, interlayering of clay structures, and their growth patterns. Figure 5b is an HRTEM of illite crystallites prepared perpendicular to the beam showing 10-Å bands of stacked illite layers (8). Care must be taken when using HRTEM to avoid long exposure times of clays to the electron beam, which is converted to thermal energy and can lead to fuzzy images.

Atomic force microscopy (AFM) allows the observation of individual atoms on mineral surfaces and their interactions. Fluid cells allow one to follow the progress of reactions between minerals and solutions in real time. Minerals with layered structures usually display good cleavage planes, which makes them easy to study by AFM. Many materials have in fact been imaged on mica because of the excellent qualities of their freshly cleaved surfaces (20). This is not a necessary condition, however, and studies on ultrafine clay particles also have been carried out (21). AFM studies of various clay mineral surfaces themselves are available, including illite and montmorillonite (22), lizardite, muscovite, and clinochlore (23), and chlorite (24). Occelli et al. have several papers discussing AFM of pillared clays (25). Finally, the molecular scale structure of cations, surfactants, and polymers on various clay surfaces has also been examined with these scanning force techniques (26), and the reader is referred to these references for detailed information. Lindgreen (27) has recently published a study of electrical conduction in micas by AFM combined with scanning tunneling microscopy.

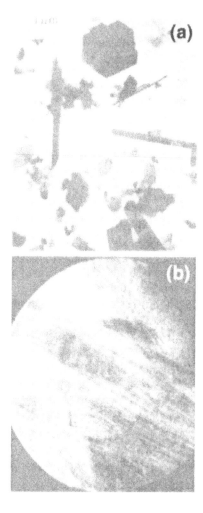

FIGURE 5 (a) TEM of illite crystal hexagons and illite/smectite mixed clay laths; (b) HRTEM of illite crystallites prepared perpendicular to the beam, showing 10-Å bands of stacked illite layers. (Reproduced with permission of Kluwer Academic Publishers from Ref. 8.)

D. Infrared Methods

Much of the power of infrared (IR) techniques does not apply to clay minerals unless they are complexed with organic molecules, because IR is less sensitive to differences in clay structures than methods such as XRD and NMR. Only broad absorbances are observed for the Si–O and Si–O–Si tetrahedral stretching vibrations near $1000-1100\ cm^{-1}$, Al–O stretching near $600\ cm^{-1}$, and octahedral Fe, Al, or Mg bending frequencies near $800-900\ cm^{-1}$ (28). One of the most useful IR regions for pure clay minerals contains the OH vibrational modes from 3500 to $4000\ cm^{-1}$. Here the bonding energies of structural hydroxyl groups, and therefore octahedral site occupancy, are easily determined. Samples must first be heated to $>400°C$ and kept dry to remove any associated water (surface, zeolitic, interlayer), which would interfere with the signal. Extensive references for clay mineral IR spectra and ancillary discussion are available in the literature (29–32).

In absorption studies of organo-clay complexes, adsorbed organic compound spectra are compared with those of the neat organic compound (or preferably a dilute solution of it in inert solvent). Differences yield the type of interactions between the clay minerals and the adsorbed species. Furthermore, spectral changes during thermal treatment with evolution of adsorbed water gives information on the fine structure of the complex. Many studies have been published that make possible the identification of bond types formed between organic functional groups and active clay surface sites, including their bond strength. The amount of adsorbed organic should be high enough to allow accurate IR determinations; therefore it is a method mostly applied to expanding clay minerals of high adsorptive capacity. Since smectites fall into this category, most IR studies have been done with them; the most investigated smectite is montmorillonite. Far fewer studies concern vermiculite and sepiolite–palygorskite, but they do exist. The organo-clay complexes formed with kaolinite are of a special class, and some information on these systems will be discussed in Chapter 4 with respect to Raman spectroscopy. A comprehensive review concerning IR and thermo-IR studies of organo-clay complexes has been published very recently (33), which includes sample preparation and experimental information, adsorption of amines (aliphatic, ammonium, di- and polyamines, anilines, benzidines, and N-ring compounds such as pyridine), hydroxylic compounds (alcohols, phenols), carboxylic acids (benzoic acid, fatty acids, stearic acid), amino acids, carbonyl compounds (e.g., amides), nitro compounds, nitriles, phosphorous and sulfur compounds, aromatic-transition metal π-complexes, and cationic dyes.

E. Nuclear Magnetic Resonance Methods

Nuclear magnetic resonance (NMR) spectroscopy has been widely used for the characterization of clay minerals and their organic interactions, providing both structural and dynamic information. NMR is sensitive to the local environment of atoms, and knowledge about crystallographic sites, coordination number, and tetrahedral sheet distortions has been gleaned from solid state ^{29}Si and ^{27}Al spectra of clays (34). Similar studies are also useful in monitoring the crystallization mechanisms of clays (35). Chapter 6 presents some NMR results of pillared clays that were done to examine inorganic pillar formation within clay interlayers. Several interlayer cations have been studied with respect to their distribution, hydration, and coordination by ^7Li (36), ^{113}Cd (37), ^{23}Na [35b,38], and ^{133}Cs [35b,39] NMR. Water and cation structure along with dynamic information at gel–clay interfaces has been examined by multinuclear magnetic resonances of aqueous solutions (40). ^1H, ^{13}C, and ^{31}P NMR studies provide information on the orientation, interaction, and mobility of organic species sorbed within interlayers (27,41). With respect to the last, a review was recently published concerning NMR of organo-clay complexes (42). Provided in this review are theoretical aspects (including line broadening in solids), and discussion of complexes with organic cations, organic molecule sorption on 2:1 and 1:1 minerals, and clay-organic grafting. In a review, Goodman and Chudek (43) discuss high-resolution NMR of clay minerals, the interactions of water with clay surfaces, and the structural distribution of paramagnetic ions.

V. SURFACE PROPERTIES, INCLUDING SURFACE ACIDITY

A. Surface Area

Among the most important parameters of clays with respect to catalytic applications are their surface area (including textural porosity) and surface acidity. When surface-area-determinative guest molecules are nonpolar, such as N_2, they do not usually penetrate the interlayer regions of typical inorganic ion-exchanged forms (such as Na^+) of well-outgassed smectites and vermiculites. Slightly larger ions, such as Cs^+ and NH_4^+, will, by virtue of their larger size, keep the basal planes far enough apart to allow some limited penetration. To calculate the specific surface area of smectites, the clay is considered to be completely dissociated into single TOT layers. The "interior" specific surface area of the oxygen cleavage planes is then 750–800 m^2/g (11). For many interlayer reactions, this large surface area value appears to be real. The exterior surface area for most natural smectites is less than 20% of this value. Vermiculites have an interior surface area that is also 750 m^2/g; however, the exterior area is a much smaller percentage than it is for smectites (<10 m^2/g). For comparison, in kaolinite the specific surface area is only about 10 m^2/g.

The most common procedure for determining the surface area of a powder is to derive the amount of adsorbed inert gas (typically nitrogen) at monolayer

TABLE 3 N_2 BET Surface Areas of Various Clay Minerals

Clay	Outgassing conditions	S. A., m^2/g
Kaolinite[a,b]	200°C, overnight, $<10^{-2}$ torr	8.75
Na,Ca-montmorillonite[a,c]	same	31.0
Ca-montmorillonite[a,d]	same	80.2
Ca-montmorillonite[a,e]	same	93.9
Na-hectorite[a,f]	same	64.3
Laponite[g]	105°C, overnight, 10^{-3} torr	360
Sepiolite[h]	96°C, 3 h	378
Palygorskite[h]	95°C, <70 h	192

[a] 3-point method of Thomas within Ref. 44 (p. 208).
[b] KGa-1, well crystallized.
[c] SWy-1, Wyoming.
[d] STx-1, Texas.
[e] SAz-1, Arizona (Cheto).
[f] SHCa-1, California.
[g] Synthetic hectorite; 6-point method of NPL within Ref. 44 (p. 212).
[h] Ref. 4 (p. 421).

coverage from a BET plot of adsorption isotherm data (see Sec. V. B), using the known cross-sectional area per molecule. Polar molecule adsorption methods (using water or glycerol, for example, which are able to penetrate the interlayers) provide the total internal and external surface area value. When this data is combined with N_2 BET, external and internal surface areas can both be determined. Theory, a discussion of standards, and data for a variety of clays are provided in Ref. 44. It is pointed out here that outgassing conditions are very important and should be kept consistent from sample to sample. Table 3 contains N_2 BET surface area data for a variety of clay minerals. Note the high values for laponite and sepiolite–palygorskite. For laponite (360 m^2/g), this is due to a very low degree of layer association, for it is well established as primarily a delaminated clay. The sorption is also substantial for sepiolite at 378 m^2/g, in this case because of the very fine fibrous nature of its crystalline morphology. At higher temperatures, sepiolites and palygorskites are known to undergo a crystal folding that decreases the surface areas to about 120–160 m^2/g (4).

B. N_2 Isotherms

Reference (4) provides detailed information about the sorption of polar and nonpolar sorbates (both gases and liquids) on many different types of clays, current to 1978. Rutherford et al. reported adsorption isotherms as recently as 1997 of various ion-exchanged montmorillonites (45). They found that the lateral dimensions of the elemental silicate sheets (aspect ratio) has a great effect on microporosity and surface area and that both values also increase with cation size as: Na $<$ Ca $<$ K $<$ Cs $<$ TMA (tetramethylammonium). Reference (46) contains indepth information up to 1987 specifically about water sorption isotherms on various clay types. Since many catalytic scientists use nitrogen surface area measurements (see preceding section), the behavior of N_2 isotherms will be the focus here. Figure 6 shows the adsorption–desorption N_2 isotherms for representative clays. These include natural TMA- and Ca-montmorillonites (47), Na-hectorite, sepiolite (48), and two synthetic hectorites. Natural smectite clays display type IV isotherms. The most recent IUPAC Commission report on gas physisorption explains that the characteristic feature of a type IV isotherm is its hysteresis loop (49). The point at low P/P_o values where the almost linear middle section of the isotherm begins is normally, it is explained, where monolayer coverage is complete and multilayer adsorption begins. Typically a large uptake occurs close to saturation pressures due to substantial vapor condensation on external surfaces (50). In many cases the hysteresis is due to capillary condensation in mesopores, with a mesoporous vertical rise seen in adsorption, limited uptake at high P/P_o values, and either H1 or H2 hysteresis loops. However, for clays, H3 or H4 hysteresis loops are observed. Furthermore, there is neither a limiting uptake at high P/P_o values nor a definitive mesoporous region. Therefore the porous network is defined differently.

Sepiolite (Fig. 6b) shows a classic H3 hysteresis loop, albeit a small one (49,51). The behavior is virtually identical to other published reports showing slight (52) to no (53) hysteresis for various sepiolite minerals. The steep adsorption branch at high partial pressures indicates the large external surface area for this fibrous clay. In H3 hysteresis loops, liquid nitrogen evaporation commences immediately upon pressure reduction. According to percolation theory, which is a method used to determine the connectivity of porous solids, the loop closes when all of the pores below their condensation pressures have access to a vapor-filled pore in the percolation cluster (51). According to the IUPAC Commission, this behavior is observed for systems containing ''aggregates of platelike particles giving rise to slit-shaped pores'' (49). Clearly the fibrous nature of sepiolite does not lend itself to this explanation, but the hysteresis demonstrated is very slight to negligible.

The Ca- and TMA-montmorillonites (Fig. 6a) and Na-hectorite (Fig. 6c) display H4 hysteresis loops. These are very similar to H3 loops, except pore size distribution is shifted to smaller pore sizes, and indeed the assignment of ''aggregates of platelike particles giving rise to slit-shaped pores'' applies to the smectites. However, when percolation theory is applied assuming certain parameters, the results are clearly unphysical (see Ref. 51 for details). In all of the smectite clay isotherms, the desorption branch shows an inflection ''knee'' at about 0.45–0.5 P/P_o. This inflection has been observed for many different

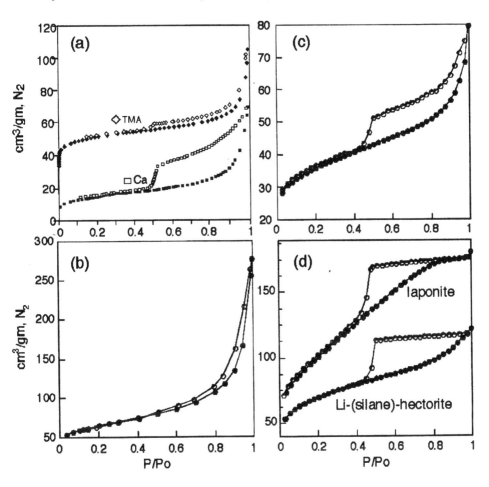

FIGURE 6 N_2 sorption isotherms for (a) TMA- and Ca-montmorillonite. (Adapted from Ref. 47, courtesy of Academic Press.) (b) An Italian sepiolite. (From Ref. 48.) (c) Natural SHCa-1 Na-hectorite and (d) synthetic laponite and Li-(silane)-hectorites. (c and d from Ref. 35a, courtesy of the Royal Society of Chemistry.) Closed symbols are adsorption, open symbols are desorption.

types of layered materials when using nitrogen as the sorbent gas (nitrogen boils at $P/P_0 \approx 0.42$), and it is therefore ignored. The phenomenon arises due to the complexity of capillary condensation in pore networks with pore-blocking effects, and it is sometimes called the *tensile strength artifact* (49). Seaton (51) has determined that the spinodal pressures (the point at which liquid in the pore reaches its limit of intrinsic stability and nitrogen vaporizes spontaneously) for a wide range of pore sizes fall within a narrow range. Therefore, the closure of H4 loops occur at almost exactly the same pressure, which is in fact the tensile strength artifact. According to Sing et al., H4 hysteresis arises in systems containing microporous slit-shaped pores (49).

Interestingly, the synthetic hectorites laponite and a Li-(silane)-hectorite (so called because it is derived from tetraethoxysilane) (35b) display hysteresis that is more similar to H2 (Fig. 6d) than the H4 loop of natural hectorite. The extreme linearity of the desorption isotherm at high partial pressures, a universal feature of H2 hysteresis loops, is interpreted as due to decompression of liquid nitrogen within the overall micro- and mesoporous network. According to Sing et al., H2 hysteresis is observed for "mesoporous networks of broad structure," although it most often occurs when the adsorption branch indicates mesoporosity (steep rise at higher P/P_0 values) (49). According to percolation theory, these clays at this point are below their so-called percolation threshold (51). The tensile strength artifact for N_2 then arises (at the "knee"), where the metastable liquid nitrogen vaporizes from the pores and the porous network has reached its percolation threshold. Why the synthetic hectorites display this behavior while natural hectorites do not may have something to do with the adsorption behavior at very high partial pressures. Natural hectorite displays proportionally more external surface area with its larger increase in adsorption toward saturation pressure than the lower-aspect-ratio synthetic hectorites. It is known that the aspect ratio of these clays increases as: laponite < Li-(silane)-hectorite < SHCa (35a).

The type IV isotherms displayed by most clays are, in general, amenable to surface area measurements using BET analysis, as long as attention is paid to certain constants in the BET equation (49). Furthermore, only the adsorption branch of the isotherm is reliably used to measure pore size distributions, which in most cases is quite broad. Finally, for the interested reader, guidelines for the characterization of porous solids containing both definitions and appropriate method selection are provided by Rouquerol et al. (54).

C. Cation Exchange Capacity

Cation exchange capacity (cec) is defined as the amount of exchangeable cations that a clay mineral can adsorb at a specific pH. This is a measurement of the total negative charges present, and it includes (1) isomorphous substitution within the lattice, (2) broken bonds at edges and external surfaces, and (3) the dissociation

TABLE 4 Ranges of Cation Exchange
Capacities for Clay Minerals

Mineral	cec (mEq/100 g)
Kaolinite	3–15
Halloysite	5–50
Illite	10–40
Chlorite	10–40
Montmorillonite	60–150
Vermiculite	100–150

Source: Ref. 55.

of accessible hydroxyl groups (55). Negative charges from the latter two sources are pH dependent. A comparison of cec values for a variety of clay minerals is provided in Table 4. There are several methods available for cec determinations, and no one general method can be reliably employed for all clay minerals. One method that is used more effectively for many clay minerals is based on saturation, primarily with barium as the index cation. Experimentally advantageous in this method is that, for most minerals, the measurement of cec is not dependent on the pH of saturation. A detailed experimental description of ammonium, sodium, and barium saturation is provided in Ref. 53. For the interested reader, a thorough discussion of the cation exchange equilibria in clays, including thermodynamics and isotherms, is available in Ref. 56.

D. Surface Acidity

The inherent surface acidity of materials such as clays leads to the carbonium ion mechanisms necessary in cracking catalysis. Contributors to acidic sites are water molecules that are coordinated to exchangeable cations, hydroxyl groups at crystal edges, hydronium ions, structural aluminum or magnesium that is coordinately unsaturated, and the bare exchangeable cations themselves (especially transition metal ions). The first three are Bronsted acid sites and the latter two are Lewis acid sites. A recent and detailed discussion of the origin of these two types of acid sites is provided in Ref. 11, including acidity and basicity of the hydroxyl and oxygen clay lattice planes, the broken bond surface, and the interlayer space.

The discussion here will pertain primarily to smectites clays, since they comprise the bulk of catalytic applications. Both Lewis and Bronsted acidity are dependent upon a clay's hydration state, and therefore the activity may vary throughout the course of a reaction. Strong Bronsted acidity in smectites is derived primarily from the dissociation of water that coordinates exchangeable cations. Hydrogen ion dissociation from hydration water under the polarizing effect of a metal ion occurs as follows:

$$[M(OH_2)_x]^{n+} \rightarrow [M(OH)(OH_2)_{x-1}]^{(n-1)+} + H^+$$

where M is the cation of $n+$ charge and x is the number of coordinating water molecules. Protons are much more mobile in the interlayer than in bulk water because the dielectric constant of interlayer water is less than that of bulk water. In fact, it is reported that the degree of dissociation of water is 10^7 times higher in the interlayer space than in bulk water (57). Acid strengths of some common ions decrease in the order of polarization (decreasing size and increasing charge) as: Fe^{3+}, Al^{3+}, Fe^{2+}, Mg^{2+}, Ca^{2+}, Ba^{2+}, Li^+, Na^+, K^+ (11).

Since acid strength increases with the polarization of cations, as the amount of water decreases the polarization (and therefore the ability to donate protons) increases. Fully dehydrated interlayer cations then behave as Lewis acids. Surface Si–OH–Al groups are stronger Bronsted acid sites than either Si–OH–Si or Al–OH–Al groups, and upon drying they convert to Lewis acid sites. In terms of adsorbing polar organic molecules, smectites are excellent because the exchangeable cation and hydration level play an important role. Strong bases such as ammonia and aliphatic amines are protonated in the interlayer. Weakly basic aromatic amines such as aniline create organic base–water–metal cation species via hydrogen bonds with bridging water molecules, which acts as a proton donor to the organic. If the adsorbed organics are proton donors, they can react with basic sites, including atoms in the oxygen plane and negative dipoles of water molecules in cation hydration spheres. Indoles, phenols, and fatty acids fall into this category. In the latter two examples, it is water that acts as the proton acceptor. When water is evolved during adsorption, a direct ion–dipole interaction occurs in the interlayer between the exchangeable cation and the polar adsorbate. If the exchangeable cation is a transition metal (Lewis acid) and the adsorbate a strong electron donor, they form a coordination d complex in the interlayer, thus stabilizing the adsorption process (11).

Yariv et al. have pointed out that interlayer water does not behave like a continuum of bulk water (58). Rather, they propose that three separate zones exist with distinguishable water structures. Zone A_m contains the ordered water of the ion's hydration sphere, zone A_o comprises the ordered water in the solid–liquid boundary at the clay oxygen plane, and zone B is the disordered water zone that separates the previous two. The disorganized water molecules of zone B are actually more active than the structurally organized zone A water molecules. The relative degrees of surface acidity then depend strongly on the interplay between exchangeable cation and water in zones A_m and B. This behavior can be separated into cations that do and do not form stable hydrates.

For stable hydrates, the strength of zone B water depends on exchangeable cation polarization power. The greater this property (e.g., Al^{3+}), the more acidic the associated hydration shell (A_m) and the greater is the extension of zone B. Many of the properties of Al-smectites, for example, result from the high activity of zone B water (58). For unstable hydrates displaying high acidity, the B zone is very large. The degree of water structure breaking increases with cation size; for example, in Cs-montmorillonite the large size of Cs^+ breaks the hydrophobic structure of zone A_o. Furthermore, the polarization of Cs (large size, low charge) is so low that zone A_m does not even form. The unexpectedly high surface acidity of Cs-montmorillonite, then, results from the acidic characteristics of zone B water. In a similar fashion but to a lesser extent, Rb- and K-smectites also display unexpectedly high surface acidities. Since there are no zone A_m water molecules, the Cs^+ ions are embedded within two parallel ditrigonal cavities of the flat oxygen planes. This acts to increase the stacking order of the smectite sheets.

In 1950 Walling defined *solid surface acidity* as "the ability of the surface to convert an adsorbed neutral base to its conjugate acid," and he developed a method to measure Bronsted acidity using indicators with a range of pK_a values (59). Iso-octane was used to eliminate effects due to water. Benesi extended the method to a wider range of pK_a values using Hammett indicator dyes and determined semiquantitative measurements of the acid strengths of various cation-exchanged clays (kaolinite and montmorillonite) (60).

This was soon supplanted by a more powerful spectroscopic method, wherein the IR behavior of chemisorbed amines onto acidic surfaces was found to distinguish the relative amounts of Lewis and Bronsted sites. The first report, for ammonia on a silica-alumina cracking catalyst, occurred in 1954 (61). Bands for NH_3 and NH_4^+ were observed; upon addition of water, the NH_4^+ bands increased at the expense of NH_3 bands. These results indicated two types of acid sites: (1) NH_3 chemisorbed by coordinate bond formation between the Lewis base (NH_3) and a Lewis acid site, and (2) transfer of a proton from a Bronsted site to the base forming NH_4^+ bound via coulombic forces.

The probe molecule pyridine has often been used more recently in the study of clay surface acidity by IR methods. Figure 7 shows data of pyridine chemisorption on a synthetic mica–montmorillonite catalyst as an example (62,63) of the types of bands of interest. The spectra are interpreted as showing chemisorption of pyridine at both protic and aprotic sites. The clay was first heated for 15 h at 650°C under vacuum and then cooled and spectrum A taken. Note that there are no bands in the 1400–1700 cm^{-1} region, the residual hydroxyl near 3450 cm^{-1}, and edge silanol hydroxyl at 3747 cm^{-1}. Pyridine vapor was then chemisorbed and spectrum B taken. The bands at 1456 cm^{-1} and 1547 cm^{-1} are assigned to Lewis and Bronsted sites, respectively. Since the 3747 cm^{-1} edge silanol band decreases, it is assumed that these protons are involved in the mechanism. Lewis sites predominated under these particular conditions.

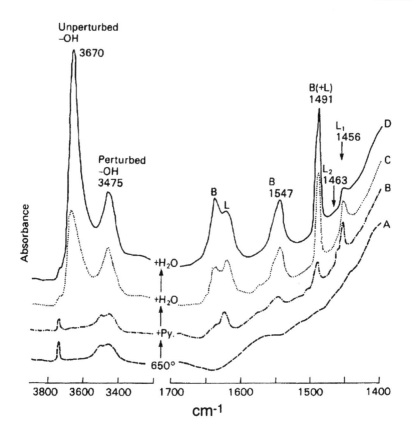

FIGURE 7 IR spectra of pyridine chemisorption on synthetic mica–montmorillonite showing the transformation of pyridine to pyridinium ion by interaction with water. L, B = Lewis and Bronsted acid sites (63). (Reproduced with permission from the Mineralogical Society of Great Britain and Ireland.)

Incremental amounts of water were then dosed into the system and spectra C and D taken. Much of the chemisorbed pyridine was transformed into pyridinium ions (compare $1456 \, \text{cm}^{-1}$ and $1547 \, \text{cm}^{-1}$ intensities) along with a sharp increase in the structural hydroxyl (3475 and $3670 \, \text{cm}^{-1}$). These changes were proposed to be due to water combining with a Lewis site to regenerate hydroxyl along with pyridinium.

Pyridine, a base of moderate base strength, is used to distinguish between not only Lewis and Bronsted acid sites on solid surfaces, but also strong and weak Bronsted acids (33). When it is protonated to pyridinium ($C_5H_5NH^+$), the diagnostic N^+-H deformation band appears in the region 1535–$1550 \, \text{cm}^{-1}$, and the skeletal modes shift to 1485 (strong), 1610 (medium), and 1640 (medium) cm^{-1}. When pyridine coordinates Lewis acid sites, diagnostic bands appear at 1455–1470 (strong), 1490 (medium), 1585 (very weak), and 1620–1634 (medium) cm^{-1}. Weak Bronsted acids form hydrogen bonds by donating protons to the pyridine nitrogen. This is shown by small ring vibration shifts at 1440–1447 and 1580–1600 cm^{-1}. A detailed review of pyridine chemisorption IR data (as well as many other amines) on different cation-exchanged smectites and pillared clays is provided in Ref. 33.

VI. CATALYSIS

Catalytic cracking of crude oil to high-octane gasoline and other usable fuels is a fundamental refining process of significant economic performance. The subject of catalytic cracking in the petroleum industry was reviewed comprehensively by Hettinger in 1991 (64). The use of clays as acidic cracking catalysts during crude oil refining for gasoline production was first disclosed in 1923, a process that became commercially feasible in 1938 when catalyst regeneration problems were solved (63). While clay catalysts were supplanted by synthetic zeolites in this industry in the 1960s, three recent extensive reviews of catalysis using clays demonstrate that the relevance of this topic has not diminished over the years. In 2002 Heller-Kellai published "Clay catalysis in reactions of organic matter" (65), in 1999 Vaccari published "Clays and catalysis: a promising future" (66), and in 1993 the book *Organic Chemistry Using Clays*, by Balogh and Laszlo, was published (67). Pertinent to catalytic reactions of clays are their complexes and fundamental interactions with organic molecules. Excellent reviews on this topic (68), including this book in Chapter 3, and also on the hydrophobicity (organophilicity) of organo-clays (69) have been published very recently as well. It becomes redundant to repeat this information in its entirety here. Rather, the highlights of catalytic cracking reactions as well as a few other specialty catalytic reactions using clays (especially as they relate to environmentally friendly technologies) will be the focus of this section.

A. Clays in Hydrocarbon Cracking Catalysts

The cracking (rupture of C–C bonds) of hydrocarbons can occur by either a thermal or a catalytic mechanism, but catalytic cracking requires an acidic catalyst. The first step in acidic cracking is carbocation formation, followed by fragmentation, disproportionation, and isomerization. Simple olefins are protonated in mineral acid solutions, but protonation of aromatics is more difficult and requires a strong acid. Either Lewis or Bronsted acids can protonate branched alkanes, but severe conditions are necessary to protonate *n*-alkanes.

Before the advent of zeolites, the preparation of a suitable catalyst was accomplished by the partial degradation of montmorillonite, halloysite, or kaolinite by mineral acids. This chemistry and the properties of the resultant catalysts have been reviewed by many (61,70). Acid pretreatments are done either by mineral acid washes, which exchange the interlayer cations with protons, or by heating clay-mineral acid suspensions (e.g., in 30% H_2SO_4 at 95°C) (71). Optimum conditions vary from one clay to another, depending on the chemical composition, hydration level, exchangeable cation, and reaction conditions. Some commercially available acid-treated montmorillonites that are widely used today as industrial acid catalysts or hydrocarbon cracking adsorbents are K10®, KSF® (Sud-Chemie, Fluka Chemical), F-13® and F-20® (Engelhard). Of these, K10 is most often used in published reports. Lewis acids such as inorganic salt catalysts or reagents can be added to them as well.

The current status of clay use in industrial cracking catalysts is provided in Ref. 65. This includes kaolinite, which is calcined and acid treated and acts as a partially active support in "semisynthetic" silica-alumina gel formulations. In addition, sepiolite is used as a V- and Ni-heavy metal trap, for these metals are present in oil feeds and can cause either overcracking or coking. Sepiolite has furthermore been found to act as an effective catalyst for asphaltene conversion in hydrotreating, after dispersal of metal oxides such as MoO_3 on its surface (72). As far as new directions in cracking reactions, Macedo et al. (72a) have shown that modified kaolinite (calcined, acid treated) can catalyze cumene cracking via a synergism between strong Lewis and Bronsted sites. These are attributed to penta- and tetra-coordinated Al, respectively, which results in superacidity. Pillared clays (Chapter 6) also offer novel cracking activities.

The relative degrees of importance of clay Bronsted and Lewis acid sites in catalytic cracking are discussed by Rupert et al. (63). In addition, an interesting phenomenon reported by Heller-Kellai (65) discusses the possible role of clay volatiles and condensates liberated via thermal activation in such reactions. The catalysis discussed thus far depends on direct contact between a clay surface and reacting molecules. However, Heller-Kellai argues that at elevated temperatures direct contact is not necessary. Rather, some of the volatiles released upon heating clays can act as powerful acid catalysts. Figure 8 shows the volatiles released, in addition to water vapor, from a montmorillonite heated at 390°C. Amounts and compositions will vary from clay to clay and also with respect to time and temperature. Therefore catalytic effects are expected to vary with heating conditions and type of clay. Significant cracking of n-alkanes by clay volatiles has been reported at mild conditions of just 160°C (73). According to one classification, the clay is then behaving as a superacid (74). The amount of evolved volatiles is very small and therefore the yields of catalytic reactions are very low. However, in clay-promoted reactions such as n-alkane cracking, clay volatiles are the only active components at mild temperatures. With clays that are activated upon heating, the contribution of surface catalysis dominates as the temperature is raised

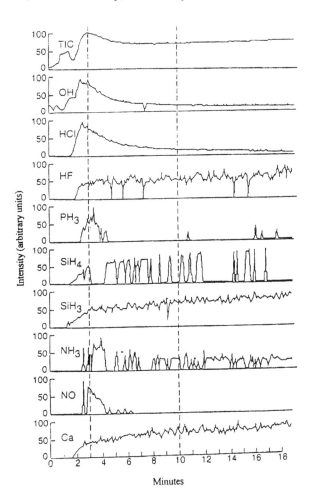

FIGURE 8 Single ion reconstruction traces (top trace = total ion count) of volatiles released from Camp Berteau montmorillonite at 390°C. Note the different volatile compositions at different times, e.g., 3 and 10 minutes (dashed lines). (Reprinted from Ref. 65.)

and obscures the minor effect of the volatiles. Interestingly, the activity of clay volatiles is preserved upon condensation for periods up to several years (65).

B. Clays as Specialty Catalysts

Many of the most useful applications of clays lie in organic/fine chemistry, because of such features as their use in catalytic amounts (unlike $AlCl_3$, for example), the ease of catalyst recovery and regeneration, simple experimental setups, mild experimental conditions, and yield and/or selectivity enhancements. These features lend clays especially attractive to more environmentally friendly technologies (66). As a demonstration of the versatility of clay minerals in organic chemistry, the types of reactions discussed in Balogh and Laszlo (67) are: electrophilic aromatic, addition, elimination, oxidation, dehydrogenation, aromatization, hydrogenation, hydrogenolysis, cyclization, Diels-Alder, isomerization, dimerization, oxidative dimerization, rearrangements, condensation, thermal and hydrolytic decomposition, reactions of carbonyls, carboxylic acids and derivatives, and amino acid and peptide formation under prebiotic conditions.

Often it is various acid-treated clays that are exploited in the foregoing reactions. For example, the relative degrees of Bronsted and Lewis acidity contributions have been studied with an acid-treated montmorillonite exchanged with Al^{3+}, Fe^{3+}, Cu^{2+}, Zn^{2+}, Ni^{2+}, Co^{2+}, and Na^+ by Brown and Rhodes (75). Two reactions were used: (a) Bronsted acid–catalyzed rearrangement of α-pinene to camphene and (b) Lewis acid–catalyzed rearrangement of camphene HCl to isobornyl chloride. Surface acidities were first measured using ammonia microcalorimetry and pyridine IR. Maximum Bronsted and Lewis acidities occurred at thermal activations of 150°C and 250–300°C, respectively. Clays and modified clays are also effective supports for inorganic salts and reagents, such as Lewis acids ($ZnCl_2$, $AlCl_3$, $FeCl_3$) and Fe(III) and Cu(II) nitrates. The acronyms Clayzic, Kaozic, and Japzic are K10-, kaolinite-, or Japanese acid clay–supported $ZnCl_2$, respectively, and Clayfen and Claycop are Fe(III) and Cu(II) nitrates on K10.

Vaccari (66) explains that acid-treatment conditions to maximize catalytic activity depend on the precise reaction of interest. Reactions between polar molecules require mildly treated clays to capitalize on the large number of acid sites available in the internal surface. In contrast, nonpolar molecules react only at the surface-accessible external face and edge sites of the clay platelets. These latter reactions require more severely treated clays, and activity depends on the interplay between surface area and cec. In a swelling solvent, maximum activity can be achieved simply by exchanging with acidic cations such as Al^{3+} without any acid treatment.

Solid acid catalysts also have been demonstrated in vapor-phase synthesis for the advantages (over liquid phase) of safety, continuous production, and simpler operation. For example, the synthesis of alkylquinolines over K10 from ethylene glycol and ethylaniline has been reported as a "green," or environmentally friendly, process (76). In another example using modified clays, Friedel-Crafts acylations are possible. Table 5 provides a comparison of exchanged clays and metal chloride–impregnated clays for the benzoylation of mesitylene. Cornelis and Laszlo cite the commercial Envirocats® catalysts as advantageous Friedel-Crafts clay catalysts for their high selectivity, yields, and conversions, their

TABLE 5 Benzoylation of Mesitylene by
Benzoyl Chloride Using Impregnated and
Exchanged Clay Catalysts at 433 K

Catalyst	Reaction time (min)	Yield (%)
Kaozic-533[a]	60	74
Japzic-533	60	100
Japzic-673	15	100
Clayzic-533	10	100
Al(III)-K10	15	98
Ti(IV)-K10	15	100
Cr(III)-K10	15	94
Fe(III)-K10	15	98
Co(II)-K10	15	100
Cu(II)-K10	15	84
Zn(II)-K10	15	85

[a] activation temperature (K).
Source: Adapted from Ref. 66.

improved reaction times, and the 2000-fold reduction in catalyst amount needed (77).

Vaccari details other reactions as well, including the loading of elements such as Pt, Pd, Ru, Rh, Ni, and Cu to create hydrogenation catalysts (where the lamellar structure appears to add size and shape selectivity), Diels-Alder reactions, phenol nitration, and triphase catalysis (66). Pillaring clays has offered many new possibilities, especially with respect to shape selectivity, as is discussed in Chapter 6.

C. NMR Studies of Clay Catalysis

The power of NMR spectroscopy is well recognized in the study of organic-supported zeolite catalytic transformations (see, for example, Chapters 6 and 16 of the first volume in this series, the *Handbook of Zeolite Science and Technology*), and especially useful are those concerning in situ organic reactions. Its use has been relatively rare in analogous clay-catalyzed organic reactions, however. One of the first examples followed the transformation of isobutene to *t*-butanol in the galleries of a synthetic Al^{3+}-hectorite (78). The Bronsted acid for this proton-catalyzed reaction is provided by the acidic nature of Al^{3+}-coordinated water. ^{13}C NMR showed two peaks for the two carbon atom types in the *t*-butanol $(CH_3)_3-COH$ product. Other workers have utilized ^{13}C NMR to follow the poly-

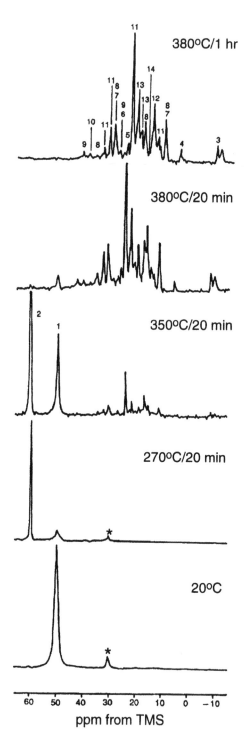

merization of styrene via a thermal-initiated radical mechanism accompanied by the Bronsted-catalyzed cationic mechanism. It was found that increasing temperature and the clay content enhanced the latter mechanism (79). Another example followed the oxygen–methyl bond cleavage of anisoles on various clays and pillared clays (montmorillonites and hectorites) by solid-state ^{13}C MAS NMR (80). These studies also demonstrated that the organics in the clay–anisole complexes were fairly mobile prior to reaction, whereas the mobility decreased markedly after reaction, as evidenced by an increase in line width and the inability of the molecules to cross-polarize. In a final example of using in situ solid-state ^{13}C NMR, the catalytic conversion of methanol on pillared clays was monitored (81). SiO_2/TiO_2 sol-pillared saponites gave a wider variety of hydrocarbon products and were more reactive than similar montmorillonites, which was attributed in part to their larger surface area. The dominant aliphatics were n-pentane, n-hexane, and n-heptane after heating at 350°C. Figure 9 shows the ^{13}C MAS NMR spectra for a saponite catalyst after various thermal treatments. The initial strong line at 50 ppm for the methyl group of MeOH was considerably broader in the montmorillonites, indicating that the motion of MeOH in these clays is more restricted than in saponites. Several hydrocarbons are found in both clays after heating at 350°C, but others are found only for saponites such as butane, isohexane, isoheptane, and neopentane. Aromatic carbons are observed for all clays, but the resonances are weak, broad, and difficult to assign. Furthermore, two types of methane resonances are observed in saponite (peak #3 at about − 10 ppm)—a narrow peak for mobile methane molecules and a broad peak for strongly adsorbed methane. In montmorillonite, only the sharp resonance is observed.

VII. FINAL COMMENT

In conclusion, a word should be said about the versatility of clay minerals in applications beyond that of catalysis. Hundreds of millions of tons find commercial applications in fields as diverse as ceramics and building materials, paper coatings and fillers, drilling muds, foundry molds, pharmaceuticals, skin care

---

FIGURE 9 Aliphatic ^{13}C MAS NMR region with proton decoupling of air-dried, SiO_2/TiO_2-sol pillared saponite after adsorption of methanol and various thermal treatments. Asterisks denote spinning side bands. The peak assignments correspond to: 1 = MeOH, 2 = DME, 3 = methane, 4 = ethane, 5 = iso- and n-butane, 6 = neopentane, 7 = isopentane, 8 = 3-methylpentane, 9 = isohexane, 10 = isoheptane, 11 = n-pentane, n-hexane, n-heptane, 12, 13 = aromatics (methylbenzenes, toluene, xylenes), 14 = hexamethylbenzene. Some CO is observed at 350°C and above (81). (Reprinted in part with permission from Ref. 81. Copyright 1995 American Chemical Society.)

products, kitty litter, and laundry detergents. In terms of catalytic, adsorbent, and ion-exchange applications, environmental issues continually press for more and more environmentally friendly solid acid catalysts (over liquid acids, for example), and future work in optimization of such materials can be expected to continue well into a bright future.

ACKNOWLEDGMENTS

The writing of this chapter was sponsored by the U.S. Dept. of Energy, Office of Basic Energy Sciences, Division of Chemical Sciences, Geosciences, and Biosciences, under contract #W-31-109-ENG-38.

REFERENCES

1. S Guggenheim, RT Martin. Clays Clay Miner 1995; 43:255–256.
2a. RE Grim. Clay Mineralogy. 2nd ed. New York: McGraw-Hill, 1968.
2b. RE Grim. Applied Clay Mineralogy. New York: McGraw-Hill, 1962.
3. CE Weaver, LD Pollard. The Chemistry of Clay Minerals. New York: Elsevier Scientific, 1973.
4. RM Barrer. Zeolites and Clay Minerals as Sorbents and Molecular Sieves. New York: Academic Press, 1978:407–486.
5. GW Brindley, G Brown, eds. Crystal Structures of Clay Minerals and Their X-Ray Identification. London: Mineralogical Society, 1980.
6. ACD Newman, ed. Chemistry of Clays and Clay Minerals. New York: Wiley Interscience, 1987.
7. SW Bailey, ed. Reviews in Mineralogy. Hydrous Phyllosilicates (exclusive of micas). Vol. 19. Chelsea. MI: Mineralogical Society of America, 1988.
8. B Velde. Introduction to Clay Minerals. New York: Chapman & Hall, 1992.
9. DM Moore, RC Reynolds. X-Ray Diffraction and the Identification and Analysis of Clay Minerals. New York: Oxford University Press, 1997.
10. S Yariv, H Cross, eds. Organo-Clay Complexes and Interactions. New York: Marcel Dekker, 2002.
11. S Yariv, KH Michaelian. In: S Yariv, H Cross, eds. Organo-Clay Complexes and Interactions. New York: Marcel Dekker, 2002:1–38.
12. ACD Newman, G Brown. In: ACD Newman, ed. Chemistry of Clays and Clay Minerals. New York: Wiley-Interscience, 1987:1–128.
13. SW Bailey. In: GW Brindley, G Brown, eds. Crystal Structures of Clay Minerals and Their X-Ray Identification. London: Mineralogical Society, 1980:1–124.
14. WD Keller. In: AG Cairns-Smith, H Hartman, eds. Clay Minerals and the Origin of Life. Cambridge: Great Britain: University Press, 1986:14–15.
15. SW Bailey, GW Brindley, DS Fanning, H Kodama, RT Martin. Clays Clay Miner 1984; 32:239–240.
16. B Ransom, HC Helgeson. Clays Clay Miner 1989; 37:189–191.

17. M Rieder, G Cavazzini, YS D'yakonov, VA Frank-Kamenetskii, G Gottardi, S Guggenheim, PV Koval, G Muller, AMR Neiva, EW Radoslovich, J-L Robert, FP Sassi, H Takeda, Z Weiss, DR Wones. Clays Clay Miner 1998; 46:586–595.
18. A Mottana, FP Sassi, JB Thompson, S Guggenheim. Micas: Crystal Chemistry and Metamorphic Petrology. Washington DC: Mineralogical Society of America, 2002.
19. A Langier-Kuzniarowa. In:. S Yariv, H Cross, eds. Organo-Clay Complexes and Interactions. New York: Marcel Dekker, 2002:273–344.
20a. H Egawa, K Furusawa. Langmuir 1999; 15:1660–1666.
20b. AG Liu, RC Wu, E Eschenazi, K Papadopoulos. Coll Surf A 2000; 174:245–252.
21. H Lindgreen, J Garnes, PL Hansen, F Besenbacher, E Laegsgaard, I Stensgaard, SAC Gould, PK Hansma. Amer Miner 1991; 76:1218–1222.
22. H Hartman, G Sposito, A Yang, S Manne, SAC Gould, PK Hansma. Clays Clay Miner 1990; 38:337–342.
23. FJ Wicks, K Kjoller, RK Eby, FC Hawthorne, GS Henderson, GA Vrdoljak. Can Miner 1993; 31:541–550.
24a. GS Henderson, GA Vrdoljak, RK Eby, FJ Wicks, AL Rachlin. Coll Surf A 1994; 87:197–212.
24b. GA Vrdoljak, GS Henderson, JJ Fawcett, FJ Wicks. Amer Miner 1994; 79:107–112.
24c. GA Vrdoljak, GS Henderson. Coll Surf A 1994; 87:187–196.
25a. ML Occelli, SAC Gould. J Catal 2001; 198:41–46.
25b. ML Occelli, SAC Gould, JM Tsai, B Drake. J Molec Catal A 1995; 100:161–166.
25c. ML Occelli, SAC Gould, B Drake. Micropor Mater 1994; 2:205–215.
25d. ML Occelli, B Drake, SAC Gould. J Catal 1993; 142:337–348.
26a. JC Schulz, GG Warr. Langmuir 2000; 16:2995–2996.
26b. HN Patrick, GG Warr, S Manne, IA Askay. Langmuir 1999; 15:1685–1692.
26c. TL Porter, MP Eastman, ME Hagerman, JL Attuso, ED Bain. J Vac Sci Technol A 1996; 14:1488–1493.
26d. X Xiao, J Hu, DH Charych, M Salmeron. Langmuir 1996; 12:235–237.
26e. S Nishimura, S Biggs, PJ Scales, TW Healy, K Tsunematsu, T Tateyama. Lamgmuir 1994; 10:4554–4559.
27. H Lindgreen. Clay Miner 2000; 35:643–652.
28. FJ Berry, MHB Hayes, SL Jones. Inorg Chim Acta 1990; 178:203–208.
29. VC Farmer. In: H VanOlphen, JJ Fripiat, eds. Data Handbook for Clay Materials and other Non-Metallic Minerals. New York: Pergamon Press, 1979:285–338.
30. JJ Fripiat. In: JJ Fripiat, ed. Advanced Techniques for Clay Mineral Analysis. New York: Elsevier, 1982:191–210.
31. HW van der Marel, H. Beutelspacher, eds. Atlas of Infrared Spectroscopy and Clay Minerals and Their Admixtures. New York: Elsevier, 1976.
32. JD Russell, AR Fraser. In: MJ Wilson, ed. Clay Mineralogy: Spectroscopic and Chemical Determinative Methods. New York: Chapman & Hall, 1994:11–67.
33. S Yariv. In: S Yariv, H Cross, eds. Organo-Clay Complexes and Interactions. New York: Marcel Dekker, 2002:345–462.
34a. MR Weir, WX Kuang, GA Facey, C Detellier. Clays Clay Miner 2002; 50:240–247.
34b. SK Lausen, H Lindgreen, HJ Jakobsen, NC Nielsen. Amer Miner 1999; 84:1433–1438.

34c. J Sanz, JL Robert. Phys Chem Minerals 1992; 19:39–45.

34d. HD Morris, S bank, PD Ellis. J Phys Chem 1990; 94:3121–3129.

34e. S Komarneni, CA Fyfe, GJ Kennedy, H Strobl. J Am Ceram Soc 1986; 69:C45–47.

35a. KA Carrado, R Csencsits, P Thiyagarajan, S Seifert, SM Macha, JS Harwood. J Mater Chem 2002; 12:3228–3237.

35b. KA Carrado, L Xu, DM Gregory, K Song, S Seifert, RE Botto. Chem Mater 2000; 12:3052–3059.

36. J Grandjean, JL Robert. J Magn Reson 1999; 138:43–47.

37a. DJ Sullivan, JS Shore, JA Rice. Amer Miner 2000; 85:1022–1029.

37b. V Laperche, JF Lambert, R Prost, JJ Fripiat. J Phys Chem 1990; 94:8821–8831.

37c. D Tinet, AM Faugere, R Prost. J Phys Chem 95:8804–8807.

38. JT Kloprogge, JBH Jansen, RD Schuiling, JW Geus. Clays Clay Miner 1992; 40: 561–566.

39. CA Weiss, RJ Kirtkpatrick, SP Altaner. Amer Miner 1990; 75:970–982.

40a. Y Nakashima. Amer Miner 2001; 86:132–138.

40b. Y Nakashima. Clays Clay Miner 2000; 48:603–609.

40c. C Gevers, J Grandjean. J Coll Interf Sci 2001; 236:290–294.

40d. J Grandjean. J Coll Interf Sci 2001; 239:27–32.

40e. M Letellier. Magn Reson Imag 1998; 16:505–510.

40f. J Grandjean, RL Robert. J Coll Interf Sci 1997; 187:267–273.

41a. LQ Wang, J Liu, GJ Exarhos, KY Flanigen, R Bordia. J Phys Chem B 2000; 104: 2810–2816.

41b. S Hayashi, E Akiba. Chem Phys Lett 1994; 226:495–500.

42. J Sanz, JM Serratosa. In: S Yariv, H Cross, eds. Organo-Clay Complexes and Interactions. New York: Marcel Dekker, 2002:223–272.

43. BA Goodman, JA Chudek. In: MJ Wilson, ed. Clay Mineralogy: Spectroscopic and Chemical Determinative Methods. New York: Chapman & Hall, 1994:121–172.

44. H van Olphen, JJ Fripiat, eds. Data Handbook for Clay Minerals and Other Nonmetallic Minerals. New York: Pergamon Press, 1979:203–216.

45. DW Rutherford, CT Chiou, DD Eberl. Clays Clay Miner 1997; 45:534–543.

46. ACD Newman. In: ACD Newman, ed. Chemistry of Clays and Clay Minerals. New York: Wiley Interscience, 1987:237–274.

47. JF Lee, CK Lee, LC Juang. J Coll Interf Sci 1999; 217:172–176.

48. KA Carrado, G Sandi, H Joachin. Unpublished data.

49. KSW Sing, DH Everett, RAW Haul, L Moscou, RA Pierotti, J Rouquerol, T Siemieniewska. Pure Appl Chem 1985; 57:603–619.

50. DW Rutherford, CT Chiou, DD Eberl. Clays Clay Mins 1997; 45:534–543.

51. NA Seaton. Chem Eng Sci 1991; 46:1895–1909.

52. AJ Dandy, MS Nadiye-Tabbiruka. Clays Clay Miner 1982; 30:347–352.

53. N Tosi-Pellenq, Y Grillet, J Rouquerol, P Llewellyn. Thermochim Acta 1992; 204: 79–88.

54. J Rouquerol, D Avnir, DH Everett, C Fairbridge, M Haynes, N. Pernicone, JDF Ramsay, KSW Sing, KK Unger. In:. J Rouquerol, F Rodriquez-Reinoso, KSW Sing, KK Unger, eds. Characterization of Porous Solids III. New York: Elsevier, 1994: 1–9.

55. DC Bain, BFL Smith. In: MJ Wilson, ed. Clay Mineralogy: Spectroscopic and Chemical Determinative Methods. New York: Chapman & Hall, 1994:300–.
56. H Laudelout. In: ACD Newman, ed. Chemistry of Clays and Clay Minerals. New York: Wiley-Interscience, 1987:225–236.
57. R Touillaux, P Salvador, C Vandermeersche, JJ Fripiat. Isr J Chem 1968; 6:337–348.
58. S Yariv. In: ME Schrader, G Loeb, eds. Modern Approaches to Wettability. New York: Plenum Press, 1992:279–326.
59. C Walling. J Am Chem Soc 1950; 72:1164–1168.
60. HA Benesi. J Am Chem Soc 1956; 78:5490–5494.
61. JE Mapes, RP Eischens. J Phys Chem 1954; 58:1059–1062.
62. AC Wright, WT Granquist, JV Kennedy. J Catal 1972; 25:65–80.
63. JP Rupert, WT Granquist, TJ Pinnavaia. In: ACD Newman, ed. Chemistry of Clays and Clay Minerals. New York: Wiley Interscience, 1987:275–318.
64. WP Hettinger. Appl Clay Sci 1991; 5:445–468.
65. L Heller-Kallai. In: S Yariv, H Cross, eds. Organo-Clay Complexes and Interactions. New York: Marcel Dekker, 2002:567–613.
66. A Vaccari. Appl Clay Sci 1999; 14:161–198.
67. M Balogh, P. Laszlo. Organic Chemistry Using Clays. Berlin: Springer-Verlag, 1993.
68. S. Yariv. In: S Yariv, H Cross, eds. Organo-Clay Complexes and Interactions. New York: Marcel Dekker, 2002:39–111.
69. RF Giese, CJ van Oss. In: S Yariv, H Cross, eds. Organo-Clay Complexes and Interactions. New York: Marcel Dekker, 2002:175–192.
70a. See, for example: LB Ryland, MW Tamele, JN Wilson. In: PH Emmett, ed. Catalysis. New York: Reinhold, 1960:1–93.
70b. TH Milliken, AG Obland, GA Mills. Clays Clay Miner 1955; 1:314–326.
71. CN Rhodes, DR Brown. Catal Lett 1994; 24:285–291.
72. S Inoue, T Takatsuka, Y Wada, S Nakaata, T Ono. Catal Today 1998; 43:225–232.
72a. JCD Macedo, CJA Mota, SMC de Menezes, V Camorim. Appl Clay Sci 1994; 8: 321–330.
73. L Heller-Kellai, I Miloslavski, Z Aizenshtat. Clays Clay Miner 1989; 37:446–450.
74. LD Field. In: JA Davies, ed. Selective Hydrocarbon Activation. New York: VCH, 1990:241–264.
75. DR Brown, CN Rhodes. Catal Lett 1997; 45:35–40.
76. M Campanati, P Savini, A Tagliani, A Vaccari, O Piccolo. Catal Lett 1997; 47: 247–250.
77. A Cornelis, P Laszlo. Synlett 1994; 3:155–161.
78a. DTB Tennakoon, R Schloegl, T Rayment, J Klinowski, W Jones, JM Thomas. Clay Miner 1983; 18:357–371.
78b. DT Tennakoon, W Jones, JM Thomas, T Rayment, J Klinowski. Mol Cryst Liq Cryst 1983; 93:147–155.
79. D Njopwouo, G Roques, R Wandji. Clay Miner 1988; 23:35–43.
80. KA Carrado, R Hayatsu, RE Botto, RE Winans. Clays Clay Miner 1990; 38:250–256.
81. H He, L Zhang, J Klinowski, ML Occelli. J Phys Chem 1995; 99:6980–6985.

2

Molecular Modeling of Clay Mineral Structure and Surface Chemistry

Sung-Ho Park and Garrison Sposito
Lawrence Berkeley National Laboratory
Berkeley, California, U.S.A.

I. INTRODUCTION

Various methods in computational chemistry have been applied to study clay minerals since the early 1970s. The molecular modeling of clay mineral structure and surface chemistry that we will discuss here is mainly about computational chemistry applied to the clay mineral alone or in contact with a specific aqueous solution.

Because of their ubiquitous presence in natural materials and their strong surface reactions with cations and organic molecules, clay minerals are involved in many environmentally important phenomena (1). A nanoscale particle size with large specific surface area (ca. 750 m^2g^{-1}) is typical for clay minerals. Therefore, cation exchange and swelling processes occur readily, governed by the electrical double layer in the interlayers of the hydrated clay mineral.

To provide molecular-scale detail about these surface reactions, four types of computer simulation for layer-type silicates (clay minerals) are utilized: quantum mechanics (QM), molecular mechanics (MM), molecular dynamics (MD), and Monte Carlo (MC) (Fig. 1). Molecular modeling of a clay mineral and its hydrates is a nontrivial task because of the relatively large cell size required for simulation (up to hundreds of atoms for a solid clay mineral surface, plus layers of water molecules, which also require up to hundreds of atoms) and complex

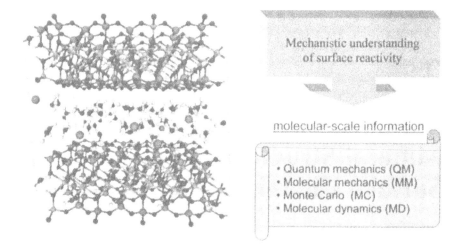

FIGURE 1   Computational surface geochemistry of clay minerals.

structure (structural features are discussed in the next section). However, recent major developments in both computational codes and computing power using the parallel architecture of supercomputers allows one to model many-atom systems with much more detail. It is now starting to be possible to perform QM as well as MM at the state-of-the-art level of atomistic molecular simulation.

Clay minerals are isostructural with the mica, but with a random pattern of isomorphic substitutions within their tetrahedral or octahedral sheets, depending on the type of clay. We shall discuss the structural aspects of clay mineral hydrates in the first section of this chapter. Then classical simulations with Monte Carlo (MC) and molecular dynamics (MD) methods will be discussed, together with their application to clay mineral structure, surfacial water structure, adsorption of ions, adsorption of organic and inorganic compounds, and clay nanocomposites.

Quantum mechanical calculations on clay minerals are essential in computational geochemistry today. They can serve to provide the interparticle potential for the clay–water–adsorbate system for MC and/or MD simulations. They can also determine the clay structure without experimental parameters. To this end, the recent development in first-principles calculations using density functional theory (DFT) will be discussed and compared with semiempirical ab initio and Hartree–Fock methods. Recent success in determining 2:1 clay mineral equilibrium structures (both neutral and charged clays) will be discussed and compared with available experimental data. The most challenging and desired simulations of the clay–water interface at an ab initio level will be discussed in terms of both energy minimization and future first-principles molecular dynamics. The quantum mechanical approach will be the most important field of computational geochemistry to understand the molecular mechanisms by which clay minerals react with adsorptives (cation, organic and inorganic compounds, and water) in aqueous environments. First-principles molecular dynamics simulations using density functional theory can provide valuable insight into clay minerals electronic properties and structure as well as behavior at surfaces and interfaces.

Recent synergetic studies between computational simulations and spectroscopic techniques will be also discussed. Experimental techniques (FTIR spectroscopy, NMR, neutron scattering, neutron diffraction, and X-ray absorption spectroscopy) will also be discussed. Computer simulations often are able to provide an excellent interpretation of what is obtained from spectroscopy. Examples of this interplay between experiment and simulation are listed. In presenting the results of computer simulations, visualization of the molecular structure or electronic structure and animations of the MD trajectories are often used. Some of these also are discussed, in terms of technical aspects and software available for visualization. One of the most up-to-date visualization techniques, virtual reality (VR), is also discussed in the context of computational chemistry.

If computational power permits, first-principles molecular modeling such as ab initio MD can provide the most accurate information without adjustable parameters. However, even the most powerful computer system available today is limited to first-principles solution of molecular information with a certain size. As a result, the hybrid methods such as QM/MM or ONIOM* (where different levels of theory for one calculation are involved) are often used (Fig. 1). As an example of future prospects, molecular design and characterization of nanocomposites using computer simulation are also briefly mentioned. Atomistic modeling of clay nanocomposites is a very promising field in clay mineral materials science.

II. STRUCTURAL FEATURES OF CLAY MINERAL HYDRATES

In order to model clay mineral structure and understand its surface chemistry, it is important to understand some of the basic features of clay mineral structures. We are dealing with the simulation of a system where solid clay mineral structure is interfaced with an aqueous system. Therefore it is essential to understand the structure of the clay mineral itself and its hydrates before any computer modeling effort.

---

* ONIOM (our own *n*-layered integrated molecular orbital and molecular mechanics) approach is discussed in section VII of this review.

A. Structure of Clay Minerals

The scientific importance of clay minerals is due to their large surface area and their characteristic structure. This layered structure is a key feature used to distinguish among different types of clay minerals: (a) The 1:1 type, in which one tetrahedral SiO_4 layer shares corners with an octahedral sheet of, e.g., $AlO_2(OH)_4$. The thickness of this two-sheet unit is about 7 Å. (b) The 2:1 type, consisting of two sheets of SiO_4-tetrahedra, all tetrahedra sharing corners with each other and with an octahedral sheet of e.g., $AlO_4(OH)_2$, situated in between. The thickness of this three-sheet unit, to be referred to as *platelet*, is about 10 Å. In both types,

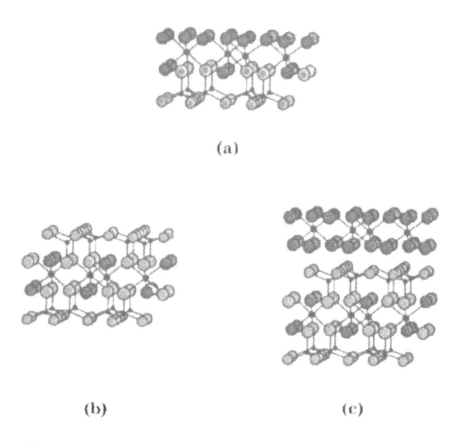

(a)

(b)   (c)

FIGURE 2 Clay layer types (a) 1:1 layer type, e.g., kaolinite; (b) 2:1 layer type, e.g., montmorillonite; (c) 2:1 layer with hydroxide layer, e.g., chlorite Dark shaded spheres are the hydroxides and light shaded spheres are oxygens. The small dots indicate silicon ions (Si^{4+}) and the large dots indicate metal ions (aluminum: Al^{3+}, Mg^{2+}, or Fe^{2+}).

aluminum ions in the octahedral sheet may be substituted by magnesium ions. This layered structure explains why the clay minerals occur in plate-shaped two-dimensional crystals. One way to visualize this extremely thin clay layer is to imagine the area-to-mass ratio of these plates as hundreds of square meters per gram (2). (c) The 2:1 layer with hydroxide interlayer.

With respect to the different kinds of isomorphic cation substitution, clay minerals are categorized into five subgroups. The layer types are shown in Figure 2, and the five groups are shown in Table 1. The main types of 2:1 layers are discussed in Chapter 1 of this book. Here, three soil clay mineral groups with this structure in the 2:1 layer type are listed: illite, vermiculite, and smectite. Among the smectites, those in which the substitution of Al for Si exceeds that of Fe(II) or Mg for Al are called *beidellite*, and those in which the reverse is true are called *montmorillonite* (Fig. 3). The sample chemical formula in Table 1 for smectite represents montmorillonite. In any of these 2:1 minerals, the layer charge is balanced by cations that reside on or in the cavities of the basal plane of the oxygen atoms of the tetrahedral sheet. These interlayer cations are represented by M in the chemical formula (Table 1). The 2:1 layer type with hydroxide interlayer is represented by dioctahedral chlorite in soil clays (3).

B. Clay Mineral Hydrates

In clay minerals, Si, Al, Fe, Mn, and some organic macromolecules are responsible for most wettability properties. Wettability of clay minerals is a crucial prob-

TABLE 1 Clay Mineral Groups

Layer structure	Clay group	Layer charge[a] $x-1.0$ (e)	Basal Spacing[b] (Å)	CEC[c] mEq/ 100 g	Chemical formula[a]
1:1	Kaolins (kaolinite)	<0.01	7.14	1–10	$[Si_4]Al_4O_{10}(OH)_8 \cdot nH_2O$ ($n = 0$ or 4)
2:1	Smectite (montmorillonite)	0.5–1.2	12.4–17	80–120	$M_x[Si_8]Al_{3.2}Fe_{0.2}Mg_{0.6}O_{20}(OH)_4$
	Vermiculites	1.2–1.8	9.3–14	120–150	$M_x[Si_7Al]Al_3Fe_{0.5}Mg_{0.5}O_{20}(OH)_4$
	Illite (hydrous micas)	1.4–2.0	10.0	~30	$M_x[Si_{6.8}Al_{1.2}]Al_3Fe_{0.25}Mg_{0.75}O_{20}(OH)_4$
2:1 with hydroxide	Chlorite	Variable	14.0	10–40	$(Al(OH)_{2.55})_4[Si_{6.8}Al_{1.2}]Al_{3.4}Mg_{0.6}O_{20}(OH)_4$

[a] Ref. 3.
[b] Ref. 4.
[c] Cation exchange capacity.
Source: Ref. 5.

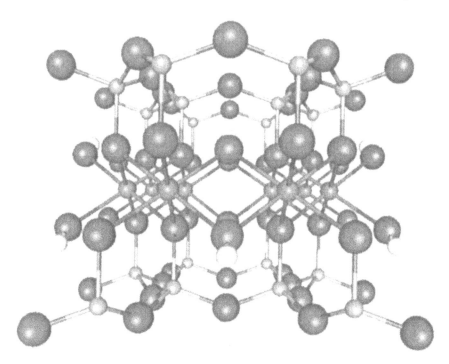

FIGURE 3 Perspective visualization of montmorillonite (2:1 smectite). Tetrahedral SiO_4 network, octahedral network composed of AlO or Al(OH), and the orientation of structural hydroxyl groups are visualized well in this perspective view. The hexagonal cavity in the surface as well as the cavity in the octahedral sheet of this 2:1 clay mineral are also well shown.

lem, faced in various industrial technologies, such as metallurgy (molding sands), drilling fluids, painting, oil refining and decolorization, water clarification, catalysis, and production of ceramics, papers, adhesives, plastics, rubber, medicines, cosmetics, etc. (6). For montmorillonite clay, which is a 2:1 TOT (tetrahedra-octahedra-tetrahedra) type of clay, crystalline clay swelling was attributed to the interlayer water content. This interlayer space of a TOT clay mineral is the space between two parallel silicate layers, sandwiched by two oxygen planes of siloxane groups. The clay swelling in the interlayer space is the result of (a) thermal motion of water molecules in the environment of the mineral, (b) electrostatic attraction forces between water molecules and the exchangeable cationic species, and (c) attraction and dispersion forces between the TOT layers (6). The geometric organization of different types of smectite was related to clay swelling using TEM analysis by Hetzel et al. (7). Significant uptake of water content causes expansion of the clay crystal along the c axis (which defines the direction between two parallel silicate layers of clay), and this expansion can be monitored by the use of adsorption isotherms. Interlayer water swelling is measured via the X-ray diffraction (XRD) technique in terms of the c-spacing (8).

C. Difficulties in Finding the Structure of Mineral Hydrates from Experiments, and the Needs for Molecular Modeling

Difficulties in the structural study of clay mineral hydrates are twofold. First is the difficulty associated with the clay mineral structure itself; the other has to do with difficulties in locating water molecules near the clay mineral surface (Fig. 4). For example the clay mineral montmorillonite is not suitable for the usual XRD method of structural analysis because it is not possible to prepare a single-crystal sample of this 2:1 clay mineral. It is, therefore, often compared with pyrophyllite because of structural similarities. The problem of locating molecules is not an easy one to resolve. One cannot obtain positional information about interlayer water using XRD. Neutron scattering and neutron diffraction are powerful techniques to provide detailed structural information on an aqueous system near a clay mineral surface. Recently, isotopic substituted neutron diffraction techniques were used to find the distribution or organization of clay interlayer water (9,10). However, this method also has a limitation, because the structural information obtained from this method is statistically averaged and, therefore, a nonaveraged picture of the water structure at molecular level is missing. Depend-

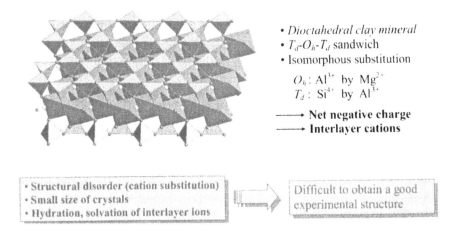

- *Dioctahedral clay mineral*
- T_d-O_h-T_d sandwich
- Isomorphous substitution

    O_h: Al^{3+} by Mg^{2+}
    T_d: Si^{4+} by Al^{3+}

——→ **Net negative charge**
——→ **Interlayer cations**

- Structural disorder (cation substitution)
- Small size of crystals
- Hydration, solvation of interlayer ions

Difficult to obtain a good experimental structure

FIGURE 4   Why simulate montmorillonite?

ing on the type of modeling technique, we are able to obtain a detailed molecular structure of both clay mineral and the hydration water at the clay–water interface.

III. MOLECULAR MODELING METHODS

We shall look into four types of molecular modeling method used to describe clay mineral structure and surface chemistry: quantum mechanics (QM), molecular mechanics (MM), molecular dynamics (MD), and Monte Carlo (MC) methods. If you are more interested in clay mineral applications, you can read section IV before reading this section.

In quantum mechanics, we shall discuss the Hartree–Fock (HF) method, the semiempirical method, and, most importantly, the density functional method, which is used most often for large condensed-matter systems and can be easily applied to clay minerals. Then we shall discuss various types of classical methods based on molecular mechanics or statistical mechanics, such as molecular mechanics, molecular dynamics, and Monte Carlo simulations. Among these methods, Monte Carlo and molecular dynamics have been used many times to study clay mineral and its hydrates. More detail on these methods will be discussed in the next section.

A. Quantum Mechanics

Quantum mechanical calculations view a molecule as a collection of point nuclei and electrons with fixed masses and charges. The energy terms include the kinetic energy of each particle and the coulombic energies between the particles (repulsion between nuclei, attraction between the nucleus and an electron, and repulsion between electrons). Here we will review some basic equations of quantum mechanics to understand quantum mechanical methods, such as Hartree–Fock, semiempirical, and density functional theory methods, that have been most widely used for the clay mineral modeling where simulation size is greater than the molecular cluster with several atoms.

1. Hartree–Fock (HF) Method

The Hartree–Fock method is a molecular orbital theory method that is central to chemistry (11). The detailed theory of the classical Hartree–Fock type of molecular orbital calculation may be found in reviews (12,13) and books (11,14,15). Most of the equations and theoretical background are taken and adapted from Szabo and Ostlund (11).

The time-independent Schrödinger equation used in quantum mechanical calculations is

$$H\Psi(r, R) = E\Psi(r, R) \tag{1}$$

where E is the total energy, Ψ is the molecular wave function, from which all chemical properties can be calculated, r and R are position vectors for electrons and nuclei (see Fig. 5), and H is the molecular Hamiltonian:

$$H(r,R) = -\sum_{i}^{n}\frac{1}{2}\nabla_i^2 - \sum_{\alpha}^{A}\frac{1}{2M_\alpha}\nabla_\alpha^2 - \sum_{i}^{n}\sum_{\alpha}^{A}\frac{Z_\alpha}{|r_i - R_\alpha|}$$
$$+ \sum_{i}^{n}\sum_{j>i}^{n}\frac{1}{|r_i - r_j|} + \sum_{\alpha}^{A}\sum_{\beta>\alpha}^{A}\frac{Z_\alpha Z_\beta}{|R_\alpha - R_\beta|} \qquad (2)$$

In Eq. (2.2), M_α is the mass ratio of nucleus A to an electron, Z_α is the atomic number of nucleus α, and ∇_i^2 and ∇_α^2 are the Laplacian operator applied to the ith electron and the αth nucleus in Cartesian coordinates.

$$\nabla_i^2 = \frac{\partial^2}{\partial x_i^2} + \frac{\partial^2}{\partial y_i^2} + \frac{\partial^2}{\partial z_i^2}; \quad \nabla_\alpha^2 = \frac{\partial^2}{\partial x_\alpha^2} + \frac{\partial^2}{\partial y_\alpha^2} + \frac{\partial^2}{\partial z_\alpha^2} \qquad (3)$$

The first term in Eq. (2) is for the electronic kinetic energy; the second term is for the nuclear kinetic energy; the third term represents the coulombic interaction

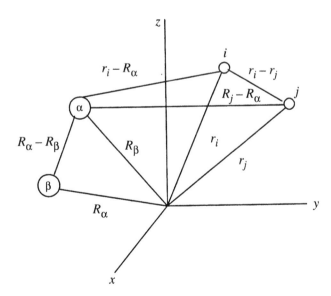

FIGURE 5 A molecular coordinate system: α and β are nuclei, while i and j are electrons.

between electrons and nuclei; the fourth and fifth terms are the operators for the electronic repulsion and nuclear repulsion, respectively.

The Schrödinger equation, shown in Eq. (3), is further simplified by the Born–Oppenheimer approximation (14), by which electrons are considered to be moving in the field of fixed nuclei because nuclei are much heavier than electrons. Therefore the second term (nuclear kinetic energy term) in Eq. (3) can be neglected for the fixed position of the nuclei in this approximation. The fifth term (nuclear repulsion) is constant and has no effect on the operator eigenfunctions. If we don't include this nonelectron term, the remaining Hamiltonian has only electronic terms describing the motion of n electrons in the field of A point charges.

$$H_{\text{elec}}(r, R) = -\sum_i^n \frac{1}{2}\nabla_i^2 - \sum_i^n \sum_\alpha^A \frac{Z_\alpha}{|r_i - R_\alpha|} + \sum_i^n \sum_{j>i}^n \frac{1}{|r_i - r_j|} \qquad (4)$$

The Schrödinger equation with electronic Hamiltonian is

$$H_{\text{elec}}\,\Psi_{\text{elec}} = E_{\text{elec}}\,\Psi_{\text{elec}} \qquad (5)$$

Instead of finding the exact solution of Ψ_{elec} by solving Eq. (5) with functions that depend explicitly on the positions of all electrons (this is not practical to use systems with more than a few electrons), we use one-electron functions $\Psi_i(r_i)$, which only depend on the position of one of the electrons.

The electronic Schrödinger equation with one electron wave function is

$$F(i)\,\Psi(r_i) = E\,\Psi(r_i) \qquad (6)$$

where $F(i)$ is called the Fock operator, which is an effective one-electron operator:

$$F(i) = -\frac{1}{2}\nabla_i^2 - \sum_\alpha^A \frac{Z_\alpha}{|r_i - R_\alpha|} + v^{HF}(r_i) \qquad (7)$$

where v^{HF} is called the Hartree–Fock potential, which is the "mean field" seen by the ith electron. The Hartree–Fock equation shown in Eq. (6) is nonlinear (the mean-field potential will change as the coefficient of all other electrons change). We can solve the Hartree–Fock equation iteratively by the procedure called the *self-consistent field* (SCF) method. This SCF procedure to obtain the solution of the trial wavefunction given by the Slater determinant is often explained by application of the variational principle. The variational principle states that the expected value of the total energy [Eq. (8)] we obtain with our trial wavefunction will always be greater than the true total energy:

$$\langle E \rangle = \frac{\int \Psi^* H \Psi\, dr}{\int \Psi^* \Psi\, dr} \qquad (8)$$

Therefore, we can obtain the best approximate wavefunction by minimizing our total energy with respect to our trial wavefunction.

The HF method is a mean-field method where all the electron–electron interactions are approximated with the Hartree–Fock potential. Therefore, the instantaneous electron–electron interaction is missing in this method. Ab initio calculations, including electron–electron correlation such as the MP2[*] method, can improve this problem. Here we will not discuss on MP2 or any other electron–electron correlation method in this review. The electron–electron correlation methods are not suitable for clay mineral surface modeling mainly due to their computational expense (computational time is proportional to M^5 for MP2, where M is the molecular size) (15).

2. Basis Set

Ab initio quantum mechanics is the most accurate calculation among many molecular orbital calculations. The accuracy depends on the type of the basis set used. Here we will discuss some of the most frequently used basis sets. The terminology and descriptions are taken mostly from Hehr et al. (16).

The contracted Gaussian type of orbital uses the linear combination of Gaussian functions to replace the Slater[†] function. STO-3G is the basis set where three Gaussian functions are used. Basis sets are usually described by the number of Gaussians per contracted Gaussian-type orbital (CGTO), and the number of CGTOs for each atomic orbital in an occupied shell (i.e., $1s$, $2s$, $2p_x$, $2p_y$, $2p_z$ for all second-row atoms). A "minimal" basis set contains only one basis function per atomic orbital (AO). A double-zeta basis set has two CGTOs per AO. The Pople family of basis sets includes the set of STO-nG (n is the number of Gaussian functions) minimal basis sets and a large number of split-valence basis sets that have one basis function for the core orbitals and two or more CGTOs for the valence orbitals. The standard Pople nomenclature is L-$M_1M_2M_3$G, where L is the number of primitive functions in the single core basis function, the number of M_i indicate how many separate basis functions describe the valence region, and the value of M_i indicates the number of primitive functions in that basis function. For example, in the 6-311G basis set for oxygen, the core $1s$ orbital is described by the basis function consisting of six primitive Gaussians. There are three of each of the valence $2s$ and $2p$ orbitals, the first containing three primitives and the next two consisting of only one Gaussian function each. The polarization of functions on the nonhydrogen atoms is indicated by adding a single "*" to

---

* Second-order Møller–Plesset (or many-body) perturbation theory (C Møller, MS Plesset. Phys Rev 46:618, 1934).
† Slater-type atomic orbitals (STOs) have exponential radial parts, and their convergence is not as good as Gaussian functions (JC Slater. Phys Rev 36:57, 1930).

TABLE 2　Commonly Used Pople Basis Sets and the Number of Functions for the Second Row Atoms (Li-Ne)

Basis set	Number of functions	Basis set	Number of functions	Basis set	Number of functions
STO-3G	5	6-31G	9	6-311G	13
3-21G	9	6-31G*	15	6-311G*	18
4-31G	9	6-31+G*	19	6-311+G*	22

The 6-31G* basis set usually includes 6d-type polarization functions; others include 5d functions.

the description or "**" if the hydrogens have p-type basis functions. Similarly the presence of diffuse functions is indicated by a "+" or "++" before the "G" (e.g., 6–31 + +G**). Commonly used Pople basis sets and the number of functions for the second row atoms (Li–Ne) are shown in Table 2.

3.　Semiempirical Methods

The need for faster calculations for large molecules gave birth to the semiempirical quantum mechanical methods, where only the valence electrons are taken into account. The decrease of the number of two-electron integrals shows a big improvement in terms of convergence speed. The inner electrons are considered to be less important for chemical phenomena, and they are therefore parameterized empirically. The amount of neglect of the diatomic differential overlap integrals is the major discrimination between most semiempirical methods. The complete neglect of differential overlap (CNDO) and intermediate neglect of differential overlap (INDO) methods started an era of development of approximate molecular orbital methods (17). The modified neglect of diatomic overlap (MNDO) method (18), AM1 (19), PM3 (20,21), and SAM (22) are all semiempirical molecular orbital calculation methods. Detailed theory of semiempirical methods and their practical use in chemical structure and energy calculations is reviewed by Clark (23).

4.　Density Functional Theory (DFT) Methods

The traditional method to study the electronic properties of molecules is based on describing the motion of individual electrons. The 1998 Nobel Chemistry Laureate, Walter Kohn, showed that it is not necessary to consider the motion of each individual electron. In other words, the system is characterized by its electron density rather than by a wave function. Due to its simplicity, we can study a very-large-molecule problem, such as the mechanism of enzymatic reaction (24). Density functional theory is currently the theoretical basis of most computer simulations of solids (25).

The basic theory and applications of DFT have been explored in textbooks (26–29) and reviews (30–32). The DFT method is based on the Hohenberg–Kohn theorem (33), which states that the ground-state energy of a molecular system is just a functional of the electronic density (ρ), so, in principle, one needs only a knowledge of the density to calculate all the properties of the system. The electron density (ρ) is a measurable quantity by X-ray scattering. Recently, the DFT method has been used to make a direct comparison with experimental X-ray data in terms of electron density (34). While in ab initio MO theory one needs a wave function that is a function of $3N$ coordinates (N is the number of the electrons), in DFT theory one only needs ρ, which is already a three-dimensional function. Kohn and Sham (KS) developed a computational approach that made the theory useful (35). According to the HK theorem, one can calculate the total energy as the sum of three terms, kinetic energy, attraction between nuclei and electrons, and electron–electron repulsion:

$$E[\rho] = T[\rho] + V_{ne}[\rho] + V_{ee}[\rho] \tag{9}$$

In density functional theory, the energy is written in terms of the functional of electron density, while it is expressed in terms of the wavefunction in the Schrödinger equation. As in HF theory, the electron–electron term, $V_{ee}[\rho]$, can be divided into a coulombic part and an exchange-correlation part, $J(\rho)$ and $E_{xc}(\rho)$, respectively. Although it is possible to prove that the exchange-correlation functional is unique, the explicit form of this functional is unfortunately not known. However, in the orbital-based formulation of DFT by Kohn and Sham (35), reasonable approximations to the density functional have been developed. The electron density can be written in terms of a one-electron orbital:

$$\rho(\vec{r}) = \sum_{i=1}^{N} |\psi_i(\vec{r})|^2 \tag{10}$$

where Ψ_i are one-electron (Kohn–Sham) orbitals. These Kohn–Sham orbitals are obtained by using a self-consistent approach to solve the Kohn–Sham equations. That is, an initial guess of the density is made and then used to obtain improved orbitals, which give an improved density, and so on.

For practical calculations, approximations of the exchange-correlation energy functional are used. The exchange-correlation energy functionals are often categorized into three generations (36,30). The simplest way to obtain the exchange and correlation contribution is to use the local density approximation (LDA) proposed by Kohn and Sham (35). This method is based on the uniform-electron-gas model, in which the electron density is constant throughout space. This method has been widely used for band calculations in the field of solid-

TABLE 3   Some of the Most Widely Used DFT Functionals

Acronym	Name (people who parameterized)	Type of functional
LDA (S-VWN)	Perdew–Zunger parameterization	Local density approximation
GGA (PW91)	Perdew and Wang (Ref, 40)	Generalized gradient approximation
B3-LYP (Becke3LYP)	Becke 3 term with Lee, Yang, Parr exchange (Ref, 39)	Hybrid

state physics. The structural parameters, such as lattice parameters and bond length computed using LDA, are known to underestimate the experimental values by 1–2% (37). The next generation of functionals, GGA, makes use of both the density and its gradients to correct the LDA for the variation of electron density with position. The letters GGA stand for *generalized gradient approximation*. It is often also called *gradient-corrected functional* or *nonlocal functional* (38). Some commonly used gradient-corrected correlation functionals include the LYP functional (39), the Perdew–Wang 1991 (PW91) parameter-free correlation functional (40), and the Perdew–Burke–Ernzerhof (PBE) exchange and correlation functional (41). The third generation of exchange-correlation functionals are post-GGA functionals such as "hybrid" functionals. Hybrid functionals are energy functionals that contain both the DFT type (LDA or GGA) and the Hartree–Fock type calculated from the orbitals (30). The most popular hybrid exchange-correlation functional at the moment is B3LYP, available for the first time in the Gaussian (42) package. Some of the most widely used DFT functionals just discussed are listed in Table 3.

B. Molecular Mechanics (MM)

There is a tendency to use the term *molecular mechanics* (MM) as opposite to quantum mechanics, therefore including all classical dynamics methods, such as energy minimization, Monte Carlo, and molecular dynamics. Sometimes it is used to describe only the energy minimization method using empirical force field (potential); or it is even used to refer to the quantum mechanical method specifically, emphasizing its use for molecular motion. Nevertheless, we will focus on the second definition of molecular mechanics (43–45). In this approach, a molecule is viewed as a collection of particles (atoms) held together by simple harmonic or elastic forces. Such forces are defined in terms of potential energy functions of the internal coordinates of the molecule and make up what is termed the *molecular force field* (44).

The results of a molecular mechanics (and also molecular dynamics) calculation depend directly on the force field. The quality of the force field describing the energetics of the system is greatly important. This is because the most important and fundamental part of MM computation is the calculation of the potential energy for a given configuration of atoms. The calculation of this energy, along with its first and second derivatives with respect to the atomic coordinates, yields the information necessary for minimization of the total energy. The force field is composed of a functional form of this expression and the parameters needed to fit the potential energy surface (46,47). It is important to understand that the force field, both the functional form and the parameters themselves, represents the single largest approximation in molecular mechanics. In an MM calculation with no experimental restraints, any results depend critically on the details of the force field function and the simulation parameters. If one is refining the force field with respect to an extensive NMR data set, then presumably the experimental data are the most important factor determining the final structures.

The Energy Expression

Molecular mechanics is based on a "ball-and-stick" picture of a molecule, occasionally with some classical electrostatistics. Neither explicit consideration of electrons nor the quantum mechanical treatment of potential energy is made. (In quantum mechanics the potential energy is represented as a sum of the nuclear repulsion energy and the electronic energy obtained from an approximate solution to the Schrödinger equation.) The potential energy in this classical MM model is written as a superposition of various two-body, three-body, and four-body interactions. The potential energy is expressed as a sum of valence (or bonded), cross-valence, and nonbonded interactions:

$$E_{total} = E_{bonded} + E_{cross\text{-}term} + E_{nonbonded} \tag{11}$$

where the valence (bonded) interactions consist of bond stretching (E_R), bond angle bending (E_θ), dihedral angle torsion (E_ϕ), and inversion (E_ω) terms. The nonbonded interactions consist of van der Waals (E_{vdw}) terms and electrostatic (E_{el}) terms. The valence cross-term set generally consists of bond–angle and angle–angle terms. Cross-terms are used to improve the accuracy of modeling mechanical properties.

Since the mid-1990s, several force fields have been developed. AMBER, CHARMm, DREIDING, MM2, MM3, and universal force field (UFF) are some of these. AMBER and CHARMm have been used most widely, especially in the biological and biochemical fields, while MM2 (48) and MM3 (49) are used mostly for small molecules (e.g., hydrocarbons). A recent development is universal force field, by Rappé and coworkers (50). This general-purpose force field has been parameterized for the full periodic table. The form of the expression of the potential energy in UFF is given by following equation:

$$E = E_R + E_\theta + E_\phi + E_\omega + E_{vdw} + E_{el} \tag{12}$$

where the name of each potential energy term at the right-hand side of the equation is as just explained. All of the mathematical representations (equations) of these potential energy terms are found in the original paper (50).

C. Energy Minimization

Because the concept of *structure* is central to chemistry, finding the energy minima of complex, multidimensional surfaces itself has been one of the most extensively studied fields. By simply minimizing energy, stable conformations can be identified. This is known as a classical treatment of molecular geometry. In the real (quantum mechanical, finite-temperature statistical mechanical) world, molecules are never at rest and molecular structure is a dynamic concept involving oscillations of the nuclear positions about equilibrium geometries (51). Minimization of a molecule is done in two steps. First, an equation describing the energy of the system as a function of its coordinates must be defined and evaluated for a given conformation. Target functions may be constructed that include external restraining terms to bias the minimization, in addition to the energy terms. Next, the conformation is adjusted to lower the value of the target function. Rarely, a minimum may be found after one adjustment, or it may, rather, require many thousands of iterations, depending on the nature of the algorithm, the form of the target function, and the size of the molecule. The efficiency of the minimization is therefore judged by both the time needed to evaluate the target function and the number of structural adjustments (iterations) needed to converge on the minimum.

The potential energy calculated by summing the energies of various interactions is a numerical value for a single conformation. This number can be used to evaluate a particular conformation, but it may not be a useful measure of a conformation because it can be dominated by a few bad interactions. For instance, a large molecule with an excellent conformation for nearly all atoms can have a large overall energy because of a single bad interaction, for instance, two atoms too near each other in space, and have a huge van der Waals repulsion energy. It is often preferable to carry out energy minimization on a conformation to find the best nearby conformation. Energy minimization is usually performed by gradient optimization: Atoms are moved so as to reduce the net forces on them. The minimized structure has small forces on each atom and therefore serves as an excellent starting point for molecular dynamics simulations. The conjugate gradient algorithm (52) is often used for MM minimization study. There are other methods available to find a global minimum conformation of a molecule, such as the steepest descent (53), Newton–Raphson methods (a quadratically convergent method for finding a root of an equation, and its variants), and the quasi-Newton–Raphson method. Detailed treatments of the Newton methods can be found in the literature (53–55).

D. Molecular Dynamics (MD)

The molecular dynamics method was first introduced in the late 1950s by Alder and Wainwright (56,57) to study the interactions of hard spheres. In 1964, Rahman carried out the first simulation using a realistic potential for liquid argon (58). In 1974, the first molecular dynamics simulation of a realistic system was done by Rahman and Stillinger in their simulation of bulk liquid water (59).

In MD, the motion of particles in a system is simulated as they react to forces caused by interactions with other particles in the system, whereas MC stochastic treatment neglects these interactions. Therefore, this dynamic feature of MD allows us to study time-dependent processes occurring in chemistry. Molecular dynamics calculations evaluate the forces acting on each particle and use these to determine the accelerations these particles undergo. Particle velocities are initially determined by a random distribution calibrated to give a Maxwell–Boltzmann distribution at a given simulation temperature, and then the following velocities are updated according to the calculated accelerations. The trajectory of each particle (each atom in the molecule) is then obtained by solving classical Newtonian dynamics over time for the system. If there are N atoms in the molecule and each atom is denoted by subscript i, then the forces acting on atom i are

$$F_i(t) = m_i \frac{\partial^2 r_i(t)}{\partial t^2}, \quad i = 1, N \; N \equiv (\text{number of atoms}) \tag{13}$$

where $F_i(t)$ is the force on atom i at time t, $r_i(t)$ is the position of atom i at time t, and m_i is the atomic mass. The force on atom i at time t is also defined as the negative gradient of the potential energy function:

$$F_i = -\frac{\partial}{\partial r_i} V(r_1, r_2, \cdots, r_N) \tag{14}$$

where $V(r)$ is the potential energy at position r, which is a function of the atomic positions ($3N$) of all the atoms in the system. The analytical solution of the integration of this equation of motion is not trivial, and many algorithms have been introduced to solve the motion. Verlet, leapfrog, and Beeman's algorithm are the most widely used integration schemes in MD. While Beeman's algorithm provides the most accurate expression for the velocities with better energy conservation, the calculation is computationally more expensive (60).

While integrating the classical (Newtonian) equations of motion provides information regarding the constant-energy surface, one may wish to explore the equilibrium thermodynamic properties of a system. If a microscopic dynamic variable A takes on values $A(t_n)$ along the trajectory at the time step t_n, then the time average

$$A = \lim_{t \to \infty} \frac{1}{t} \int_0^t A(t_n) \, dt \tag{15}$$

will give the measured thermodynamic value for the selected variable. This dynamics variable can be any function of the coordinates and momenta of the particles of the system. Through time averaging one can compute properties such as internal energy, kinetic energy, and pressure, regardless of the type of ensemble employed. However, properties like specific heats, compressibilities, and elastic constants depend upon the type of ensemble used. Therefore, it is important to choose a correct ensemble to perform the dynamics simulations. There are constant-NVT (canonical; constant number of particles, volume, and temperature) dynamics, constant-NVE (microcanonical; constant number of particles, volume, and energy) dynamics, and constant-NPT (constant number of particles, pressure, temperature), dynamics and grand canonical ensemble (constant μVT, where μ is the chemical potential).

For the canonical dynamics simulation, the temperature (T) is held constant by coupling to a thermal bath. Nosé (61) and Hoover (62) suggested different methods of thermal coupling for canonical dynamics. Canonical dynamics using Hoover's heat bath gives the trajectory of particles in real time, while the molecular dynamics based on Nosé's bath does not give the trajectory of particles in real time due to its time scaling method. Therefore, for a real-time evaluation of the system, Hoover's heat bath should be used for the canonical MD.

The treatment of long-range nonbonded interactions is important, especially when we have many atoms in a widely spread lattice or a three-dimensional glob, such as a big protein molecule or clay mineral. The computational cost of calculating every nonbonded (van der Waals and electrostatic) interaction is very high in terms of time. The system that has 350 atoms inside exceeds 60,000 interaction pairs. Therefore the accuracy of the results in MD will depend greatly on the quality of the approximation used.

If one uses cutoff distances in MD for a system, one may acquire undesired border artifacts in which the energy jumps abnormally at the border when an atom moves in and out of this cutoff range. Inclusion and exclusion of the non-bonded energy term with respect to the cutoff distance are no longer valid. The direct method with fixed cutoff distance is also very slow for large cutoff distances. An improvement can be made to reduce the large jump of the direct method with the cutoff distance. In this method, called the *spline- switching method*, the energy is multiplied by a spline function. Instead of a fixed single cutoff, two different parameters, called spline-on and spline-off distances, are used to define the cutoff. Within the spline-on/spline-off range, the nonbonded interaction energy is reduced according to the spline function, while it is ignored beyond the spline-off distance. The narrower the spline-on/spline-off range, the faster the interaction energy converges. One may want to extend the spline-on/ spline-off range to increase the accuracy of the computation. However, the computational time will simultaneously increase dramatically.

Ewald summation is one of the procedures developed to solve the problems just mentioned. While VDW has rapid potential drop across certain interatomic distances due to its 6–12 exponential function, the electrostatic interaction's convergence over the interatomic distance variation is very slow due to its $1/r$ dependency. The use of a two-step summation (one in real space and one in reciprocal space) for the periodic system will give a more accurate value for the electrostatic interactions (63). One summation is carried out in reciprocal space; the other is carried out in real space. Based on Ewald's formulation, the simple lattice sum can be reformulated to give absolutely convergent summations that define the principal value of the electrostatic potential, called the *intrinsic potential*. Given the periodicity present in both crystal calculations and in dynamics simulations using periodic boundary conditions, the Ewald formulation becomes well suited for the calculation of electrostatic energy and force.

E. Monte Carlo Simulation

The method of Monte Carlo simulation is often called the Metropolis method, since it was introduced by Metropolis and coworkers (64). Monte Carlo techniques in general provide data on equilibrium properties only, whereas MD gives nonequilibrium properties, such as transport properties, as well as equilibrium properties.

As indicated by its name, MC is a stochastic method that uses random numbers to generate a sample population of the system from which one can determine the properties of interest. The Metropolis algorithm allows one to calculate expectation value of property F from a canonical ensemble* using the equation

$$<F> = \frac{\int F e^{-E/k_B T} \, dq \, dp}{\int e^{-E/k_B T} \, dq \, dp}$$

(16)

---

* Also called as NVT ensemble because the number of particles, the volume, and the temperature of the system are constant.

where $<F>$ is the expectation value of property F, k_B is the Boltzmann constant, T is the temperature of the system, $dq\ dp$ is a volume element in phase space,* and E is the energy of the system. This integral is not trivial to solve analytically. A numerical simulation with a sufficiently large sample is required. The F calculated from the simulation with N_c, the number of total sample configurations considered, is written as

$$
F = \frac{\displaystyle\sum_{c=1}^{N_c} F_c e^{-E_c/k_B T}}{\displaystyle\sum_{c=1}^{N_c} e^{-E_c/k_B T}}
$$

(17)

where F_c and E_c are the property and the energy at configuration c, respectively. The complex analytical integration over the phase space is now replaced by the numerical summation over a large number configuration sets.

A particle selected at random is moved in a random way within the prescribed limits, and it will have new configuration. According to the Metropolis algorithm, this configuration is selected depending on the Boltzmann factor $e^{-Ec/kBT}$ before and after the move. If the new configuration, $E_c + I$, is smaller than that of E_c, the new configuration is accepted. Otherwise, the acceptance of the new configuration is controlled by selecting a random number i $(0 < i < 1)$ and comparing it in the following way:

1. If $e^{-(Ec+1-Ec)kBT} < i$, accept the new configuration.
2. If $e^{-(Ec+1-Ec)kBT} \geq i$, keep the original configuration and start a new selection of an atom randomly for another displacement in the space and calculate $E_{c+i} - E_c$. Then step (1) is repeated.

Over the course of perhaps several hundred thousand to several million attempted steps, a large number of energetically accessible configurations of the system are acquired. This collection of energetically accessible configurations, or the *states*, is called an *ensemble*. Thermodynamic and other properties of the system can be computed as an ensemble average during the simulation. Some properties converge more rapidly than others. For example, average internal energy may converge relatively quickly, but heat capacity may require a much large ensemble sample to compute reliably.

The random nature of Monte Carlo simulation makes it a useful tool for sampling conformational space. Although the efficiency of sampling conforma-

---

* Phase space in canonical dynamics is composed of q_x, q_y, q_z, p_x, p_y, and p_z considering heat transfer and momentum change over the 3D space.

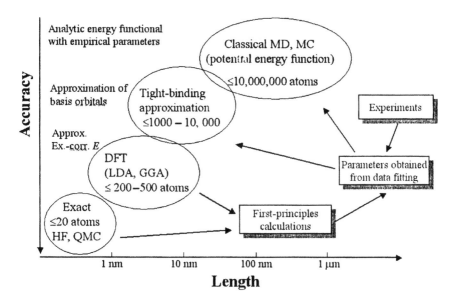

FIGURE 6 Computational methods for clay mineral structure and surface chemistry.

tional space is not as great as MD, the power of MC is in the fact that it can search much more space in a stochastic manner, which is impossible in the case of MD simulation. For instance, the dihedral probability grid Monte Carlo method can rotate a dihedral angle in a single step, without regard to an energy barrier that might prevent the same rotation in molecular dynamics (65). Monte Carlo simulation is good for coarse-grained sampling of conformational space, while the MD technique with force field minimization is a good tool for the complementary searching for the optimized conformation (Fig. 6).

IV. COMPUTATIONAL STUDIES OF CLAY MINERAL STRUCTURE AND THE CLAY–WATER INTERFACE

For the application of molecular modeling methods discussed in Sec. II, we shall go over the computational studies achieved in two different ways. First, we shall discuss structural studies in which the main result was the optimized structure of the clay mineral, either in hydrated or dehydrated form. Therefore, the emphasis is on the clay mineral network structure instead of the interlayer or adsorbate structure (exchangeable ions, organic/inorganic compounds, or water). Next, we shall review the investigations in which the main interest was in the structure of the interlayer with respect to the (hydrophobic or hydrophilic) clay surface. This

will include the interlayer water structure, the adsorbate configuration, ion solvation structure, and any polymer or complex compound at the clay–water interface.
Clay nanocomposite simulations therefore will be discussed in this section, too.

A. Clay Mineral Structure

Both quantum mechanical and molecular mechanical methods have been used
for clay mineral structure determinations, while quantum mechanical simulations
have been the majority of the studies. This is due mainly to the difficulties in
finding accurate semiempirical parameters (force field or potential) for layer structures where both an octahedral (O_h) sheet and a tetrahedral sheet (T_d) are found
in its structure. Historically, ab initio methods, such as extended Hückel methods
and molecular orbital (MO) methods with a Gaussian-type basis set (such as
6–31G), were used on the cluster model representing the clay mineral layer
structure in the early days. More recently DFT methods on either the cluster
model or a unit cell with periodic boundary conditions have been used. Due to
the recent development of both hardware and software, the quantum mechanical
method using density functional theory is more frequently used. The DFT method
with pseudopotentials can handle complex systems having many atoms with periodic boundary conditions. This electronic structure calculation using ab initio
methods is exciting in the field of geochemistry, where many spectroscopic data
are available that in turn can be benefited by using advanced theoretical interpretations. Both quantum mechanical (QM) and molecular mechanical (MM or MD)
studies on the clay mineral structure and closely related silicate material are
summarized in Table 4. We will discuss the computational methods in relation
to spectroscopic interpretation in the following section.

Electronic structure calculation using ab initio molecular orbital calculation
were performed by Gibbs and coworkers (66–69) for the model structure representing tetrahedral hexagonal rings of silicates and siloxanes found in clay mineral
surface. They used the cluster model of (SiO_4^{4-}) tetrahedral structure and obtained
optimized structure in terms of bond lengths and bond angles. The geometries
and charge distributions obtained were compared with experimental values for
silicates and siloxanes. Quantum mechanical calculations of this kind was performed later by Teppen et al. (70) by studying several aluminum and silicon oxide
clusters using density functional theory calculation to investigate the structure and
charge distribution of these model systems representing smectite clays.

In the field of mineralogy, the first ab initio calculations based on density
functional theory were performed on cordierite (magnesium aluminum silicate;
$Mg_2Al_4Si_5O_{18}$) by Winkler et al. (71). The DFT methods based on local density
approximation (LDA) have been performed for hydrated Mg end-member cordierite to elucidate the location, orientation, and total energy of hydration in the
ground state. The calculations shoed that the energetically most stable orientation

TABLE 4 Computational Studies on Clay Mineral Structure Determination

Model	Method/basis set	Aim	Refs.
Cluster model of $(SiO_4{}^{4-})$ tetrahedral structure	SCF/STO-3G	Optimized structure of silicate tetrahedra in terms of bond lengths and bond angles	66–68
$(OH)_3M_2Si_9O_{28}$ (M: O_h metals) with 17 H's for dangling bonds	Extended Hückel	Favored isomorphic substitution of Al^{3+} by Na^+, K^+, Mg^{2+}, Ca^{2+}, Fe^{2+}, and Fe^{3+} in the octahedral layer	113
$H_6Si_2O_7$	SCF/6-31G	Geometry and Si–O–Si force field	114
SiO_4 silicate cluster	SCF/6-31G**	Optimized structure of silicate tetrahedra in terms of bond lengths and bond angles	69, 115, 116
$Al_2(OH)_2$ $(Si_2O_5)_2$, $MgAl(OH)_2(Si_2O_5)_2$ for pyrophyllite and celadonite	Extended Hückel, tight binding (TB) method	Effect of isomorphous substitution on bonding and charge distribution	117
Al_2Si_2 O_9 H_4 (kaolinite)	Periodic ab initio HF using SCF/STO-3G, 6-21G	Geometry, charge density	118
$H_3SiOSiH_3$, $Si(OH)_4$, $H_6Si_2O_7$	SCF/MC6-311G and MP2/MC6-311G**	Geometry optimization, effect of basis set size and electron correlation	119
$AlSiO_7H_7$, $AlSi_2O_{10}H_9$, $AlSi_2$ $O_{10}H_9$, $AlSi_3$ $O_{13}H_{11}$, $AlSi_3$ $O_{13}H_{11}$, $AlSi_3O_{12}H_9$, $Al_2Si_2O_{12}H_{10}$, $AlSi_5O_{18}H_{13}$, $Al_3Si_3O_{18}H_{15}$, $Al_2Si_6O_{20}H_{10}$, $Al_4Si_4O_{20}H_{12}$, $Al_2Si_{10}O_{30}H_{14}$	SCF/DZP and SCF/TZP	Development force field for a luminosilicates	120
$Mg_3Si_4O_{10}(OH)_2$, $Al_2Si_4O_{10}(OH)_2$, talc and pyrophyllite	DFT/LDA	Geometry of clay minerals, interaction with H_2O	74
$Si(OH)_4$, $(OH)_3SiOSi(OH)_3$, $Al(OH)_4{}^-$, $H_3SiOHAlH_3$, $(OH)_3SiOAl(OH)_3{}^-$, $(OH)_3SiOHAl(OH)_3$	SCF/3-21G* and DZP/MP2	Development forcefield for aluminosilicates	121, 122

(*Continued*)

TABLE 4 *Continued*

Model	Method/basis set	Aim	Refs.
$KAlSiO_4$	Energy minimization using empirical potential, DFT/ norm-conserving	Enthalpy of disorder for Al/Si exchange in T_d	123
Gibbsite, kaolinite, pyrophyllite	MM/MD, cff91 force field	Force field development, geometry optimization of clay minerals	124
$Al_2Si_2O_5(OH_4)$; kaolinite	DFT/LDA	Geometry optimization	72
$[Al(OH)_4]^-$, $Al(OH)_3)(H_2O)_3$, $[Al(OH)_2(H_2O)_2]_2$- $(OH)_2$, $[Al(OH)_3$-O- $Si(OH)_3]^-$, $Al(OH)_2(H_2O)_3$-O- $Si(OH)_3$	HF/6-311G**, B3LYP/6-311G**, MP2/6-311G**	Geometry, atomic charge for clay mineral potential	125
$Al(OH)_6^{3-}$	SCF/6-311G**	Force field for AlO_6 octahedra for kaolinite	77
$[Al_2O_8(OH)_2]^{12-}$ as a pyrophyllite cluster model, Al substituted by Fe or Mg	HF/LANL2DZ, HF/ 6-31+G*	Geometry, isomorphous substitution effect of Al by Mg and Fe	78
$NaSi_8Al_3O_{20}(OH)_4$ and $NaSi_8Al_3O_{20}(OH)_4$ with H_2O	DFT/LDA	Geometry, effect of cation on the water orientation near clay surface	126
$Al_2Si_2O_5(OH)_4$ for dickite and kaolinite	DFT/GGA-USP, ab initio MD	Geometry, IR band calculation (structural OH stretching)	127–129
Smectites/illites where Si^{-4+}, Al^{3+}, Mg^{2+}, Fe^{3+} content varies in both T_d and O_h	MM lattice minimization (based on empirical potentials)	Geometry (cis- /trans-) cis-/trans- vacant ratio, OH angle and length	130
Pyrophyllite, muscovite, margarite, beidellite, montmorillonite, illites	MM lattice minimization (based on empirical potentials)	Geometry, effect of isomorphic substitution	131
Vermiculite unit cell: $Mg_3Si_4O_{10}(OH)_2$	SCF/6-311G**	Geometry, clay force field including MgO_6	132
$Si_6Al_6O_{36}H_{30}$ (kaolinite)	ONIOM (B3LYP/SVP:PM3)	Geometry optimization, H_2O adsorption, acetate adsorption	79
Montmorillonite/ beidellite cluster	DFT/BLYP, DNP	Sorption of organic molecules (dioxin, furan)	133

of the H_2O molecule in alkali-free Mg end-member cordierite is with the proton–proton vector parallel to [001]. They found that the water molecule in cordierite is nearly undistorted and very weak hydrogen bonded, which is in contrast to the calorimetric study but consistent with spectroscopic (IR) and quasi-elastic neutron-scattering experiments. Hobbs et al. (72) also used the same technique based on local density functional theory to calculate and minimize the total energy to obtain the optimized geometry of all atomic coordinates for two proposed kaolinite crystal structures. All calculations were performed using published unit-cell parameters. Inner- and interlayer H atom positions agree well with those determined by Bish (73) from neutron diffraction data and confirm a unit cell with *C1* symmetry. More importantly, ab initio total energy minimization study on talc and pyrophyllite, which are uncharged structural analogs of smectites, was conducted by Bridgeman et al. (74). They found the calculated atomic coordinates were in good agreement with experiment. All of these studies suffered from the limitation that they did not optimize unit-cell parameters in response to internal stress. This means that the structural parameters reported were calculated at some undetermined, nonzero pressure. Bond angles are particularly sensitive to pressure because of the relative weakness of the forces involved as compared to bond lengths.

Molecular modeling using multiple scattering formalism (75) provides XAFS observables (see also Sec. V.A). Therefore, one can simulate the XAFS from candidate structures and compare with experimental results to find which structure is correct. The theoretically derived (suggested) clay mineral structures can now be confirmed by experiments.

Ebina et al. (76) performed comparative study of XPS X-ray photoelectron spectroscopy (XPS) and DFT calculations for five phyllosilicates: dioctahedral smectite, mica, magnesium chlorite, saponite, and hectorite. They concluded that the trends of distribution of Al in phyllosilicates can be determined qualitatively using this method. These results are very encouraging with respect to clay minerals due to the lack of available experimental data. Therefore, DFT calculations can be used as a reliable method to determine the distribution of tetrahedral and octahedral Al in clay minerals.

Vibrational spectra and structure of clay mineral kaolinite have been investigated by the interplay between spectroscopy (IR/Raman) and quantum mechanical calculations (HF and MP2 calculations with 6–311G** basis sets) (77). Experimental and theoretical methods proved to be valuable combination tools for exploring the structure and dynamics of complex clay mineral structure. Their force field (potential) developed is also expected to be helpful in predictions and in interpretation of vibrational spectra of other clay minerals. Their quantum mechanically derived force-field parameters can be applied to MD simulations and vibrational normal mode analysis.

Sainz-Diaz et al. (78) studied the isomorphous substitution effect on the vibration frequencies of hydroxyl groups in clay octahedral sheet by using ab initio calculations on cluster model of the clay mineral. They were not able to obtain calculated values of the OH vibration frequency of clay minerals matching experimental results because their simulation didn't include the hydrogen bonding between structural OH and apical oxygens. However, they claimed that their model represents a relative behavior with respect to the isomorphous substitution of octahedral sheet cations similar to the experimental data.

Tunega et al. (79) showed that systematic DFT calculations are successful for the study of the adsorption of water, acetate anion, and acetic acid on the tetrahedral and octahedral surfaces of a single kaolinite layer. They found the water molecules and the acetic acid are only weakly bound to the basal oxygen atoms on the tetrahedral side. They also found that the octahedral surface is more highly energetic than the tetrahedral one and is more attractive for polar and/or negatively charged species; therefore there are more possibilities to form hydrogen bonds with adsorbates than the tetrahedral side due to the existence of hydroxyl groups.

Refson et al. (80) showed that current DFT methods with parallel algorithms performed on massively parallel supercomputers now make it possible to engage accurate ab initio structure calculations on pyrophyllite. They allowed full structural relaxation; therefore, the results are appropriate for comparison to ambient-pressure crystallographic data. Their simulation results gave an excellent account of bond lengths and angles when compared to single-crystal X-ray diffraction data for pyrophyllite. The effect of charge substitution and interlayer cation on the montmorillonite structure, especially its hydroxyl groups, was studied systematically (81). These quantum mechanical calculations of clay mineral structure, which require no adjustable parameters, gave excellent results for both equilibrium structures and total energies. Experimentally unavailable but important features such as OH orientation were determined for pyrophyllite (26°) and montmorillonite. The determination of theoretical bulk structure of montmorillonite is also crucial because of the lack of quality single crystals (see Sec. II.C).

B. Clay Mineral Hydrates (Clay–Water Interface)

Adsorption of nucleotides on homoionic bentonite clays was modeled using intermediate neglect of differential overlap (INDO) semiempirical quantum mechanical calculation by Liebmann and coworkers (82). They characterized the binding of five different nucleotides to Zn^{2+} and Mg^{2+} when adsorbed in the interlayer space of hydrated bentonite (mainly Na-montmorillonite). The energetics of complex formation was obtained as well as cation–nucleotide–water (hydrating cation) binding pattern. The preference of the bindings of cation-exchanged montmorillonite among nucleotides was studied. These results implicated direct cation–nucleotide complex formation in the interlayer space of hydrated clay, which could then be involved in subsequent polymerization.

A macroscopic energy balance model for crystalline swelling of 2:1 phyllosilicates was devised by Laird (83). The method uses a balance among the potential energies of attraction, repulsion, and resistance. He predicted the basal spacing of the clay swelling as a function of interlayer cation and layer charge. However, his modeling results did not predict the experimental basal spacing. This discrepancy was attributed to the wrong assumption of his continuous and irreversible crystalline swelling.

Chatterjee and coworkers (84) used quantum mechanical calculations at both the semiempirical (MNDO-modified neglect of differential overlap) and first-principles (DFT) level. Cluster models of montmorillonite were used to study the interaction of one water molecule near the clay surface, and they found that lower energy obtained is when hydrogens of the water molecule point toward the clay surface. The effect of tetrahedral substitution on the clay–water interaction has been studied, and they concluded that tetrahedral substitution greatly modifies the position of the energy levels, reducing the gap between HOMO[*] and LUMO,[†] stabilizing the energy of smectite–water interaction.

1. Monte Carlo Simulations on the Clay–Water Interface

Monte Carlo (MC) simulations on the clay–water interface system have been studied by many researchers worldwide since the mid 1990s. The use of MC for clay–water simulation was popular because the stochastic Metropolis algorithm has been used successfully for many other liquid-state physical chemistry (63). This method has strong advantages in sampling conformational surface, and this advantageous technique has been used to find the clay–adsorbate–water equilibrium structure.

One of the earliest applications of MC on the clay system was done by Skipper and coworkers (85). They performed computer calculations of water–clay interactions using atomic pair potentials (86). They used an MCY (Matsuoka–Clementi–Yoshimine) model (87) for water–water interactions. They also proposed as the energetically preferable site for water molecules to reside the top of the interlayer hydroxyl groups located inside the hexagonal holes of siloxane surface. Their use of MC in the study of the interaction between clay, water, and interlayer cations to find a starting point of the clay swelling properties enabled them to calculate the interlayer spacing of hydrated Na and Mg smectites. They proposed an interlayer cation solvation model with water-forming octahedral complexes. Interlayer swelling was attributed to solvation of interlayer cations,

---

* HOMO: highest occupied molecular orbital
† LUMO: lowest unoccupied molecular orbital

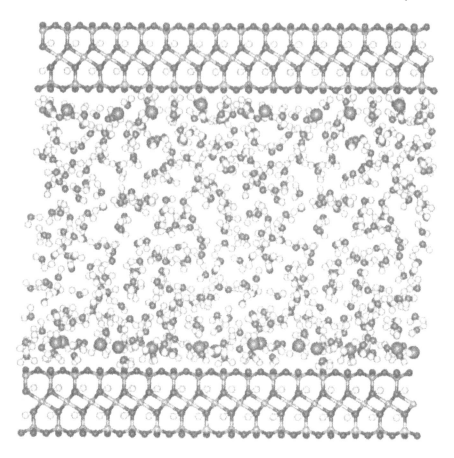

FIGURE 7   (a) Mica–water interface—Monte Carlo simulation. (From Ref. 149.)

which they believe hold the clay layers apart. They determined the calculated equilibrium layer spacings for hydrated Na and Mg smectites to be 14.2 Å and 14.7 Å, respectively, which showed relatively good agreement with the experimental X-ray diffraction results, 14.5 Å and 15.1 Å for Chambers montmorillonite at 79% relative humidity (85).

Alfred Delville (88) also performed Monte Carlo simulations independently to study the clay–water interface and published his results in the same year as Skipper and coworkers. However, he used the water–water potential parameters

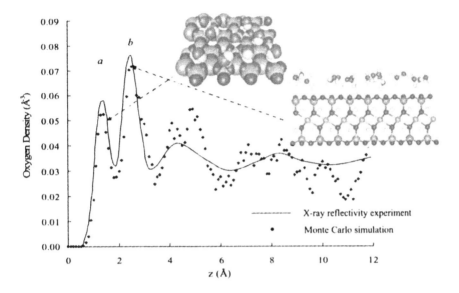

FIGURE 7   (b) Mica–water interface—Monte Carlo vs. X-ray reflectivity. (From Ref. 149.)

from the TIP4P* model of Jorgensen and coworkers (89) instead of the MCY model used by Skipper. He used a modified intermediate neglect of differential overlap (MINDO) quantum mechanical calculation to determine the parameters for clay–water and clay–cation energies. The phenomenon of clay swelling was again attributed to the solvation of the interlayer cations. The mean water organization of hydrated sodium montmorillonite and the description of the immersion enthalpy were reported to be in a good correlation with the experimental results. Since then, Delville has performed a series of MC computations to study the clay–water system, including a wettability study of clay minerals in relation to their surface charge and chemical composition (90–92). He found no structural liquid film in contact with the surface in the case of kaolinite, while he found a dense and structured liquid film for potassium–mica and sodium–montmorillonite. He concluded that his molecular simulations and the description of the clay–water interface give clear evidence of the influence of the surface electric

---

* In this model, the water molecule is rigid and has four intermolecular interaction sites.

charge and chemical composition on the content and the organization of its hydrating film.

Skipper and coworkers continued to develop the model for the clay hydration system (93). Both the effects of simulation cell size and shape and the effects of potential function model were thoroughly studied to prepare a solid methodology of modeling the clay-swelling system using Monte Carlo simulations. They then performed MC with their clay–water system in the form of monolayer hydrate model with the previously developed methodology (94). They found that water molecules are induced to interact with the siloxane surface oxygen atoms through hydrogen bonding as the percentage of tetrahedral layer charge increases. They also found that sodium counterions are induced to form inner-sphere surface complexes. From their simulation, they suggested the need for new, careful diffraction study on a series of monolayer hydrates of montmorillonite where the charge of the interlayer region and isomorphic substitution at the tetrahedral portion of the clay can be systematically varied.

A study of adsorption/desorption hysteresis loops in the swelling process previously observed experimentally by Fu and coworkers (95) was also tackled by Monte Carlo simulation (96). They conducted a series of simulations in which the interlayer water content is increased systematically from 0 to 300 mg/g of clay. Then the calculated clay layer spacing values as a function of water content were compared with experimental data. They claimed that their simulation established, for the first time, the true equilibrium clay layer spacings of the system.

The role of potassium as a clay-swelling inhibitor was studied using MC simulation modeling by Boek and coworkers (96). The driving force of their attention on this matter is due to the previously mentioned property of clay to fall apart during oil well drilling through shales.[*] They performed a series of MC simulations on different types of montmorillonite that contain sodium, lithium, or potassium cation in the interlayer. Their simulations showed that a water content increase in the interlayer region resulted in the induced hydration of Na^+ and Li^+, which in turn increased the interlayer spacing. However, K^+-montmorillonite didn't show an increase in the interlayer spacing. They concluded that K^+ ions migrate to and remain bound to the clay surface, and this hydrophobic effect of K^+ reduces the tendency of K^+-saturated clays to expand, explaining the shale-swelling inhibition properties of K^+ ions.

Sato and coworkers, in Japan, also studied the same montmorillonite clay system using MC (97–99). They attempted the modeling of the clay system in order to find the theoretical support for their group's earlier finding, "chirality effects of the clay," (100–103) in which the bound chelates interact stereoselec-

---

[*] In petroleum engineering, borehole instability is associated with the uptake of water by smectites from the drilling fluid.

tively in the interlayer space of a clay and that the interaction is affected drastically by slight changes in the ligand structures. The free energy of binding, the roles of the upper and lower clay sheets in determining the orientation of a adsorbates, and the effects of water on the binding were studied. They claimed that their simulations of cationic metal complexes as binding guest of the interlayer surface of montmorillonite clay showed a more detailed picture about the bound structure of the chelate than the experimental results. They also performed MC simulation for intercalation of the same metal complexes on a different clay substrate, saponite. Their MC simulation predicted that both the racemic mixture[*] and the pure enantiomer[†] of $[M(phen)_3]^{2+}$ form a monomolecular layer when they are adsorbed within the cation exchange capacity (CEC) of a model tetrahedral sheet (104).

2. Molecular Mechanics and Molecular Dynamics

While numerous works of Monte Carlo simulations were carried out for the study of the clay–water system, especially in terms of clay swelling and interlayer cation solvation, molecular dynamics (MD) has only recently been used. While MC has an advantage in conformation sampling in the entire system in equilibrium by its stochastic method, MD is more suited to the dynamical or diffusion properties of the chemical system. Here, several examples of using MD techniques on clay minerals are reviewed for understanding the current activities in this field.

Keldsen and coworkers (105) used the MD technique to calculate the enthalpies of adsorption of series of hydrocarbons on a smectite clay. The rigid model of clay structure was used, while the hydrocarbons were allowed to move freely during the course of simulations. Then, two different force fields, MM2 by Allinger (106) and Hopfinger's force field (107), were used to calculate the enthalpy of adsorption from both molecular mechanics (energy minimization) and molecular dynamics. The average error of less than 2.0 kcal/mol between the theoretical and experimental data was reported from the molecular dynamics using Hopfinger's force field.

The stereoselectivity of the clay mineral surface previously proposed by Yamagishi and coworkers from their experiments (100–103) was chosen for the molecular mechanics study by Breu and coworkers (108,109). They performed a lattice energy minimization technique with periodic boundary condition on a monolayer of tris(2,2'-bipyridine) ruthenium(2+) complexes[‡] and tris(1,10-

---

* Racemic mixture: a compound that is a mixture of equal quantities stereoisomers of the same compound and therefore is optically inactive.
† Enantiomer: one of an isomeric pair of crystalline forms or compounds whose molecules are nonsuperimposable mirror images.
‡ $[Ru(bpy)_3]^{2+}$

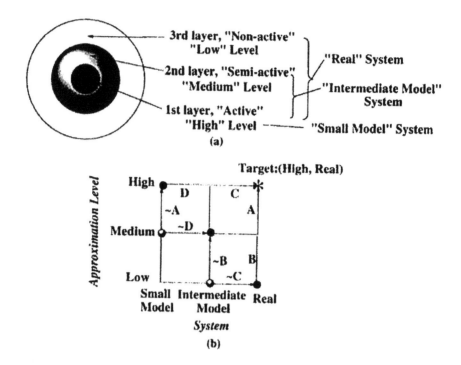

FIGURE 8 Concept of the ONIOM method. (From Ref. 167.)

phenanthroline) ruthenium(2 +) complexes.[*] They proposed that the stereoselectivity of the clay surface is the result of the lateral interactions between the guest
complexes as modified by the corrugation of the silicate layer of the clay. Different two-dimensional arrangements of ruthenium cation complexes were obtained
for enantiomeric and racemic forms. They also reported smaller interlayer spacing
from their simulation and claimed that this result is due to a 0 K temperature
treatment.

Combination of both molecular dynamics and Monte Carlo simulations
were used to study the mechanism of swelling of sodium–montmorillonite clay
by Karaborni and coworkers (110). Their molecular simulations were performed
in the grand-canonical ensemble, that is, at constant chemical potential, volume,
and temperature (μ, V, and T, respectively). In their simulation, the distance

---

[*] $[Ru(phen)_3]^{2+}$

between the clay layers was fixed after each incremental increase in the interlayer distance, and the system was allowed to take or reject water molecules until equilibrium was reached. Their simulation showed that there are four stable states at basal spacings of 9.7, 12.0, 15.5, and 18.3 Å. They claimed that the amount of swelling in terms of basal spacing showed good agreement with the experimental data.

One of the recent successes in computer simulation of clay mineral hydrates is that the structure of interlayer water in terms of quantitatively obtainable content called *total radial distribution function* was achieved by both Monte Carlo simulations and neutron diffraction methods (111,112). The accuracy of the method and the predicting power of the molecular simulation of this complex solid–liquid interface was first confirmed for the interlayer structure of 2:1 clay mineral smectite. Computational studies on the clay mineral–water interface using quantum mechanics, molecular mechanics, Monte Carlo, and molecular dynamics simulations are summarized in Table 5.

V. PERSPECTIVE ON COMPUTATIONAL CLAY MINERAL CHEMISTRY

So far we have discussed the structural features of clay minerals and their hydrates along with computational techniques to investigate these structural characteristics. Examples of the computational studies on structure of clay mineral and its hydrates were also provided where the emphasis was on the application of the molecular modeling techniques. Experimental techniques have been developed to elucidate these important structures of clay mineral and its hydrates. Due to the development of the computational techniques and their advantage of separating detailed molecular structure, the results of computational molecular modeling now often challenge us to interpret or validate the spectroscopic methods applied on these systems. Instead of rivaling between theoretical and experimental methods, the cooperative relationship between computational molecular modeling and spectroscopic methods will benefit both sides to understand the structure of clay mineral and its hydrates.

Another power of the computational molecular modeling is again due to its molecular detailed results on structures. All of these molecular simulations also provide the three-dimensional positions (coordinates) of all atoms. These results can easily be viewed with molecular graphics software developed for biological or materials science. Often these commercially available computational packages themselves or academic utility codes written for simulation analysis can produce simulated spectroscopic results according to given structural information. These predicted spectra obtained from theoretical methods (either potential based or ab initio) can be compared directly with the experimentally observed spectra. The visualization of the structures during the course of simulations can also help

TABLE 5 Computational Studies on the Clay Mineral–Water Interface

Model	Method (basis set, potential, MD length/MC steps, etc.)	Aim	Refs.
256 H_2O molecules between two parallel uncharged layers of oxygen atoms; 15.5 × 17.9-Å cluster; d_{int} = 33 Å	ST2 water and surface oxygens 5.5 ps MD	Structural and dynamics effect of uncharged silicate surface to surfacial water.	134
31.08 × 35.88-Å MMT unit, 14 Na^+, 476 water for MC	MINDO calculation of MMT geometry and atomic charge, MCY water, grand canonical ensemble MC (μPT).	Interlayer water orientation, immersion enthalpy, clay swelling	88
$Al_2Si_4O_{10}(OH)_2$ clay unit, 4 × 2 cell (18.28 × 21.12-Å) MMT unit, 4Mg^{2+} or 8Na^+, 64 water for Mg and Na smectite	MCY water, parameterization for clay–water–cation system (SRM), NPT ensemble with 1.0 MPa 1,250,000 MC steps.	Transferable potential development for hydrated smectite, interlayer structure of water and cations	85
$O_{28}Si_8Al_4(OH)_4$; mica, 31.08 × 35.88-Å MMT unit, 2–3 layers of water	GCMC, MCY water	Layering of interlayer water molecules as a function of thickness (neutral or charged mica)	91
$O_{28}Si_8Al_4(OH)_4$; 31.08 × 35.88-Å MMT unit; kaolinite and mica	GCMC, TIP4P water	Effect of clay surface charge on water structure	92
$M_{0.75}[Si_{7.75}Al_{0.25}](Al_{3.5}Mg_{0.5})$ $O_{20}(OH)_4$; 21.12 × 18.28-Å MMT, M = Li^+, Na^+, and K^+, varying water contents, (0–96 water molecules)	TIP4P water, NPT ensemble MC with 100 kPa,	Interlayer water structure, hydration of interlayer cations	96, 135

System	Method	Purpose	Ref.
21.12 × 18.28-Å MMT; montmorillonite, vermiculite	MCY and TIP4P water, NPT ensemble MC with 100 kPa,	Interlayer molecular structure (water, cation) of monolayer hydrates	93, 94
21.12 × 18.28-Å Na^+-MMT; 1, 2, 3 layers of water (32, 64, 96 H_2O)	MCY water, NPT-MC (100 kPa) followed by NVE-MD	Interlayer molecular structure (water, cation) of Na-MMT hydrates	136
21.12 × 18.28-Å MMT, random insertio n of water	MCY water, Grand canonical (μVT) MC/MD, orientational-bias MC	Mechanism of swelling of Na-MMT, stable states of MMT hydrate.	110
$Mg_3Si_4O_{10}(OH)_2$, 4 × 8 talc unit cells, 256 water	TIP4P water, NPT-MC (1 atm), 2,500,000 MC steps	Surfacial water structure near talc	137
21.12 × 18.28-Å Li^+-MMT; 1, 2, 3 layers of water	MCY water, NPT-MC (100 kPa) followed by NVE-MD	Interlayer structure of Li-MMT hydrates	138
21.12 × 18.28-Å K^+-MMT; 1, 2, 3 layers of water	MCY water, NPT-MC (100 kPa) followed by NVE-MD	Interlayer structure of K-MMT hydrates	139
21.12 × 18.28-Å Li^+-smectite unit (hectorite, beidellite, montmorillonite) at low water content ($H_2O/Li = 3$)	MCY water, NPT-MC (100 kPa) followed by NVE-MD	Interlayer structure	140
Unit cell = $Si_4Al_4O_{10}(OH)_8$, 15.447 × 17.868-Å kaolinite	SPC water, NVE-MD (22.98 ps)	Structure and short time dynamics of interlayer water in kaolinite with 8.5- and 10-Å interlayer spacing	141
21.12 × 18.28-Å MMT; $Cs_{0.75}{}^+[Si_8](Al_{3.25}Mg_{0.75})O_{20}(OH)_4$	SPC/E water, grand canonical ensemble MD	Clay mineral swelling and hydration	142

(Continued)

Park and Sposito

TABLE 5 *Continued*

Model	Method (basis set, potential, MD length/MC steps, etc.)	Aim	Refs.
21.12×18.28-Å Na^+-MMT; $M_{0.75}$ $^+[Si_8](Al_{3.25}Mg_{0.75})O_{20}(OH)_4$ M = Na^+, Cs^+, Sr^{2+}	SPC/E water, NPT-MD	Clay mineral swelling and hydration	143
2×2 crystallographic unit cells of hydrocalumite	SPC water, CVFF potential for ion–ion and ion–water interactions, NPT-MD	Dynamics of Cl and H_2O in the Ca-aluminate hydrate hydrocalumite (Friedel's salt) interlayer	144
25.81×26.70-Å kaolinite, 820 water molecules, Na^+, Cl^-	TIP3P water, NVT-MD, all atoms (including clay) allowed to move	Dynamics of clay surfacial water	145
Mg-beidellite, 21.12×18.28-Å; $Mg_{0.375}$ $[Si_{7.25}Al_{0.75}](Al_4)O_{20}(OH)_4$	MCY water, NPT-MC, NPH- MD	Water diffusion and interlayer structure of Mg-smectite hydrates	146
21.12×18.28 Å MMT, Li^+, Na^+, K^+	GCMC-MD, MCY water	Adsorption isotherms of water, interlayer structure	147
21.12×18.28-Å Ca^{2+} -MMT; $Ca_{0.375}[Si_{7.75}Al_{0.25}](Al_{3.5}Mg_{0.5})$ $O_{20}(OH)_4$	TIP4P water, NPT-MC	Interlayer structure of Ca- montmorillonite hydrate up to three water layers	148
Hydrated muscovite, $K_2Al_4(Al_2Si_6)O_{20}(OH)_4$, 256 water molecules	MCY water, NPT-MC	Detailed molecular structure of hydration water on micaceous	149

ps = picosecond = 1.0×10^{-12} s; MMT = montmorillonite; ST2 = Stillinger potential 2 (59).
NPH-MD = the layer spacing remains a dynamical quantity and the method of Parrinello and Rahman was used to solve the resulting equation of motion, which replaces the usual Newton–Euler equations; GCMC-MD = MD simulation with MC attempts to insert or delete a water molecule at specific frequency (e.g., each third) of MD time step.

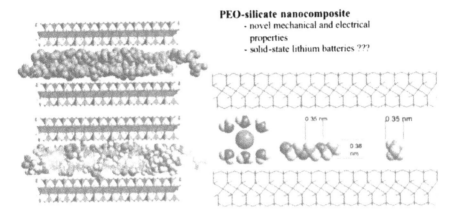

PEO-silicate nanocomposite
- novel mechanical and electrical properties
- solid-state lithium batteries ???

FIGURE 9 Clay–polymer nanocomposite application. (From Ref. 174.)

the direction or termination of the entire simulations. Instead of looking at radial distribution function or density profile results either from simulations or spectroscopic results, the direct visualization of the system can help us to understand the structure in much better way. Layers of water structures, interlayer cation solvation structure, water adsorption to clay mineral surface, cation adsorption to surface, and the protonation or deprotonation of clay interlayer aquatic system (in first-principles methods only) and the newly designed clay nanocomposite structures are also great examples where the visualization can aid the understanding of structure of clay mineral and its hydrates. We shall first discuss the examples where the molecular modeling was used to interpret or validate spectroscopic results of clay mineral or its hydrate structures. Then the visualization of the simulation results will be discussed.

A. Interpretation of Spectra

Spectroscopic methods for hydration of ions were reviewed for structural aspects and dynamic aspects of ionic hydration by Ohtaki and Radnai (150). They discussed X-ray diffraction, neutron diffraction, electron diffraction, small-angle X-ray (SAXS) and neutron-scattering (SANS), quasi-elastic neutron-scattering (QENS) methods, extended X-ray absorption fine structure (EXAFS), X-ray absorption near-edge structure (XANES) spectroscopies, nuclear magnetic resonance (NMR), Mössbauer, infrared (IR), Raman, and Raleigh–Brillouin spectroscopies. The clay interlayer molecular modeling where clay surface is interfaced with aqueous solution also includes ions that are also solvated by interlayer water.

Therefore, knowledge of these scattering and spectroscopic methods applied to ionic solutions is quite useful for clay–water–ion interface simulations. They discussed the utility of the computational results directly comparable to the scattering and spectroscopic methods. These includes radial distribution function comparison (MD or MC with X-ray or neutron diffraction methods), angular distribution of water molecules around ions (MD or MC with neutron diffraction measurements), self-diffusion coefficient of water and ions (MD with quasi-elastic neutron scattering and NMR measurements), and intramolecular vibrations (flexible potential MD with IR or Raman spectroscopic data).

An excellent review with compilation of spectroscopic methods with theoretical insights were previously done for the smectite–water system, where they concentrated their study on the structure of the water in the clay–water system (151). Due to the recent development of computational simulations, the same topic of interlayer surface structure was revisited, and this time the computer simulation part was embedded to support the conceptual understanding of the molecular aspect of clay–water interface molecular structure (1).

Interpretation of vibration spectra using molecular orbital theory calculations was reviewed and compiled by Kubicki (152). Use of attenuated total reflectance Fourier transform infrared spectroscopy (ATR-FTIR) to obtain vibrational frequencies of the organic ligands on the mineral surfaces and in solution was explained. Then the theoretically obtained frequencies from quantum mechanical calculations were used to predict vibrational frequencies for similar organic–mineral complexes. Frequencies were obtained via force constant analyses of the minimum energy structures (see Ref. 153 for further discussion on obtaining vibrational frequency), and scaling of calculations to account for anharmonic and basis set effects was also explained. Kubicki (152) pointed out that even the most accurate representation of the molecular structure and force constant will cause the calculated value to have a positive deviation from experiment. This is mainly due to the fact that experimental vibrational frequencies have some anharmonic component, unlike the theoretical frequencies calculated from the HF method, which is a purely harmonic one. As a practical solution to this problem, the method by Pople and coworkers (154) was introduced. Briefly, one can plot experimental vibrational frequencies for the series of the molecules and then use the slope of the experimental plot to "correct" frequencies calculated with a given method for comparison with experiment (well-known scaling constants for vibrational frequencies, e.g., 0.89 in Hartree–Fock theory) (154).

The interplay between molecular simulations and neutron-scattering experiments on supercritical or ambient water was realized by Chialvo et al. (155). They claimed that the excellent agreement between two techniques is an indication of the increasing reliability of the intermolecular potential models and the accuracy of the simulation results giving us greater confidence in our abilities to measure and predict the microstructural properties of water at all condition.

The solvation of cation and electrical double layer structure near clay surface was studied by neutron diffraction methods (156–158). The interplay between molecular simulations and neutron diffraction techniques also has been also applied to this clay mineral–water–cation interface system. Park and Sposito (112) simulated the total radial distribution function (TRDF) of interlayer water from Na-, Li-, and K-montmorillonite hydrates as a physical quantity from molecular simulations. They obtained TRDF values from Monte Carlo simulations and directly compared with previously obtained $^1H/^2D$ isotopic difference neutron diffraction results (9,10).

X-ray absorption spectra were simulated by FEFF* calculations (75). This ab initio self-consistent real-space multiple-scattering code calculates extended X-ray-absorption fine structure (EXAFS), full multiple scattering calculations of X-ray absorption near-edge structure (XANES), and projected local densities of states (LDOS). This method can be used for clay–water interface, where one can simulate X-ray absorption spectroscopy and then compare with available experimental results. McCarthy et al. (159) have performed molecular dynamics simulations for dehydrated and hydrated Na^+-MgO [(100) plane] to predict extended X-ray absorption fine structure (EXAFS) spectra. Their MD-driven EXAFS calculations were carried out in the following six steps: (a) molecular mechanics (MM) simulations on clusters to obtain the energy-minimized structure; (b) constant-energy (NVE) molecular dynamics simulations with the same potentials as used in MM; (c) calculations of the scattering potentials using the ab initio high-order multiple-scattering EXAFS code, FEFF6 (160); (d) enumeration of the multiple-scattering paths for a given cluster; (e) calculation of effective scattering amplitudes for each scattering path; and (f) calculation of the EXAFS spectrum. McCarthy et al. (159) observed a significant enhancement of the EXAFS signal from the hydrated Na^+-MgO interface in comparison with the dehydrated interface, attributing this to the effect of water oxygens acting as efficient photoelectron scatterers about the absorber atom (Na). Contributions to the EXAFS signal from water oxygens were separated from those from the MgO surface oxygens by using deconvolution of Fourier transform spectra obtained by calculating each scattering path amplitude independently. Comparisons with calculations on Na^+ in aqueous solution were also made to distinguish between cations at the MgO interface and in solution. McCarthy et al. (159) also pointed out the possible applications of this technique, using MD along with the multiple-scattering calculations, to more complex, naturally occurring interfaces.

---

* FEFF stands for the effective scattering amplitude f_{eff}, an important ingredient in the theory of EXAFS (extended X-ray absorption fine structure) FEFF Project, Physics Department, University of Washington. See their World Wide Web address at: http://leonardo.phys.washington.edu/feff.

Table 6 lists some examples of molecular modeling techniques that have been applied to the interpretation of spectra.

B. Scientific Visualization

Visualization of results from the molecular simulations discussed in the previous sections is in a sense considered as important as the computational techniques used for the simulations. Therefore, the visualization methods and the software for this purpose have also been developed for many years. Visualizations we discuss here can be one of the following:

1. Structure viewing (either molecular or electronic structure)
2. Animation movie
3. Virtual reality presentation

Visualization of the lowest-energy structure is often accomplished by showing the snapshot of the lowest energy structure coordinates using molecular viewer software. Most modern software packages now include rendering capability with many editing features, with built-in file conversion or equipped with file format import/export functions. There are viewing softwares—in many cases freely available, solely made for molecular viewing purpose. These require the 3-D coordinates information from the computer simulations. The coordinates obtained from a specific molecular simulation (MC, MD, QM, or MM) can be imported to the viewing software if the software permits that file format. Otherwise, we can convert the file format produced from a simulation to the file format compatible to the viewing software. One useful software (UNIX-based) for this purpose is "BABEL."* BABEL is a program designed to interconvert a number of file formats currently used in molecular modeling. The program is available for Unix (AIX, Ultrix, Sun-OS, Convex, SGI, Cray, Linux) and MS-DOS and on Macs. The software (either commercial or freeware/shareware) for viewing molecular (or electronic) structures are listed in Table 7. Different software has different emphasis in visualization, depending on the scientific purpose. For example, there are viewers better suited for biological molecules, while there are viewers for organic or materials science results. Clay mineral simulation results visualization that we discuss here may well work with more materials science–based software, where the software is often capable of generating surface structures or two-dimensional or three-dimensional mineral structures.

Unlike molecular structure viewing, the animation requires more than just the final 3D coordinates (e.g., for the lowest-energy structure). Most of the molec-

---

* Babel is available via anonymous ftp from ccl.osc.edu in the directory pub/chemistry/software/ UNIX/babel.

TABLE 6 Molecular Simulations Applied to the Interpretation of Spectra

Model system	Method (expt./theory)	Physical property compared (interpreted)	Ref.
Salicylic acid adsorbed onto illite clay	ATR-FTIR/QM HF 3-G**, HF 6-311+G**	Correlation between calculated frequencies from QM vs. measured ATR-FTIR frequencies	161
Phyllosilicates, including smectite, mica, chlorite, saponite, and hectorite	X-ray photoelectron spectroscopy (XPS)/DFT	Calculated vibrational spectra from DOS (density of state) was used as an interplay with XPS results to measure the qualitative distribution of Al in the phyllosilicates.	76
Silicates and aluminosilicates models	NMR/QM HF 6-31G*, 6-311 (2d,p)	Calculated 23Na NMR deshielding calculation vs. experimental NMR shift	162
Carboxylic acids adsorbed onto mineral surfaces (illite, kaolinite, montmorillonite)	ATR-FTIR/QM HF 3-21G**	Calculated vibrational frequencies vs. measured ATR-FTIR frequencies	163
Li-, Na-, and K-montmorillonite	MC (MCY water)	Total radial distribution function from MC vs. 1H/2D isotopic difference neutron diffraction results	112
2:1 phyllosilicates	FT-IR, NMR/MC (inverse Monte Carlo)	Cation distributions obtained from MC compared with proportion of cation pairs determined by FT-IR and NMR	131
Vermiculite	IR, Raman/MD (HF 3-21G*/MP2 6-311G** used for force-field development	IR calculated from Fourier transformation of the autocorrelation function of the total dipole moment; Raman spectra calculated from Fourier transformation of the autocorrelation function of the polarizability tensor	132
Mica (muscovite)	MC (MCY water)	Water O density distribution from MC vs. X-ray reflectivity mearsurements	149

TABLE 7   Visualization Software

Visualization software	Features	Company/Web site/(Ref.)
Material Studio (PC version) Cerius2 (Unix version)	Both Material Studio and Cerius2 are complete simulation packages. They include variety of classical and quantum mechanical simulation codes with GUI (graphical user interface). Many visualization features are included, such as viewer, 3-D sketcher, and animation movie.	Accelrys http://www.accelrys.com
VMD (Visual Molecular Dynamics)	VMD is a molecular visualization program for displaying, animating, and analyzing large (biomolecular) systems using 3-D graphics and built-in scripting	Theoretical Biophysics Group http://www.ks.uiuc.edu/ Research/vmd (165)
DS Viewer (formerly WebLab Viewer)	Easy-to-use visualization tools. It shares results with computational packages or by including chemical information in documents, spreadsheets, etc. It also helps one to evaluate geometry and understand chemistry.	Accelrys http://www.accelrys.com/d studio/ds_viewer
AVS/5	Interactive tool for scientific visualization, often used for molecular structure visualization	Advance Visual Systems http://www.avs.com/soft ware/soft_t/avs5.html (166)
Chemscape Chime	It can rotate a molecular image within a Netscape page. Multiple molecules can be displayed within a single Web page. It also supports scripts (using a superset of the RasMol scripting language).	MDL Information Systems, Inc. http://www.mdli.com/ products/chemscape.html
RasMol	Easy-to-use visualization tool for molecules, and one can prepare publication-quality images (freeware).	http://www.openrasmol.org
MAGE (Windows, Mac, Unix)	It offers over a thousand excellent tutorials on molecular structures. It is a freeware with powerful macromolecular visualization.	http://www.umass.edu/ microbio/rasmol/ mage.htm
RM Scene Graph	An intuitive and easy-to-use viewer (renderer) for describing scenes (or molecules and atoms) using OpenGL. More than one camera and viewports can be used to achieve the multiple views.	R3Vis Corp. http://www.r3vis.com

ular dynamics simulation codes have a feature to output the trajectory files that describe the motion of all or part of the molecules simulated. Therefore, one can visualize this motion by showing the anim.rion, a movie with many frames, from this trajectory output file. This can be either classical MD or ab initio MD results. Many commercial software programs are capable of making animation movies from the trajectory output file generated within the molecular simulation software package. There is also software developed to produce an animation movie from the MD trajectory outputs elsewhere. See Table 8 for animation software for MD trajectory visualization.

Virtual reality (VR) visualization is the most up-to-date technology through which one can quite thoroughly examine the simulation results (molecular structure, electronic structure, animation, etc.). The VR technique creates fully immersive environments where the modeler (here, the clay mineral modeler) is surrounded by data objects in 3D space. One can actually feel oneself sitting (or standing) within the system (not in front of the system) of the molecular or electronic structure and feel the surrounding environment and almost touch each individual part (atomic, surface, bonds, chain, and so on). VR thus gives insights impossible to achieve by any other means and is not merely a 3D visualization or animation described by VRML (virtual reality modeling language),* which also can be used to visualize molecular structures. Most VR software requires special hardware for input and output. The basic components of full effect VR are (a) a high-performance computer graphic system, (b) a head-tracked display that presents the virtual world from the user's current head position, (c) three-dimensional input devices that allow the user to provide input directly in three dimensions. The virtual reality technique, with descriptions of its approaches, 3D database interfaces, document navigation, data mining, and creative chemical visualization, has been reviewed (164). One can apply the virtual reality technique to the molecular simulation results of clay mineral structure or the clay mineral–water interface to examine the structural detail and possibly obtain further insight from the fully immersive virtual environment.

VI. SUMMARY AND FUTURE PROSPECTS

So far we have discussed molecular modeling of clay minerals and their surface chemistry (either clay mineral's structural features or the clay mineral–water interface), different molecular modeling techniques (either classical mechanics or quantum mechanics in general), the interplay between spectroscopic methods and molecular simulations, and the visualization techniques of the results from

---

* VRML is a portable, platform-independent, and flexible file format for the transport of 3D graphical information. It can transport static scenes, animation, and multimedia.

computer simulations. Examples of methods, applications, and visualization software are provided for readers of interest who want to participate or understand the computer simulations as a theoretical tool for this important geochemical field.

Computer simulations of clay mineral surface chemistry can provide valuable insight, often allowing faster and more elaborate tests of theory than would be possible by experiment. They also allow a much more sophisticated analysis of experiment—such as the direct interplay between spectroscopy and simulations—than theory alone could provide or can further describe or visualize the system that has never been possible with experimental techniques.

Thanks to both hardware and software development, the most advanced form of molecular modeling, requiring no adjustable parameter, is becoming more popular in the field of computational chemistry. Clay mineral simulation will also benefit from this powerful technique, which is becoming more and more possible for large molecular systems such as the clay mineral–water interface. Recent applications of density-functional-theory-based ab initio molecular dynamics in chemically relevant systems are reviewed (32). Recent advances in the development of efficient numerical algorithms for the prediction of spectroscopic properties are also highlighted. Another recent approach to involving the application of first principles to large systems is ONIOM (our own n-layered integrated molecular orbital and molecular mechanics) approach (167) or the QM/MM approach (168). ONIOM is an attractive alternative to conventional ab initio MD, which is computationally expensive for large systems (167). ONIOM combines different levels of theory for one calculation, ranging from an MM level of theory for describing the steric and electrostatic effects of the exterior part of the system, to some intermediate levels of MO methods describing the electronic effects of functional groups or ligands close to the center of action, and finally to a highly accurate method to deal with the electron correlation at a very high level of theory on the most important action center of the system. One can possibly apply this multilayered simulation technique to the clay mineral–water system, where we can deal with different parts of the system with different levels of theoretical approach. Although the hybrid QM and MM treatments (QM/MM), where only a limited core part of the system is treated quantum mechanically, is getting more attention, there are still a number of open issues, such as how best to merge the QM and MM regions, what parameters to use for the empirical potential, and what level of treatment to use in the QM region (such as specially parameterized semiempirical treatments, ab initio)

The field of clay–polymer nanocomposite technology is attracting a great amount of attention (169). Among clay minerals, the 2:1 natural clay mineral montmorillonite has been used for clay nanocomposite applications. The structure of nanocomposites, the dynamics of confined polymer clay nanocomposites using NMR and computer simulations, rheology of clay nanocomposites have been

TABLE 8 Molecular Simulation Studies on Clay–Polymer Nanocomposites

Model	Method (basis set, potential, MD length/MC steps, etc.)	Aim	Ref.
Li^+-montmorillonite/PEO nanocomposite; Na^+-montmorillonite/PEO* nanocomposite	GCMC† simulation with MCY water model and potentials previously used for hydrated MMT (Skipper et al., 1995) and PEO–salt systems (Smith et al., 1993). NVT-MD simulations up to 1 ns.	Detailed structure of intercalated PEO nanocomposites; mobility of Li^+ ions; role of cation's hydration shells in cation conductivity	174
Li^+-montmorillonite/PEO nanocomposite vs. PEO polymers in bulk	MD simulations with the same potentials used by Ref. 174.	Comparison of structure and Li^+ diffusion between bulk Li^+/PEO and clay hybrid Li^+/PEO nanocomposite system.	175
Na-MMT with methanal (CH_2O) and ethylenediamine ($NH_2CH_2CH_2NH_2$) MMT–$MgAl_3(OH)_4(AlSi_7O_{20})$	Plane-wave DFT/GGA	Investigate potential acid sites of montmorillonite for catalyzing polymerization of methanal with ethylenediamine	176
MMT–$Si_{24}(Al_{10}Mg_2)O_{60}(OH)_{12}$ 10 selected ammonium ions (quats); two amino acids; nylon 6,6	MD simulations with Dreiding 2.21 force field; 200 ps; charge equilibration method; 8.5-Å cut-off	Predicting binding energy between nylon 6,6 and a clay platelet according to adsorbed hydrocarbon chain length, volume, and polarity of quats	177
Li^+-montmorillonite/PEO nanocomposite vs. PEO polymers in bulk	MD simulations with the potentials used by Ref. 174.	To trace how the confinement of PEO in MMT affects the segmental dynamics of PEO	178

* PEO: poly(ethylene oxide)
† GCMC: grand canonical Monte Carlo in which chemical

reviewed in the materials science field (170). Excellent reviews on polymer layered clay nanocomposites are also available (171–173). These reviews aim at reporting on recent developments in syntheses, properties, and future applications of polymer–layered silicate (smectite) nanocomposites. Also in this handbook, Ruiz-Hitzky et al. discuss clay–polymer complexes or nanocomposites in detail (see Chapter 3). Despite many experimental or synthetic studies of clay–polymer nanocomposite materials (see Chapter 3 for examples), molecular modeling (or even computer modeling in general) of clay nanocomposite is still under development. There are only handfuls of examples of molecular modeling (the methods discussed in Section III) where clay mineral (montmorillonite) and polymer intercalates have been used to model clay–polymer nanocomposites. Table 8 summarizes these molecular simulations applied to clay–polymer nanocomposites. Molecular modeling of clay–polymer nanocomposites can be a great aid to provide structure–property models that are essential for the further development of clay–polymer composite materials.

REFERENCES

1. G Sposito, NT Skipper, R Sutton, SH Park, AK Soper, JA Greathouse. Proc. Natl. Acad. Sci. USA 1999; 96:3358.
2. GH Bolt, MGM Bruggenwert. Soil Chemistry: A. Basic Elements. Amsterdam: Elsevier Scientific, 1978.
3. G Sposito. The Chemistry of Soils. New York: Oxford University Press, 1989.
4. KH Tan. Principles of Soil Chemistry. New York: Marcel Dekker, 1993.
5. H Bohn, B McNeal, G O'Connor. Soil Chemistry. New York: Wiley, 1979.
6. ME Schrader, GI Loeb. Modern Approaches to Wettability: Theory and Applications. New York: Plenum Press, 1992.
7. F Hetzel, D Tessier, AM Jaunet, H Doner. Clay Clay Min 1994; 42:242.
8. C Shang, ML Thompson, DA Laird. Clay Clay Min 1995; 43:128.
9. DH Powell, HE Fischer, NT Skipper. J. Phys. Chem. B 1998; 102:10899.
10. DH Powell, K Tongkhao, SJ Kennedy, PG Slade. Physica B 1998; 241–243:387.
11. A Szabo, NS Ostlund. Modern Quantum Chemistry. New York: McGraw-Hill, 1989.
12. MC Zerner. Semiempirical molecular orbital methods In: KB Lipkowitz, DB Boyd, eds. Reviews in Computational Chemistry. New York: VCH, 1991.
13. J Cioslowski. Ab initio calculations on large molecules: methodology and applications. In: KB Lipkowitz, DB Boyd, eds. Reviews in Computational Chemistry. New York: VCH, 1993.
14. M Born, JR Oppenheimer. Ann. Physik 1927; 84:457.
15. A Luchow, JB Anderson. Annu. Rev. Phys. Chem 2000; 51:501.
16. WJ Hehre, L Radom, PvR Schleyer, JA Pople. Ab Initio Molecular Orbital Theory. New York: Wiley, 1986.
17. JA Pople, DL Beveridge. Approximate Molecular Orbital Theory. New York: McGraw-Hill, 1970.

18. MJS Dewar, W Thiel. J. Am. Chem. Soc 1977; 99:4899.
19. MJS Dewar, EG Zoebisch, EF Healy, JJP Stewart. J. Am. Chem. Soc 1985; 107: 3902.
20. JJP Stewart. J. Comput. Chem 1989; 10:209.
21. JJP Stewart. J. Comput. Chem 1989; 10:221.
22. MJS Dewar, CX Jie, JG Yu. Tetrahedron 1993; 49:5003.
23. T Clark. A Handbook of Computational Chemistry: A Practical Guide to Chemical Structure and Energy Calculations. New York: Wiley, 1985.
24. M Boero, K Terakura, M Tateno. J. Am. Chem. Soc 2002; 124:8949.
25. B Winkler. Z. Kristall 1999; 214:506.
26. RG Parr, W Yang. Density Functional Theory of Atoms and Molecules. Oxford: Oxford University Press, 1989.
27. RM Dreizler, EKU Gross. Density functional theory. In: EKU Gross, RM Dreizler, eds. NATO ASI Series B. New York: Plenum Press, 1995.
28. J Seminario, P Politzer, eds. Modern Density Functional Theory—A Tool for Chemistry (Theoretical and Computational Chemistry Vol. 2). Amsterdam: Elsevier, 1995.
29. J Seminario, ed. Recent Developments and Applications of Modern Density Functional Theory (Theoretical and Computational Chemistry Vol. 4). Amsterdam: Elsevier, 1996.
30. H Chermette. Coord. Chem. Rev 1998; 180:699.
31. FQ Ban, KN Rankin, JW Gauld, RJ Boyd. Theor. Chem. Acc 2002; 108:1.
32. JS Tse. Annu. Rev. Phys. Chem 2002; 53:249.
33. P Hohenberg, W Kohn. Phys. Rev. B 1964; 136:B864.
34. M Krack, A Gambirasio, M Parrinello. J. Chem. Phys 2002; 117:9409.
35. W Kohn, LJ Sham. Phys. Rev 1965; 140:1133.
36. T Ziegler. Can. J. Chem.-Rev. Can. Chim 1995; 73:743.
37. NJ Ramer, AM Rappe. Phys. Rev. B 1999; 59:12471.
38. IN Levine. Quantum Chemistry. New York: Prentice-Hall, 2000.
39. C Lee, W Yang, RG Parr. Phys. Rev. B 1988; 37:785.
40. JP Perdew, Y Wang, E Engel. Phys. Rev. Lett 1991; 66:508.
41. JP Perdew, M Ernzerhof, K Burke. J. Chem. Phys 1996; 105:9982.
42. Gaussian 94. Pittsburgh, PA: Gaussian, Inc., 1995.
43. JE Williams, PJ Stang, PV Schleyer. Annu. Rev. Phys. Chem 1968; 19:531.
44. EM Engler, JD Andose, PV Schlever. J. Am. Chem. Soc 1973; 95:8005.
45. NL Allinger, JT Sprague. J. Am. Chem. Soc 1973; 95:3893.
46. O Ermer. Struct. Bond 1976; 27:161.
47. AT Hagler. Theoretical Simulation of Conformation, Energetics, and Dynamics of Peptides. New York: Academic Press, 1985.
48. NL Allinger. J. Chem. Soc.Am 1977; 99:8127.
49. NL Allinger, YH Yuh, JH Lii. J. Am. Chem. Soc 1989; 111:8551.
50. AK Rappe, CJ Casewit, KS Colwell, WA Goddard, WM Skiff. J. Am. Chem. Soc 1992; 114:10024.
51. RJ Bartlett, JF Stanton. Application of Post-Hartree–Fock Methods: A Tutorial. New York: VCH, 1994.

52. R Fletcher, CM Reeves. Comput. J 1964; 7:149.
53. J Dennis, RB Schnabel. Numerical Methods for Unconstrained Optimization and Nonlinear Equations. Englewood Cliffs, NJ: Prentice-Hall, 1983.
54. R Fletcher. Practical Methods of Optimization. Tiptree. Essex. UK: Wiley, 1987.
55. PE Gill, M W Gill. Practical Optimization. New York: Academic Press, 1983.
56. BJ Alder, TE Wainwright. J. Chem. Phys 1957; 27:1208.
57. BJ Alder, TE Wainwright. J. Chem. Phys 1959; 31:459.
58. A Rahman. Phys. Rev 1964; 136:A405.
59. F Stillinger, A Rahman. J. Chem. Phys 1974; 60:1545.
60. PM Rodger. Mol. Simulat 1989; 3:263.
61. S Nosé. Mol. Phys 1984; 52:255.
62. WG Hoover. Phys. Rev. A 1985; 31:1695.
63. MP Allen, DJ Tildesley. Computer Simulation of Liquids. New York: Oxford University Press, 1989.
64. N Metropolis, AW Rosenbluth, MN Rosenbluth, AH Teller, E Teller. J. Chem. Phys 1953; 21:1087.
65. AM Mathiowetz. Ph.D. dissertation, California Institute of Technology. 1993.
66. MD Newton, GV Gibbs. Phys. Chem. Miner 1980; 6:221.
67. GV Gibbs. Am. Miner 1982; 67:421.
68. BC Chakoumakos, RJ Hill, GV Gibbs. Am. Miner 1981; 66:1237.
69. GV Gibbs, LW Finger, MB Boisen. Phys. Chem. Miner 1987; 14:327.
70. BJ Teppen, CH Yu, DM Miller, L Schafer. J. Comput. Chem 1998; 19:144.
71. B Winkler, V Milman, MC Payne. Am. Miner 1994; 79:200.
72. JD Hobbs, RT Cygan, KL Nag, PA Schultz, MP Sears. Am. Miner 1997; 82:657.
73. DL Bish. Clay Clay Min 1993; 41:738.
74. CH Bridgeman, AD Buckingham, NT Skipper, MC Payne. Mol. Phys 1996; 89: 879.
75. JJ Rehr, JM Deleon, SI Zabinsky, RC Albers. J. Am. Chem. Soc 1991; 113:5135.
76. T Ebina, T Iwasaki, A Chatterjee, M Katagiri, GD Stucky. J. Phys. Chem. B 1997; 101:1125.
77. D Bougeard, KS Smirnov, E Geidel. J. Phys. Chem. B 2000; 104:9210.
78. CI Sainz-Diaz, V Timon, V Botella, A Hernandez-Laguna. Am. Miner 2000; 85: 1038.
79. D Tunega, G Haberhauer, MH Gerzabek, H Lischka. Langmuir 2002; 18:139.
80. K Refson, S-H Park, G Sposito. J. Phys. Chem. B. In press 2003; 107.
81. SH Park, K Refson, G Sposito. J. Phys. Chem. B. Submitted. 2003.
82. P Liebmann, G Loew, S Burt, J Lawless, RD MacElroy. Inorg. Chem 1982; 21: 1586.
83. DA Laird. Clay Clay Min 1996; 44:553.
84. A Chatterjee, T Iwasaki, T Ebina, H Hayashi. Appl. Surf. Sci 1997; 121:167.
85. NT Skipper, K Refson, JDC McConnell. J. Chem. Phys 1991; 94:7434.
86. NT Skipper, K Refson, JDC McConnell. Clay Min 1989; 24:411.
87. O Matsuoka, E Clementi, M Yoshimine. J. Chem. Phys 1976; 64:1351.
88. A Delville. Langmuir 1991; 7:547.
89. WL Jorgensen, J Chandrasekhar, JD Madura, RW Impey, ML Klein. J. Chem. Phys 1983; 79:926.

90. A Delville. Langmuir 1992; 8:1796.
91. A Delville. J. Phys. Chem 1993; 97:9703.
92. A Delville. J. Phys. Chem 1995; 99:2033.
93. NT Skipper, F-RC Chang, G Sposito. Clay Clay Min 1995; 43:285.
94. NT Skipper, G Sposito, F-RC Chang. Clay Clay Min 1995; 43:294.
95. MH Fu, ZZ Zhang, PF Low. Clay Clay Min 1990; 38:485.
96. ES Boek, PV Coveney, NT Skipper. J. Am. Chem. Soc 1995; 117:12608.
97. H Sato, A Yamagishi, S Kato. J. Phys. Chem 1992; 96:9377.
98. H Sato, A Yamagishi, S Kato. J. Am. Chem. Soc 1992; 114:10933.
99. H Sato, A Yamagishi, S Kato. J. Phys. Chem 1992; 96:9382.
100. A Yamagishi, M Soma. J. Am. Chem. Soc 1981; 103:4640.
101. A Yamagishi. J. Phys. Chem 1982; 86:2472.
102. A Yamagishi. J. Chem. Soc.-Dalton Trans 1983:679.
103. A Yamagishi. Inorg. Chem 1985; 24:1689.
104. H Sato, A Yamagishi, K Naka, S Kato. J. Phys. Chem 1996; 100:1711.
105. GL Keldsen, JB Nicholas, KA Carrado, RE Winans. J. Phys. Chem 1994; 98:279.
106. U Burkert, NL Allinger. Molecular Mechanics. Washington, DC: American Chemical Society, 1982.
107. A Hopfinger. B. Am. Phys. Soc 1973; 18:372.
108. J Breu, CRA Catlow. Inorg. Chem 1995; 34:4504.
109. J Breu, N Raj, CRA Catlow. J. Chem. Soc.-Dalton Trans 1999:835.
110. S Karaborni, B Smit, W Heidug, J Urai, E vanOort. Science 1996; 271:1102.
111. G Sposito, S-H Park, R Sutton. Clay Clay Min 1999; 47:192.
112. SH Park, G Sposito. J. Phys. Chem. B 2000; 104:4642.
113. S Aronowitz, L Coyne, J Lawless, J Rishpon. Inorg. Chem 1982; 21:3589.
114. M O'Keeffe, PF McMillan. J. Phys. Chem 1986; 90:541.
115. J Sauer. Chem. Phys. Lett 1983; 97:275.
116. AC Hess, PF McMillan, M Okeeffe. J. Phys. Chem 1986; 90:5661.
117. WF Bleam, R Hoffmann. Inorg. Chem 1988; 27:3180.
118. AC Hess, VR Saunders. J. Phys. Chem 1992; 96:4367.
119. BJ Teppen, DM Miller, SQ Newton, L Schafer. J. Phys. Chem 1994; 98:12545.
120. JR Hill, J Sauer. J. Phys. Chem 1995; 99:9536.
121. VA Ermoshin, KS Smirnov, D Bougeard. Chem. Phys 1996; 209:41.
122. VA Ermoshin, KS Smirnov, D Bougeard. J. Mol. Struct 1997; 410:371.
123. JDC McConnell, A DeVita, SD Kenny, V Heine. Phys. Chem. Miner 1997; 25:15.
124. BJ Teppen, K Rasmussen, PM Bertsch, DM Miller, L Schafer. J. Phys. Chem. B 1997; 101:1579.
125. BJ Teppen, CH Yu, SQ Newton, DM Miller, L Schafer. J. Mol. Struct 1998; 445:65.
126. A Chatterjee, T Iwasaki, T Ebina. J. Phys. Chem. A 2000; 104:8216.
127. L Benco, D Tunega, J Hafner, H Lischka. J. Phys. Chem. B 2001; 105:10812.
128. L Benco, D Tunega, J Hafner, H Lischka. Am. Miner 2001; 86:1057.
129. L Benco, D Tunega, J Hafner, H Lischka. Chem. Phys. Lett 2001; 333:479.
130. CI Sainz-Diaz, A Hernandez-Laguna, MT Dove. Phys. Chem. Miner 2001; 28:322.
131. CI Sainz-Diaz, J Cuadros, A Hernandez-Laguna. Phys. Chem. Miner 2001; 28:445.

132. M Arab, D Bougeard, KS Smirnov. Phys. Chem. Chem. Phys 2002; 4:1957.
133. A Chatterjee, T Iwasaki, T Ebina. J. Phys. Chem. A 2002; 106:641.
134. PF Low, JH Cushman, DJ Diestler, DJ Mulla. J. Colloid Interf. Sci 1984; 100:576.
135. ES Boek, PV Coveney, NT Skipper. Langmuir 1995; 11:4629.
136. F-RC Chang, NT Skipper, G Sposito. Langmuir 1995; 11:2734.
137. CH Bridgeman, NT Skipper. J. Phys.-Condes. Matter 1997; 9:4081.
138. F-RC Chang, NT Skipper, G Sposito. Langmuir 1997; 13:2074.
139. F-RC Chang, NT Skipper, G Sposito. Langmuir 1998; 14:1201.
140. J Greathouse, G Sposito. J. Phys. Chem. B 1998; 102:2406.
141. KS Smirnov, D Bougeard. J. Phys. Chem. B 1999; 103:5266.
142. RM Shroll, DE Smith. J. Chem. Phys 1999; 111:9025.
143. DA Young, DE Smith. J. Phys. Chem. B 2000; 104:9163.
144. A Kalinichev, RJ Kirkpatrick, RT Cygan. Am. Miner 2000; 85:1046.
145. MR Warne, NL Allan, T Cosgrove. PCCP Phys. Chem. Chem. Phys 2000; 2:3663.
146. JA Greathouse, K Refson, G Sposito. J. Am. Chem. Soc 2000; 122:11459.
147. EJM Hensen, TJ Tambach, A Bliek, B Smit. J. Chem. Phys 2001; 115:3322.
148. JA Greathouse, EW Storm. Mol. Simul 2002; 28:633.
149. SH Park, G Sposito. Phys. Rev. Lett 2002; 89:085501.
150. H Ohtaki, T Radnai. Chem. Rev 1993; 93:1157.
151. G Sposito, R Prost. Chem. Rev 1982; 82:553.
152. JD Kubicki. Interpretation of vibrational spectra using molecular orbital theory calculations. In: RT Cygan, JD Kubicki, eds. Reviews in Mineralogy and Geochemistry. Washington DC: Geochemical Society and Mineralogical Society of America, 2001.
153. AC Lasaga, GV Gibbs. Phys. Chem. Miner 1988; 16:29.
154. JA Pople, HB Schlegel, R Krishnan, DJ Defrees, JS Binkley, MJ Frisch, RA Whiteside, RF Hout, WJ Hehre. Int. J. Quantum Chem 1981:269.
155. AA Chialvo, PT Cummings, JM Simonson, RE Mesmer, HD Cochran. Ind. Eng. Chem. Res 1998; 37:3021.
156. NT Skipper, AK Soper, MV Smalley. J. Phys. Chem 1994; 98:942.
157. NT Skipper, MV Smalley, GD Williams, AK Soper, CH Thompson. J. Phys. Chem 1995; 99:14201.
158. GD Williams, AK Soper, NT Skipper, MV Smalley. J. Phys. Chem. B 1998; 102:8946.
159. MI McCarthy, GK Schenter, MR ChaconTaylor, JJ Rehr, GE Brown. Phys. Rev. B 1997; 56:9925.
160. SI Zabinsky, JJ Rehr, A Ankudinov, RC Albers, MJ Eller. Phys. Rev. B 1995; 52:2995.
161. JD Kubicki, MJ Itoh, LM Schroeter, SE Apitz. Environ. Sci. Technol 1997; 31:1151.
162. JA Tossell. Phys. Chem. Miner 1999; 27:70.
163. JD Kubicki, LM Schroeter, MJ Itoh, BN Nguyen, SE Apitz. Geochim. Cosmochim. Acta 1999; 63:2709.
164. WD Ihlenfeldt. J. Mol. Model 1997; 3:386.
165. W Humphrey, A Dalke, K Schulten. J. Mol. Graph 1996; 14:33.

166. BS Duncan, TJ Macke, AJ Olson. J. Mol. Graph 1995; 13:385.

167. M Svensson, S Humbel, RDJ Froese, T Matsubara, S Sieber, K Morokuma. J. Phys. Chem 1996; 100:19357.

168. M Freindorf, JL Gao. J. Comput. Chem 1996; 17:386.

169. SD Burnside, EP Giannelis. Chem. Mater 1995; 7:1597.

170. EP Giannelis, R Krishnamoorti, E Manias. Adv. Polym. Sci 1999; 138:107.

171. M Alexandre, P Dubois. Mater. Sci. Eng. R-Rep 2000; 28:1.

172. TJ Pinnavaia, GW Beall, eds. Polymer-Clay Nanocomposites. New York: Wiley, 2000.

173. KA Carrado. Polymer-clay nanocomposites. In: G Shonaike, S Advani, eds. Advanced Polymeric Materials: Structure–Property Relationships. Boca Raton. FL: CRC Press, 2003.

174. E Hackett, E Manias, EP Giannelis. Chem. Mater 2000; 12:2161.

175. V Kuppa, E Manias. Chem. Mater 2002; 14:2171.

176. S Stackhouse, PV Coveney, E Sandre. J. Am. Chem. Soc 2001; 123:11764.

177. G Tanaka, LA Goettler. Polymer 2002; 43:541.

178. V Kuppa, E Manias. J. Chem. Phys 2003; 118:3421.

3

Clay–Organic Interactions: Organoclay Complexes and Polymer–Clay Nanocomposites

Eduardo Ruiz-Hitzky, Pilar Aranda, and José María Serratosa
Consejo Superior de Investigaciones Científicas (CSIC)
Cantoblanco
Madrid, Spain

I. INTRODUCTION: BACKGROUND, HISTORY, AND SIGNIFICANCE OF CLAY–ORGANIC INTERACTIONS

Clay minerals are silicates that, due to their unique characteristics (layer or fibrous structures, ion exchange capacity, variability in chemical composition and in electrical charge of particles, etc.), present special physicochemical behaviors that determine their surface properties and, therefore, their interactions with organic substances. One of the most salient features of these interactions is that sorption of molecules of different functionality affects not only the external surface but also, in most cases, the intracrystalline region of the silicate structure. Moreover, such interactions involve a wide variety of bonding mechanisms and energies (Table 1) resulting in organic–inorganic materials of variable stability. Due to the peculiarities just mentioned and the existence of well-crystallized species amenable to detailed structural analyses, clay minerals constitute appropriate model systems for the study of physical and chemical phenomena pertinent to adsorption and reactivity in constrained spaces. These processes have special characteristics different from those on open surfaces or in solution. The knowl-

TABLE 1   Mechanisms of Clay–Organic Interactions

Nature of the interactions	Characteristics
Electrostatic	Ion exchange of interlayer metal cations with organic cations
van der Waals forces Hydrogen bonding and water bridges Ion dipole and coordination Proton transfer Electron transfer	Adsorption of neutral molecules by interactions with external or internal (intracrystalline region of silicates) surfaces
Covalent bonding	Grafting reactions of organic groups

edge already acquired on clay–organic interactions has been extended to other constrained systems, such as inorganic or organic layered compounds and zeolites. In fact, many of the studies published recently concerning adsorption of organic species on layered compounds have their antecedents in studies made some years ago on clay minerals.

The importance of clay–organic interactions also arises because such compounds are used in industrial applications, such as additives, fillers, rheological agents, and specific sorbents, and, recently, in technological applications as advanced materials. Moreover, the clay–organic interaction is of crucial importance regarding agricultural production, the origin and exploitation of oil resources, and the origin of life on our planet. The current relevance and vitality of the field can be inferred from the exponential increase in publications appearing during the last few decades in scientific journals of general scope and in journals specializing in clays, colloids, and materials topics and also from the overwhelming number of patents devoted to applications on this subject. In addition to this book, other useful reviews and books have been published (1–10). The aim of this chapter is to review and summarize the main mechanisms governing clay–organic interactions, from discrete molecules or cations to large entities, such as polymers. Only a few detailed examples, with emphasis on our own experience, will be presented in trying to group these interactions into several representative systems. In addition, the characterization of organic–inorganic solids derived from various common clays, as well as a discussion of current research and trends in the development of new hybrid compounds, will be addressed. Finally, a general overview about the perspectives for application of these hybrids compounds as advanced materials will be presented. Milestones concerning clay–organic interactions and relevant research achievements are collected in Table 2, as a sort of historical synopsis.

TABLE 2 Milestones in Clay–Organic Interactions

Date	Milestones	Examples	Authors (Ref.)
8th century AD	Clay–dye hybrid compounds	Palygorskite/indigo mixtures (Maya blue)	Cited by van Olphen in 1966 (74)
1911	Catalytic transformation of organic compounds activated by clays	Pinene to camphene over palygorskite	Montaland (245)
1939–41	Intercalation of organic cations in smectites	Montmorillonite/aliphatic and aromatic ammonium cations	Gieseking (11) and Hendricks (12)
1944	Intercalation of neutral species in clays	Montmorillonite/glycerol	MacEwan (246)
1945	Structural features by XRD, one-dimensional Fourier analysis of clay–organic compounds	Montmorillonite/diamines and glycols	Bradley (41)
1949	Application of thermal analysis in organoclay characterization	DTA of montmorillonite/alkylammonium	Jordan (32) and Allaway (247)
1954	Selective sorption of hydrocarbons by palygorskite and sepiolite	Palygorskite/n-heptane + iso-octane	Barrer et al. (68)
1961	Intercalation of salts in kaolinite	Kaolinite/K-acetate	Wada (63)
1961	Intercalation of neutral molecules in kaolinite	Kaolinite/urea	Weiss (248)
1961	UV-v is application to study clay–organic systems	Montmorillonite/benzidine	H. Hasegawa (249)
1961	Polymer–clay intercalation compounds	Montmorillonite/ polyacrylonitrile	Blumstein (109)
1965	Orientation of organic molecules in the interlayer space by IR spectroscopy	Montmorillonite and vermiculite/pyridine and pyridinium	Serratosa (25,26)
1968	Organic derivatives of clays through covalent bonds (grafting)	Vinyl derivatives of chrysotile	Fripiat and Mendelovici (84)
1969	π Bonds in clay–aromatic compounds	Cu-montmorillonite/benzene	Doner and Mortland (54)
1974	Organic reactions in the interlayer space of clays	Vermiculite/L-ornithine peptide formation	Rausell and Fornes (21)

(*Continued*)

TABLE 2 *Continued*

Date	Milestones	Examples	Authors (Ref.)
1974	Intracrystalline sorption of organic compounds in sepiolite	Sepiolite/hexane	Serna and Fernandez-Alvarez (250)
1976	Organic pillared clays	Montmorillonite/ diprotonated-triethylenediamine	Mortland and Berkheiser (213)
1980	Application of ^{13}C-NMR to characterize clay–organic systems	Ag-hectorite/benzene	Resing et al. (251)
1983	Catalysts based on organometallic–clay complexes	Montmorillonite/ $[Rh(PPh_3)_3]^+$	Pinnavaia (219)
1984	LMMS application to characterize clay–organic systems	Sepiolite/organosilanes	De Waele et al. (252)
1985	Photostabilization of coadsorbed labile bioactive species	Montmorillonite/methyl green/bioresmethrin	Margulies et al. (226)
1989	Microwave activation of organic reactions on clay–adsorbed compounds	Rearrangement of pinacol to pinacolone intercalated in montmorillonite	Gutierrez et al. (137)
1990	Ion-conducting polymers—2D intercalated materials	Montmorillonite/PEO	Ruiz-Hitzky and Aranda (122)
1993	Polymer melt intercalation in organo-smectites	Montmorillonite/ polystyrene	Vaia et al. (134)
1993	Grafting of organic groups in the interlayer space of kaolinite	Methoxy derivatives of kaolinite	Tunney and Detellier (108)
1998	Templated synthesis of polymer–clay nanocomposites	Synthetic fluoro-smectite/ polyvinylpyrrolidone	Carrado and Xu (111)

II. INTERACTION OF CLAYS WITH ORGANIC CATIONS

Gieseking (11) and Hendricks (12) showed for the first time that inorganic exchangeable cations of clay minerals can be replaced by organic cations through ion exchange reactions in aqueous solution of organic-base salts, such as hydrochlorides. The reaction may be expressed as in Eq. (1). X-ray diffraction was used to demonstrate that penetration of the organic cations into the interlayer space occurred and, therefore, that true organic–inorganic compounds were obtained.

$$RH^+ + M^+\text{-clay} \rightleftarrows RH^+\text{-clay} + M^+ \tag{1}$$

(R = organic base capable of protonation; M^+ = exchangeable inorganic cation.)

Since then, numerous organic salts of clay minerals have been prepared and studied. In most cases, the organic bases are aliphatic or aromatic amines and nitrogenated heterocycles or macrocycles. Among these species are cationic dyes (methylene blue, acridine orange, etc.) and others that have biological interest (aminoacids, purine bases, or nucleosides) (Table 3). The bonding mechanism between the organic cation and the charged clay layers is essentially electrostatic, but other noncoulombic forces may also contribute to the adsorption. In particular, van der Waals attractions between the organic species and the silicate surface, as well as between adjacent organic species themselves, adds to the adsorption forces. These become progressively more significant as the molecular weight of the organic species increases. Adsorption isotherms show that, for large cations, van der Waals forces dominate, since the principal interaction is between the

TABLE 3 Interactions of Clay Minerals with Organic Cations

Ionic species	Characteristic examples	Authors (Ref.)
Alkyl- and aryl-ammonium cations	$C_nH_{2n+1}NH_3^+$/smectites and vermiculites	Gieseking (11), Weiss (27), Lagaly (28), Johns and Sen Gupta (17), Martín-Rubí et al. (19)
Protonated nitrogenated heterocycles	Pyridinium/montmorillonite and vermiculite chlorobenzene on pyridinium/montmorillonite	Serratosa (26,30)
Aminoacids	Complexes with montmorillonite and vermiculite	Garret and Walker (253), Rausell-Colom and Salvador (254), Rausell-Colom and Fornés (21)
Purines and nucleosides	Adenine and adenosine/montmorillonite	Hendricks (12), Lailach and Brindley (24)
Alkaloids	Codeine/montmorillonite	Hendricks (12), Weiss (27)
Cationic dyes	Methylene blue/montmorillonite	Gieseking (11), Bergman and O'Konski (221)
Pesticides	Diquat and paraquat/montmorillonite and vermiculite	Weber et al. (256), Weed and Weber (257)
Surfactants	Complexes with smectites	Barrer (258)

adsorbed organic species themselves rather than between organic species and the clay surface (13,14).

Fixation is favored if the organic cations contain radicals capable of interacting by H bonding with the surface oxygens of the silicate. Since these are arranged in ditrigonal six-membered rings, the presence of groups with trigonal symmetry, i.e., $-NH_3^+$ groups in alkylammonium ions, favors reaction (15). When the charge on the silicate is tetrahedrally located, the combined effect of electrostatic and H-bond interactions leads to *keying* of the alkylammonium ion $-NH_3^+$ groups into the surface cavities (16–19). The possibility of $-CH_2-$ or terminal $-CH_3$ groups being H bonded to the surface oxygens has also been investigated, and IR spectroscopy shows evidence in favor of the existence of such bonds. In the case of alkylammonium complexes of vermiculite, IR spectra show a splitting of the symmetric deformation vibration band of $-CH_3$ at 1380 cm^{-1}, with a new, perturbed, dichroic component appearing at 1395 cm^{-1} (20). This is interpreted as terminal $-CH_3$ groups with their C_3 axes perpendicular to the surface, and indicates weak interactions between these groups and the silicate oxygen atoms.

Hydrogen bonding of $-NH_3^+$ groups to surface oxygens may be prevented if the base itself contains functional groups acting as electron donors. If they also carry a negative charge, then coulombic and/or H-bond interactions may be preferentially directed toward those groups rather than to the surface. This is the case for L-ornithine cations adsorbed on vermiculite, where the presence of a charged carboxyl group causes both $-NH_3^+$ groups to be located away from the surface and directed toward the carboxyls of neighboring cations (21).

The adsorption of organic cations is also dependent upon solution pH. At a given pH, the concentration of cations in solution relative to the concentration of uncharged molecules is dependent on the pK value of the base. If the pH is adjusted to be equal to the pK value, then the ratio of cation to free base is equal to unity, and cation exchange may be accompanied by adsorption of neutral molecules of the same organic species. For cations to be the dominant species in solution, the pH should be at least one or two units lower than the pK. If too acidic, the adsorption may be hindered due to competition with H$^+$ ions or with metal cations released from the silicate lattice by acid attack. Adsorption will depend, too, on the solubility of the base in water that, in turn, may be pH dependent (22).

The dependence of organic base adsorption on pH is illustrated by the case of benzidine. This base is adsorbed from aqueous solutions on Ca- and Na-montmorillonite as a mixture of monovalent and divalent cations at pH $\approx$ 3 and as neutral molecules when the pH is higher. Aniline is also adsorbed as a mixture of monovalent cations and neutral molecules at pH $=$ 3.2. Amounts adsorbed are in excess of the exchange capacity of the clay, and in both cases intermolecular association by hydrogen bonds (see Fig. 1) is assumed to exist between cations and molecules in the interlayer space (23).

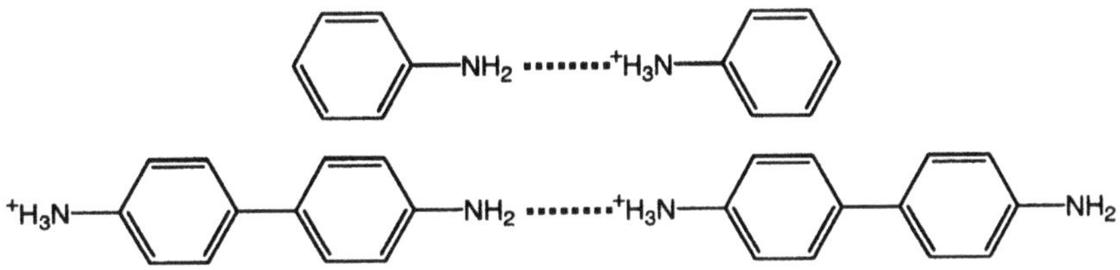

FIGURE 1 Intermolecular association by hydrogen bonds of aniline and benzidine.

The role played by intermolecular H bonding in the simultaneous adsorption of organic molecules and cations is more strikingly illustrated by the specific coadsorption of purines and pyrimidines on montmorillonite. Thymine and uracil are adsorbed appreciably if solutions also contain adenine, which is coadsorbed. Coadsorption is attributed to H bonding between molecules (see Fig. 2) in solution, rather than to adsorption of the purine cations and subsequent uptake of pyrimidine molecules by the complex (24).

The arrangement of organic cations within the interlayer space of layer silicates depends, essentially, on three factors: the size of the organic cation, the charge density, and the location of charge (tetrahedral or octahedral) in the silicate layers. The influence of these factors is illustrated by complexes of smectites and vermiculites with butylammonium (19) and pyridinium ions (25,26). In the case of butylammonium, three arrangements have been identified, depending upon the layer charge and the location of charge between three clays: Wyoming-montmorillonite (half-unit-cell charge of 0.33, octahedrally located), Beni-Buxera vermiculite (half-unit-cell charge of 0.72, tetrahedrally located), and Llano vermiculite

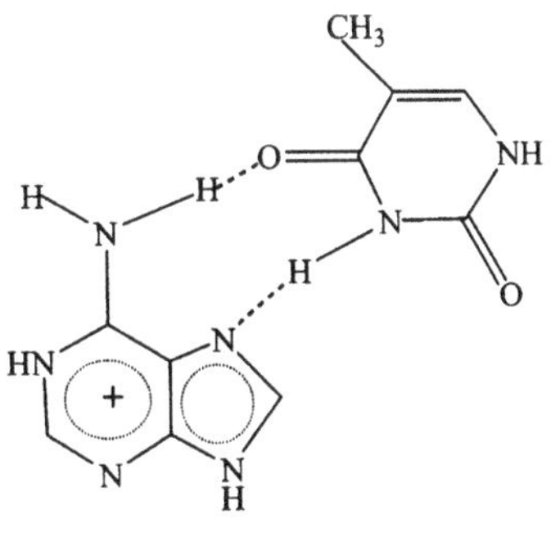

FIGURE 2 Possible coadsorption complex of adenine–thymine. (Based on Ref. 24.)

(half-unit-cell charge of 0.95, tetrahedrally located). With montmorillonite, where the charge density is low and predominantly octahedral, organic cations lie parallel to the silicate layers with an all-trans conformation, as in Figure 3a. Both the basal spacing of the complex (d_{001} = 1.35 nm) and evidence from IR (nondichroic, nonperturbed δ_{NH3} at 1500 cm^{-1}) are consistent with this disposition.

In contrast, in the complexes with the two vermiculites, the δ_{NH3} band is both dichroic and shifted toward higher frequencies (1532 and 1572 cm^{-1}), indicating that the C_3 axes of NH_3^+ end groups are perpendicular to the layers and that NH_3 groups are hydrogen bonded to the surface oxygens. However, the spacing (d_{001} = 1.47 nm) for the high-charge Llano vermiculite complex corresponds to molecules tilted 55° to the surface in an all-trans conformation (Fig. 3c). For the medium-charge Beni-Buxera vermiculite complex, the spacing (d_{001} = 1.32 nm) corresponds to molecules lying flat on the surface (Fig. 3b) by means of a conformational change via a rotation of 120° around the C_1—C_2 bond. These different arrangements result from differences in charge density and charge location in the substrates. The tetrahedral location of charge in vermiculites causes the NH_3^+ end groups to be keyed into the ditrigonal cavities. The surface charge density of the layers will determine the conformation that the aliphatic chains will adopt. Molecules with conformations such as those in Figure 3b occupy an area of 0.30 nm^2, which is just smaller than the area available in Beni-Buxera vermiculite (0.35 nm^2 per unit charge). For Llano vermiculite (available area of 0.28 nm^2 per unit charge), steric hindrance prevents such an interlayer arrangement, and aliphatic chains stand up in an all-trans conformation, as in Figure 3c.

Similarly, pyridinium ions, $C_5H_5NH^+$, adopt different orientations in the interlayer space of layer silicates, depending on the layer charge of the substrate, as shown by IR and X-ray diffraction results (25,26). Thus, in montmorillonite, pyridinium ions are arranged with their planes parallel to the silicate layers; in vermiculite, the plane and C_2 axis of the pyridinium ion are essentially perpendicular to the layers. In the latter case, monodimensional X-ray Fourier analysis confirms the orientation obtained from IR spectra (Fig. 4).

The arrangement of long-chain alkylammonium cations in the interlayer space of 2:1 clay minerals has been studied extensively, and excellent reviews on the subject have been published (27–29). The arrangement adopted by the intercalated alkylammonium ions depends, as mentioned earlier, on the charge of the silicate and on the length of the alkyl chain (Fig. 5). Alkylammonium ions adsorbed on smectites are typically arranged as monolayers (d_{001} = 1.35 nm) or bilayers (d_{001} = 1.74 nm). But for high charged layer silicates, such as vermiculites [where the equivalent surface area, i.e., the area available on the silicate layer per unit charge (nm^2/unit charge) is smaller], the alkylammonium ions form paraffin-type structures. Their arrangements are also influenced by the possibility of conformational changes through rotations around C—C bonds (kinks). This

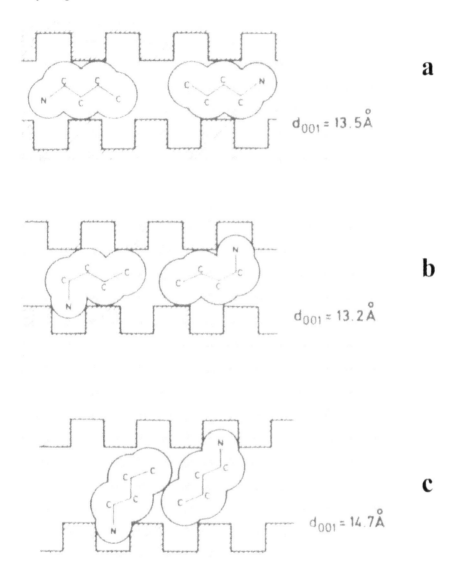

a

$d_{001} = 13.5\,\overset{\circ}{A}$

b

$d_{001} = 13.2\,\overset{\circ}{A}$

c

$d_{001} = 14.7\,\overset{\circ}{A}$

FIGURE 3 Arrangement of butylammonium cations in the interlayer space of (a) Wyoming montmorillonite, (b) medium-charge Beni-Buxera vermiculite, and (c) high-charge Llano vermiculite. (After Ref. 243. Reprinted with permission from Elsevier, copyright 1984.)

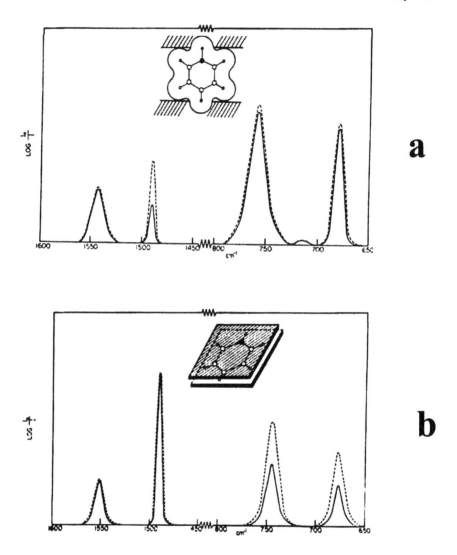

FIGURE 4 Orientation of pyridinium ions in the interlayer space of (a) vermiculite and (b) montmorillonite from IR spectra registered at two incidence angles (_____ 0° and – 40°). In vermiculite (a), the IR absorption band at 1491 cm^{-1} (vibration of symmetry class A1 with dipole change along the C_2 axis) shows a clear directional dependence, in montmorillonite (b), the bands that present a clear dichroism are those at 748 and 677 cm^{-1} (out-of-plane vibrations of symmetry class B_2). (After Ref. 26.)

occurs for pseudo-trimolecular arrangements, where some chain ends are shifted above one another, and the spacing (ca. 2.2 nm) becomes the thickness of three alkyl chains (Fig. 5c).

Clay complexes with organic cations may absorb neutral organic molecules in the interlayer space. This process is accompanied by a separation of the silicate layers and, generally, with a change of orientation of the organic cation. Thus, adsorption of benzene or chlorobenzene in pyridinium–montmorillonite changes the disposition of the pyridinium ion from parallel to normal to the silicate layers, and an increase of d_{001} from 1.25 to 1.50 nm is observed. IR studies of the dichroism of specific IR absorption bands show that the pyridinium cations have their N—H groups directed to the layer surface (C_2 axis perpendicular to the layers). For chlorobenzene, the molecules also adopt a perpendicular orientation but with the C—Cl bond axis (C_2 axis) parallel to the layers (30).

When interlayer cations are long alkylammonium ions and the adsorbed molecules are polar alkyl compounds, considerably more swelling takes place. Clay organic derivatives of this type are plentiful and are extensively documented in the literature (27,31). In this case, strong van der Waals interactions between the cation alkyl chains and the molecules cause them to be densely packed in bimolecular layers between the silicate plates, either with their longitudinal axes

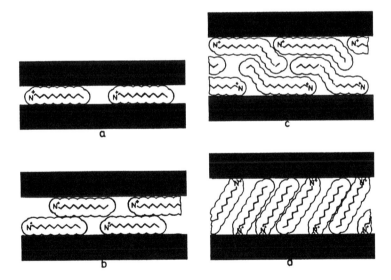

FIGURE 5 Arrangements of the alkyl chains in alkylammonium-exchanged 2:1 clay minerals: (a) monolayers; (b) bilayers; (c) pseudotrimolecular layers of chains lying flat on the silicate surfaces; and (d) paraffin-type structure. (After Ref. 28. Reprinted with permission from Elsevier, copyright 1986.)

perpendicular to the silicate layers (montmorillonite) or with the longitudinal axes inclined 54° to the plates (vermiculite). The charged groups of the organic cations and the polar groups of the organic molecules are in contact with the silicate surface, probably keyed in the ditrigonal cavities of the tetrahedral sheets. In many cases, the alkyl chains forming the film are not fully stretched, but they undergo conformational changes by rotations around the C—C bonds (*kinks*). These kinks were first proposed to occur as defects and structural elements in crystals of the paraffins and polymers. When the bimolecular clay complexes are heated, phase transitions are observed that have been explained as nucleation of kink-block defects in the interlayer organic films. Such defects may propagate cooperatively so that, if a kink is formed in a chain, steric effects cause a kink to be formed on the neighboring chain and the reaction continues through the organic film in an ordered fashion (Fig. 6).

Complexes of smectites with long-chain alkylammonium cations have the property of dispersing in organic liquids, forming thixotropic gels with high liquid content (32,33). For the development of good organophilic properties, alkyl chains of 12 carbon atoms or more are required. Maximum swelling occurs with organic liquids that combine a high polarity with high organophilic characteristics, such as benzonitrile and nitrobenzene. With nonpolar hydrocarbons, swelling is enhanced if small amounts of polar compounds, such as alcohols, esters, or aldehydes, are dissolved in them. Organophilic clays have been used as gas chromatographic stationary phases, in the·manufacture of lubricating greases and paints, and in the preparation of oil-base drilling fluids, among other applications (see Section VI). In 2000, organophilic clays were used in the preparation of pesticide formulations in order to reduce the volatility and leachability of nonpolar pesticide molecules, such as alachlor and metalachlor, and consequently extending their biological activity in the field (34).

In inverse relation to the development of organophilic properties, the adsorption of water reduces as the size of the organic cation increases. However, hydration takes place in complexes with organic cations containing functional groups capable of interacting with water molecules by hydrogen bonding. Hydrates of L-ornithine–vermiculite complexes having d_{001} spacings as high as 4.2 nm have been reported by Mifsud and coworkers (35). Structural analyses of these hydrates indicate that part of the water is associated with the silicate surface, with the remainder in association with the functional groups of the amino acid cations (21). Furthermore, complexes of vermiculite with short-chain alkylammonium ions (propyl- or butyl-) swell in water or in dilute solutions of the alkylammonium salts, forming coherent gel structures in which the silicate layers are separated by several tens of microns (36,37). Vermiculites saturated with certain amino acids in cationic form swell in water in the same manner, and these systems have been used as model for the study of *double-layer* interaction (38–40). From these studies, it is deduced that the theory of Derjaguin–Landau–Verwey–Overbeek (DLVO theory) is operative in these systems.

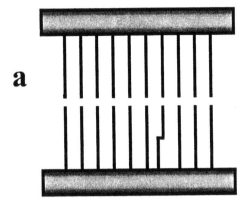

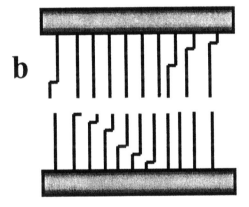

FIGURE 6   Conformation of alkyl chains in a bimolecular film: (a) all-trans conformation with an isolated kink and (b) nucleation of kink-block defects. (Based on Ref. 31.)

III. ADSORPTION OF NEUTRAL ORGANIC MOLECULES ON CLAY MINERALS

A. Intercalation of Organic Molecules in 2:1 Phyllosilicates

The formation of organic–inorganic compounds by intercalation of organic neutral molecules in 2D solids was accomplished for the first time in clay minerals by Bradley (41) and MacEwan (42,43). Since these early reports, this type of compound has been studied extensively (see Table 4). Neutral molecules penetrate

TABLE 4 Interaction of Clay Minerals with Neutral Organic Molecules

Neutral species	Characteristic examples	Predominant mechanisms	Authors (Ref.)
Aliphatic hydrocarbons	Alkanes/sepiolite	van der Waals	Barrer et al. (68)
Aromatic hydrocarbons	Benzene/ Cu-montmorillonite	Electron transfer	Doner and Mortland (54)
Hydroxy compounds (alcohols, polyols, phenols)	Ethylene glycol/ Ca-montmorillonite	Hydrogen bonding/ water bridges	Bradley (41)
Carbonyl compounds	acetone/ Ca-montmorillonite	Hydrogen bonding/ water bridges	Glaeser (259)
Carboxylic acids	Benzoic acid/ Ca-montmorillonite	Hydrogen bonding/ water bridges	Yariv et al. (260)
Sulfoxides	DMSO/kaolinite	Hydrogen bonding	Olejnik et al. (261)
Amines	Alkylamines/M^{n+}- montmorillonites	Proton transfer, ion dipole, and/or coordination*	Fripiat et al. (62)
Amides	DMF/kaolinite	Hydrogen bonding	Cruz et al. (262)
Urea	Urea/montmorillonite	Hydrogen bonding	Mortland (263)
Nitriles	Benzonitrile/ Ca-montmorillonite	Water bridges/ion dipole	Serratosa (26)
N-Heterocyclic compounds	Pyridine/M^{n+}- montmorillonites	Water bridges, proton transfer or coordination*	Farmer and Mortland (59)
Macrocyclic compounds	Crown ethers/ montmorillonite	Ion dipole	Ruiz-Hitzky and Casal (44)

*Depending on the nature of the M^{n+} exchangeable cations.

into the interlayer space of clays when the energy released in the adsorption process is sufficient to overcome the attraction between layers.

Possible adsorption sites in the clay structure include the exchangeable metal ions, with which sorbate molecules may form coordination compounds, and surface oxygens of the tetrahedral sheets, which may act as proton acceptors for the formation of H bonds with molecules containing —OH or —NH groups. These adsorption mechanisms may act simultaneously, but their relative contribution to the adsorption process will depend on the nature of sorbate molecules and on the type of exchangeable cation.

Earlier studies had attributed a predominant role to the interaction of sorbate molecules with the silicate surfaces. But after IR spectroscopic techniques became widespread, the relevance of the interactions with the exchangeable cations was recognized. Hydrogen-bond interactions between surface oxygens of the silicate layers and functional NH or OH groups of the adsorbed organic molecules are only significantly strong when the layer charge is tetrahedrally located. Therefore, this contribution is relevant for silicates with tetrahedral charge (beidellites, saponites, vermiculites) saturated with cations of low solvation energy and for organic molecules with multiple groups capable of forming hydrogen bonds (OH, NH). Hydrogen bonding to surface oxygens is more evident in the case of alkyl-ammonium cations (see Section II). Also, van der Waals attractions between molecules and the mineral substrate contributes to the adsorption process, but the significance is secondary except for organic compounds of large molecular weight.

Adsorption of organic molecules is also influenced by the clay hydration level. When interlayer water is present, the cohesion forces between layers are considerably reduced, consequently, penetration of the sorbate molecules is facilitated. Organic molecules, then, compete with water for coordination sites around the exchangeable cations and, depending on the relative values of the hydration and solvation energies of these, they will do one of the following:

1. Replace water and become coordinated directly to the inorganic sites.
2. Occupy sites in a second sphere of coordination around the cation, bonded via bridging water molecules.
3. Accept a proton from the water of coordination around the cations or from the cations themselves if the clay is saturated with H^+ or NH_4^+.

Direct coordination to the exchangeable cations has been recognized in complexes of smectites and vermiculites with a wide variety of organic compounds, such as alcohols, amines, nitriles, urea and amides, and pyridine and pesticide molecules. Complexes could be formed with alkaline and alkaline earth cations or with transition metal ions, and excellent reviews have been published on this matter (3–5,7,8).

For non–transition metal ions, the size and charge of the cations determines the stoichiometry and the arrangement of adsorbed molecules in the interlayer space. Examples that illustrate this are complexes formed by homoionic smectites and vermiculites with macrocyclic organic ligands such as crown ethers and cryptands (44,45). When these compounds are adsorbed by smectites and vermiculites from methanol solution, the interlayer water is excluded and the compounds are coordinated directly to the interlayer cations in monolayer or bilayer complexes, depending on the size of the cations. If the ratio r_c/r_i (r_c = radius of the macrocyclic cavity and r_i = cationic radius) is greater than 1, then the cation is occluded into the cavity of the cyclic ligand, consequently, one ligand molecule is adsorbed per cation. One-layer or two-layer complexes will result (Fig. 7), depending on the area of the equivalent charge in the silicate surface relative to the projected area of the ligand molecule. When the ratio $r_c/r_i < 1$, then each

ligand/cation arrangement	r_c/r_i ratio	$\Delta d_L(\text{Å})$	characteristic examples
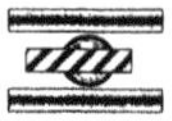	>1	~4	15C6/Na-mont.; 18C6/Ba-mont
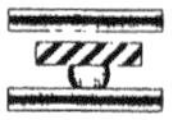	≤1	~4	12C4,15C5,18C6/K- and NH$_4$-mont.
	<1	~8	12C4/Na-mont. ; 15C5/Ba-mont
	<1	6-7	12C4/Sr-mont. ; 12C4/Ba-mont
	>1	~6	18C6/Na-mont.

FIGURE 7 Proposed interlayer arrangements of exchangeable cations and associated macrocyclic ligands in complexes with montmorillonite and vermiculite. r_c = radius of the macrocyclic cavity; r_i = radius of cations; Δd_L = interlayer distance.

cation is generally sandwiched between two cyclic ligands and two-layered complexes result (Fig. 7). However, in the case of K^+ or NH_4^+, one-layer complexes are formed in which the cations, due to their sizes, are coordinated to the oxygens of the cyclic ligand from one side and to the oxygens of the ditrigonal cavities on the silicate surface from the other. Further, for the complex of 18C6 crown ether with NH_4^+-montmorillonite, the IR spectrum indicates a change in NH_4^+ symmetry from T_d to C_{3v}, suggesting that NH_4^+ is hydrogen bonded to the crown oxygens (46) (Fig. 8).

Laser microprobe mass spectrometry (LMMS) confirms the existence of true interlayer complexes between cations and macrocyclic ligands (47). This technique, which provides mass spectra of fragmented solids after irradiation with a laser beam, was first used to characterize the intercalation compounds of crown ether– and cryptand–smectite complexes (47,48). One of the most interesting results derived from LMMS of these materials is the ability to corroborate the macrocycle-interlayer cation complexation, such as cryptand C(222)-Na-montmorillonite vs. Cu-montmorillonite. Alkaline–cryptand complexes are clearly assigned since the m/e values correspond to the sum of both the sodium and cryptand atomic mass (i.e., 399 Daltons for $Na^+/C(222)$). Transition metal

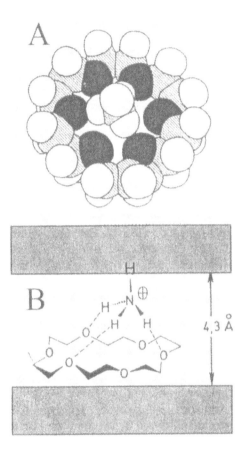

FIGURE 8 Molecular model of $18C_6/NH_4^+$-montmorillonite complex showing the cation–ligand interaction (A) and the arrangement in the interlayer space of the silicate (B). (Adapted from Ref. 46. Reproduced by permission of The Royal Society of Chemistry.)

ions do not form complexes; the macrocycle is protonated due to the acidic environment ensured by such cations (i.e., $m/e = 377$ Daltons, corresponding to $C(222)\text{-}H^+$) (47). The high affinity of macrocycles towards interlayer cations has also been seen by enthalpy measurements using adsorption microcalorimetry (49).

Nitrogenated macrocycles based on the cationic form of hemin can also be intercalated in layer silicates. Thus, different porphyrin derivatives, such as phenyl- and pyridyl-porphyrins, metalloporphyrins, chlorophylin, and related ma-

crocycles of synthetic or natural origin, have been intercalated in smectites saturated with different inorganic cations (50,51). Since this type of ligand forms
complexes with transition metal cations, the stability and spectroscopic properties
of porphyrins with Fe(II), Mn(II), Cu(II), Fe(III), Sn(IV), and other cations have
been investigated, particularly in relation with the ability of the macrocycle to
be protonated in the interlayer space of clays. In fact, it has been speculated that
the prebiotic synthesis of porphyrins in the primitive Earth may have contributed
to the evolution of photosynthesis and respiration in living systems (52).

In general, organic complexes formed within the interlayer space of clay
minerals have their counterparts in solution. However, in certain cases, the combined effect of the silicate-layer electric field and of steric restrictions can modify
the stability of the complexes to the extent that certain complexes can be formed
only in the interlayers. Numerous examples exist showing that the interlayer space
provides a special environment in which unique chemistry occurs. Of particular
relevance are the studies of Cremers and coworkers (53) in which free energies
of formation are determined for Cu-, Ni-, Zn,-, Cd-, and Hg-montmorillonite
complexes with polyamines and for Ag-montmorillonite with thiourea. Their
results show that thermodynamic constants for the complexes in interlayer are
two to three orders of magnitude higher than those for the same complexes in
solution. Anther example of stabilization occurs for aromatic compounds via
donation of π electrons. This was shown by Donner and Mortland (54) and
Mortland and Pinnavaia (55) for complexes of Cu^{2+}-smectites with arenes, which
are not formed in solution. For the formation of these complexes, the water of
hydration must be removed from the Cu-montmorillonite. Two kinds of complexes are obtained with benzene, depending upon the extent of dehydration.
Type I complexes (green) are formed by exposing the clay to benzene vapor in
a desiccator with P_2O_5; benzene is truly coordinated to the cation through the π
electron system, and the aromaticity of the molecule is preserved. Type II complexes (red) are formed by further dehydration of the former. Evidence from IR
and ESR spectroscopy indicates that the benzene ring no longer retains its D_{6h}
symmetry, but rather as part of a radical cation represented tentatively as

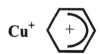

Arene complexes are also formed with other Cu^{2+}- or Ag^+-saturated smectites.
Charge location (tetrahedral or octahedral) has an influence on the formation of
these compounds, with complexation easier in octahedrally charged smectites.
This is because water is more firmly bound and more difficult to eliminate in
tetrahedrally charged smectites; thus, aromatic molecules will have greater difficulty entering into direct coordination with the cations (56). Also, arene com-

plexes are formed more easily with Ag-nontronite than with Cu-nontronite, due to the lower hydration energy of Ag^+ ions.

In arene complexes prepared as oriented aggregates, it is possible to determine the arrangement of the interlayer organic molecules by NMR spectroscopy. One interesting study of this kind is the case of benzene sorbed on Ag-hectorite (57). The ^{13}C NMR spectra were recorded at different temperatures for various orientations of the clay aggregates with respect to the applied magnetic field (B_0). From interpretation of the NMR spectra, the more realistic model (Fig. 9) is one in which the benzene molecules stand nearly perpendicular to the layers; a slight tilt of about 15° of the C_6 axis out of the (**a,b**) plane of clay platelets is necessary to reproduce the experimental NMR patterns. The benzene molecules are submitted to a double rotation: one around the C_6 axis and the other around that normal to the clay layers. Only the latter motion is completely quenched at 77 K. Formation of π-coordinated complexes and the resulting possible disruption of the bond resonance in the benzene ring suggest important new pathways for organic synthesis via these complexes (58).

The second association mechanism of neutral organic molecules to exchangeable cations is through bridging water molecules. This type was first demonstrated for pyridine sorbed on montmorillonite (59), and it has been shown

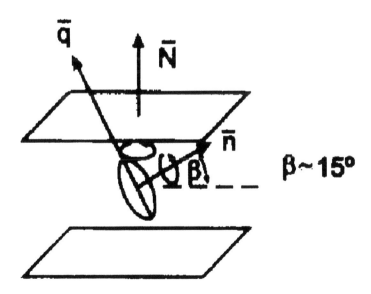

FIGURE 9 Model deduced from the ^{13}C NMR spectrum of benzene/Ag-hectorite complex for the average orientation of C_6H_6 molecules between the clay layer. (After Ref. 251. With kind permission of Kluwer Academic Publishers.)

since to occur in many other systems. Displacement of IR absorption bands from both water and adsorbed organic molecules provide evidence for this type of complex. One illustrative example is the montmorillonite–benzonitrile system (Fig. 10) studied in detail by Serratosa (60). The benzonitrile molecules are arranged with the plane of the benzene ring at a large tilt angle to the silicate layers (basal spacing of approximately 1.5 nm) and with the principal C_2 axis parallel to the layers. The IR absorption frequency of the O—H_a bond depends on the polarizing power of the interlayer cation; for alkaline earth cations it is 3405 cm^{-1} (Ba^{2+}), 3390 cm^{-1} (Ca^{2+}), and 3348 cm^{-1} (Mg^{2+}). Hydrogen bonds displace the frequency of the IR —C≡N stretching band, but its position is not affected by the cations. When the samples are evacuated and heated (≈80°C), bridged water molecules are lost and benzonitrile is coordinated directly to the cations. This is indicated by the position of the $\nu_{C≡N}$ stretching band, which is sensitive to the metal ion present: 2240 cm^{-1} for Ba^{2+}, 2249 cm^{-1} for Ca^{2+}, and 2261 cm^{-1} for Mg^{2+}. In the Mg complex, some water is still present after prolonged evacuation at 80–100°C. This residual water presents a relatively sharp

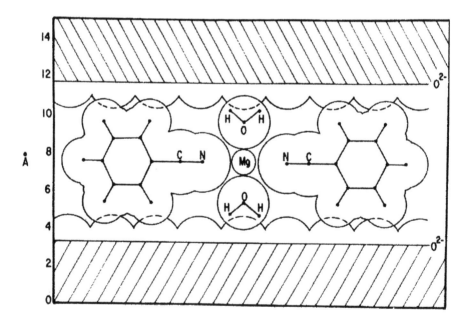

FIGURE 10 Arrangement of interlayer species in a benozonitrile/Mg-montmorillonite complex. The Mg^{2+} ions are midway between the layers and coordinated to four benzonitrile molecules (nearly perpendicular to the layers) and two water molecules. (After Ref. 60.)

IR absorption at 1650 cm^{-1}, corresponding to the deformation vibration of H_2O, and is sensitive to the orientation of the clay film with respect to the incident radiation. These several pieces of evidence permit construction of a model in which the Mg^{2+} ions are coordinated with six ligands: four benzonitrile molecules arranged as a paddle wheel and two water molecules situated above and below at the two remaining corners of the octahedron (Fig. 10).

The third mechanism of adsorption concerns protonation of organic bases when adsorbed as neutral molecules into layer silicates. Sources of protons are (a) hydrogen ions in NH_4^+ or short-chain alkylammonium–saturated clays or (b) water molecules coordinated to the exchangeable cations. By far the most common mechanism is the transfer of protons from interlayer water to adsorbed organic molecules. Metal ions containing water ligands are acids of measurable strength whose ionization constants (or pK) have been determined in water (61). The acidity of interlayer water coordinated to exchangeable cations is significant, although pK values are not necessarily the same as in solution. In fact, the acidity of interlayer-coordinated water can be greater than in solution, depending on the nature of the metal ion. Evidence for the formation of protonated organic bases is obtained by IR, since the absorption bands for neutral molecules and protonated species are well separated. The phenomenon was observed for the first time by Fripiat and coworkers for the adsorption of short aliphatic amines by montmorillonite (62).

B. Intercalation Processes in Kaolin Minerals

With the exception of halloysite (42), kaolin minerals were considered as nonexpanding layer silicates until the work of Wada (63), who showed that kaolinite could be expanded by repeated grinding with K-acetate and with other salts of organic acids of low molecular weight. The resulting kaolinite complexes were called *intersalation* compounds, but it was found later that many nonsaline organic substances (urea, formamide, dimethylsulfoxide, and others) also penetrate between kaolinite layers, whereupon the term *intercalation* became accepted.

Intercalation compounds have been classified in two main groups according to the method of preparation (64): (a) compounds prepared by direct reaction of mineral with organic substance, and (b) compounds prepared only by replacement of a substance previously intercalated by direct reaction, type d. Molecules that readily form type d complexes are diverse:

1. Metal salts of low-molecular-weight organic acids of (acetates, cyanoacetates, and propionates) with large monovalent cations of low hydration energy
2. Compounds with a strong tendency for hydrogen-bond formation, such as urea, formamide, acetamide, and imidazole

3. Molecules having a high dipole moment or with mesomeric structures, such as dimethylsulphoxide and pyridine-N-oxide

4. Molecules combining two or more of the preceding characteristics, such as ammonium acetate, N-methylacetamide, and the potassium salt of picolinic acid-N-oxide.

These compounds intercalate directly from the liquid (hydrazine), from the melt (acetamide), or from concentrated aqueous solutions (10 M solution of urea).

The mechanism of intercalation has not been well established, but the process may be envisaged as resulting from the tendency of the dipolar kaolinite layers to become solvated with molecules. It is evident that hydrogen bonds between the surface hydroxyls of kaolinite and the intercalated molecules, as demonstrated by IR spectroscopy, contribute to the intercalation process, but calculation of the energies involved indicates that other factors contribute to the energy balance (65). Other postulates are that a decrease of the electrostatic attraction between the layers is caused by a higher dielectric constant in the interlayer volume after intercalation or that compensation of the internal dipole moment of the kaolinite layers by the dipole moment of the intercalated species occurs.

The arrangement of the organic molecules in the interlayer space of kaolinite has been inferred, as for smectites, from measurements of the basal spacing d_{001} by XRD and from the displacement of specific IR absorption bands. Lately, the use of other techniques has given more precise knowledge about the arrangement and mobility of the intercalated organic molecules. One illustrative example is the study by Hayashi (66) of the kaolinite–dimethylsulfoxide complex by ^{13}C, ^{1}H, and ^{2}H NMR at 170–380 K (Fig. 11). Below 300 K, all interlayer DMSO molecules are equivalent with one of the methyl groups keyed into the ditrigonal holes of the silicate sheet. Above 320 K, the keyed methyl group of some of the interlayer DMSO molecules is released from the trapped holes, and two types of DMSO molecules coexist in the interlayer space. This process is complete at about 415 K, when all the interlayer DMSO molecules are essentially free and not keyed into the kaolinite layers. The DMSO methyl groups undergo free rotation around their C_3 axis over the temperature range studied; the C_3 axis is fixed at low temperature (about 160 K), but at higher temperatures the methyl groups initiate a wobbling motion, whose amplitude increases with temperature. The DMSO molecules released from the ditrigonal holes undergo an anisotropic rotation of the entire molecule.

C. Adsorption on Palygorskite and Sepiolite

Palygorskite and sepiolite are porous microcrystalline solids that have high adsorptive capacity, from which derives multiple industrial applications. Their peculiar crystal structure consists of talclike ribbons parallel to the fiber axis, alter-

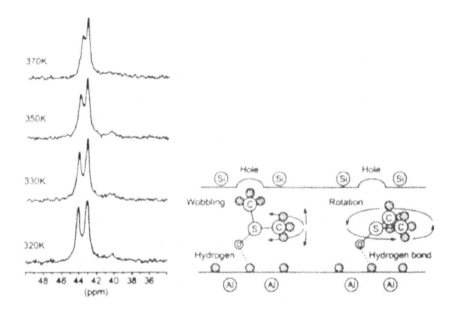

FIGURE 11 (left) ^{13}C CP/MAS-NMR spectra of kaolinite-DMSO intercalate at different temperatures; (right) schematics of the arrangement and motion of DMSO guest molecules. (After Ref. 66.)

nating with tunnels (or channels at the external surfaces) along the fiber axis (Fig. 12). The width of the ribbons and, consequently, the width of the tunnels (channels) are different in the two minerals. In palygoskite the cross section of the tunnels is 0.64×0.37 nm^2-while in sepiolite the cross section is 1.06×0.37 nm^2. The existence of tunnels and channels makes them accessible for molecular sorption, an aspect that has been in discussion for many years. Controversy arose, at the beginning, from adsorption studies that indicated a large discrepancy between calculated and experimental surface areas. Later, IR helped confirm the penetration of adsorbed molecules in the tunnels by the shift of IR absorption bands associated with H_2O coordinated to Mg^{2+} ions at the border of the ribbons inside the tunnels. For example, Serna and VanScoyoc (67) clearly demonstrated the penetration of short-chain alcohols (methanol and ethanol) in the tunnels of sepiolite. From these earlier studies, it was generally accepted that the main factor for accessibility was polarity rather than molecular size and that only small polar molecules, such as NH_3, CH_3OH, CH_3—CH_2OH, acetone, and ethylene glycol, would have access to the intracrystalline tunnels. These molecules can, in certain cases, replace the water coordinated to the Mg^{2+} at the border of the ribbons.

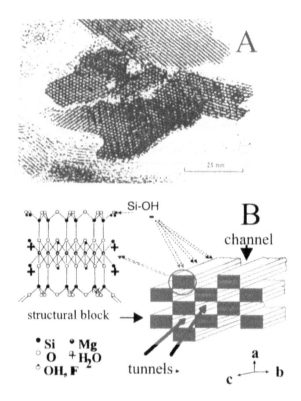

FIGURE 12 TEM image of the cross section of a sepiolite fiber (A) and structural model of this silicate (B), showing the tunnels and channels available for interaction with organic species. (Adapted from Refs. 244 and 224.)

The accessibility of tunnels to the nonpolar gas molecules used in surface area determinations, although questioned in some earlier works (68), has now been clearly demonstrated for sepiolite by Ruiz-Hitzky (69). This was accomplished by using the Howarth–Kawazoe analysis (70) of argon adsorption isotherms. For samples previously degassed at 120°C, the analysis shows the predominance of pores with a diameter of 0.6–0.7 nm. These pores disappear when sepiolite is heated at 350°C under dynamic vacuum due to loss of coordinated water and folding of the structure (71).

Other organic molecules such as pyridine also have access to the intracrystalline tunnels of sepiolite, as demonstrated clearly by Inagaki (72a), Kuang (72b), and Ruiz-Hitzky (69). The latter has recently reviewed the subject of accessibility and concluded that, for sepiolite, larger molecules (Fig. 13), such as methylene blue (molecular dimensions $1.7 \times 0.76 \times 0.32\ nm^3$), acridine orange, thioflavine, and poly(ethyelene glycol), can also penetrate the tunnels. In the case of acridine orange, Ridler and Jennings (73) demonstrated the highly specific directional binding of the dye into sepiolite, the molecules being arranged with their long axes parallel to the sepiolite fiber axis. Further, natural indigo, with a planar molecular configuration, is also firmly adsorbed within sepiolite and palygorskite tunnels. It has been suggested that Maya Blue, a pigment used in Indian pottery and murals during the pre-Spanish period, is an indigo–palygorskite complex (74). The color cannot be extracted with disolvents, resists hot concentrated mineral acids, and persists upon heating to about 250°C. Analogous stable pigments have been prepared from palygorskite or sepiolite with indigo (74,75). In both cases, stability is achieved by heating the freshly prepared complex for several days to $\approx 100°C$.

Apart from the accessibility of intracrystalline tunnels, it is evident that the existence of external channels on sepiolite and palygorskite also contribute to adsorption of organic sorbates. Such sorbates have molecular sizes and shapes that closely match the dimensions of the open channels at the exposed crystal faces. This was first demonstrated (76) by percolating binary mixtures of saturated hydrocarbons, e.g., long-chain n-paraffins and naphthenes dissolved in pentane, through palygorskite columns. Extremely good separations of the two components were obtained, the long-chain n-paraffins being the most retained by the clay. Barrer and coworkers (68) found that free energy, heat, and entropy changes for the n-paraffins (n-pentane) sorbed on palygorskite are greater than for the adsorption of branched-chain paraffins (isopentane or neopentane) and that the former are adsorbed in larger amounts than the latter. Sepiolite shows similar behavior, but its selectivity is different from that of palygorskite; sepiolite sorbs n- and isopentane equally strongly but neo-pentane is less strongly adsorbed. In sepiolite the channel width is larger than in palygorskite, and it appears that the former sorbates can be accommodated into the channels at the external surface (or eventually into the tunnels), but not neo-pentane.

IV. INTERACTIONS THROUGH COVALENT BONDING: ORGANIC DERIVATIVES OF CLAYS

Organic derivatives of clays are formed by grafting organic groups onto the silicate surfaces through covalent bonds. The first such attempts were reported by Deuel in 1952 (77), who claimed the formation of clay derivatives upon reacting diazomethane with montmorillonite. However, these results were critically examined (78) because Deuel explained the esterification using the highly suspect montmorillonite structure proposed by Edelman and Favejee, which had $\equiv Si$—OH silanol groups in the interlayer space (79). Diazomethane was successfully used later by Hermosín and Cornejo (80) in reactions with sepiolite and

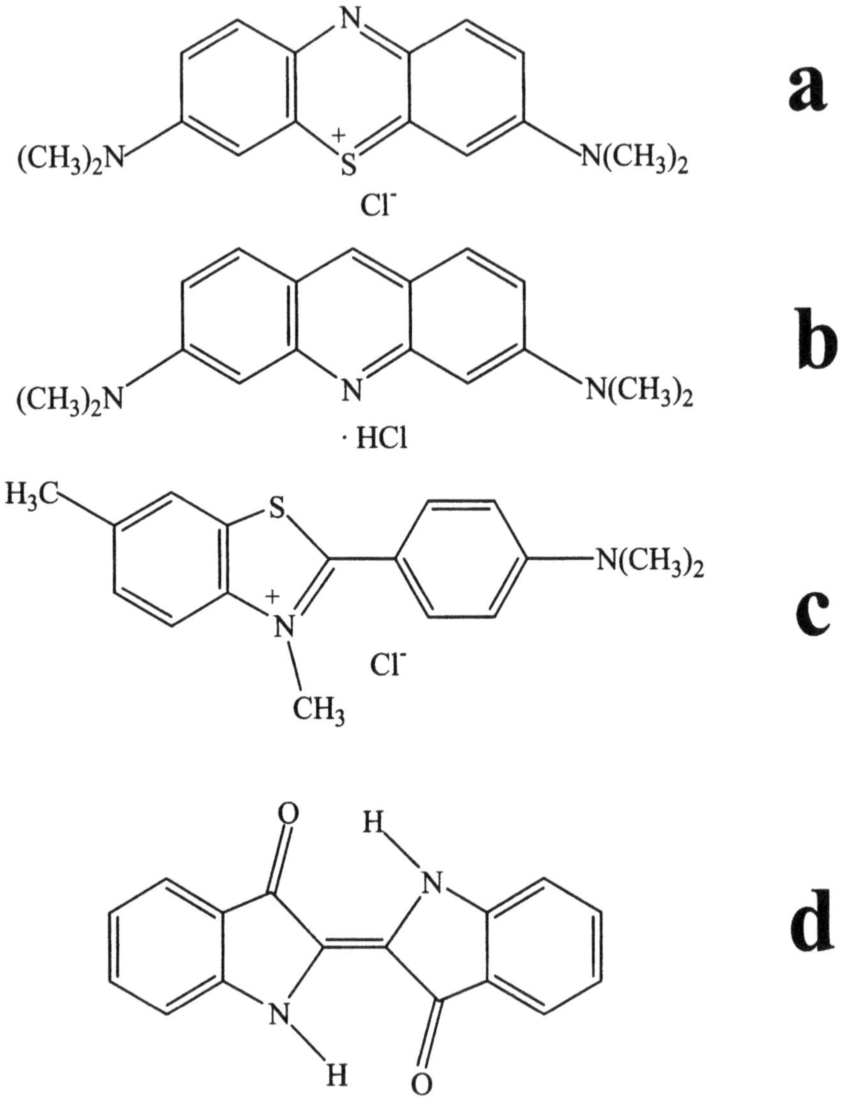

FIGURE 13 Structures of the larger molecules (dyes) adsorbed on sepiolite or palygor-kite: (a) methylene blue, (b) acridine orange, (c) thioflavine-T, (d) indigo.

palygorskite, which are known to contain available surface silanol groups (71). These allow methylation as shown in Eq. (2):

$$\equiv\!\text{Si-OH} + CH_2N_2 \rightarrow \equiv\!\text{Si-O-CH}_3 + N_2 \tag{2}$$

In any event, attachment to the silicate surface through covalent $\equiv\!\text{Si—O—C—}$ bonds is of low stability since they are very sensitive to water molecules. Stable hybrid organic–inorganic compounds are obtained by the reaction of silicates with functionalized organosilanes. The chemistry arises because of the reactivity of organosilanes containing $\equiv\!\text{Si—X}$ groups (X = OR, Cl). These compounds are able to interact with surface silanol groups and yield very stable siloxane bridges ($\equiv\!\text{Si—O—Si}\!\equiv$), as in Eq. (3):

$$[\text{surface}]\text{Si-OH} + \text{X-Si}[R_1R_2R_3] \rightarrow [\text{surface}]\text{Si-O-Si}[R_1R_2R_3] + XH \tag{3}$$

In general, clays have a low content of silanol groups, and these are located exclusively at the external surfaces, i.e., at the edges of the tetrahedral layers. Thus, the direct reaction of layered silicates with organosilanes in either the vapor phase or in aprotic solvent solutions gives organic derivatives where the organic groups are attached to the external surface. Due to the interlayer water content of 2:1 charged silicates, such as vermiculite, rapid hydrolysis of the X functions of the silane takes place, giving organosiloxanes as side products (81). Intermediate [R$_3$SiOH] organosilanol species are thought to be involved in such processes:

$$R_3\text{Si-X} + H_2O \rightarrow [R_3\text{Si-OH}] \rightarrow R_3\text{Si-O-SiR}_3 \tag{4}$$

The silanol groups necessary to react with the organosilane reagents could be produced by acid treatment of clay minerals. Lentz (82) first developed an efficient procedure for the preparation of organic derivatives of silicates based on simultaneous hydrolysis of both the silicate (e.g., olivine) and the organosilane (e.g., hexamethyldisiloxane). This procedure, known as *cohydrolysis* (83), was subsequently applied to prepare organic derivatives of silicates such as chrysotile, vermiculite, and sepiolite (84–87). Octahedral cations (e.g., Mg^{2+}) belonging to the silicate are extracted by acid (e.g., HCl), giving *fresh* silanol groups [Eq. (5)] able to react with the organosilanes [Eq. (3)]:

$$\text{Si-O-Mg-O-Si} + [H^+/H_2O] \rightarrow [\text{surface}]\text{-Si-OH} + Mg^{2+} \tag{5}$$

In these cases, the synthesis of organic derivatives of clays is characterized by significant structural and textural alteration of the silicate. The complete extraction of octahedral cations leads to amorphous silica that is subsequently grafted by the organosilanes. The resulting organosilicic compounds show textural characteristics derived from the original silicate. For instance, organic derivatives of sepiolite obtained by cohydrolysis with alkyl- and alkenyl-silanes maintain the microfibrous morphology typical of the parent clay (86). The presence of grafted species has been proven via addition of OsO_4 [Eq. (6)] to unsaturated groups of vinylsilanes, their localization and distribution on the mineral surface being observed as dark dots of nanometer size by TEM (88).

$$[\text{sepiolite}] \equiv \text{Si-O-CH=CH}_2 + \text{OsO}_4 \longrightarrow [\text{sepiolite}] \equiv \underset{\underset{\underset{\underset{O \quad O}{\diagdown \diagup}}{\overset{\diagdown \diagup}{Os}}}{\overset{|\quad\;\;|}{O \quad O}}{\text{Si-O-CH-CH}_2}$$

$$(6)$$

Layered polysilicates such as magadiite and kenyaite contain hydrated Na^+ cations in their interlayer space (89). Such cations are easily replaced by protons after treatment with low concentrated acid solutions (90), giving layered silicic acids, such as the so-called H-magadiite. Therefore, this type of solid has the unique property of containing silanol groups inside the interlayer region (91). Therefore, the grafting of internal silanol groups is feasible by using polar molecules. DMSO and NMF are two molecules able to expand the silica layers, allowing access of organosilanes and facilitating corresponding intracrystalline grafting reactions (Fig. 14) (92,93). More recently, alkylammonium-exchanged versions have been successfully used as intermediate compounds in the synthesis of organic derivatives of magadiite and other alkaline layered silicates (94,95).

As previously mentioned, sepiolite contains many available silanol groups that are a significant portion of their large external surface (71,96). It is through these that this clay is able to yield organic derivatives by direct reaction with organosilanes, either in the vapor phase or in organic solvents (97). Grafting of sepiolite using alkylchlorosilanes, such as trimethylchlorosilane, creates organophilic and hydrophobic organic–inorganic materials, whereas the use of silanes containing functional groups (benzyl, amino, mercapto, etc.) strongly modifies the chemical reactivity of the mineral. For instance, the grafting of aryl species allows the introduction of sulphonic acid groups that confer strong acidic behavior (98) useful for heterogeneous catalysis (99). In addition, the grafting of unsaturated groups such as methylvinyldichlorosilane provides organic derivatives that are able to copolymerize with unsaturated monomers or polymers (100). The resulting nanocomposites have the distinction that the polymer is covalently attached to the mineral, which could be of great interest in the improvement of mechanical and rheological properties.

In addition to organosilanes, other reagents are also able to give organic derivatives by grafting reaction with $\equiv$Si—OH groups (101,102,104,105). In this way, isocyanates ($R-N=C=O$) interacts with sepiolite, giving the corresponding

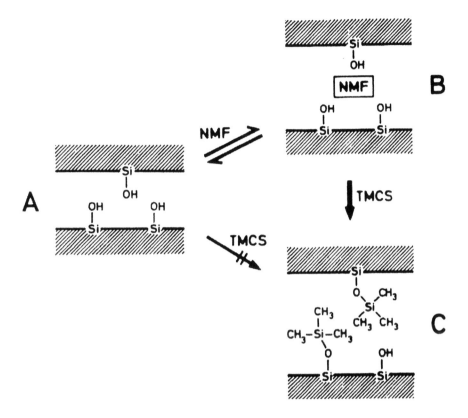

FIGURE 14 Intracrystalline grafting reactions of organosilanes to the interlayer region of a layered silicic acid after treatment with DMSO or NMF molecules, which expand the silica layers. (From Ref. 93. Copyright Spinger-Verlag.)

organic–inorganic materials, in which aliphatic or aryl groups remain attached to the mineral surface through silyl urethane ($\equiv$Si—O—CO—NH—R) bridges (101).

$$\equiv\text{Si-OH} + \text{R-N}=\text{C}=\text{O} \rightarrow \equiv\text{Si-O-CO-NH-R} \qquad (7)$$

The inductive effect of alkyl groups in alkyl isocyanates, such as *n*-butyl isocyanate, stabilizes the silylurethane bonds. This improves thermal and hydrolytic resistance over aryl isocyanates. In this context, phenyl isocyanate reacts with sepiolite, giving large amounts of diphenyl urea, whereas butyl isocyanate gives stable grafted compounds, producing low amounts of secondary products (dibutyl urea) (101). Because diisocyanates (e.g., hexamethylene diisocyanate and 2,4-

toluene diisocyanate) yield organic derivatives containing excess $-C = N = O$ groups, further reactions with polyols and polyesthers could be useful for preparing nanocomposites based on silicate-polycondensate materials (102). The reaction of isocyanates with vermiculite has also been reported, although this clay has a low amount of silanol groups (only on the external surface of particles) (103).

When molecules containing epoxy groups are used as organic reagents, interaction with the silanol groups causes the rings to open. Amorphous silica and sepiolite react in either the vapor phase or in organic solvents with 1,2-epoxides, giving the organic derivatives shown in Eq. (8) (104,105):

$$\equiv Si-OH + CH_2CH-R \longrightarrow \equiv Si-O-CH-R$$

with the epoxide group $\overset{\diagdown}{O}\diagup$ on the left and CH_2OH on the right.

$$(8)$$

In this case, the organic groups (R) remain covalently attached to the surface through $\equiv Si-O-C$ bonds, and their stability depends on the nature of the R substituent (104). Possible secondary reactions include rearrangements of the epoxides, giving carbonyl compounds, which is also enhanced in the case of R = aryl groups. For instance, epoxystyrene reacts with sepiolite and other clays, giving large amounts of phenylacetaldehyde via catalytic processes involving silico-alumina species present on mineral surfaces (106). In contrast, epoxides containing alkyl substituents give very stable derivatives, which are of interest as fillers for elastomer and thermoplastic reinforcement. This is the case of allyl-glicidyl ether, a reagent containing unsaturated groups able to give further copolymerization reactions with different monomers or polymers (107).

The interlayer aluminol groups (Al-OH) present in the internal surfaces of kaolinite can be grafted by reaction with alkanols and polyols. Tunney and Detellier (108), using previously expanded kaolinite with DMSO or NMF, obtain organic derivatives of this silicate trough $Al-O-C$ bonds. Surprisingly, some of these hybrid materials (e.g., methoxy-kaolinite) show high thermal stability (>350°C) as well as elevated resistance to water hydrolysis.

V. POLYMER–CLAY INTERACTIONS: NANOCOMPOSITES

The adsorption of different types of macromolecules on clays and other soil components was commonly known and systematically studied in the 1950s, but it was only in 1961 that Blumstein (109) reported for the first time the possibility of preparing well-defined polymer—clay materials. In 1979, Theng (6) edited the first monograph entirely focused on providing state-of-the-art knowledge about polymer–clay interactions, because at that time it was a topic of enormous and still-expanding interest. Theng's book was directed mainly at soil scientists and agronomists interested in liquid-phase interactions, and it scarcely mentioned features regarding the resulting organic–inorganic compounds, except for clay–biopolymer complexes. The main interest of the former studies was in rheological properties of polymer/clay suspensions. From these studies it was deduced that the predominance of a specific interaction mechanism (hydrogen bonding, electrostatic, coordination, etc.) depends on: (a) the nature of the polymer (charge, molecular weight, hydrophobicity, etc.), (b) the type of solvent (water, polar or nonpolar organic liquids), and (c) the type of clay mineral (charged, ion exchange capacity, swelling ability, etc.). The most important conclusions were that the existence of water molecules, present in the solvent or associated with the cations, plays an important role in the process and that the main driving force to produce the polymer–clay complexes is the gain of entropy.

Blumstein (109) demonstrated the possibility of forming polymer/clay intercalation compounds by in situ polymerization of various monomers, such as acrylonitrile, vinyl acetate, and methyl metacrylate, previously intercalated into smectites in the presence of a initiator. The resulting products are stable organic–inorganic materials that nowadays are known as *polymer–clay nanocomposites* (PCNs) (109). Some years later, Friedlander (110) demonstrated that layered silicates containing transition metal ions with redox properties as exchangeable cations could also induce the polymerization of unsaturated monomers, giving materials where the polymer remains strongly associated with the mineral substrate. From these pioneering works to the present, the studies on intercalation of organic polymers into different 2D host lattices have increased exponentially, being at the present very intense. The resulting materials are considered as nanocomposites, with interactions at the atomic level between the inorganic hosts and the polymer guests. The term *nanocomposites* includes structural architectures ranging from well-ordered and stacked multilayers to completely delaminated materials, as indicated by several authors (111,112). Some researchers recognize only the latter as true nanocomposites, i.e., fully delaminated polymer–clay derivatives (113). As in conventional composites, there are structural and functional nanocomposites. The main interest of the first group concerns mechanical or rheological properties, such as nylon–clay nanocomposites (114), whereas the latter group refers to nanostructured materials showing a specific chemical or physical behavior, such as conducting nanocomposites (115). In this section we will briefly discuss several aspects related to the procedures used to prepare PCNs and the ways to produce host–guest compatibility.

A classification of methods for preparing polymer-layered inorganic solid compounds takes into account the main processes involved in the synthesis of the final hybrid material. The nature of both the 2D host solid and the guest polymer determines the pathway applicable to obtain a particular nanocomposite and, in certain cases, is decisive in the behavior of the resulting material. In general, the synthetic routes commonly employed have been classified into four groups:

Direct intercalation of the polymer by (a) adsorption from solutions, (b) ion exchange reactions, or (c) polymer melt intercalation.

In situ intercalative polymerization in the interlayer spaces, which requires the previous intercalation of the monomer/s and a second step to induce their polymerization—usually activated by a thermal treatment or by addition of a catalyst.

Delamination and entrapping-restacking is a method that profits from the possibility of swelling the host solids by separation of their layers using an appropriated solvent or some specific treatment and then adding the monomer or the polymer to be intercalated.

Templating synthesis is a method that can be used when the layered host matrix could be synthesized in presence of the polymer that acts as a template agent remaining between the layers of the assembled solid.

The election of one or another method depends on the characteristics of both host and guest species (Table 5). For example, templating synthesis is a common accepted way for preparing polymer-LDH (layered double hydroxide) materials because LDHs themselves are predominantly synthetic. Therefore, this method has also been extrapolated to polymer–synthetic clay nanocomposites. An important factor that must be considered is the necessity of compatibility between the organic macromolecule and the interlayer space of the layered solid. Thus, due to the organophillic character of most polymers, it is usually necessary to modify such regions to ensure affinity between polymer and inorganic substrate. For clays this problem is in general satisfactorily resolved by previous exchange with alkyl-ammonium cations. Here we will not consider separately the delamination and entrapping-restacking routes, because they can be viewed as specific cases of the others.

A. Direct Intercalation

Although rarely applied, positively charged polymers, for instance, β-dimethyl-aminoethlylmethacrylate hydroacetate (DMAEM) (116) and other polycations (117), can be intercalated into clays by a cation exchange procedure. In these cases, the organic–inorganic interaction is established trough coulombic forces between the cationic groups belonging to the polymer and the negatively charged clay surface. As early as 1952, Ruehrwein and Ward (116) observed that DMAEM penetrates the interlayer space of Na-montomorillonite by exchanging Na^+ ions. The partial ion exchange reaction gives rise to Na^+-rich and polycation-rich layers that are randomly interstratified within a single crystal. The maximum layer expansion is about 0.5 nm, corresponding to the presence of a single layer

TABLE 5 Routes of Preparation for Polymer–Clay Nanocomposites and Selected Representative Examples

Preparation method	Requirements	Examples and characteristics	Ref.
Direct intercalation based on:			
Cation exchange	Applicable to polycations	Intercalation of DMAEM into montmorillonite; coulombic interaction	116
Adsorption from solutions	Polar soluble polymers	Intercalation of PEO into smectites from water[a] or polar solvents (acetonitrile)[b]; water bridges and/or ion–dipole interactions	[a] 121 [b] 122
Adsorption from melts	Polymers stable beyond its melting point	Intercalation of PEO melted by thermal heating[c] or microwave irradiation[d]	[c] 133 [d] 135
In situ intercalative polymerization	Step 1: Intercalated monomers as ions or neutral molecules	Intercalation of aniline into Cu-montmorillonite; Cu^{2+} centers induce polymerization	149
	Step 2: Induction of polymerization: active sites, thermal treatment, additives (promoters, monomers,…)	Intercalation of anilinium into Na-montmorillonite; further polymerization induced by treatment with $(NH_4)_2S_2O_8$	152
		Preparation of nylon-6/montmorillonite nanocomposites from protonated ω-aminoacid, ϵ-caprolactam and thermal treatment	153
Templating synthesis of clay in presence of the polymer	Applicable only to synthetic clays and polymers both compatible and stable in the reaction media	Synthetic hectorite/PAN nanocomposites	111
Two step methods	Step 1: Modification of either the clay or the polymer	Nanocomposites prepared from modified clays: methacrylates/vinyl–vermiculite,[a] PAN/organo-vermiculites,[b,c]	[a] 100 [b] 165 [c] 166
	Step 2: direct intercalation or in situ polymerization	PS/montmorillonites[d,f]	[d] 134 [f] 168
		Nanocomposites based on modified polymers: PVB15C5/mica,[g] PEG-PE block copolymers/montmorillonite,[h] ammonium-terminated PS/synthetic mica[i]	[g] 169 [h] 171 [i] 172

of organic molecules adsorbed in a flat, extended conformation. Later studies have demonstrated that polycation adsorption processes are essentially irreversible, since desorption requires the simultaneous desorption of all the ionic sites in a polymer chain and the diffusion away from the clay surface (6,117). Parameters such as molecular weight and charge density of the polymer, clay concentration, purity and nature of exchangeable cations, as well as other external factors, such as ionic strength and pH, govern the process and consequently may lead to the formation of different polymer–clay complexes. The potential interest of polycation–clay nanocomposites as sorbents or catalysts may attract further investigations of these not well-understood materials.

Direct intercalation is also the common method for intercalating a large variety of neutral, polar, water-soluble polymers, such as poly(vinyl alcohol) (PVA) (118–120), poly(ethylene glycol) (PEG) (121), poly(ethylene oxide) (PEO) (122), and poly(N-vinyl pyrrolidone) (PVP) (123). The driving force for forming these compounds is, in addition to entropic effects, ion–dipole interactions between the interlayer cations and the polar units (for instance, oxygen or nitrogen atoms) in the polymer. The presence of water molecules, associated mainly with the interlayer cations, or the nature of certain interlayer cations, such as ammonium or ammonium-derivative cations, may induce the existence of other types of polymer–clay interaction mechanisms, such as water bridging and hydrogen bonding, respectively. In the same way, the oxygen atoms in the silicate surface could induce the formation of hydrogen bonds with hydrogen atoms from the aliphatic chain.

A representative and typical example of polymer–clay nanocomposites prepared by direct adsorption is the intercalation of polyoxyethylenes into smectites. Direct intercalation of PEGs of different molecular weights into Ca-montmorillonite leads to the formation of different phases, in which the amount adsorbed and the basal spacings depend on the length of the polymer chain (121). The use of water as solvent and the high hydration energy of calcium ions ensure the presence of a large amount of water in the interlayer region, facilitating the formation of water bridges between the polar units of the polymer and the interlayer cations.

When PEO is intercalated from acetonitrile solutions into montmorillonite and hectorite (122,124–126), the resulting nanocomposites present characteristics that depend on the nature of the interlayer cation in a similar way to that observed for intercalation of crown ethers into smectites (44,46,49,127,128). Such behavior implies the existence of ion–dipole interactions between the oxygen atoms of the polymer and certain interlayer cations, as occurs for PEO-salt complexes (129). The fact that the synthesis is carried out in nonaqueous solutions determines that the polymer may replace the hydration shell usually accompanying the interlayer cations, as shown by IR spectroscopy (124). The effectiveness of such a process is related to the hydration energy of the cation. When this energy is high (calcium or aluminum), the hydration shell is maintained and water bridges are formed instead (124,125). In these cases, interstratified materials or mixtures of two distinct phases are obtained. However, for clays exchanged with cations of low hydration energies, such as sodium and barium, adsorption isotherms are L-type, and a single phase with a basal spacing increase of about 0.8 nm is obtained (124). In these cases, IR and NMR evidence supports a model in which the helical structure of the polymer is preserved (Fig. 15). Thus, PEO-hectorite materials containing Na^+, K^+, or Ba^{2+} show a single peak at about 70 ppm in ^{13}C-NMR spectra, close to the chemical shift observed for pure PEO, suggesting similar conformations in the intercalated and nonintercalated polymer, i.e., preservation of the helix (125). The ^{23}Na-NMR spectrum of PEO/Na-hectorite clearly shows that the Na^+ ions are perturbed by the polymer (124) and also that the polymer provides an environment with similarity either to the natural hydration shell or, more likely, to crown-ether derivatives (49).

Nevertheless, the formation of true PEO-cation complexes in the interlayer space of clays has been the object of some controversy (130,131). In fact, when intercalation is performed from PEO/water solutions, different phases are formed, according to Parfitt and Greenland, for PEG/Ca-montmorillonite (121). Also, Wu and Lerner (130) claimed the formation of two types of intercalation compounds, one layer (d_L = 1.36 nm) and two layers (d_L = 1.77 nm), in which the polymer chains adopt a planar zigzag conformation, similar to the well-known ethylene glycol/montmorillonite complex (132). To resolve all these apparently contradictory results, one might consider the fact that the polymer helix could be distorted by interactions with both the cations and the silicate surfaces. But there are also other factors influencing the process. As pointed out by Bradley in 1945 (41), the interaction between hydrogen atoms of an aliphatic chain with oxygens of the silicate surface involves energy comparable in magnitude to that of the

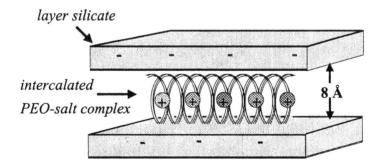

FIGURE 15 Structural model representing the intercalation of PEO in a homoionic smectite, from acetonitrile solutions. (After Ref. 122.)

O—H⋯O bonds of a polyglycol–water system. It is clear that for similar clays, the intercalation procedure will affect strongly the balance between both types of interactions and, consequently, the characteristics of the final products. In the presence of water, which could also swell the clay, it will be energetically more favorable to have the polymer helix distorted as it interacts with the clay surface than to replace the hydration shell of cations. For adsorption from nonaqueous solvents, the energy required to change from the helix to zigzag polymer conformation is greater than the energy necessary to replace the solvation shell of the cation. Therefore, PEO will interact mainly with the cations and preserve most of its helical conformation.

The attractive ion-conducting behavior of PEO-clay nanocomposites, which could be of potential interest for applications as solid electrolytes (122), led Vaia and coworkers (133) to utilize a new method of synthesis profiting from the fact that certain polymers can melt without chemical degradation (134). Insertion of PEO into alkaline metal homoionic montmorillonites is reached after 6 h of treatment at 80°C of a PEO-clay mixture. The chosen temperature is ≈10°C above the melting point of PEO, ensuring sufficient mobility of polymer chains to migrate toward the intracrystalline region of the clay.

The effectiveness of melt intercalation can be improved by using microwave (MW) irradiation in the so-called "MW-assisted blending intercalation" procedure (115). In this procedure, water molecules associated with the interlayer cations absorb MW energy and, consequently, accelerate the heating of the polymer–clay mixture. Thus, MW irradiation applied to PEO-montmorillonite preparations produces the rapid melting of the polymer that diffuses through the interlayer space of the phyllosilicates as water molecules escape (135,136). This procedure is also interesting because it saves time and energy compared to conventional thermal heating, as is true for numerous organic–inorganic processes carried out under MW irradiation (dry media reactions, molecular intercalations, etc.) (137–140).

It is interesting to note that PEO-clay nanocomposites prepared following the melt-intercalation procedure show isotropic behavior such as ion conductivity (133,135). In these materials, the elimination of the hydration shell from the cations occurs in a nonequilibrium process that may induce different type of polymer–clay interactions together with the gluing of particles by a polymer chain interacting with surfaces of different particles.

B. In Situ Intercalative Polymerization

This method consists of the intercalation of precursor species (monomers) followed by polycondensation in the intracrystalline region of the clay. Blumstein in 1961 (109) reported formation of polymers after interlayer adsorption of monomers such as acrylonitrile and methyl-methacrylate into Na-montmorillonite using benzoyl peroxide as a catalyst. Kanatzidis (141), who in 1987 described the preparation of polypyrrole/FeOCl nanocomposites, coined the term *in situ intercalative polymerization*. In FeOCl, as in Cu-smectites (142), pyrrole is intercalated and then polymerized by the action of transition metal ions with redox properties present in the inorganic matrix (either in the structure or as exchangeable cations). In situ polymerization has been extensively applied to the preparation of a large variety of nanocomposite materials based on conducting and nonconducting polymers intercalated in FeOCl, clays, and other layered solids (V_2O_5 xerogel, LDHs, layered phosphates, etc.), as has been illustrated in several review articles (115,143,144).

The formation of polyaniline (PANI) between the layers of montmorillonite (Fig. 16) is an illustrative example of the in situ polymerization procedure. As already postulated in the 1960s by several authors, montmorillonite-adsorbed aniline spontaneously oxidizes and polymerizes by the catalytic action of Cu^{2+} ions (145,146) or by atmospheric oxygen (147,148). Evidence of such reactions and mechanisms were first advanced by Cloos and coworkers (149,150) 20 years

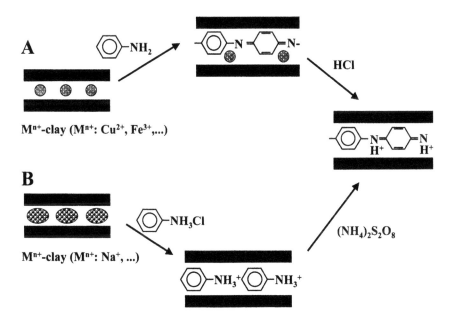

FIGURE 16 Schematic representation of two routes to the preparation of PANI-clay nanocomposites by in situ polymerization: (A) adsorption of neutral aniline in Cu^{2+}- or Fe^{3+}-exchanged clays; (B) ion exchange of interlayer cations with anilinium cations and subsequent polymerization with an oxidant agent (e.g., $(NH_4)_2S_2O_8$).

later. These authors found that the adsorption of aniline in homoionic montmoril-lonites is determined by the conjunction of several factors: (a) basicity of the solute, (b) complexing ability (Brönsted and Lewis acidity) of the exchangeable cation, and (c) polarity and Lewis basicity of the solvent. These characteristics are responsible for the adsorption mechanism of the monomer (protonation, indirect coordination through water bridges, direct coordination to the exchangeable cat-ion, anilinium-aniline association). The presence of ions with redox properties in an acid environment determines feasibility to form the polymer. Thus, Cu-montmorillonite induces the formation of an aniline radical and the further poly-merization of these species in the interlayer region of the clay (149). Although the resulting PANI was formed in one of the insulating forms, these nanocompos-ites could be doped in a further step to yield the conducting emeraldine salt form (Fig. 16A) by exposure to HCl vapors (151). Typical values of the electrical conductivity in the plane defined by the silicate layers are around 10^{-3} S/cm.

As presented schematically in Figure 16, a second way to prepare poly-mer–clay nanocomposites via in situ polymerization consists of intercalation of the monomer (or a precursor of the monomer) in the form of a cation and then later addition of an initiator to induce/polymerization. Thus, the direct exchange of the interlayer cations of smectites by anilinium cations, followed by oxidation with $(NH_4)_2S_2O_8$, could be an alternative procedure to reach the formation of PANI/clay nanocomposites. In this case, the experimental conditions allow the direct formation of PANI as a conducting emeraldine salt (152).

A third route to induce in situ formation of the polymer is exemplified by the nylon–clay nanocomposites of Fukushima and coworkers at the Toyota Central Research Laboratories (114,153). Their formation requires two steps, each of which involves addition of a monomer required for the condensation reaction. Thus, a protonated ω-aminoacid (e.g., $^+NH_3\text{-}(CH_2)_n\text{-}COOH$) with $n = 2\text{–}8$, 11, 12, 18) is first intercalated; and in a second step, the reaction, thermally activated, with ε-caprolactam generates the nylon-6 nanocomposite (114,154). The interest-ing mechanical properties of the resulting materials led to their rapid industrial commercialization and probably account in large part for the increasing attention on polymer–clay nanocomposites at present.

Following the appropriate pathway, in situ polymerization has allowed the preparation of a large variety of polymer–clay nanocomposites with interesting functional and/or mechanical properties. For instance, a doped PPy-synthetic hec-torite nanocomposite exhibits conductivity from about 10^{-5} to 10^{-2} S/cm (155). In general, monomers showing affinity to be adsorbed by smectites, e.g., hydro-philic species, can produce intracrystalline homocondensations. Other monomers, such as acrylonitrile, are also easily intercalated in smectites, because such mole-cules are directly associated to the interlayer cations M^{n+} (M^{n+} = Li^+, Na^+, etc.) through $—C{\equiv}N{\cdots}M^{n+}$ ion–dipole interactions (156–158). The action of γ-irradiation (156) or thermal (158) treatments can induce the polymerization reaction to give polyacrylonitrile (PAN), as has been clearly shown via IR, through evolution of monomer cyclization. Heating of PAN-clay nanocomposites is of great interest for generating carbon–clay nanocomposites that can be used as components of electrodes for electrochemical devices (159). Finally, it is interesting to remark that the preparation by in situ polymerization of polymer–clay nanocomposites from monomers of low polarity, such as styrene, requires other strategies to make compatible the inorganic host, as will be explained in (Section V.D).

C. Template Synthesis of the Clay in the Presence of a Polymer

This procedure is based on the synthesis of the inorganic host in the presence of a polymer that acts as a templating agent for the growing of the solid, in a way similar to alkylammonium ions or surfactants that are used in the formation of certain zeolites or mesoporous silica materials (MCMs), respectively (10,160). This procedure has often been employed for the preparation of nanocomposites derived from synthetic inorganic host solids, such as layered double hydroxides (LDHs) (161). In the case of clays, this route is relatively new and was applied for the first time by Carrado and coworkers (111,162,163) to obtain different polymer–hectorite nanocomposites. These nanocomposites are formed by in situ hydrothermal crystallization of gels prepared from a mixture of silica sol, magnesium hydroxide, LiF, and a selected water-soluble polymer (PANI, PAN, PVP, PDDA, etc.), which are incorporated into the solution in different proportions (Fig. 17). The synthetic clay formed under such conditions is a semicrystalline layered fluorohectorite according to the XRD patterns (111). These authors observed that in the case of neutral polymers, it is possible to increase the polymer loading with a corresponding increase of the basal spacing until nearly complete delamination occurs. For polymers containing ionizable groups, such as PDDA (polydimethyldiallylammonium chloride), the limit in the incorporation of the polymer is determined by the neutralization of the charged sites on the silicate lattice with the cationic sites in the polymer chain.

   Although it is applicable only to certain systems, one of the major interests of this synthetic approach is the possibility to prepare nanocomposites with higher polymer/clay ratios than those prepared by other methods (e.g., in situ polymerization). This is the case for PANI-clay nanocomposites, which are essentially semi-delaminated materials that form well-dispersed systems in water without the loss of the polymer to solution (162). Also, nanocomposites obtained by this route have been used for the preparation of porous materials in which the polymer is removed by calcination (e.g., PVP at 500°C, air). The loss of the polymer produces the rearrangement of clay layers, giving porous solids with mesopores that are in the range 4–10 nm, depending on the molecular weight of the polymer (163,164).

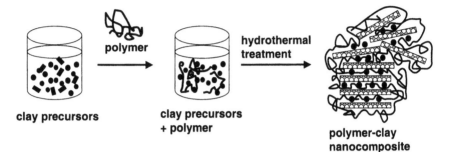

FIGURE 17 Schematic representation of the templating synthesis pathway employed to prepare polymer–clay nanocomposites that generate the clay under hydrothermal conditions in the presence of the polymer.

D. Two-Step Methods

The preparation of polymer-clay nanocomposites with polymers of low polarity requires the application of strategies involving either the modification of the interlayer space of the clay to produce an organophillic environment or the inclusion of functional groups in the polymer (or the monomer precursor) to facilitate intercalation in the clay.

1. Modification of the Clay

This can be achieved following two approaches: (a) using coupling agents such as organosilanes to graft specific organic functions through the silanol groups on the clay surface, and (b) producing organoclays by intercalation of alkylammonium cations. In the second step, the polymer or the monomer is incorporated by application of direct intercalation or in situ polymerization methods to produce the desired nanocomposite.

   a. Minerals, such as chrysotile, sepiolite, and vermiculite, that are modified by the grafting of vinylsilanes (85) (following the methods of cohydrolysis described earlier (Section IV)) can be either copolymerized with unsaturated monomers or blended with an organic unsaturated polymer; the latter is followed by a cure with peroxide at high temperature. Examples include nanocomposites obtained by the reaction of crysotile– and vermiculite–vinyl derivatives with unsaturated monomers, such as methylacrylate, methylmethacrylate, n-butylacrylate, n-dodecylacrylate, vinylacetate, and styrene (100). The resulting materials show extraordinary improvement in mechanical properties, especially elastic modulus.

b. Organoclays of organophillic character may allow the inclusion of monomers capable of in situ polymerization as well as the direct intercalation of polymers from solution or melts. For instance, PAN, which does not intercalate in vermiculites, can be incorporated in the interlayer space of the alkylammonium-exchanged version (165,166). Alternatively, vermiculite/PAN nanocomposites arise from adsorption of the monomer (acrylonitrile), followed by the action of a radical initiator such as benzoyl peroxide at 50°C over 24 h (166). Interest in PAN-clay systems lies primarily in their use as a carbon–clay material that can be transformed at high temperature to sialons via carboreduction (165).

Low-polarity polymers such as polystyrene (PS) can be intercalated into organoclays following two procedures: (a) direct melt intercalation and (b) copolymerization of styrene with an organoclay containing unsaturated bonds. Materials resulting from alkylammonium–montmorillonite and PS were the first reported example of polymer–clay nanocomposites prepared by melt intercalation (134). Typical experimental conditions require heating in vacuum at 165°C for not less than 24 h. TEM images of the resulting PS-organoclay nanocomposites show a heterogeneous microstructure composed of well-ordered intercalated material and a semiexfoliated phase, indicating the complexity of the mechanism governing the intercalation–delamination process (167). PS-organoclay materials can be also prepared by reacting an alkenylammonium–clay, such as $(CH_2 = CH—C_6H_4)CH_2(CH_3)_3N^+$–montmorillonite, with styrene monomer (168). The PS attaches to the clay through the exchangeable cations, and excess polymer is extracted with organic solvents. However, it should be pointed out that neither of these two approaches leads to homogeneous PS-clay nanocomposites.

2. Modification of the Polymer

The incorporation of modified polymers into the clay, is the other way to produce compatibility that leads to obtain nanocomposites. The procedure uses polymers with grafted functionalities that can interact either with the hydrophilic surface of a clay or with the exchangeable cations. For instance, adsorption of polymers with crown-ether functions (e.g., poly(4'-vinylbenzo-15-crown-5, PVB15C5) has been observed on mica (169). In this case, the crown-ether groups coordinate the surface cations and thereby the polymer is attached to the silicate. Surprisingly, vinyl–crown-ether monomers (e.g., 4'-vinylbenzo-15-crown-5, VB15C5) do not show the same tendency. This observation, and the fact that polymer adsorption increases with chain length, suggests the existence of additional effects, such as interactions between polymer chains and loop formation (which is also observed in polymers adsorbed on clays (6) that stabilize PVB15C5 on the solid).

The propensity for oxyethylene groups to interact with interlayer cations has also been used to produce the intercalation of poly(methyl methacrylate) (PMMA) and polyethylene (PE). For example, PEO-PMMA (170) or PEG-PE (171) block copolymers have been intercalated in montmorillonite, resulting in blend nanocomposites with interesting mechanical and ion-conductivity properties.

Other routes involve the synthesis of more complicated polymers, such as that used by Hoffmann and coworkers (172) for the preparation of PS with ionizable sites. A PS with terminal ammonium groups was intercalated in a synthetic clay by an exchange reaction, resulting in a nanocomposites in which the organophillic polymer is bonded through electrostatic forces to the silicate. The resulting nanocomposites can be completely exfoliated by melting with PS and show remarkable elastic properties.

E. Nanocomposites Derived from Sepiolite and Kaolinite

In Section IV, the ability of sepiolite to be coupled to polymers through covalent bonds using organosilanes was highlighted. In these polymer–clay systems, only the external surface of the silicate is involved. However, the accessibility of geometrically compatible molecules to the structural tunnels of sepiolite (see Sec. III.C) allows the preparation of nanocomposites with polymers incorporated inside the sepiolite structure. Small monomers, such as isoprene and styrene, penetrate the tunnels and are subsequently polymerized using mild conditions initiated solely by the surface acidity of the mineral itself (173). In addition, large macromolecules of, for example, PEO can themselves slowly penetrate into the sepiolite tunnels, irreversibly replacing the zeolitic water (69). This result opens the possibility of inserting other polymers into sepiolite, especially if they are linear and polar.

Attempts to prepare polymer–kaolinite nanocomposites by direct intercalation or in situ polymerization have thus far been unsuccessful (174). However, following the procedures used to expand the kaolinite layers as described in Section III.B, i.e., using DMSO or other polar molecules, the interlayer space of this aluminosilicate could be made accessible to polar polymers, such as polyethylene glycols (175). In these nanocomposites, the Δd_L values range between 0.37 and 0.41 nm, indicating that PEGs are intercalated and form a monolayer of polyoxyethylene chains in planar zigzag conformation. IR data indicate that at least some of the PEG (O—CH$_2$CH$_2$—O) groups are in a trans conformation. Moreover, ^{13}C DD/MAS NMR spectra show that the polymer is intercalated intact and is more constrained in the interlayer space of kaolinite than it is in bulk. This conformation is consistent with hydrogen-bond interactions between oxygens of the polymer and OH of the mineral surface. A similar situation for PEO-smectite intercalations when water molecules are retained by the interlayer cations has been described earlier (Sec. V.A). Distortion of the helical structure of PEGs by OH internal surface groups also occurs in modified polyoxyethylenes intercalated in layered double hydroxides (LDHs) (176).

VI. USES OF CLAY–ORGANIC COMPOUNDS: PERSPECTIVES TOWARD NEW APPLICATIONS AS ADVANCED MATERIALS

Hybrid organic–clay materials, including polymer–clay composites, are used in many industrial applications due to their organophillic character and reproducible thixotropic behavior. Production of commercial organoclays, such as quaternary alkyl- and aryl-ammonium salts, amounts to more than 25,000 tons/year (177). For example, Bentone® 27 and Bentone® 34 (Rheox Inc. Elementis Specialties) are used as additives in interior and exterior paints, in other coatings of asphaltic composition, in mineral oils to produce greases of different viscosities, in cosmetics such as facial masks, skin cleaners, and shampoos, in water- and oil-based printing inks, adhesives, drilling fluids, sealants, etc. (178). To a lesser extent, sepiolite treated with surfactants (rheological grade, Pangel) (179) is used in similar industrial applications, receiving special interest due to its gelling efficiency in low-polarity solvents. Some clays are used directly as fillers of commercial polymers, but others are subjected to treatments that modify the nature of the surface and enhance compatibility between the inorganic layer charge and the organic matrix. For example, coupling agents based on functionalized silanes (e.g., containing mercapto groups) are useful for conditioning kaolinites to be used as reinforcements in rubber and plastics. Industrial use of smectites in these applications is favored if the clay mineral is previously saturated with organic cations (methylene blue or malachite green). This precludes sorption of accelerators and other additives needed for rubber processing (180).

Innovative applications are possible because the mechanisms governing interactions between clays and organic compounds, their microstructural organization, and the resulting synergy are fairly well understood. The ability to introduce suitable functionality into clays by rational use of organic or organometallic reagents allows for controlled surface modification, giving nanostructured solids based on clays (see Sec. IV). Chemical reactivity is therefore changed, and the added functional groups allow for further chemistry that is typical of organic reagents. Also, the resulting nanocomposites have physical characteristics related to mechanical, surface, optical, magnetic, electronic, or electrochemical properties that are inherent to the organic species involved (181).

A. Selective Adsorption and Separation Processes Using Organoclays

Organoclays based on the interlayer adsorption of quaternary ammonium species are receiving increasing attention due to their ability to selectively adsorb neutral or ionic species (182–184). Intercalated organic cations, such as tetramethyl- and tetraethyl-ammonium (TMA$^+$ and TEA$^+$), act as pillars that separate the silicate layers along the c axis of smectites and vermiculites, creating galleries that can be filled by small nonpolar molecules, such as benzene and toluene (182). As such, these organoclays act as molecular sieves. The design of chemical sensors based on this selectivity has been reported by Yan and Bein, showing a clear discrimination based on molecular size (185). Thin films of TMA-hectorite deposited on QCM (quartz crystal microbalance) piezoelectric sensors displayed discrimination between benzene and cyclohexane (Fig. 18).

Removal of inorganic and organic pollutants from air, water, and soils can be carried out using long-chain alkylammonium–exchanged smectites and other organoclays. Such materials adsorb not only cationic radionuclides such as ^{134}Cs$^+$ and ^{85}Sr^{2+}, but also long-lived radioiodide (^{125}I$^-$), which is present in water primarily in anionic form (186–189). Hexadecylpyridinium–bentonite is effective for the removal of iodine, whereas TMA-clay is completely ineffective. The ability to adsorb I$^-$ species could be attributed to an inversion of the micellar charge, which is produced by the adsorption of the long-chain alkylammonium in excess of the cation exchange capacity of the clay mineral. Nonpolar organic pollutants can also be removed by organoclays, including carcinogenic polycyclic aromatic hydrocarbons (naphthalene, phenanthrene, etc.) that are encapsulated by dimethylbenzylstearyl ammonium–exchanged bentonites (190).

These organoclays are also receiving attention as gas chromatography packing materials (191–193); e.g., TMA-smectites are used for separating light hydrocarbons (193). Other important uses concern the development of improved formulations of pesticides (34,194–198). Here, organosmectites and organosepiolites decrease the tendency of volatile herbicides, such as alachlor and metalachlor, to evaporate (34).

Grafting of —SH (*thiol* or *mercapto*) groups using appropriate organosilanes (e.g., 3-mercaptopropyltrimetoxysilane, 3-MPTMS) modifies the reactivity of clay surfaces to allow the uptake of heavy metals from aqueous solutions. Mercier and Detellier (199) report the grafting of 3-MPTMS onto the interlayer surface of montmorillonite; the resulting material (so-called *thiomont*) adsorbs Pb^{2+} and Hg^{2+} ions and, to a lesser extent, Cd^{2+} and Zn^{2+}. However, the bonding mechanism between the clay surface and the organosilane is not clear, because montmorillonite does not contain accessible intracrystalline hydroxyls. Condensed organopolysiloxanes may form and act as a coating, with some points attached to the ≡Si—OH groups located at the edges of the silicate microcrystals. Conversely, the grafting of 3-MPTMS onto sepiolite, with its high content of silanol groups, gives organosepiolite derivatives that are able to complex Cd^{2+} ions via thiol–cation interactions (200).

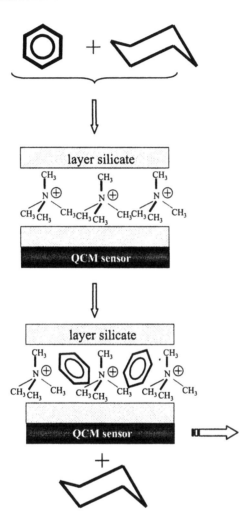

FIGURE 18 Molecular discrimination of benzene over cyclohexane from selective sorption in a TMA-hectorite film deposited on a QCM sensor. (Based on Ref. 185.)

B. Membranes and Electroactive Sensors Based on Organoclays

Membranes containing macrocyclic compounds can be very efficient in the separation and identification of ions and, therefore, in the preparation of ion-selective sensors (201). The immobilization of crown ethers and cryptands by intercalation on 2:1 phyllosilicates (45) has been used in the preparation of composite membranes made up of sandwich-like materials encapsulated into thin polybutadiene coatings to improve their mechanical properties (202). The entrapped macrocyclic compounds modulate the transport properties of cations in aqueous solution, and this can be used for their individual discrimination and recognition. Although the mechanism controlling the ion selectivity of these composite membranes is still not clear, development of ion-selective sensors based on the coating of electrodes by this class of membranes can be envisaged.

The use of organoclays to entrap electroactive chemical species or enzymes offers an excellent opportunity to develop efficient electrochemical sensors and biosensors (203–205). Montmorillonite-CTAB has been used to prepare modified electrodes, based on the affinity showed by these organoclays to uptake electroactive anions such as $Fe(CN)_6^{4-}$, $Mo(CN)_8^{4-}$, and $Fe(C_2O_4)_3^{3-}$ (203). The peak potentials of such species observed by cyclic voltammetry can be used to analyze these compounds in aqueous solutions. To explain the driving forces governing the entrapment of anionic electroactive species by the organoclay, inversion of the particle charge when the incorporated alkylammonium long-chain species exceeds the CEC must occur.

The immobilization of urease and glucose oxidase enzymes using a Laponite® (synthetic hectorite)-poly(pyrrole-pyridinium) film, which confers electronic conductivity to this hybrid system, is the basis for the development of conductimetric biosensors (204). Use of these materials in microelectrode arrays made possible the design of microbiosensors for the detection of glucose. It can be expected that the future involvement of conducting nanocomposites based on polymer–clay materials will be an important aspect of designing new enzyme-based microsensors.

Organoclays derived from montmorillonite and synthetic fluorohectorite exchanged with $(C_{16}H_{33}NMe)^+$, $(C_{18}H_{37}NMe)^+$, and $(C_{16}H_{33}PBu)^+$ (Me and Bu: methyl and butyl groups, respectively) have been used to prepare conventional CPEs (carbon paste electrodes) in which the clay fraction amounts to about 5 wt% (206). Curiously, only the so-called *mixed-ion heterostructure* based on the fluorohectorite/[$(C_{16}H_{33}PBu^+ + Na^+)$] compound, is especially active in a test gauging electrochemical response toward certain contaminants in aqueous solution. From the oxidation signals observed via differential pulse voltammetry, such apparently mixed-ion materials are effective for analyzing the molecular pollutants 2,4-diclorophenoxyacetic (the basis of the herbicide 2,4-D) and 2,4-

dichlorophenol. The electrochemical response was attributed to the ability of this unique clay intercalate to achieve electrical neutrality upon oxidation of the dichlorophenoxy moiety (to quinone) through the rapid release of sodium ions (206). Carbon paste electrodes are known to degrade with time; therefore, more robust electrodes for organic species determinations based on organoclays must be designed using epoxy resins or sol-gel hybrid materials obtained from organo-alkoxysilanes. An alternative method to developing biosensors with improving mechanical resistance is based on the treatment of Laponite with alkyltrialkoxysi-lanes containing amino functions. According to Labbé et al. (204), the resulting compounds consist of oligomeric organosiloxanes (organosilsesquioxanes) that are intercalated into the layered silicate, conferring both mesoporosity and ion exchange capacity. Although these intercalated materials are disordered, they can be regarded as promising materials for novel sensor devices. Amperometric biosensors for glucose determination have been successfully prepared using these compounds. In addition, anionic electroactive species such as $Fe(CN)_6^{3-}$, $Mo(CN)_8^{4-}$, and $SiW_{12}O_{40}^{4-}$ were incorporated into the organic derivative of the clay, exhibiting electroactivity that it is of interest for analytical purposes. In addition, these systems are good candidates for electrocatalysts because they can form Pt nanoparticles from $PtCl_4^{2-}$ species.

C. Organoclays as Supports for Organic Reactions and Heterogeneous Catalysts

Anionic activation reactions, e.g., replacement of the halogen atom in alkylbro-mides by reaction with alkaline salts of different anions, usually take place in the presence of organic solvents using phase-transfer catalysts. They are based on the complexing ability of crown ether macrocyclic ligands or on the presence of surfactants (207). As proposed by Regen (208), the *triphase catalysis* process is facilitated by appropriate solids that allow for the displacement of substituted groups and the partition of reagents and products with different solubilities. Immobilization of such macrocycles by grafting on silica surfaces allows, under mild conditions, the alkylation (209) of acetate and similar reactions, with the advantage of efficient recuperation of the transfer agent:

$$n\text{-}C_8H_{17}Br + CH_3COOK \rightarrow KBr + n\text{-}C_8H_{17}COOCH_3 \ (93\%) \qquad (9)$$

The facile alkylation of anions such as thiocyanide by n-alkylbromides in almost quantitative yield has been also successfully carried out using organoclays, as was first reported by Lin and Pinnavaia (210). These authors point out the importance of the alkyl chain arrangement in the interlayer space, which is imposed by the microstructure of the host layered solid—in particular the location of charge within the silicate and the electrical charge density of the clay. Different bilayer arrangements of methyltrioctylammonium–hectorites favor different anionic activation reactions (210). Other hybrid organic–inorganic systems, such as benzoate-LDHs, also activate this type of reaction, operating in this case without organic solvents, i.e., under so-called *dry media conditions* (211,212). These reactions are especially interesting when they are promoted by microwave irradiation, as initially reported by Gutierrez et al. (137,138). Thus, the usefulness of microwave-assisted reactions involving organoclays as triphase catalysts is suggested. This process enhances efficiency, saves time, and avoids the use of organic solvents.

Another significant interest of organoclay systems as catalysts is related to the acidic character of certain intercalation compounds. In montmorillonite/TED-$2H^+$ (diprotonated triethylene diamine, i.e., 1,4-diazabicyclo [2,2,2] octane), the diprotonated diamine acts as a pillar facilitating the uptake of small molecules and also confers the acidic environment necessary for conversions of organic groups such as nitriles (e.g., acetonitrile) to amides (213). Few nitriles, such as trichloromethylacetonitrile, can be directly hydrolyzed by the untreated layered silicates (214). Strong acidic behavior can also be introduced by grafting of arylsilanes such as $Cl_2Si(CH_3)(CH_2)_2Ph$ on sepiolite, which gives very stable organic derivatives that can be sulfonated to produce arylsulfonic groups (98,215). Since these groups contain H^+ species, this type of material can be regarded as a catalyst for alcohol dehydrogenation. Here the selectivity toward diethylether is determined by both experimental conditions and the textural properties of the organic derivative. Beckman rearrangement reactions are also efficiently catalyzed by these sulphoaryl derivatives (99).

Another class of heterogeneous catalyst based on organic derivatives of sepiolite contains transition metals anchored to the clay surface through covalent bonds. Osmium species were obtained by direct reaction of OsO_4 with vinyl groups, previously grafted on sepiolite by reaction of methylvinyldichlorosilane (216) [Eq. 6]. After controlled hydrogenation (217), the resulting Os derivatives act as effective oxidation catalysts for the production of free hydrogen from water photodecomposition (218). Another approach to preparing transition metal clay systems with catalytic activity is in situ formation or ion exchange of ligand–metal complexes. An illustrative example is $[Rh(PPh_3)_3]^+$-smectites, which demonstrate the making of a heterogeneous catalyst from one conventionally used in homogeneous conditions (219). The resulting materials are efficient in the hydrogenation of terminal olefins, and they also avoid the competitive isomerization that takes place when the reactions are carried out in homogeneous media. Yet another advantage is the easy recuperation of the catalysts, which allows their reutilization for several cycles without appreciable loss of activity. This is a significant improvement with respect to the homogeneous catalysis process.

D. Clay–Dye Complexes and Photoactive Materials and Devices

Clays were used in antiquity as a host to entrap organic dyes. Maya Blue, a stained clay that was prepared in America several centuries ago, consists of indigo pigment sequestered by palygorskite (74). The special ability of this mineral, observed also for the similar mineral sepiolite, to trap molecular species within intracrystalline tunnels is the basis for preparing such a clay–dye complex (75,69). Usually, cationic dyes, such as methylene blue, methyl green, crystal violet, safranin T, thioflavine T, acrydine orange, and rhodamine 6G, are ion exchanged readily into smectites and vermiculites (198). One of the most interesting features of such complexes is their metachromasy effect, i.e., the decrease in intensity of absorption bands assigned to isolated species (*monomers*), with a concomitant intensity increase at higher energies assigned to the aggregation of organic species (220). This effect is observed for methylene blue intercalated in clays such as montmorillonite (221). Self-aggregation produces dimers, trimers, and higher aggregates that are usually in a planar conformation, with the aromatic rings lying parallel to the silicate layers (222). Clays of high charge density, such as the synthetic clay fluortaenolite, cause a nearly perpendicular orientation of dye species in the interlayer space (223).

Cationic dyes are also adsorbed at the external surface of sepiolite, giving rise to different types of aggregates, depending upon the initial concentration of dye (224,225). These sepiolite–dye complexes have been investigated for their possible use in the photostabilization of coadsorbed labile bioactive species (197), similar to the cationic dye–montmorillonite compounds studied by Margulies et al. (226). In some cases, the dimensions of the dye molecules are compatible with the topology of the structural tunnels of sepiolite, allowing their inclusion into the interior of the silicate microcrystal, where they remain strongly associated to the host clay (69,75). The confinement of photoactive species in such structural tunnels is of potential interest in the design of materials for microlaser applications, similar to the encapsulation of cationic dyes along the c-axis in AlPO single crystals (227).

Sequestration of optically active species (or other functional molecules) is now possible using organoclays organized as hollow microspheres. Muthusamy and coworkers (228) in 2002 reported the synthesis of an organoclay named Mg-phyllo(hexadecyl)silicate (MgHD) from hexadecyltrimetoxysilane and $MgCl_2$ in a water–oil microemulsion, producing spheroids with diameters in the range of 5–25 microns. These hybrid materials are poorly crystallized spheres, torus-like particles, or broken spheres (Fig. 19), with large central cavities surrounded by spongelike thin walls of the organoclay. Interestingly, these synthetic procedures are applicable for preparing microspheres of organoclays containing dye molecules that are covalently attached or occluded in their cavities (228).

Organoclays based on quaternary ammonium compounds are able to adsorb azobenzene derivatives, which confer to the final product attractive photoresponse properties (229). The azobenzene reversibly changes from trans to cis conformation upon UV irradiation or alternatively under mild heat treatment. Interestingly, the variation of the basal spacing shows that the photoresponse is stable for more than 15 cycles (229).

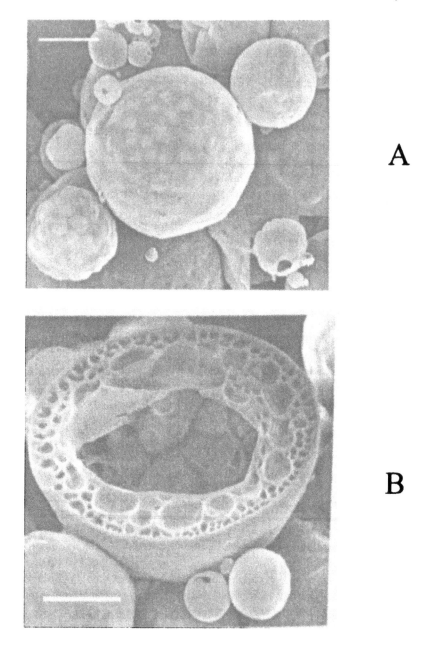

FIGURE 19 SEM image of intact (A) and broken (B) hollow spheroids of synthetic Mg-phyllo(hexadecyl)silicate prepared by microemulsion methods. Scale bar represents 20 microns. (From Ref. 228.)

Other materials derived from clay–organic systems include small semiconductor nanoparticles (CdS, for instance), which are extremely interesting in terms of new nanoelectronic and optoelectronic devices. The functionalization of sepiolite using 3-MPTMS (see Sec. VI.A) allows complexation of Cd^{2+} ions by thiol–cation interactions (200). These materials are potentially very attractive for their quantum dot properties, because it can be expected that treatment with H_2S gas or Na_2S solutions could produce semiconducting CdS particles of nanosized dimensions. In addition, such systems would be of interest in terms of photocatalytic activity, such as for the generation of H_2 from isopropanol, as was observed for MCM mesoporous silica that was functionalized in a similar way (230).

E. Organoclays as Ionic and Electronic Conductors

Much effort is being devoted to preparing new solids with ionic, electronic, or mixed electrical behaviors as novel devices. The combination of conducting polymers with layer silicates offers one avenue for this. The intercalation of cation complexing agents based on oxyethylene compounds, such as crown ethers and PEO (44,122,45) (see Sec. V.A), allows for modulation of the ionic mobility of interlayer cations, such as Li^+ and Na^+, and therefore control of the ionic conductivity (124,231,232). PEO-clay nanocomposites show improved characteristics compared to conventional PEO-salt complexes. For example, the thermal stability raises to 250–300°C from 70–100°C (233). These anisotropic solid electrolytes present ionic conductivity values in the range of 10^{-5}–10^{-4} S/cm at elevated temperature (>200°C), i.e., several orders of magnitude higher than the original smectites but still far from typical PEO-salt complexes (10^{-3}–10^{-4} S/cm at room temperature), making less probable their use in electronic or electrochemical devices. Moreover, the conductivity of Li^+- and Na-montmorillonite/PEO complexes was measured in a direction parallel to the (a,b) plane of the silicate (layer plane), while values in a direction perpendicular to the clay layers are 10^2–10^3 lower (115). One exception to this behavior is the unique PEO-synthetic hectorite nanocomposite developed by Sandi and coworkers (234). A membrane of this material shows high conductivity (4.87×10^{-3} S/cm) at low temperatures and a transference number near unity (0.95). These remarkable numbers are speculated to arise from the inclusion of small silica nanoparticles encapsulated during the synthesis of the clays. Transport numbers practically equal to unity, deduced from electrical polarization measurements in solid state, have been reported by Aranda and coworkers for PEO-montmorillonite nanocomposites prepared by the MW-assisted blending-intercalation method (136).

Polymer–clay nanocomposites using PANI and PPy exhibit better electrical behavior than PEO-clay systems, but in these cases the conductivity is electronic. These nanocomposites are prepared, as discussed earlier (Sec. V.B), by interlamellar adsorption of aniline or pyrrole in smectites saturated with transition metal cations (115). Although the so-called *intercalative polymerization* of these nitrogenated bases initially gives insulating materials, they can be easily transformed into conductive solids. Aniline is intercalated in Cu-fluorohectorite and spontaneously forms the nonconducting emeraldine base form, as shown by electronic absorption spectra. After exposure to HCl vapors, the changes observed in the spectra indicate the formation of lattice polarons and, consequently, a significant increase in the electrical conductivity to 5×10^{-2} S/cm (151). Similarly, PPy-clay systems, when doped with I_2, display electronic conductivity that increases from 2×10^{-5} to 1.2×10^{-2} S/cm (155). These conductivity values are lower than for pure PANI or PPy, however (115,143,144).

In order for this type of material to be a potential candidate as electrode materials (batteries and electrochemical sensors or electrocatalysts), the incorporation of suitable electroactive elements, such as V or Mn, would be helpful. This would provide a certain number of centers with higher oxidation states useful for redox processes.

F. Polymer–Clay Nanocomposites: Mechanical Properties

Additional important characteristics of polymer–clay nanocomposites (PCNs), such as their fire-retarding effect, resistance to ignition, gas impermeability, as well as improved mechanical properties, are also of great interest for new industrial applications (235,236). Among the estimated markets of nanocomposites for the coming years, automobile, building, and food packaging are the sectors that are expected to receive the most emphasis, with a predicted commercial demand in 2009 of 156,500, 68,000, and 161,000 tons, respectively (237). These values represent a revolutionary challenge for novel applications of clay derivatives, attracting influential companies such as Toyota, Nanocor, Southern Clay Products, Süd Chemie, Dow Chemical, and others.

Some PCNs have excellent mechanical behavior, such as high Young's modulus (tensile modulus) and good flexural, elasticity, or impact properties. In addition to the nature of the constituents, i.e., clay, polymer, quaternary ammonium species, and other conditioning compounds, the mechanical properties vary with factors such as clay content and mode of preparation. Consider the nylon 6–clay nanocomposite, which was the first PCN to receive commercial interest. This material was initially investigated by Toyota, and it remains probably the most studied system to date. Nylon and other polyamide resins (usually reinforced with glass fiber) are currently used in the automobile industry as a conventional composite material. Different procedures have been applied to prepare nylon–clay nanocomposites, including melt intercalation (using organoclays, such as smectites exchanged with octadecylammonium) and in situ intercalative polymerization of ε-caprolactam (in Na^+— or CO_2H—$(CH_2)_{11}$—NH_3^+-exchanged smectites), using different amounts and types of clay (montmorillonite, hectorite,

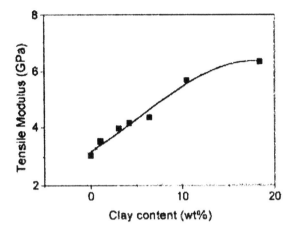

FIGURE 20 Evolution of the tensile modulus with the clay content in nylon 6–montmorillonite nanocomposites prepared by melt intercalation using octadecylammonium-exchanged smectite. (After Ref. 240. Reprinted by permission of John Wiley & Sons, Inc.)

saponite) (235,238). The amounts of smectite needed to obtain materials with suitable mechanical properties are relatively low (ca. 5%). The Young's modulus measured at room temperature of commercial nylon 6 increases from 1.11 to 1.87 GPa when incorporated as a nanocomposite with montmorillonite exchanged with CO_2H—$(CH_2)_{11}$—NH_3^+ (239). An increase of tensile modulus with clay content is clearly observed for nylon 6–montmorillonite nanocomposites prepared by melt intercalation using octadecylammonium–smectite (Fig. 20) (240). Table 6 summarizes the mechanical properties of nylon-PCNs and pure nylon-6; only the Izod impact strength is moderately reduced in the nanocomposites (from 20.6 to 18.1 J/m), while the other values increase significantly. The improved mechanical

TABLE 6 Mechanical Properties, Measured at Room Temperature, of Nylon 6–Montmorillonite Nanocomposites (Clay Content 4.7%)

Mechanical properties	Nylon 6	Nanocomposite nylon-6–montmorillonite (GPa)
Tensile modulus	1.11 GPa	1.87 GPa
Flexural modulus	89.3 MPa	143 MPa
Izod impact	20.6 J/m	18.1 J/m

Source: Ref. 235.

properties, together with the enhanced heat distortion temperature (from 65°C to 150°C) in the nanocomposites prepared by melt intercalation, allow their use as belt covers in cars. Nanocomposites based on smectites and magadiite layered solids combined with epoxy resins and polyurethanes display a 2–3× increase in Young's modulus (241,242).

Future work in this field includes alternative uses of polymers and copolymers combined with other modified clays (kaolinite, palygorskite, and sepiolite) and the use of reagents with different coupling ability. The use of organosilanes remains an interesting way for obtaining nanocomposites with good mechanical behavior, as reported for vermiculite vinyl derivatives copolymerized with unsaturated monomers such as methylacrylate, methylmethacrylate, n-butylacrylate, n-dodecylacrylate, vinylacetate, and styrene (100). The resulting materials show an increase in the isochronal modulus (Fig. 21) by a factor of close to 100, which is by far the highest value yet observed within polymer–clay nanocomposites. Therefore, it would be of great interest to revisit the grafting procedures presented in Section IV because they would provide new functionalized organomineral products.

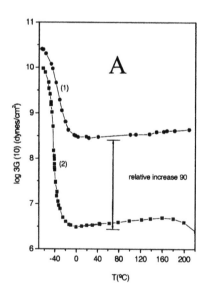 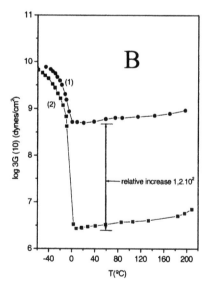

FIGURE 21   Isochronal modulus of vermiculite vinyl derivatives copolymerized with unsaturated monomers such as *n*-butylacrylate (A) and *n*-dodecylacrylate (B). (After Ref. 100. Reprinted by permission of John Wiley & Sons, Inc.)

ABBREVIATIONS

2D	solid of bidimensional structure
3-MPTMS	3-mercaptopropyltrimetoxysilane
18C6	18-crown-6
15C5	15-crown-5
ALPO	porous aluminophosphate
C(222)	cryptand (2,2,2)
CEC	cation exchange capacity
CPE	carbon paste electrode
CTAB	cetyltrimethylammonium bromide
DIQUAT	1,1′-ethylene-2,2′-dipyridilium dibromide
d_L and d_{001}	basal space distance
DMAEM	β-dimethyl-aminoethlylmethacrylate hidroacetate
DMF	dimethyl formamide
DMSO	dimethylsulfoxide
LDH	layered double hydroxide
LMMS	Laser microprobe mass spectrometry
MB	methylene blue
MCM	hexagonally arranged mesoporus silica
MgHD	Mg-phyllo(hexadecyl)silicate
MW	microwave irradiation
NFM	N-methylformamide
PAN	polyacrylonitrile
PANI	polyaniline
PARAQUAT	1,1′-ethylene-4,4′-dipyridilium dichloride
PCNs	polymer clay nanocomposites
PDDA	poly(dimethyldiallyl-ammonium)
PE	polyethylene
PEG	poly(ethylene glycol)
PEG-PE	poly(ethylene glycol)-polyethylene block copolymers
PEO	poly(ethylene oxide)
PMMA	poly(methylmethacrylate)
PPy	polypyrrole
PS	polystyrene
PVA	poly(vinyl alcohol)
PVB15C5	poly(4′-vinylbenzo-15-crown-5)
PVP	poly(N-vinyl pyrrolidone)
QCM	quartz crystal microbalance
TEA	tetraethylammonium
TED	triethylene diamine (1,4-diazabicyclo [2,2,2] octane)
TMA	tetramethylammonium
VB15C5	4′-vinylbenzo-15-crown-5

REFERENCES

1. DJ Greenland. Soils Fertilizers 1965; 28:415–425.
2. GW Brindley. In: JM Serratosa, ed. Proc Reunión Hispano-Belga de Minerales de la Arcilla. Madrid: Consejo Superior de Investigaciones Científicas, 1970:55–66.
3. MM Mortland. Adv Agronom 1970; 22:75–117.
4. JA Rausell-Colom, JM Serratosa. Reactions of clays with organic substances. In: ACD Newman, ed. Chemistry of Clays and Clay Minerals. London: Mineralogical Society, 1987:371–422.
5. BKG Theng. The Chemistry of Clay–Organic Reactions. London: Adam Hilger, 1974.
6. BKG Theng. Formation and Properties of Clay–Polymer Complexes. New York: Elsevier, 1979.
7. G Lagaly. In: K Jasmund, G Lagaly, eds. Tonminerale und Tone. Darmstadt. Germany: Steinkopff Verlag, 1993:89–167.
8. S Yariv, H Cross. Organoclay Complexes and Interactions. New York: Marcel Dekker, 2002.
9. TJ Pinnavaia, GW Beall. Polymer–Clay Nanocomposites. West Sussex. UK: Wiley, 2000.
10. JM Thomas, WJ Thomas. Principles and Practices of Heterogeneous Catalysis. Weinheim. Germany: VCH, 1997.
11. JE Gieseking. Soil Sci 1939; 47:1–13.
12. SB Hendricks. J Phys Chem 1941; 45:65–81.
13. DJ Greenland, SP Quirk. Clays Clay Miner 1962; 9:484–499.
14. BKG Theng. Organic complexes of montmorillonite. PhD dissertation, University of Adelaida, Australia. 1964.
15. A Weiss, E Michel, AL Weiss. In: D Hazdi, ed. Wassertoffbrukenbindungen. Ein- und Zweidimensionale Innerkristalline Quellungsvorgänge. Hydrogen Bonding (a Symposium). London: Pergamon Press, 1958:495–508.
16a. GF Walker. In: ITh Rosenqvist, ed, Proc Internat Clay Conference, Stockholm, 1963, Pergamon Press, Oxford, 1963.
16b. GF Walker. Clay Miner 1967; 7:129–143.
17. WD Johns, PK Sen Gupta. Amer Miner 1967; 52:1706–1724.
18. JM Serratosa, WD Johns, A Shimoyama. Clays Clay Miner 1970; 18:107–113.
19. JA Martín-Rubí, JA Rausell-Colom, JM Serratosa. Clays Clay Miner 1974; 22:87–90.
20. T González-Carreño, JA Rausell-Colom, JM Serratosa. In: ITh Rosenqvist, ed, Proc III European Clay, Conference Oslo, Nordic Society for Clay Research, Oslo, 1977.
21. JA Rausell-Colom, V Fornes. Amer Miner 1974; 59:790–798.
22. GW Brindley, A Tsunashima. Clays Clay Miner 1972; 20:233–240.
23. T Furukawa, GW Brindley. Clays Clay Miner 1973; 21:279–288.
24. GE Lailach, GW Brindley. Clays Clay Miner 1969; 17:95–100.
25. JM Serratosa. Nature (London) 1965; 208:679–681.
26. JM Serratosa. Clays Clay Miner 1966; 14:385–391.
27. A Weiss. Angew Chem Int Ed Engl 1963; 2:134–143.
28. G Lagaly. Solid State Ionics 1986; 22:43–51.

29. M Ogawa, K Kuroda. Bull Chem Soc Jpn 1997; 70:2593–2618.
30. JM Serratosa. Clays Clay Miner 1968; 16:93–97.
31. G Lagaly. Angew Chem Int Ed Eng 1976; 45:575–586.
32. JW Jordan. Mineralog Mag 1949; 28:598–605.
33. JW Jordan. J Phys Colloid Chem 1949; 53:294–306.
34. Y El Nahal, S Nir, C Serban, O Rabinovitch, B Rubin. J Agric Food Chem 2000; 48:4791–4801.
35. A Mifsud, V Fornés, JA Rausell. In: JM Serratosa, ed. Proc Reunión Hispano-Belga de Minerales de la Arcilla. Madrid: Consejo Superior de Investigaciones Científicas, 1970:121–127.
36. WG Garret, GF Walker. Clays Clay Miner 1962; 9:557–567.
37. JA Rausell-Colom. Trans Faraday Soc 1964; 60:190–201.
38. JA Rausell-Colom, P Salvador. Clay Miner 1971; 9:193–208.
39. JA Rausell-Colom, J Sáez-Auñón, CH Pons. Clay Miner 1989; 24:459–478.
40. J Sáez-Auñón. Estudio de la gelificación del sistema vermiculita-ornitina en solución salina mediante análisis de la difusión de rayos-X a pequeños ángulos. PhD dissertation, Complutense University. 1990.
41. WF Bradley. J Amer Chem Soc 1945; 67:975–981.
42. DMC MacEwan. Faraday Soc Trans 1948; 44:349–367.
43. DMC MacEwan. Nature 1946; 157:169–160.
44. E Ruiz-Hitzky, B Casal. Nature 1978; 276:596–597.
45. E Ruiz-Hitzky, B Casal, P Aranda, JC Galván. Rev Inorg Chem 2001; 21:125–159.
46. B Casal, E Ruiz-Hitzky, JM Serratosa, JJ Fripiat. J Chem Soc Faraday Trans I 1984; 80:2225–2232.
47. B Casal, E Ruiz-Hitzky, L Van Vaeck, FC Adams. J Incl Phenom 1988; 6:107–118.
48. JK de Waele, H Wouters, L Van Vaeck, FC Adams, E Ruiz-Hitzky. Anal Chim Acta 1987; 195:331–336.
49. P Aranda, B Casal, JJ Fripiat, E Ruiz-Hitzky. Langmuir 1994; 10:1207–1212.
50. SS Cady, TJ Pinnavaia. Inorg Chem 1978:1501–1507.
51. F Obrecht, MI Cruz, JJ Fripiat. J Colloid Interf Sci 1978; 66:43–54.
52. AG Cairns-Smith. Genetic Takeover and the Mineral Origins of Life. Cambridge: Cambridge University Press, 1982:335–339.
53a. J Pleysier, A Cremers. J Chem Soc Faraday I 1975; 71:256–264.
53b. A Maes, P Peigneur, A Cremers. In: SW Bailey, ed, Proc Internat Clay Conference, Mexico, 1975, Applied Publishing, Wilmette, IL, 1976.
53c. A Maes, P Peigneur, A Cremers. J Chem Soc Faraday I 1978; 74:182–189.
53d. P Peigneur, A Maes, A Cremers. In: MM Mortland, VC Farmer, eds, Proc Internat Clay Conference, Oxford, 1978, Elsevier, Amsterdam, 1979.
54. HF Donner, MM Mortland. Science 1969; 166:1406–1407.
55. MM Mortland, TJ Pinnavaia. Nature (Phys Sci) 1971; 229:75–77.
56. DM Clementz, MM Mortland. Clays Clay Miner 1972; 20:181–186.
57. HA Resing, D Slotfeldt-Ellingsen, AN Garroway, DC Weber, TH Pinnavaia, K Unger. In: JP Fraissard, MA Resing, eds, Magnetic Resonance in Colloid and Interface Science. Proc NATO Adv Sudy Inst, Menton, France, 1979, Reidel, Dordrecht, 1980.

58. TJ Pinnavaia. In: GV Smith, ed. Catalysis in Organic Synthesis. New York: Academic Press, 1977:131–138.
59. VC Farmer, MM Mortland. J Chem Soc A 1966:344–351.
60. JM Serratosa. Amer Miner 1968; 53:1244–1251.
61. JP Hunt. Metal Ions in Aqueous Solution. New York: Benjamin, 1963.
62. JJ Fripiat, A Servias, A Leonard. Bull Soc Chim France 1962:635–644.
63. K Wada. Amer Miner 1961; 46:79–91.
64. A Weiss, W Thielepape, H Orth. In: A Weiss, L Heller, eds, Proc Internat Clay Conference, Jerusalem, 1966, Israel Program for Scientific Translations, Jerusalem, 1966.
65. M Cruz, H Jacobs, JJ Fripiat. In: JM Serratosa, ed, Proc Internat Clay Conf, Madrid, Divisioón de Ciencias CSIC, Madrid, 1973.
66. S Hayashi. Clays Clay Miner 1997; 45:724–732.
67. JC Serna, GE VanScoyoc. In: MM Mortland, VC Farmer, eds, Proc Internat Clay Conference, Oxford, 1978, Elsevier, Amsterdam, 1979.
68a. RM Barrer, N Mackenzie. J Phys Chem 1954; 58:560–568.
68b. RM Barrer, N Mackenzie, DM MacLeod. J Phys Chem 1954; 58:568–572.
69. E Ruiz-Hitzky. J Mater Chem 2001; 11:86–91.
70. G Howarth, K Kawazoe. J Chem Eng Jpn 1983; 16:470–475.
71. JL Ahlrichs, CJ Serna, JM Serratosa. Clays Clay Miner 1975; 23:119–124.
72a. S Inagaki, Y Fukushima, H Dot, O Kamigaito. Clay Miner 1990; 25:99–105.
72b. W Kuang, GA Facey, C Detellier, B Casal, JM Serratosa, E Rulz-Hitzky. Chem Mater (in press).
73. PS Ridler, BR Jennings. Clay Miner 1980; 15:121–133.
74. H van Olphen. Science 1966; 154:645–646.
75. B Hubbard, W Kuang, A Moser, GA Facey, C Detellier. Clays Clay Miner 2003; 51:318–326.
76. GW Nederbragt, JJ De Jong. Rec Trav Chim 1946:832–834.
77. H Deuel. Clay Miner Bull 1952; 1:205–212.
78. G Brown, R Green-Kelley, K Norrish. Clay Miner Bull 1952; 1:214–220.
79. CH Edelman, SCL Favejee. Z Kristallogr 1940; 102:417–431.
80. MC Hermosín, J Cornejo. Clays Clay Miner 1986; 34:591–596.
81. F Aragón de la Cruz, F Esteban, C Vitón. In: JM Serratosa, ed, Proc Internat Clay Conf, Madrid, División de Ciencias CSIC, Madrid, 1973.
82. CW Lentz. Inorg Chem 1964; 3:574–579.
83. E Ruiz-Hitzky, A Van Meerbeek. Coll Polym Sci 1978; 256:135–139.
84. JJ Fripiat, E Mendelovici. Bull Soc Chim Fr 1968:483–492.
85. L Zapata, J Castelein, JP Mercier, JJ Fripiat. Bull Soc Chim France 1972:54–63.
86. E Ruiz-Hitzky, JJ Fripiat. Bull Soc Chim France 1976:1341–1348.
87. A Van Meerbeek, E Ruiz-Hitzky. Coll Polym Sci 1979; 257:178–181.
88. J Barrios-Neira, L Rodrique, E Ruiz-Hitzky. J Microsc Spec Electron 1974; 20:295–298.
89a. HP Eugster. Science 1967; 157:1177–1180.
89b. G Lagaly, K Beneke, A Weiss. Naturforsch 1973; 28b:234–238.
90. JM Rojo, E Ruiz-Hitzky, J Sanz. Inorg Chem 1988; 27:2785–2790.

91. JM Rojo, E Ruiz-Hitzky, J Sanz, JM Serratosa. Rev Chim Miner 1983; 20:807–816.
92. E Ruiz-Hitzky, JM Rojo. Nature 1980; 287:28–30.
93. E Ruiz-Hitzky, JM Rojo, G Lagaly. Coll Polym Sci 1985; 263:1025–1030.
94. M Ogawa, S Okutomo, K Kuroda. J Am Chem Soc 1998; 120:7361–7362.
95. M Ogawa, M Miyoshi, K Kuroda. Chem Mater 1998; 10:3787–3789.
96. JM Serratosa. In: MM Mortland, VC Farmer, eds, Intern Clay Conf 1978. Proc Internat Clay Conference, Oxford, 1978, Elsevier, Amsterdam, 1979.
97. E Ruiz-Hitzky, JJ Fripiat. Clays Clay Miner 1976; 24:25–30.
98. AJ Aznar, E Ruiz-Hitzky. Mol Cryst Liq Cryst Inc Nonlin Opt 1988; 161:459–469.
99. E Gutierrez, AJ Aznar, E Ruiz-Hitzky. In: M Guisnet, J Barrault, C Bouchoule, D Duprez, G Pérot, R Maurel, C Montassier, eds. Heterogeneous Catalysis and Fine Chemicals II. Amsterdam: Elsevier, 1991:539–547.
100. L Zapata, A Van Meerbeek, JJ Fripiat, M della Faille, M Van Russelt, JP Mercier. J Polymer Sci Symp 1973; 42:257–272.
101. MN Fernández-Hernández, E Ruiz-Hitzky. Clay Miner 1979; 14:295–305.
102. E Ruiz-Hitzky, MN Fernández-Hernández, JM Serratosa. Spanish Patent No. 479518, 1979.
103. B Siffert, H Biava. Clays Clay Miner 1976; 24:303–311.
104. B Casal, E Ruiz-Hitzky. In: ITh Rosenqvist, ed, Proc III European Clay Conference, Nordic Society for Clay Research, Oslo, 1977.
105. B Casal, E Ruiz-Hitzky. An Quím 1984; 80:315–320.
106. E Ruiz-Hitzky, B Casal. J Catalysis 1985; 92:291–295.
107. B Casal, E Ruiz-Hitzky, JM Serratosa. Spanish Patent No. 480550, 1980.
108a. JJ Tunney, C Detellier. Chem Mater 1993; 5:747–748.
108b. JJ Tunney, C Detellier. J Mater Chem 1996; 6:1679–1685.
109. A Blumstein. Bull Soc Chim France 1961:899–905.
110. HZ Friedlander. ACS Div Polym Chem Reprints 1963; 4:300–306.
111. KA Carrado, L Xu. Chem Mater 1998; 10:1440–1445.
112. E Ruiz-Hitzky. Organic–Inorganic Materials: From Intercalations to Devices. Chapter 2. In: P Gómez-Romero, C Sánchez, eds. Funtional Hybrid Materials: Wiley-VCH (in press).
113. G Lagaly. Apl Clay Sci 1999; 15:1–9.
114. Y Fukushima, A Okada, M Kawasumi, T Kurauchi, O Kamigaito. Clay Miner 1988; 23:27–34.
115. E Ruiz-Hitzky, P Aranda. Electroactive Polymers Intercalated in Clays and Related Solids. Chapter 2. In: TJ Pinnavaia, GW Beall, eds. Polymer–Clay Nanocomposites. West Sussex. UK: Wiley, 2000:19–46.
116. RA Ruehrwein, DW Ward. Soil Sci 1952; 73:485–492.
117. C Breen. Appl Clay Sci 1999; 15:187–219.
118. DJ Greenland. J Colloid Sci 1963; 18:647–664.
119. G Lagaly. Smectitic clays as ionic macromolecules. In: AD Wilson, HJ Prosser, eds. Developments in Ionic Polymers. Vol. 2. London: Elsevier, 1986:77–140.
120. N Ogata, S Kawakage, T Ogihara. J Appl Polum Sci 1997; 66:573–581.
121a. RL Parfitt, DJ Greenland. Clay Miner 1970; 8:305–315.
121b. RL Parfitt, DJ Greenland. Clay Miner 1970; 8:317–323.

122. E Ruiz-Hitzky, P Aranda. Adv Mater 1990; 2:545–547.
123. R Levy, CW Francis. J Colloid Interface Sci 1975; 50:442–450.
124. P Aranda, E Ruiz-Hitzky. Chem Mater 1992; 4:1395–1403.
125. P Aranda, E Ruiz-Hitzky. Acta Polymer 1994; 45:59–67.
126. P Aranda, E Ruiz-Hitzky. Appl Clay Sci 1999; 15:119–135.
127. E Ruiz-Hitzky, B Casal. In: R Setton, ed. Chemical Reactions in Organic and Inorganic Constrained Systems. NATO-ASI Series C. Vol. 165. Dordrecht: Reidel, 1986:179–189.
128. B Casal, P Aranda, J Sanz, E Ruiz-Hitzky. Clay Miner 1994; 29:191–203.
129. JMG Cowie, SH Cree. Annu Rev Phys Chem 1989; 40:85–113.
130. J Wu, MM Lerner. Chem Mater 1993; 5:835–838.
131. MM Döeff, JS Reed. Solid State Ionics 1999; 113–115:109–115.
132. RC Reynolds. Amer Miner 1965; 50:990–1001.
133. RA Vaia, S Vasudevan, W Krawiec, LG Scanlon, EP Giannelis. Adv Mater 1995; 7:154–156.
134. RA Vaia, H Ishii, EP Giannelis. Chem Mater 1993; 5:1694–1696.
135. P Aranda, JC Galván, E Ruiz-Hitzky. In: RM Laine, C Sanchez, CJ Brinker, E Giannelis, eds. Organic/Inorganic Hybrid Materials. MRS Symposium Proceedings. Vol. 519. Warrendale. PA: Materials Research Society, 1998:375–380.
136. P Aranda, Y Mosqueda, E Pérez-Cappe, E Ruiz-Hitzky. J Polym Sci B Polym Phys 2003; 41:3249–3263.
137. E Gutierrez, A Loupy, G Bram, E Ruiz-Hitzky. Tetrahedron Lett 1989; 30:945–948.
138. G Bram, A Loupy, M Majdoub, E Gutierrez, E Ruiz-Hitzky. Tetrahedron 1990; 46:5167–5176.
139. DR Baghurst, DMP Mingos. Chem Soc Rev 1991; 20:1–47.
140. MM Kingston, SJ Haswell. Microwave-Enhanced Chemistry. Washington, DC: American Chemical Society, 1997.
141. MG Kanatzidis, LM Tonge, TJ Marks, HO Marcy, CR Kannewurf. J Am Chem Soc 1987; 109:3797–3799.
142. D Vande Poel. Adsorption de Composés Aromatiques sur la Montmorillonite Saturée en Cu(II). PhD dissertation, Catholic University of Louvain. 1975.
143. E Ruiz-Hitzky. Adv Mater 1993; 5:334–340.
144. E Ruiz-Hitzky, P Aranda. Ans Quim Int Ed 1997; 93:197–212.
145. S Yariv, L Heller, Z Sofer, W Bodenheimer. Israel J Chem 1968; 6:741–756.
146. L Heller, S Yariv. In: L Heller, ed, Proc Internat Clay Conf, Tokyo, 1969, Israel University Press, Jerusalem, 1969.
147. A Weiss. Clays Clay Miner 1963; 10:191–224.
148. T Furukawa, GW Brindley. Clays Clay Miner 1973; 21:279–288.
149. P Cloos, A Moreale, C Braers, C Badot. Clay Miner 1979; 14:307–321.
150. A Moreale, P Cloos, C Badot. Clay Miner 1985; 20:29–37.
151. V Mehrota, EP Giannelis. Solid State Commun 1991; 77:155–158.
152. T-C Chang, S-Y Ho, K-J Chao. J Chin Chem Soc 1992; 39:209–212.
153. Y Fukushima, S Inagaki. J Incl Phenom 1987; 5:473–482.
154. A Usuki, Y Kojima, M kawasumi, A Okada, Y Fukushima, T Kurauchi, O kamigaito. J Mater Res 1993; 8:1179–1184.

155. V Mehrota, EP Giannelis. Solid State Ionics 1992; 51:115–122.
156. J Sanz, JM Serratosa. Nuclear magnetic resonance spectroscopy of organoclay complexes. Chapter 6. In: S Yariv, H Cross, eds. Organoclay Complexes and Interactions. New York: Marcel Dekker, 2002:223–272.
157. R Blumstein, A Blumstein, KK Parikh. Appl Polym Sci 1974; 25:81–88.
158. F Bergaya, F Kooli. Clay Miner 1991; 26:33–41.
159. L Ducleaux, E Frackowiak, T Cibinski, R Benoit, F Beguin. Mol Cryst Liq Cryst 2000; 340:449–454.
160. JS Beck, JC Vartuli, WJ Roth, ME Leonowicz, CT Kresge, KD Schmitt, C T-W Chu, DH Olson, EW Sheppard, SB McCullen, JB Higgins, JL Schlenker. J Am Chem Soc 1992; 114:10834–10843.
161. F Leroux, JP Besse. Chem Mater 2001; 13:3507–3515.
162. KA Carrado, L Xu, S Seifert, R Csencsits. Polymer–clay nanocomposites from synthetic polymer-silicate gels. Chapter 3. In: TJ Pinnavaia, GW Beall, eds. Polymer–Clay Nanocomposites. West Sussex. UK: Wiley, 2000:47–93.
163. KA Carrado. App Clay Sci 2000; 17:1–23.
164. KA Carrado, L Xu. Microp Mesop Mater 1999; 27:87–94.
165. MA Aviles, A Justo, PJ Sánchez-Soto, JL Pérez-Rodríguez. J Mater Chem 1993; 3:223–224.
166. JL Pérez-Rodríguez, C Maqueda. In: S Yariv, H Cross, eds. Organoclay Complexes and Interactions. New York: Marcel Dekker, 2002:157–159.
167. RA Vaia, KD Jandt, EJ Kramer, EP Giannelis. Chem Mater 1996; 8:2628–2635.
168. AS Moet, A Akelah. Mater Lett 1993; 18:97–102.
169. E Herzog, W Caseri, UW Suter. Colloid Polymer Sci 1994; 272:986–990.
170. W Chen, Q Xu, RZ Yuan. J Mater Sci Lett 1999; 18:711–713.
171. B Liao, M Song, H Liang, Y Pang. Polymer 2001; 42:10007–10011.
172. B Hoffmann, C Dietrich, R Thomann, C Friedrich, R Mülhaupt. Macromol Rapid Commun 2000; 21:57–61.
173. S Inagaki, Y Fukushima, M Miyata. Res Chem Intermed 1995; 21:167–180.
174. J Sanz, JM Serratosa. Nuclear magnetic resonance spectroscopy of organoclay complexes. Chapter 6. In: S Yariv, H Cross, eds. Organoclay Complexes and Interactions. New York: Marcel Dekker, 2002:259.
175. JJ Tuney, C Detellier. Chem Mater 1996; 8:927–935.
176. F Leroux, P Aranda, JP Besse, E Ruiz-Hitzky. Eur J Inorg Chem 2003:1242–1251.
177. HH Murray. In: H Kodama, AR Mermut, J Kenneth Torrance, eds, Proc 11th Internat Clay Conf, Ottawa, 1997, ICC97 Organizing Committee, Ottawa, 1997.
178. Web site: www.rheox.com.
179. A Alvarez, J Santaren, R Perez-Castell, B Casal, E Ruiz-Hitzky, P Levitz, JJ Fripiat. In: LG Schultz, H van Olphen, FA Mumpton, eds, Proc Internat Clay Conference, Denver, 1985, Clay Minerals Society, Bloomington, IN, 1987.
180. S Yariv. In: S Yariv, H Cross, eds. Organoclay Complexes and Interactions. New York: Marcel Dekker, 2002:498.
181. E Ruiz-Hitzky. Chem Record 2003; 3:88–100.
182. RM Barrer. Clays Clay Miner 1989; 37:385–395.
183. WF Jaynes, GF Vance. Clays Clay Miner 1999; 47:358–365.

184. S Yariv. In: S Yariv, H Cross, eds. Organoclay Complexes and Interactions, Marcel Dekker. 2002:88–89.
185. Y Yan, T Bein. Chem Mater 1993; 5:905–907.
186. J Bors. Radiochim Acta 1990; 51:139–143.
187. J Bors, A Gorny, S Dultz. Radiochim Acta 1994; 66/67:309–313.
188. J Bors, A Gorny, S Dultz. Clay Miner 1997; 32:21–28.
189. J Bors, S Dultz, A Gorny. In: H Kodama, AR Mermut, J Kenneth Torrance, eds, Proc 11th Internat Clay Conf, Ottawa, 1997, ICC97 Organizing Committee, Ottawa, 1997.
190. A Wefer-Roehl, KA Czurda. In: H Kodama, AR Mermut, J Kenneth Torrance, eds, Proc 11th Internat Clay Conf, Ottawa, 1997, ICC97 Organizing Committee. 1997: 123–128.
191. T González-Carreno, JA MartinRubi. J Chromatogr 1977; 133:184–189.
192. A Yamagishi. In: LG Schultz, H van Olphen, FA Mumpton, eds, Proc Internat Clay Conference, Denver, 1985, Clay Minerals Society, Bloomington, In, 1987.
193. HB Lao, C Detellier. Clays Clay Miner 1994; 42:477–481.
194. WF Jaynes, SA Boyd. Soil Sci Soc Amer J 1991; 55:43–48.
195. S Wu, G Sheng, SA Boyd. Adv Agron 1997; 59:25–62.
196. G Sheng, X Wang, S Wu, SA Boyd. J Environ Qual 1998; 27:806–814.
197. B Casal, J Merino, JM Serratosa, E Ruiz-Hitzky. Appl Clay Sci 2001; 18:245–254.
198. S Yariv. Introduction to organoclay complexes and interactions. Chap 2. In: S Yariv, H Cross, eds. Organoclay Complexes and Interactions. New York: Marcel Dekker, 2002:39–111.
199. L Mercier, C Detellier. Environ Sci Technol 1995; 29:1318–1323.
200. R Celis, MC Herrmosin, J Cornejo. In: J Zapatero, AJ Ramírez, MV Moya, eds. Integración Ciencia-Tecnologia de las Arcillas en el Contexto Tecnológico-Social del Nuevo Milenio. Malaga: Sociedad Española de Arcillas & Diputación Provincial de Málaga, 2000:103–111.
201. JH Fendler. Membrane Mimetic Chemistry, Wiley. 1982:184.
202. P Aranda, JC Galván, B Casal, E Ruiz-Hitzky. Coll Polym Sci 1994; 272:712–720.
203. P Falaras, D Petridis. J Electroanal Chem 1992; 337:229–239.
204. L Coche-Guérente, V Desprez, P Labbé. J Electroanal Chem 1998; 458:73–86.
205. A Senillou, N Jaffrezic, C Martelet, S Cosnier. Anal Chim Acta 1999; 401:117–124.
206. D Ozkan, K Kerman, B Meric, P Kara, H Demirkan, M Polverejan, TJ Pinnavaia, M Ozsoz. Chem Mater 2002; 14:1755–1761.
207. LF Lindoy. The Chemistry of Macrocyclic Ligand Complexes. Cambridge: Cambridge University Press, 1989:109.
208. SL Regen. J Am Chem Soc 1976; 98:6270–6274.
209. E Blasius, KP Janzen. Israel J Chem 1985; 26:25–34.
210. CL Lin, TJ Pinnavaia. Chem Mater 1991; 3:213–215.
211. AL Garcia-Ponce, V Prevot, B Casal, E Ruiz-Hitzky. New J Chem 2000; 24: 119–121.
212. V Prevot, B Casal, E Ruiz-Hitzky. J Mater Chem 2001; 11:554–560.
213. MM Mortland, V Berkheiser. Clays Clay Miner 1976; 24:60–63.
214. A Sánchez, A Hidalgo, JM Serratosa. In: JM Serratosa, ed. Proc Internat Clay Conf, Madrid, División de Ciencias CSIC, Madrid, 1973.

215. AJ Aznar, J Sanz, E Ruiz-Hitzky. Colloid Polym Sci 1992; 270:165–176.
216. J Barrios-Neira, L Rodrique, E Ruiz-Hitzky. J Microsc Spec Electron 1974; 20: 295–298.
217. J Barrios, G Poncelet, JJ Fripiat. J Catal 1981; 63:362–370.
218. B Casal, F Bergaya, D Challal, JJ Fripiat, E Ruiz-Hitzky, H Van Damme. J Mol Catalysis 1985; 33:83–86.
219. TJ Pinnavaia. Science 1983; 220:365–371.
220. RA Schoonheydt. In: JJ Fripiat, ed. Advanced Techniques for Clay Mineral Analysis. Amsterdam: Elsevier, 1981:163–189.
221. K Bergman, CT O'Konski. J Phys Chem 1963; 67:2169–2177.
222. J Cenens, R Schoonheydt. Clays Clay Miner 1988; 36:214–224.
223. T Fujita, N Iyi, T Kosugi, A Ando, T Deguchi, T Sota. Clays Clay Miner 1997; 45:77–84.
224. AJ Aznar, B Casal, E Ruiz-Hitzky, I López-Arbeloa, F López-Arbeloa, J Santarén, A Álvarez. Clay Miner 1992; 27:101–108.
225. G Rytwo, S Nir, L Margulies, B Casal, J Merino, E Ruiz-Hitzky, JM Serratosa. Clays Clay Miner 1998; 46:340–348.
226. L Margulies, H Rozen, E Cohen. Nature 1985; 315:658–659.
227. I Braun, G Ihlein, F Laeri, JU Nöckel, G Schulz-Ekloff, F Schüth, U Vietze, O Weiss, D Wöhrle. Appl Phys B 2000; 70:335–344.
228. E Muthusamy, D Walsh, S Mann. Adv Mater 2002; 14:969–972.
229. T Fujita, N Iyi, Z Klapyta. Mat Res Bull 1998; 33:1693–1701.
230. H Wellmann, J Rathousky, M Wark, A Zukal, G Schulz-Ekloff. Microporous Mesoporous Mater 2001; 44:419–425.
231. E Ruiz-Hitzky, P Aranda, B Casal, JC Galván. Adv Mater 1995; 7:180–184.
232. P Aranda, JC Galván, B Casal, E Ruiz-Hitzky. Electrochim Acta 1992; 37: 1573–1577.
233. E Ruiz-Hitzky, P Aranda, E Pérez-Cappe, A Villanueva, Y Mosqueda Laffita. Rev Cubana Quím 2000; 12:58–63.
234a. G Sandi, H Joachin, R Kizilel, S Seifert, KA Carrado. Chemistry Materials 2003; 15:838–843.
234b. G Sandi, KA Carrado, H Joachin, W Lu, J Prakash. J. Power Sources 2003, in press.
234c. LJ Smith, J-M Zanotti, G Sandí, KA Carrado, P Porion, A Delville, DL Price, ML Saboungi. In: P Knauth, J-M Tarascon, E Traversa, HL Tuller, eds. Solid-State Ionics-2002 (MRS Symp. roc., Vol. 756), 2003, in press.
235. E Ruiz-Hitzky, A Van Meerbeeck. Polymer–clay nanocomposites. In: F Bergaya, BKG Theng, G Lagaly, eds. Handbook of Clay Science: Elsevier (in press).
236. M Alexandre, Ph Dubois. Mater Sci Eng 2000; 28:1–63.
237. J Quarmley, A Rossi. Industrial Minerals (January). 2001:47–53.
238. M Biswas, SS Ray. Polymer 1998; 39:6423–6428.
239. Y Kojima, A Usuki, M Kawasumi, A Okada, Y Fukushima, T Kurauchi, O Kamigaito. J Mater Res 1993; 8:1185–1189.
240. LM Liu, ZN Qi, XG Zhu. J Appl Polym Sci 1999; 71:1133–1138.
241. T Lan, PD Kaviratna, TJ Pinnavaia. Chem Mater 1995; 7:2144–2150.

242. Z Wang, TJ Pinnavaia. Chem Mater 1998; 10:3769–3771.
243. JM Serratosa, JA Rausell-Colom, J Sanz. J Mol Catal 1984; 27:225–234.
244. M Rautureau, C Tchoubar. Clays Clay Min 1976; 24:43–49.
245. L Montaland. Process for Converting Pinene into Camphene. U.S. Pat. No. 999,667.
246. DMC MacEwan. Nature 1944; 154:577–578.
247. WH Allaway. Soil Sci Soc Amer Proc 1949; 13:183–188.
248. A Weiss. Angew Chem 1961; 73:736.
249. H Hasegawa. J Phys Chem 1961; 65:292–296.
250. CJ Serna, T Fernández-Álvarez. An Quím 1974; 70:760–764.
251. HA Resing, D Slotfeldt-Ellingsen, AN Garroway, DC Weber, TJ Pinnavaia, K Unger. In: JP Fraissard, MA Resing, eds. Proc NATO Adv Study Inst, Menton, 1979. Dordrecht. Netherlands: Reidel, 1980:239–258.
252. JK de Waele, FC Adams, B Casal, E Ruiz-Hitzky. Mikrochim Acta III 1984: 117–128.
253. WG Garret, GF Walker. Nature 1961; 191:1389–1390.
254. JA Rausell-Colom, P Salvador. Clay Miner 1971; 9:139–149.
255. GE Lailach, TD Thompson, GW Bradley. Clays Clay Miner 1961; 16:285–293.
256. JB Weber, PW Perry, RP Upchurch. Soil Sci Amer Proc 1965; 29:678–687.
257. SB Weed, JB Weber. J Amer Miner 1968; 53:478–490.
258. RM Barrer. Zeolites and Clay Minerals as Sorbents and Molecular Sieves. London: Academic Press, 1978:453–475.
259. R Glaeser. Clay Miner Bull 1948; 1:88–90.
260. S Yariv, JD Russell, VC Farmer. Isr J Chem 1966; 4:201–213.
261. S Olejnik, LAG Aylmore, AM Posner, JP Quirk. J Phys Chem 1968; 72:241–249.
262. M Cruz, A Laycock, JL White. In: L Heller, ed, Proc Internat Clay Conf, Tokyo, 1969, Israel University Press, Jerusalem, 1969.
263. MM Mortland. Clay Miner 1966; 6:143–156.

4

Sorption of Nitroaromatic Compounds on Clay Surfaces

Cliff T. Johnston
Purdue University
West Lafayette, Indiana, U.S.A.

Stephen A. Boyd and Brian J. Teppen
Michigan State University
East Lansing, Michigan, U.S.A.

Guangyao Sheng
University of Arkansas
Fayetteville, Arkansas, U.S.A.

I. INTRODUCTION

Nitroaromatic compounds (NACs) are widely used as pesticides, explosives, solvents, and intermediates in chemical syntheses. NACs and their degradation products pose a potential threat to ecological and human health because they are toxic environmental contaminants commonly found in soil and subsurface environments at elevated concentrations. At a site in Texas, for example, soil concentrations of NAC contaminants in excess of 75,000 mg/kg have been reported and have been detected at depths of greater than 30 meters at concentrations above human health exposure limits (1). In general, NACs degrade relatively slowly in the environment (2,3). Recent evidence has shown that certain NACs have a high affinity for certain types of clay minerals, and this may contribute to their recalcitrant behavior (4–11). This chapter will examine the chemical mechanisms that govern NAC–clay interactions from two perspectives. First, we will examine

155

the properties of the clay surface that contribute to NAC sorption. Second, the chemical and physical properties of the NAC solute that influence sorption will be examined.

NACs that show a high affinity for clay minerals generally have two molecular attributes: (a) the presence of two or more $-NO_2$ groups and (b) the lack of a bulky substituent (5,9). Examples include 1,3,5-trinitrobenzene (1,3,5-TNB) and 6-methyl-2,4-dinitrophenol (DNOC) with adsorption coefficient K_d values of 159,000 and 213,000 L kg^{-1}, respectively (10). These values are several orders of magnitude higher than those of similar organic compounds. For example, 2-nitrotoluene and nitrobenzene have K_d values of just 4.6 and 7.2 L kg^{-1}, respectively (5). The anomalously high K_d values for compounds like 1,3,5-TNB (12) indicate that site-specific interactions occur between these types of NAC molecules and the clay surface. The apparent high degree of specificity of NACs for sorption sites on clay minerals compared to other nonionic organic solutes is somewhat surprising because clay minerals have been generally viewed as having a low sorptive potential for nonionic organic compounds like NACs in the presence of water (13). Expandable clay minerals are strongly hydrophilic, with a presumed low affinity for neutral organic compounds. More recently, however, the hydrophobic–hydrophilic character of smectite surfaces has been reexamined and *portions* of the smectite surface have been shown to have partial hydrophobic character (14). The higher-than-expected sorption of NACs and related compounds on clay minerals is shedding new light on the nature of the clay surface and the role these surfaces play in attenuation of different types of nonionic organic solutes, including NACs.

The focus of this chapter is on abiotic interactions of NACs with clay minerals. These interactions have direct relevance to the broader topics of the fate and transport of NACs in soil and subsurface environments (15–20). The processes governing the release and transformation of NACs into soils, sediments, and the aquatic environment are not well understood. As stated earlier, NACs are recalcitrant to microbial degradation, and this may be linked, in part, to their high affinity for mineral substrates. Sorption reactions of NACs on clay surfaces have also been implicated in biogeochemical processes controlling the microbially mediated transformations of NACs (3,21–23), and there is supporting evidence that NACs that are strongly sorbed are resistant to microbial degradation (19,24). In addition, understanding the surface chemistry of smectite-NAC interactions may help in the design of more selective and efficient catalysts for the removal of NACs from contaminated waste streams (25). On a more fundamental level, there has been considerable interest recently in the design and fabrication of hybrid clay films consisting of layers of smectite and organic molecules (26–32) for applications involving optical devices, nonlinear optical materials, catalysts, and sensors (33). For these applications, NACs are being used to modify the photophysical and photochemical properties of clay minerals (34).

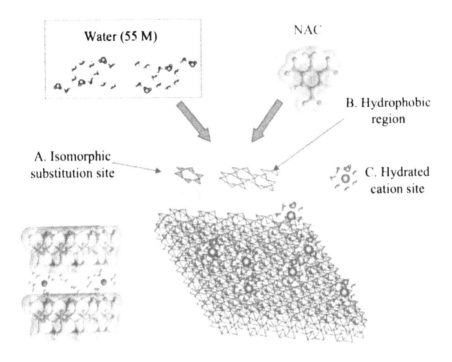

FIGURE 1   Potential surface interactions of an NAC molecule sorbed to a clay particle.

A conceptual illustration of an NAC molecule interacting with a clay particle is shown in Figure 1. At the top of the diagram two solution components
are shown, the dissolved NAC and water. The dominant, and often neglected,
component in sorption studies is water, which is present at a concentration of
about 55 M (moles dm^{-3}). For comparison, the aqueous solubility of 2,4,6-
trinitrotoluene (TNT) is $\sim 6.6 \times 10^{-4}$ M (moles dm^{-3}) at 25°C (35). The interaction of water with clay minerals plays a critical role in controlling the interaction
of organic solutes with clay minerals. Mineral surfaces that are strongly hydrophilic will have a low affinity for nonionic organic solutes such as NACs due to
preferential adsorption of water. However, sorption of such sparingly water-soluble solutes is favorable when the surface has some hydrophobic character. Many
organic solutes have limited solubility in aqueous solution, and their sorption is
entropy driven. Depending on the type of NAC and the properties of the clay
surface, NACs may be strongly or weakly bound to the surface, excluded from
the solid–water interface, and/or undergo degradation. These types of interactions
strongly influence the ultimate fate and transport of NACs in soil and subsurface

environments. Microbial degradation of NACs, for example, is presumed to be controlled by the amount of NAC in the soil solution (18). Thus, strong retention of NACs by clay minerals will directly impact their bioavailability to microorganisms as well as to plants and other species.

II. ACTIVE SITES ON CLAY MINERALS

Four different types of active sites can be used to represent the reactive features on clay minerals implicated in NAC sorption mechanisms. Analogous to describing the structure and function of biological macromolecules, such as proteins and enzymes, active sites can also be used to describe the surface chemistry of clay minerals (36–38). Although a broader array of sites are present on clay minerals (37), this subset of active sites is thought to represent the dominant source of reactive sites involved in NAC attenuation. The term *active site* implies a process wherein a surface chemical reaction of interest is promoted by a molecular-scale feature on the surface of a clay mineral. Active sites are described with respect to their location (edge vs. basal surface), the geometric arrangement of surface atoms, chemical composition, and accessibility. An in-depth understanding of the surface reactivity of clay minerals toward NACs requires knowledge about the crystal structure of the clay mineral and the defects that are often associated with these sites. Several excellent reviews on the crystallography and mineralogy of clay minerals have been published (39–42). Sorption of NACs on clay minerals generally involves more than one type of surface site, and the division and assignment of sites is somewhat artificial. It is hoped that the conceptual framework of active sites on clay minerals will promote increased understanding about the interaction of NACs with clay minerals.

Active sites involved in NAC sorption on clay minerals can be divided into two broad categories of surface sites: polar and nonpolar. *Polar sites* originate from isomorphic substitution sites and from broken edges. *Nonpolar sites*, on the other hand, occur on the surfaces of neutral mineral surfaces. Current evidence would suggest that NAC accessibility to both types of surface sites is a prerequisite to sorption.

A. Neutral Siloxane Surface

The least reactive type of surface site found on clay minerals is the neutral siloxane surface. Features associated with this type of site include no charge, no permanent dipole moment, and weak interactions with the hydrogen-bonded network of water molecules. Hydrophobic surfaces occur on 2:1 phyllosilicates, where there is no isomorphic substitution (e.g., talc and pyrophyllite), and also on the siloxane side of the kaolinite 1:1 TO unit. There is currently some question about the structural microscopic hydrophilicity of talclike surfaces (43); however, from a macroscopic point of view, it is clearly hydrophobic (44). The neutral siloxane surface is thought to function as a weak Lewis base (45); therefore, the electron-donating ability of this site is limited. Neutral siloxane surfaces have a *low* affinity for water (43,46). Solute interactions with this type of surface are restricted to weak, nonspecific surface interactions resulting from dipole–dipole and dipole–polarization interactions, also called *London forces*.

The neutral siloxane surface has been invoked in NAC sorption mechanisms (8–11) and in the sorption of related compounds, such as atrazine (47,48). An illustration of a hydrated smectite surface is shown in Figure 1. The surface of the smectite particle is represented by isomorphic substitution sites that, depending on the surface charge density (SCD) of the clay, are separated by distances of 1–2 nm. K-SWy-2 smectite with a specific surface area of 800 m^2 g^{-1} and a cation exchange capacity of 80 $cmol_c$ kg^{-1}, for example, has a density of charge sites of 0.6 sites/nm^2, or 0.096 coulombs/m^2 (assuming that all of the charge originates from isomorphic substitution). The corresponding area per site is 1.6 nm^2, and the distance between charge sites for the clay is ~1.3 nm (Fig. 2) (36). However,

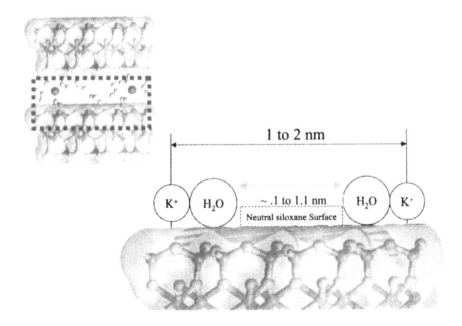

FIGURE 2 Expanded side view of a smectite particle showing the approximate distance between isomorphic substitution sites and the approximate sizes of K^+ ions (van der Waals radius is 0.13) and water molecules (van der Waals radius of 0.15 nm) surrounding the K^+ ions.

guest species in smectite interlayers often contact the two opposing surfaces (11). In this case, the approximate charge density is 1.2 sites/nm^2, and the distance between charge sites is 0.91 nm. The region between the isomorphic substitution sites, labeled site B in Figure 1, is the neutral portion of the siloxane surface (also shown in Fig. 2). Sorption of NACs on clay minerals is assumed stabilized by these types of neutral sites, because this surface can favorably interact with the neutral, planar NAC molecule. Accessibility of organic solutes to these neutral hydrophobic sites is inversely related to the surface charge density of the clay (48,49). As shown in Figure 2, if the isomorphic substitution sites are separated by distances of 1–2 nm (depending on the surface charge density), the corresponding size domain of the neutral siloxane surface is 0.1–1.1 nm (space between hydrated cations), assuming one hydration sphere of water molecules, as shown. If the organic solute can "fit" into this region, then the sorbed NAC molecule is stabilized. Although not highly energetic, this type of site is thought to play an important role in governing the sorption of neutral organic compounds on clay minerals, including NACs.

Evidence supporting the hypothesis that the smectite surface has a partially hydrophobic nature is provided from several previous studies. The first comes from a sorption study of nonpolar organic solutes (e.g., benzene, toluene, *p*-xylene) from aqueous suspension on modified smectites (14,49). On naturally occurring smectites with Na$^+$, K$^+$, Ca^{2+}, or Mg^{2+} ions, little if any sorption of these types of nonpolar solutes occurs. In contrast, appreciable sorption occurs when the inorganic exchangeable cations are replaced by the organic trimethyl-phenylammonium (TMPA) cation (49). The enthalpy of hydration of this organic cation is considerably less than that of the alkali and alkaline earth cations. Consequently, fewer water molecules surround the trimethylphenylammonium cation in the interlayer, leaving a portion of the neutral siloxane surface exposed and accessible to organic solutes. Sorption of nonpolar organic solutes on this type of modified clay indicates that the neutral siloxane surface has some hydrophobic character. In addition, nonpolar organic solute sorption greatly increases with a reduction of the surface charge density (49). These observations indicate that the strongly hydrophilic nature of smectites results from the exchangeable cations. When the hydrophilic sites, or pillars, are no longer present, the surface takes on a moderately hydrophobic character. Nonpolar organic solutes are also strongly sorbed on clay minerals from the vapor phase at low water contents (50–53). At low water content, organic solutes in the vapor phase can gain access to the clay surface when they do not have to compete with water molecules.

Support for the hydrophobic character of the siloxane surface is further provided by a sorption study of atrazine from aqueous suspension on a series of Ca^{2+}-exchanged smectites (48). A total of 13 Ca^{2+}-exchanged smectites were used to represent a wide range of cation exchange capacities and surface charge densities. Atrazine is a weak organic base, and its molecular structure has both polar and nonpolar regions (the inset of an electrostatic potential map of atrazine is shown in Fig. 3). As illustrated in Figure 3, sorption ranged from 0% to $\sim 100\%$ of the added atrazine on the different Ca-exchanged smectites. This variation was surprising, since all of the sorbents were Ca-exchanged smectites. The amount of atrazine sorbed by the clays from aqueous suspension was determined mainly by the surface charge density of the smectite, with Freundlich sorption coefficients ranging from <0.01 L kg^{-1} for the high surface charge density Otay montmorillonite sample to 1330 L kg^{-1} for the Panther Creek Beidellite sample. Surface charge density (and/or the cation exchange capacity) of the clay was the most important determinant, with increased atrazine sorption occurring on the smectites with the lowest surface charge densities (47,48). A similar study of atrazine sorption on K-exchanged smectites was reported recently (8). Greater atrazine sorption occurred on the K-exchanged smectites compared to the Ca smectites studied by Laird (48). This was attributed to the lower hydration requirements of the K$^+$ ion compared to Ca^{2+}. Also, NAC sorption was shown (11) to increase as the surface charge density of a smectite is reduced (see later).

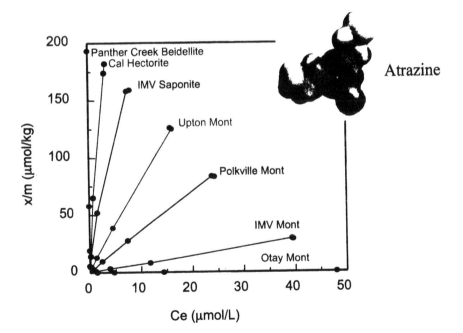

FIGURE 3   Atrazine sorption on seven different types of Ca^{2+}-exchanged smectites. The equilibrium solution concentration of atrazine (Ce) is plotted on the x-axis, expressed as μmoles L^{-1}. (Adapted from Ref. 48.)

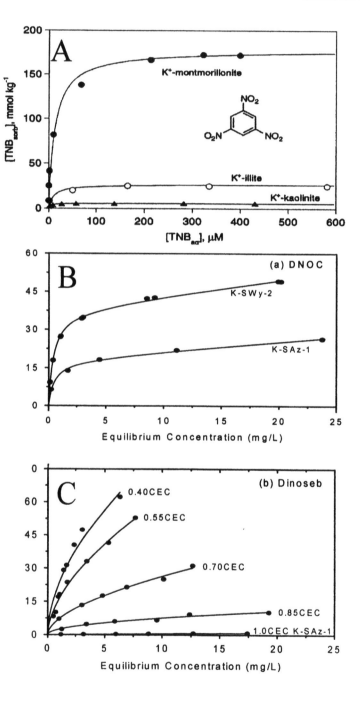

B. Exchangeable Cations

NAC sorption has been reported on a variety of clay minerals, including smectite, kaolinite, and illite (5). As shown in Figure 4A, significantly more sorption occurs on smecite than on kaolinite or illite (4,5). This is attributed to the larger specific surface area of smectite, to isomorphic substitution sites, and to site-specific interactions (37). Most of the cation exchange capacity of smectites results from isomorphic cationic substitution (40). These sites are characterized by a permanent negative charge and are referred to as *constant-charge sites*. Isomorphic substitution can occur in either the octahedral or tetrahedral sheets of 2:1 phyllosilicates, and many clay minerals are characterized by varying degrees of substitution in both sheets (40). In the case of substitution of Mg^{2+} for Al^{3+} in the octahedral sheet, the charge is delocalized over a relatively large region thought to be about nine oxygen atoms on the siloxane surface, as shown by the larger dashed circle in Figure 5 (36,54,55). Substitution of Al^{3+} for Si^{4+} in the tetrahedral sheet results in a more localized charge distribution, represented by the smaller circle in Figure 5. The topography of the siloxane surface, showing the hexagonal network of siloxane ditrigonal cavities, is also shown in Figure 5. This representation of the clay surface is useful because it probably most accurately depicts the surface topography that a sorbed NAC molecule "sees."

In the case of expandable clay minerals such as smectite and vermiculite, clusters of water molecules surround the exchangeable cations. Exchangeable cations play an important role in determining the sorption potential for NACs (10,11). Hydration of these inorganic cations creates a hydrophilic environment at the clay surfaces and interlayers. As the hydration enthalpy of the cation increases (Table 1), the strength of the interaction between the cation and the water molecules surrounding the cation is increased. Cation hydration causes separation of the clay layers, a phenomenon referred to as *swelling*. If the cation exchange capacity of the clay is above a threshold value of ~150 cmol$_c$ kg^{-1}, however, attractive forces between the cations and the clay surface become too large, which limits the degree of swelling and, therefore, NAC accessibility to the interlamellar region. In addition, for these high-charge clays, the distance between the hydrated exchangeable cations (i.e., the sizes of the hydrophobic sites) is minimized.

◄───

FIGURE 4 Representative nitroaromatic sorption isotherms. (A) Sorption isotherm of 1,3,5-trinitrobenzene on K-exchanged smectite, illite, and kaolinite. (B) Sorption of DNOC on K-exchanged SWy-2 smectite (low surface charge density) and SAz-1 smectite (high surface charge density). (C) Sorption isotherms of DINOSEB on K-SAz-1 and a series of charge-reduced SAz-1 smectites. The degree of charge reduction is shown on each curve, ranging from no charge reduction (1.0) to 60% charge reduction (0.4).

Siloxane ditrigonal cavity

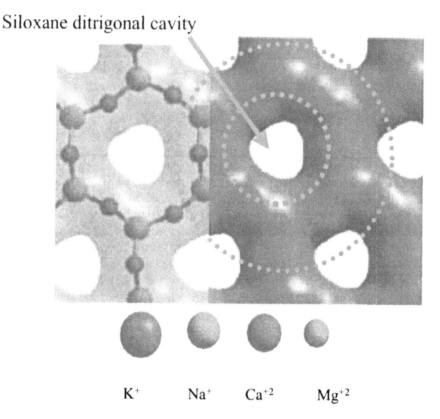

K$^+$ Na$^+$ Ca^{+2} Mg^{+2}

FIGURE 5   A 1.6-nm by 1.6-nm portion of the siloxane surface is shown in two representations. Ball-and-stick representation is shown on the left, total electron density is shown on the right. For comparison, the sizes of the Na$^+$, K$^+$, Ca^{2+}, and Mg^{2+} are shown. (Ionic radii obtained from Ref. 86.)

C.   Water Molecules Surrounding Exchangeable Cations

Depending on the hydration energy of the exchangeable cation (Table 1), a varying number of water molecules will be strongly associated with these interlayer cations. Of importance to NAC sorption, with cations of lower hydration energies (Cs$^+$ and K$^+$, for example), less energy is required to displace water molecules from the cation. This is illustrated by the sorption isotherms of water on smectites exchanged with four different cations, as shown in Figure 6. Note that significantly less water uptake occurs on the Cs-smectite compared to Na- or Ca-smectites. Water molecules surrounding exchangeable cations have chemical and phys-

TABLE 1 Selected Properties of Metal Ions

	Effective ionic radius† (re_{ff}) nm	Ionic potential‡ Z^2/re_{ff}	Enthalpy of hydration§ kJ mol^{-1}	Hydrolysis¶ constant pK$_h$
Ca^{2+}	0.126	2.15	−1669	12.7
Mg^{2+}	0.086	2.8	−1998	11.4
K$^+$	0.165	0.45	−360	
Na$^+$	0.132	0.5	−444	14.5
Cs$^+$	0.188	0.4	−315	
Cu^{2+}	0.087	2.6	−2174	7.5
Fe^{3+}	0.069	6.3	−4491	2.2
Fe^{2+}	0.075	2.5	−2009	
Co^{2+}	0.079	2.6	−2106	9.6
Zn^{2+}	0.088	2.6	−2106	9.6
Cd^{2+}	0.124	2.3	−1882	11.7
Pb^{2+}	0.143	2	−1556	7.8
Al^{3+}	0.067	6.9	−4774	5.1
Li$^+$	0.090	0.7	−559	13.8

† Ref. 85.
‡ Ref. 86.
§ Ref. 87.
¶ Ref. 88.

ical properties that are distinct from those of bulk water because of their close proximity to the ions. Their mobility is more restricted due to polarization effects by the cation (56–59) and, depending on the nature of the interlayer cation, can be more acidic than that of bulk water (60,61). The acidic nature of interlayer water may be especially important in the sorption of ionizable NACs, such as 2,4-dinitro-6-methyphenol (DNOC), which is sorbed to an appreciable extent at pH values above its pK$_a$ value. In this case, the surface acidity of the smectite (61) shifts the equilibrium of DNOC in aqueous solution toward that of the neutral species, which is more strongly sorbed than the ionized form (11). Thus, sorption of ionizable NACs, such as DNOC and DINOSEB, is influenced by pH (11). DNOC is a weak organic acid, with a pK$_a$ value of 4.35, and DNOC sorption from aqueous solution on smectite decreases with increasing pH. The observation that with increasing pH adsorption becomes lower indicates that DNOC is primarily adsorbed as the neutral species. In its ionized form the negatively charged clay surface minimizes DNOC sorption, but not completely. The interlamellar region of clay surfaces is known to be more acidic than bulk solution, and this

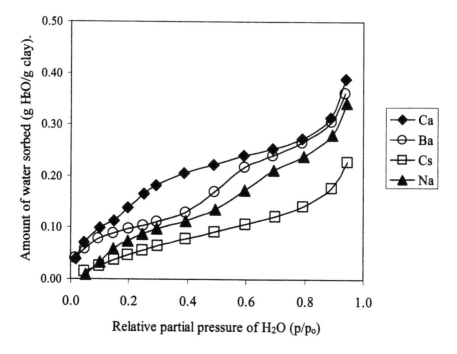

FIGURE 6   Water vapor sorption isotherms on smectites exchanged with Ca^{2+}, Ba^{2+}, Cs^+, and Na^+ ions. (Adapted from Ref. 89.)

may account for the appreciable sorption of DNOC at pH $>$ pK_a. The acidic nature of smectite surfaces is strongly dependent upon pH and water content.

For neutral NACs, exchangeable cations influence NAC sorption in at least two ways. Probably the most direct role is in determining the size of the sorptive domains. If the cations are too strongly hydrated (e.g., Mg^{2+}), then little sorption can occur because the surface is largely obscured by waters of hydration. Cation hydration also influences interlayer distances. Optimal interlayer spacings may exist wherein sufficient separation allows access while simultaneously minimizing contact of the organic solute with water (9). In addition, the hydrated cations have a partial positive charge, and these charged, hydrated pillars can interact favorably with negatively charged functional groups, such as the —NO_2 group of sorbed NACs. Molecular properties of NACs that contribute to their sorption by clay minerals will be covered in the following section.

III.   NITROAROMATIC COMPOUND SORPTION ON CLAY MINERALS

A relatively large number of NACs have been used in sorption studies on clay minerals, representing a broad range of chemical and physical properties (see

Table 2). All of the compounds listed in Table 2A and shown in Table 2B have been used in clay sorption studies, and they represent a broad range of aqueous solubilities, polarities, sizes, and Hammett constants. As will be discussed later, sorption is strongly dependent on the number and ring location of the nitro substituents, as well as other functional groups.

A. Influence of Clay Type

The sorption of 1,3,5-trinitrobenzene (TNB) from aqueous suspension on kaolinite, illite, and smectite is shown in Figure 4A (5). Considerably more sorption of 1,3,5-TNB occurs on the K-montmorillonite compared to kaolinite or illite. Related sorption studies of NACs on clay minerals indicate that NAC sorption on smectites occurs in the interlamellar regions. Internal sorption sites are not available on kaolinite or illite; as a consequence, considerably less sorption occurs. Smectites, in particular, show a high affinity for NACs from aqueous suspension.

Within the group of expandable clay minerals known as smectites (2:1 phyllosilicates with layer charge (χ) values of 0.3–0.6), large variations in NAC sorption have been shown to occur. This can be illustrated by comparing the sorption isotherm of 6-methyl-2,4-dinitrophenol (DNOC) on SWy-2 smectite, a low-charge smectite, to the isotherm obtained using SAz-1 smectite, a high-surface-charge-density (SCD) smectite (Figure 4b) (11). Considerably more DNOC sorption occurs on the low-SCD clay (SWy-2). This is attributed to (a) favorable interactions between the nitro group and the exchangeable cation, and (b) the neutral portion of the NAC molecule becoming stabilized by the clay surface through a combination of van der Waals interactions and electrostatic attraction forces between the hydrated cation and the —NO_2 group, which carries a partial negative charge. On the higher-SCD clay (SAz-1), the latter surface sites are less available, due to shrinkage of adsorption domains, and therefore sorption is diminished.

The interplay between different surface sites is also shown by the dramatic sorption increase of 4,6-dinitro-6-sec-butyl phenol (DINOSEB) on a high-SCD clay (SAz-1) upon systematically reducing its SCD (Fig. 4C) (11). Sorption of DINOSEB from aqueous solution is shown on a series of smectites that represented varying degrees of charge reduction. The starting clay was of high surface charge density (the Cheto SAz-1 smectite from Arizona). This clay has a CEC of 120 $cmol_c$ kg^{-1} and is represented by the label 1.0 CEC; there was very little sorption of DINOSEB on this sample. Heating the clay in the presence of a Li or Mg salt can artificially reduce the surface charge density of smectites. These cations are small enough to penetrate into the clay lattice and locally satisfy the charge deficit, thereby reducing the cation exchange capacity of the sample (62,63). Sorption was greatly enhanced as the degree of charge reduction increased from 15% (denoted 0.85 CEC) to 60% (0.4 CEC) (11), consistent with

TABLE 2A K_d Sorption Coefficients for Selected NACs with Kaolinite and Montmorillonite

Compound	Abbreviation	K_d (L kg^{-1}) (Cs$^+$–kaol)	K_d (L kg^{-1}) (K$^+$–mont.)
2-Nitrophenol	2-NP	35	45
3-Methyl-2-nitrophenol	3-Me-2NP	8.9	
4-Methyl-2-nitrophenol	4-Me-2NP	130	
5-Methyl-2-nitrophenol	5-Me-2NP	450	
6-Methyl-2-nitrophenol	6-Me-2NP	120	
4-sec-Butyl-2-nitrophenol	4-sBu-2NP	4.7	
4-Methoxy-2-nitrophenol	4-OMe-2NP	1,100	
4-Chloro-2-nitrophenol	4-Cl-2NP	120	
5-Fluoro-2-nitrophenol	5-F-2NP	220	
4-(Trifluoromethyl)-2-nitrophenol	4-CF3-2NP	33	
2,4-Dinitrophenol	4-NO2-2NP	>9,000	
2,5-Dinitrophenol	5-NO2-2NP	>6,000	
2,6-Dinitrophenol	6-NO2-2NP	>7,000	
3,4-Dinitrophenol	3-NO2-4NP	8.3	
3-Nitrophenol	3-NP	23	
4-Carboxy-3-nitrophenol	4-COOH-3NP	1300	
4-Nitrophenol	4NP	34	
2-Methyl-4-nitrophenol	2-Me-4NP	67	
3-Methyl-4-nitrophenol	3-Me-4NP	409	
6-Methyl-2,4-dinitrophenol	DNOC	18,000	37,000
4-Methyl2-6,-dinitrophenol	2,6-DNOC		19,000
6-sec-Butyl-2,4-dinitrophenol	DINOSEB	54	64
2,4-Dinitro-6-sec-butylphenylmethylether	DINOSEB-ME		17
2,4-Dinitro-6-sec-butylphenyl acetate	DINOSEB-AC		2.9
6-tert-Butyl-2,4-dinitrophenol	2,4-DINO-6-TERB (DINOTERB)	18	28
4-tert-Butyl-2,6-dinitrophenol	2,6-DINO-4-TERB	8.6	
Nitrobenzene	NB	3.5	7.2
2-Nitrotoluene	2-Me-NB (2-NT)	3.4	4.6
3-Nitrotoluene	3-Me-NB (3-NT)	18	21
4-Nitrotoluene	4-Me-NB (4-NT)	54	45
2,4-Dinitrotoluene	2,4-DNT		7,400
2,6-Dinitrotoluene	2,6-DNT		125
2,4,6-Trinitrotoluene	TNT		21,500
2-Amino-4,6-dinitrotoluene	2-A-4,6-DNT		2,900
4-Amino-2,6-dinitrotoluene	4-A-2,6-DNT		125
2,6-Diamino-4-nitrotoluene	2,6-DA-4-NT		11
2,4-Diamino-6-nitrotoluene	2,4-DA-6-NT		3.5
4-Ethylnitrobenzene	4-Et-NB	62	
4-n-Butylnitrobenzene	4-n-But-NB	75	
2-Nitrobiphenyl	2-Phen-NB	3.6	
3-Nitrobiphenyl	3-Phen-NB	58	
2-Chloronitrobenzene	2-Cl-NB	6	
3-Chloronitrobenzene	3-Cl-NB	24	
4-Chloronitrobenzene	4-Cl-NB	44	

(*Continued*)

TABLE 2A *Continued*

Compound	Abbreviation	K_d (L kg^{-1}) (Cs+–kaol)	K_d (L kg^{-1}) (K+–mont.)
4-Bromonitrobenzene	4-Br-NB	52	
1,2-Dinitrobenzene	2-NOZ-NB (1,2-DNB)	1.7	4.2
1,3-Dinitrobenzene	3-NOz-NB (1,3-DNB)	1800	4500
1,4-Dinitrobenzene	4-NO2-NB (1,4-DNB)	4000	3100
1,3,5-Trinitrobenzene	TNB		>60000
4-Nitroanisole	4-OCHy NB	340	
4-Cyanonitrobenzene	4-CN-NB	520	
4-Nitrobenzaldehyde	4-CHO-NB	730	
4-Acetylnitrobenzene	4-CHsCO-NB	2100	
1,3-Dinitronaphthalene	1,3-DNN	>4000	
1,5-Dinitronaphthalene	1,5-DNN	>5000	
1,8-Dinitronaphthalene	1,8-DNN	48	
2-Nitroaniline	2-NA		8.4
3-Nitroaniline	3-NA		3.5
4-Nitroaniline	4-NA		13.5
Nitrocyclohexane	NCH	<0.1	
1,2-Dihydroxybenzene	catechol	25	
4-Methylphenol	4-Me-Ph	0.9	
2-Chlorophenol	2-Cl-Ph	<0.1	
3-Chlorophenol	3-Cl-Ph	0.3	
4-Chlorophenol	4-Cl-Ph	<0.1	
1,3-Dichlorobenzene	1,3-DCB	1.4	
1,3,5-Trichlorobenzene	1,3,5-TCB	1.3	
1,2,3,4-Tetrachlorobenzene	1,2,3,4-TeCB	5.4	
1,2,3,5-Tetrachlorobenzene	1,2,3,5-TeCB	7.2	
N,2,4,6-Tetranitro-*N*-methylaniline	TETRYL		5.8
1,3,5-Trinitro-hexahydro-1,3,5-triazine	RDX		1.2
O,O-Diethyl-*O*-(4-nitrophenyl) monothiophosphate	Parathion		9.6
O,O-Dimethyl-*O*-(4-nitrophenyl) monothiophosphate	Me-parathion		6.2
(Trifluoromethyl)-2,6-dinitro-*N,N'*- dipropylaniline	Trifluraline		8.1
(methylsulfonyl)-2,6-dinitro-*N,N'*- dipropylaniline	Nitraline		3

Source: Refs. 4–5.

Table 2B   Structures of Selected Nitroaromatic Compounds

the atrazine results (48). The lack of DINOSEB sorption on smectites with a high CEC indicates that these surfaces are strongly hydrophilic; therefore, water is more difficult to displace as the hydration spheres of the cations overlap one another and the hydrophobic sites are effectively blocked. Considerably greater sorption occurs on smectites with lower CECs or clays that have had their surface charge density reduced artificially (64).

Sorption isotherms based on macroscopic observations are not capable of providing detailed molecular-level insight into the chemical mechanisms underlying the sorption behavior. In order to obtain this type of information, spectroscopic, structural, and computational data and methods are required. X-ray diffraction methods were combined with sorption and spectroscopic data to study the interaction of DNOC with smectite (11). The d-spacing of the DNOC-smectite complex in the presence and absence of water and varying amounts of adsorbed DNOC are shown in Figure 7. In the air-dried DNOC-smectite complex, the d-

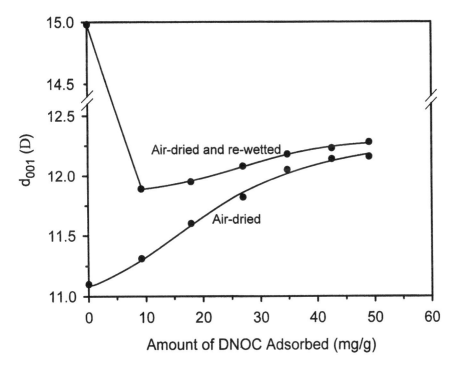

FIGURE 7 Basal spacings of DNOC-adsorbed K-SWy-2 clays obtained from XRD patterns in the presence and absence of water. The d_{001} spacing of the DNOC-SWy-2 complexes are plotted on the y-axis in angstroms. (From Ref. 11.)

spacing gradually increased from 1.1 nm to 1.22 nm with increasing surface coverage of DNOC, thus demonstrating intercalation of the NAC into smectite interlayers. Of interest here is the behavior when the NAC-smectite complex is exposed to water. As shown by the top curve in Figure 7, swelling of the air-dried clay (no DNOC) causes an increase in the interlayer spacing from ~1.1 nm to ~1.5 nm, consistent with prior studies (65). Interestingly, the presence of sorbed DNOC restricts swelling of the rewetted clay. The interlayer spacings for the air-dried and rewetted clays are nearly equal (~1.22 nm) at the higher DNOC loadings. The presence of DNOC causes the monolayer structure to be favored, even in cases where the smectite might swell further in the absence of DNOC (e.g., K-SWy-2). The ~1.2-nm spacing appears optimal for DNOC sorption. This may be understood as a result of the ~0.35-nm van der Waals thickness of planar NAC molecules (8). The structure of pyrophyllite (66) has a d_{001}-spacing of 0.92 nm, representative of the pure interaction between siloxane surfaces. Adding the thickness of NACs to that value yields 1.27 nm for the idealized d_{001}-spacing of a 2:1 clay intercalated with NAC. There is another way to estimate this distance. Quantum chemical optimizations of 1,3,5-trinitrobenzene on one siloxane surface (67,68) yielded 0.32 nm for the optimal distance between the basal plane of O and the NAC ring plane. In the structure of pyrophyllite (66), the basal planes are ~0.27 nm apart, so they would have to be expanded another 0.37 nm in order for an NAC to be 0.32 nm from each basal plane; this would correspond to a d_{001}-spacing of 1.29 nm. The two estimates are perhaps somewhat larger than the observed d_{001}-spacings of 1.22 nm because the experimental peaks are somewhat broad and contain contributions from both expanded and unexpanded smectite layers. The d-spacing of K-exchanged smectites at high humidity (p/po ~1.0) is 1.2–1.25 nm and is referred to as a monolayer structure (one layer of water in the interlamellar region of the clay). In the case of low-SCD smectites (e.g., SWy-1), the d-spacing is increased to 1.5 nm (two layers of water) and the corresponding spacing for Ca- and Ba-exchanged SWy-2 is greater than 1.5 nm in aqueous suspensions.

With an appreciable amount of DNOC sorbed (~45 mg DNOC/g of clay), the d-spacing is ~1.2 nm. The only possible configuration for this complex is for the sorbed DNOC molecules to lay flat in the interlamellar region, with the DNOC molecule in contact with both sides of the opposing siloxane surfaces. If the sorbed DNOC molecules were oriented on their side or on end, the d-spacing would be larger than the observed 1.2-nm value. Removal of DNOC from water may provide sufficient energy to prevent K-smectite from swelling beyond ~1.2–1.25 nm, since the free energy of hydration of many small organic molecules is in the range of 10–30 kJ/mol. In contrast, the amount of energy required to compress the interlayers of smectites exchanged with divalent cations (e.g., Ca and Ba) is too large, and dehydration of the sorbed NAC molecule does not occur. Cs-smectites have a strong tendency to equilibrate with ~1.25-nm spacings at 100% relative humidity and, due to the low hydration energy of Cs^+, rarely expand further. For these clays, there is no energy penalty for compressing the layers. Therefore, Cs-SWy-2 is an even more effective adsorbent of DNOC than K-SWy-2.

Polarized infrared spectroscopy can provide a direct means of determining the molecular orientation of sorbed species on oriented self-supporting clay films (11). If the orientation of the induced dipole moment for a given vibrational mode is known, then the orientation of the sorbed species can be determined by measuring the dichroic ratio of the band (69). Highly oriented self-supporting films of smectite can be made because of the high aspect ratio of the individual smectite particles (70,71). Using the linear dichroism methods developed by Margulies and coworkers (71), the orientation of sorbed DNOC was reported in 2002 (11). This is shown by the polarized FTIR data obtained for DNOC sorbed on K-SWy-1 clay in Figure 8. The vibrational bands having greatest intensity parallel to the clay surface are represented by the dashed line and those having greater intensity perpendicular to the clay surface by the solid line in both plots. The bands shown in the left plot correspond to the $\nu_{sym}(NO)$ bands and those in the right plot to the out-of-plane $—NO_2$ deformation mode. The dashed line shows the vibrational modes obtained when the IR beam is normal to surface of the oriented clay film. In this configuration, the vibrational bands parallel to the clay film (i.e., siloxane surface) have maximum intensity. The solid line corresponds to a clay film tilted at an angle of $\sim 45°$. The increase in intensities of the $\nu_{sym}(NO)$ bands at 1350 and 1330 cm^{-1} relative to the out-of-plane spectra (solid line) indicates that the molecular plane of the $—NO_2$ functional group is oriented parallel to the clay surface. Quantum chemical and structural studies have confirmed that the $—NO_2$ groups and the aromatic ring are essentially coplanar (11). The opposite behavior is observed for *out-of-plane* vibrational modes. This is shown by the increased intensity of the out-of-plane modes ($—NO_2$ deformation) shown on the right side of Figure 8. These observations provide direct evidence that the sorbed DNOC molecules are laying nearly flat with respect to the clay surface, at least in air-dried clay films.

B. Influence of Exchangeable Cation

The interaction of NACs with clay minerals is strongly influenced by the nature of the exchangeable cation (4,10,11). The amount of DNOC sorbed by smectites exchanged with different cations is shown in Figure 9. For the alkali metal and alkaline earth cation studies, the amount of DNOC sorbed increased with increasing ion size within each series: ($Cs^+ > K^+ > Na^+$) and ($Ba^{2+} > Ca^{2+}$). This trend is inversely related to the hydration enthalpies for these cations, supporting the hypothesis that NAC sorption is enhanced for exchangeable cations that have low hydration enthalpies (Table 1). The high sorption of DNOC on Cs^+- and K^+-

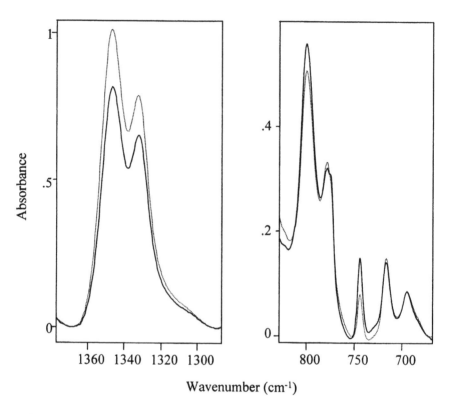

FIGURE 8 Linear dichroism FTIR spectra of a self-supporting clay film of DNOC sorbed to K-SWy-2 montmorillonite. The clay film was tilted at an angle of 45° about its vertical axis with respect to the incident IR beam in the sample compartment. The spectrum represented by the dashed line was obtained with an IR polarizer along the vertical axis. The solid line spectrum was obtained with the IR polarizer at 90° (along the horizontal axis). In this configuration (solid line), the intensities of the out-of-plane vibrational modes will have maximum intensity. (From Ref. 82.)

exchanged SWy-2 smectites is attributed, in part, to the low hydration energies of these cations. The modest hydration energies of these cations mean that homoionic clays tend to be hydrated by only a monolayer of water (in the case of Cs^+) or that there is little penalty for restricting swelling to a monolayer (in the case of K^+). If the clay interlayer, in turn, contains only a monolayer of water, then the sorbed NAC molecules will be almost removed from water because the NAC thickness of ~0.35 nm is very close to the thickness of the water monolayer. This can stabilize the sorbed NAC in two ways: (a) removing the low-solubility

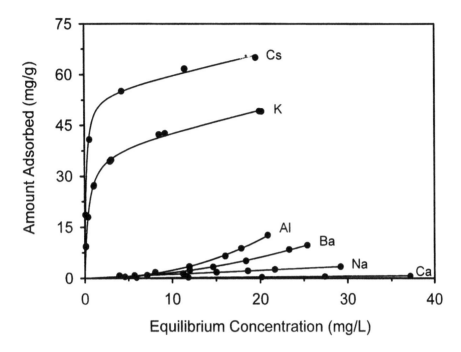

FIGURE 9   Effect of exchangeable cations on DNOC adsorption by K-SWy-2. The bands shown in the left plot correspond to the $\nu_{sym}(NO)$ bands and those in the right plot to the out-of-plane —NO_2 deformation mode. (From Ref. 11.)

NAC from water is energetically favorable because the NAC is hydrophobic and the solvation energy is positive; (b) the two largest surfaces of the NAC both interact with the most hydrophobic regions of the smectite surface, which should be modestly favorable for the system energy.

Site-specific interactions between the —NO_2 groups of sorbed NAC molecules in the interlamellar region of smectites and exchangeable cations were first proposed to account for NAC sorption by Yariv and coworkers over 30 years ago (72,73). They found that the vibrational bands of the sorbed NACs were shifted relative to the neat compound. The spectral trends in this study were not conclusively related to the nature of the exchangeable cation because of differences in water content (73). More recently, evidence against site-specific interactions was reported by Weissmahr and coworkers based on spectral *similarities* between NACs sorbed on K^+- and Cs^+-exchanged clays obtained using UV-visible, infrared, and NMR methods (6). They reported spectral shifts in the ATR-FTIR spectra of sorbed NACs relative to the NAC in aqueous solution. However,

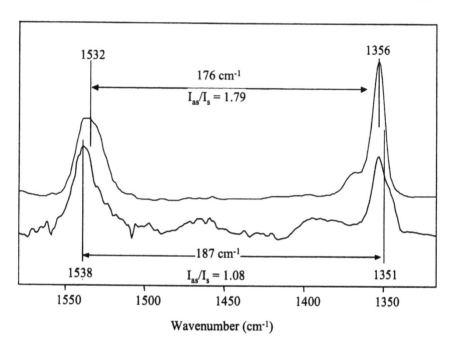

FIGURE 10 FTIR spectra of 1,3-dinitrobenzene sorbed to K-SWy-1 montmorillonite (upper trace) and 1,3-dinitrobenzene in aqueous solution (lower trace). The splitting between the $\nu_{asym}(NO)$ and $\nu_{sym}(NO)$ bands is shown, as well as the intensity ratio of I_{as}/I_s. (From Ref. 9.)

they interpreted the spectral shifts as being relatively minor. In addition, they attributed the spectral shifts reported in earlier IR studies, in part, to the use of nonaqueous solvents (72,73). In an attempt to clarify the chemical mechanisms responsible for NAC sorption, they combined sorption and spectroscopic data. They found that IR spectra of NAC sorbed to K- and Cs-exchanged smectites were similar, showing only minor shifts from the NAC in aqueous solution. They concluded that cations did not play a significant role and attributed the high affinity of certain NACs for smectites to the formation of an electron–donor–acceptor complex with the clay surface (6). Unfortunately, the exchangeable cations used in this study, K^+ and Cs^+, have similar hydration enthalpies (Table 1) (6). Furthermore, one could argue that the observed spectral perturbations, although small, are significant in the context of prior NAC vibrational studies (74).

In the past few years, sorption, spectroscopic, and computational methods have been combined to elucidate the chemical mechanisms behind the high affinity of certain NACs for clay surfaces (8–11). The vibrational modes associated with the —NO_2 groups of NACs can be used as molecular probes of site-specific interactions involving cations in the interlamellar environment of clay minerals. The movement of the nitrogen and oxygen atoms in the —NO_2 group create a large induced dipole moment; consequently, the vibrational bands associated with the stretching and bending vibrations of these groups have high molar absorptivity values (i.e., they are strong bands in the IR spectrum). The positions of the symmetric and asymmetric —NO_2 bands of nitroaromatic compounds sorbed to a SWy-1 smectite exchanged with different cations are influenced by the nature of the cation (Figs. 10–12). The positions and intensities of both the asymmetric and symmetric NO_2 stretching bands of DNOC are influenced by the nature of the exchangeable cation, as shown in Figure 12. In general, the splitting between the asymmetric and symmetric NO_2 stretching bands is decreased upon NAC complexation to clay minerals with weakly hydrated cations (Fig. 10). As the ionic potential of the cations is *decreased*, the splitting between the $\nu_{asym}(NO)$ and $\nu_{sym}(NO)$ bands is decreased from 200 cm^{-1} to 187 cm^{-1}, indicative of a *stronger* interaction (Figure 11A). For cations with ionic potentials greater than 3, NAC solutes cannot effectively compete with water molecules for coordination sites around the cation. Similar cation-induced perturbations of the $\nu_{sym}(NO)$ and

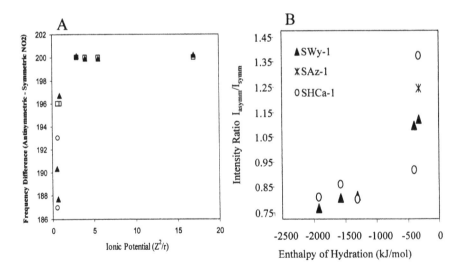

FIGURE 11   (A) Effect of ionic potential on the frequency difference between the $\nu_{asym}(NO)$ and $\nu_{sym}(NO)$ bands of 1,3,5-trinitrobenzene sorbed to SAz-1, SWy-1, and SHCa-1 smectite (cations included are Mg, Ca, Ba, Na, and Cs). (B) Intensity ratio of the $\nu_{asym}(NO)$ band/$\nu_{sym}(NO)$ bands as a function of the enthalpy of hydration of the cation in the series Mg, Ca, Ba, Na, K, and Cs. (From Ref. 10.)

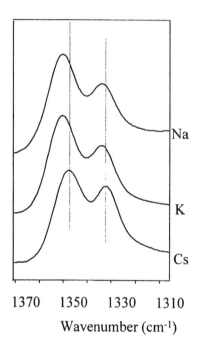 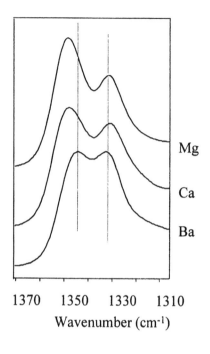

1370   1350   1330   1310   1370   1350   1330   1310
Wavenumber (cm⁻¹)   Wavenumber (cm⁻¹)

FIGURE 12   Influence of the exchangeable cation on the NO_2 stretching bands of DNOC sorbed to SWy-1 smectite. FTIR spectra of DNOC in the 1370- to 1320-cm⁻¹ region showing the $\nu_s b_2$ and $\nu_s a_1$ modes of DNOC sorbed to SWy-2 exchanged with different exchangeable cations.

$\nu_{asym}(NO)$ bands occurred on three different smectites (hectorite, SWy-1, and SAz-1), which indicated that the interaction is associated primarily with the hydrated cations and not directly with the smectite surface. In general, the vibrational modes of the NO_2 groups are relatively insensitive to interactions involving NO_2 groups (74,75). In the context of prior vibrational studies of NACs, the shift in band positions of the $\nu(NO)$ bands that were observed upon adsorption of TNB by smectite represent some of the largest shifts in vibrational studies of NACs that have been reported (74–81).

The cation-induced shifts in position can be correlated with a number of properties of the exchangeable cations. In addition to the hydration enthalpies, for example, the differences in frequency between the asymmetric and symmetric NO_2 stretching bands of 1,3,5-trinitrobenzene are related to the ionic potential of exchangeable cation. Similar cation-induced shifts have been reported in 2002 (82) for DNOC sorbed to smectite (Fig. 12). The vibrational modes of the symmetric and asymmetric stretching vibrations of DNOC are more complex than those of 1,3,5-TNB because DNOC has a lower symmetry than that of TNB and the two —NO_2 groups of DNOC are in different environments. Despite these differences, similar shifts were found for both the alkali and alkaline earth cations.

The relative intensities of the $\nu_{asym}(NO)$ and $\nu_{sym}(NO)$ bands (I_{as}/I_s) are also influenced by the nature of the exchangeable cation. This is shown in Figure 11B, where the ratio of the integrated areas of the asymmetric to the symmetric $\nu(NO)$ bands is plotted as a function of the enthalpy of hydration of the exchangeable cation. As the enthalpy of cation hydration decreases, the cations can interact more directly with the —NO_2 substituents, as reflected in the change in positions and relative intensities of the $\nu_{asym}(NO)$ and $\nu_{sym}(NO)$ bands. In a related study of nitrotoluene sorption to Si- and Ti-oxides (74), the ratio of band intensities (I_{as}/I_s) was influenced by the oxide surface. In the case of titania, the (I_{as}/I_s) ratio decreased to values less than 1.0, which was attributed to compression of the O—N—O angle. For silica, the ratio was greater than 1.0 and was attributed to hydrogen-bonding interactions. This study is relevant insofar as it shows that the relative intensities of the $\nu(NO)$ bands are sensitive to differing modes of surface interactions.

Classical molecular dynamics simulations of NACs in smectite interlayers predicted that inner-sphere K^+–O_2N— complexes would readily form between NACs and interlayer K^+ (9,11). Therefore, in order to help rationalize the observed FTIR spectral shifts in terms of the proposed complexation reactions between —NO_2 groups and interlayer cations, quantum calculations of some simple models (Fig. 1) for inner-sphere NAC-K^+ interactions were performed; they are summarized in Table 3. The quantum results follow the experimental trends for complexed versus uncomplexed —NO_2 stretching frequencies. That is, complexation by K^+ results in a red shift (i.e., to lower frequency) of the $\nu_{asym}(NO)$ band and a simultaneous blue shift of the $\nu_{sym}(NO)$ band. Taken together, the splitting between the two frequencies decreases upon complexation. The K-O_2N interac-

TABLE 3 Key Vibrational Frequencies and Geometries Obtained from Quantum Calculations of NACs and NACs Complexed to K^+ Ions

Molecule	Asymm. NO_2 frequency	Symm. NO_2 frequency	Frequency splitting	Average N—O bond length	C—N bond length body length	Average O—K distance
1,3-DNB	1537,1532	1355,1345	185	1.222	1.485	—
K_2-1,3-DNB	1499,1496	1357,1344	147	1.229	1.469	2.92
1,3,5-TNB	1538	1346	192	1.220	1.488	—
K_3-1,3,5-TNB	1517	1348	169	1.223	1.484	3.18

tion causes the N—O bonds to be somewhat weakened, while the C—N bonds are somewhat strengthened, relative to isolated, uncomplexed NACs (Table 3). This causes the N—O stretching components of normal modes in the complexes to decrease in frequency and the C—N components to increase in frequency compared to those of the uncomplexed NAC molecule. On this basis, complexation of K^+ results in a shift to lower frequency for the asymmetric —NO_2 modes because they are dominated almost purely by the weakened N—O stretching components (83). At the same time, the potential energy distribution of the symmetric —NO_2 modes is comprised of ~1/3 C–N stretching components and ~2/3 N—O stretching components (83), so strengthening of the C—N bond upon complexation tends to induce a shift to slightly higher frequency (which is somewhat offset by the N—O bond weakening). Thus, the shifts observed experimentally (Figs. 10–12) for the vibrational peaks of NACs interacting with smectites can be rationalized in terms of inner-sphere complexation between —NO_2 groups and interlayer K^+. By implication, the shift is not observed for Na^+, Mg^{2+}, or Ca^{2+} because their stronger hydration precludes inner-sphere complexation by —NO_2 groups. When adsorbed to clays saturated by these cations, the FTIR spectra of TNB are very similar to those in aqueous solution. These data provide molecular-level support for the hypothesis that SNB sorption on smectites is controlled, in part, by interactions with exchangeable cations.

It seems reasonable to attribute the strong NAC sorption observed for K- and Cs-smectites to specific complex formation, because experimental estimates for adsorption enthalpies are approximately 40 kJ/mol (4,6). While complex formation between NAC rings and the siloxane surface itself has been postulated (4), Johnston et al. (82) (10,11) have followed the FTIR and computational results to stress complexation interactions between NACs and interlayer cations as an alternative explanation for those cases in which strong NAC adsorption is observed. Other quantum mechanical investigations (84) have found interaction energies near 40 kJ/mol for small NACs interacting with molecular fragments of clay basal surfaces. They attributed these strong attractive energies to ~50% dispersion interactions (67) and ~50% electrostatic interactions with neutral clay fragments (68). Despite the electrostatic contributions, these studies discounted the formation of specific chemical bonds (68) or electron donor–acceptor complexes (67) between the NAC and the basal clay surface. Alternatively, interactions of nitrobenzene with one partially hydrated Na^+ cation atop a clay fragment were also shown to give rise to a 49-kJ/mol interaction energy (68). Taken together, these studies would imply that both NAC-cation complexation (especially when more than one NAC group can form complexes) and NAC dispersion/electrostatic interactions with clay basal surfaces by direct molecular contact should help to stabilize adsorption of NACs by smectites.

Perhaps the most compelling evidence that cations can have a significant influence on NAC sorption by smectites comes from a recent field study (17).

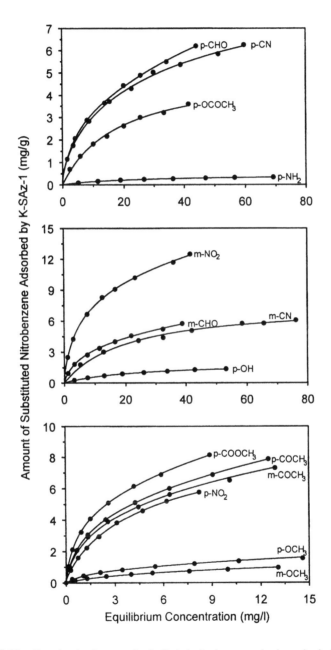

FIGURE 13   Sorption isotherms of substituted nitrobenzenes by homoionic K-smectite (K-SAz-1) from water. (From Ref. 9.)

A two-step field test was used to examine the mobility of 4-nitroluene. A solution containing both 4-nitrotoluene and KCl was injected into a sandy aquifer and monitored at an extraction well. A solution containing $CaCl_2$ was subsequently injected into the aquifer, and mobility of the NAC was found to increase. In the presence of KCl, NAC mobility was attenuated by K-exchanged minerals present. Upon introduction of $CaCl_2$, ion exchange of K^+ ions by Ca^{2+} ions was proposed to account for the increased mobility of the NAC. This study provided field-scale evidence that exchangeable cations play a definite role in NAC mobility in soils and subsurface environments.

C. Influence of NAC Properties

NAC sorption on smectites is strongly influenced by the properties of the NAC itself. This is shown in Figure 13, where the sorption isotherms of 14 NACs on K-SAz-1 smectite from aqueous suspension are plotted (9). Freundlich adsorption coefficients obtained from these isotherms are listed in Table 4. NACs with substituents such as $-NO_2$, $-CN$, $-COCH_3$, and $-COOCH_3$ have high affinities for smectite surfaces, whereas those with $-NH_2$, $-OH$, and $-OCH_3$ substituents have lower sorption coefficients (9). The substituents $-NO_2$, $-CN$, $-COCH_3$, and $-COOCH_3$ are strong electron-withdrawing groups, and $-NH_2$ and $-OH$ are electron-releasing groups. Quantitative relationships between

TABLE 4 Physicochemical Properties of Substituted Nitrobenzenes, the Freundlich Adsorption Coefficients (K_f), and the Hammett Constants of Substituents (σ)

Substituent	Water solubility $S_w(mg/L)^a$	K_f (l/g)	σ^b
p-NH$_2$	728	0.0499	−0.66
p-OH	1.16×10^4	0.182	−0.36
p-OCH$_3$	590	0.452	−0.12
m-OCH$_3$	428	0.244	0.10
p-CHO	2320	1.05	0.22
p-OCOCH$_3$	—	0.533	0.31
m-CHO	—	1.02	0.36
m-COCH$_3$	750	2.51	0.36
p-COOCH$_3$	539	3.28	0.44
p-COCH$_3$	1300	2.62	0.52
m-CN	1710	1.03	0.62
p-CN	1650	1.16	0.63
m-NO$_2$	533	2.87	0.71
p-NO$_2$	69	2.87	0.78

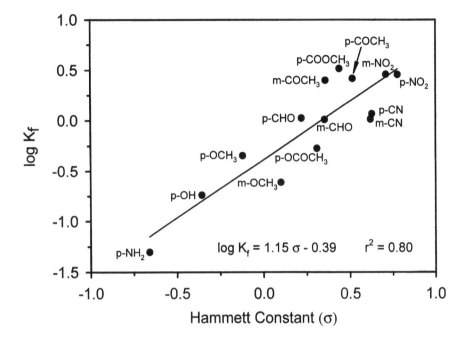

FIGURE 14 Freundlich adsorption coefficients (K_f) of substituted nitrobenzenes plotted as their logarithm versus the Hammett constants of substituents. (From Ref. 9.)

structure and reactivity have been developed; one such relationship is given by the Hammett constant (Table 4). Figure 14 shows the logarithm of the Freundlich sorption coefficients plotted as a function of the Hammett constant (Table 4). As shown, the strong positive correlation is evidence that NAC sorption by smectite is sensitive to electronic effects and favored by electron-withdrawing groups such as —NO$_2$ and —CN (9). These results are consistent with the proposed role of NACs as electron acceptors (4), but they do not provide unequivocal evidence for the proposed electron donor–acceptor mechanism.

Quantum calculations (9) were used to calculate the electron density on the ring and on the —NO$_2$ group (Fig. 15). Despite the large change in the Hammett constant, the total charge on the ring carbon atoms showed little variation. In contrast, a significant change was observed in the charge on the —NO$_2$ group from electron-withdrawing to electron-donating groups. The lack of change in ring electron density implied that electron density donated by substituents flows through to the nitro group and that the ring itself was relatively unaffected (9). These results are consistent with other recent quantum results (67) asserting

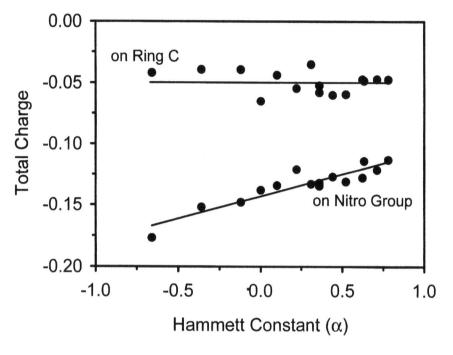

FIGURE 15 Relationships between the Hammett constant and electronic properties of substituted nitrobenzenes.

that donor–acceptor interactions should not play a significant role in complexation of nitroaromatic compounds by siloxane surfaces.

In light of hypotheses advanced earlier for NACs such as TNB, DNB, and DNOC (i.e., that —NO_2 groups form complexes with certain exchangeable cations), it seems plausible that the sorption of other NACs, too, should be dependent on the abilities of all substituents to participate in complexation with interlayer cations. That is, it seems reasonable that —CN, —$COCH_3$, and —$COOCH_3$ would form complexes with interlayer K^+ (although these have not yet been demonstrated), while —NH_2 and —OCH_3 substituents would be less likely to form them. In general, NAC sorption to smectite increases as solubility decreases. This is a smaller effect compared to the importance of interlayer complex formation, because otherwise poorly soluble aromatics such as toluene would adsorb more strongly to smectites. However, in the case of similar NACs with substituents capable of forming complexes, removal of poorly soluble NACs from aqueous solution would be expected to contribute favorably to the energy of adsorption. In the case of adsorption to K^+-smectites in which d_{001}-spacings are near

1.25 nm, the NAC interacts very little with water and is almost dehydrated. Using the solubility of each pure-phase NAC as a crude estimate for its solvation energy, the contributions to adsorption energies from desolvating the NACs in Table 4 would range from 6 kJ/mol (for p-nitrophenol) to 19 kJ/mol (for p-DNB). These energies are not adequate to drive adsorption processes, but they may become a source of significant differences when the other system forces are in balance.

In summary, the potential for adsorption of NACs on smectites is determined largely by the additive interactions of the —NO$_2$ groups and the secondary substituent with interlayer cations. A conceptual illustration of a NAC molecule interacting with a hydrated clay surface is shown in Figure 16. Recent spectroscopic studies of NAC-clay complexes have provided unambiguous evidence that the —NO$_2$ groups interact with exchangeable cations or with water molecules surrounding these cations. In addition, polarized spectroscopic measurements combined with X-ray diffraction show that sorbed NACs are oriented "flat" on the clay surface, as shown in Figure 16. Water molecules compete with NACs

NAC oriented flat on the
smectite surface

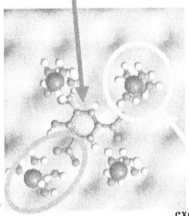

Water competes
for sites around
exchangeable cations

NO$_2$ group of
NAC interacts
with exchangeable
cations

FIGURE 16   Illustration of sorbed DNOC molecule laying flat on the siloxane surface showing the siloxane surface and water molecules surrounding the exchangeable cations. The dimensions of the DNOC molecule and the distance between the exchangeable cations are drawn to scale. (Adapted from Refs. 82 and 90.)

for coordination sites around the exchangeable ions, and this is dictated by the hydration energy, charge, and size of the cations as well as by the molecular properties of the NAC itself. The bulk adsorption data, FTIR spectral trends, and computational studies are self-consistent with the hypotheses that (a) NO_2 groups effectively complex certain cations (e.g., Cs^+ or K^+) both directly and through the intermediation of water, (b) secondary functional groups that interact favorably with K^+ (e.g. $—COOCH_3$, $—CN$, $—NO_2$) cause comparatively higher sorption, and those unable to complex K^+ as strongly ($—H$, $—NH_2$, $—OCH_3$) reduce adsorption, and (c) the resultant overall adsorption is tempered by the aqueous solubility of the NAC.

ACKNOWLEDGMENTS

The authors would like to thank Max Mortland for helpful discussions related to this review.

REFERENCES

1. Mason and Hanger Corporation/Battelle Pantex. Risk Reduction Rule Guidance for Pantex Plant RCRA Facility Investigations. Final, 1996.
2. A Esteve-Nunez, A Caballero, JL Ramos. Microbiology and Molecular Biology Reviews 2001; 65:335.
3. N Hannink, SJ Rosser, CE French, A Basran, JAH Murray, S Nicklin, NC Bruce. Nature Biotechnology 2001; 19:1168–1172.
4. SB Haderlein, RP Schwarzenbach. Env. Sci. Tech 1993; 27:316–326.
5. SB Haderlein, KW Weissmahr, RP Schwarzenbach. Env. Sci. Tech 1996; 30: 612–622.
6. KW Weissmahr, SB Haderlein, RP Schwarzenbach, R Hany, R Muesch. Env. Sci. Tech 1997; 31:240–247.
7. KW Weissmahr, SB Haderlein, RP Schwarzenbach. Soil Sci. Soc. Am. J 1998; 62: 369–378.
8. GY Sheng, CT Johnston, BJ Teppen, SA Boyd. J. Agri. Food Chem 2001; 49: 2899–2907.
9. SA Boyd, G Sheng, BJ Teppen, CT Johnston. Env. Sci. Tech 2001; 35:4227–4234.
10. CT Johnston, MFD Oliveira, G Sheng, SA Boyd. Env. Sci. Tech 2001; 35: 4767–4772.
11. G Sheng, CT Johnston, BJ Teppen, SA Boyd. Clays Clay Minerals 2002; 50:25–34.
12. MM Mortland. Adv. Ag 1970; 22:75–117.
13. CT Chiou, LJ Peters, VH Freed. Science 1979; 206:831–832.
14. WF Jaynes, SA Boyd. J. Air Waste Man. Asso 1990; 40:1649–1653.
15. MDR Pizzigallo, P Ruggiero, M Spagnuolo. Fresenius Envir. Bull 1998; 7:552–557.
16. MM Broholm, N Tuxen, K Rugge, PL Bjerg. Env. Sci. Tech 2001; 35:4789–4797.
17. KW Weissmahr, M Hildenbrand, RP Schwarzenbach, SB Haderlein. Env. Sci. Tech 1999; 33:2593–2600.

18. SF Nishino, JC Spain, H Lenke, HJ Knackmuss. Env. Sci. Tech 1999; 33:1060–1064.
19. C Achtnich, H Lenke, U Klaus, M Spiteller, HJ Knackmuss. Env. Sci. Tech 2000; 34:3698–3704.
20. TW Sheremata, A Halasz, L Paquet, S Thiboutot, G Ampleman, J Hawari. Env. Sci. Tech 2001; 35:1037–1040.
21. K Rugge, TB Hofstetter, SB Haderlein, PL Bjerg, S Knudsen, C Zraunig, H Mosbaek, TH Christensen. Env. Sci. Tech 1998; 32:23–31.
22. GR Johnson, BF Smets, JC Spain. Appl. Env. Microbio 2001; 67:5460–5466.
23. BT Oh, CL Just, PJJ Alvarez. Env. Sci. Tech 2001; 35:4341–4346.
24. GR Lotufo, JD Farrar, LS Inouye, TS Bridges, DB Ringelberg. Env. Toxicol. Chem 2001; 20:1762–1771.
25. OV Makarova, T Rajh, MC Thurnauer, A Martin, PA Kemme, D Cropek. Env. Sci. Tech 2000; 34:4797–4803.
26. NA Kotov, FC Meldrum, JH Fendler, E Tombacz, I Dekany. Langmuir 1994; 10: 3797–3804.
27. B van Duffel, RA Schoonheydt, CPM Grim, FC De Schryver. Langmuir 1999; 15: 7520–7529.
28. K Inukai, Y Hotta, S Tomura, M Takahashi, A Yamagishi. Langmuir 2000; 16: 7679–7684.
29. Y Umemura, A Yamagishi, R Schoonheydt, A Persoons, F De Schryver. Thin Solid Films 2001; 388:5–8.
30. Y Umemura, A Yamagishi, R Schoonheydt, A Persoons, F De Schryver. Langmuir 2001; 17:449–455.
31. S Takahashi, M Taniguchi, K Omote, N Wakabayashi, R Tanaka, A Yamagishi. Chem. Phys. Lett 2002; 352:213–219.
32. Y Umemura, A Yamagishi, R Schoonheydt, A Persoons, F De Schryver. J. Am. Chem. Soc 2002; 124:992–997.
33. K Inukai, Y Hotta, M Taniguchi, S Tomura, A Yamagishi. J. Chem. Soc. Chem. Comm 1994:959.
34. M Ogawa, K Kuroda. Chem. Rev 1995; 95:399–438.
35. JM Phelan, JL Barnett. J. Chem. Eng. Data 2001; 46:375–376.
36. CT Johnston, E Tombacz. In: J. B. Dixon, D. G. Schulze, eds. Soil Mineralogy with Environmental Applications. Madison, WI: Soil Science Society of America, 2002: 37.
37. CT Johnston. In: B. Sawhney, ed. Organic Pollutants in the Environment. Boulder, CO: Clay Minerals Society, 1996:1.
38. A Yamagishi. In: K. Tamaru, ed. Dynamic Processes on Solid Surfaces. New York: Plenum Press, 1993:307.
39. SW Bailey. Micas. Reviews in Mineralogy Vol. 13., Mineralogical Society of America,. 1984.
40. SW Bailey. Hydrous Phyllosilicates (Exclusive of Micas). Reviews in Mineralogy Vol. 19. Washington, DC: Mineralogical Society of America, 1988.
41. DL Bish, GD Guthrie. Mineralogy of clay and zeolite dusts (exclusive of 1:1 layer silicates). In: Health Effects of Mineral Dusts, GD Guthrie, BT Mossman, eds. Washington, DC: Minerlogical Society of America, 1994:139–184.

42. GW Brindley, G Brown. Crystal Structures of Clay Minerals and Their X-Ray Identification. London: Mineralogical Society, 1980.
43. LJ Michot, F Villieras, M Francois, J Yvon, R LeDred, JM Cases. Langmuir 1994; 10:3765–3773.
44. H Malandrini, F Clauss, S Partyka, JM Douillard. J. Colloid Interface Sci 1997; 194: 183–193.
45. G Sposito. The Surface Chemistry of Soils. New York: Oxford University Press, 1984.
46. KHL Nulens, H Toufar, GOA Janssens, RA Schoonheydt, CT Johnston. The latest frontiers of clay chemistry: proceedings of the Sapporo conference on the chemistry of clays and clay minerals (Sapporo, Japan, 1996). A Yamagishi, A. Aramata, M. Taniguchi, eds. Sendai: The smectite forum of Japan, 1998:116.
47. DA Laird, PD Fleming. Env. Toxicol. Chem 1999; 18: 1668–1672.
48. DA Laird, E Barriuso, RH Dowdy, WC Koskinen. Soil Sci. Soc. Am. J 1992; 56: 62–67.
49. WF Jaynes, SA Boyd. Clay. Clay Miner 1991; 39:428–436.
50. MM Mortland, TJ Pinnavaia. Nature Physical Sci 1971; 229:75–77.
51. TJ Pinnavaia, MM Mortland. J. Phys. Chem 1971; 75(26):3957–3962.
52. CT Johnston, T Tipton, DA Stone, C Erickson, SL Trabue. Langmuir 1991; 7: 289–296.
53. T Tipton, CT Johnston, SL Trabue, C Erickson, DA Stone. Rev. Sci. Inst 1993; 64: 1091–1092.
54. WF Bleam. Clay. Clay Miner 1990; 38:527–536.
55. WF Bleam, R Hoffmann. Inorg. Chem 1988; 27:3180–3186.
56. C Poinsignon, JM Cases, JJ Fripiat. J. Phys. Chem 1978; 82:1855–1860.
57. G Sposito, R Prost. Chem. Rev 1982; 82:553–573.
58. CT Johnston, G Sposito, C Erickson. Clay. Clay Miner 1992; 40:722–730.
59. W Xu, CT Johnston, P Parker, SF Agnew. Clay. Clay Miner 2000; 48:120–131.
60. J Helsen. J. Chem. Edu 1982; 59:1063–1065.
61. MM Mortland, KV Raman. Clay. Clay Miner 1968; 16:393–398.
62. GW Brindley, G Ertem. Clay. Clay Miner 1971; 19:399–404.
63. WF Jaynes, SJ Traina, JM Bigham, CT Johnston. Clay. Clay Miner 1992; 40: 397–405.
64. SA Boyd, WF Jaynes. Layer charge characteristics of 2:1 silicate clay minerals. AR Mermut, ed. Boulder, CO: Clay Minerals Society, 1994:48.
65. DC MacEwan, MJ Wilson. In: GW Brindley, G Brown, eds. Crystal Structures of Clay Minerals and Their X-Ray Identification. London: Mineralogical Society, 1980: 197.
66. JH Lee, S Guggenheim. Am. Mineralogist 1981; 66:350–357.
67. A Pelmenschikov, J Leszczynski. J. Phys. Chem. B 1999; 103:6886–6890.
68. L Gorb, JD Gu, D Leszczynska, J Leszczynski. Phys. Chem. Chem. Phys. A 2000; 2:5007–5012.
69. JM Serratosa. Clay. Clay Miner 1967; 16:93–97.
70. CT Johnston, GS Premachandra. Langmuir 2001; 17:3712–3718.
71. L Margulies, H Rozen, A Banin. Clay. Clay Miner 1988; 36:476–479.

72. S Yariv, JD Russell, VC Farmer. Isr. J. Chem 1966; 4:201–213.
73. S Saltzman, S Yariv. Soil Sci. Soc. Am. J 1975; 39:474–479.
74. I Ahmad, TJ Dines, CH Rochester, JA Anderson. J. Chem. Soc. Faraday Trans 1996; 92:3225–3231.
75. RA Nyquist, SE Settineri. Appl. Spec 1990; 44:1552–1557.
76. CP Conduit. J. Chem. Soc 1959:3273–3277.
77. F Borek. Naturwissenschaften 1963:471–472.
78. NBH Jonathan. J. Mol. Spectrosc 1960; 4:75–83.
79. W Baitinger, PvR Schleyer, TSSR Murty, L Robinson. Tetrahedron 1964; 20: 1635–1647.
80. JHS Green, HA Lauwers. Spectrochim. Acta 1971; 27A:817–824.
81. T Urbanski, U Dabrowska. Bulletin De L'Academie Polonaise des Sciences 1959; 7:235–237.
82. CT Johnston, G Sheng, BJ Teppen, SA Boyd, MF De Oliveira. Env. Sci. Tech 2002; 36:5067–5074.
83. GN Andreev, B Schrader, H Takahashi, D Bougeard, IN Juchnovski. Can. J. Spec 1984; 29:145–147.
84. A Pelmenschikov, J Leszczynski. J. Phys. Chem. B 1999; 103:6886–6890.
85. RD Shannon. Acta. Cryst 1976; A32:751.
86. JE Huheey. Inorganic Chemistry. Principles of Structure and Reactivity. New York: Harper and Row, 1978.
87. HL Friedman, CV Krishnan. Water: A Comprehensive Treatise. Volume 3: Aqueous Solutions of Simple Electrolytes. F. Franks, ed. New York: Plenum Press, 1973:1.
88. KB Yatsimirksii, VP Vasil'ev. Instability Constants of Complex Compounds. Elmsford, NY: Pergamon Press, 1960.
89. RW Mooney, AG Keenan, LA Wood. J. Am. Chem. Soc 1952; 74:1371–1374.
90. KW Weissmahr, SB Haderlein, RP Schwarzenbach. Env. Sci. Tech 1997; 31: 240–247.

5

Photoprocesses in Clay–Organic Complexes

Makoto Ogawa
Waseda University
Tokyo, Japan

I. INTRODUCTION

Photophysical and photochemical reactions in heterogeneous media may differ significantly from analogous reactions in a homogeneous solution (1–4). The important role of reaction media in control rates, product distributions, and stereochemistry has long been recognized. Therefore, the study of the photoprocesses of organic and inorganic photoactive species in restricted geometry is a growing new field that yields a wide variety of useful applications in such areas as reaction media for controlled photochemical reactions and molecular devices for optics. For this purpose, nanomaterials with an ordered structure have an advantage, in that the properties of immobilized species can be discussed on the basis of their defined nanoscopic structures (5). Their structure–property relationships will provide indispensable information on designing materials with novel chemical, physical, and mechanical properties. In other words, one can control the attractive properties, such as photochromic and photocatalytic behavior, by organizing photoactive species into matrices with appropriate geometry and chemical environments.

In addition, spectroscopic properties of the adsorbed species, which are very sensitive to the environment, have given insight to the microscopic structures of the host–guest systems where conventional instrumental analysis does not have access (5–9). By utilizing photoprocesses of adsorbed photoactive species, one can obtain microscopic information such as the distribution and mobility of the guest species.

Among many ordered or constrained systems utilized to organize reactants, layered materials offer a two-dimensional expandable interlayer space for organizing guest species (10–12). The motivation to study intercalation reactions arises because the optical and electronic properties of both guest and host can be altered by the reactions (13,14). The microscopic structure can be tailored by selecting and designing both the guests and hosts and also by coadsorption. From X-ray diffraction studies, interlayer distances are measured and the orientation of the intercalated species are estimated. Moreover, some materials have been processed into single crystals or oriented films, in which case microscopic anisotropy can be converted into a macroscopic property. Such structural features make it possible to discuss structure–property relationships.

In this chapter, studies regarding the photofunctions of clay–organic intercalation compounds are summarized. Some of these studies are for the purpose of characterizing the properties of host materials and host–guest systems, and others are for the purpose of contributing to future practical applications. The well-defined layered structures as well as the ability to accommodate guest species on the surface of the layers are very useful for organizing photoactive species to evaluate and control the photofunctions. Attention will be focused mainly on the role of the layered structure on the organization of photoactive species and on the photofunctions of intercalation compounds in connection with microscopic structures.

II. SYNTHESES OF DYE–CLAY INTERCALATION COMPOUNDS

A. Host–Guest Systems

Smectite clay minerals, which is a group of 2:1 clays consisting of negatively charged silicate layers and readily exchangeable interlayer cations (see Chapter 1), has widely been used as the host materials (15–17). Isomorphous substitution of metal cations with similar size and lower valency generates a net negative charge for layers. To compensate for the negative charge, metal cations occupy the interlayer space. The amount as well as the site of the isomorphous substitution influences the surface and colloidal properties of smectites. Synthetic analogs of smectites, e.g., hectorite (Laponite, Laporte Ind. Co.) (18), saponite (Sumecton SA, Kunimine Ind. Co.) (19), and swelling mica (sodium-fluor-tetrasilicic mica, TSM, Topy Ind. Co.) (20–21), have attracted increasing interest, since natural clay minerals contain impurities. Amounts of impurities depend on the source of the clay minerals, and they occur both within the structure and on the surface.

The mechanism of the intercalation of guest species into clays has been described in previous chapters (see Chapters 1 and 2). The mechanism of the intercalation can be classified into two broad types (15). One involves cation exchange with interlayer exchangeable cations; the other concerns adsorption of polar molecules by ion–dipole interactions with interlayer cations and/or hydrogen bonding with the surface oxygen atom of the silicate sheets. When transition metal ions are substituted for interlayer Na ions, charge transfer between the interlayer transition metal ions to guest species can be a driving force for the intercalation (22–27). These reactions have been carried out by solid–liquid, solid–gas and solid–solid reactions (28–31). One of the characteristic features of smectites is the possible surface modification by means of ion exchange reactions. Nanoporous pillared smectites have been synthesized using inorganic particles and small organic cations as pillars (see Chapter 7) (32). Organophilic modification has been conducted by cation exchange with cationic surfactant (33–36), and organic dyes have been introduced into such modified clays (Fig. 1).

Layered alkali silicates are also capable of incorporating guest species in the interlayer space to form intercalation compounds (37). Among them, magadiite (ideal formula $Na_2Si_{14}O_{29} \cdot nH_2O$) has been used as a host for organoammonium ions (38) and polar molecule (39). The preparation of organosilane grafted (40–44) and pillared (45,46) derivatives has been reported previously. Compared with smectites, magadiite possesses unique properties for organizing guest species: (a) The density of the cation exchange sites on the layer surface is higher than that of smectite, and (b) it can be conveniently prepared by hydrothermal synthesis. However, the intercalation of guest species is not as easy as it is for smectites. Therefore, studies on the preparation of intercalation compounds from magadiite have been limited. In order to introduce bulky organic species, organoammonium-exchanged forms, which are prepared by conventional ion exchange reactions in aqueous media, are used as the intermediates.

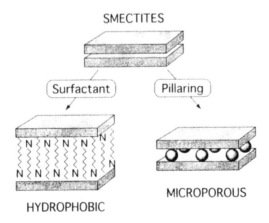

FIGURE 1 Schematic model of possible surface modifications of smectite clays.

Layered double hydroxides (LDHs) are synthetic minerals with positively charged brucite-type layers of mixed-metal hydroxides (see Chapter 9 for details) (47). Exchangeable anions located at the interlayer spaces compensate for the positive charge of the brucite-type layers. The chemical composition of the LDHs is generally expressed as $[M(II)_{1-x}M^1(III)_x(OH)_2][A^{n-}_{x/n}]^{x-}$, where $M(II)$ = Mg, Co, Ni, etc., $M(III)$ = Al, Cr, Fe, etc., and A is an interlayer exchangeable anion such as CO_3^{2-} or Cl^-. Anionic species have been introduced into the interlayer spaces of LDHs by three methods. Anion exchange reactions using an aqueous solution of guest species has been investigated widely. Compared with the cation exchange of smectites, the ion exchange reaction for LDHs is more difficult because of their high selectivity to carbonate anions and their large anion exchange capacity. Therefore, CO_2 should be excluded during the sample preparation. Intercalation compounds have also been prepared via direct synthesis in which an LDH phase precipitates in the presence of a guest species (48). The treatment of a mixed-metal oxide solid solution, obtained by the heat treatment of LDH-carbonate, with an aqueous solution of guest results in its reconstitution into an LDH intercalation compound (49).

In addition to the clay–organic systems, dye-intercalated layered solids including zirconium phosphates and transition metal oxides will be discussed (50–51). Due to the variation of layer charge density, particle morphology, and electronic properties, host–guest systems with unique microstructures and properties have been obtained.

B. Forms of Intercalation Compounds

For the evaluation of the photoprocesses, samples have been obtained as powders, suspensions, and thin films. One of the most unique and attractive properties of smectites is their swelling property (32). By dispersing in water, smectites form a stable thixotropic gel or suspension. When the suspension is evaporated on a flat plate, the clay particles pile up, with their *ab* plane parallel to the substrate, to form a film. Thin films of clay–dye intercalation compounds have been prepared by processing the presynthesized intercalation compounds into films or by the intercalation of dyes into clay films (52–62). Surfactant-intercalated smectites exhibit swelling behavior in organic solvents, while tetramethylammonium-smectites swell in water (11). By casting clay–dye suspensions in appropriate solvents, thin films can be made. Self-supporting films are also available (52–54). The preparation of thin films by the Langmuir–Blodgett technique from exfoliated platelets of surfactant-intercalated clays in organic solvent has also been reported (55,56). Inorganic–organic multilayered films have also been prepared via alternate adsorption of a cationic species and the anionic sheet of an exfoliated layered solid (57–62). These films make it possible to investigate various physicochemical properties, such as photochemical and barrier behavior.

III. PHOTOPHYSICAL PROCESSES OF ADSORBED DYES

A. Location of the Dyes

The change in the absorption spectra upon the adsorption on clays has been used to probe the clay surfaces (63). Metachromasy is a deviation from Beer's law, occurring in dye molecules that tend to adhere to each other. As a result, gradual replacement of the principal band in the visible region by a band with a shorter (or longer) wavelength occurs. From the change in the absorption spectra, adsorption behavior such as dye aggregation and the interactions of dye molecules with the surface has been investigated. Photoluminescence is also a powerful tool that provides information on the compositional, structural, and dynamic characteristics of the media (7,8). There are two commonly encountered kinds of luminescence: fluorescence and phosphorescence. Fluorescence is generally associated with the emission of light accompanying the transition of excited singlet states to the ground state, while phosphorescence is generally associated with the emission of light from the lowest triplet state to the ground state. Luminescence parameters are sensitive to changes in the microenvironment of the probe, so a luminescence probe in different microenvironments will display distinct luminescence properties. Emission spectra, luminescence decay, quantum efficiency of luminescence, polarization of luminescence (the orientation of the emitted light vector relative to the orientation of the absorbed light vector), and excitation spectra all have been employed to obtain information on microenvironments. Quenching, sensitization, and energy transfer, processes that have been observed by adding a second component or by changing the microenvironment, also provide information, including especially dynamics. The absorption and luminescence properties of clay–organic systems have been extensively investigated, and ideas on the location, aggregation (or isolation), and mobilities of the dyes on the clay surfaces have been put forth.

1. Luminescence of Ruthenium Poly(pyridine) Complexes

Tris(2,2'-bipyridine)ruthenium(II) (abbreviated as $[Ru(bpy)_3]^{2+}$ is the probe molecule used most extensively, due to, among other properties, its unique combination of chemical stability, long excited-state lifetime, and redox properties (64–66). The absorption spectrum of $[Ru(bpy)_3]^{2+}$ in aqueous solution has a metal–ligand (d-π) charge transfer (MLCT) band around 460 nm and a π–π^* transition for the ligands around 300 nm. The luminescence of $[Ru(bpy)_3]^{2+}$ on ion exchange resins (67), micelles (68), cellulose (69), porous Vycor glass (70), silica gels (71–73), zeolites (74), and layered materials (10,11) has been investigated, and valuable information on microscopic structures has been obtained. Studies on the photoprocesses of $[Ru(bpy)_3]^{2+}$ on layered systems with well-defined separation between an excited probe molecule and a quencher can provide insights into the distance dependence of electron transfer.

The cation exchange ability of smectites is isutilized to incorporate $[Ru(bpy)_3]^{2+}$ cations. The luminescence probe studies have been carried out mainly in colloidal clay systems in which an aqueous solution of $[Ru(bpy)_3]^{2+}$ is mixed with an aqueous suspension of smectites. In some cases, a species that quenches the excited state of $[Ru(bpy)_3]^{2+}$ is added (80,82). The concentrations of hosts, $[Ru(bpy)_3]^{2+}$, and quenchers in the suspensions have been varied in order to see the dynamics of the reactions. The photoprocesses of the $[Ru(bpy)_3]^{2+}$-smectite systems have revealed that the steady-state and time-resolved luminescence properties of $[Ru(bpy)_3]^{2+}$ are not simple, and various interpretations of the unusual properties have been proposed (80,83,91).

X-ray diffraction studies of metal tris(2,2'-bipyridine) complex–smectite intercalation compounds, prepared by conventional ion exchange, show basal spacings of ca. 1.8 nm when the complexes substitute the interlayer exchangeable cation quantitatively (75). Considering the size and shape of the complexes, the intercalated tris(2,2'-bipyridine) complexes are arranged as monomolecular coverage of the silicate sheets, with their threefold axis perpendicular to the silicate sheets. "Intersalation" is also possible (75,76).

In the absorption spectra of $[Ru(bpy)_3]^{2+}$ adsorbed on smectites, the MLCT band occurs near 460 nm, which is red-shifted from aqueous $[Ru(bpy)_3]^{2+}$ solution (77). From the observed spectral shifts of both MLCT and $\pi-\pi^*$ bipyridine bands, as well as Raman and XPS studies, it has been proposed that covalently hydrated or slightly distorted bipyridine ligands due to steric constraints are formed when $[Ru(bpy)_3]^{2+}$ is adsorbed on montmorillonite. The formation of partially oxidized $[Ru(bpy)_3]^{2+}$ has also been reported (78). Judging from the surface charge density of the silicate sheets and the size of $[Ru(bpy)_3]^{2+}$, these ions were arranged in close contact with each other in the interlayer space when cation exchange was quantitative. This close proximity may cause the distortion of bipyridine ligands. The spectral shifts in the absorption and luminescence spectra have been observed even when the loaded amounts of $[Ru(bpy)_3]^{2+}$ are far below the cation exchange capacity of smectites.

It has been reported that the montmorillonite interlayer segregates $[Ru(bpy)_3]^{2+}$ from exchangeable Na^+ ions, resulting in high local concentrations of the complex ions in the interlayer space even when the added concentration of $[Ru(bpy)_3]^{2+}$ is only 1–2% of the cation exchange capacity (77). (See Fig. 2.) Very efficient self-quenching due to the high local concentration has been observed. When $[Zn(bpy)_3]^{2+}$ was coadsorbed with $[Ru(bpy)_3]^{2+}$ on hectorite, the effective self-quenching rate was dramatically reduced, presumably due to dilution of $[Ru(bpy)_3]^{2+}$ in the interlayer. One important observation in segregation behavior is that MV^{2+} does not quench the excited state of $[Ru(bpy)_3]^{2+}$ even though such quenching is readily observed for the neutral viologen, propyl viologen sulfonate. The results were attributed to ion segregation; i.e., MV^{2+} and $[Ru(bpy)_3]^{2+}$ are located in different regions of the clay interlayers. The

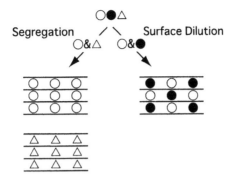

FIGURE 2 Schematic drawing for segregation and surface dilution.

origin of the ion segregation process is unclear. Nonuniform charge distribution among different layers or weak interactions between $[Ru(bpy)_3]^{2+}$ may be involved.

Steady-state and time-resolved studies of the excited properties of the $[Ru(bpy)_3]^{2+}$ adsorbed on a variety of clay minerals have been carried out by some research groups in order to understand the adsorbed states of $[Ru(bpy)_3]^{2+}$ more clearly (77–87). Habti et al. have reported nonexponential decay of the excited state of $[Ru(bpy)_3]^{2+}$ adsorbed on a variety of clay minerals with different iron contents (78). From the effects of iron content on the decay profiles, they point out the quenching effect of the irons within the lattice of the minerals and the essentially immobile character of adsorbed $[Ru(bpy)_3]^{2+}$ on the microsecond time scale. Each adsorbed probe ion is able to interact with a very limited number of neighboring quencher ions around the adsorption sites. The total quenching probability for a particular probe is determined by the quencher concentration in the solid and by the number of solid particles in contact with the probe. They have also mentioned that the degree of swelling affects the quenching.

DellaGuardia and Thomas reported similar decay profiles for $[Ru(bpy)_3]^{2+}$ bound to montmorillonite and attributed the components to two species, one at the outer surface and edges and one between the silicate sheets (intercalated) (79). Schoonheydt et al. also pointed out that $[Ru(bpy)_3]^{2+}$ is adsorbed on the external (edge sites) and interlamellar surfaces (planar sites), even at low loadings of $[Ru(bpy)_3]^{2+}$ (80). This was first proposed by Turro et al. (81). The occupancy of edge sites with respect to planar sites increases with decreasing particle size of the clay minerals. Nakamura and Thomas also reported two different adsorption sites for $[Ru(bpy)_3]^{2+}$ on laponite (82). At low concentrations of laponite, $[Ru(bpy)_3]^{2+}$ is adsorbed on outer layers and is in contact with the aqueous phase.

$[Ru(bpy)_3]^{2+}$ is incorporated in the interlayer space at high concentrations of laponite.

Turro et al. reported that the addition of $MgCl_2$ and tetrabutylammonium chloride leads to an increase in the internal surface area due to flocculation and formation of clay aggregates and leads to an increase in the contribution of the interlayer species (81). From time-resolved luminescence and excited-state resonance Raman studies, they concluded that the effects of laponite on the excited state of $[Ru(bpy)_3]^{2+}$ are due primarily to surface–probe interactions.

The time-resolved luminescence of $[Ru(bpy)_3]^{2+}$ has been deconvoluted into two components, and these have two different emission maxima (83). This observation suggests the existence of two distinct adsorption sites for $[Ru(bpy)_3]^{2+}$ on laponite. At one type of the adsorption zone the water is held rigidly and presents an environment for the $[Ru(bpy)_3]^{2+}$ probe that results in a short lifetime and a blue-shifted emission. The second absorption zone involves a stronger interaction directly with the clay surface such that the photophysics are less influenced by the nature of the surrounding water. This explanation was supported by a direct comparison of photophysical properties of the luminescence probe on the clay and in ice at $-20°C$.

Cationic (e.g., cupric ion) and ionically neutral (e.g., dimethylaniline and nitrobenzene) quenchers have been added in the colloidal clay system to probe the dynamic behavior of adsorbed species on clay surfaces (82). Quenchers as exchangeable cations behaved similar to those in solution. The quenching rate constant for O_2 on the clay was smaller by about two orders of magnitude than for those in solution. In addition, the effect of M^{3+} ions was stronger when they were present as substitution ions in the lattice than when they were exchangeable ions. This is due to the difference in distance between the species and in ion mobility. Copper(II) was adsorbed strongly, which simplified the kinetics. Stern–Volmer type of kinetics were observed, and a quenching rate constant was obtained that was lower than that in aqueous solution; this provides an estimate of the degree of movement of cupric ions on the clay surfaces. Dimethylaniline and nitrobenzene are weakly adsorbed on a clay around the $[Ru(bpy)_3]^{2+}$, in a zonelike effect, rather than being adsorbed randomly throughout the system.

From these studies, the following general conclusions for the adsorption of $[Ru(bpy)_3]^{2+}$ on smectites have been derived.

1. There are more than two adsorption sites with different chemical environments on smectites; one is exposed externally (outer surface) and the other is confined in the sheets (intercalated).
2. There are strong interactions between the surface of the silicate sheets and $[Ru(bpy)_3]^{2+}$ as well as between adjacent $[Ru(bpy)_3]^{2+}$ complexes.
3. Naturally occurring clay minerals contain impurities, such as Fe^{3+},

that quench the excited state of adsorbed species through an energy transfer mechanism.

Yamagishi and Soma have shown that optical isomers of ruthenium and iron polypyridine and phen (phen = 1,10-phenanthroline) chelate complex ions have different adsorption behaviors on montmorillonite (87,88). When a racemic mixture of ferric tris phen complex ions ($[Fe(phen)_3]^{2+}$) is adsorbed by a clay, the ions are adsorbed as a racemic pair unit rather than in a random distribution of optical isomers. The adsorption of a racemic mixture occurs in two successive steps: The initial adsorption occurs within the CEC limit, and the succeeding adsorption exceeds the CEC limit. When a tris(phen) metal chelate is placed on a silicate sheet of montmorillonite with its threefold symmetry axis perpendicular to the silicate surface, the three bottom coordinated phenanthroline ligands make an equilateral triangle, with each side approximately 6.5 Å. Since this distance is close to the distance between the centers of neighboring SiO_4 hexagonal holes (5.5 Å) on a silicate sheet, it is possible for the bound chelate to have the three ligand hydrogen atoms buried into the silicate surface. As a result, it is suggested that the chelate is fixed on the surface with a definite orientation. From molecular model considerations, it has been concluded that racemic adsorption by metal chelates on a solid surface can be achieved under the following conditions.

1. A bound chelate should be adsorbed to such a density that it interacts with its neighbors stereochemically. In the case of cation exchange by a clay, a molecule carrying two positive charges is required to have a molecular radius larger than 5 Å. Otherwise, it does not interact sterically with its neighbors.
2. A surface should be two dimensional and capable of fixing a bound chelate in a definite orientation. Such a strict stereochemical restriction precludes stacking of a chelate into molecular aggregates.

Joshi and Ghosh utilized luminescence probes for the study of racemic adsorption in clay-$[Ru(phen)_3]^{2+}$ systems and concluded that layered clays promote recognition between optical antipodes of cationic poly(pyridine) metal complexes (89,90). The absorption and emission spectra and time-resolved luminescence decay profiles of enantiomeric and racemic $[Ru(phen)_3]^{2+}$ adsorbed on hectorite at ca. 3.5% loading indicate differences in the binding modes of the two forms. The quenching effect of coadsorbed $[Co(bpy)_3]^{2+/3+}$, $[Ni(bpy)_3]^{2+}$, and $[Rh(bpy)_3]^{3+}$ were also discussed. From these data they concluded that such chiral interactions, which are partly responsible for the spectral shifts of the racemic complexes upon adsorption, occur spontaneously on clay and are not a steric constraint imposed on the system at high packing density. Spontaneous chiral interactions, which presumably occur via π overlap of partially oriented chelates, also account for many of the intriguing absorption and emission spectral results of $[Ru(phen)_3]^{2+}$ in the presence of $[M(bpy)_3]^{n+}$ coadsorbates. Such interactions have been thought to be the driving force for their optical resolution.

The isolation of $[Ru(bpy)_3]^{2+}$ in interlayer spaces is a topic of interest for excited-state $[Ru(bpy)_3]^{2+}$ applications. The isolation of $[Ru(bpy)_3]^{2+}$ has been achieved by the cointercalation of photoinactive species, such as water-soluble polymers (91,92). Ogawa et al. have prepared the ruthenium-tris(bipyridine)-fluor-tetrasilicic mica(TSM)-poly(vinyl pyrrolidone)(PVP) intercalation compounds with variable concentration of $[Ru(bpy)_3]^{2+}$ by cation exchange of sodium ions in a preformed TSM-PVP intercalation compound with $[Ru(bpy)_3]^{2+}$ (91,92). PVP was added in order to avoid aggregation of $[Ru(bpy)_3]^{2+}$, and the compounds show unique photoluminescence properties. The luminescence maxima of intercalated $[Ru(bpy)_3]^{2+}$ shifted gradually toward blue with decreased $[Ru(bpy)_3]^{2+}$ loading. This reflects the change in the polarity and/or rigidity of the microenvironment of $[Ru(bpy)_3]^{2+}$ that is caused by cointercalated polar PVP. Moreover, $[Ru(bpy)_3]^{2+}$ was isolated effectively to suppress self-quenching, even at its highest concentration loading. It was suggested that cointercalated PVP was forced to surround $[Ru(bpy)_3]^{2+}$ in close contact in the sterically limited interlayer spaces, as schematically shown in Figure 3. X-ray diffraction studies supported the conclusion. When $[Ru(bpy)_3]^{2+}$ was intercalated in TSM without PVP, the diffraction peaks split into two, suggesting segregation.

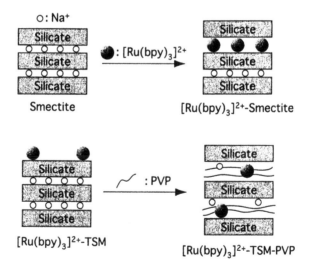

FIGURE 3   Schematic structure of (top left) Na-TSM, (top right) TSM-PVP intercalation compound, (bottom left) $[Ru(bpy)_3]^{2+}$-TSM intercalation compound, and (bottom right) $[Ru(bpy)_3]^{2+}$-TSM-PVP intercalation compound. (From Ref. 91. Copyright 1993 American Chemical Society.)

Luminescence due to the MLCT transition of $[Ru(bpy)_3]^{2+}$ was observed after intercalation of PVP, and the luminescence maximum shifted toward shorter wavelength. Similar luminescence blue shifts have been observed for $[Ru(bpy)_3]^{2+}$-TSM-PVP intercalation compounds with different $[Ru(bpy)_3]^{2+}$ loadings. Blue shifts of MLCT luminescence have been observed in various $[Ru(bpy)_3]^{2+}$-heterogeneous systems, including colloidal silica (Wheeler and Thomas) (73), and silicate and aluminosilicate gels from sol-to-gel conversion through gel aging and drying (71). It was speculated that an increased possibility for relaxation in a more fluid state caused the spectral red shifts. In the present $[Ru(bpy)_3]^{2+}$-TSM-PVP intercalation compounds, the cointercalated PVP is thought to surround $[Ru(bpy)_3]^{2+}$ to cause the spectral blue shifts.

Another observation is a luminescence shift during hydration/dehydration. When $[Ru(bpy)_3]^{2+}$(1 mmol)-TSM-PVP was stored at relatively high humidity, the basal spacings increased to 2.58 nm from 2.32 nm (92). This shows the intercalation of water molecules into the interlayer space of TSM to alter the interlayer microenvironment from $[Ru(bpy)_3]^{2+}$-PVP to $[Ru(bpy)_3]^{2+}$-PVP-water. Since PVP is a hydrophilic polymer, water molecules can easily be intercalated. The MLCT luminescence band shifted toward longer-wavelength regions upon hydration from 588 nm to 598 nm. The amount of adsorbed water was determined by thermogravimetry (TG) to be 2 and 10 mass % for the dehydrated and the hydrated samples, respectively. The relationship between basal spacings and MLCT luminescence maxima are shown in Figure 4.

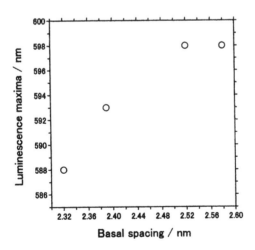

FIGURE 4 Relationship between basal spacings and MLCT luminescence maxima of adsorbed $[Ru(bpy)_3]^{2+}$ on smectite clay.

The variation of luminescence maxima was plotted against the calculated average distance between adsorbed [Ru(bpy)$_3$]$^{2+}$ molecules in the interlayer space in Figure 5 (92). The average distance between adsorbed [Ru(bpy)$_3$]$^{2+}$ molecules (see Fig. 6) was determined by using the gallery height, the ideal surface area (estimated at 700 m^2/g clay (32), which is calculated based on the surface area of each cell), and the composition. The luminescence maxima shifts toward shorter wavelengths with decreasing [Ru(bpy)$_3$]$^{2+}$ concentration when the [Ru(bpy)$_3$]$^{2+}$-[Ru(bpy)$_3$]$^{2+}$ distance is shorter than 6 nm. On the other hand, the luminescence maxima does not change when the average [Ru(bpy)$_3$]$^{2+}$-[Ru(bpy)$_3$]$^{2+}$ distance is longer than 6 nm. These observations indicate that the intermolecular interactions between adjacent [Ru(bpy)$_3$]$^{2+}$ are a possible factor for the observed luminescence shifts, in addition to host–guest interactions.

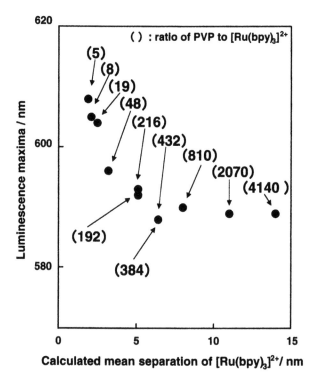

FIGURE 5 Luminescence maxima as a function of the average distance of the intercalated [Ru(bpy)$_3$]$^{2+}$ in the interlayer space of TSM. (From Ref 92. Copyright 2000 American Chemical Society.)

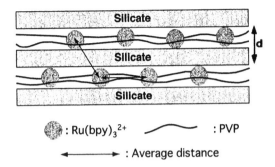

FIGURE 6 Schematic drawing for the $[Ru(bpy)_3]^{2+}$-TSM-PVP intercalation compounds.

The concentration-quenching radius was reported to be 1.3 nm for $[Ru(bpy)_3]^{2+}$ synthesized within zeolite Y cages (93). Here the adhorbed sites of $[Ru(bpy)_3]^{2+}$ are determined by the crystalline structures of host zeolites. In amorphous materials, the distribution can be controlled by changing the concentration of dispersed $[Ru(bpy)_3]^{2+}$. Nagai et al. reported the distance-dependent concentration quenching of $[Ru(bpy)_3]^{2+}$ dispersed in a polydimethylsiloxane film (94). Considering the excluded volume of $[Ru(bpy)_3]^{2+}$, the quenching radius was determined to be 1.1 nm. Luminescence shifts as a function of $[Ru(bpy)_3]^{2+}$ concentration in $[Ru(bpy)_3]^{2+}$-TSM-PVP are thought to reflect a larger separation. Although a more detailed explanation is difficult at present, the luminescence shifts do reflect the unique nature of the two-dimensional nanospace for guest organization. This idea was supported by the luminescence lifetimes (92). The lifetimes for $[Ru(bpy)_3]^{2+}$-TSM-PVP are longer than those for the $[Ru(bpy)_3]^{2+}$-TSM system, confirming that the coadsorbed PVP suppresses the self-quenching of $[Ru(bpy)_3]^{2+}$.

The saponite-dodecyldimethylamine N-oxide (abbreviated as C12AO) intercalation compound is a novel surfactant-modified clay (54). Since C12AO is nonionic, the intercalation behavior is expected to be different from those of cationic surfactants. Moreover, these intercalation compounds have been obtained as thin films with good optical quality. The coadsorption of C12AO with $[Ru(bpy)_3]^{2+}$ on saponite has been investigated (54). Upon coadsorption of C12AO, the luminescence maximum shifts to shorter wavelengths, as shown in Figure 7. Since the luminescence lifetime (ca. 1 μs) of $[Ru(bpy)_3]^{2+}$ did not change, self-quenching due to $Ru(bpy)_3]^{2+}$ aggregation is negligible. Accordingly, the luminescence shift was ascribed to the change in the environment of molecularly dispersed $Ru(bpy)_3]^{2+}$ on the saponite surface. The ordered microstructures of intercalated C12AO may provide a rigid environment for

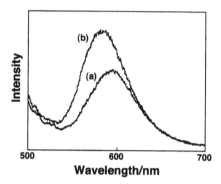

FIGURE 7   Luminescence spectra of $[Ru(bpy)_3]^{2+}$-Na-saponite intercalation compounds (a) before and (b) after intercalation of C12AO surfactant. The adsorbed amounts of $[Ru(bpy)_3]^{2+}$ and C12AO are 1 and 400 mmol/100 g clay, respectively. (From Ref 54. Copyright 1998 American Chemical Society.)

$[Ru(bpy)_3]^{2+}$ in the interlayer space. The polarity of the hydrophilic head group of C12AO may also affect the excited state of $Ru(bpy)_3]^{2+}$

Intercalation of ruthenium poly(pyridine) complexes into other layered materials has been reported (95–106). The intercalation is not as facile as in smectite systems, which is partly due to the higher charge densities of these host materials. Consequently, quantitative ion exchange of ruthenium complexes with the interlayer cations is difficult. Synthetic efforts have been made to introduce ruthenium polypyridine chelate complexes into magadiite, zirconium phosphate and phosphonates, LDH, $MnPS_3$, and a transition metal oxide and to control the adsorption states.

Intercalation of $[Ru(bpy)_3]^{2+}$ into the layered silicate magadiite (ideal formula $Na_2Si_{14}O_{29} \cdot nH_2O$) was investigated (95). Since the complex cation did not intercalate by a direct ion exchange reaction with the interlayer sodium ions in an aqueous medium, a crown ether, 15-crown-5, was added to promote the reaction. Two possible reaction mechanisms were proposed, as schematically shown in Figure 8; one is a two-step reaction where 15-crown-5 first intercalates and then exchanges with $[Ru(bpy)_3]^{2+}$. The other is a one-step reaction where 15-crown-5 forms complex ions with sodium ions deintercalated from the interlayer space simultaneously and $[Ru(bpy)_3]^{2+}$ is intercalated into the interlayer space to compensate the charge balance. Alkylammonium–magadiite has also been used as an intermediate for the introduction of $[Ru(bpy)_3]^{2+}$ (96).

Three types of $[Ru(bpy)_3]^{2+}$-zirconium phosphates have been prepared by three different methods (98). By ion exchange or impregnation of $[Ru(bpy)_3]^{2+}$, the complex ions are adsorbed only on the external surface. When a mixture of

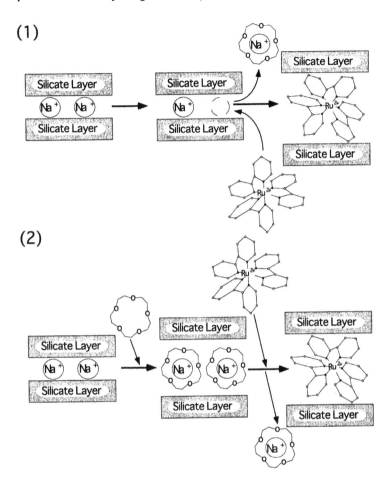

FIGURE 8 Schematic drawing of possible cation exchange reactions in magadiite in the presence of 15-crown-5: (1) one-step process and (2) two-step process, as described in the text.

zirconium oxy chloride($ZrOCl_2 \cdot 8H_2O$), H_3PO_4 and $[Ru(bpy)_3]^{2+}$ Cl_2 is refluxed, a compound that emits at 590 nm, thought to be $Ru(bpy)_2(bpyH^+)^{3+}$, is formed. Vliers et al. have also prepared stable aqueous suspensions of zirconium phosphate and hexylammonium-zirconium phosphate containing $[Ru(bpy)_3]^{2+}$ (99). In the case of zirconium phosphate, there is no interlamellar adsorption of Ru(II) and the cations are located on the external surface. On the other hand, in the case of hexylammonium–zirconium phosphate, $[Ru(bpy)_3]^{2+}$ was adsorbed in both the interlamellar space and the external surface. By quenching with $Fe(CN)_6^{3-}$, it has reportedly been possible to distinguish between $[Ru(bpy)_3]^{2+}$ on the external surface and in the interlamellar space.

The photophysics of $[Ru(bpy)_3]^{2+}$ has been used to investigate the chemical microenvironment within a sulfonated derivative of layered zirconium phosphate sulfophenylphosphonates (abbreviated as ZrPS) (100,101). Similar spectral shifts have been observed for $[Ru(bpy)_3]^{2+}$ in other chemical microenvironments (e.g., micelles, clays), in theory the result of interactions between phenyl rings of the host and bpy rings of neighboring complexes. Movement of $[Ru(bpy)_3]^{2+}$ in the interlayer space of ZrPS is restrained to some extent, but diffusion leading to dynamic quenching reaction can occur. A model combining diffusional quenching and sphere of action quenching accounts for the quenching of $[Ru(bpy)_3]^{2+}$ by MV^{2+} in ZrPS (101).

The photoprocesses of ruthenium tris(4,7-diphenyl-1,10-phenanthrolinedisulfonate) $(Ru(BPS)_3^{4-})$ intercalated in a layered double hydroxide (LDH) has been studied (102). The compound was prepared by coprecipitation of Mg/Al LDH from an aqueous NaOH solution containing the metal chloride salts and $Ru(BPS)_3^{4-}$ with hydrothermal treatment. Multiexponential decay profiles were observed. From the effects of "surface dilution" by $Zn(BPS)_3^{4-}$ on luminescence, the decay profile has been attributed to self-quenching processes similar to those for $[Ru(bpy)_3]^{2+}$-smectite systems.

Intercalation of $[Ru(bpy)_3]^{2+}$ into $MnPS_3$ (103–105) and niobate $(K_4Nb_6O_{17}\cdot3H_2O)$ (106,107) has been reported as well. From spectroscopic studies of the $[Ru(bpy)_3]^{2+}$-MnPS3 system, it was concluded that intercalated $[Ru(bpy)_3]^{2+}$ is bound weakly. Nakato et al. have prepared two types of intercalation compounds of $K_4Nb_6O_{17}$ with $[Ru(bpy)_3]^{2+}$ Two types of alkylammonium-intercalated $K_4Nb_6O_{17}$ were used as intermediates. One compound involves $[Ru(bpy)_3]^{2+}$ in only interlayer I, and the other involves Ru(II) in both interlayers I and II (106). Since the electronic properties of the hosts (both niobate and $MnPS_3$) are quite different from those of smectites, one may observe unique photoprocesses. Possible surface dilution effects on the photophysics of $[Ru(bpy)_3]^{2+}$ have been reported for zirconium phosphate (108) and $K_4Nb_6O_{17}$(107) systems.

2. Luminescence of Pyrenes

Aromatic hydrocarbons such as pyrene have also been employed as a luminescence probe of polarity and microviscosity in a variety of organized assemblies (109). Pyrene is a good excimer-forming probe due to the long lifetime of fluorescence and formation of excited-state dimers (excimers) at low concentration. Figure 9 shows an example pyrene luminescence spectrum. The ratio of excimer to monomer fluorescence intensity is often utilized as a measure of pyrene mobility and proximity. The vibronic fine structure of the pyrene monomer is sensitive

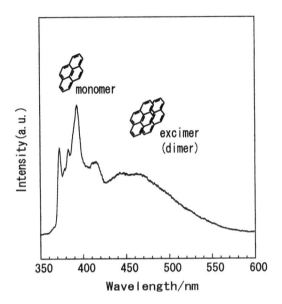

FIGURE 9   A typical pyrene luminescence spectrum.

to its surrounding polarity (110,111). Cationic pyrene derivatives (1-pyrenyl)tri-methylammonium (PN$^+$), [3-(1-pyrenyl)propyl]trimethylammonium (P3N$^+$), [4-(1-pyrenyl)butyl]trimethylammonium (P4N$^+$), and [8-(1-pyrenyl)octyl]tri-methylammonium (P8N$^+$) ions have been studied with respect to the anionic surface of silicate sheets (112–116).

DellaGuardia and Thomas studied the emission properties of P4N$^+$ ad-sorbed on colloidal montmorillonite and found that excimers formed even at low concentrations of P4N (112). This indicates a clustering of adsorbed P4N$^+$ on the surface. Coadsorption of C$_{16}$3C$_1$N$^+$ (the amount is 20% of CEC) was carried out by adding an aqueous solution of C$_{16}$3C$_1$N$^+$ to a colloidal montmorillonite suspension containing P4N$^+$ (the amount is 0.5% of CEC) (113). Upon addition of hexadecyltrimethylammonium ions, the P4N$^+$ can be dispersed over the silicate surface. This causes emission from the pyrene monomer to increase and the excimer emission intensity to decrease. Excited state P4N$^+$ exhibited a double-exponential decay, suggesting that two different environments exist, though their origin is unclear. The quenching study demonstrated that surfactants reduced the accessibility of quenchers into montmorillonite. When quenchers are adsorbed on montmorillonite, the quenching rate constant is reduced by at least one order of magnitude compared to a homogeneous solution.

The fluorescence quenching of P4N$^+$ adsorbed on colloidal laponite by coadsorbed C$_n$Py$^+$ ions showed unusual behavior (113). Increasing the quencher concentration led to an efficient quenching of P4N$^+$ fluorescence, but increasing the quencher concentration further produced a reverse effect, whereby the fluorescence started to recover, only to be followed by a smaller degree of quenching. The alkyl chain length of alkylpyridinium ions affects the degree of recovery. It has been suggested therefore that geometry, or arrangements of reactants, affects the quenching of excited adsorbed chromophores on colloidal clay.

Viane et al. studied the adsorption of P3N$^+$ bromide on hectorite, barasym, and laponite by an ion exchange mechanism (114,115). Even at very low concentrations, a 480-nm emission was observed due to ground-state interactions that result in efficient excimer formation. This indicates a clustering of adsorbed P3N$^+$ on the silicate surface. The 480-nm emission was suppressed by detergent molecules or calcium ions. In the former case, detergent molecules solubilized P3N$^+$ on the surface of laponite, as was observed in the P4N$^+$-montmorillonite system discussed earlier. On the other hand, mobility of P3N$^+$ was restricted in the latter case due to ordering of silicate particles in suspension. Nonhomogeneous distributions of P3N$^+$ ions and a minimization of the contact surface between hydrophobic pyrene derivatives and the surrounding water phase were proposed to explain the excimer formation. In order to see which was the major determining factor, suspensions in different solvents were studied in their subsequent work (115). In that study, they used three pyrene derivatives, PN$^+$, P3N$^+$, and P8N$^+$ bromides. In nonaqueous suspension, no excimer emission was observed. Since efficient intermolecular excimer formation in aqueous suspension had been ascribed to cluster formation, the absence of excimer emission indicated the absence of clusters on the clay surface when suspended in nonaqueous media. These observations suggested that the distribution of adsorbed ions is determined by the surrounding medium and not by the distribution of negative adsorption sites. Further, the bonding between positive probe and negative site on the clay surface is not strong enough to inhibit diffusion of the adsorbed ions.

Adsorption and binding behavior of 9-anthracenemethylammonium ions (abbreviated as AMAC) and P4N$^+$ on zirconium phosphate in the presence of butylamine hydrochloride has been studied (117,118). AMAC tends to aggregate while P4N$^+$ gives monomer emission at low surface coverage. Excimer formation of P4N$^+$ depends on phosphate concentration, additives, and the type of probe. However, the reason for differences in the binding behavior of AMAC and P4N$^+$ has not been clarified.

Pyrene luminescence also has been utilized to probe surfactant-modified smectites. Pyrene and anthracene themselves are poorly adsorbed on the hydrophilic surface of smectites, but they are readily adsorbed by organoammonium-intercalated clays. Intercalation of pyrene, anthracene, and azobenzenes into long-chain quaternary alkylammonium-smectites was successfully achieved by solid–solid reactions (119–121). This is accomplished by grinding both species together with a mortar and a pestle at room temperature. Hydrophobic interactions between the guest species and organoammonium ions are thought to be the driving force for intercalation. Upon solid-state reaction of octadecyltrimethylammonium($C_{18}3C_1N^+$)–montmorillonite and pyrene, a new $d(001)$ diffraction peak at ca. 3.7 nm appears, and the intensity of the $d(001)$ diffraction peak due to unreacted $C_{18}3C_1N^+$–montmorillonite decreases. Changes in the XRD of dioctadecyldimethylammonium($2C_{18}2C_1N^+$)–montmorillonite after reaction with pyrene is different. The basal spacings increased gradually from 3.0 to 3.8 nm as a function of the amount of anthracene. The arene molecules are solubilized at extremely high concentrations in alkylammonium–montmorillonites while retaining the ordered lamellar structure. In fluorescence spectra of these pyrene intercalated compounds, monomer fluorescence with vibrational structure was observed around 400 nm, together with a broad peak due to excimer emission at 500 nm. The ratio of monomer to excimer for the $2C_{18}2C_1N^+$–montmorillonite system is three times higher than that for the $C_{18}3C_1N^+$ system, suggesting that the adsorbed pyrene molecules are more isolated in the interlayer space of the $2C_{18}2C_1N^+$-montmorillonite than in $C_{18}3C_1N^+$-montmorillonite.

Similar results were observed when fluorotetrasilicic mica (TSM) was used as the host material (121). The basal spacings and concentration of adsorbed organoammonium cations in $2C_{18}2C_1N^+$- and $C_{18}3C_1N^+$-TSMs were similar to those of the corresponding organoammonium–montmorillonites, suggesting similar arrangements of intercalated quaternary ammonium ions in the interlayer spaces. The changes in the X-ray diffraction patterns upon reaction with anthracene or pyrene were also similar to those observed for the montmorillonite systems. Spectroscopic properties of the intercalated pyrene molecules showed a similar changes as well. The ratio of monomer to excimer intensity for $2C_{18}2C_1N^+$-TSM was twice that of $C_{18}3C_1N^+$-TSM, suggesting that the adsorbed pyrene molecules are more isolated in the interlayer space of $2C_{18}2C_1N^+$-TSM than $C_{18}3C_1N^+$-TSM.

In order to elucidate differences in adsorption states, a synthetic saponite with a CEC of 71 mEq./100 g clay was used. This is a lower layer charge density than montmorillonite. The $2C_{18}2C_1N^+$-saponite displayed a smaller basal spacing of 2.2 nm, with intercalated ions possibly arranged in a pseudo-trimolecular layer similar to that for $C_{18}3C_1N^+$-montmorillonite. When pyrene was intercalated, change in the fluorescence spectra as a function of concentration was similar to that for $C_{18}3C_1N^+$-montmorillonite. The fluorescence spectrum showed a tendency for pyrene molecules to form excimers (or dimers) similar to those for the $C_{18}3C_1N^+$-montmorillonite and TSM systems.

It was proposed that these spectral differences are due to different geometrical arrangements of the interlayer alkylammonium ions (121); see Figure 10. If the cations are arranged parallel to the silicate sheets, adsorbed arenes tend to

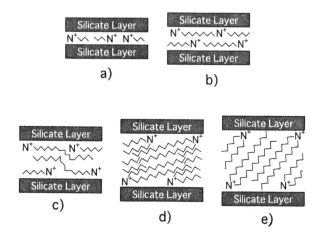

FIGURE 10 Possible geometrical arrangements of intercalated alkylammonium ions in smectite interlayers: (a) monolayer, (b) bilayer, and (c) pseudo-trimolecular layer; paraffin-type arrangements with (d) monolayers and (e) bilayers.

aggregate. Arene molecules intercalated into alkylammonium-smectites with a paraffin-type arrangement tend to solubilize the hydrocarbons molecularly. In other words, one can create various reaction environments by selecting hosts with various layer charge densities and guests of different size. This idea was supported by a luminescence probe study using anthracene carboxylic acid (122).

The phase transitions of dialkyldimethylammonium ($2C_n2C_1N^+$, where n denotes the carbon number in the alkyl chain)-silicates have been investigated via pyrene fluorescence temperature dependence and photochromism. The gel-to-liquid crystal-phase transition has been determined using 1,3-di(1-pyrenyl)pro-pane and pyrene (123) 1,3-Di(1-pyrenyl)propane can be a better probe than pyrene for microviscosity studies. Because it has two pyrene moieties per molecule, intramolecular excimers form at extremely low concentrations. Films of $2C_{18}2C_1N^+$- and $2C_{12}2C_1N^+$-bentonite containing pyrene or 1,3-di(1-pyrenyl)-propane have been prepared by casting their suspensions in chloroform on In-doped SnO_2 electrodes, and luminescence has been examined at a temperature range from 0 to 75°C (124). The relative intensity of the excimer peak increased gradually with increasing temperature, suggesting a lower viscosity at tempera-tures above the phase transition. Such discontinuities have been observed for $2C_{12}2C_1N^+$- and $2C_{18}2C_1N^+$-clay films. The phase transition temperature has been estimated as 7.4 and 53°C for the $2C_{12}2C_1N^+$- and $2C_{18}2C_1N^+$-clay films, respectively. These temperatures are in agreement with those measured by DSC. In the gel state, surfactant hydrocarbon chains are all in the trans conformation, causing rigidity. In the liquid crystalline state, the surfactant alkyl chains contain a fraction of C–C gauch conformation bonds to produce kinks, which produces more fluidity. The enhanced excimer peak of 1,3-di(1-pyrenyl)propane reflects the increase in the mobility of the alkyl chains. These results are consistent with increased diffusion rates of solutes in $2C_n2C_1N^+$-clay films above the phase transition temperature (124,125). Such phase transitions have been known to affect the permeability of a luminescent probe as well as catalytic reactivity (124,125). An important aspect of $2C_n2C_1N^+$-clay films is the change in the states of guest species after soaking in water. The probes are initially solubilized in a polar region close to the head groups of the surfactant. This region is thought to be relatively rigid, and this might explain the observed large microviscosities. Upon soaking in water, the probe microenvironment slowly becomes less polar and much less viscous.

Nakamura and Thomas have studied a stable aqueous suspension of the hexadecyltrimethyl–ammonium chloride ($C_{16}3C_1NCl$)/laponite system using pyrene as a probe (126). Their suspension contained 2 mmol of $C_{16}3C_1NCl$ and 1 g of laponite, which was able to dissolve 0.1 mmol pyrene. The kinetics of pyrene quenching and pyrene excimer formation reactions suggests that the $C_{16}3C_1NCl$ forms a cluster-like double layer on the clay surface. Since the amount of $C_{16}3C_1NCl$ used is twice the CEC of laponite, the effects of Cl^- on the photophysics of adsorbed pyrene cannot be excluded.

DellaGuardia and Thomas reported the incorporation of 1-dodecanol and pyrene in the interlamellar space of montmorillonite (interlayer spacing of 13 Å) (127). Upon suspending the powders in water, a fraction of 1-dodecanol and pyrene is released into the aqueous phase, forming dodecanol micelles into which released pyrene is solubilized. Weiderrecht et al. showed that pyrene was effectively adsorbed onto a pillared montmorillonite from solutions to form dimers (or excimers). On the other hand, similar polycyclic hydrocarbons, such as naphthalene and perylene, were not adsorbed into the gallery regions. The affinity of pyrene to form dimers is thought to be critical to explain this difference in adsorption behavior (128).

3. Other Dyes

In addition to ruthenium(II) polypridine complexes and pyrene, other cationic dyes have been examined using absorption and luminescence characteristics upon intercalation. Adsorption of cationic dyes such as methylene blue (129–136), pyronine Y, (137), rhodamines (138–145), and coumarines (146) on smectites in colloidal suspension has been documented. Adsorption behavior of the dye molecules has been investigated from changes in visible absorption spectra. The concentrations of dyes and hosts, the relative amounts of dye to hosts, and the nature of the hosts all affect the states of the adsorbed dyes.

Cenes and Schoonheydt studied the adsorption of methylene blue on various clay minerals and concluded that dye aggregation was responsible for the metachromatic effects (132). The distribution of the dye ions over the surface was determined not only by the electrostatic interaction between the negatively charged exchange sites and methylene blue, but also by dye–dye interactions. The effects of hosts on the adsorption states of pyronin Y were observed by Grauer et al (137). In montmorillonite, adsorption led to metachromasy of the dye and the appearance of a new band at a shorter wavelength than the original band (480 and 545 nm, respectively), even at very small coverages. In laponite, on the other hand, no metachromasy was observed with small amounts of dye. In montmorillonite, the organic cation was oriented with the plane of the rings parallel to the silicate layer and π interaction between the oxygen plane of the silicate layer and the aromatic dye gave rise to the metachromasy. In laponite, the plane of the aromatic ring was tilted relative to the silicate layer, so π interactions between the oxygen plane and the aromatic dye did not occur. They ascribed the different behavior to the difference in the location of the isomorphous substitution. The negative charge of laponite results from octahedral substitution, whereas tetrahedral substitution occurs in montmorillonite. Such a substitution is thought to lead to increasing basic strength of the oxygen plane in montmorillonite compared to laponite.

Organic laser dyes have found an increasing variety of applications in spectroscopy, optics, and lasers. One of the key problems in their investigation and application is their fixation into matrices, because the spectral characteristics are largely affected by the nature of matrices. Endo et al. have studied the fluorescence properties of rhodamine 590 and 640 adsorbed on montmorillonite and rhodamine 590, pyronine Y, and coumarine dyes on saponite (139,146,147). These dyes are known to show high fluorescence quantum efficiency and are used as tunable laser dyes. The fluorescent properties of the resulting intercalation compounds have been discussed on the basis of the dye arrangements in the interlayer space. Surface modification by surfactant (148) and pillaring (149) with alumina are shown to be an effective way to isolate cationic dyes, tetrakis-(1-methyl-4-pyridyl)porphyrin and 7-diethylamino-4-methylcoumarin. Luminescence intensity of the coumarin-pillared clay composite was reported to be six times greater than that of coumarin–clay composite.

Porphyrins and phthalocyanines are well known not only for biological but also for catalytic, conductive, and photoactive properties. There has been an increased interest in the organization of porphyrins and metalloporphyrins into well-defined molecular architecture because of the possibility of controlling their desirable properties. Along this line, the intercalation of porphyrins (148,150–160) and phthalocyanines (161–171) into layered solids has been reported. Since porphyrins undergo reversible protonation–deprotonation reactions, depending on the acidity of the environment, they can be used as a probe for the interlayer environment.

Cady and Pinnavaia reported the reactions of *meso*-tetraphenylporphyrin ($TPPH_2$) with interlayer exchangeable cations of montmorillonite. The acidity of the interlayer cations affects the adsorption state of $TPPH_2$ (150). Strongly acidic hydrated Fe^{3+} and VO^{2+} ions react quantitatively with the free base porphyrins to afford the protonated porphyrin cations, which form monolayers in the interlayer space of the silicate sheets. On the other hand, weakly acidic hydrated Na^+ and Mg^{2+} ions afford only a trace amount of $TPPH_42^+$ $TPPH_2$ reacts with (n-$C_3H_7)_4N^+$, Co^{2+}, Cu^{2+}, and Zn^{2+} ions to give mainly metalloporphyrin in solution and a hydronium-exchanged form of the montmorillonite. They have also reported the formation of porphyrin from aldehyde and pyrrole catalyzed by the Bronsted acidity of hydrated cations. Van Damme et al. reported the reaction of $TPPH_2$ and *meso*-tetra(4-pyridyl)porphyrin with montmorillonite (151). There is an equilibrium between various species on montmorillonite that is a function of the hydration state of the surface. Carrado and Winans investigated ion exchange of two water-soluble meso-substituted porphyrins (tetrakis(1-methyl-4-pyridiniumyl)porphyrin and tetrakis(N,N,N,-trimethyl-4-anilinumyl)porphyrin) and the corresponding metalloporphyrins Fe(II), Fe(III), and Co(II) with montmorillonite, hectorite, and fluorhectorite and proposed similar effects of the exchangeable cations on their adsorption state (152).

The metallation–demetallation reaction of tin tetra(4-pyridyl) porphyrin in Na-hectorite has been investigated (153). UV-vis and luminescence spectra revealed that the adsorbed complex demetallated to the tetra(4-pyridyl) porphyrin dication upon dehydration. This process was reversible, however, indicating that the Sn4$^+$ ion remains in the vicinity of the porphyrin upon demetallation.

Kuykendall and Thomas studied the photophysics of tetrakis(N-methylpyridyl)porphyrin adsorbed on colloidal smectite and proposed that the photophysics can be a convenient method to monitor the degree of dispersion or extent of deflocculation of a clay dispersion to produce primary particles or single sheets (156).

B. Dye Aggregates

Dye aggregation has been a topic of interest because it is thought to be a key issue in natural and artificial photosynthetic systems as well as in photonic materials such as nonlinear optical materials. The adsorpion of a cationic cyanine dye, 1,1'-diethyl-2,2'-cyanine (pseudoisocyanine, abbreviated as PIC, whose molecular structure is shown in Fig. 11) on smectites (montmorillonite and saponite) was reported (172,173). It is well known that the spectral properties of cyanine dyes are strongly affected by aggregation (174,175). These aggregates (so-called

n = 0,1, 2 for dye-1, -2, -3   Et = ethyl

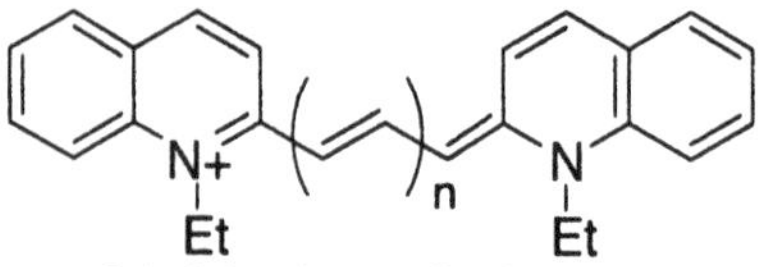

FIGURE 11   Molecular structures of cationic cyanine dyes.

J- and H-aggregates) have attracted much attention for the spectral sensitization of photographic processes and useful optical properties (176,177).

The adsorption and aggregation of PIC cations on montmorillonite and saponite have been investigated in aqueous suspensions and cast films. The absorption spectrum of an aqueous mixture of montmorillonite and PICBr is shown in Figure 12b. (The absorption spectrum of a 5×10^{-6} M PICBr aqueous solution is shown in Figure 12a, in which the absorption band of monomeric PIC appears at around 520 nm.) In the absorption spectrum, a new absorption band appeared at 567 nm, which is red-shifted relative to the monomer absorption at 522 nm. The emission spectrum of the aqueous mixture is shown in Figure 13a. (The excitation wavelength is 520 nm.) The luminescence spectrum consists of an intense resonance band at around 570 nm. These absorption and luminescence bands are characteristic of PIC J-aggregates (174–177). Therefore, PIC forms J-aggregates on the surface of montmorillonite in aqueous solution. However, spectroscopic features of PIC were different when synthetic saponite was used. The absorption band due to monomeric PIC Br red-shifted slightly upon addition of saponite, showing that the PIC cations distributed molecularly in the aqueous mixture with possible interaction with silicate surface.

In contrast to the observations for aqueous suspensions, the cast films of PIC/clays give different results. J-aggregates are absent when the amount of PICBr is 2.5 mmol/100 g clay. On the other hand, when the amount of PIC is increased to 50 mmol of PIC/100 g clay, the band due to monomeric PIC disappears and the J-band appears at around 570 nm, irrespective of clay type. The basal spacings of the PIC/saponite films were ca. 1.3 and 1.7 nm for 2.5 and 50 mmol of PIC/100 g saponite, respectively. The PIC cations distribute molecularly on the surface of clays when the loading amount is low. Higher dye loadings result in their aggregation within the interlayer space.

Since spectroscopic features vary with the type of clay, it is apparent that different clays have the ability to organize the dye molecules in different ways. The change in the state of adsorbed cyanine dyes during evaporation of solvent has also been revealed. The optical properties of J-aggregates formed on layered silicates may vary because of their stable and confined microstrutucre. For exam-

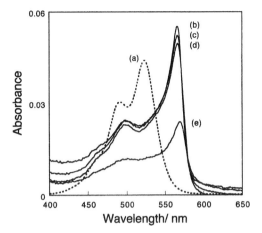

FIGURE 12   Absorption spectra of (a) 5×10^{-6} M PICBr aqueous solution and aqueous mixtures containing (b) 20, (c) 10, (d) 5, and (e) 1 mg of montmorillonite and 100 ml of 5×10^{-6} M PICBr.

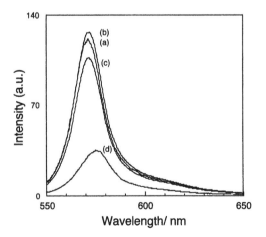

FIGURE 13   Luminescence spectra of aqueous mixtures containing (a) 20, (b) 10, (c) 5, and (d) 1 mg of montmorillonite and 100 ml of 5×10^{-6} M PICBr aqueous solution. The excitation wavelength is 520 nm.

ple, the aggregation of a stilbazolium cation on clay has led to optical second harmonic generation (178).

C. Orientation of Intercalated Dye

The orientation of intercalated dye molecules has been investigated primarily from geometrical constraints using the size of the guest and the gallery height of the intercalation compound (179,180). One-dimensional Fourier analysis is a powerful tool to evaluate the microstructures of intercalation compounds. The orientation of cationic dyes such as rhodamines and oxazines in smectites and taeniolite and that of anionic dyes such as (4-phenylazo-phenyl)acetic acid in an LDH have been derived in this way (181,182).

Spectroscopic (UV-is and ESR) anisotropy has been applied to determine the orientation of guest species in oriented films. The molecular orientation of Cu(II) tetrakis-(1-methyl-4-pyridiniumyl)porphyrin (CuTMPyP) in the interlayer space of Na hectorite and Li fluorhectorite has been studied by X-ray diffraction and anisotropic ESR spectroscopy (154). In the fluorhectorite, the long axis of the porphyrin cross section becomes inclined to the silicate sheet with a tilt angle of $35°$. Differences in the arrangements of the intercalated porphyrin were attributed to the different charge densities of the host materials. The intercalation and arrangement of Co(II)tetrakis(N-methyl-pyridiniumyl)porphyrin (CoTMPyP) into the interlayer space of hectorite and fluorohectorite has been studied by X-ray diffraction and EPR spectroscopy (155). The orientations of the intercalated porphyrines were reportedly affected by the layer charge density. When intercalated on the low-charge-density hectorite, CoTMPyP orients with the molecular plane parallel to the silicate layers and has no water in axial coordination site of Co(II). In contrast, CoTMPyP orients with its ring at $27°$ to the silicate layer, with water molecules coordinated to Co(II) in the high-charge-density fluorohectorite. Vacuum dehydration decreases the basal spacing of CoTMPyP-fluorohectorite from 19.6 to 17.6 Å, causing a rearrangement of the intercalated CoTMPyP into a staggered bilayer with no axial water bound to Co(II).

Nakato et al. have reported the intercalation of a free-base porphyrin, 5,10,15,20-tetrakis(1-methyl-4-pyridiniumyl)-porphyrin, into layered tetratitanic acid (158). In order to insert the bulky ions into the interlayer space, n-propylammonium-$H_x Ti_4 O_9$ was used as an intermediate. Judging from the observed d-values, which indicated the expansion of the interlayer space by 1.05 nm, and the spectroscopic characteristics, the intercalated porphyrin ions were present as the monomeric free-base ions and were inclined with a tilt angle of about $35°$. Luminescence decay profiles revealed self-quenching due to interactions between porphyrin ions.

Aminophenyl(TAPP)- and pyridinium(TMPyP)-substituted porphyrins were intercalated into a-zirconium hydrogen phosphate by exchanging the porphyrins into the p-methoxyaniline(PMA) preintercalated compound (Zr(O$_3$POH)-$_2$·2PMA) (160). Powder X-ray diffraction patterns and EPR spectra of the uniaxially ordered thin films suggest that the p-TAPP derivatives consisted of a monomolecular porphyrin layer in which the heme planes were tilted nearly 45° relative to the host lamellae, whereas o-TAPP derivatives assembled predominantly into a porphyrin bilayer in which the heme macrocycles lay parallel with the host sheets.

The intercalation of 5,10,15,20-tetra(4-sulfonatophenyl)porphine into the interlayer space of a layered double hydroxide has been accomplished by anion exchange (159). The powder X-ray diffraction pattern showed that the intercalated porphyrin anions are arranged with their molecular planes perpendicular to the host lamellae.

The intercalation and orientation of phthalocyanines in anionic and cationic layered materials have been investigated by means of X-ray diffraction and EPR. The orientation of the intercalated species was found to depend on the layer charge density of the host materials (152,155). Carrado et al. reported the incorporation of phthalocyanines and metal phthalocyanines into the interlayer space of anionic (Mg–Al LDH) and cationic (hectorite) clays via ion exchange or in situ crystallization of the synthetic host layers (167). The phthalocyanines are oriented parallel to the silicate layer of hectorite and perpendicular to the anionic layer of LDH, in correlation with their hosts' layer charge density. EPR studies on the Co(II)tetrasulphophthalocyanine(CoPcTs)-Mg/Al LDH system indicated the existence of Co with water in axial coordination sites. Outgassing the sample results in the disappearance of the Co(II) spectrum, and the original spectrum has been restored upon rewetting the sample with water. However, the orientation of the phthalocyanines ring, which was determined by XRD, could not be verified because attempts to prepare well-oriented films of the CoPcTs-LDH were unsuccessful.

Thus, intercalation of porphyrins and phthalocyanines into layered solids results in the organization of macrocycles with variable arrangements. The charge density of the host layer, the charge and molecular size of the guest porphines and phthalocyanines, etc. are the factors to control their states. Oriented films of the intercalation compounds are very important for the determination of the orientation of porphyrins in the interlayer spaces. Although porphines and phthalocyanines are well known for their biological, catalytic, conductive, and photoactive properties, only a limited number of studies have been made to investigate the properties of their intercalation compounds. 2C$_{12}$2C$_1$N–montmorillonite–metal phthalocyanine intercalation compounds (168,169) and a CoPcTs-Mg-Al LDH intercalation compound (170) have been prepared and their catalytic activities have been reported. Systematic studies on the structure–property relationships of these systems can provide indispensable information of both practical and scientific importance.

Amphiphilic cationic azobenzene derivatives (Fig. 14) have been used as the guest species for the intercalation into layered silicates (magadiite and montmorillonite) (183,184). The dye orientation in the interlayer spaces has been discussed from the spectral shifts and the gallery heights of the products. During their molecular assembly, the chromophore interacts to give aggregated states, and the dye–dye interactions cause both bathochromic and hypsochromic spectral shifts, depending on the microstructures (185). The spectral shifts reflect the orientation of the dipoles in the aggregates; smaller spectral red shifts are expected for the aggregates with larger tilt angles of the dipoles, as schematically shown in Figure 15. Depending on the layer charge density (cation exchange capacity) of host materials and the molecular structures of the amphiphilic azo dyes, aggregates (J- and H-aggregates) with different microstructures (tilt angle) formed in the interlayer spaces of layered silicates are summarized in Table 1.

Nonlinear optical properties are also affected by dye molecule orientation. Nonlinear optics comprises the interaction of light with matter to produce a new light field that is different in wavelength or phase (186,187). Examples of nonlinear optical phenomena are the ability to alter the frequency (or wavelength) of light and to amplify one source of light with another, switch it, or alter its transmission characteristics throughout the medium, depending on its intensity. Since nonlinear optical processes provide key functions for photonics, recent activity in many laboratories has been directed toward understanding and enhancing second- and third-order nonlinear effects. Second-harmonic generation (SHG) is a nonlinear optical process that converts an input optical wave into an outwave of twice the input frequency. Large molecular hyperpolarizabilities of certain organic materials lead to anomalously large optical nonlinearities. Major research

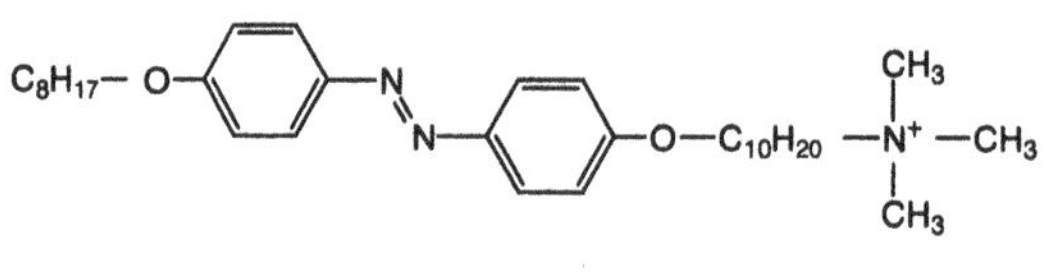

π-(ω-trimethylammoniodecyloxy)-π'-(octyloxy)azobenzene cation

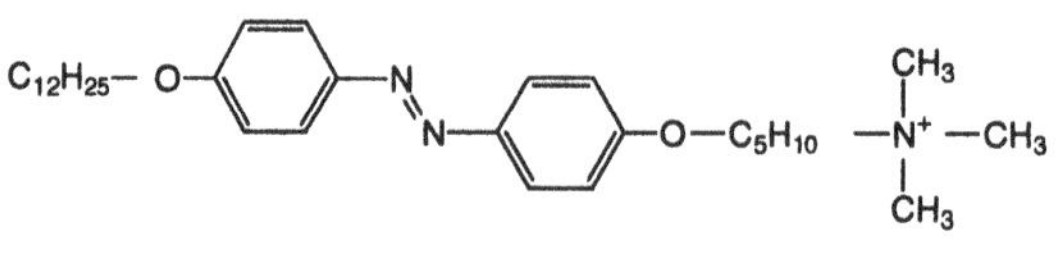

π-(ω-trimethylammonioheptyloxy)-π'-(dodecyloxy)azobenzene cation

FIGURE 14   Molecular structures of amphiphilic cationic azobenzene.

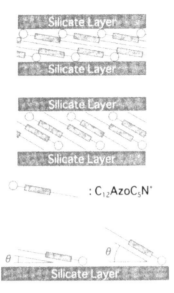

: $C_{12}AzoC_5N^{\cdot}$

FIGURE 15 Proposed microstructures of the $C_{12}AzoC_5N^+$-montmorillonite intercalation compound.

Table 1 Basal Spacings and Gel-to-Liquid-Crystalline Phase-Transition Temperatures for $2C_{18}\,2C_1\,N^+$-Silicates

Host	Phase-transition temperature	Basal spacing	Technique	Ref.
Montmorillonite	327	4.24	Thermal decoloration of photomerocyanine to spiropyran	210
Montmorillonite	327	4.83	Permeation of fluorescence probe and DSC	124
Bentonite	326	4.3	Pyrene luminescence	123
Bentonite	327	Not given	Electrochemical reduction of trichloroacetic acid	125
TSM	328	3.4	Photochromism of azobenzene	213

efforts are directed toward (a) identifying new molecules possessing large nonlinear polarizability and (b) controlling molecular orientation on a microscopic level. For purpose (b), intercalation compounds have potential applicability because intercalated organic species can take unique arrangements in the interlayer space to lead to the nonlinear effects.

p-Nitroaniline possesses high hyperpolarizability but crystallizes in a centrosymmetric manner so that it cannot show any SHG. The intercalation compounds mentioned earlier have been obtained as powders, and the measurements of SHG intensity have been carried out by the Kurtz–Perry powder technique (188). For the detailed characterization of optical properties of the host–guest systems as well as their practical applications as photonics devices, such materials must be available as single crystals or thin films. One of the possible resolutions for the problem is self-assembled multilayers grown on substrates such as organosilicon-derived materials (189) and zirconium phosphonate–derived materials (190). Although they are not intercalation compounds, their structures are composed of alternating inorganic and organic layers similar to intercalation compounds. Compared with Langmuir–Blodgett films, which are studied most extensively in this research field, the inorganic–organic composite systems show better stability. Therefore, they show promise for the future application of photonic materials.

Cooper and Dutta have prepared intercalation compounds that exhibit SHG by the intercalation of 4-nitrohippuric acid ($NO_2C_6H_4CONHCH_2COOH$) into a layered lithium aluminate ($LiAl_2(OH)_6{}^+$) (191). (Lithium aluminate is an LDH consisting of layers of Li and Al surrounded by hydroxyl groups, and the positive charge on the layer is neutralized by exchangeable anions.) From the observed basal spacing of the product (the interlayer distance was 18 Å), the guest species are thought to be stacked perpendicular to the host layer. It has been proposed that the 4-nitrohippuric acid molecules are held in the interlayers as carboxylic acid and the process is not via ion exchange. The orientation of the 4-nitrohippuric acid molecules in the interlayers leads to an ordered arrangement of dipoles, and the material exhibits frequency-doubling characteristics, generating 532-nm radiation from incident 1064-nm radiation. SHG from kaolinite-p-nitroaniline intercalation compounds has also been reported (192).

The intercalation compounds 4-[-(4-dimethylaminophenyl)ethenyl]-1-methylpyridinium ($DAMS^+$)-MPS_3, where M is either Mn^{2+} or Cd^{2+}, were prepared by an ion exchange process (193,194). The intercalation compounds ($Cd_{0.86}PS_3(DAMS)_{0.28}$ and $Mn_{0.86}PS_3(DAMS)_{0.28}$) exhibited a significantly large SHG activity (194). This was obtained by direct reaction between MPS_3 powder and an ethanolic solution of DAMS iodide at 130°C in the presence of pyridinium chloride. In this reaction, pyridinium chloride played a role in generating in situ an intermediate pyridinium intercalated compound that underwent rapid exchange with the $DAMS^+$. Although IR and XRD showed that these compounds were structurally similar irrespective of the reaction conditions, the compounds obtained by the direct reaction were much more crystalline and did not contain residual solvents. Minor changes had a dramatic influence on the NLO properties, however; for example, $Mn_{0.86}PS_3(DAMS)_{0.28}$ exhibited an efficiency 300 times that of urea, and the $Cd_{0.86}PS_3(DAMS)_{0.28}$ reached 750 times the efficiency of urea. Additionally, no significant decay of the NLO signal over a period of several months was observed, which is a crucial point in the search for new SHG materials based on organic NLO chlomophores. The high intensity of SHG excluded the possibility of surface effects proposed in their first report and indicated that the NLO chromophore was spontaneously well aligned between the interlayers. It should be noted that the intercalation did not cause significant dilution of the chromophore: There is one DAMS species per 525 $Å^3$ in the tosylate salt, while there is one DAMS per 740 $Å^3$ of the intercalation compounds. This ability to incorporate guest species at such high concentration is one of the most attractive features of the intercalation process.

Thin films of a TMA-saponite-p-nitroaniline (p-NA) intercalation compound showed optical second-harmonic generation (195). The compound was prepared by a solid–gas reaction between a TMA-saponite film and p-NA vapor. In order to align p-NA dipoles in the TMA-saponite, an external electric field was applied to the host during the intercalation of p-NA. No SHG was observed for the compound prepared without an electric field. This indicates that the applied electric field caused a noncentrosymmetric alignment of the p-NA in the interlayer micropore of the TMA-saponite. The proposed alignment of p-NA dipoles in the TMA-saponite is schematically shown in Figure 16. The TMA-saponite film is composed of oriented particles of TMA-saponite with their ab plane parallel to the substrate in which microscopic anisotropy can be directly converted into a macroscopic one. Moreover, the film shows transparency in the visible region and possesses micropores in the interlayer space. Because of the matching between the size and geometry of the micropore of the TMA-saponite and those of p-NA molecules, the incorporation into the TMA-saponite is a novel way to create self-assembled aggregates of p-NA with noncentrosymmetry.

In addition to the out-of-plane anisotropy of guest species, in-plane anisotropy of guest species is worth investigating for the construction of intercalation compounds with three-dimensional anisotropy. Recently, we have reported the adsorption of cationic cyanine dyes (Fig. 11) on $K_4Nb_6O_{17}$ single crystals to create unidirectional orientation of the dyes by means of the in-plane anisotropy of $K_4Nb_6O_{17}$ (196). $K_4Nb_6O_{17}$ has an anisotropic surface with regularly arranged NbO^- groups that act as cation-exchangeable sites, as shown in Figure 17 (197,198). The in-plane anisotropy was evaluated by polarized absorption spectra. The polarized absorption spectra (Fig. 18) of PIC/$K_4Nb_6O_{17}$ showed the in-plane anisotropy of the adsorbed cyanine dyes. It is thought that the difference in the periodic distances between the a and c axes of $K_4Nb_6O_{17}$ matches the distance

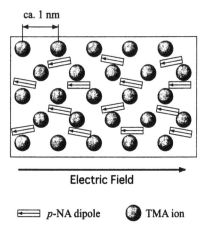

FIGURE 16 Schematic structure of TMA-saponite-*p*-nitroaniline intercalation compounds prepared with application of an electric field during intercalation.

of the positive charges in the cyanine dye aggregates to cause unidirectional orientation of the adsorbed dyes.

In addition to measuring solid samples by the methods just described, the orientation of dyes in suspension has been investigated using electric dichroism (199–205). Electric linear dichroism (ELD) measures the change in the absorption of incident light linearly polarized parallel and perpendicular to the applied electric field direction, as schematically shown in Figure 19. This technique has been applied extensively for dye–polyelectrolyte solutions. For cationic dye–clay systems, the roll and tilt angles of dyes at the surface of clays have been derived by detailed analysis of the ELD. It should be noted that electric dichroism can be performed only in solution of low ionic conductivity. Intercalation compounds

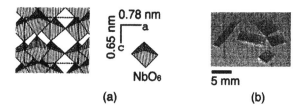

FIGURE 17 (a) Surface structure of $K_4Nb_6O_{17}$ cleaved from interlayer I. (b) Photograph of single crystal of $K_4Nb_6O_{17}$.

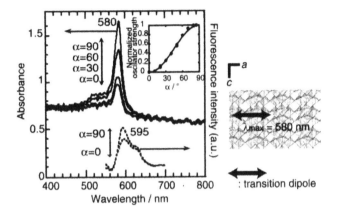

FIGURE 18 Polarized absorption spectra of $PIC/K_4Nb_6O_{17}$.

of 4-chlorostilbene-4′-carboxylate/LDH and 2- and 4-[4-(dimethylamino)styryl]-1-ethylpyridinium/saponite were suspended in N,N-dimethylformamide (DMF) to examine the ELD.

For the N-alkylated acridine orange cation on colloidally dispersed mont-morillonite (199) and zirconium phosphate (200) in aqueous systems, the orienta-

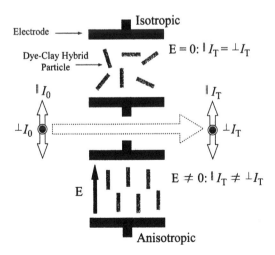

FIGURE 19 Principles of electric dichroism in clay–dye systems.

tion and mobility of the dye were dependent on alkyl-chain length. The steric hindrance and hydrophobic interaction among the alkyl chains of adjacent dyes are thought to be involved.

By analyzing the ELD spectra of 2- and 4-[4-(dimethylamino)styryl]-1-ethylpyridinium cations on saponite suspended in DMF (Fig. 20), the tilt and roll angles of the dye were found to be controlled by the amount of the intercalated dye as well as the molecular structure (the position of cationic site within the dye) (103). The adsorption models of 4-[4-(dimethylamino)styryl]-1-ethylpyridinium cations on solid saponite are shown in Figure 21. Depending on the coverage, the orientation of 4-[4-(dimethylamino)styryl]-1-ethylpyridinium varied, as shown in Figure 21. The important role of the orientation of the stilbazolium cations will be discussed in a subsequent section.

IV. PHOTOCHEMCIAL REACTIONS IN CLAY–ORGANIC SYSTEMS

A. Photochromism

A wide variety of photochemical reactions in clay–organic systems has been investigated. Among photochemical reactions studied for dye–clay systems, photochromism predominates. Photochromism, which deals with photochemical reactions that are thermally or photochemically reversible, has received considerable attention because of its actual and potential applications and for its paramount importance in biological phenomena (206). Studies concerning photochromic reaction behaviors and their controls in solid media have significance for practical applications, such as in optical recording, as well as in probing host–guest systems. Accordingly, the intercalation of photochromic dyes into the interlayer space of layered solids has been investigated (207–218).

Takagi et al. reported the intercalation of 1',3',3'-trimethylspiro[2H-1-benzopyran-2,2'–indoline] (H-SP) and its 6-nitro (NO$_2$-SP) and 6-nitro-8-(pyridinium)-methyl (Py$^+$-SP) derivatives into montmorillonite; their photochromic behavior (the photochromic reaction of spiropyran is shown in Fig. 22) has been studied for colloidal systems (208). The effects of intercalation on the rate of thermal coloration and decoloration have been compared with those in other systems, such as colloidal silica, hexadecyltrimethylammonium bromide, and sodium dodecylsulfate (SDS) micelles.

Py$^+$-SP was intercalated into montmorillonite quantitatively as an equilibrium mixture with the corresponding merocyanine (MC) with the ratio of Py$^+$-SP:Py$^+$-MC of 35:65 and exhibited reversed photochromism. It is known that thermal equilibria between SP and MC are dependent on the polarity of surroundings; MC becomes the major product with increasing polarity. Therefore, the observed reverse photochromism has been explained in terms of the polar inter-

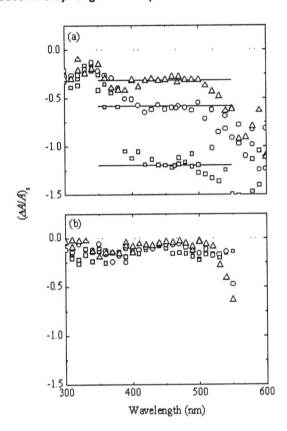

FIGURE 20   ELD spectra of 2- and 4-[4-(dimethylamino)styryl]-1-ethylpyridinium cations on saponite suspended in DMF. [Dye]:[saponite] = 0.1 (squares), 0.5 (circles), and 1.0 (triangles).

layer of montmorillonite. The thermal isomerization of Py^+-SP intercalated in aqueous colloidal montmorillonite exhibited a linear combination of two components of first-order kinetics, indicating the presence of two adsorption environments. It has been suggested that one is due to molecularly separated species and the other is due to aggregated species.

In contrast, a preferential adsorption as SP was observed when $C_{16}3C_1N^+$ was coadsorbed with Py^+-SP, H-SP, and NO_2-SP, and normal photochromism has been observed in these systems. Single first-order kinetics has been observed for the Py^+-SP-$C_{16}3C_1N^+$-montmorillonite system. The effects of coadsorbing $C_{16}3C_1N^+$ on the photochromic behavior showed that the $C_{16}3C_1N^+$ surrounds Py^+-SP to create a hydrophobic environment.

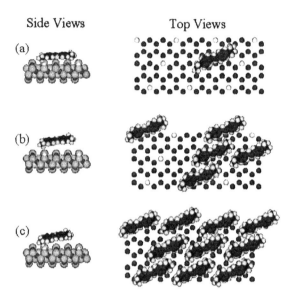

FIGURE 21   Adsorption model of 4-[4-(dimethylamino)styryl]-1-ethylpyridinium cations on saponite. [Dye]:[saponite] = 0.1 (a), 0.5 (b), and 1.0 (c).

The photochromism of a cationic diarylethene, 1,2-bis(2-methyl-3-thiophenyl)perfluorocyclo-pentene bearing two pyridinium substituents at each thiophenyl ring (Fig. 23), intercalated in montmorillonite was reported (209). The product was prepared as oriented films by casting, so the dye orientation was deduced from the basal spacing and spectroscopic behavior was determined with polarized light. The photochromic reaction was efficient and smooth, while the efficiency decreased upon repeated irradiation. The decrease was attributed to the formation

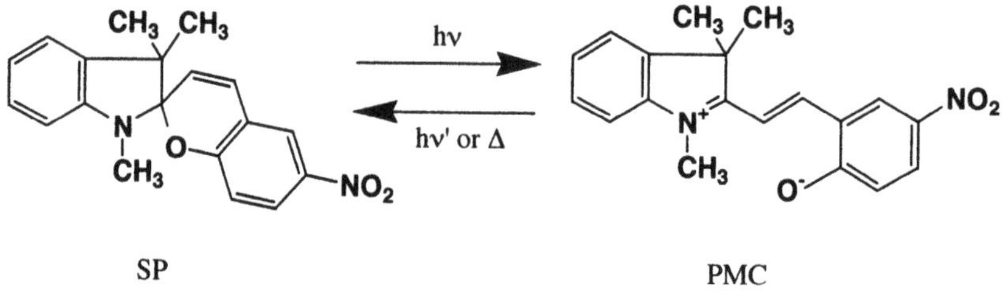

SP   PMC

FIGURE 22   Isomerization of spiropyrane to merocyanine.

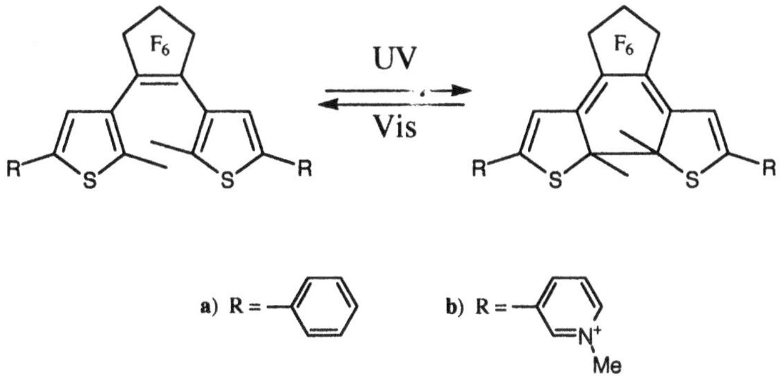

FIGURE 23   Molecular structures and photochemical reactions of diarylethenes.

of photoinactive species. The degradation was successfully suppressed by the coadsorption of dodecylpyridinium cations. Presynthesized long-chain organoammonium-modified silicates have been used for the introduction of spiropyrans (210–211) and azobenzenes (212,213,217,218). The role of the surfactant is not only for producing hydrophobic interlayer spaces, but also for controlling the states of the adsorbed dyes.

    Seki and Ichimura have investigated the thermal isomerization kinetics of photoinduced merocyanine (MC) to spiropyran (SP) in solid films having multibilayer structures, which consists of ion complexes between ammonium bilayers forming amphiphiles ($2C_{18}2C_1N^+$) and polyanions (montmorillonite and poly(-styrene sulfonate), abbreviated as PSS) (210). 1,3,3-Trimethyl-6'-nitrospiro[indo-line-2,2'-2'H-benzopyran](SP) was incorporated into thin films of the polyion complexes, which were prepared by casting the chloroform solution or suspension of the polyion complexes.

    X-ray diffraction studies showed that the cast films were composed of a multilamellar bilayer structure whose membrane plane was preferentially oriented parallel to the film surface. Endothermic peaks were observed in the differential scanning calorimetry (DSC) curves of the films; the observed phase-transition temperature was 54.5 and 48.5°C for the $2C_{18}2C_1N^+$-montmorillonite and $2C_{18}2C_1N^+$-PSS films, respectively. Both DSC and XRD results indicated that the $2C_{18}2C_1N^+$-montmorillonite film had a more ordered structure than $2C_{18}2C_1N^+$-PSS film. Since annealing the film at 60–70°C at a relative humidity of ca. 100% for a few hours resulted in the well-structured $2C_{18}2C_1N^+$-PSS film, the photochromic behavior was investigated for the annealed films. The difference in the film structure influences the kinetic properties of the thermal decaying of MC embedded in the films.

The incorporated SP exhibited photochromism in both of the immobilized bilayer complexes with montmorillonite and PSS. Kinetic measurements of the thermal isomerization (decoloration) were carried out for the annealed film. The decoloration reaction rate is dependent on the mobility of the surroundings and, in polymer matrices, is influenced by the glass transition. It was found that the reaction rates abruptly increased near the gel-to-liquid-crystal phase-transition temperature (54°C) of the immobilized bilayer due to increased matrix mobility in this system. The film prepared with montmorillonite gives more homogeneous reaction environments for the chromophore than those with the linear polymer (PSS). This leads to drastic changes in the reaction rate at the crystal-to-liquid-crystal phase transition of the bilayer, showing the effect of the phase transition of immobilized bilayers to be more pronounced than that of the glass transition of amorphous polymer matrices.

The formation of stable H (parallel type) and J (head-to-tail type) aggregates of photo-merocyanines has been suggested in a photochromism study of a series of 1'-alkyl-3',3'-dimethyl-6-nitro-8-alkanoyloxymethylspiro($2H$-1-benzopyran-2,2'-indoline) derivatives with different lengths of alkyl chains in $2C_{12}2C_1N^+$-montmorillonite (211). A cast film consisting of the SP incorporated into a bilayer intercalated into a clay was prepared on a glass plate by slowly evaporating a solution of the SP and $2C_{12}2C_1^+$-montmorillonite. When longer R^1 alkyl chains were introduced, new, very sharp absorption peaks at around 500 nm appeared in addition to the absorption at 570 nm due to the monomeric photomerocyanine (abbreviated as PMC) upon UV light irradiation. A new, sharp absorption band appeared at a longer wavelength region around 610 nm when SPs bearing longer alkyl chains were exposed to UV light. These new absorption bands are attributed to aggregates of PMCs that are reported to form occasionally in organized molecular assemblies (211). The absorption bands around 500 and 610 nm have been ascribed to H and J-aggregates of PMC, respectively. A high activation energy and highly positive activation entropy for J and H-aggregates of PMCs, which directly correlate with the thermal stability of these aggregates (slow decoloration), have been observed.

Azobenzenes are well-known photochromic dyes with trans–cis isomerization (Fig. 24), and their photochromic reactions in various heterogeneous systems have been reported. Photochemical isomerization of azobenzenes intercalated in the hydrophobic interlayer space of alkylammonium–montmorillonite and TSMs has been investigated (212,213). Intercalation compounds were obtained by a solid–solid reaction between the organophilic silicates and azobenzene. The intercalated azobenzene showed reversible trans-to-cis photoisomerization upon UV irradiation and subsequent thermal treatment (or visible light irradiation). The hydrophobic interlayer spaces of the alkylammonium silicates were proved to serve as media for the immobilization and photochemical isomerization of the azo dye.

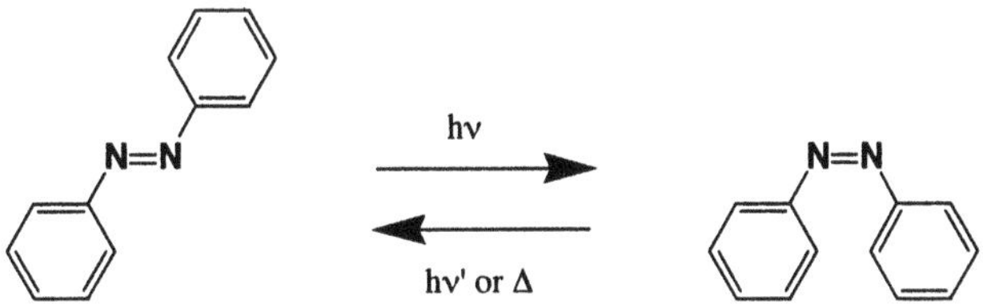

FIGURE 24 Photoisomerization of azobenzene.

$2C_n2C_1N^+$-TSM-azobenzene intercalation compounds were also synthesized as films by casting the suspension of the $2C_n2C_1N^+$-TSM and azobenzene in organic solvents on a quartz substrate (213). The intercalated azobenzene exhibits reversible photochromic reactions. The fraction of the photochemically formed cis-isomer in photostationary states depends on the reaction temperature, suggesting that a change in the states of the interlayer $2C_n2C_1N^+$ occurs. The transition temperatures estimated from the photochemistry of azobenzene were in good agreement with values determined by other techniques (123,124,210).

It should be noted here that the variation of the d-values for the $2C_n2C_1N^+$-smectites are summarized in Table 2 (123–125,210,213). The difference in the basal spacings corresponds to the difference in the orientation of the alkyl chains. It has been known that excess guest species can be accommodated in the interlayer spaces of smectites as a salt; this phenomenon is refered to as *intersalation*. The large d-values reported in the literature may be due to the intersalation as well as to the difference in the surface layer charge density. Although this does not cause the change in the phase-transition temperatures, intersalation might affect the photoprocesses of the guest species.

Recently, the introduction of retinal, which is the chromophore of rhodopsin, into a surfactant-(dimethyoctadecyllammonium)-modified clay was investi-

Table 2 Interlayer Spacings and Absorption Maxima of Intercalation Compounds

Sample	Gallery height (nm)	Absorption maximum (nm)	Type
$C_{12}AzoC_5N^+$- montmorillonite	1.4	385	J-aggregate bilayer
$C_8AzoC_{10}N^+$- montmorillonite	1.5	373	J-aggregate monolayer
$C_{12}AzoC_5N^+$-magadiite	2.8	360	J-aggregate bilayer
$C_8AzoC_{10}N^+$-magadiite	3.2	320	H-aggregate

gated in order to mimic the properties of rhodopsin (214,215). The spectroscopic and photochemical properties of retinal in vitro are of interest in studying the primary chemical process of vision and in developing novel photoresponsive materials. The modified clay interlayer offers environments for retinal similar to rhodopsin in two respects: color regulation and efficient isomerization at a cryogenic temperature. Protein environments have the ability to tune the color of retinal Shiff bases; however, the color regulation in artificial systems has not been satisfactory. In rhodopsins, a retinal molecule forms a Schiff base linkage with a lysine residue and the retinal Shiff base is protonated. It was proposed that a proton was supplied from dimethyoctaecyllammonium to retinal Shiff base, as schematically shown in Figure 25. The trans–cis isomerization of a protonated retinal Shiff base occurs even at 77 K, as revealed by visible and infrared spectroscopy. (The infrared difference spectra before and after illumination is shown in Fig. 26.) The efficient isomerization of retinal at 77 K was worth mentioning as a successful mimic of the primary photochemical reaction in rhodopsin. However, azobenzene isomerization in dialkyldimethylammonium-TSM was reportedly suppressed at lower temperature. The difference is worth investigating using similar materials (clays, surfactants, and chromophore). The use of the intercalation compounds containing retinal in the field of sensors is a worthwhile field of new research.

Amphiphilic cationic azobenzene derivatives (scheme Fig. 14) have been used as the guest species for intercalation into the layered silicates magadiite and montmorillonite (182–185). The azobenzene chromophore photoisomerized effectively in the interlayer space of silicates, despite the fact that the azobenzene

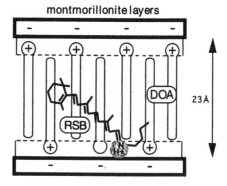

FIGURE 25 Schematic drawing of the protonation of intercalated all-trans retinal Shiff base in dimethyloctadecylammonium–montmorillonite.

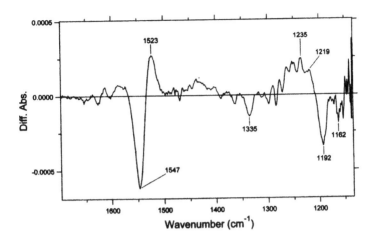

FIGURE 26 Infrared difference spectra after the illumination of intercalated all-trans retinal Shiff base in dimethyloctadecylammonium–montmorillonite.

chromophore is aggregated there. Motivated by the success of effective photo-isomerization of azo dyes in the interlayer spaces, we expect development of their photoresponsive properties. One notable example has been achieved in the system of a cationic azo dye, p-[2-(2-hydroxyethyldimethylammonio)ethoxy]azo-benzene bromide (Fig. 27) and magadiite (216). The dye consists of a photoiso-merizable azobenzene unit and a cationic dimethyl-hydroxyethylammonioethoxy group. XRD and elemental analysis results indicated the formation of a π-(ω-dimethyl-hydroxyethylammonioethoxy)-azobenzene magadiite intercalation compound. The spectral properties as well as XRD results revealed that the ad-sorbed azo dye cations form head-to-head aggregates in the interlayer space, as schematically shown in Figure 28. The change in the X-ray diffraction pattern of the film upon photochemical reaction is shown in Figure 29. The basal spacing changed after UV irradiation from 2.69 to 2.75 nm, and the value came back to 2.69 nm upon visible light irradiation. The reversible change in the basal spacing has been observed repeatedly, as shown in Figure 29B. Considering the change in the absorption spectrum, the trans-azo dye cations are thought to form a densely packed aggregate in the interlayer space. Upon UV irradiation, half of the trans-form isomerized to cis form and coexists with trans form in a same interlayer, as suggested by the single-phase X-ray diffraction patterns. The relative contribu-tion of cis form and trans form generally depends on the environmental factor as well as irradiation conditions (light source, temperature, and so on). It is postu-

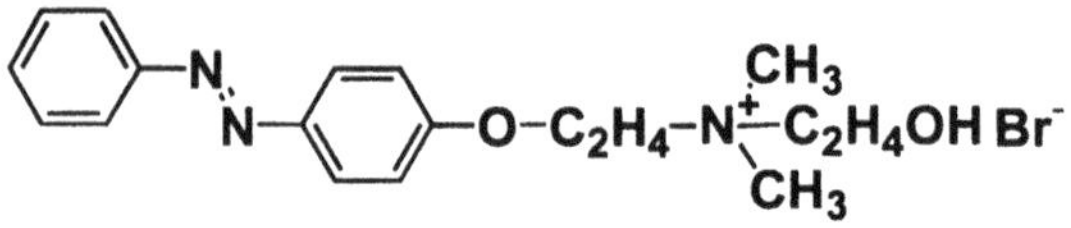

FIGURE 27 Molecular structure of π-(ω-dimethyl-hydroxyethylammonioethoxy)-azo-benzene.

lated that a densely packed aggregate is difficult to form in the interlayer space of magadiite at the photostationary state due to the geometric difference of two isomers, and this causes the change in the basal spacings. This demonstrates a type of photomechanical effect.

The change in the basal spacings triggered by the photochemical reactions of the intercalated azobenzenes has been observed for organophilic clay–azobenzene intercalation compounds (217,218). However, the location of the intercalated azobenzene is difficult to determine in those systems, since the dyes were solubilized in the hydrophobic interlayer spaces. In the present system, the photoisomerization of a certain portion of the intercalated dye is thought to induce the change in the microstructure of the intercalation compound. The dye used in the present study does not contain flexible units such as long alkyl chains, and therefore the photoisomerization induced a change in the microstructure detectable by XRD.

B. Photochemical Cycloaddition

The interlayer spaces of clay minerals have been shown to provide a stable and characteristic reaction field suitable for spatially controlled photochemical reactions. Regioselective photocycloaddition of stilbazolium cations, intercalated in the interlayer space of saponite, have been reported (219–225). There are four possible photochemical reaction paths of the stilbazolium ion. Upon irradiation of UV light to a stilbazolium–saponite suspension, *syn* head-to-tail dimers were predominantly formed at the expense of cis–trans isomerization, which is a predominate path in homogeneous solution (Fig. 30). The selective formation of head-to-tail dimers suggests that the intercalation occurs in an antiparallel fashion, as is shown in Figure 31. Since the dimer yields were only slightly dependent on the loading amount of guest ions, stilbazolium ions were adsorbed inhomogeneously to form aggregates with antiparallel alternative orientation even at very low loading (e.g., 1% of CEC) This aggregation was supported by the fluorescence spectrum of the dye adsorbed on saponite, in which excimer fluorescence was observed at 490–515 nm at the expense of the monomer fluorescence at ca. 385–450 nm. The selective formation of *syn* head-to-tail dimers indicates the

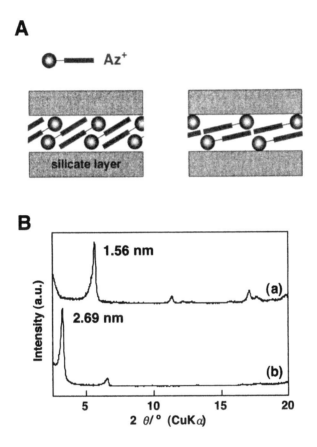

FIGURE 28 (A) X-ray powder diffraction patterns of magadiite and AZ^+-magadiite. (B) Proposed microstructures of the cationic azobenzene–magadiite.

formation of aggregates owing to hydrophobic interaction between the adsorbate ions.

Changes in the aggregation state of γ-stilbazolium [4-(2-phenylvinyl)pyridinium] ion on saponite by coadsorption of alkylammonium ions (C_nN^+) has been investigated (222). Figure 32 shows the effect of C_8N^+ on photoreactivity of preintercalated stilbazolium ions on clay. On coadsorbing C_nN^+ longer than the stilbazolium ion, the major photoreaction was changed from cyclodimerization to E–Z isomerization and the excimer emission of the intercalated stilbazolium ions was dramatically reduced. The formation of dimers or clusters of adsorbent were simulated by a Monte Carlo method for a two-dimensional lattice.

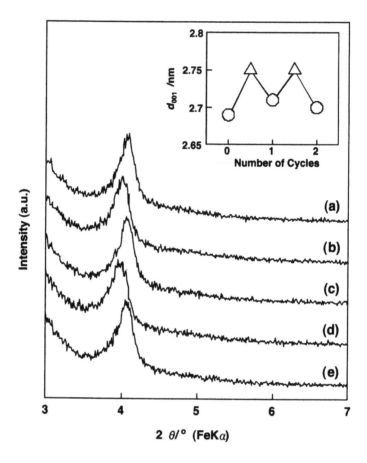

FIGURE 29 Reversible change in the XRD pattern of the cationic azobenzene–magadiite. *Inset*: Change in the basal spacing upon photochemical reactions.

Photochemical cycloaddition for several unsaturated carboxylates has been studied in the presence of hydrotalcite (LDH) (223–225). In addition to the antiparallel packing of the guest, the intermolecular distances of two double bonds of adjacent carboxylates were found to affect the stereoselectivity of the photochemical reactions. While cinnamate yielded head-to-head dimers exclusively, stilbenecaroboxylates (Fig. 33) led to a significant amount of head-to-tail dimer in addition to a head-to-head dimer. This difference was explained by molecular packing of the dye anions in the interlayer space of LDH. The proposed molecular packing of stilbenecarboxylate and *p*-phenylcinnamate ions in the interlayer space of Mg/Al LDH is shown schematically in Figure 34.

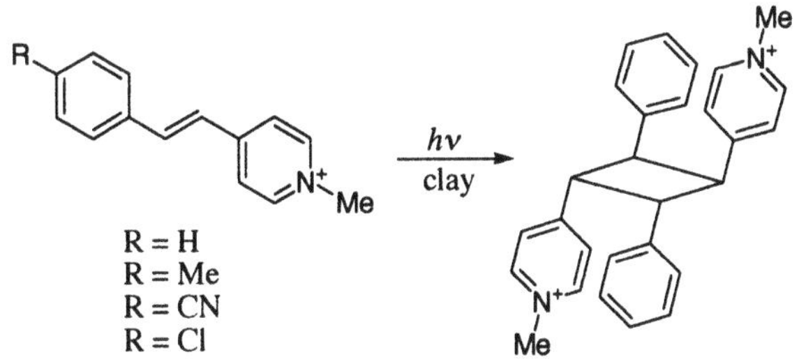

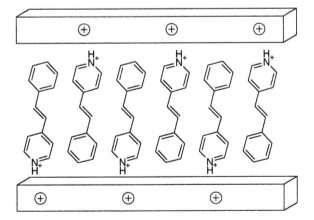

FIGURE 30 Formation of syn head-to-tail dimmers of stilbazolium cations in the presence of clay.

Similar to the effects of organoammonium ions on the photochemistry of stilbazolium ions in smectite, the addition of sodium p-phenethylbenzoate, a photochemically inactive coadsorbate, affected significantly the product distribution in the photolysis of p-(2-phenylethenyl)benzoate intercalated in hydrotalcite. This series of investigations shows that the organization of organic species into the interlayer space of layered materials is a way of crystal engineering in which reactions can be controlled. In other words, relating selectivity of reactions and interlayer spacing is a method of probing the geometric relationships between intercalated species.

FIGURE 31 Schematic representation of alkene packing in the interlayer space of saponite.

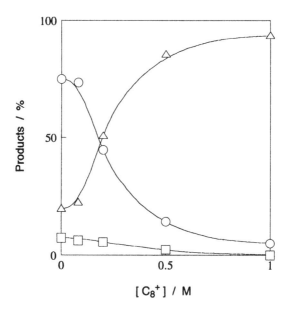

FIGURE 32 Effect of *n*-octylammonium ions on photoactivity of preintercalated g-stilbazolium ion in saponite: –○–, *syn* head-to-tail dimer; –□–, *syn* head-to-head dimer (3); –Δ–, Z-isomer. Intercalated γ-stilbazolium–saponite complex was suspended in 5 mL of aqueous solution of *n*-octyammonium ions.

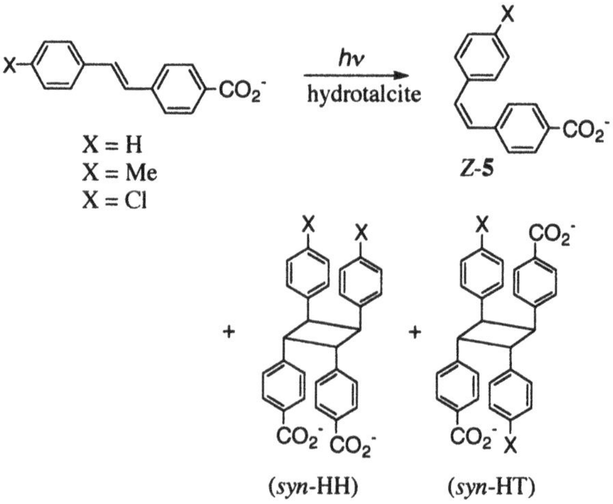

X = H
X = Me
X = Cl

Z-5

(*syn*-HH)   (*syn*-HT)

FIGURE 33 Photochemical reaction paths of stilbenecarboxylates.

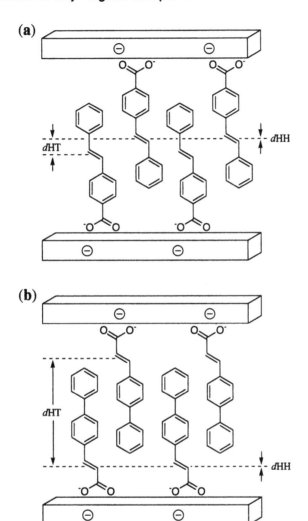

FIGURE 34 Simplified drawing of antiparallel packing of (a) *p*-phenyl-cinnamate and (b) stilbene carboxylate ions in LDH.

C. Photochemical Hole Burning

Photochemical hole burning (PHB) is the site-selective and persistent photo-bleaching of an inhomogeneously broadened absorption band, induced by reso-nant laser light irradiation at cryogenic temperatures (226). PHB has attracted increasing attention due in part to its possible applicability to high-density fre-quency-domain optical storage, in which more than 10^3 times more storage den-sity than present optical disk systems would in principle be available. There remain, however, many obstacles to overcome in materials optimization as well as in storage system and architecture. Among many efforts, a search for new materials is important because the hole-formation processes depend significantly on the nature of host–guest systems. For the formation of persistent holes, the existence of both solid matrix and a photoreactive molecule is essential. Since hole formation depends significantly on the nature of host–guest systems, it can be both used as and probed by high-resolution solid-state spectroscopy. On this basis, two intercalation compounds that exhibit PHB reactions have been prepared (157,227). In both systems, saponite was used as the host material. The photoac-tive centers are 1,4-dihydroxyanthraquinone (abbreviated as DAQ) and cationic porphines, both of which are typical PHB dyes.

Ogawa et al. (227) have prepared the TMA-saponite-DAQ intercalation compound and investigated its PHB reaction to show the merits of an ordered matrix for a PHB material. DAQ is one of the molecules most extensively used as a PHB probe, and its hole formation has been observed in numerous amorphous matrices. The PHB reaction of DAQ is likely due to the breakage of internal hydrogen bond(s) and the subsequent formation of external hydrogen bond(s) to proton acceptor(s) within a matrix (Fig. 35). To be a molecularly dispersed system is another basic prerequisite for composing efficient PHB materials, in order to avoid line broadening due to energy transfer. For this purpose, saponite was modified by pillaring with TMA ions to obtain independent micropores in which DAQ molecules were incorporated at a monomolecular level without aggregation.

The schematic structure of the TMA-saponite is shown in Fig. 36. The basal spacing of the TMA-saponite is 1.43 nm, with a TMA-pillared interlayer spacing of 0.47 nm. Taking into account the molecular size and shape of DAQ and the geometry of the micropore of TMA-saponite, DAQ was intercalated with the molecular plane nearly perpendicular to the silicate sheet.

A persistent spectral zero-phonon hole was obtained at liquid helium tem-peratures by Kr^+ laser light irradiation (520.8 nm). In spite of the high concentra-tion of DAQ (ca. 1.5 mol kg^{-1}), a narrow hole with an initial width of 0.25 cm^{-1} (4.6 K) was obtained. This width is narrower than those (e.g., 0.4–0.8 cm^{-1}) of DAQ doped in ordinary polymers and organic glasses (e.g., PMMA, ethanol/methanol mixed glass) obtained under similar experimental conditions. The narrow line width is desirable for the application to optical recording, since

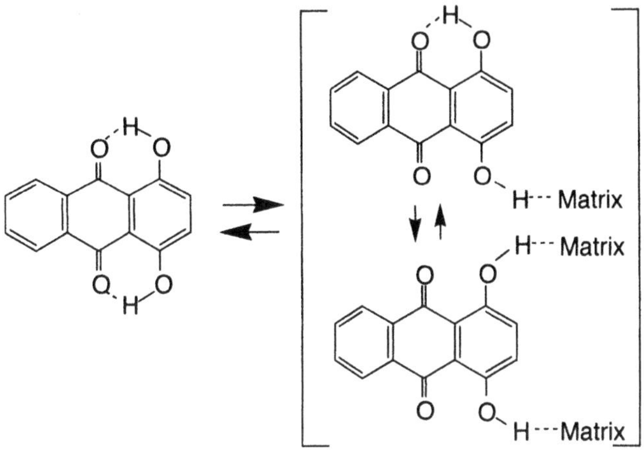

FIGURE 35 Photochemical reactions of 1,4 dihydroxyanthraquinone.

the number of holes to be made in an inhomogeneously broadened band increases. This width may be related mainly to the dephasing, but contributions from spectral diffusion cannot be neglected. On the other hand, a broad pseudo-phonon side-hole, whose shift from the zero-phonon hole is 25 cm^{-1}, appears only after irradiation stronger than 1500 mJ cm^{-2}. In Table 3, hole width, homogeneous width $\Delta\omega_i$, and burning efficiency obtained for the 520-nm band are summarized.

The burning efficiency is about 8.3×10^{-4} at the initial stage of burning, estimated from the temporal evolution of the area of zero-phonon hole. This value is similar to or slightly higher than the typical one, 1×10^{-4}, observed in ordinary dispersed cases. Taking into account the high concentration of DAQ (ca. 0.15 mol kg^{-1} corresponding to ca. 3 mol L^{-1}) in the present system, no distinct decrease in the yield also supports the monomolecular dispersion of DAQ

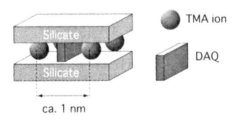

FIGURE 36 Schematic drawing for the TMA-saponite-DAQ intercalation compound.

Table 3 Hole Width, Inhomogeneous Width, and Burning Yield Obtained for
520-nm Band of TMA-saponite-DAQ System for Wet and Dried Conditions (See Text)
and for 545-nm Band for Wet Condition Respectively

Sample	$\Gamma_h(cm^{-1})$	$\Delta\omega_i(cm^{-1})$	Φ	$\lambda_B(nm)$	Comment
Wet	0.25	930	8.3×10^{-4}	520.8	1.3 mJ/cm^2
					1.4%
Dry	0.38	940	4.5×10^{-4}	520.8	1.3 mJ/cm^2
					0.5%
Wet	1.38	1040	8.4×10^{-5}	547	1.5 mJ/cm^2
					0.2%

Γ_h: holewidth; $\Delta\omega_i$; inhomogeneous width; Φ: burning yield; λ_B: burning wavelength;
Comment: The laser fluence and relative hole depth where the parameters are determined. Burning was
carried out at 4.6 K.

within the interlayer space of saponite. This property is most noteworthy from
the viewpoint of the fabrication of recording media, because the ability for such
high dopant concentrations is desirable for preparing materials as thin films. At
such high concentrations, the broadening of holes and the decrease in the burning
efficiency are usually inevitable in amorphous matrices prepared with conventional procedures. The microporous structure of the TMA-saponite intercalation
compound therefore apparently leads to some desirable characteristics regarding
hole formation.

D. Photoinduced Electron/Energy Transfer

Photoinduced electron transfer in the interlayer space of clays has been reported.
Protection of pesticides from photodegradation by energy transfer at the surface
of clay is a notable example (228,229). Synthetic nonhalogenated pyrethroid
bioresmethrin (Fig. 37) is a powerful and safe contact insecticide; however, rapid
photodecomposition limits its agricultural use. Therefore, attempts have been
made to stabilize photolabile bioresmethrin. When bioresmethrin was adsorbed
on montmorillonite with methylgreen (Fig. 38), the degradation was significantly
suppressed. After ion exchange with methylgreen, the interlayer space becomes
hydrophobic and accommodates bioresmethrin. Desorption of bioresmethrin is
then also possible, as required for insecticidal activity. The interactions between
methylgreen and bioresmethrin played an important role in the stabilization. The
efficiency of the stabilization depends on the structure of the pesticide molecule
and on the spatial distribution of dye and pesticide. The stabilization of a photola-

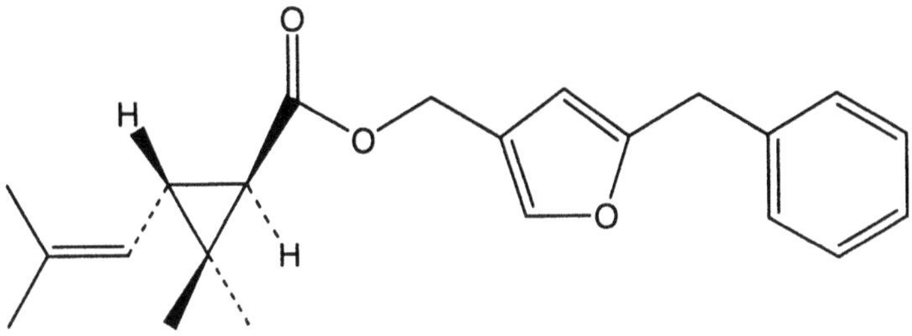

FIGURE 37 Molecular structure of bioresmethrin.

bile herbicide, trifluralin, was also achieved by using a sepiolite clay modified with a cationic dye, thioflavine (230).

Energy transfer from rhodamin 6G to cationc dye acceptors such as thionin and crystal violet was investigated in aqueous laponite suspension (231). A concentrating effect was demonstrated by electronic electron transfer. All preparation methods showed rapid microscopic and macroscopic equilibration of the systems. On the other hand, time-dependent spectral changes were observed for aqueous smectite suspensions containing methylene blue and acridine orange (232). Sensitized photoisomerization of cis-stilbazolium ion by $[Ru(bpy)_3]^{2+}$ was studied in saponite (233). The reaction yield was much higher than that in homogeneous

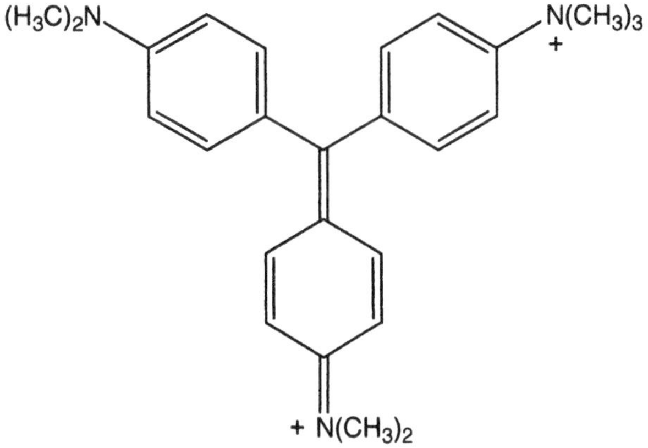

FIGURE 38 Molecular structure of methylgreen.

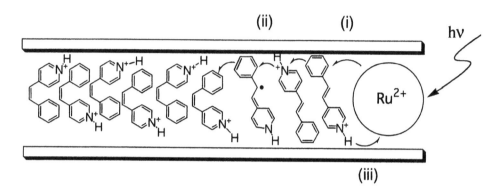

FIGURE 39 Schematic representation of the sensitized isomerization of cis-stilbazolium cation in the interlayer space of saponite.

solution, indicating aggregation of the stilbazolium ions in the interlayer space. The proposed mechanism for the sensitized isomerization is shown in Figure 39.

Viologens are photoreduced reversibly in the presence of an electron donor to form blue radical cations (234). The color development by electrochemical reduction and UV light irradiation is shown in Figure 40. Miyata et al. reported the photochromism of viologens (1,1'-dialkyl-4,4'-bipyridinium ions) intercalated in the interlayer space of montmorillonite with cointercalating poly(vinyl pyrrolidone) (PVP) (235). Viologen dications were intercalated in the interlayer space of montmorillonite-PVP intercalation compound by cation exchange. Photochemical studies were carried out for self-supporting films of viologen–montmorillonite-PVP by irradiation with Hg lamp. Upon irradiation, viologen radical cations formed to show blue color and characteristic absorption bands at 610 and 400 nm. Reversible color development and color fading were observed. In this system, cointercalated PVP was assumed to act as an electron donor for the reduction of the viologens. The color fading required a longer period than that in a pure PVP matrix. Since the color-fading process in the PVP matrix was an oxidation caused by oxygen in air, the slow color-fading reaction observed for the viologen–montmorillonite-PVP system was explained by the prevention of contact between the viologen radical cations and the oxidizing agent.

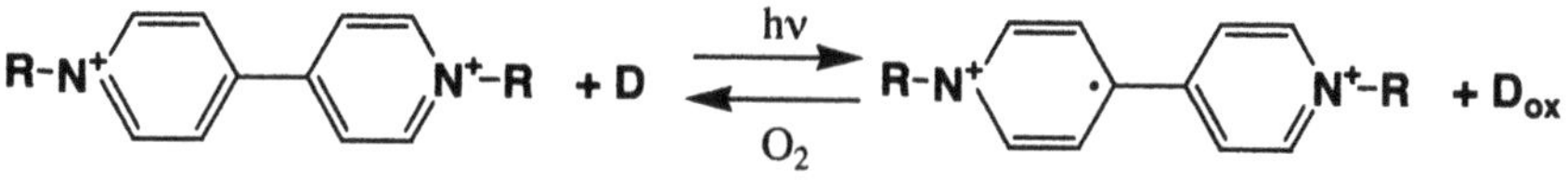

$$R-N^+ \underset{O_2}{\overset{h\nu}{\rightleftarrows}} R-N^+ \cdots N^+-R + D_{ox}$$

FIGURE 40 Photochromism of methylviologen.

Thompson and his coworkers have reported efficient photoinduced charge separation in layered zirconium viologen phosphonate compounds in both powdered and thin-film samples (236,237). In order to overcome the scattering problems associated with powder, transparent multilayer thin films of ZrPV(X) were grown directly onto fused silica substrates from aqueous solution. The sequential growth method has been applied in the preparation of ZrPV(X) films. In this case, fused-silica slides are treated with $(et)_3SiCH_2CH_2CH_2NH_2$, followed by treatment with $POCl_3$. This procedure leads to a phosphonate-rich surface suitable for treatment with $ZrOCl_2$. The slides are then allowed to react with $H_2O_3P–CH_2CH_2$-bipyridinium-$CH_2CH_2–PO_3H_2X_2$.

Photolysis of $Zr(O_3PCH_2CH_2(bipyridinium)CH_2CH_2PO)X_2$ (X = Cl, Br, I), ZrPV(X), resulted in the formation of blue radical cations of viologen that are stable in air. The photoreduction of viologen in these thin-film samples was very efficient (quantum yields = 0.15), showing simple isobestic behavior in the electronic spectra. Contrary to bulk solids, photoreduced thin films are very air sensitive. The mechanism for the formation of charge-separated states in these materials involves both irreversible and reversible components. An irreversible component is proposed to involve hydrogen atom abstraction by photochemically formed halide radicals, followed by structural rearrangements. Optimization of the reversible process may make it possible to use these materials for efficient conversion and storage of photochemical energy.

Energy transfer among naphthyl-, anthryl-, and pyrenylalkylammonium bound to zirconium phosphate and photoinduced electron transfer from 5,10,15,20-tetrakis(4-phosphonophenyl)porphyrin to N,N'-bis(3-phosphonopropyl)-4-4'-bipyridinium organized in Zr-based self-assembled multilayers have been reported (238–240). The layered structure plays an important role in organizing reactants to control the reaction.

The photochromic behavior of methylviolgen (MV^{2+}) intercalated into a series of layered transition metal oxides has been reported (241). $K_2Ti_4O_9$ (242), $HTiNbO_5$ (243), $K_4Nb_6O_{17}$ (244,245), HNb_3O_8 (245), and $HA_2Nb_3O_{10}$ (A = Ca, Sr) (246) were used as host materials and MV^{2+} was intercalated by cation exchange with interlayer cations. The photochemical studies were conducted for powdered samples by irradiation with a Hg lamp, and the reactions were monitored by diffuse reflectance spectra. Semiconducting host layers acted as electron donors for the reduction of viologen to form radical cations of the intercalated MV^{2+} in the interlayer space. The stability of the photochemically formed blue radical cations has been discussed with respect to their microscopic structures.

The photochemistry of intercalation compounds formed between layered niobates $K_4Nb_6O_{17}$ and HNb_3O_8 with MV^{2+} can be controlled by changing the interlayer structures (245). Two types of MV^{2+}-intercalated compounds with different structures have been prepared for each host. In the $K_4Nb_6O_{17}$ system, two intercalation compounds were obtained by changing the reaction conditions.

In both of the intercalation compounds, MV^{2+} is located only in interlayer I. HNb_3O_8 also gave two different intercalation compounds: One was prepared by the direct reaction of HNb_3O_8 with MV^{2+} and the other was obtained by using propylammonium-exchanged HNb_3O_8 as an intermediate. In the latter compound, propylammonium ions and MV^{2+} were located in the same layer. All the intercalation compounds formed $MV^{\cdot+}$ in the interlayers by host–guest electron transfer. The presence of cointercalated K^+ and propylammonium ion in the $K_4Nb_6O_{17}$ and HNb_3O_8 systems, respectively, significantly affected the decay of $MV^{\cdot+}$. This difference was explained by guest–guest interactions, which varied with the cointercalation of photoinactive guests (K^+ and propylammonium ion).

E. Photocatalysis

There are many reactions promoted by light-activated solids that are not consumed in the overall reactions. Such solids are called *photocatalysts*. The construction of highly organized systems is very important to obtain higher efficiency and selectivity in photocatalytic reactions.

Among photocatalytic reactions, the photocatalytic splitting of water with transition metal complexes as sensitizers has attracted the most attention. Numerous homogeneous and heterogeneous systems are currently being investigated. Heterogeneous and microheterogeneous systems offer the possibility of avoiding back reactions, of stabilizing high-energy intermediates, and of catalyzing difficult redox reactions. Catalytic oxidation or reduction of water has been achieved with colloidal metal or oxide particles in numerous systems. In order to achieve the separation of H_2- and O_2-producing systems at a molecular level, the heterogeneous character of solid surfaces has been investigated. Along this line, the photooxidation of water catalyzed by metal complexes within the adsorbed layers of clays has been investigated (247–257). The possible role of electrostatic effects between negatively charged clay minerals and photocatalysts for photoreduction of water has been reported.

Newsham et al. have reported the excited state properties of trans-dioxorhenium(V) ions immobilized in the intercrystalline environments of three complex-layered oxides (255). Intercalation of *trans*-$ReO_2(py)_4^+$ in hectorite and fluorohectorite and *trans*-$ReO_2(CN)_4^{3-}$ in Mg/Al LDH were conducted. Despite structurally similar intercalated *trans*-ReO_2 cores, steady-state and time resolved luminescence experiments have revealed that the three intercalation compounds are quite distinct: $ReO_2(py)_4$-hectorite is highly emissive with a luminescence spectrum similar to that of the native ion; $ReO_2(py)_4$-fluorohectorite exhibits broad and featureless emission whose intensity is attenuated by a factor of 50 relative to that of the $ReO_2(py)_4$-hectorite; no luminescence has been observed for the $ReO_2(CN)_4$-hydrotalcite. From the excited-state properties and the X-ray diffraction results, the arrangements of the intercalated oxoions have been proposed. It is suggested that these different arrangements are responsible for the differences in the excited-state properties. Though the catalytic properties have not been reported, the observed differences in the excited state properties suggest the important role of supports.

Uranyl-exchanged smectites have been used to photo-oxidize alcohols to ketones (256). Interlayer uranyl ions have been thought to be catalytically active sites. The selectivity for ketone production by uranyl clay is less than that by uranyl-loaded zeolites, and the stability of the clay photocatalysts is not as high as that of the zeolite catalysts. However, unusual radical coupling products can be formed on the uranyl–clay catalysts. One important observation is that the excitation wavelength can be significantly changed by changing the hosts.

Uranyl-containing pillared clay has also been used as a photocatalyst (257). Uranyl ions in the clay pillars produce more selective catalysts than materials obtained by inserting uranyl ions in the interlamellar space of the clay. A comparison of the photocatalytic properties of uranyl–clays, uranyl-containing pillared clay, and uranyl–zeolites showed that photochemical selectivity was a function of pore size and surface area.

The photophysical and photochemical properties of small semiconductor crystallites have been a subject of interest. The small size of these particles endows them with unusual optical properties due to quantum size effects that may find application in photocatalysts and nonlinear optical materials. Crystallites with nanometer sizes are inherently unstable due to their high surface tensions, and the crystallites tend to grow larger under conventional nucleation conditions. Thus, terminating or stabilizing reagents, or a matrix, are indispensable for the preparation of clusters with controlled size. Among possible solid matrices (258), semiconductor quantum particles formed with one constrained dimension and two free dimensions are likely to be highly anisotropic, potentially giving rise to unusual photophysical properties. On this basis, layered materials have been utilized as a support, with the resulting materials characterized by optical properties and TEM observations. There are two possible approaches: One is the incorporation of semiconductor colloidal particles, and the other is the formation of the clusters in the interlamellar space.

The incorporation of TiO_2 into montmorillonite has been reported by two different methods (259–264). One is the oxidation of the intercalated Ti(III) and the other is the direct intercalation of TiO_2 colloid. Yoneyama et al. studied the photocatalytic activities of microcrystalline TiO_2 incorporated in the interlayer space of montmorillonite (259,260). The X-ray diffraction pattern showed that the TiO_2 pillar has a 15-Å height. The pillared TiO_2 showed a ca. 0.58-eV blue shift in its absorption and fluorescence spectra, compared with TiO_2 particles. The excited electronic states of the pillared TiO_2 were determined to be 0.36 eV more negative than those of a TiO_2 powder. It was found that the TiO_2/clay system exhibited higher photocatalytic activity than TiO_2 powder for decomposition of

2-propanol to give acetone and hydrogen and for *n*-carboxylic acids with up to eight carbons (from acetic acid to capric acid), which yielded the corresponding alkanes and carbon dioxides. They explained that the higher excited electronic state contributes to the higher activity. One important observation in this study is that the pillared TiO_2 exhibited lower catalytic activities for decomposition of capric acid, which has 10 carbons. It was thought that the space of the TiO_2-montmorillonite interlayers is too narrow to accommodate large molecules, resulting in the poor photocatalytic activity. Thus, the incorporation of TiO_2 small particles into montmorillonite led to the enhancement of the photocatalytic activities as well as guest selectivity in the reaction due to its defined pore size.

The photocatalytic activity of TiO_2-montmorillonite in the reduction of MV^{2+} with oxidation of triethanolamine (261) and the oxidation of several aliphatic alcohols have also been reported (259). The photodegradation of dichloromethane, which is not readily degraded or hydrolyzed in an aquatic environment, to hydrochloric acid and carbon dioxide using titanium-exchanged montmorillonite, titania-pillared montmorillonite, and titanium-aluminum polymeric cation pillared montmorillonite has been reported (261).

The photocatalytic decomposition of acetic acid, propionic acid, and *n*-butyric acid on iron oxide incorporated in the interlayer space of montmorillonite has been studied (263,264). TEM observation showed an interlayer spacing of 0.66 nm, which is in agreement with X-ray diffraction analysis. The iron oxide prepared in montmorillonite had a greater bandgap and exhibited remarkable photocatalytic activities, being in marked contrast to a ferric oxide bulk powder. These observations indicate the formation of size-quantized iron oxide particles on montmorillonite. The catalytic activities were reported to be greatly dependent on solution pH.

The preparation of CdS and CdS-ZnS in the interlayer space of montmorillonite (262b,265), laponite (266–268), and Mg/Al LDH (269) has been reported. In the case of montmorillonite and laponite, the particles have been prepared by ion exchange with Cd^{2+} ions and subsequent treatment with H_2S. For the LDH system, cadmium ethylenediamine tetraacetic acid was introduced into the interlayer space, and the compound was allowed to react with a Na_2S solution. The interlayer spacing of the CdS-incorporated materials has been determined to be ca. 0.5, 0.2, and 0.3 nm for montmorillonite, laponite, and hydrotalcite, respectively. Spectral shifts in visible absorption spectra suggest that the CdS particles were confined to some extent. Quantum-size ZnSe, PbS, CdS, and CdSe particles with a diameter of 3–5 nm in the interlamellar region of the layered host material $Zr(O_3PCH_2CO_2H)_2$ has been prepared by reaction of H_2S or H_2Se with M(II) $[Zr(O_3PCH_2CO_2)_2]$ (270). The semiconductor particles exhibit quantum size effects in their electronic absorption spectra relative to bulk semiconductors. The correlation between the size and the magnitude of blue shifts agrees closely with that reported in the literature for well-characterized "capped" semiconductor particles.

The evolution of hydrogen was observed when an aqueous suspension containing chlorophyll a adsorbed hectorite-PVP, MV^{2+}, 2-mercaptoethanol, and hydrogenase was illuminated (271). MV^{2+} and 2-mercaptoethanol were thought to be an electron carrier and electron donor, respectively. The chlorophyll a was effectively stabilized by the adsorption onto hectorite-PVP intercalation compound.

Domen et al. studied the photocatalytic activities of several layered metal oxide semiconductors, mainly $K_4Nb_6O_{17}$, for water cleavage by bandgap excitation (272). Motivated by the success, a wide variety of photocatalysts based on the layered transition metal oxides have been investigated (273–280). The unique structure has successfully been utilized for efficient photocatalytic reactions. In some cases, metal or oxide clusters have been deposited on $K_4Nb_6O_{17}$.

V. POSSIBLE PRACTICAL APPLICATIONS OF DYE–CLAY HYBRIDS

As already mentioned, a wide variety of photochemical and photophysical properties of clay–dye systems have been reported. Motivated by the progress in controlling photoprocesses by organizing species in the interlayer space of layered materials has led researchers to fabricate intercalation compounds as thin films. For such purposes, films with precisely controlled thickness and the multilayered structures with alternate heteroaggregates are required. Accordingly, the layer-by-layer deposition technique and the LB method have been conducted.

Coumarin–clay intercalation compounds have been deposited on solid substrates with poly(diallyldimethylammonium) chloride (Mw = 200,000–350,000) (281,282). The absorbance and luminescence intensity increased linearly with thickness (deposition cycle). More simple and versatile ways to fabricate multilayered thin films were reported by Fan et al., using a bicationic sexithiophene derivative (283). Linear relationships between deposition cycle and the absorbance of the films were observed, indicating the sequential adsorption of negatively charged clay plates and the cationic sexithiophene derivative. Nonlinear optical properties (second-harmonic generation, SHG) of multilayered films of clay/(PDDA)/4-{4-[N-allyl-N-methylamino]phenylazo}benzenesulfonic acid praed by the laye-by-layer deposition technique have been investigated (284). Nonlinear optical properties were explained by the noncentrosymmetric organization of 4-{4-[N-allyl-N-methylamino]-phenylazo}benzenesulfonate anions in the films. The substrate type (pretreatment with a silane coupling reagent), the PDDA concentration and the clay type were found to be factors for optimizing SHG.

Similar multilayered films containing clay plates and dyes have been prepared by a Langmuir–Blodgett technique. LB films of [Ru(II)(phen)$_2$(dcC12-

bpy)] (phen = 1,10-phenanthroline, dcC12-bpy = 4,4′-carboxy-2,2′-bipyridyl didodecyl ester) hybridized with clays were prepared and the SHG was investigated (285). No increase in the SHG singnals from the films as a function of the layer number was observed, while the films modified with octadecylammonium chloride showed a quadratic relation between the SHG signal and the layer number. This difference indicated that the noncentrosymmetric orientation of the Ru(II) complex in the films was achieved by the addition of octadecylammonium chloride. LB films containing two kinds of amphiphilic ruthenium complexes, [Ru(III)(acac)$_2$(acacC12)](acac = 2,4-pentanedionate; acac12 = 5-dodecyl-2,4-pentanedionate) and [Ru(II)(phen)$_2$(dC18-bpy)](ClO$_4$)$_2$ (phen = 1,10-phenanthroline, dC18-bpy = 4,4′-dioctadecyl-2,2′-bipyridyl) as electron acceptor and donor, respectively, was prepared with and without clay (286). The photoinduced electron transfer was suppressed, suggesting that the clay platelets acted as an efficient barrier for photoinduced electron transfer.

Electroluminescence from clay/polymer intercalation compound has been reported (287). Electroluminescent poly[2-methoxy-5-(2′-ethyl-hydroxy)-1,4-phenylenevinylene] was complexed with dimethyl dehydrogenated-tallow ammonium–montmorillonite. The product was fabricated as thin films by spin coating the suspension on ITO-coated glass substrate and subsequent deposition of Al on the films as cathodes. Complexation with clay led to improved photostability and photoluminescence output. The absorption characteristics of dyes also have been utilized to prepare clay–dye hybrids as possible ultraviolet radiation collectors (288–290).

Very recently, the application of dye–clay intercalation compounds for the removal of toxic substances from water has been examined (291–293). Organic modification of clays with aliphatic or simple aromatic ammonium or pyridinium ions has been conducted to control the adsorptive properties of dyes for this purpose (294,295). Such dye–clay intercalation compounds therefore can be considered as a new class of adsorbents with additional functions, such as sensing and photoresponsive adsorptive properties.

The controlled photochemical reactions, reaction dynamics, and selectivity effects reported thus far may lead researchers to fabricate devices involving dye–clay intercalation compounds for such applications as optical recording. Intercalation has already been applied for the stabilization of dyes on carbonless copying paper (296) and for thermal dye transfer printing (297,298). In thermal dye transfer printing, the dyes are expected to migrate from an ink layer (a color ribbon) to clay dispersed in a receiver layer, as schematically shown in Figure 41. These researchers used tetra-n-decylammonium- or dioctadecyldimethylammonium–montmorillonite as the dye-receiving layer for cationic dye (Rhodamine 6G and an oxazine dye) fixation. Image fixation occurred by cation exchange mechanisms such that the counteranions of the dye in the ink layer and the

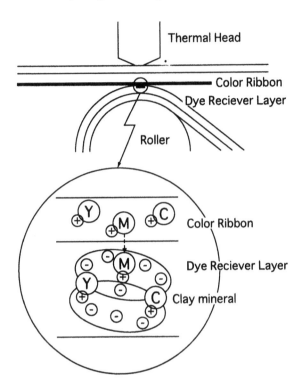

FIGURE 41   Schematic drawing of the fixation of cationic dyes in montmorillonite for thermal transfer printing.

interlayer cation of clays in the receiving layer affects the printing speed. The stability of the color image was also satisfactory.

Another novel recording system has been proposed wherein electrochemically induced acids and bases promote the alternate intercalation of fluoran dye and tetra-*n*-decylammonium ions, along with a color change of the dye, in the acetone–clay suspension (299). Designing more electrochemically sensitive recording layers and suppressing any decomposition will be required to improve the performance.

VI.   CONCLUSIONS AND FUTURE PERSPECTIVES

From all of the studies concerning the photoprocesses of ionic species on clays, the following general ideas can be derived:

1. Cationic dyes preferentially occupy the cation exchangeable site on the smectite surfaces. The intercalation of ionic dyes into other layered solids with high surface charge density requires optimized reaction conditions, such as the use of an intermediate or careful pH adjustment.
2. In order to avoid flocculation of clay particles, the dye amounts should be very low compared to the cation exchange capacity.
3. Naturally occurring clay minerals contain impurities such as Fe^{3+} that quench the excited state of adsorbed dye species through energy transfer mechanisms.
4. Dyes tend to aggregate on silicate surfaces (both internal and external), even at very low concentration loadings.
5. The charge density of the host layers and the size and shape of the guest species significantly affect the orientation of the guest species in the interlayer space.
6. The adsorption of two different species often results in segregration, while appropriate additives can be coadsorbed to alter the spatial distribution of the dyes.

Host–guest interactions (e.g., electrostatic and interactions between p-electron and surface oxygen layers) as well as guest–guest interactions (e.g., hydrophobicity) are important factors in determining states of the intercalated dyes. Aggregation of guest species in the interlayer space does not originate from inhomogeneous charge distribution on the host surfaces, but rather from guest–guest interactions. Appropriate coadsobates, such as surfactants, can alter the states (aggregated vs. isolated) of photoactive species. The rotation and diffusion of guest species depends on host–guest interactions, guest–guest interactions, coadsorbates, etc. in the host–guest systems.

For nonionic dyes, their surface interactions, such as ion–dipole and hydrogen bonding, can significantly alter the photophysical and photochemical properties of the dyes. On the other hand, organically modified clays accommodate a variety of nonionic dyes without significant modification of their properties. While the amounts of dye intercalated cannot be controlled (or evaluated) precisely, the photoprocesses of such dyes can yield important microscopic information. Hierarchical control from micro- to macroscopic structures is a key issue for the practical application of intercalation compounds. Recent developments regarding controlled macroscopic forms of intercalation compounds, such as particle shape (300,301), and the characterization of suspensions (302) represent milestones for the application of intercalation compounds.

REFERENCES

1. M Anpo, T Matsuura. Photochemistry on solid surfaces. Studies in Surface Science and Catalysis 47. Amsterdam: Elsevier, 1989.

2. M. Anpo. Surface Photochemistry. Chichester: Wiley Interscience, 1997.
3. J Klafter, JM Drake. Molecular Dynamics in Restricted Geometries. New York: Wiley Interscience, 1989.
4. V Ramamurthy. Photochemistry in Organized and Constrained Media. New York: VCH, 1991.
5. G Alberti, T Bein. Comprehensive supramolecular chemistry. Vol. 7. Oxford: Pergamon Press, 1996.
6. JK Thomas. J Phys Chem. 1987; 91:267.
7. JK Thomas. Chem Rev. 1993; 93:301.
8. NJ Turro, M Grätzel, AM Braun. Angew Chem Int Ed Engl. 1980; 19:675.
9. V Ramamurthy. Tetrahedron. 1986; 42:5753.
10. JK Thomas. Acc Chem Res. 1988; 21:275.
11. M Ogawa, K Kuroda. Chem Rev. 1995; 95:399–438.
12. T Shichi, K Takagi. In: V Ramamurthy, KS Schenze, eds. Solid State and Surface Photochemistry. Vol. 5. New York: Marcel Dekker, 2000:31–110.
13. MS Whittingham, AJ Jacobson. Intercalation Chemistry. New York: Academic Press, 1982.
14. W Müller-Warmuth, R Schöllhorn. Progress in Intercalation Research. Dordrecht: Kluwer Academic, 1994.
15. BKG Theng. The Chemistry of Clay–Organic Reactions. London: Adam Hilger, 1974.
16. RE Grim, Clay Mineralogy. New York: McGraw-Hill, 1953.
17. H Van Olphen. An Introduction to Clay Colloid Chemistry. 2nd ed. New York: Wiley-Intersciecnce, 1977.
18. DW Thompson, T Butterworth. J Colloid Interface Sci. 1992; 151:236.
19. M Ogawa, Y Nagafusa, K Kuroda, C Kato. Appl Clay Sci. 1992; 7:291.
20. K Kitajima, N Daimon. Nippon Kagaku Kaishi 685:1974.
21. M Soma, A Tanaka, H Seyama, S Hayashi, K Hayamizu. Clay Sci. 1990; 8:1.
22. HE Doner, MM Mortland. Science. 1969; 66:1406.
23. SA Boyd, MM Mortland. Nature. 1985; 316:532.
24. Y Soma, M Soma, I Harada. Chem Phys Lett. 1983; 94:475.
25. Y Soma, M Soma, I Harada. J Phys Chem. 1984; 88:3034.
26. A Moreale, P Cloos, C Badot. Clay Miner. 1985; 20:29.
27. M Soma, Y Soma. Chem Lett. 1988:405–408.
28. M Ogawa, K Kuroda, C Kato. Chem Lett. 1989:1659–1662.
29. M Ogawa, T Hirata, K Kuroda, C Kato. Chem Lett. 1992:365–368.
30. M Ogawa, T Handa, K Kuroda, C Kato. Chem.Lett. 1990:71–74.
31. M Ogawa, T Hashizume, K Kuroda, C Kato. Inorg Chem. 1991; 30:584.
32. IV Mitchell. Pillared Layered Structure. London: Elsevier, 1990.
33. M Ogawa, K Kuroda. Bull Chem Soc Jpn. 1997; 70:2593.
34. G Lagaly. Clay Miner. 1981; 16:1.
35. G Lagaly. Solid State Ionics. 1986; 22:43.
36. G Lagaly, K Beneke. Colloid Poly Sci. 1991; 269:1198.
37. G Lagaly. Adv Colloid Interface Sci. 1979; 11:105.
38. G Lagaly, K Beneke, A Weiss. Am Miner. 1975; 60:650.

39. G Lagaly, K Beneke, A Weiss. Am Miner. 1975; 60:642.
40. E Ruiz-Hitzky, M Rojo. Nature. 1980; 287:28.
41. E Ruiz-Hitzky, M Rojo, G Lagaly. Colloid Polym Sci. 1985; 263:1025.
42. M Ogawa, S Okutomo, K Kuroda. J Am Chem Soc. 1998; 120:7361–7362.
43. M Ogawa, M Miyoshi, K Kuroda. Chem Mater. 1998; 10:3787.
44. K Isoda, K Kuroda, M Ogawa. Chem Mater. 2000; 12:1702–1707.
45. ME Landis, BA Aufdembrink, P Chu, ID Johnson, GW Kirker, MK Rubin. J Am Chem Soc. 1991; 113:3189.
46. JS Dailey, TJ Pinnavaia. Chem Mater. 1992; 4:855.
47a. WT Reichle. Chemtech. 1986; 16:58–63.
47b. F Tsifiro, A Vaccari. In: G Alberti, T Bein, eds. Comprehensive Supramolecular Chemistry. Vol. 7. Oxford: Pergamon Press, 1996:251–292.
47c. V Rives, MA Ulibarri. Coord Chem Rev. 1999; 1999:61–120.
48. IY Park, K Kuroda, C Kato. J Chem Soc Dalton Trans. 1990:3071.
49. M Chibwe, TJ Pinnavaia. J Chem Soc Chem Commun. 1993:278.
50. A Clearfield, U Constantino. In: G Alberti, T Bein, eds. Comprehensive Supramolecular Chemistry. Vol. 7. Oxford: Pergamon Press, 1996:151–188.
51. B Raveau. Rev Inorg Chem. 1987; 9:37.
52. M Ogawa, M Takahashi, C Kato, K Kuroda. J Mater Chem. 1994; 4:519.
53. M Isayama, K Sakata, T Kunitake. Chem Lett 1993:1283.
54. M Ogawa, N Kanaoka, K Kuroda. Langmuir. 1998; 14:6969–6973.
55. K Inukai, Y Hotta, M Taniguchi, S Tomura, A Yamagishi. J Chem Soc Chem Commun. 1994:959.
56. Y Hotta, M Taniguchi, K Inukai, A Yamagishi. Clay Miner. 1997; 32:79.
57. ER Kleinfeld, GS Ferguson. Science. 1994; 265:370.
58. ER Kleinfeld, GS Ferguson. Chem Mater. 1996; 8:1575.
59. Y Lvov, K Ariga, I Ichinose, T Kunitake. Langmuir. 1996; 12:3038.
60. SW Keller, H-N Kim, TE Mallouk. J Am Chem Soc. 1994; 116:8817.
61. T Sasaki, M Watanabe, H Hashizume, H Yamada, H Nakazawa. Chem Commun. 1996:229.
62. T Sasaki, S Nakano, S Yamauchi, M Watanabe. Chem Mater. 1997; 9:602.
63. H Zollinger. Color Chemistry. Weinheim: VCH-Verlag, 1987.
64. K Kalyanasundaram. Photochemistry of Polypyridine and Porphyrin Complexes. London: Academic Press, 1992.
65. J Juris, V Balzani, F Barigelletti, S Campagna, P Belser, A Von Zelewsky. Coord Chem Rev. 1988; 84:85.
66. K Kalyanasundaram. Coord Chem Rev. 1982; 46:159.
67. AT Thornton, GS Laurence. J Chem Soc Chem Commun. 1978:408. PC Lee, D Meisel. J Am Chem Soc. 1980; 102:5477. Y Kurimura, M Nagashima, K Takato, E Tsuchida, M Kaneko, A Yamada. J Phys Chem. 1982; 86:2432. T Miyashita, Y Arito, M Matsuda. Macromolecules. 1991; 24:872.
68. JT Kunjappu, P Somasundaran, NJ Turro. J Phys Chem. 1990; 94:8464.
69. BH Milosauljevic, JK Thomas. Macromolecules. 1984; 17:2244. BH Milosauljevic, JK Thomas. Chem Phys Lett. 1985; 114:133.
70. HD Gafney. Coord Chem Rev. 1990; 104:113.

71. P Innocenzi, H Kozuka, T Yoko. J Phys Chem B. 1997; 101:2285.
72. K Matsui, K Sasaki, N Takahashi. Langmuir. 1991; 7:2866.
73. J Wheeler, JK Thomas. J Phys Chem. 1982; 86:4540.
74. TII Kim, TE Mallouk. J Phys Chem. 1992; 96:2879.
75. MF Traynor, MM Mortland, TJ Pinnavaia. Clays Clay Miner. 1978; 26:318.
76. M Borah, JN Ganguli, DK Dutta. J Colloid Interface Sci. 2001; 233:171–179.
77. PK Ghosh, AJ Bard. J Phys Chem. 1984; 88:5519.
78. A Habti, D Keravis, P Levitz, H Van Damme. J Chem Soc Faraday Trans 2. 1984; 80:67.
79. RA DellaGuardia, JK Thomas. J Phys Chem. 1983; 87:990.
80. RA Schoonheydt, P de Pauw, D Vliers, FC de Schryver. J Phys Chem. 1984; 88: 5113.
81. NJ Turro, CV Kumar, Z Grauer, JK Barton. Langmuir. 1987; 3:1056.
82. T Nakamura, JK Thomas. Langmuir. 1985; 1:567.
83. VG Kuykendall, JK Thomas. J Phys Chem. 1990; 94:4224.
84. D Krenske, S Abdo, H Van Damme, M Cruz, JJ Fripiat. J Phys Chem. 1980; 84: 2447.
85. A Awaluddin, RN DeGuzman, CV Kumar, SL Suib, SL Burkett, ME Davis. J Phys Chem. 1995; 99:9886.
86. S Abdo, P Canesson, M Cruz, JJ Fripiat, H Van Damme. J Phys Chem. 1981; 85: 797.
87. A Yamagishi. J Coord Chem. 1987; 16:131.
88. A Yamagishi, M Soma. J Am Chem Soc. 1981; 103:4640.
89. V Joshi, PK Ghosh. J Am Chem Soc. 1989; 111:5604.
90. V Joshi, PK Ghosh. J Chem Soc Chem Commun. 1987:789.
91. M Ogawa, M Inagaki, N Kodama, K Kuroda, C Kato. J Phys Chem. 1993; 97: 3819.
92. M Ogawa, M Tsujimura, K Kuroda. Langmuir. 2000; 16:4202–4206.
93. W Turbeville, DS Robins, PK Dutta. J Phys Chem. 1992; 96:5024.
94. K Nagai, N Takamiya, M Kaneko. J Photochem Photobiol A Chem. 1994; 84:271.
95. M Ogawa, N Maeda. Clay Miner. 1998; 33:643–650.
96. M Ogawa, T Takizawa. J Phys Chem B. 1999; 103:5005–5009.
97. RC Yeates, SM Kuznicki, LB LLoyd, EM Eyring. J Inorg Nucl Chem. 1981; 43: 2355.
98. DP Vliers, RA Schoonheydt, FC de Schrijver. J Chem Soc Faraday Trans 1. 1985; 81:2009.
99. DP Vliers, D Collin, RA Schoonheydt, FC de Schryver. Langmuir. 1986; 2:165.
100. JL Colón, C-Y Yang, A Clearfield, CR Martin. J Phys Chem. 1988; 92:5777.
101. JL Colón, C-Y Yang, A Clearfield, CR Martin. J Phys Chem. 1990; 94:874.
102. EP Giannelis, DG Nocera, TJ Pinnavaia. Inorg Chem. 1987; 26:203.
103. R Clement. J Am Chem Soc. 1981; 103:6998.
104. O Poizat, C Sourisseau. J Phys Chem. 1984; 88:3007.
105. R Jakubiak, AH Francis. J Phys Chem. 1996; 100:362.
106. T Nakato, D Sakamoto, K Kuroda, C Kato. Bull Chem Soc Jpn. 1992; 65:322.
107. T Nakato, K Kusunoki, K Yoshizawa, K Kuroda, M Kaneko. J Phys Chem. 1995; 99:17896.

108. CV Kumar, ZJ Williams. J Phys Chem. 1995; 99:17632.
109. JB Birks. Photophysics of Aromatic Molecules. London: Wiley Interscience, 1970.
110. A Nakajima. Bull Chem Soc Jpn. 1971; 44:3272.
111. K Kalyanasundaram, JK Thomas. J Am Chem Soc. 1977; 99:2039.
112. RA DellaGuardia, JK Thomas. J Phys Chem. 1983; 87:3550.
113. T Nakamura, JK Thomas. J Phys Chem. 1986; 90:641.
114. K Viane, J Caigui, RA Schoonheydt, FC De Schryver. Langmuir. 1987; 3:107.
115. K Viane, RA Schoonheydt, M Cruzen, B Kunyima, FC De Schryver. Langmuir. 1988; 4:749.
116. B Kunyima, K Viane, MM Hassan Khalil, RA Schoonheydt, M Cruzen, FC De Schryver. Langmuir. 1990; 6:482.
117. CV Kumar, EH Asuncion, G Rosenthal. Microporous Mater. 1993; 1:123.
118. CV Kumar, EH Asuncion, G Rosenthal. Microporous Mater. 1993; 1:299.
119. M Ogawa, H Shirai, K Kuroda, C Kato. Clays Clay Miner. 1992; 40:485.
120. M Ogawa, T Aono, K Kuroda, C Kato. Langmuir. 1993; 9:1529.
121. M Ogawa, T Wada, K Kuroda. Langmuir. 1995; 11:4598.
122. I Momiji, C Yoza, K Matsui. J Phys Chem B. 2000; 104:1552–1555.
123. MF Ahmadi, JF Rusling. Langmuir. 1995; 11:94.
124. Y Okahata, A Shimizu. Langmuir. 1989; 5:954.
125. N Hu, JF Rusling. Anal Chem. 1991; 63:2163.
126. T Nakamura, JK Thomas. Langmuir. 1987; 3:234.
127. RA DellaGuardia, JK Thomas. J Phys Chem. 1984; 88:964.
128. GP Wiederrecht, G Sandi, KA Carrado, S Seifert. Chem Mater. 2001; 13:4233.
129. PT Hang, GW Brindley. Clays Clay Miner, S Yariv, D Lurie. Isr J Chem. 1971; 9:537.
130. RK Taylor. J Chem Tech Biotechnol. 195; 35A.
131. D Saehr, RL Dred, DCM Hoffman. Clay Miner. 1978; 13:411.
132. J Cenes, RA Schoonheydt. Clays Clay Miner. 1988; 36:214.
133. RA Schoonheydt, L Heughebaert. Clay Miner. 1992; 27:91.
134. C Breen, B Rock. Clay Miner. 1994; 29:179.
135. J Bujdak, P Komadel. J Phys Chem B. 1997; 101:9065.
136. KY Jacobs, RA Schoonheydt. Langmuir. 2001; 17:5150.
137. Z Grauer, GL Grauer, D Avnir, S Yariv. J Chem Soc Faraday Trans 1. 1987; 83:1685.
138. GL Grauer, D Avnir, S Yariv. Can J Chem. 1984; 62:1889.
139. T Endo, N Nakada, T Sato, M Shimada. J Phys Chem Solids. 1988; 49:1423.
140. Z Grauer, D Avnir, S Yariv. Can J Chem. 1984; 62:1889.
141. Z Grauer, AB Malter, S Yariv, D Avnir. Colloids Surfaces. 1987; 25:41.
142. MJT Estévez, FL Arberoa, TL Arberoa, IL Arberoa, RA Schoonheydt. Clay Miner. 1994; 29:105.
143. MJT Estévez, FL Arbeloa, TL Arbeloa, IL Arbeloa. Langmuir. 1994; 9:3629.
144. F López Arbeloa, MJ Tapia Estévez, T López Arbeloa, I López Arbeloa. Langmuir. 1995; 11:3211–3217.
145a. R Chaudhuri, F López Arbeloa, I López Arbeloa. Langmuir. 2000; 16:1285–1291.
145b. F López Arbeloa, V Martínez Martínez, J Bañuelos Prieto, I López Arbeloa. Langmuir. 2002; 18:2658–2664.

146. T Endo, T Sato, M Shimada. J Phys Chem Solids. 1986; 47:799.
147. T Endo, N Nakada, T Sato, M Shimada. J Phys Chem Solids. 1989; 50:133.
148. S Takagi, T Shimada, T Yui, H Inoue. Chem Lett. 2001:128–129.
149. P Wlodarczyk, S Komarneni, R Roy, WB White. J Mater Chem. 1996; 6: 1967–1969.
150. SS Cady, TJ Pinnavaia. Inorg Chem. 1978; 17:1501.
151. H Van Damme, M Crepsin, F Obrecht, MI Cruz, JJ Fripiat. J Colloid Interface Sci. 1978; 66:43.
152. KA Carrado, RE Winans. Chem Mater. 1990; 2:328.
153. S Abdo, MI Cruz, JJ Fripiat. Clays Clay Miner. 1980; 28:125.
154. EP Giannelis. Chem Mater. 1990; 2:627.
155. L Ukrainczyk, M Chibwe, TJ Pinnavaia, SA Boyd. J Phys Chem. 1994; 98:2668.
156. VG Kuykendall, JK Thomas. Langmuir. 1990; 6:1350.
157. K Sakoda, K Kominami. Chem Phys Lett. 1993; 216:270.
158. T Nakato, Y Iwata, K Kuroda, M Kaneko, C Kato. J Chem Soc Dalton Trans. 1993:1405.
159. IY Park, K Kuroda, C Kato. Chem Lett. 1989:2057.
160. RM Kim, JE Pillion, DA Burwell, JT Groves, ME Thompson. Inorg Chem. 1993; 32:4509.
161. KA Carrado, P Thiyagarajan, RE Wianes, RE Botto. Inorg Chem. 1991; 30:794.
162. L Garillon, F Bedioui, J Devynck, P Battioni, L Barloy, D Mansuy. J Electroanal Chem. 1991; 303:283.
163. L Garillon, F Bedioui, J Devynck, P Battioni. J Electroanal Chem. 1993; 347:435.
164. F Bergaya, H VanDamme. Geochim Cosmochim Acta. 1982; 46:349.
165. H Kameyama, H Suzuki, A Amano. Chem Lett. 1988:1117.
166. L Barloy, JP Lallier, D Mansuy, Y Piffard, M Tournox, JB Valim, W Jones. New J Chem. 1992; 16:71.
167. KA Carrado, JE Forman, RE Botto, RE Winans. Chem Mater. 1993; 5:472.
168. JF Rusling, MF Ahmadi, N Hu. Langmuir. 1992; 8:2455.
169. N Hu, JF Rusling. Anal Chem. 1991; 63:2163.
170. M Chibwe, TJ Pinnavaia. J Chem Soc Chem Commun. 1993:278.
171. L Grigoryan, K Yakushi, C-J Liu, S Takano, M Wakata, H Yamauchi. Physica C. 1993; 218:153.
172. M Ogawa, R Kawai, K Kuroda. J Phys Chem. 1996; 100:16218.
173. M Iwasaki, M Kita, K Itoh, A Kohno, K Fukunishi. Clays Clay Miner. 2000; 48: 392–399.
174. EE Jelley. Nature. 1936; 138:1009.
175. G Scheibe. Angew Chem. 1936; 49:563.
176. AH Herz. Adv Colloid Interface Sci. 1977; 8:237.
177. D Möbius. Adv Mater. 1995; 7:437.
178. T Coradin, K Nakatani, I Ledoux, J Zyss, R Clément. J Mater Chem. 1997; 7: 853–854.
179. N Iyi, K Kurashima, T Fujita. Chem Mater. 2002; 14:583–589.
180. U Costantino, N Coletti, M Nocchetti, GG Aloisi, F Elisei, L Latterini. Langmuir. 2000; 16:10351–10358.

181. T Fujita, N Iyi, T Kosugi, A Ando, T Deguchi, T Sota. Clays Clay Miner. 1997; 45:77–84.
182a. M Ogawa. Chem Mater. 1996; 8:1347.
182b. M Ogawa, A Ishikawa. J Mater Chem. 1998; 8:463.
183. M Ogawa, M Yamamoto, K Kuroda. Clay Miner. 2001; 36:263.
184. M Ogawa, R Goto, N Kakegawa. Clay Sci. 2000; 11:231–242.
185. M Shimomura, S Aiba, N Tajima, N Inoue, K Okuyama. Langmuir. 1995; 11:969.
186. PN Prasad, DJ Williams. Introduction to Nonlinear Optical Effects in Molecules and Polymers. New York: Wiley Interscience, 1991.
187. RW Munn, CN Ironside. Principles and Applications of Nonlinear Optical Materials. London: CRC, 1993.
188. SK Kurtz, TT Perry. J Appl Phys. 1968; 39:3798.
189. DQ Li, MA Ratner, TJ Marks, C Zhang, J Yang, GK Wong. J Am Chem Soc. 1990; 112:7389.
190. HE Katz, G Scheller, TM Putvinski, ML Schilling, WL Wilson, CED Chidsey. Science. 1991; 254:1485.
191. S Cooper, PK Dutta. J Phys Chem. 1990; 94:114.
192. K Kuroda, K Hiraguri, Y Komori, Y Sugahara, H Mouri, Y Uesu. Chem Comm. 1999:2253–2254.
193. PG Lacroix, AVV Lemarinier, R Clément, K Nakatani, JA Delaire. J Mater Chem. 1993; 3:499.
194. PG Lacroix, R Clément, K Nakatani, J Zyss, I Ledoux. Science. 1994; 263:658.
195. M Ogawa, M Takahashi, K Kuroda. Chem Mater. 1994; 6:715.
196. N Miyamoto, K Kuroda, M Ogawa. J Am Chem Soc. 2001; 123:6949.
197. K Nassau, JW Shiever, JL Bernstein. J Electrochem Soc. 1969; 116:348.
198. M Gasperin, MTL Bihan. J Solid State Chem. 1982; 43:346.
199. A Yamagishi, M Soma. J Phys Chem. 1981; 85:3090.
200. M Taniguchi, A Yamagishi, T Iwamoto. J Phys Chem. 1990; 94:2534.
201. Z Zernia, D Gill, S Yariv. Langmuir. 1994; 10:3988–3993.
202. K Yamaoka, R Sasai. J Colloid Interface Sci. 2000; 225:82–93.
203. R Sasai, T Shichi, K Gekko, K Takagi. Bull Chem Soc Jpn. 2000; 73:1925–1931.
204. R Sasai, T Shin'ya, T Shichi, K Takagi, K Gekko. Langmuir. 1999; 15:413.
205. S Holzheu, H Hoffmann. J Colloid Interface Sci. 2002; 245:16–23.
206. H Dürr, H Bouas-Laurent. Photochromism Molecules and Systems. Amsterdam: Elsevier, 1990.
207. JM Adams, AJ Gabbutt. J Incl Phenom. 1990; 9:63.
208. K Takagi, T Kurematsu, Y Sawaki. J Chem Soc Perkin Trans. 1991; 2:1517.
209. R Sasai, H Ogiso, I Shindachi, T Shichi, K Takagi. Tetrahedron. 2000; 56: 6979–6984.
210. T Seki, K Ichimura. Macromolecules. 1990; 23:31.
211. H Tomioka, T Itoh. J Chem Soc Chem Commun 532:1991.
212. M Ogawa, H Kimura, K Kuroda, C Kato. Clay Sci. 1996; 10:57.
213. M Ogawa, M Hama, K Kuroda. Clay Miner. 1999; 34:213.
214. M Sasaki, T Fukuhara. Photochem Photobiol. 1997; 66:716–718.
215. H Kandori, T Ichioka, M Sasaki. Chem Phys Lett. 2002; 354:251–255.

216. M Ogawa, T Ishii, N Miyamoto, K Kuroda. Adv Mater. 2001; 13:1107.
217. M Ogawa, K Fujii, K Kuroda, C Kato. Mater Res Soc Symp Proc. 1991; 233:89.
218. T Fujita, N Iyi, Z Klapyta. Mater Res Bull. 1998; 33:1693–1701.
219. K Takagi, H Usami, H Fukaya, Y Sawaki. J Chem Soc Chem Commun. 1989: 1174.
220. H Usami, K Takagi, Y Sawaki. J Chem Soc Perkin Trans. 1990; 2:1723.
221. H Usami, K Takagi, Y Sawaki. J Chem Soc Faraday Trans. 1992; 88:77.
222. H Usami, K Takagi, Y Sawaki. Bull Chem Soc Jpn. 1991; 64:3395.
223. K Takagi, T Shichi, H Usami, Y Sawaki. J Am Chem Soc. 1993; 115:4339.
224. T Shichi, K Takagi, Y Sawaki. Chem Lett. 1996:781.
225. T Shichi, K Takagi, Y Sawaki. Chem Commun. 1996:2027.
226. WE Moerner. Persistent Spectral Hole Burning: Science and Applications. Berlin: Springer-Verlag, 1988.
227. M Ogawa, T Handa, K Kuroda, C Kato, T Tani. J Phys Chem. 1992; 96:8116.
228a. L Margulies, H Rozen, E Cohen. Nature. 1985; 315:658–659.
228b. L Margulies, H Rozen, E Cohen. Clays Clay Miner. 1988; 36:159.
229. L Margulies, H Rozen. J Mol Struct. 1986; 141:219.
230. B Casal, J Merino, JM Serratosa, E Ruiz-Hitzky. Appl Clay Sci. 2001; 18:245–254.
231. D Avnir, Z Grauer, S Yariv, D Huppert, D Rojanski. Nouv J Chim. 1986; 10:153.
232. APP Cione, MG Neumann, F Gessner. J Colloid Interface Sci. 1998; 198:106–112.
233. H Usami, T Nakamura, T Makino, H Fujimatsu, S Ogasawara. J Chem Soc Faraday Trans. 1998; 94:83–87.
234. PMS Monk. The Viologens. Chichester: J Wiley, 1998.
235. H Miyata, Y Sugahara, K Kuroda, C Kato. J Chem Soc Faraday Trans 1. 1987; 83:1851.
236. LA Vermeulen, ME Thompson. Nature. 1992; 358:656.
237. LA Vermeulen, J Snover, LS Sapochak, ME Thompson. J Am Chem Soc. 1993; 115:11767.
238. J Snover, ME Thompson. J Am Chem Soc. 1994; 116:765.
239. SB Ungashe, WL Wilson, HE Katz, GR Scheller, TM Putvinsky. J Am Chem Soc. 1992; 114:8717.
240. CV Kumar, A Chadhari, GL Rosenthal. J Am Chem Soc. 1994; 116:403.
241. T Nakato, K Kuroda. Eur J Solid State Chem. 1995; 32:809–818.
242. H Miyata, Y Sugahara, K Kuroda, C Kato. J Chem Soc Faraday Trans 1. 1988; 84:2677.
243. T Nakato, H Miyata, K Kuroda, C Kato. React Solids. 1988; 6:231.
244. T Nakato, K Kuroda, C Kato. J Chem Soc Chem Commun. 1989:1144.
245. T Nakato, K Kuroda, C Kato. Chem Mater. 1992; 4:128.
246. T Nakato, K Ito, K Kuroda, C Kato. Microporous Materials. 1993; 1:283.
247. B Casal, E Ruiz-Hitzky, F Bergaya, D Challal, J Fripiat, H Van Damme. J Mol Catal. 1985; 33:83.
248. H Nijs, M Cruz, J Fripiat, H Van Damme. J Chem Soc Chem Commun. 1981: 1026.
249. H Nijs, M Cruz, J Fripiat, H Van Damme. J Phys Chem. 1983; 87:1279.
250. C Detellier, G Villemure. Inorg Chim Acta. 1984; 86:L19.

251. C Detellier, G Villemure, H Kodama. Can J Chem. 1984; 63:1139.
252. G Villemure, G Bazan, H Kodama, AG Szabo, C Detellier. Appl Clay Sci. 1987; 2:241.
253. H Van Damme, H Nijs, JJ Fripiat. J Mol Catal. 1984; 27:123.
254. H Van Damme, F Bergaya, A Habti, JJ Fripiat. J Mol Catal. 1983; 21:223.
255. MD Newsham, EP Giannelis, TJ Pinnavaia, DG Nocera. J Am Chem Soc. 1988; 110:3885.
256. SL Suib, KA Carrado. Inorg Chem. 1985; 24:863.
257. SL Suib, JF Tanguay, ML Occelli. J Am Chem Soc. 1986; 108:6972.
258a. JH Fendler. Chem Rev. 1987; 87:877.
258b. For example, M Krishran, JR White, MA Fox, AJ Bard. J Am Chem Soc. 1983; 105:7002.
258c. BH Milosauljevic, J Kuczynski, JK Thomas. J Phys Chem. 1984; 88:980.
258d. J Kuczynski, JK Thomas. J Phys Chem. 1985; 89:2720.
258e. N Herron, Y Wang, ME Eddy, GD Stucky, DE Cox, K Moller, T Bein. J Am Chem Soc. 1989; 111:530.
259. H Yoneyama, S Nippa. Chem Lett. 1988:1807–1808.
260. H Yoneyama, S Saga, S Yamanaka. J Phys Chem. 1989; 93:4833.
261. JF Tanguay, SL Suib, RW Coughlin. J Catal. 1989; 117:335.
262a. F-RF Fan, H-Y Liu, AJ Bard. J Phys Chem. 1985; 89:4418.
262b. I Dekany, L Turi, Z Kiraly. App Clay Sci. 1999; 15:221–239.
263. H Miyoshi, H Mori, H Yoneyama. Langmuir. 1991; 7:503.
264. H Miyoshi, H Yoneyama. J Chem Soc Faraday Trans 1. 1989; 85:1873.
265. O Enea, AJ Bard. J Phys Chem. 1986; 90:301.
266. RD Stramel, T Nakamura, JK Thomas. Chem Phys Lett. 1986; 130:423.
267. RD Stramel, T Nakamura, JK Thomas. J Chem Soc Faraday Trans 1. 1988; 84: 1287.
268. X Liu, JK Thomas. J Colloid Interface Sci. 1989; 129:476.
269. T Sato, H Okuyama, T Endo, M Shimada. React Solids. 1990; 8:63.
270. G Cao, LK Rasenberg, CM Nunn, TE Mallouk. Chem Mater. 1991; 3:149.
271. T Itoh, A Ishii, Y Kodera, A Matsushima, M Hiroto, H Nishimura, T Tsuzuki, T Kamachi, I Okura, Y Inada. Bioconjugate Chem. 1998; 9:409–412.
272. K Domen, A Kudo, A Shinozaki, A Tanaka, K Maruya, T Onishi. J Chem Soc Chem Commun. 1986:356.
273. K Domen, A Kudo, M Shibata, A Tanaka, K Maruya, T Onishi. J Chem Soc Chem Commun. 1986:1706.
274. A Kudo, A Tanaka, K Domen, K Maruya, K Aika, T Onishi. J Catal. 1988; 111: 67.
275. A Kudo, K Sayama, K Asakura, K Domen, K Maruya, T Onishi. J Catal. 1989; 120:337.
276. A Kameyama, K Domen, K Maruya, T Endo, T Onishi. J Mol Catal. 1990; 58: 205.
277. K Sayama, A Tanaka, K Domen K Maruya, T Onishi. J Phys Chem. 1991; 95: 1345.
278. T Sekine, J Yoshimura, A Tanaka, K Domen, K Maruya, T Onishi. Bull Chem Soc Jpn. 1990; 63:2107.

279. YI Kim, SJ Atherton, ES Brigham, TE Mallouk. J Phys Chem. 1993; 97:11802.
280. YI Kim, S Salim, MJ Huq, TE Mallouk. J Am Chem Soc. 1991; 113:9561.
281. DW Kim, A Blumstein, J Kumar, SK Tripathy. Chem Mater. 2001; 13:243–246.
282. DW Kim, A Blumstein, SK Tripathy. Chem Mater. 2001; 13:1916–1922.
283. X Fan, J Locklin, JH Youk, W Blanton, C Xia, R Advincula. Chem Mater. 2002; 14:2184–2191.
284. B van Duffel, T Verbiest, S van Elshocht, A Persoons, FCD Schryver, RA Schoonheydt. Langmuir. 2001; 17:1243–1249.
285. Y Umemura, A Yamagishi, R Schoonheydt, A Persoons, FD Schryver. J Am Chem Soc. 2002; 124:992–997.
286. K Inukai, Y Hotta, S Tomura, M Takahashi, A Yamagishi. Langmuir. 2000; 16: 7679–7684.
287. TW Lee, OO Park, JJ Kim, JM Hong, YC Kim. Chem Mater. 2001; 13:2217–2222.
288. C Del Hoyo, MA Vicente, V Rives. Clay Miner. 2001; 36:541–546.
289. C Del Hoyo, V Rives, MA Vicente. Clay Miner. 1998; 33:467–474.
290. MA Vicente, M Sánchez-Camazano, MJ Sánchez-Martín, M DelArco, C Martín, V Rives, J Vicente-Hernández. Clays Clay Miner. 1989; 37:157–163.
291. M Borisover, ER Graber, F Bercovich, Z Gerstl. Chemosphere. 2001; 44: 1033–1040.
292. M Ogawa, T Morita, T Okada. Submitted.
293a. T Okada, M Ogawa. Chem Lett. 2002:632–633.
293b. T Okada, M Ogawa. Chem Commun. 2003:1378–1379.
294. RM Barrer. Zeolites and Clay Minerals as Sorbents and Molecular Sieves. London: Academic Press, 1978.
295a. MM Mortland, S Shaobai, SA Boyd. Clays Clay Miner. 1986; 34:581.
295b. SA Boyd, JF Lee, MM Mortland. Nature. 1988; 333:345.
295c. SA Boyd, S Shaobai, JF Lee, MM Mortland. Clays Clay Miner. 1988; 36:125.
296. R Fahn, K Fenderl. Clay Miner. 1993; 18:447.
297. K Ito, N Zhou, K Fukunishi. J Imaging Sci Technol. 1994; 38:575–579.
298. K Ito, M Kuwabara, K Fukunishi, Y Fujiwara. J Imaging Sci Technol. 2000; 40: 275–280.
299. K Ito, K Fukunishi. Chem Lett. 1997:357–358.
300. U Constantino, F Marmottini, M Nocchetti, R Vivalni. Eur J Inorg Chem. 1998: 1439–1446.
301. M Ogawa, H Kaiho. Langmuir. 2002; 18:4240–4242.
302. JCP Gabriel, F Camerel, BJ Lemaire, H Desvaux, P Davidson, P Batail. Nature. 2001; 413:504–508.

6

Pillared Clays and Porous Clay Heterostructures

Pegie Cool and Etienne F. Vansant
University of Antwerp
Antwerp, Belgium

I. INTRODUCTION

After about 50 years of intensive research on the development of inorganic molecular sieves, the porous solids have now taken a firm position in materials science, finding important applications in the field of adsorption and catalysis. The longest-known molecular sieves are the natural zeolites, with a three-dimensional open-structure framework built up of very small micropores (pore diameter < 2 nm). After the oil crisis of 1973, the main objective was to obtain materials with larger pores allowing more bulky molecules in their porous network. It was then that scientists started to create porosity in the interlayer space of layered clays, developing the first pillared clays with pores in the larger microporous region. Since the mid-1990s, however, there has been an industrial need for materials with even larger mesopores (pore diameter > 2 nm) for the processing of heavy oil fractions and the synthesis of pharmaceuticals, fine chemicals, and bio-compounds. Therefore, based on these changing requirements of the catalytic industry, much effort has been expended in producing mesoporous solids. In 1995, then, the first clay-derived mesoporous material (named *porous clay heterostructure*) was developed based on a templated technique. Both pillared clay and porous clay heterostructures will be discussed in this chapter, with the focus on their synthesis and applications.

II. PILLARED CLAYS

Pillaring is the process by which an inorganic layered compound is transformed into a thermally stable micro- and mesoporous material with retention of the layer structure. The pillared material is a pillared layered solid, or PLS. In the case of clay minerals, one obtains a pillared clay mineral. A PLS is a subgroup of the group of intercalation compounds, which meet the following criteria (1):

1. The minimum increase of basal spacing is equal to the diameter of the N_2 molecule, commonly used to measure surface areas and pore volumes (0.315–0.353 nm).
2. The layers are separated vertically and do not collapse upon removal of the solvent.
3. The pillaring agent has molecular dimensions and is laterally spaced on a molecular-length scale.
4. The interlamellar space is porous. The minimum size of the pore opening is the diameter of the N_2 molecules; there is no maximum limit to the size of the pores.

These four criteria mean that a PLS has chemical and thermal stability, that there is a molecular distribution of pillars in the interlamellar region, but that ordering of the pillars is not required. The elemental platelets, or lamellae, of the crystalline layered solid must be ordered so as to give an X-ray diffraction pattern, which allows the determination of the d_{001} spacing. Finally, the interlamellar region of the PLS must be accessible to molecules at least as large as N_2.

A. Al-Pillared Smectite Clays

The preparation of pillared clay minerals involves the following steps: (a) swelling of the clay mineral in water; (b) ion exchange of the pillaring agent or its precursor; (c) washing of the product; (d) calcination. The chemistry of each step has been very difficult to study because, even for the most commonly used pillaring cation, Al^{3+}, the chemistry is not fully understood, nor is anything known about the distribution of the pillars in the interlamellar region and their shapes. Butruille and Pinnavaia (2) reviewed the alumina pillared clays up to 1996, while Gil and Gandia (3) discussed the pillaring with solutions containing two or more cations, one being Al^{3+}. Both review papers stress the catalytic applications of the pillared materials. Here, we discuss the chemistry of each step in the pillaring process, with Al^{3+} as an example.

1. Solution Chemistry of Al^{3+}

Al^{3+} readily hydrolyzes in water to form a variety of products (4). One of them is the Keggin ion $[Al_{13}O_4(OH)_{24}(H_2O)_{12}]^{7+}$, Al_{13}^{7+}, which is stable in a broad

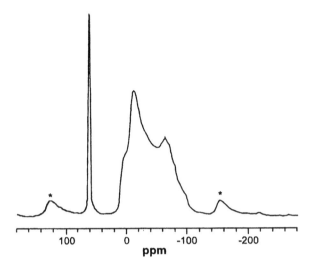

FIGURE 1 The ^{27}Al MAS-NMR spectrum of the $Al_{13}{}^{7+}$-sulphate salt. Spinning side bands are indicated by an asterisk. (From Ref. 5.)

range of conditions: $[Al^{3+}] < 0.01M$, $[OH^-]/[Al^{3+}] < 2.4$ and pH < 5. It consists of a central AlO_4 tetrahedron with every O connected to three $Al(OH)_4(H_2O)_2$ octahedra. In solution the ^{27}Al NMR spectrum consists of a single sharp line at 60 ppm. The Keggin ion can be precipitated as a sulphate salt, and, in the solid state, the ^{27}Al NMR spectrum contains the sharp tetrahedral line at 60 ppm and a complex line system of octahedral Al between 20 and -100 ppm (Fig. 1).

Further hydrolysis of $Al_{13}{}^{7+}$ gives dimers, trimers, polymers, and finally pseudoboehmite, which precipitates out of the solution. All these species can be distinguished on the basis of their Al–OH vibrations, as given in Table 1 (5).

Table 1 AlOH Vibrations of Oligomeric Al

Species	δ(OH), cm^{-1}	γ(OH), cm^{-1}
$Al_{13}{}^{7+}$	980	780
Associated Al_{13}	960–980	720–730
Polymeric Al_{13}	880–960	—
Pseudoboehmite	1070–1080	740–760

Source: Ref. 5.

Most of the solutions used for pillaring contain a variety of Al^{3+} species, with Al_{13}^{7+} as the majority species. These solutions are slightly acidic. Thus, several species of Al^{3+} and H^+ might be exchanged simultaneously. The relative amount of each species on the clay surface will depend on their relative amounts in solution, the selectivity of each species in the ion exchange reaction, and its chemical stability at the clay surface. To understand the process fundamentally, one species has to be selected and studied in detail. This is what has been done with the Al_{13}^{7+} Keggin ion (6). The isotherm of saponite is shown in Figure 2. It confirms that the reaction is an ion exchange reaction:

$$\overline{7Na^+} + Al_{13}^{7+} \rightarrow \overline{Al_{13}^{7+}} + 7Na^+ \tag{1}$$

There is a quantitative uptake of Al^{3+} up to almost 2 mmol/g and a release of 1.5 mol/g of Na^+. The numbers indicate that the expected ratio [Al]/[Na] = 13/7 is not attained. More Na^+ is released in solution than required for a stoichiometric ion exchange reaction. This can be ascribed either to exchange of protons

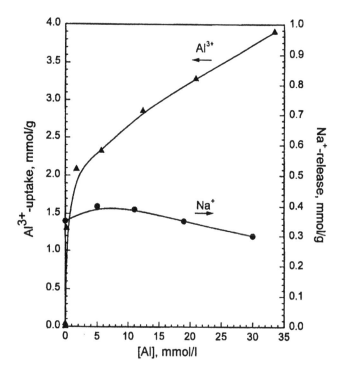

FIGURE 2 Uptake of Al and release of Na upon ion exchange of Al_{13}^{7+} with saponite. (From Ref. 5.)

from the slightly acidic solution or to hydrolysis of the Keggin ion in contact with the clay surface:

$$[Al_{13}O_4(OH)_{24}(H_2O)_{12}]^{7+} + xH_2O \leftrightarrow [Al_{13}O_4(OH)_{24+x}(H_2O)_{12-x}]^{7-x} + xH_3O^+ \tag{2}$$

In the second stage of the isotherm, the amount of Al taken up increases proportionally with the amount present in solution, but the amount of Na^+ released decreases. This indicates that the ion exchange is partial and that part of the $Al_{13}{}^{7+}$ is precipitated on the clay. The relative amount of precipitated $Al_{13}{}^{7+}$ increases with the amount present in the exchange solution.

2. The Pillaring Process

After ion exchange, washing or dialysis is a necessary step to obtain a pillared clay mineral. Three regimes can be distinguished: (a) If the loading of Al is less than 1.4 mmol/g, the d_{001} spacing is 1.25 nm and does not jump to 1.80–1.90 nm upon washing; (b) at Al loadings above 2.5 mmol/g, the d_{001} spacing is 1.80–1.90 nm before washing and does not change upon washing; (c) at the intermediate Al loadings, the d_{001} spacing is 1.25 nm before washing and increases to 1.80–1.90 nm upon washing. Most of the preparations of pillared clay minerals have these intermediate loadings, and washing is required to obtain a pillared product.

The washing process releases Cl^-, Al^{3+}, and Na^+ into solution, especially with the first washing. At the same time, the pH of the clay suspensions in the dialysis bags increases from about 4.2 to 5–5.5 for beiddelite, montmorillonite, and saponite and to 7–8 for hectorite and laponite (7). These data suggest a complex chemistry accompanying washing or dialysis. The presence of Cl^- indicates either that Cl^- substitutes for OH^- in the Keggin ion or that Al-chloro complexes are adsorbed, which hydrolyze upon washing:

$$Al-Cl + 2H_2O \rightarrow Al-OH + H_3O^+ + Cl^- \tag{3}$$

and the proton exchanges for Na^+. The increase of the pH of the suspension in the dialysis tube and the concommitant (small or negligible) increase of the pH in the surrounding solution also indicate release of protons due to hydrolysis. Aceman et al. (7) suggest that the Keggin ion is adsorbed on the external surface of the clay particles, hydrolyzes, and breaks up into smaller oligomers. These Al-oligomers diffuse in the interlamellar space together with Cl anions and hydrolyze to form the pillaring Al species. Whatever the exact chemistry, the experimental data indicate that pillaring is a complex process involving Al Keggin ions and other Cl-containing Al species in solution. Once adsorbed, these species undergo hydrolysis reactions, and washing or dialysis is necessary to obtain the pillared product with the characteristic 1.80–1.90-nm spacing. At the same time, the d_{001} line sharpens, which is indicative of an increase in the degree of ordering in the pillared product.

The final step in the pillaring process is calcination. This converts the pillaring species into an Al oxide pillar with retention of the d_{001} spacing. The thermal stability of this spacing depends on the crystallinity of the product obtained after washing: The sharper the d_{001} line, the more thermally stable is the pillared clay mineral. In any case, the structure of the oxidic cluster is unknown. In the starting Keggin ion, the ratio octahedral Al:tetrahedral Al is 12:1, and it is unlikely that this ratio exists in the cluster. A detailed NMR study of the pillaring of an Al-free smectite such as hectorite would shed some light on this problem. The formation of a chemical bond between the cluster and the Si-tetrahedra of the clay mineral has been discussed on the basis of NMR studies. Several authors have invoked such a Si–O–Al bond with inversion of the Si-tetrahedron for smectites with isomorphic substitution in the tetrahedral layer (8).

For industrial production of pillared clays, the pillaring process has to be optimized and scaled up. Sonication of the clay suspension, acid and base treatments, and competitive exchange are some of the techniques found in the literature (9). In laboratory preparations, large amounts of water are needed. Thus pillaring of concentrated aqueous smectite suspensions has been tried in order to find better conditions for scale-up (10).

In conclusion, pillaring of smectites is a well-established process under laboratory conditions. There is room for further investigation, mainly in two areas: determination of the structure of the oxidic Al cluster and optimization of the pillaring conditions for scale-up. An essential condition for pillaring is the swelling of the smectites in aqueous suspension. Swelling of clay minerals with a much higher charge density than that of smectites does not occur spontaneously in aqueous suspensions. The pillaring of these types of clays is discussed in the next section.

B. Al-Pillared Clays Derived from High-Charge-Density Clays

Phlogopites are trioctahedral 2:1 sheet silicates with all the possible octahedral positions occupied mostly by divalent cations. The numerous octahedral and tetrahedral layer substitutions are responsible for the high net negative layer charge of micas. The end-member phlogopite has the following general formula: $K_2Mg_6(Si_6Al_2)O_{20}(OH,F)_4$. The potassium ions, the dominant charge-compensating interlayer cations, are located between unit layers, with adjacent layers being stacked in such a way that the potassium ion is equidistant from 12 oxygens, six of each tetrahedral layer (11). In their original state, natural micas do not swell in the presence of water because the hydration energy of the interlayer potassium ions is insufficient to overcome the cooperative structural forces at the coherent edges of a cleavage surface (11) and, hence, ion exchange does not occur. Two obstacles have to be overcome in order to pillar natural micas: (a) the high selectivity of mica for the potassium ions makes it very difficult to exchange other cations, and (b) even a Na-exchanged mica, when brought in contact with an Al-pillaring solution, selectively retains Al^{3+} instead of the bulkier Al_{13}^{7+} species from the pillaring solution (12).

Vermiculites may be considered as swelling trioctahedral micas with substitutions of Al for Si in the tetrahedral layers and of Fe– and Al– for Mg in the octahedral layers. They constitute intermediate minerals in the natural weathering sequence of micas to smectites (13), with a negative layer charge density between that of micas and smectites, and where hydrated cations, most often Mg^{2+} and Ca^{2+}, have replaced the charge-neutralizing K^+ ions of the parent mica, providing partial swelling properties. Straight pillaring of vermiculite suspensions with Al_{13}-containing solutions gives materials with interlayer spacings of about 1.4 nm at room temperature (14–16), namely, half the gallery height (or interlayer free spacing) commonly achieved in 1.8-nm Al-pillared smectites (0.4 nm vs. about 0.8 nm (17,18), a failure attributed to the location of the negative charge on the basal oxygens of the tetrahedral layers, preventing the selective intercalation of the bulkier Keggin-type Al_{13}^{7+} cations (16). A preliminary treatment of vermiculite with an ornithine solution prior to pillaring treatment results in a mixture of mostly nonpillared material and a small pillared fraction (19).

Successful pillaring of vermiculites and phlogopites has recently been achieved after a preliminary treatment aiming at a reduction of the layer charge density of the minerals. This treatment is comparable with a controlled "accelerated weathering" process, which mimics the natural alteration sequence: micas → vermiculites → smectites. Bringing the charge-reduced minerals in contact with the pillaring solution results in 1.8-nm pillared materials (20,21). The higher structural stability of vermiculites and micas to temperature compared with smectites is of considerable interest in obtaining pillared materials with high resistance to thermal treatments, a weakness shared by all the smectite-based pillared materials. From the catalytic point of view, the higher number of Al-for-Si tetrahedral substitutions compared with, e.g., saponites and beidellites makes those minerals very attractive because it may be anticipated that more Brønsted acid sites can be generated upon pillaring.

1. Pillaring Process

Pillaring of vermiculites and phlogopites was possible only after a treatment aiming at the charge reduction of the minerals. In setting up the pillaring method, three requirements need to be considered: (a) There should be no need for an additional grinding treatment of the mineral; (b) the mineral dispersions should not be too diluted, and (c): the method should be applicable to vermiculites and phlogopites from different deposits.

The conditioning treatment consists of the following four successive steps:

1. Acid leaching with diluted nitric acid in controlled conditions
2. Calcination in air at 873 K for 4 h
3. Acid leaching with a diluted solution of a complexing acid
4. Saturation of the residual exchange positions with sodium (or calcium) ions

At the end of this "conditioning" treatment, the overall negative charge of the minerals is reduced by about one-third. Charge reduction occurs mainly in steps 1 and 2, whereas step 3 is carried out in order to eliminate extraframework elements produced in the preceding steps, which block the exchange sites in the interlamellar space. This treatment is not indispensable; but when carried out, it significantly improves the characteristics of the final product. Thorough Na (or Ca) exchange is essential for the quality of the pillared materials. The charge-reduced vermiculites and phlogopites are then brought in contact with the Al-pillaring solution, as in the case of the pillaring of smectites. Details on the operating variables of the conditioning treatment and pillaring process of these minerals can be found in Refs. 20 and 21.

2. Interlayer Spacings and Thermal Stability

Adequately Al-pillared smectites are characterized by an interlayer distance of about 1.8 nm (or gallery height of about 0.8 nm), which does not significantly decrease after a calcination at temperatures of 673–773 K. Typical XRD patterns illustrative of Al-pillared vermiculite and Al-pillared phlogopite are shown in Figure 3. Both materials exhibit narrow peaks with interlayer distance of 1.88–1.9 nm after heating at 473 K. Spacings of 1.83–1.84 nm are found after heating at 773 K, with a slight decrease after a calcination at 973 K and 1073 K (1.7–1.71 nm). The maintenance of the spacings at those high temperatures is in contrast to equivalent materials obtained with smectites, which generally collapse at lower temperatures owing to the lower intrinsic stability of the clay itself. Framework dehydroxylation of smectites generally occurs at around 873–923 K, compared with 1073 K and for vermiculite and 1373 K for phlogopite, a difference reflected in the thermal stability of the corresponding pillared forms. It has been shown that the structural stability is lowered upon pillaring, as a result of the protolysis of some bonds in the octahedral layers and, consequently, a weakening the thermal framework stability (20,21).

C. Synthesis of Mesoporous Clay–Derived Materials

Different strategies are reported in the literature to obtain mesoporous materials out of a layered silicate: (a) hydrothermal pillaring with rare earth–Al pillar moieties (22); (b) adsorption of colloidal sols (23); (c) gallery templated polymer-

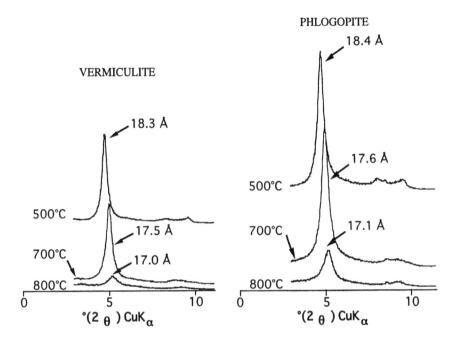

FIGURE 3   X-ray diffraction patterns of Al-pillared vermiculite and phlogopite after calcination at 773, 973, and 1073 K.

ization of e.g., Si(OEt)$_4$ (24,25); (d) topochemical transformation of layered silicates in the presence of surfactants (26,27).

In the first method, commercial, partially neutralized Al^{3+} solutions are modified by addition of La^{3+} or Ce^{3+}, and the pillaring process of a smectite clay mineral is performed under reflux or under hydrothermal conditions. A material is obtained that is, at least partially, expanded to a d_{001} spacing of 2.6 nm. The structure and composition of the pillaring oligomer is unknown, nor is it known to what extent the La^{3+} or Ce^{3+} cations are incorporated into the pillar. In view of the pH conditions of the pillaring process, only traces are expected to be incorporated in the pillar; some may be ion exchanged or adsorbed on the pillar. Anyhow, the data of Table 2 clearly show the increase in surface area and pore volume with respect to the regularly pillared materials.

When sols of TiO$_2$–SiO$_2$, SiO$_2$–Al$_2$O$_3$, or ZrO$_2$–SiO$_2$ are used in combination with supercritical drying, pillared smectites are obtained that are almost X-ray amorphous (23). The materials can be seen as expanded clay structures with a 5- to 10-fold increase of pore volume, due mainly to the effect of supercritical drying (Table 2). When imogolite, a tubular aluminosilicate, is used as pillaring

Table 2 Mesoporous Pillared Layered Silicates

Material	d_{001} (nm)	Surface area BET ($m^2\,g^{-1}$)	Pore volume (cc g^{-1}) Meso	Micro
La, Al-PM	2.6	493	0.13	0.15
Ce, Al-PB	2.57	431	0.23	—
La, Al-PB-AD	2.54	426	0.24	—
Ti, Si-PB-AD	—	403	0.50	—
Ti, Si-PB-SCD	—	501	4.8	—
Zr, Si-PB-AD	—	354	0.54	—
Zr, Si-PB-SCD	—	521	5.0	—
Ti, Si-PB-AD	—	125	0.03	—
Ti, Si-PB-SCD	—	467	1.5	—
Imogolite-PM	—	480	0.20	—
FSM	3.5–4.5	900–1200	0.7–1.3	—

PM: pillared montmorillonite; PB: pillared beidellite; AD: air dried; SCD: supercritically dried; FSM: folded sheet material.

agent, two types of micropores are formed—intratubular and intertubular (Fig. 4)—with a very small mesoporosity (28).

True mesoporous clay derivatives can be obtained via two routes, resulting in different structures, depending on the clay dimensions. The first one, proposed by Pinnavaia's group, describes the synthesis of porous clay heterostructures, or PCHs (24). These materials will be discussed in detail in Section III.

An alternative synthesis route has been developed by Kuroda and coworkers (27), who modified monolayered silicates such as kanemite with cationic surfactants. Under controlled pH conditions, the intercalated surfactants form micelles and the thin kanemite sheets undergo a topotactic transformation by folding around the micelles (Fig. 5). After calcination, the organics are removed, resulting in a highly porous mesoporous material. They are named folded sheet mesoporous (FSM) solids, after their mechanism of formation. The mesopore size can be tuned by a variation of the chain length of the used surfactant.

D. Porosity Characteristics of Pillared Clays

The introduction of micropores in clays results in a significant increase in the total specific surface area. Surface area and porosity are important characteristics of porous pillared layered solids for applications in the fields of adsorption and catalysis. They are crucial criteria in heterogeneous catalysis since they determine the accessibility of the active sites and are therefore related to the catalytic activity. The pore architecture of a porous solid controls transport phenomena and governs

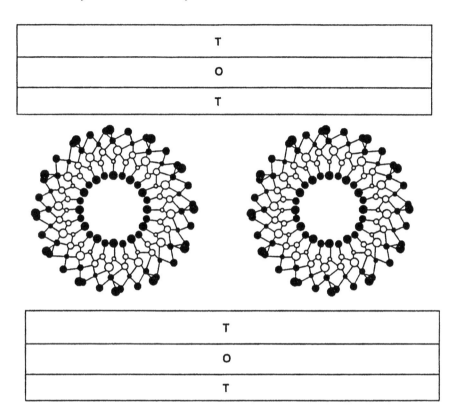

FIGURE 4 Cross-sectional view of Na^+ montmorillonite intercalated with a monolayer of imogolite $(SiAl_2O_3(OH)_4)$. (From Ref. 5.)

shape selectivity. Thus properties such as pore volume and pore size distribution are essential parameters to be analyzed.

1. Surface Area and Microporosity

The surface area of a solid is commonly derived from its nitrogen adsorption/desorption isotherm at 77 K. Brunauer et al. have defined five different types of isotherms (29). Type I isotherms are characteristic of microporous adsorbents, while type IV isotherms are typical of mesoporous solids. A significant increase in volume adsorbed in the low P/P_0 region in type IV isotherms indicates the presence of micropores associated with mesopores. In case of pillared clays, the uptake of N_2 in the low-pressure region is largely due to the high number of micropores that have been generated by the pillaring process. The shape of the

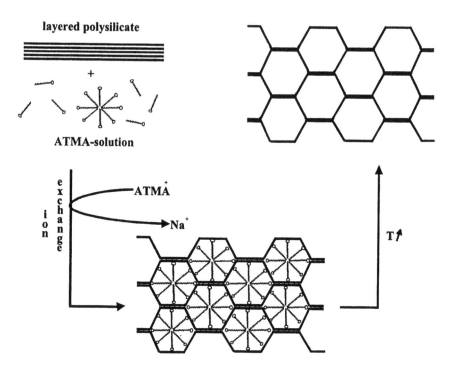

FIGURE 5   Formation mechanism of mesoporous FSM materials.

isotherms of pillared clays is dependent on the pillar species introduced (30) (see Fig. 6).

The isotherms of Al-PILC and Zr-PILC are of type I, indicating the small dimensions of the pillars. The shape of the adsorption isotherm of Fe-PILC resembles a type II isotherm (nonporous). The isotherm of Ti-PILC is close to type IV, but with a distortion in the relative pressure range 0.05–0.5 followed by capillary condensation. The distortion is due to the presence of pores on the border region between micro- and mesopores (1.5–3.5 nm). Isotherms of intermediate shape can also be observed, for example, alumina PILC doped with lanthania (LaAl-PILC) (31). For some PILC materials, the presence of a large hysteresis loop in the isotherm might be observed, indicating the importance of mesopores in the structure. From the adsorption isotherm data, several parameters describing the porosity can be quantitatively derived.

In the case of PILCs, three main methods have been applied to calculate the surface area based on the isotherm data: the Langmuir method, the BET method, and the t-plot or α_s-plot (t-plot method: see Sec. II.D.2).

Vads/g (ccSTP/g)

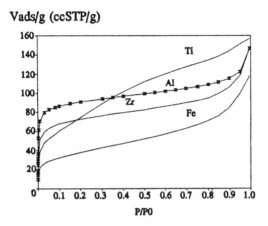

FIGURE 6 N_2 adsorption isotherms at 77 K obtained for montmorillonite pillared with different pillaring species (Al, Zr, Fe, Ti). (From Ref. 30.)

Physical adsorption in microporous solids shows type I isotherms because the micropores limit the adsorption to a few molecular layers. Using a kinetic approach, Langmuir described the type I isotherm, considering that adsorption was limited a monolayer (32). This approach assumes that the adsorption energy is constant and is independent of the fraction of the surface occupied by the adsorbed molecules.

Brunauer, Emmett, and Teller (BET) extended Langmuir's theory to multilayer adsorption (33). The BET theory assumes that the uppermost molecules in the adsorbed stacks are in dynamic equilibrium with the vapor and that the adsorbed molecules, not in direct interaction with the surface, are all equivalent to the liquid state. Discussions on the validity of the assumptions, as well as a complete description of the BET theory, may be found in the literature (34–37). For PILCs, the Langmuir surface area is always higher than the BET surface area, about 30–40% (Table 3). It is difficult to determine which one is closest to the real situation, though the reliability of specific surface areas can be checked by setting an upper limit for the monolayer capacity of porous solids (38). Since clay sheets have a silica surface, a monolayer of adsorbed N_2 on a nonporous silica surface is completed at a P/P_0 of 0.09, according to Gregg and Sing (35). Since the adsorption force in pores should be stronger than on an open surface, the monolayer capacity of a PILC should be lower than its uptake at $P/P_0 = 0.09$. The BET and Langmuir monolayer capacities of PILCs can be compared now to their adsorption at $P/P_0 = 0.09$. Zhu and Vansant performed these calculations on a series of different Al-PILC samples (Table 3) and concluded that the

Langmuir surface areas are too high and that the BET data are lower than in the real situation (38). For these Al-PILCs, after the filling of pores of about 0.8 nm, the adsorption changes slightly with the increase in P/P_0. In this case, the slope of the BET plot will increase steeply,

$$s = 1/[V(1 - P/P_0)] \qquad (4)$$

resulting in an underestimation of the BET surface area. Therefore, the deviation is caused by the space restriction of the very fine pores.

In order to obtain the micropore volume and the external surface area of microporous solids, De Boer et al. (39) have developed the *t-method*. This method is based on the comparison of adsorption isotherm data of a porous sample and of a nonporous sample of identical chemical composition and surface character (reference isotherm). When the adsorbed N_2 volume is plotted against the statistical thickness t (in nanometers) of the adsorbed N_2 layer (V_{ads} vs. t), a linear relation can be obtained. If both reference and sample isotherms are identical, as is the case for nonporous solids, a straight line passing through or close to the origin is obtained (Fig. 7). Horizontal departures from the straight line indicate the presence of micropores.

The microporous volume is obtained from a straight line extrapolated to a positive intercept on the ordinate. The estimation of t should be based on the reference isotherm, which yields the same BET constant (C_{BET}) as that of the material tested (40). The choice of the standard t function and the part of the V–t curve used for linear fitting drastically affect the values of the external surface and micropore volume. To avoid the foregoing complications, Sing (41) introduced and developed the α_s method, in which the normalized adsorption, α_s (= $n/n_{0.4}$), is derived from the isotherm of a reference material by using the amount adsorbed at a relative pressure of 0.4 ($n_{0.4}$) as the normalization factor. Lecloux

Table 3   BET and Langmuir Surface Areas and Monolayer Capacities of PILCs, in Comparison with Their Adsorption at P/P_0 of 0.09 ($V_{0.09}$)

Sample	SA_{BET} (m^2 g^{-1})	SA_L (m^2 g^{-1})	$V_{BET\text{-monolayer}}$ (cc STP g^{-1})	$V_{L\text{-monolayer}}$ (cc STP g^{-1})	$V_{0.09}$ (cc STP g^{-1})
Al-PILCs					
Al-PILC1	223.9	295.7	55.1	68.0	63.1
Al-PILC2	284.0	377.1	71.0	86.6	80.2
Al-PILC3	337.9	447.9	77.6	102.9	93.5
Fe-PILC	129.4	178.0	29.7	40.9	29.5
Ti-PILC	251.5	348.0	57.8	91.2	54.4

Source: Ref. 38.

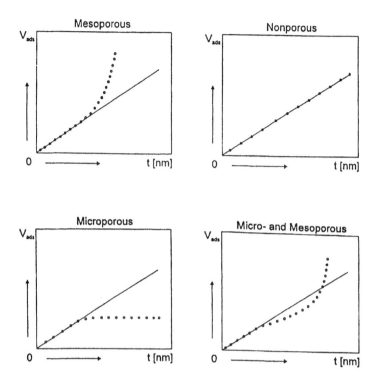

FIGURE 7 *t*-Plots of mesoporous, nonporous, microporous, and combined micro-meso-porous solids.

and Pirard (42) have shown that the α_s- and *t*-methods, when based on the same standard isotherm, differ by only a constant factor and must give the same estimation of the micropore volume and external surface. Table 4 gives an overview of the micropore parameters derived from t-plots of PILCs.

2. Textural Characteristics of Al-Pillared Clays

For Al-pillared smectites, micropore volumes and specific surface areas in the ranges of 0.06–0.130 cm^3 g^{-1} and 250–350 m^2 g^{-1} are obtained, respectively calculated with the *t*-plot and BET method (43). Figure 8 clearly illustrates the development of the microporosity resulting from the Al pillaring of vermiculite and phlogopite. Nitrogen adsorption–desorption isotherms of the starting minerals (1), after the conditioning treatment (prior to Al pillaring) (2), and of pillared vermiculite and phlogopite calcined at 773 K (3) and 973 K (4) are shown.

Table 4   Micropore Parameters of Pillared Clays Derived from t-Plots

Sample	SA_{ext} $(m^2\ g^{-1})$	$SA_{micropore}$ $(m^2\ g^{-1})$	$V_{micropore}$ $(cc\ g^{-1})$	$d_{micropore}$ (nm)
Al-PILC	57.8	223	0.092	0.82
Ti-PILC	70.0	193	0.133	1.38
Zr-PILC	40	177	0.078	0.88
LaAl-PILC	37	185	0.082	0.89

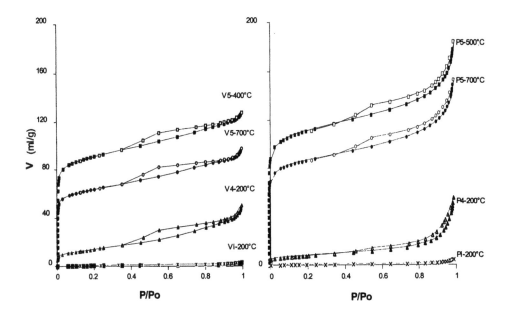

FIGURE 8   Nitrogen adsorption–desorption isotherms at 77 K of (1) initial vermiculite (left) and phlogopite (right); after the conditioning treatment (2), and of the Al-pillared forms after calcination at 773 K (3) and 973 K (4).

Typical values of the specific surface areas and micropore volumes of Al-pillared phlogopite and vermiculite calcined at 773 K are 322 m^2 g^{-1} and 0.099 cm^3 g^{-1} for a pillared vermiculite (20) and 315 m^2 g^{-1} and 0.102 cm^3 g^{-1} for pillared phlogopite (21).

3. Microscopic Techniques as Tool for the Characterization of Porosity

Transmission electron microscopy and scanning electron microscopy (TEM/SEM) can be used to characterize the micro- and mesoporosity of pillared clays, though they are not routine techniques. Basal spacings observed using TEM/SEM agree well with X-ray diffraction (XRD) results. Electron microscopy can provide information on the stacking and ordering of clay sheets and, as a consequence, also on the porosity, since the contribution of micro- and mesoporosity in the PILC structure is greatly determined by the orientation of the clay sheets. For pillared montmorillonite, for instance, there is a characteristic long-range face-to-face stacking of silicate layers, resulting in a large microporosity (44). Delaminated laponite clays, on the other hand, are characterized by short stackings of 10–15 layers. Face-to-edge and edge-to-edge orientation of these smaller aggregates generate the meso-/macropores present in delaminated structures (44). Not only the clay ordering but also the existence of a possible external phase after pillaring or of nonintercalated clay plates can be elucidated using the TEM technique. In the case of Ti-pillared montmorillonite, black dots in the TEM picture clearly indicate the presence of an external TiO_2 phase as particles of approximately 3 nm (45). For Al-pillared montmorillonite, a more homogeneous structure is depicted in the image without evidence for an external phase (45). Ahenach et al. studied the influence of water concentration on the pillaring of montmorillonite with an organosilane as pillaring precursor (46). The TEM images proved that with increasing H_2O concentration to enhance the hydrolysis of the silane, particles of a certain size (appearing as black spots in the image) are formed at the external surface of the clay. In the field of porous pillared clay membranes, Vercauteren et al. (47) showed that SEM can be very useful to obtain information on the Al-pillared clay top layer deposited on an Al_2O_3 support membrane. Smooth and uniform top layers, which are very thin, can be seen in the SEM pictures, and no cracks are present.

Since the invention of the atomic force microscope (AFM) by Binning et al. (48), which can generate atomic-scale images of materials, this new tool has often been used in combination with electron microscopy to examine the surface and porosity of pillared clays. The principle of this technique is based on scanning the surface of a sample with a very sharp tip, brought within close proximity of the sample, to map the contours of the surface. Hartman et al. (49) demonstrated the ability of the AFM to image molecular-scale features of montmorillonite and illite. Occelli et al. (50,51) conducted a profound characterization of Al-pillared montmorillonite with AFM. The AFM imaged the clay surface structure and provided detail of the sorption of polyoxycations in pillared clay catalysts. Atomic-scale-resolution images of the clay surfaces before and after pillaring consist of hexagonal arrays of bright spots (representing the three basal oxygens of a SiO_4 unit). The average nearest-neighbor distance (the spot-to-spot distance) before and after pillaring with Al_{13} ions could be calculated, and it was consistently observed that pillared montmorillonites possess larger nearest-neighbor distances than the parent materials. The pillaring reactions seem to increase the center-to-center distance between bright spots (from 0.515 ± 0.05 nm to 0.545 ± 0.05 nm), so it is assumed that by replacing the parent-clay charge compensating cations with Al_{13} ions in the interlamellar space, stretching of the SiO_4 layers occurs. With AFM, direct observation of Al_2O_3 clusters (the pillars) between the layers is not possible, though the expansion of the clay layers as a result of the pillaring can be visualized. The mean basal spacing, as calculated from the distribution of expanded layers, is 0.90 ± 0.01 nm. This result is in good agreement with the basal spacing obtained by XRD (0.85 nm).

E. Surface Acidic Properties of Pillared Clays

Pillaring also confers acidic properties to the resulting materials. In Al-pillared smectites, Lewis acidity has generally been associated with the aluminum pillars, whereas the origin of the Brønsted acidity (protons) is more diverse. Protons can be formed at the pillaring step, as a result of polymerization processes of the pillaring species occurring in the interlayer space (18); protons are also produced at the calcination step as a result of the transformation of the hydrated pillar precursors to oxidic pillars (52). In the case of Al-pillared clays obtained from smectites with Al-for-Si substitutions (saponites, beidellites), Brønsted acidity has also been associated with Si—OH—Al groups resulting from the proton attack of Si—O—Al linkages of the tetrahedral layers. Such acid groups give rise to a new OH stretching band at 3440 cm^{-1} in Al-pillared beidellite and at 3595 cm^{-1} in Al-pillared saponite. These bands disappear upon pyridine adsorption and are restored after desorption of the base (53–55). This feature differs from pillared clays without tetrahedral substitutions, where such acid groups are absent.

Two main methods have been employed to characterize the acidity of solids: temperature-programmed desorption of ammonia and infrared spectroscopy of adsorbed bases. Temperature-programmed desorption (TPD) over acid solids previously exposed to ammonia allows one to quantify the acid content and provides information on the strength of the acid sites. However, the distinction between Lewis and Brønsted acidities is not straightforward and is often hampered by physisorbed ammonia. The validity of this method has recently been questioned (56). Infrared spectroscopy of adsorbed pyridine and ammonia have been most extensively used to investigate the acid properties of solids including the pillared clays. This approach allows one to distinguish and quantify the species that interact with Lewis and Brønsted acid sites, owing to the fact that both types of interaction give rise to distinct IR bands. The quantification of the Lewis and Brønsted acid contents ($q_{B,L}$) commonly is based on the integrated intensities of the IR bands at 1448 cm^{-1} characteristic of pyridine in interaction with Lewis sites (Py–L) and at 1545 cm^{-1} for pyridine adsorbed on Brønsted sites (Py–B), by means of the following equation:

$$q_{B,L} = (A_I \, \pi D^2)(4w\varepsilon_{B,L})^{-1} \tag{5}$$

where A_I, D, w, and $\varepsilon_{B,L}$ represent the integrated band area of Py–B and Py–L, the diameter of the wafer (cm), the sample weight, and the extinction coefficient of Py–B (1.67 ± 0.12 cm μmol^{-1}), and Py–L (2.22 ± 0.21 cm μmol^{-1}) (57), respectively. After outgassing under vacuum at 373 K, values of 51 (Py–B) and 254 (Py–L) μmol g^{-1} were found for Al-pillared phlogopite and of 39 (Py–B) and 183 (Py–L) μmol g^{-1} for Al-pillared vermiculite. After outgassing at 473 K, these values decreased to 29 (Py–B) and 155 (Py–L) μmol g^{-1} for the pillared mica and to 22 (Py–B) and 112 (Py–L) μmol g^{-1} for the pillared vermiculite. Using a similar method, an Al-pillared saponite had 15 (Py–B) and 122 (Py–L) μmol g^{-1} after outgassing at 473 K (58). Both Al-pillared vermiculite and phlogopite have more Brønsted acid sites than Al-pillared saponite, in line with prediction. The pillared mica contains more B and L acid sites than the pillared vermiculite, with the latter one and Al-pillared saponite having similar L acid content. Figure 9 compares the variation of the normalized Py–B and Py–L band areas against outgassing temperature relative to the values found after an outgassing in vacuum at 373 K for Al-pillared vermiculite and phlogopite. Both materials have B and L sites with similar acid strength. NH$_3$-TPD runs over Al-pillared phlogopite gave a B + L acid content of 290 μmol g^{-1}, which is in fair agreement with the values obtained from pyridine adsorption (305 μmol g^{-1}). For comparison, the acid contents (B and L) established from NH$_3$-TPD measurements over different Al-pillared smectites were in the range 110–150 μmol g^{-1} for Al-pillared montmorillonites and between 180 and 280 μmol g^{-1} for Al-pillared materials prepared with saponites from different deposits (43).

F. Applications in the Field of Adsorption and Catalysis

1. Adsorption

In recent years a number of investigations have been carried out to study adsorption in pillared clays, almost all of which are experimental in nature. Despite their many potential applications and the existence of valuable information that can be obtained through extensive experimental efforts, very few fundamental theoretical studies have been undertaken so far to investigate transport and adsorp-

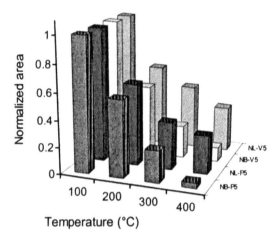

FIGURE 9 Relative variation of the normalized band areas of the Py–B (at 1545 cm^{-1}) and Py–L (1448 cm^{-1}) bands versus outgassing temperature for Al-pillared vermiculite (V5) and Al-pillared phlogopite (P5).

tion processes in PILCs. A few computer simulation studies have been carried out (59–63), while fundamental molecular simulations of adsorption and diffusion in model pillared clays have been performed by the group of Yi et al (64–66).

Adsorption studies dealing with liquid systems clearly show the potential use of PILCs as adsorbents for environmental applications. One example is the uptake of toxicants such as chlorophenols on pillared and delaminated clay structures (67). Al-delaminated laponite is more effective in adsorbing pentachlorophenol (PCP) than the Al-pillared montmorillonite. At an equilibrium time = 24 h, an equilibrium pH = 4.7, and an initial PCP concentration of 38 μmol L^{-1}, the maximal adsorption of PCP was 27 μmol g^{-1} on Al-delaminated laponite and 12 μmol g^{-1} on Al-pillared montmorillonite. The binding of PCP onto the substrate is attributed to interactions between the PCP molecule and the immobilized Al$_2$O$_3$ species. The greater adsorption capacity for the delaminated structure arises from the greater dispersion and availability of the Al oxide aggregates in the clay.

In order to adsorb organics selectively from aqueous solution, it is very important that the adsorbents are hydrophobic. Therefore, Shu et al. (68) investigated the adsorption of the same organic toxicants on a hydrophobic surfactant (Tergitol)-modified Zr-pillared montmorillonite. The surfactant-modified external surfaces have a high affinity for the organics, and this affinity is related to the surfactant loading. An adsorption capacity of phenol equal to 0.8 mmol g^{-1} has been reached at an equilibrium concentration of 7 mg mL^{-1}

Porous pillared clays can also be used for the removal of color-generating compounds such as "humic" and "fulvic" acids from water (69). The adsorption of "humic acid" from both distilled and tap water was studied, with a maximum adsorption capacity of 23.4 mg g^{-1} "humic acid" out of a tap water solution with an initial concentration of 160 mg L^{-1} on Al-PILC thermally treated at 453 K.

For the adsorption of inorganic gases on pillared clays, studies of the affinity, capacity, and selectivity of PILCs toward inorganic gases were first described by Barrer (70,71). In these studies, Co(en)$_3$-pillared fluorhectorites (en = ethylenediamine) were used to adsorb H$_2$, D$_2$, Ne, N$_2$, O$_2$, Ar, Kr, and CO$_2$. Low-temperature-uptake rates showed the potential molecular sieving effects of the microporous materials. Gameson et al. (72) investigated low-temperature adsorption of Xe and Kr on Al-PILC, which resulted in a model for the coverage mechanism in the pores.

Most work on the adsorption of gases at ambient temperature on various PILCs was performed by Baksh and Yang, who proposed that the pores in pillared clays consist of narrow and tall paths limited by the interpillar spacing rather than the interlayer free spacing (73,74). For these slit-shaped micropores, the electrostatic fields of top and bottom clay layers overlap in the middle of the pore and result in a very high interaction field. In a first study, Zr-PILCs were used for testing their adsorption properties (N$_2$, O$_2$, hexane, benzene) (73). The equilibrium isotherms for N$_2$ and O$_2$ at 298 K on Zr-PILC are shown in Figure 10, together with comparison data of 5A zeolite, which is widely used for air separation. The N$_2$/O$_2$ equilibrium selectivity on Zr-PILC is high but still smaller than that of zeolite 5A (approximately 3), as a result of the limiting interpillar spacing in Zr-PILC, not allowing multilayer buildup.

The same authors tested the adsorption of O$_2$, N$_2$, CH$_4$, CO$_2$, SO$_2$, and NO on five different pillared clays (Zr–, Al–, Cr–, Fe–, and Ti-PILCs) (74). The adsorption isotherms at 298 K showed that the PILC materials exhibit a high and selective adsorption toward the gases. The adsorption data were used to determine diffusion constants and potential energy profiles in the slit-shaped pores of the PILCs. The equilibrium selectivity of CH$_4$/N$_2$ on Al-PILC is greater than 5.0 (at 1013 hPa), exceeding all known sorbents by a large margin (Fig. 11). In addition, high SO$_2$/CO$_2$ equilibrium selectivities are observed on the pillared clays.

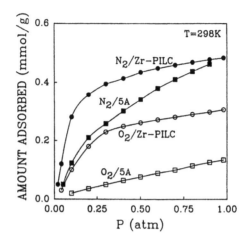

FIGURE 10   Equilibrium isotherms of N_2 and O_2 on Zr-PILC compared with 5A zeolite. (From Ref. 73.)

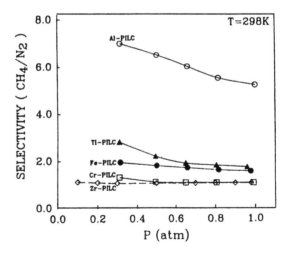

FIGURE 11 CH_4/N_2 equilibrium selectivities at 298 K on Zr-, Al-, Cr-, Fe-, and Ti-PILCs. (From Ref. 74.)

A systematic investigation was undertaken in 1997 to tailor the micropore dimensions of pillared clays for enhanced gas adsorption (75). Two clays with different CECs are used for Al pillaring, namely, Arizona montmorillonite (CEC = 1.4 meq g^{-1}) and Wyoming montmorillonite (CEC = 0.76 meq g^{-1}), resulting in a different pillar density and interpillar distance. It is shown that the CH$_4$ adsorption on the Arizona pillared clay can be nearly doubled by the smaller interpillar spacing, due to the back-to-back overlapping potential in the micropores (Table 5). The N$_2$ adsorption is not significantly influenced because of the low polarizability of N$_2$. The CH$_4$/N$_2$ selectivity ratio for the Al-pillared Arizona montmorillonite calcined at 873 K is 2.35, which is adequate for separation by the pressure swing adsorption process (76).

In order to obtain a porous pillared material with improved adsorption properties, some additional modification techniques on PILCs, both during or after the synthesis, have been developed. The incorporation of metals in the pillars, when performed during the synthesis, results in the formation of mixed oxide-pillars. Heylen et al. (77) observed an enhanced adsorption at 273 K of cyclohexane, CCl$_4$, and CO$_2$ on Fe-/Cr-pillared montmorillonite compared to the pure Fe-pillared montmorillonite. By the synthesis of mixed oxide-pillars, specific adsorption sites are created in the PILC, exerting a positive influence on the adsorption capacity and selectivity towards gases.

A way to increase the porosity of pillared clays and to modify their adsorption behavior is the preadsorption of organic templates prior to the ion exchange with the pillaring precursors. n-Alkylammonium ions have therefore been pre-exchanged on the clay in an amount equal to a fraction of the CEC. As a result, the pillar density decreases, since part of the interlayer space is occupied by the templates, which are removed only in the final calcination step. It is shown that this indirect pillaring procedure induces an increase in the adsorption capacity toward inorganic gases. Heylen et al. (78,79) investigated the influence of the preadsorption of butylammonium ions on Fe- and Al-pillared montmorillonite (in an amount = ½ of the CEC). The surface area and micropore volume of Fe-PILC was 2.5 times higher, compared to the unmodified Fe-PILC (Table 6).

Table 5   Adsorption Properties of Al-Pillared Clay Minerals

	Amount adsorbed (mmol g^{-1})	
Type of PILC (calcined at 873 K)	CH$_4$	N$_2$
Arizona montmorillonite	0.087	0.037
Wyoming montmorillonite	0.048	0.032

Source: Ref. 75.

Cool et al. (80,81) performed a templated synthesis of Zr-pillared laponite using ethylenediamine in an amount exceeding the CEC of laponite. Here, the amines not only have a positive influence on the pillar distribution, but also favor the parallel orientation of the clay sheets, resulting in a more homogeneously pillared structure with increased microporosity. Thus, on laponite, the main function of the template is not to block exchange sites, but to influence the stacking of clay layers. The porosity and adsorption characteristics of the different pillared clays are summarized in Table 6.

The unmodified and modified pillared clays, synthesized using amines or with mixed pillars, have an intermediate hydrophylic–hydrophobic character and thus have some potential for the removal of chlorinated hydrocarbons. In Figure 12, the adsorption isotherms of some hydrocarbons on different pillared clays are compared to the adsorption isotherm on Na-montmorillonite (79). In all cases, the adsorption capacity of the mixed Fe-/Zr-PILC is a factor 4 higher than that of the pure Fe-PILC. Besides the adsorption capacity, the isotherm type also changes.

Another technique for modifying PILCs, found to be useful to improve the adsorption properties, is the incorporation of specific cations in the porous structure of pillared clays, serving as specific adsorption sites in certain applications. It is known that the cation exchange capacity (CEC) of montmorillonite greatly decreases after pillaring with alumina pillars, due to the nonexchangeable H^+ cations in the clay structure that are formed during the calcination. It is, however, possible to restore the CEC of a PILC in two ways (82,83).

The direct ion exchange procedure involves the exchange of a desired cation or anion from an alkaline or acidified salt solution, respectively. This is possible because of the amphoteric character of the hydroxyl groups, present on the pillars

Table 6 Porosity and Adsorption Characteristics (N_2 and O_2; at 273 K on Al- and Zr-PILC and at 194 K on Fe-PILC) of Pillared Clays, Prepared Without and With Templates

Type of PILC	SA_{BET} $(m^2 \, g^{-1})$	MicroPV $(cc \, g^{-1})$	Amount adsorbed (mmol g^{-1}) N_2	O_2
Al-montmorillonite	340	0.112	0.058	0.057
BuA-Al-montmorillonite	361	0.131	0.110	0.105
Fe-montmorillonite	134	0.037	0.000	0.030
BuA-Fe-montmorillonite	233	0.121	0.230	0.171
Zr-laponite	425	0.227	0.203	0.175
Etdiam-Zr-laponite	482	0.340	0.286	0.219

BuA = butylammonium; Etdiam = ethylenediamine.
Source: Refs. 78 and 81.

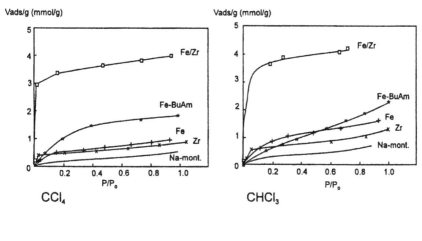

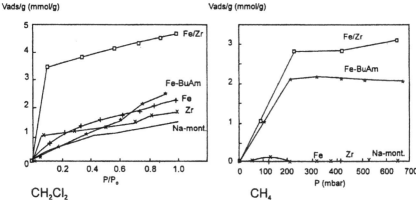

FIGURE 12 Adsorption isotherms at 273 K of chlorinated hydrocarbons (CCl_4, $CHCl_3$, CH_2Cl_2, CH_4) on different pillared clays and on the parent Na-montmorillonite. (From Ref. 79.)

and on the clay layer edges (84,85). At low pH the —OH groups protonate and act as anion exchangers, while at high pH they deprotonate and become cation exchangers.

The indirect ion exchange method consists first of a modification of the PILC with ammonia. The reaction of the gas with the H^+ ions present in the clay structure after calcination results in the formation of NH_4^+ ions, which can subsequently be exchanged for any other desired cation from a salt solution. These ion modifications result in the introduction of cations in the PILC pores and alter the adsorption properties of the substrate.

Molinard and Vansant (86,87) prepared a series of cation- (Mg^{2+}, Ca^{2+}, Sr^{2+}, Ba^{2+}) -modified Al-PILCs and tested the adsorption capacity/selectivity for the air components N_2 and O_2 (Fig. 13). The introduction of cations has a clear positive influence on the adsorption behavior of the pillared clays. Compared to oxygen, nitrogen has a higher quadrupole moment, resulting in additional interaction with the adsorption sites on the pillared clay surface. The difference in affinity between O_2 and N_2 is most pronounced on Sr^{2+}–Al-PILC.

The adsorption isotherms of nitrogen and oxygen on Al-PILC and Sr^{2+}-Al-PILC above 273 K are presented in Figure 14. For the original Al-PILC, there is an equal uptake of both gases over the entire pressure range; after the introduction of cationic adsorption sites (Sr^{2+}) in the PILC structure, the N_2 capacity at 507 hPa is doubled. As a result, the selectivity ratio N_2/O_2 can be significantly increased by ion modification.

In order to study the influence of the amount of exchanged cations on the gas adsorption properties of the porous substrate, two different PILCs have been prepared with a different Ca^{2+} loading. The N_2/O_2 ratios of the Ca^{2+}-Al-PILCs at 273 K are determined as a function of the pressure (Fig. 15). Both PILCs have a higher preference for nitrogen than the parent Al-PILC ($N_2/O_2 \approx 1$), and the PILC with the highest cation amount (2.23 wt%) exhibits a better N_2/O_2 selectivity. Compared to the Sr^{2+}-exchanged PILCs, the capacities obtained on the Ca^{2+}-Al-PILCs are higher (88).

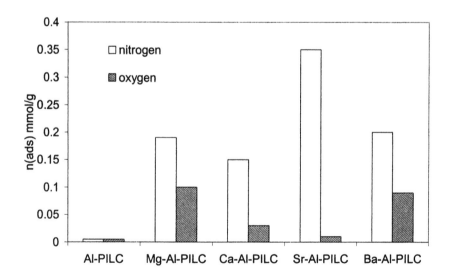

FIGURE 13 Comparison of the adsorption capacity for N_2 and O_2 on cation-modified Al-PILCs ($T = 194$ K; $p = 51332.67$ Pa). (From Ref. 86.)

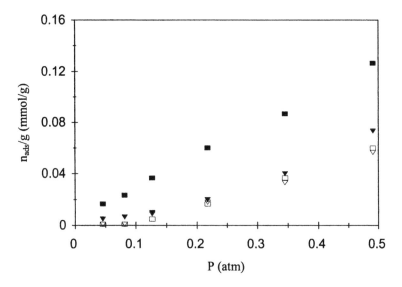

FIGURE 14 Adsorption isotherms at 273 K on Al-PILC and Sr-Al-PILC. (From Ref. 88.)

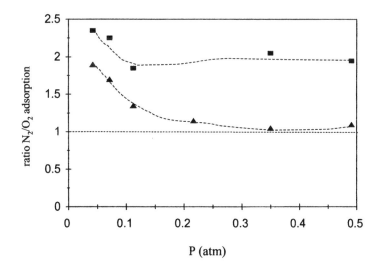

FIGURE 15 Ratio of N_2/O_2 adsorption of two Ca-Al-PILCs, containing 1.26 wt% Ca^{2+} and 2.23 wt% Ca^{2+}. (From Ref. 88.)

In a 1998 review by Zhu and Lu (89), the many different techniques for cation doping on PILCs, described throughout the literature in order to tailor the pore structure and properties of pillared clays, were discussed. The adsorption of water, air components, and organic vapors on cation-doped pillared clays are studied. It is shown that by the presence of cations on the PILC surface, their hydrophilicity can be enhanced. By comparing water adsorption on Al-PILC before and after loading with Ca^{2+} ions, it becomes clear that introducing cations enhances the water sorption, particularly at low vapor pressures (90). Moreover, there is the possibility to adjust the shape of the water isotherm by loading various amounts of cations. This technique will be very useful in developing PILC-based dessicants and adsorbents for dehumidification and cooling applications.

Adsorption of organic molecules on pillared clays was also investigated (89). An Al-PILC with slit-shaped pores of ~0.8 nm was modified with various amounts of Na^+ ions, in order to fine-tune the size of the interlayer micropores. The adsorption results of p- and m-xylene on the Na-doped samples at 298 K are presented in Figure 16. When the Na^+ content is low, the PILC shows almost no selectivity toward the different isomers. As the amount of Na^+ increases, the uptake of both p- and m-isomer decreases. A sudden drop is observed for the m-isomer at a Na^+ loading of only 0.066 mmol g^{-1}. The uptake ratio of p- over m-xylene reaches a maximum of ca. 2 here. Since they are slightly larger molecular entities, m-isomers are more subjected to sterical hindrance than p-isomers; at this cation loading, quite large portions of the micropores are no longer accessible to m-isomers. The uptake ratio decreases further with increased cation loading, since both isomers are excluded from the smaller micropores then. The results indicate that the modification of micropores with cations is a simple and powerful strategy to tailor the adsorption properties of the pillared clay structure. Adjusting the amount of doped ions is of crucial importance to fine-tune the opening size of the narrow micropores.

2. Catalysis

Pillared clays with pillars based on Al and also different elements (e.g., Cr, Fe) or with mixed pillars have been extensively investigated in numerous catalytic reactions of hydrocarbons, principally in proton-catalyzed reactions. Catalytic aspects have been reviewed in several articles (3,91,92). Comparison of the catalytic performances of acid solids with potential catalytic applications preferably with a reference catalyst is a straightforward way to obtain a preliminary information on their efficiency. With this approach, the activity of Al-pillared materials prepared with different smectites and pillaring methods have been compared, using the hydroisomerization of linear paraffins over Pt-impregnated pillared samples. In these bifunctional catalysts, the metal function is necessary to dehydrogenate the paraffin and rehydrogenate the branched olefins, and the protons aid the formation of carbenium ions, which isomerize via protonated cyclopropane

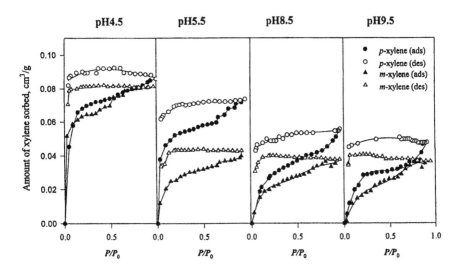

FIGURE 16   Adsorption–desorption isotherms of *p*- and *m*-xylene on Al-PILC doped with various amounts of Na$^+$ ions. The pH value of the dispersion for loading the Na$^+$ into the PILC structure is given on top of the plots. (From Ref. 89.)

structures (93). In typical reactions carried out at increasing temperatures, the isomers are first formed in yields that increase with temperature to reach a maximum and decrease at still higher temperatures. At a given temperature, mainly depending on the paraffin chain length and catalyst acidity, the isomers start to undergo β-scission, giving lower alkanes. At still higher temperatures, cyclization occurs. In such a reaction, the best catalysts, of course, are those exhibiting high selectivities to the isomers at high conversions. Comparative results of heptane, octane, and decane hydroisomerization obtained over Al-pillared forms of montmorillonite, hectorite, beidellite, and saponite have been reported in several studies (43,94–99). These studies have shown, without exception, that the Al-pillared forms of smectites with Al-for-Si substitutions in the tetrahedral layers (beidellite, saponites) were significantly more efficient catalysts than clays without such substitutions (montmorillonites, hectorite), independent of the paraffin used. The higher activity of Al-pillared beidellites and saponites has been attributed to the higher strength of the Si–OH–Al acid sites of the tetrahedral layers compared with Al-pillared montmorillonite and hectorite, where such sites are absent. Illustrative conversion-time curves of octane hydroisomerization obtained over Al-pillared vermiculite and phlogopite are shown in Figure 17.

As seen, total conversion is achieved at about 573 K. The yields of C8 isomers reach a maximum of 80% at 503 K for the pillared phlogopite and at

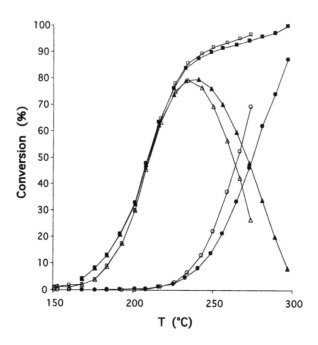

FIGURE 17 Variation of octane conversion (squares), yields of C8 isomers (triangles) and cracking products (circles) versus reaction temperature for Al-pillared vermiculite (filled symbols) and Al-pillared phlogopite (open symbols).

513 K for Al-pillared vermiculite, for overall conversions of 85% and 88% and selectivities to C8 isomers of 94% and 91%, respectively. Under similar conditions, Al-pillared montmorillonite and saponite exhibited a maximum yield of isomers of 61.5% at 574 K and 70.4% at 524 K, respectively (100). The results obtained over the pillared vermiculite and phlogopite are the best ever obtained with Al-pillared materials and most of the commercial zeolites.

III. POROUS CLAY HETEROSTRUCTURES

Since the early 1990s, extensive research has been undertaken on the synthesis of (semi)crystalline porous inorganic materials with a narrow pore size distribution in the mesoporous region (pore diameter: 2–10 nm), because these type of materials fulfill the increasing demand for adsorbents and catalytic supports with ordered large pores. Following the original patent of Mobil Oil Corporation in 1992 introducing the so-called M41S materials, the synthesis of these materials involves the use of ionic surfactants that interact with the inorganic ions to form a range of ordered mesostructures (101).

Since the porosity of pillared clays w as limited primarily to the microporous region (max. 2 nm), in 1995 researchers started to intercalate the same surfactants into layered clay hosts to perform a templated reaction in the clay interlayer region, in order to obtain true mesoporous clay materials (pore diameter > 2 nm), the so-called porous clay heterostructures, or PCHs. Because this research on PCHs is very young, many challenges still remain. In what follows, an overview of what is known about these materials in the literature will be given.

A. Synthesis Mechanism of Porous Clay Heterostructures

Whereas pillared clays are formed by the insertion of dense, metal polyoxycationic aggregates into the galleries of the layered clay host, porous clay heterostructures are formed by an intragallery in situ assembly process of surfactant-inorganic precursor nanostructures. The approach to design PCHs is based on the use of intercalated quaternary ammonium cations and neutral amines as cosurfactants to direct the clay interlamellar hydrolysis and polymerization reaction of an inorganic Siprecursor. After subsequent removal of the templates, a mesoporous silicate network remains between the clay sheets.

1. Porous Clay Heterostructures Derived from High-Charge-Density Clays

The mechanism proposed by Galarneau et al. (24) in 1995 for the formation of PCH derived from high-charge-density Li^+-fluorohectorite is presented in Figure 18. In the first step of the synthesis, long-chained surfactants $Q^+ = C_nH_{2n+1}N(CH_3)_3^+$ ($n = 10–16$) are introduced in the interlayer space of the clay by exchange with the Li^+ ions. Then, after subsequent addition of the neutral amine cosurfactant $C_nH_{2n+1}NH_2$ ($n = 6–12$), an amine-solvated bilayer structure with a thickness equivalent to the length of the long-chained quaternary cation Q^+ and the neutral amine is formed (Fig. 18A). Subsequent intercalation of TEOS takes place by a partial displacement of the neutral amine. The interactions between the surfactants and the silicate ions introduced into the interlayer space gives rise to rodlike micellar assemblies of surfactant and neutral amine surrounded by hydrated silica structures (Fig. 18B). These micelles are analogous to the spherical micelles formed during the templated synthesis of MCM-41 structures (101). Since the extragallery water concentration is very low, the base-catalyzed hydrolysis of TEOS is much faster in the interlayer region than in solution, and this minimizes the formation of extragallery silica. The fraction of the surfactant in the micelle is determined by clay layer charge matching. In the last step, a calcination is performed to remove the organics from the structure, resulting in a porous silica framework intercalated between the clay sheets (Fig.

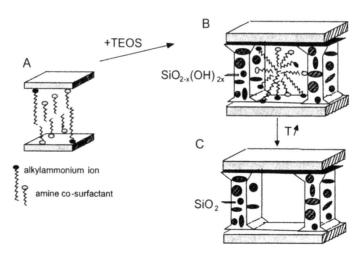

FIGURE 18 Synthesis mechanism of the formation of porous clay heterostructures. (From Ref. 24.)

18C). The resulting PCHs are characterized by nitrogen BET surface areas between 470 and 750 m^2/g and gallery heights in the range 1.5–2.4 nm, depending on the chain length of the amine (C6 or C12).

The relationships between the PCH pore size distributions (max. between 1.4 and 2.2 nm), the chain lengths of the templating surfactants (C6–C12), and the applied reaction stoichiometries (Q$^+$-clay: amine:TEOS = 1:2:15; 1:5:37.5, and 1:20:150) point out that PCH formation is based on a templating mechanism and not on a pillaring mechanism. If the pillaring mechanism was to be applied with the formation of dense laterally spaced silica aggregates in the clay gallery, there should be a pore size dependence on the clay:silica ratio, and this is not the case for PCHs. Besides, the observed pore sizes are uncharacteristic for pillaring: Pillared clays exhibit average pore sizes that are limited by the lateral separations of the pillars to values under 1 nm (102), while PCHs exhibit mesopore sizes substantially larger than this pillaring limit.

The same authors (101) also prepared other PCH solids derived from high-charge-density vermiculite, rectorite, and magadiite clays. Their layer charge per unit surface area increases in the order rectorite < fluorohectorite < magadiite < vermiculite. All obtained PCH materials have an average pore size equal to the pore size of the fluorohectorite-PCH analog. These observations are additional evidence for the intragallery templating mechanism, ruling out the pillaring mechanism, since in the case of pillaring the pore size should decrease with increasing charge density on the host layers, whereas here the observed pore sizes are invariant over the entire layer charge range.

Because of the thermal instability of the developed fluorohectorite-PCHs, Polverejan et al. (103) more recently applied the chemistry of PCH formation on synthetic smectite saponite clays with high layer charge density. For the synthetic saponites, the layer charge density is regulated by controlling the extent of silicon substitution by aluminum in the clay layers during their hydrothermal preparation. Saponite unit-cell compositions of $Na_x(Mg_6)(Si_{8-x}Al_x)O_{20}(OH)_4 \cdot nH_2O$, with $x = 1.2$, 1.5, and 1.7, were synthesized, with increasing charge densities and CEC values of 0.9, 1.0, and 1.1 meq/g. All three saponites have been used as hosts to assemble the intragallery mesostructure following the same process developed by Galarneau et al. (24). Systematically larger basal spacings are observed with increasing aluminum content or charge density (CEC) of the saponite layers. This can be explained by the more vertical orientation of the intercalated surfactant cations, resulting in an increased TEOS loading and mesostructured silica in the gallery region of the final PCH. The specific surface areas (max. 921 m²/g) and pore volumes (max. 0.44 cc/g) are larger than those obtained for fluorohectorite-PCH, but they progressively decrease with increasing aluminum loading, due to the larger amount of intragallery silica.

2. Porous Clay Heterostructures Derived from Lower-Charge-Density Clays

In the early work of Galarneau, a host clay with a high cation exchange capacity was considered essential for the PCH formation. However, more recent PCHs have been successfully synthesized from other clays of lower charge density. Cool and Ahenach reported the formation of PCHs from natural montmorillonite and synthetic saponite and laponite (25,104). The authors reported on the existence of two different heterostructures, depending on the origin of the host clay used (natural or synthetic) as well as on a combined micro- and mesoporosity for these materials.

The formation of these two different PCH structures is in relation to the different dimensions of the clays, determining the mutual interaction between plates and consequently also the organization of clay plates around the formed micelles. For the large-sized natural clay host montmorillonite, with plate size dimension 1000 nm, the micellar template formation occurs in the clay interlayer region, since the parallel or face-to-face (F-to-F) orientation between layers is favored. However, in case of the smaller-sized laponite (dimension 30 nm), the stacking of clay sheets surrounding the micelles will not necessarily occur in an F-to-F way. For this type of host, the important contribution of edge-to-face (E-to-F) and edge-to-edge (E-to-E) interactions leads to a different heterostructure. As a result of this type of stacking, the structure is built up of individual laponite plates with the growing templated silicate structure in between. After calcination, a three-dimensional porous heterostructure is obtained in which the clay sheets are linked by the hexagonal SiO_2 network. The number of E-to-F stacked layers in the structure will determine to a large extent the final micro- and mesopore volume of the PCH.

Both synthesized PCH materials exhibit supermicropores or mesopores between 1.5 and 3.0 nm and surface areas over 500 m^2/g. Figure 19 illustrates a typical N_2 adsorption–desorption isotherm at 77 K of a porous clay heterostructure, with the corresponding mesopore size distribution in the inset. The adsorption curve clearly exhibits a nearly linear portion in the partial pressure region ~0.03–0.4, indicative of supermicropores (1.5–2.0 nm) and small mesopores (2.0–3.0 nm). The increase in adsorbed volume at very small pressures (<0.03) is indicative of the filling of micropores. The mesopore size distribution shows a maximum at a pore diameter of 2.2 nm.

By a variation in the synthesis conditions of PCHs, a careful control of the pore diameter and also of the final micro/mesoporosity ratio can be established.

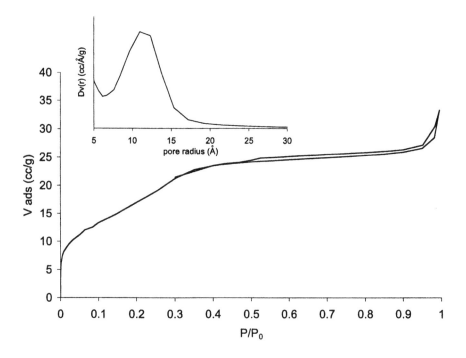

FIGURE 19   Typical N_2 adsorption–desorption isotherm at 77 K of a porous clay heterostructure, with its corresponding mesopore size distribution in the inset.

This is illustrated by the surface area and porosity data in Table 7 for PCHs derived from montmorillonite (P-Mont-H) and laponite (P-Lap-H).

The mesopore dimensions of these materials can easily be tailored by changing the chain lengths of the used surfactants and cosurfactants, since they determine the size of the formed micelles and the created pore size of the PCH after calcination. Using amines of increasing chain length ($C_6H_{13}NH_2$, $C_{12}H_{25}NH_2$), the influence on the porosity characteristics of P-Mont-H becomes clear. For this type of PCH, the contribution of micro- as well as mesopores is doubled. Second, by using the larger-sized dodecylamine instead of hexylamine, a clear shift in the P-Mont-H pore size distribution from the supermicroporous to the small mesoporous region is observed, respectively from 0.9- to 1.5-nm pore radius (Fig. 20A).

Also, the concentration ratio Q^+-montmorillonite/$C_{12}H_{25}NH_2$/TEOS used in the reaction mixture has its influence on the final pore characteristics of the PCH. A survey of the results is given in Table 7 for P-Mont-H using ratios of, respectively, 1/2/15, 1/7/50, and 1/20/150. It can be inferred that an increased concentration of cosurfactant and TEOS results in an important enhancement of the micropore volume and surface area. When this information is transformed

Table 7 Survey of the Most Important Characteristics of the Porous Montmorillonite Heterostructure (P-Mont-H) and Porous Laponite Heterostructure (P-Lap-H)

PCH	S_{BET} (m²/g)	μPV (cc/g)	MesoPV (cc/g)	Total PV (cc/g)
P-Mont-H				
co-surf., Q^+-mont/c-s/T[a]				
$C_6H_{13}NH_2$, 1/2/15	451	0.14	0.36	0.50
$C_{12}H_{25}NH_2$, 1/2/15	690	0.30	0.70	1.00
$C_{12}H_{25}NH_2$, 1/7/50	1093	0.50	0.16	0.56
$C_{12}H_{25}NH_2$, 1/20/150	1170	0.60	0.14	0.74
P-Lap-H				
co-surf., Q^+-conc.[b]				
$C_{12}H_{25}NH_2$, 100%	845	0.41	0.11	0.52
$C_{12}H_{25}NH_2$, 75%	1083	0.52	0.07	0.59
$C_{12}H_{25}NH_2$, 30%	652	0.29	0.16	0.45

BET surface area (S_{BET}), micropore volume (μPV), and mesopore volume (mesoPV). For the synthesis of P-Lap-H the molar ratio Q^+-Lap/$C_{12}H_{25}NH_2$/TEOS = 1/2/15. Mixtures are reacted for 4 h at room temperature. After reaction and air-drying, a calcination at 550°C with a heating rate of 2°C/min is performed.
[a] Co-surf.: type of cosurfactant used.
    Q^+-mont/cs/T: ratio Q^+-clay/cosurfactant/TEOS in the reaction mixture
[b] Q^+-conc.: Q^+ concentration with respect to the cation exchange capacity of laponite (0.733 meq/g).

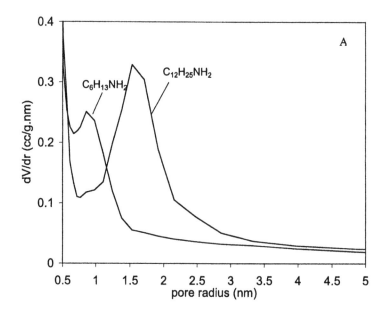

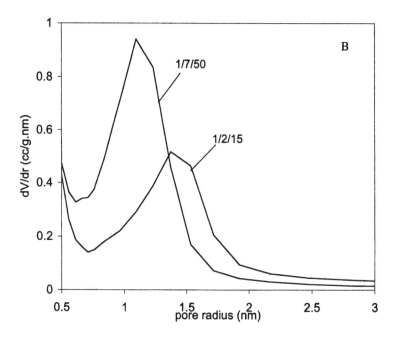

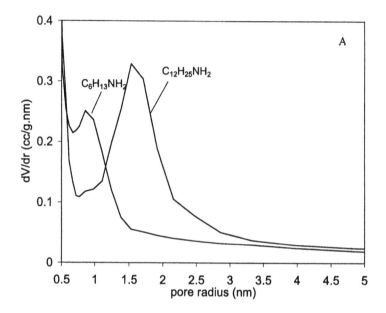

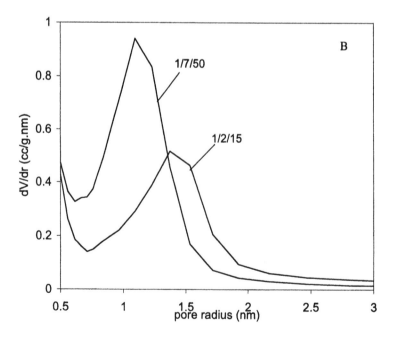

into a pore size distribution, the maximum is clearly shifted toward smaller pore radius for the concentrations 1/2/15 and 1/7/50 (Fig. 20B). The obtained shift to smaller pore size in the PSD is explained in terms of thicker Si walls present between the clay sheets due to a higher TEOS concentration, creating microporosity instead of mesoporosity.

Finally, the surfactant (Q^+) concentration is a very important parameter that can be varied during the synthesis. Therefore, in a first step $C_{16}H_{33}N(CH_3)_3^+$ has been exchanged on laponite in different amounts, respectively, 100%, 75%, and 30% of the CEC of the clay (0.733 meq/g). In the next step, micelle formation has been accomplished using $C_{12}H_{25}NH_2$ as the cosurfactant. The resulting surface area and porosity data of the corresponding P-Lap-H are summarized in Table 7, indicating that the relative contribution of the micropores is greatly enhanced when only partially exchanging Q^+(75%). In case of this partial exchange for 75% of the CEC of the clay, an optimum is found for the surface area ($S_{BET} > 1000$ m²/g) and the micropore volume ($\mu PV > 0.5$ cc/g). However, a further decrease of the Q^+ concentration to 30% of the CEC is unfavorable. Another very important feature of this partial surfactant exchange is the large residual cation exchange capacity that remains on the PCH in the form of Na^+ ions present in the structure.

The study of the extraction of templates from porous clay heterostructures instead of calcination is very interesting because of the possible degradation of the structure in this last step of the synthesis. Benjelloun et al. (105) studied the template extraction of two types of porous clay heterostructures based on montmorillonite and saponite clay hosts, since previous studies indeed pointed out that calcination may lead to local heating effects that cause a degradation of the silica framework (106). It is thus of interest to apply a "softer" extraction method for the removal of the organics. On PCHs, up to 90% of the templates can be removed via a one-step reproducible extraction procedure in methanol solvent in the presence of cations (H^+) without a subsequent calcination step for both types of PCHs. The extracted materials have a high surface area (896 m²/g for P-Mont-H and 1122 m²/g for P-Sap-H) and pore volume (0.84 cc/g for P-Mont-H and 1.13 cc/g for P-Sap-H) and are structurally very stable. When comparing the physical properties of calcined and extracted PCHs, it seems that the saponite-derived structure is more susceptible to structural collapse due to local heating effects during the calcination step, as a result of the small-sized layers' being more randomly stacked around the micellar templates. As a result, the

FIGURE 20 Pore size distributions of (A) P-Mont-H using different cosurfactant chain lengths; (B) P-Mont-H prepared with Q^+-montmorillonite/cosurfactant/TEOS concentrations of 1/2/15 and 1/7/50.

surface area and pore volume of extracted P-Sap-H is significantly larger than that of calcined P-Sap-H. In contrast, the physical properties of calcined and extracted P-Mont-H are very similar, indicating a more stable porous structure built up of large clay sheets nicely stacked in a parallel fashion. Second, by extraction it is possible to obtain slightly larger dimensions for the mesopores compared to calcination (from 2.46 to 2.74 nm for P-Mont-H and from 2.45 to 2.94 nm for P-Sap-H). Again the effect is more pronounced for saponite-derived PCH attributed to a less resistant structural stability under calcination conditions.

The same authors (107) investigated the residual cation exchange capacities of P-Mont-H and P-Sap-H. This residual CEC is the result of the liberation of protons from the surfactants during the final calcination of PCH, remaining present in the structure to balance the negative clay layer charge. When extraction is applied instead of calcination, it will be the cations present in the extraction mixture that will exchange for the surfactants and remain present in the PCH structure. Due to the inability of exchanging the small protons present in the PCH, both calcined and extracted samples in acidified methanol have been treated under an NH_3 gas flow to obtain exchangeable ammonium ions in the pore structures of the PCHs. Subsequently, these NH_4^+ ions could be exchanged for K^+ ions in a KCl solution. This results in maximal K^+ loadings or CEC-values of 0.55 mmol/g and 0.37 mmol/g for, respectively, montmorillonite- and saponite-derived PCH. This study illustrates the promising cation exchange properties of PCH solids.

B. Acidic Properties of Porous Clay Heterostructures

Silica intercalated PCHs are intrinsically acidic due to the replacement of the initial small interlayer cations (Na^+, Li^+) for surfactant cations and subsequently for protons during template removal under calcination or extraction in acidified solvent. These protons remain present in the structure in order to compensate the negative clay layer charge.

Galarneau et al. (108) were first to investigate qualitatively the acidic surface properties of fluorohectorite-PCH. For this type of PCH, Brønsted acidity results not only from protons present in the structure, but also from the hydrolysis of lattice fluorine at temperatures over 250°C, followed by proton dissociation from the resulting hydroxyl groups. In addition, the fluorohectorite layers undergo local restructuring upon calcination, giving rise to Lewis acid sites in the gallery region. Evidence for both types of acidity (Brønsted and Lewis) can be found by performing a chemisorption of the base probe molecule pyridine onto the PCH, followed by an FTIR study.

By using saponite clays with a high (tetrahedrally coordinated) Al content, the acidity of PCHs can be enhanced. Polverejan et al. (103) quantitatively determined the acidity by a temperature-programmed desorption of chemisorbed cyclohexylamine (CHA) onto PCHs derived from synthetic saponites with decreasing Si/Al content or increasing charge densities and CEC-values (0.9; 1.0 and 1.1 meq/g). The presence of both weak and strong acid sites has been proven, corresponding to desorption temperatures near 220°C and 410°C, respectively. The intensity of the first desorption peak is correlated directly with the Si/Al ratio on the clay layers and is associated with the internal gallery surfaces of the PCH. The stronger acid site at 410°C is independent of the layer composition and can be attributed to sites at the external surfaces of the PCHs. The total acidity (0.64; 0.73; 0.77 mmol CHA/g) thus increases with the saponite layer charge density, indicating that the acidity is correlated with the number of protons balancing the clay layer charge after calcination.

Another way of enhancing the acidity of PCHs is by using acid-activated clays as hosts for the synthesis of porous acid-activated clay heterostructures (PAACHs) (109). Since most of the (Brønsted) acid sites on PCHs are associated with the layers of the clay, the acidity of the host matrix can be enhanced by acid treatment prior to PCH formation. Pichowicz and Mokaya (109) used a natural montmorillonite clay on which a mild acid treatment was performed to ensure that the acid-activated clay retained a layered structure and a substantial CEC. The physical properties of the obtained PAACHs are compared with those of the PCHs in Table 8, both prepared following the procedure of Ahenach et al. (104). It is clear that the acid-activated clay–derived samples have higher surface areas and that the use of a longer-chained amine results in slightly higher pore volumes. Also, the acid contents of the samples have been compared, as determined by thermally programmed desorption of cyclohexylamine (CHA), pointing out that the total acid content (desorption between 200 and 420°C) of PAACH is at least 30% higher than that of PCH. Furthermore, the number of medium to strong acid sites (desorption between 300 and 420°C) for PAACH is almost twice that of PCH. To conclude, the higher acid content of PAACHs is due mainly to an increase in the number of medium/strong acid sites present on the clay sheets after acid treatment.

Another method to enhance the acidity of PCHs is via a postsynthesis grafting of aluminum onto the surfaces of montmorillonite- and saponite- derived PCH solids, as explored by Cool and Ahenach (25,104). They make use of an interesting feature of these materials, having a large amount of hydroxyl groups (>1 OH/nm^2) available on their surfaces after the synthesis to further react and modify the surface. For this purpose a novel modification technique, the molecular-designed dispersion (MDD) (110,111), has been applied, which allows the creation of uniform and highly dispersed supported aluminum oxide AlO$_x$ structures. The molecular-designed dispersion method consists of two steps (Fig. 21): In a first step, the aluminum acetylacetonate complex, or Al(acac)$_3$, has been anchored onto the PCH surface silanol groups. In a consecutive step, the adsorbed complex is decomposed in an oxygen-containing atmosphere at elevated tempera-

Table 8 Physical Properties and Acidity of Porous Clay Heterostructures Derived from Montmorillonite (PCH) and Acid-Activated Montmorillonite (PAACH) Clays

Samples	Surface area $(m^2 g^{-1})$	Pore volume $(cm^3 g^{-1})$	Acidity[a] $(mmol\ CHA\ g^{-1})$
PCH (C8)	795	0.75	0.54 (0.14)
PCH (C10)	782	0.82	0.56 (0.15)
PAACH (C8)	915	0.71	0.71 (0.25)
PAACH (C10)	951	0.78	0.74 (0.27)

[a] Total acid content obtained from weight loss between 200 and 420°C. Values in parentheses are the number of medium to strong acid sites obtained from weight loss between 300 and 420°C.
Source: Ref. 109.

tures, yielding aluminium oxide species at the surface of the PCH. First, the authors clarified the reaction mechanism between the Al complex and the PCH surface as a function of the initial Al complex concentration used. Following the MDD method the adsorption of the complex can occur in two distinct ways: (a) by hydrogen bonding between an acetylacetonate ligand and the surface hydroxyls and (b) by a ligand exchange mechanism, with formation of a covalent aluminum–oxygen support bonding and loss of a ligand as acetylacetonate (Hacac).

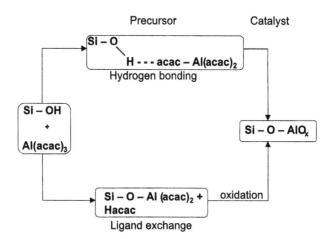

FIGURE 21   Visualization of the creation of a heterogeneous AlO$_x$ catalyst by the "molecular designed dispersion" process.

The actual reaction mechanism can be determined by calculating the ratio R:

$$R = \frac{mmol(acac)(g \text{ support})^{-1}}{mmol\ Al(g \text{ support})^{-1}} = \frac{n_{acac}}{n_{Al}} \qquad (5a)$$

with n_{acac} determined by measuring the weight loss of the modified support in an oxygen flow at 500°C. n_{Al} is obtained by chemical analysis. If the complex interacts by hydrogen bonding, the R-value should be equal to 3 (the number of acetylacetonate ligands). A lower value ($R = 2, 1$) is an indication of the ligand exchange mechanism.

For the P-Mont-H and for low concentrations of Al(acac)$_3$, R reaches 3, indicating the anchoring of the complex onto the surface —OH groups via hydrogen bonding. For P-Sap-H, however, the R-values are situated between 2 and 1.5, indicating a covalent bonding mechanism. This means that mutual differences between the two PCH supports have a direct influence on the Al-deposition mechanism: Since only saponite contains Al in its tetrahedral clay sheet, a chemical reaction between these sites and the Al complex is favored, resulting in a ligand exchange mechanism.

The acidity of the AlO$_x$-modified PCH materials can be investigated via the adsorption of the basic probe molecules NH$_3$ and CD$_3$CN. Ammonia being a strong base gives an indication of the total acidity of the PCHs. The quantitative evaluation of the acidity (adsorbed amount of NH$_3$) as a function of the degree of Al deposition on the P-Sap-H is depicted in Figure 22. It is clear that the total acidity of the PCH solid is significantly enhanced by Al grafting onto the surface.

By adsorption of a weaker base, such as acetonitrile-d_3 CD$_3$CN, one can discriminate between Brønsted and Lewis acid sites (112,113). The analysis is based on the study of the C $\equiv$ N stretching region by infrared spectroscopy. In Figure 23 the infrared spectra of Al-modified P-Sap-H before and after acetonitrile adsorption and as a function of the desorption temperature are compared (25). In the spectrum of Al-P-Sap-H before adsorption, OH vibration peaks at 3745 cm^{-1} and 3684 cm^{-1} can be distinguished of, respectively, the silica and clay surfaces. After adsorption of the probe, new bands appear in the spectrum: (a) at 2116 cm^{-1}, attributed to the stretching vibration δs (CD$_3$); (b) at 2268 cm^{-1}, corresponding to physisorbed CD$_3$CN, and (c) at 2309 cm^{-1}, due to CD$_3$CN adsorbed onto Brønsted acid sites. No evidence for Lewis acidity is found. Adsorption of the base results in a broad band between 3600 and 3300 cm^{-1}, characteristic of H-bonding interactions with the PCH support, as expected since the CD$_3$CN is a weak base and is difficult to protonate. With increasing desorption temperature, the physisorbed fraction of CD$_3$CN disappears as well as the H-bonding effects with the surface hydroxyls. However, the thermally stable Brønsted acidity remains present at the higher temperatures. This high contribu-

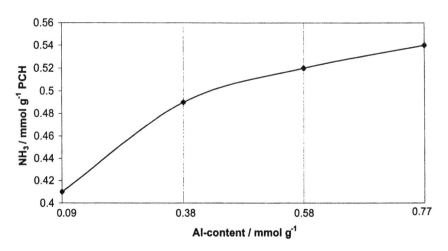

FIGURE 22 Quantitative evaluation of the acidity of Al-modified P-Sap-H as a function of the Al content by NH_3 adsorption.

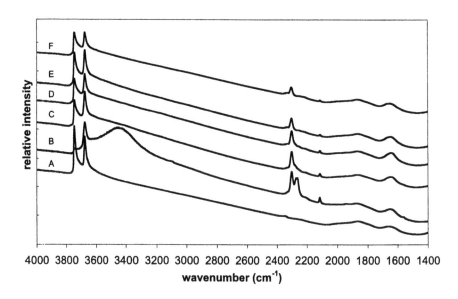

FIGURE 23 Infrared spectra of Al-modified P-Sap-H: (A) before adsorption; (B) after CD_3CN adsorption; (C) after evacuation at 25°C, (D) 60°C, (E) 120°C, (F) 150°C. (From Ref. 25.)

tion of Brønsted acid sites originates partly from the presence of protons after calcination of the PCH. Moreover, by the grafting of AlO_x species onto the PCH support, Si-(OH)-Al bonds have been created, giving rise to additional strong Brønsted acidity.

C. Catalytic and Adsorption Applications of Porous Clay Heterostructures

Because of their strong acidic properties, their unique combined micro- and meso-porosity, and their high thermal stability, PCH solids offer many interesting perspectives for applications in the field of acid catalysis.

The first report on the use of these materials, as a heterogeneous acid catalyst, was by Galarneau et al. in 1997 (108). Fluorohectorite-derived PCH was applied for the selective dehydration reaction of 2-methylbut-3-yn-2-ol (MBOH). This probe reaction is sensitive to the presence of acid–base sites on the catalyst surface (114). Acidic sites result in the dehydration of MBOH to 2-methylbut-3-yn-1-ene (Mbyne), whereas basic sites cause cleavage to acetone and acetylene. In this study, the acid catalytic activity of the PCH was compared to the activity of Li^+-fluorohectorite, alumina-pillared fluorohectorite (APF) (115), a commercially acid-treated clay (K-10), and a mesoporous pure silica MCM-41 structure. The results prove that the PCH shows a high selectivity (99.9%) to Mbyne, indicating that its surface contains mainly acidic sites. In contrast, Li^+-fluorohectorite is highly basic, affording primarily acetone (46.5%) and acetylene (36.2) as reaction products. Thus, the intercalation of a templated silica structure, followed by calcination to form a PCH, imparts significant acidic surface functionality to the structure. Further on, the study reveals that PCH is much more active (MBOH conversion 52.7%) than either MCM-41 (11%) or APF (16.4%) and almost reaches the reactivity of the K-10 montmorillonite (67.9%). This is as expected, since MCM-41 silica is only weakly acidic. And although PCH and APF share the same type of acid functionality and layered structural features, the enhanced accessibility of the framework acidic gallery sites in mesoporous PCH compared to microporous APF account for the superior catalytic activity of this PCH material.

Porous clay heterostructures, in combination with zeolite Y, are suitable catalysts for hydrocracking conversion processes, as described in the patent of the Institut Français du Petrole by Benazzi et al. (116). Since hydrocracking of heavy petroleum allows the production of lighter fractions starting from excess heavy feedstocks, this is a very important refining process. Compared to catalytic cracking, the advantage of hydrocracking is to provide middle distillates, jet fuels, and light gas-oils of very good quality. The catalysts used for hydrocracking are always of the bifunctional type that combine an acid function with a hydrogenating function. In this work, the acid function has been provided by substrates with large surface areas and important surface acidic properties (a combination of zeolite Y and porous clay heterostructures derived from 2:1 phyllosilicates containing fluorine), while the hydrogenating function is provided by one or more metals of group VIII of the periodic table (e.g., Fe, Co, Ni, Pd, Pt). The porous clay heterostructures used in the substrate of the catalyst have been pillared with either SiO_2, Al_2O_3, TiO_2, ZrO_2, or V_2O_5. Up to now, the majority of the conventional catalysts for hydrocracking consisted of weakly acidic substrates, such as amorphous silica-aluminas (116). These systems have a very good selectivity for middle distillates, but their main drawback is their low activity. By the use of mixed PCH/zeolite substrates with strong acidity, it becomes possible now to provide very active catalysts with a good selectivity for middle distillates. The feedstocks that are used in this process are heavy fractions of gas-oils consisting of at least 80% of compounds whose boiling points are between 350 and 580°C (corresponding to compounds with 15–20 carbons). Under standard operating conditions, it has been proven that the new catalysts allow selectivities for middle distillates with boiling points between 150 and 380°C that are greater than 65%, for conversion levels of more than 55% by volume. Moreover, the catalysts show a remarkable thermal stability, able to withstand temperatures of 800°C without degradation. And finally, due to the composition of the catalyst, the latter can be easily regenerated.

Polverejan et al. (117) reported on another catalytic application of mesostructured clay catalysts. Saponite-derived PCH has been used for the condensed-phase Friedel–Crafts alkylation of bulky 2,4-di-*tert*-butylphenol (DBP) with cinnamyl alcohol to produce a large flavan, namely, 6,8-di-*tert*-butyl-2,3-dihydro[4H]benzopyran. Because the molecular dimensions of DBP are very large ($1.35 \times 0.79 \times 0.49$ nm), conventional zeolites and pillared clays are not suitable to catalyze this reaction, while the mesopores of saponite PCH are large enough to allow access to the interlayer acid sites. The catalytic results obtained on saponite-PCH are compared to those on H^+-saponite, zeolite HY, and acid-treated montmorillonite K-10 (Table 9). The reaction is visualized in figure 24.

As can be deduced from Table 9, only a very little amount of flavan is obtained over HY, because of the strongly restricted diffusion of DBP through the small pores of HY. Similarly, on K-10 and H^+-saponite, the main product is dealkylated 4-*tert* butylphenol. This indicates that beside diffusion problems, the accessability of the reactants and the shape selectivity are major problems; therefore K-10 and saponite are incapable as shape-selective catalysts. However, in the case of saponite-PCH, a much higher yield of flavan (15.3%) is obtained, explained by the better accessibility of reactants and products through the mesopores of the material, proving that PCH solids are effective catalysts for the alkylation reaction.

For environmental adsorption applications, porous clay heterostructures can be functionalized with suitable surface groups in order to immobilize heavy metal

Table 9 Alkylation of 2,4-Di-*tert*-Butylphenol (DBP) with Cinnamyl Alcohol (CA)

Catalysts	Conversion (%)	Selectivity (%)*		Yield (%) of flavan
		4-*tert*-Butylphenol	Flavan	
HY	13.3	57.8	11.1	1.5
K-10	47.6	59.7	—	<1
H⁺-saponite	36.1	87.7	3.3	1.2
PCH-saponite	36.6	32.1	41.9	15.3

Reaction condition: 125 mg catalyst, 0.25 mmol DBP, 0.25 mmol cinnamyl alcohol, 12.5 mL isooctane, 60°C, 6 h.
* The total selectivity is less than 100% because part of the organic product is adsorbed onto the catalyst and is nonrecoverable.
Source: Ref. 117.

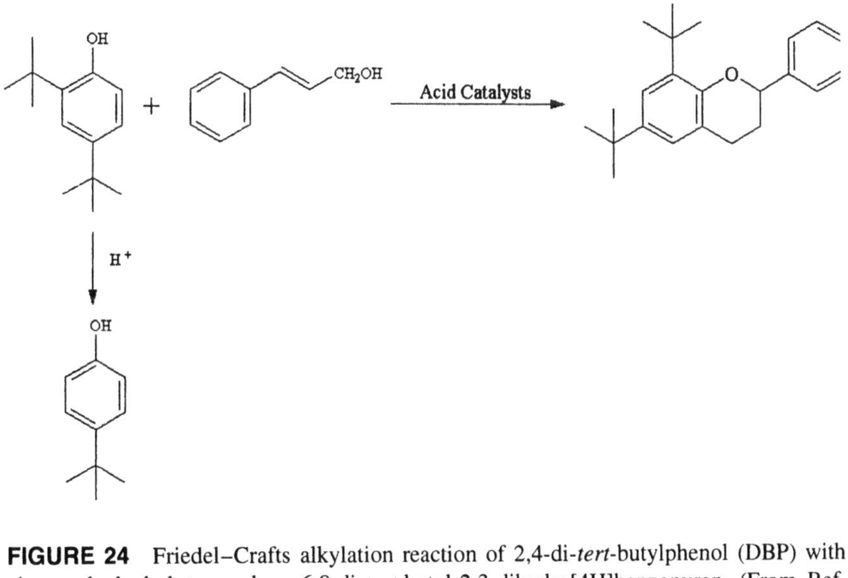

FIGURE 24 Friedel–Crafts alkylation reaction of 2,4-di-*tert*-butylphenol (DBP) with cinnamyl alcohol to produce 6,8-di-*tert*-butyl-2,3-dihydro[4H]benzopyran. (From Ref. 117.)

FIGURE 25 Grafting of mercaptopropylsilane groups to the inner and outer walls of mesostructured silica intercalated in clay. (From Ref. 118.)

pollutants. One example is a functionalized PCH adsorbent for Hg^{2+} trapping, prepared by Mercier and Pinnavaia (118) by grafting of 3-mercaptopropyltrimethoxysilane onto the intragallery silica framework walls of a porous fluorohectorite clay heterostructure. This grafting procedure of mercaptopropylsilane groups (denoted—SH) onto the silanol groups of the interlayer silica framework causes a swelling of the clay layers and thus a gallery expansion (see Fig. 25). However, the initial PCH with mesopores of 2.1 nm loses its mesoporosity after the modification, so the grafting process results in a certain constriction of the pore channels, although the modified PCH still has a surface area of 175 m^2/g and a pore volume of 0.07 cc/g. Heavy metal Hg^{2+} ions can be adsorbed on functionalized PCH up to a maximum capacity of 0.74 mmol/g, a value representing 67% of the total —SH groups present on the substrate. For initial solution concentrations under 10 ppm, the functionalized PCH is capable of removing the Hg^{2+} ions completely. But because the Hg^{2+} ions are initially quantitatively bound to the more accessible complexing sites located near the pore channel openings of the PCH and because for higher Hg^{2+} concentrations the ions must diffuse to binding sites located deeper within the clay interlayer region, the adsorption efficiency becomes limited when reaching the saturation capacity of the PCH. Anyhow, the performance of this PCH is a definite improvement for heavy metal ion binding over any other functionalized layered silicate system, such as pillared clays, due to the improved accessibility of the binding sites in the mesoporous structure.

IV. CONCLUSIONS

The research on porous layered clay structures has been abundant over the last decades. First, pillared interlayered clays have been developed, which are fine-tuned toward adsorption and catalysis purposes. Further developments in this field can be expected in the future, although a better fundamental knowledge of the pillaring process and of the distribution and geometry of the pillars in the interlamellar region is therefore required. If this chemistry were better known, researchers could start to develop pillared solids with more uniformly distributed pillars, all of the same size and shape. Pillaring could then be optimized to obtain "tailor-made" pillared clays with very narrow pore size distributions. More recently, mesoporous clay–derived layered structures (porous clay heterostructures) have been successfully developed based on a templated mechanism. Several adsorption and catalytic applications have already proven the capability of these materials to processes generating and using relatively large molecules of high value, based on their unique mesopore dimensions and high pore volumes. The main problem to be solved for the future is the high cost of these materials, compared to conventional adsorbents such as zeolites. Solutions have to be found that recover and reuse the expensive surfactants and that improve the reproducibility of the synthesis process. When this is solved, porous clay heterostructures will certainly find important applications in industrial catalytic and adsorption processes.

ACKNOWLEDGMENTS

Pegie Cool acknowledges the FWO Vlaanderen (Fund for Scientific Research—Flanders—Belgium) for financial support.

REFERENCES

1. R.A. Schoonheydt, T. Pinnavaia, G. Lagaly, N. Gangas. Pure Appl. Chem 1999; 71:2367–2371.
2. J.R. Butruille, T.J. Pinnavaia. In: J.L. Atwood, J.E.D. Davies, D.D. Macnicol, F. Vogtle, J.-M. Lehn, eds. Comprehensive Supramolecular Chemistry. In: G. Alberti, T. Bein, eds. Solid-State Supramolecular Chemistry: Two- and Three-Dimensional Solid-State Networks. Vol. 7. Oxford: Pergamon-Elsevier, 1996:219–250.
3. A. Gil, L.M. Gandia. Catal. Rev., Sci. Eng, J.T. Kloprogge, J. Porous. Mater 1998; 5:5–41.
4. V. Seefeld, R. Bertram, P. Starke, W. Gessner. Silikattechnik 1988; 39:239–241. V. Seefeld, R. Bertram, H. Gorz, W. Gessner, S. Schonherr. Z. Anor. Allg. Chem 1991; 603:129–135. V. Seefeld, R. Bertram, D. Muller, W. Gessner. Silikattechnik 1991; 42:305–308. J.T. Kloprogge, D. Seykens, J.B.H. Jansen, J. W. Geus. J. Noncrystalline Solids 1992; 142:94–102. J.T. Kloprogge, D. Seykens, J.W. Geus, J.B.H. Jansen. J. Noncrystalline Solids 1992; 142:87–93. J.T. Kloprogge, D. Seykens, J.B.H. Jansen, J.W. Geus. J. Noncrystalline Solids 1993; 152:207–211. J.T. Kloprogge, D. Seykens, J.W. Geus, J.B.H. Jansen. J. Noncrystalline Solids 1993; 160: 144–151. G. Furrer, C. Ludwig, P.W. Schindler. J. Colloid Interface Sci 1992; 149: 56–67.
5. R.A. Schoonheydt, K.Y. Jacobs. Clays from two to three dimensions. In: H. Van Bekkum, P.A. Jacobs, E.M. Flanigen, J.C. Jansen, eds. Introduction to Zeolite

Science and Practice, Studies in Surface Science and Catalysis. Vol. 137. Amsterdam: Elsevier, 2001:299.

6. R.A. Schoonheydt, H. Leeman, A. Scorpion, I. Lenotte, P.G. Grobet. Clays Clay Miner 1997; 45:518–525.

7. S. Aceman, N. Lahav, S. Yariv. Appl. Clay Sci 2000; 17:99–126.

8. D. Plee, F. Borg, L. Gatineau, J.J. Fripiat. J. Am. Chem. Soc 1985; 107:2362–2369.

9. M. De Bock, H. Nijs, P. Cool, E.F. Vansant. J. Porous Mater 1999; 6:323–333. S.P. Katdare, V. Ramaswany, A.V. Ramaswany. J. Mater. Chem 1997; 7:2197–2199. J. Ahenach, P. Cool, E.F. Vansant. Microporous Mesoporous Mater 1998; 26: 185–192.

10. L. Storaro, M. Lenarda, M. Perissinotto, V. Luchini, R. Ganzerla. Microporous Mesoporous Mater 1998; 20:317–331.

11. R.E. Grim. Clay mineralogy, New York: McGraw Hill, 1968. E. Nemecz. Clay Miner. Budapest: Akadémiai Kiado, A.C.D. Newman. Chemistry of Clays and Clay Minerals. Harlow: Longman Scientific and Technical, 1987.

12. E. Maes, L. Vielvoye, W. Stone, B. Delvaux. Eur. J. Soil Sci 1999; 50:107.

13. L.A. Douglas. In J.B. Dixon, S.B. Weed, eds. Minerals in Soil Environments. Soil Sci. Soc. of America. Madison, WI, 1977:259.

14. P.H. Hsu, T.F. Bates. Soil Sci. Soc. Am. Proc 1964; 28:763.

15. P.H. Hsu. Clays Clay Miner 1992; 40:300.

16. B. d'Espinose de la Caillerie, J.J. Fripiat. Clays Clay Miner 1991; 39:270.

17. N. Lahav, U. Shani, J. Shabtai. Clays Clay Miner 1978; 26:107.

18. A. Schutz, W.E.E. Stone, G. Poncelet, J.J. Fripiat. Clays Clay Miner 1987; 35:251.

19. L.J. Michot, D. Tracas, B.S. Lartiges, F. Lhote, C.H. Pons. Clay Miner 1994; 29: 133.

20. F.J. del Rey-Perez-Caballero, G. Poncelet. Microporous Mesoporous Mater 2000; 37:313.

21. F.J. del Rey-Perez-Caballero, G. Poncelet. Microporous Mesoporous Mater 2000; 41:169.

22. E. Booij, J.T. Kloprogge, J.A.R. van Veen. Appl. Clay Sci 1996; 11:115–162.

23. M.L. Occelli, P.A. Peaden, G.P. Ritz, P.S. Iyer, M. Yokoyama. Microporous Mater 1993; 1:99–113.

24. A. Galarneau, A. Barodawalla, T.J. Pinnavaia. Nature 1995; 374:529–531.

25. P. Cool, J. Ahenach, E.F. Vansant. In: D.D. Do, ed. Adsorption Science and Technology; Proc. 2nd Pacific Basin Conference on Adsorption Science and Technology, Brisbane, World Scientific Publishing. 2000:638–642.

26. S. Inagaki, Y. Fukushima, K. Kuroda. In: J. Weitkamp, H.G. Karge, H. Pfeifer, W. Hölderich, eds. Zeolites and Related Microporous Materials: State of the Art 1994, Studies in Surface Science and Catalysis. Vol. 84. Amsterdam: Elsevier, 1994:125–132.

27. T. Yanagisawa, T. Shimizu, K. Kuroda, C. Kato. Bull. Chem. Soc. Jpn 1990; 63: 988–992.

28. T.J. Pinnavaia, T. Kwon, S.K. Kun. In: E.G. Derouane, F. Lemos, C. Naccache, F. Ramao-Ribeiro, eds. Zeolite Microporous Solids: Synthesis, Structure and Reactivity, NATO ASI Series C. Vol. 352. Dordrecht: Kluwer, 1992:91–104.

29. S. Brunauer, L.S. Deming, E. Deming, E. Teller. J. Am. Chem. Soc 1940; 62:1723.
30. N. Maes, I. Heylen, P. Cool, E.F. Vansant. Applied Clay Science 1997; 12:43–60.
31. H-Y Zhu, N. Maes, E.F. Vansant. In: J. Rouquerol, F. Rodriguez-Reinoso, K.S.W. Sing, K.K. Unger, eds. Characterization of Porous Solids III, Studies in Surface Science and Catalysis, Elsevier. 1994; 87:457–466.
32. I. Langmuir. J. Am. Chem. Soc 1915; 40:1361.
33. S. Brunauer, P.H. Emmet, E. Teller. J. Am. Chem. Soc 1938; 60:309.
34. S. Lowell. Introduction to Powder Surface Area. Chichester: Wiley Interscience, 1979.
35. S.J. Gregg, K.S.W. Sing. Adsorption, Surface Area and Porosity. London: Academic Press, 1982.
36. R.D. Vold, M.J. Vold. Colloid and Interface Chemistry. Vol. chapter 3. Reading, MA: Addison-Wesley, 1983.
37. K.S.W. Sing, D.H. Everett, R.A.W. Haul, L. Moscow, R.A. Pierotti, J. Rouquérol, T. Siemieniewska. Pure Appl. Chem 1985; 57(4):603.
38. H-Y Zhu, E.F. Vansant. J. Porous Mater 1995; 2:107–113.
39. J.H. de Boer, B.G. Linsen, T.J. Osinga. J. Catal 1964; 4:643. J.C.P. Broekhoff, J.H. de Boer. J. Catal 1968; 10:391.
40. M.J. Remy, A.C. Vieira Coelho, G. Poncelet. Microporous Mater 1996; 7:287.
41. K.S.W. Sing. Chem. Ind 1967; 20:829.
42. A. Lecloux, J.P. Pirard. J. Colloid Interface Sci 1980; 18:355.
43. S. Moreno, R. Sun Kou, G. Poncelet. J. Phys. Chem 1997; 101:1569.
44. M.L. Occelli, D.J. Lynch, J.V. Sanders. J. Catal 1987; 197:557.
45. N. Maes, I. Heylen, P. Cool, M. De Bock, C. Vanhoof, E.F. Vansant. J. Porous Mater 1996; 3:47–59.
46. J. Ahenach, P. Cool, E. Vansant, O. Lebedev, J. Van Landuyt. Phys. Chem. Chem. Phys 1999; 1:3703–3708.
47. S. Vercauteren, K. Keizer, E.F. Vansant, J. Luyten, R. Leysen. J.Porous Mater 1998; 5:241–258.
48. G. Binning, C.F. Quate, C. Gerber. Phys. Rev. Lett 1986; 56:430.
49. H. Hartman, G. Sposito, A. Yang, S. Manne, S.A.C. Gould, P.K. Hansma. Clays Clay Miner 1990; 38(4):337.
50. M.L. Occelli, B. Drake, S.A.C. Gould. J. Catal 1993; 142:337–348.
51. M.L. Occelli, J.A. Bertrand, S.A.C. Gould, J.M. Dominguez. Microporous Mesoporous Mater 2000; 34:195–206.
52. D.E.W. Vaughan, R.J. Lussier. In: L.V. Rees, ed. Proceedings of the 5th International Conference of Zeolites. London: Heyden, 1980:94.
53. G. Poncelet, A. Schutz. In: R. Setton, ed. Chemical Reactions in Organic and Inorganic Constrained Systems. Dordrecht: Reidel, 1986:165.
54. S. Chevalier, R. Franck, H. Suquet, J.-F. Lambert, D. Barthomeuf. J. Chem. Soc. Faraday Trans 1994; 90:667.
55. S. Moreno, R. Sun Kou, G. Poncelet. J. Catal 1996; 162:198.
56. R.J. Gorte. Catal. Lett 1999; 62:1.
57. C.A. Emeis. J. Catal 1993; 141:347.
58. F. Kooli, W. Jones. J. Mater. Chem 1998; 8:2119.

59. P.A. Politowicz, J.J. Kozak. J. Phys. Chem 1988; 92:6078.
60. P.A. Politowicz, L.B.S. Leung, J.J. Kozak. J. Phys. Chem 1989; 93:923.
61. M. Sahimi. J. Chem. Phys 1990; 92:5107.
62. Z-X Cai, S.D. Mahanti, S.A. Solin, T.J. Pinnavaia. Phys. Rev. B 1990; 42:6636.
63. B.Y. Chen, H. Kim, S.D. Mahanti, T.J. Pinnavaia, Z-X Cai. J. Chem. Phys 1994; 100:3872.
64. X. Yi, S. Shing, M. Sahimi. AIChE. J 1995; 41:456.
65. X. Yi, S. Shing, M. Sahimi. Chem. Eng. Sci 1996; 51:3409.
66. X. Yi, J. Ghassemzadeh, S. Shing, M. Sahimi. J. Chem. Phys 1998; 108:5.
67. R.C. Zielke, T.J. Pinnavaia. Clays Clay Miner 1988; 36(5):403–408.
68. H.T. Shu, D. Li, A.A. Scala, Y.H. Ma. Separation Purification Technol 1997; 11: 27–36.
69. R. Wibulswas, D.A. White, R. Rautiu. Environm. Techn 1998; 19:627–632.
70. R.M. Barrer. Zeolites and Clay Minerals as Sorbents and Molecular Sieves. London: Academic Press, 1978.
71. R.M. Barrer, R.J.B. Craven. J. Chem. Soc. Faraday Trans 1992; 88(4):645–651.
72. I. Gameson, W.J. Stead, T. Rayment. J. Phys. Chem 1991; 95:1727–1730.
73. R.T. Yang, M.S.A. Baksh. AIChE J 1991; 37(5):679–686.
74. M.S.A. Baksh, R.T. Yang. AIChE J 1992; 38(9):1357–1368.
75. L.S. Cheng, R.T. Yang. Microporous Mater 1997; 8:177–186.
76. R.T. Yang. Gas Separation by Adsorption Processes. Boston: Butterworth, 1987.
77. I. Heylen, N. Maes, A. Molinard, E.F. Vansant. In: E.F. Vansant, ed. Separation Technology, Process Technology Proceedings. Vol. 11. Amsterdam: Elsevier, 1994: 355–361.
78. I. Heylen, C. Vanhoof, E.F. Vansant. Microporous Mater 1995; 5:53–60.
79. I. Heylen, E.F. Vansant. Microporous Mater 1997; 10:41–50.
80. P. Cool, E.F. Vansant. Microporous Mater 1996; 6:27–36.
81. P. Cool, E.F. Vansant. In: D.D. Do, ed. Adsorption Science and Technology; Proc. 2nd Pacific Basin Conference on Adsorption Science and Technology, Brisbane. World Scientific. 2000:633–637.
82. D.E.W. Vaughan, J.S. Magee, R.J. Lussier. Pillared Interlayered Clay Products, U.S. Patent 4,271,043, 1981.
83. A. Molinard, K.K. Peeters, N. Maes, E.F. Vansant. In: E.F. Vansant, ed. Separation Technology, Process Technology Proceedings. Vol. 11. Amsterdam: Elsevier, 1994: 445–454.
84. S. Yamanaka, P.B. Malla, S. Komarneni. J. Colloid Interface Sci 1990; 134:51–58.
85. P.B. Malla, S. Yamanaka, S. Komarneni. Solid State Ionics 1989; 32(33):354–362.
86. A. Molinard, E.F. Vansant. In: E.F. Vansant, ed. Separation Technology, Process Technology Proceedings. Vol. 11. Amsterdam: Elsevier, 1994:423–436.
87. A. Molinard, E.F. Vansant. Adsorption 1995; 1:49–59.
88. P. Cool, A. Clearfield, R.M. Crooks, E.F. Vansant. Adv Environ Res 1999; 3(2): 139–151.
89. H.Y. Zhu, G.Q. Lu. J. Porous Mater 1998; 5:227–239.
90. H.Y. Zhu, W.H. Gao, E.F. Vansant. J. Colloid Interface Sci 1995; 171:377.
91. F. Figueras. Catal. Rev. Sci. Eng 1988; 30:457.

92. J.-F. Lambert, G. Poncelet. Topics Catal 1997; 4:43.
93. J.A. Martens, P.A. Jacobs. In: J.B. Moffat, ed. Theoretical Aspects of Heterogeneous Catalysis, Van Nostrand-Reinhold. 1990:52.
94. A. Schutz, D. Plée, F. Borg, P. Jacobs, G. Poncelet, J.J. Fripiat. In: L.G. Schultz, H. van Olphen, F.A. Mumpton, eds. Proceedings Intern. Clay Conf., Denver. Clay Mineral Soc.. 1987:305.
95. A. Vieira Coelho, G. Poncelet. In: I.V. Mitchell, ed. Pillared Layered Structures: Current Trends and Applications. Dordrecat: Kluwer Academic Publishers, 1990: 185.
96. S. Moreno, E. Gutierrez, A. Alvarez, N.G. Papayannakos, G. Poncelet. Appl. Catal. A General 1997; 165:103.
97. S. Moreno, R. Sun Kou, R. Molina, G. Poncelet. J. Catal 1999; 182:174.
98. R. Molina, S. Moreno, A. Vieira Coelho, J.A. Martens, P.A. Jacobs, G. Poncelet. J. Catal 1994; 148:304.
99. R. Molina, S. Moreno, G. Poncelet. In:. A. Corma, F.V. Melo, S. Mendioroz, J.L.G. Fierro, eds. 12th International Congress on Catalysis, Proc. of the 12th ICC, Granada, Spain, July 9–14 2000, Studies in Surface Science and Catalysis. Vol. 130. Amsterdam: Elsevier, 2000:983.
100. F.J. del Rey-Perez-Caballero, M.L. Sanchez-Henao, G. Poncelet. In:. A. Corma, F.V. Melo, S. Mendioroz, J.L.G. Fierro, eds. 12th International Congress on Catalysis, Proc. of the 12th ICC, Granada, Spain, July 9–14 2000, Studies in Surface Science and Catalysis. Vol. 130. Amsterdam: Elsevier, 2000:2417.
101. C.T. Kresge, M.E. Leonowicz, W.J. Roth, J.C. Vartuli. U.S. Patent 5,098,684.
102. M.S. Baksh, E.S. Kikkindes, R.T. Yang. Ind. Eng. Chem. Res 1992; 31:2181–2189.
103. M. Polverejan, T.R. Pauly, T.J. Pinnavaia. Chem. Mater 2000; 12:2698–2704.
104. J. Ahenach, P. Cool, E.F. Vansant. Phys. Chem. Chem. Phys 2000; 2:5750–5755.
105. M. Benjelloun, P. Cool, P. Van Der Voort, E.F. Vansant. Phys. Chem. Chem. Phys 2002; 4:2818–2823.
106. R. Mokaya, W. Jones. J. Catal 1997; 172:211.
107. M. Benjelloun, P. Cool, T. Linssen, E.F. Vansant. Microporous Mesoporous Mater 2001; 49:83–94.
108. A. Galarneau, A. Barodawalla, T.J. Pinnavaia. Chem. Commun 1997:1661–1662.
109. M. Pichowicz, R. Mokaya. Chem. Comm 2001:2100–2101.
110. M.G. White. Catalysis Today 1993; 18:73.
111. P. Van Der Voort, M. Baltes, E.F. Vansant, M.G. White. Interface Sci 1997; 5: 209.
112. G. Busca. Phys. Chem. Chem. Phys 1999; 1:723.
113. A.G. Pelmenschikov, R.A. van Santen, J. Jänchen, E. Meijer. J. Phys. Chem 1993; 97:11071–11074.
114. H. Laron-Pernot, F. Luck, J.M. Popa. Appl. Catal 1991; 78:213.
115. J.-R. Butruille, L.J. Michot, T.J. Pinnavaia. J. Catal 1993; 139:664.
116. E. Benazzi, J. Brendle, R. Le Dred, J. Baron, D. Saehr, N. Georges-Marchal, S. Lacombe. U.S. Patent, 6,139,719, 2000.
117. M. Polverejan, Y. Liu, T.J. Pinnavaia. In: A. Sayari, ed. Studies in Surface Science and Catalysis. Vol. 129: Elsevier Science, 2000:401–408.
118. L. Mercier, T.J. Pinnavaia. Microporous Mesoporous Mater 1998; 20:101–106.

7

Layered α-Zirconium Phosphates and Phosphonates

Challa V. Kumar, Akhilesh Bhambhani, and Nathan Hnatiuk
University of Connecticut
Storrs, Connecticut, U.S.A.

1. INTRODUCTION

A. Zr(IV) Phosphates and Phosphonates: Layered, Porous, Inorganic Materials

Smectite-type layered inorganic materials include many solids, such as silicates, phosphates, phosphonates, phosphites, oxides, hydroxides, niobates, chalcogenides, and sulfides. These inorganic layered solids have received considerable attention in the past two to three decades due to their potential for catalytic and ion exchange applications. In addition, these materials are being explored for biological applications such as biosensors, biocatalysis, biopharmaceuticals, and protein encapsulation and in the design of inorganic enzyme mimics. Among these various layered inorganic solids, this chapter focuses on crystalline α-zirconium(IV) phosphates and phosphonates, in order to provide an overview of studies of personal interest to the authors. Selected examples of synthesis, structure, catalytic properties, and biochemical applications of Zr(IV) phosphates/phosphonates are provided.

B. Potential of Metal Phosphates and Phosphonates

Interesting applications of these layered intercalates include, but are not limited to, material design, ion exchange (1), catalysis (2), the study of quantum-sized semiconductor particles (3), assembly of molecular multilayers of controlled thickness (4), designer electrode surfaces (5), biocatalysis, biomedical applications, and the preparation of low-dimensional conducting polymers (6,7). High thermal stability, solid-state ion conductivity, high chemical stability, and the versatility to incorporate organic functional groups in the layers make these materials very attractive for practical applications. One advantage of the lamellar solids is that the interlamellar spacings can be expanded to accommodate guests of any size, from protons to proteins to protozoa, in the galleries. Metal ions, metal complexes, molecules, cyclodextrins, semiconductor particles, metal clusters, and proteins of various sizes are intercalated into the galleries of these layered materials. Formation of such intercalates can be verified using absorption, fluorescence, X-ray diffraction (XRD), circular dichroism (CD), and chemical/biological activity methods. Due to the large particle sizes (100–1000 nm), these solid materials can be separated from the unbound guest molecules by centrifugation. The binding affinity and the binding stoichiometry of the guests bound in the galleries, therefore, are readily estimated from centrifugation studies.

The intergallery spacings can be increased by pillaring, and increased spacings enhance access to the galleries for improved catalysis, conductivity, or ionexchange. Intercalation of catalytic nanoparticles in the galleries, for example, increases the layer spacings and expands the catalytic potential of the layered materials. Intercalation or covalent anchoring of enzymes in the galleries extends the catalytic applications of these materials to biochemical transformations. Due to the availability of the extensive selection of enzymes, this area of research has a high potential for biocatalytic applications, including the industrial synthesis of pharmaceuticals. Both chemical and biochemical catalytic applications of these layered materials, therefore, will be the focus of many future investigations.

Another interesting application of these layered materials has been the construction of organized, uniform, and functional supramolecular assemblies of molecules, ions, metal complexes, and proteins (8). Such molecular assemblies are important for the design of biosensors, solar energy harvesting, and photochemical applications (9). Layered zirconium phosphate/phosphonates, therefore, are versatile solids, and they have been used for the organization of a variety of guest molecules/ions at the galleries (10). Research in this area resulted in the construction and discovery of highly organized molecular systems with unique properties. The unique properties of the assemblies contrast with the properties of the individual components of the assembly, such that the whole is more than the sum of its components. The preparation, characterization, and properties of the Zr(IV) phosphate/phosphonate materials is reviewed here first, and then the catalytic, ionexchange, and biochemical as well as photochemical properties of guests intercalated in the galleries are illustrated. This chapter focuses on the topics of general interest, with some bias toward the authors' favorites; it is not intended to be exhaustive.

II. α-ZIRCONIUM PHOSPHATE (α-ZRP)

A. Preparation

The synthesis of crystalline α-Zr(IV) phosphate (α-Zr(HPO$_4$)$_2$ H$_2$O, abbreviated α-ZrP) and its structural details are closely related to the corresponding properties of the α-zirconium phosphonates. The amorphous form of ZrP is prepared by mixing an aqueous solution of zirconyl chloride (ZrOCl$_2$·8H$_2$O) with excess phosphoric acid [Eq. (1)] (11,12). The resulting white precipitate is crystallized into the layered form by extended reflux in a strong acid, such as HF, HCl, or H$_3$PO$_4$. The progress of crystallization and the formation of the layered solid is monitored in powder X-ray diffraction experiments, and the powder patterns indicate the conversion from the amorphous gel to the crystalline state (13). The degree of crystallinity varies with refluxing time, temperature, and the concentration of the acid. When HF is used, the formation of α-ZrP from the initially formed fluorozirconate is controlled by the rate of evaporation of HF; extremely slow evaporation results in the direct precipitation of the crystalline α-ZrP (14). The rate of precipitation of α-ZrP is controlled by the rate of removal of HF by passing nitrogen or water vapor over the reaction mixture; when this is very slow, highly crystalline α-ZrP is precipitated. The use of molten organic acids such as oxalic acid can dissolve the amorphous gel, and they have been used to produce pure α-ZrP as well as mixed metal phosphates (15). X-ray powder diffraction data confirm the crystalline, stacked layers of the material.

$$ZrOCl_2 + H_3PO_4 \rightarrow Zr(PO_3OH)_2 \cdot nH_2O \qquad (1)$$

Scanning electron micrographs of α-ZrP recorded in this laboratory show the stacking of uniform disks approximately 600 nm in diameter (Fig. 1). The stacks are usually upright on the SEM grid, but an occasional side view is also seen. This view clearly illustrates the layers of the materials, the uniform size of the disks, and their alignment within the stacks (16). Large α-ZrP particles of uniform size and high crystallinity are obtained by the HF method (17). The gamma form of ZrP differs significantly from the alpha phase, and current discussion will be concerned with the latter phase. A mesoporous form of the ZrP is also known, which is produced by the sol-gel method while using cetyltrimethylammonium bromide as the structure-directing reagent in the synthesis (18). Nonaqueous synthesis of ZrP in the presence of templated cations results in the formation of novel layered (19) or monodimensional (20) zirconium phosphates.

B. Structure

The crystal structure of α-ZrP indicates that two layers of phosphate anions sandwich a layer of Zr(IV) cations (21). Each Zr(IV) is coordinated to six phosphate oxygens such that each phosphate bridges three Zr(IV) ions via oxygen atoms, and

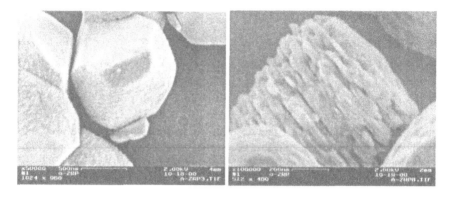

FIGURE 1 Scanning electron micrographs of crystalline α-ZrP. Stacks of the well-ordered plates are pointed toward the viewer (left), and an occasional side view is also seen (right).

the phosphates alternate above and below the plane of Zr(IV) ions. A schematic structure of α-ZrP is shown in Figure 2. The remaining OH group of the phosphate is oriented perpendicular to the α-ZrP layer, (22) and this OH group can be readily ionized to produce a lattice of negative charges in the α-ZrP galleries. Acidity of this OH group is the chemical basis for the catalytic, ion exchange, ion conductivity, and intercalation properties of α-ZrP materials. The interlayer distance, as estimated from the powder diffraction patterns, is 7.6 Å, and each phosphate occupies approximately 25 Å^2 of area in the galleries.

Hydration of the galleries and hydrogen-bonding networks within the galleries have been examined using neutron powder diffraction. The metal phosphate layer forms cavities in the framework of the galleries, and water molecules are

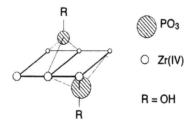

FIGURE 2 Schematic structure of α-ZrP (R = OH) consisting of a layer of metal ions sandwiched between two layers of phosphates.

located in these cavities. The HPO_4 groups form donor hydrogen bonds with these water molecules; in turn, the water molecules hydrogen bond with the oxygen atoms of the phosphate (23). No hydrogen bonding exists between the layers, and adjacent layers of the metal phosphate are held together by van der Waals interactions. Infrared and Raman studies indicate that the crystallographically distinct water molecules have significantly distinct hydrogen bonding strengths (24). A set of force field parameters for the bonds and atoms of ZrP were derived, and these parameters have been tested by calculating the crystal structures of ZrP, their energies, and characteristic vibrational frequencies (25). These calculations are in acceptable agreement with the experimental values.

Significant, attractive features of Zr(IV) phosphates and phosphonates are: (a) They are readily prepared from simple starting materials; (b) the gallery spacings can be expanded to accommodate small as well as large guests (from protons and proteins to cells) and improve the diffusion of reagents into the galleries; (c) the phosphate/phosphonate matrix does not absorb light in the visible region, although it scatters light; (d) they provide surfaces with high charge density (as high as one charge per 25 $\mathring{A}^2$), in contrast to synthetic and natural clays; and (e) the charge field of the matrix can restrict orientation/mobilities of the bound guests. Some of the exciting applications of these materials include ion exchange, solid-state ion conduction, catalysis, development of sensors, and biocatalysis.

C. Ion Exchange Properties

Ionization of the HPO_4 group provides an ion exchangeable site, and these sites are situated both inside and outside of the solid. For ion exchange purposes, binding in the galleries vs. on the outside is not distinguished (Fig. 3), but the overall binding affinity, capacity, and selectivity are of interest. There is an extensive body of literature on the ion exchange properties of α-ZrP, and only a handful of examples are chosen here to illustrate specific aspects of ion exchange (Table 1). α-ZrP matrices are rigid, robust, inert (at pH ≤ 7), and thermally stable and provide excellent materials for ion exchange applications (26). The interlayer binding sites of α-ZrP can be accessed by various cations, and this behavior has been exploited for a variety of applications, such as chromatographic separation of inorganic acids, alkaloids, (27) and metal ions (28). The exchange of ions into and out of the galleries is studied by monitoring the replacement of protons by the metal ion of interest via titration. The titration curves for the exchange of protons by sodium ions and the exchange of sodium ions by protons display hysteresis, which indicates the differences in the forward and the reverse processes (29). The hysteresis is attributed to the participation of different phases of the ion exchange material during the forward and reverse titrations. The kinetics of sodium/proton (Na^+/H^+) exchange depends on the pH of the medium, with three distinct plateau regions (pH < 7, pH = 7–9, and pH > 9); this is due to the weak

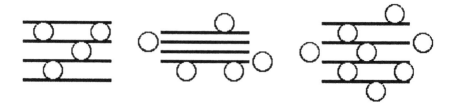

FIGURE 3 Binding of guest molecules inside, outside, and both inside and outside of α-ZrP layers.

acidity of the HPO_4 functional groups in the matrix. The rates of ion exchange also depend on the nature of the counterion, (30) and the order of increasing rates of Na^+/H^+ exchange is $I^- < Br^- < Cl^- < ClO_4^- < NO_3^- < SO_4^{-2}$. This dependence of rates on counterions indicates the participation of the counterions in the forward and reverse processes to a significant extent. In a similar process, the presence of small amounts of sodium ions also catalyzes the exchange of protons by Cs^+ and Mg^{+2}, although sodium ions do not compete in the reaction (31).

The kinetics of intercalation and deintercalation of alkali metal ions were investigated in pressure-jump experiments while monitoring the electrical conductivity of the samples (32). These studies indicate biphasic kinetics whose magnitudes are in milliseconds; the rates of the fast and slow components increased with increased concentrations of the metal ions. The forward and reverse rates depend on the interlayer distances, and the fast and slow components have been attributed to the ingress of ions into the galleries and interlayer diffusion, respectively. Similar biphasic kinetics on millisecond–second time scales were also observed in pressure-jump experiments for the deprotonation–reprotonation of α-ZrP (33). In the latter case, the slow and fast components have been attributed to deprotonation from the surface and from the interlayer regions of the solid, respectively.

In a related study, the ion exchange of potassium/proton was examined in calorimetric measurements to understand the thermodynamics of ion exchange. The K^+/H^+ ion exchange is exothermic, with an equilibrium constant of 3×10^{-5}. The corresponding enthalpy and entropy changes are 1.58 kcal mol^{-1} and -15.3 kcal $mol^{-1}K^{-1}$, respectively (34). The acetate salts of alkaline earth cations undergo facile ion exchange. Acetate was suggested to facilitate ion exchange by deprotonating HPO_4 groups and by maintaining the pH needed for metal binding (35). Another unique example concerns the ion exchange of transition metal ions, lanthanides, and actinides into α-ZrP (46,48,50). Furthermore, an important biological application of ion exchange is the removal of ammonium

Table 1 Ion Exchange Examples of Zirconium Phosphate and Its Derivatives

Structure	Ion(s) exchanged	Focus	Ref.
ZrP embedded in glass wool	Li^+, Na^+, K^+, Cs^+	Behavior	38
ZrP	Au^{+3}, Cu^{+2}, Ag^+	Separation	39
ZrP$_{crystalline}$	Sr^{+2}, Ba^{+2}	Behavior	40
ZrP$_{crystalline}$	$Mg(OH)_2$ $Ca(OH)_2$, $Sr(OH)_2$, $Ba(OH)_2$	Hydroxide	41
α-ZrP	Na^+, Mg^{+2}, Cs^+	Behavior	31
$NaZrH(PO_4)_2$ $5H_2O$	Li^+, Na^+, K^+, Cs^+	Behavior	42
ZrP$_{crystalline}$	Mn^{+2}, Co^{+2}, Ni^{+2}, Cu^{+2}, Zn^{+2}	Behavior	43
ZrP$_{gel}$	Cu^{+2}, Fe^{+2}, Fe^{+3}, Al^{+3}	Separation	44
ZrP$_{crystalline}$	Cs^+, Rb^+	Behavior	45
ZrP	Th, Pa, U, Np, Pu	Behavior	46
α-ZrP	NH_4^+	Behavior	47
α-ZrP	Mg^{+2}, Co^{+2}, Ni^{+2}, Cu^{+2}, Zn^{+2}	Behavior	48
ZrP$_{crystalline}$	Rb^+, Na^+, Ag^+, Li^+	Selectivity	49
ZrP$_{crystalline}$	Cr^{+3}, La^{+3}, Tl^{+3}	Behavior	50
ZrP$_{crystalline}$	NH_3, NH_4^+	Behavior	36
ZrP$_{crystalline}$	Mg^{+2}, Co^{+2}, Ni^{+2}, Zn^{+2}	Mechanism	51
$Zr(HPO_4)_2 \cdot H_2O$	Li^+, Na^+, K^+, Cs^+	Microcrystals	52
Zr(IV) acid phosphate	Mg^{+2}, Ca^{+2}, Ba^{+2}, Cu^{+2}, Zn^{+2}, La^{+3}	Microcrystals	53
ZrP	Alkaline metals	Acetate salts	35
ZrP	Alkaline, Cu^{+2}	Behavior	54
α-ZrP	Li^+, Na^+, K^+, Cs^+	Behavior	55
ZrP$_{crystalline}$	$Co(NH_3)_4^{+3}$	Behavior	56
ZrP	UO_2^{+2}	Behavior	57
ZrP$_{amorphous}$	Alkali metals	Thermodynamics	58
ZrP$_{crystalline}$	Alkali metals	Thermodynamics	59
α-ZrP	$Co(NH_3)_4^{+3}$	Behavior	60
α-ZrP	$Pt(NH_3)_4^{+2}$	Behavior	61
Polyacrylamide ZrP	Cs^+, Co^{+2}, Ce^{+3}	Behavior	62
α-ZrP	VO^{+2}	Behavior	63
ZrP	I^-, Br^-, Cl^-, ClO_4^-, NO_3^-, SO_4^{-2}	Behavior	30
ZrP	U	Behavior	64
α-ZrP	Fe^{+2}, Fe^{+3}	Behavior	65
ZrP$_{gel}$	Sr, Y	Behavior	66
α-ZrP	Ag^+, Cu^{+2}	Molten Salt	67

ions from blood during assisted dialysis of kidney patients. α-ZrP materials have been tested for the removal of ammonia and ammonium ions from blood by ion exchange. The exchange was slow with α-ZrP due to steric restrictions, while results from the polyhydrated dihydrogen form of α-ZrP were encouraging due to enhanced access to the galleries (36). New hybrid materials consisting of α-ZrP and organic polymers have recently been reported, and such materials can be tailored to produce novel materials for specific applications of ion exchange (37).

D. Intercalation Properties

Another aspect of binding of guests to α-ZrP involves predominant binding in the galleries by ionic, hydrophobic, as well as other interactions with the matrix. Binding on the outside vs. in the galleries is often distinguished by using a number of spectroscopic and other methods. When the binding free energy of the guests is greater than the interlayer interaction energy, guests spontaneously bind in the galleries of α-ZrP. But the small interlayer distances, as in the case of α-ZrP (7.6 Å), presents a kinetic barrier for the intercalation. Expanding the gallery spacings prior to binding or raising the temperature is necessary for the successful intercalation of large guests, such as cyclodextrins, polymers, surfactants, metal clusters, polyions, and proteins. The facile control over the spacings to accommodate the needs of guest ions of varying sizes is one of the main features of a lamellar materials (Fig. 4) (68). Short-chain surfactant molecules, for example, are introduced into the α-ZrP interlayers as spacers, and guest molecules of sizes comparable to that of the spacer are readily exchanged into the expanded galleries, while the binding of much larger guests is kinetically slow (69). By tuning the interlayer distances, therefore, a considerable amount of size selectivity for the binding can be achieved. Binding of guest ions releases the bound spacers from the galleries, contributing to the increase in the free energy of the system. Binding, therefore, is favored when the affinity of the guest is greater than that of the spacer. Appropriate choice of the spacer is essential for the successful intercalation of the desired guest ions. Control over the gallery spacings of α-ZrP is readily accomplished by the intercalation of amines of various chain lengths (RNH_2) into the galleries for the substitution by the desired guest ions. Intercalation or ion exchange may be important prerequisites for the catalytic steps, and they may be important in the release of products from the catalyst surface into solution. For these reasons, both ion exchange and intercalation properties of layered materials are important to study.

Intercalation of alkyl amines into α-ZrP has been studied in detail (Table 2). At low loadings of amine, the alkyl chains lie parallel to the α-ZrP layers, and the interlayer spacing increases from 7.6 Å for α-ZrP to about 10.4 Å for the intercalate, independent of chain length (70). At higher loadings or with longer-chain amines, the guests aggregate even at low loadings, and the alkyl

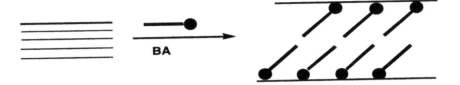

FIGURE 4   Intercalation of short-chain amines to increase the α-ZrP gallery height.

chains begin to incline with respect to the ZrP layers, at increasing angles until optimum packing is achieved. For instance, monoalkylammonium ions, such as n-butyl ammonium (BA), intercalate into the galleries of α-ZrP to form an intercalation solid (BAZrP) featuring a stable bilayer of the alkyl chains, inclined at 55–60° to the metal phosphate sheets (Fig. 4). Bilayer arrangement of the alkyl amines is confirmed by an increased interlayer distance from 7.6 Å in α-ZrP to 18.6 Å in BAZrP (71,72). The intercalation process may proceed by the moving boundary model, where the initially formed islands of amine grow with increased loadings (74).

In BAZrP, the polar ammonium groups of BA are located near the phosphate surface, and the butyl chains are packed parallel to each other while oriented at an angle to the plane of the α-ZrP layers. Electrostatic attraction between the cationic head group of BA and the negatively charged phosphate lattice provides the enthalpic propensity for the intercalation, even though the organization of the BA side chains in the galleries, as a bilayer, is not entropically favored. Favorable van der Waals interactions between adjacent BA chains, dehydration of the head groups, as well as the dehydration of the metal phosphate layer contribute to the binding thermodynamics. Entropy increases due to the loss of water from the hydrophobic surfaces of the guest molecules as well as the galleries, and this increase is an important component in the binding of hydrophobic guest ions in the ionic galleries.

Differential thermal analysis of the amine-α-ZrP intercalates indicate three temperature regions, which are assigned to the dehydration, demination, and dehydroxylation processes (73). The temperatures of dehydration and dehydroxylation steps depend on the extent of amine intercalation, and at low loading the POH groups present in the nonintercalated regions are more stable when compared to the ones in the intercalated regions.

Intercalation of α, ω-alkyl diamines (C2–C10) follows the moving boundary model, and a number of intermediate phases appear before the intercalation is complete (74). In the final phase, the diamines form a bilayer, with the chains fully extended at an angle of 58° to the α-ZrP layer. The transformation from a layered structure to a framework structure allows for the binding of short-chain

Table 2 Intercalation of Various Guests into ZrP

Structure	Intercalates	Focus	Ref.
ZrP$_{mycrocrystals}$	Methylene blue	Behavior	90
ZrP$_{crystalline}$	Pyridine, n-butylamine	Behavior	91
Zirconium hydrogen phosphate monohydrate	Alkanols, glycols	Behavior	77
Zirconium hydrogen phosphate	Cobaltocene	Behavior	87
$\alpha+\gamma-$ZrP	Pyridine	Thermal decomposition	80
α-ZrP	L-Histidine, L-lysine	Behavior	92
α-ZrP	L-Asparagine, L-alanine	Behavior	93
α-ZrP	2,2′-bipyridyl and Co(II), Ni(II), and Cu(II), 2,2′-bipyridyl complexes	Behavior	94
α-ZrP	Monoalkylamines	Behavior	95
α-ZrP	n-Alkylamines	Behavior	70
α-ZrP	Imidazole, benzimidazole, histamine, and histadine	Formation	84
ZrP	Copper(II) dithiooxamide	Formation	87
α-ZrP	2,9-Dimethyl-1, 10-phenanthroline	Behavior	96
α-ZrP	α,ω-Alkyldiamines	Formation	74
ZrP	Diamines	Behavior	75
Zr(IV) phosphate-phosphate	n-Alkylamines	Behavior	97
α-ZrP	2,2′-bipyridyl; 1,10-phenanthroline; 2,9-dimethyl-1, 10-phenanthroline	Behavior	82
α-ZrP	Palladium(II), 2,2′-bipyridyl	Behavior	98
α-ZrP	Pyridine and quinoline	Formation	81
α-ZrP	n-Alkylamine	Thermal decomposition	73
α-ZrP	Ethidium and acridinium ions	Behavior	99
α-ZrP	n-Alkylquinolinium, isoquinoliniom, and acridinium	Fluorescence	100
α-ZrP	Crystal violet	Preparation	83
α-ZrP	Ga/Cr mixed oxides	Behavior	88
ZrP	EtOH	Characterization	101
α-ZrP	TBA $^+$OH$^-$	Behavior	102
α-ZrP	Aminomethyl crowns	Behavior	86
α-ZrP	N,N'-bis(3-aminopropyl)-1,3-propanediamine	Conformational change	76
α-ZrP	EtOH	Guest/Host relations	78
α-ZrP	Fluorescein, rhodamine, cetyltrimethylammonium bromide	Donor/acceptor	103
ZrP	Crown ethers and iminodiacetates	Review	104
α-ZrP	1,ω-Alkanediols, 1-alkanols	Behavior	79

alkanols and water. Intercalation of long-chain diamines (C0–C12) into mixtures of phosphate–phosphite solids yields different results, and the phosphite regions affect the intercalation rates (75). At low loading of amine, a first phase appears where the alkyl chains are parallel to the solid; and at higher amine loadings, a second phase with the bilayer structure of the amine appears. In the case of polyamines such as N,N'-bis(3-aminopropyl)-1,3-propane-diamine, the intercalated amine assumes an extended conformation at room temperature, as indicated from ^{13}C-solid-state NMR studies, while at higher temperatures the polyamine exists in a bent form (76).

As pointed out earlier, BAZrP was used as a precursor to prepare intercalation products of several aromatic hydrophobic cations by substitution. This substitution process is facile when the size of the guest cation is comparable to the gallery spacing of BAZrP (Fig. 4). In addition to favorable electrostatic interactions with the matrix, the intercalation of hydrophobic cations is promoted by favorable entropic contributions due to the dehydration of the hydrophobic ions when they are transferred from the aqueous phase to the hydrophobic bilayers in the galleries.

Thus, several hydrophobic guest cations, such as derivatives of anthracene and pyrene, ethidium bromide, acridinium hydrochloride, rhodamine, fluorescein, chlorophylls, porphyrins, and ruthenium complexes, are intercalated into the galleries of BAZrP (Fig. 5). Depending on their size, these guests form either monolayers or bilayers in the galleries. The diffusion, orientation, and optical properties of the guests change drastically, in specific cases, upon intercalation. More detailed discussion of intercalation of hydrophobic guests into the BAZrP matrix is presented later. A strong correlation of the gallery spacings with guest size is demonstrated from X-ray diffraction data (see Table 3).

Even though electrostatic interactions are important for the intercalation of guests into galleries and binding of cations is expected to be spontaneous, the binding of neutral molecules is also observed. Intercalation of alcohols and glycols into α-ZrP is known, and X-ray powder diffraction data indicate that the alkanols are intercalated as a bilayer while the glycols form a single layer of guest molecules (77). In 1999, the intercalation of ethanol into the α-ZrP was examined using vibrational spectroscopy and molecular simulations (78). The IR and Raman spectral data indicate the preservation of the host and guest structures subsequent to intercalation. This fact reinforces the notion that the host layers are rigid. The molecular modeling allowed for a more detailed interpretation of the spectral data. Due to the weak binding of these alcohols to the matrix, they are readily exchanged for polar organic molecules, such as DMF, acetonitrile, urea, aniline, and alkylamines.

Intercalation of long-chain alkanols and α,ω-alkanediols into α-ZrP is accelerated by microwave radiation (79). The stoichiometry of the intercalate de-

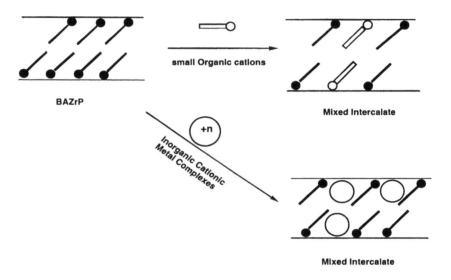

FIGURE 5 Facile substitution of *n*-butylammonium (BA) ions bound in α-ZrP galleries by hydrophobic cations or by metal complexes.

pends on the number of OH groups present in the guest, and the alkanols form a bilayer with an all-trans conformation of the alkyl chain. On heating, both the mono and the diol intercalate, the $O-C_1-C_2-C_3$ torsion angle changes from 180° to 136°, and the temperature at which this phase transition occurs depends on the chain length. Thus, as evidenced by the foregoing examples of the intercalation of alcohols, the guest molecules need not be positively charged, and intercalation is promoted by other interactions, such as hydrogen bonding and ion–dipole interactions.

Table 3 Variability of Gallery Spacing with Respect to Guest Size as Observed from X-Ray Diffraction Data

Solid	Gallery spacing (Å)
α-ZrP	7.6 (Ref. 10)
BAZrP	18.6 (Ref. 10c, 70)
NMAC/BAZrP	17.4 (Ref. 16)
AMAC/BAZrP	17.6 (Ref. 71)
$Ru(bpy)_3^{2+}$/α-ZrP	24.1

The role of water in the intercalation of guest molecules is demonstrated by intercalation studies with pyridine. The intercalation of pyridine into α-ZrP increases the water content, and the gallery spacing increases to 10.9 Å (80). Dehydration of the pyridine-ZrP intercalate occurs at 140°C, with the concomitant formation of a pyridine-free phase and a pyridine-enriched phase. In a related study, the uptake of pyridine from ethanol, chloroform, carbon tetrachloride, ethanol–water, and acetone–water solutions was observed (81). The prominent role of water in accelerating the intercalation of pyridine into α-ZrP was reported. Taken together, these studies indicate the important role of water in the intercalation of pyridine and therefore possibly other organic molecules. The role of water in the interaction between the matrix and the guest is also observed with other heterocyclic systems related to pyridine. Intercalation of 2,2'-bipyridine, 1,10-phenanothroline, and 2,9-dimethyl-1,10-phenanthroline indicate the protonation of only one of the two nitrogens, and dehydration of the materials strongly reduced the acid–base interaction between the guest and the matrix (82). The role of solvent in controlling the orientation of an intercalated guest is illustrated with crystal violet. Intercalation of crystal violet (Fig. 6) into the ethanol form of α-ZrP resulted in the expansion of interlayer spacings to 21.4 Å, and this spacing is consistent with the arrangement of the dye as a monolayer with a perpendicular orientation of its π-electron system in the galleries. The dye–dye interactions in the perpendicular orientation result in a shifting of the electronic absorption band of crystal violet to 510 nm. Drying of the sample reduced the gallery spacing to 18 Å, with a simultaneous loss of the 510-nm band, which indicates the role of water in controlling the dye–dye interactions in the galleries (83).

Intercalation of other heterocycles, such as imidazole, histamine, and histidine, was reported, and in some cases elevated temperatures were required for intercalation (84). Intercalation of other basic amino acids, such as lysine, and arginine into α-ZrP was also reported (85). Histidine and arginine indicated a single-stage binding isotherm for the intercalation, with two phases at low loadings, while lysine indicated a two stage binding curve with a gallery spacing of

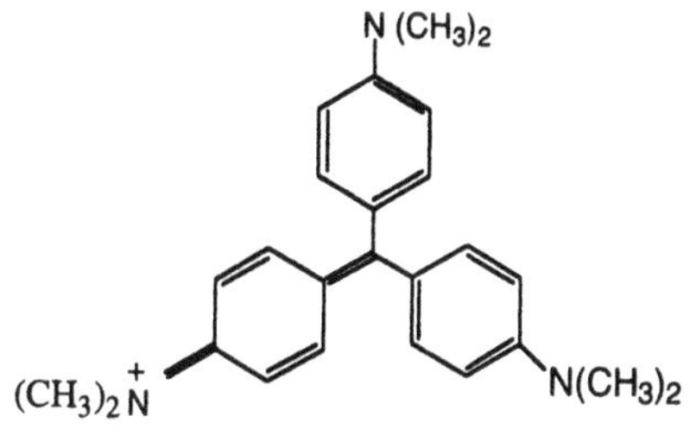

FIGURE 6   Chemical structure of crystal violet cation.

23.1 Å. The contrasting behavior between lysine and other amino acids is also noted with the gamma phase of α-ZrP (85). Another related example of biological interest is the intercalation of aminomethyl derivatives of crown ethers. Intercalation of 2-aminomethyl-12-crown-4, 15-crown-5, and 18-crown-6 was readily achieved (86). The former two crowns form intercalation compounds in two and three stages, while the latter is more complex. Protonation of the amino group of the crown by the phosphate lattice was inferred from ^{13}C solid-state NMR and IR studies.

In contrast to the direct intercalation of organic guest molecules in α-ZrP galleries, the intercalation of the metal complexes can be achieved in a number of different ways. Four of the popular routes are: (a) intercalation of the metal ion followed by intercalation of the ligand to assemble the metal complex in the galleries; (b) intercalation of the ligand followed by the metal ion; (c) direct intercalation of the metal complex; and (d) intercalation of the metal complex subsequent to preintercalation with an amine or an alkanol to open the galleries. The latter two approaches normally work better. Cobaltocene, tris(2,2'-bypyridine)Ru(II), dithiooxamideCu(II), Cr(III)acetate, and many other metal complexes have been intercalated directly into the α-ZrP galleries (87). As in the case of organic chromophores, the physical-chemical properties of the metal complexes are influenced by the rigid matrix surrounding the metal phosphate as a solid solvent. Specific examples of changes in the properties of the intercalated metal complexes and their photochemical behavior are examined in Section III.

Other examples of inorganic species intercalated in the galleries include oxides, sulfides, and gold nanoparticles. Intercalation of Ga-Cr mixed oxides into α-ZrP resulted in moderately active acid catalysts (88). Exposure of α-ZrP to gold nanoparticles stabilized with 2-(dimethylamino)ethanethiol resulted in rapid flocculation of the solid, and the incorporation of the gold particles in the galleries was investigated via powder diffraction and IR methods (89). Such facile access to nanocomposite materials is promising for the development of special catalysts for specific chemical transformations.

III. PROBING THE GALLERIES OF α-ZRP IN PHOTOPHYSICAL STUDIES

α-ZrP galleries can serve as two-dimentional microreactors, and their chemical/physical influence on reactants in terms of polarity, two-dimensional diffusion, ordering of guest molecules, guest orientation, and modified reactivity has been probed using spectroscopic methods. Diffusion in two dimensions, for instance, is quite distinct from that in homogenous solution phase. The phosphate/phosphonate galleries act as a solid solvent positioned above and below the guest molecules, and the polarizability, dipole strength, and the ability to interact with the guest molecules in a unique way make these materials especially interesting.

Application of these materials in catalysis, ionexchange, and advanced hybrid materials requires a clear understanding of the distribution, orientation, electronic properties, and host–guest interactions of molecules in the galleries. Many of these properties are investigated using fluorescent probes because the excited states of these molecules are sensitive to their microenvironment. The packing or ordering of guest molecules in α-ZrP galleries depends on their size, shape, charge density, and hydrophobicity. For example, guest molecules with polar or charged head groups at either end of an alkyl chain prefer a monolayer arrangement, while guests with a single polar head group prefer a bilayer arrangement. These arrangements can be readily distinguished in X-ray diffraction experiments.

Guest–guest interactions, in addition to the guest–host interactions, also play major role in the distribution of molecules in the galleries. While some molecules are distributed uniformly in the galleries, others prefer to aggregate and form islands. Guest–guest interactions are monitored by the observation of excimer formation, where excimers are complexes between the ground and excited states, as a function of guest loading. If the guest molecules aggregate in the galleries, for example, excimer formation is observed even at low loadings, while monomer emission will be observed if the guest molecules prefer to be distributed individually with little or no interaction between nearest neighbors. The binding affinities and stochiometries are conveniently monitored either in spectroscopic experiments or via centrifugation studies. Since phosphate/phosphonate particles are large (100–1000 nm across), the particles can be readily centrifuged to separate the bound species from the free ions/molecules. Specific examples of binding of the guest ions in the galleries are investigated in the spectroscopic experiments illustrated shortly.

A. Probing the Galleries with Organic, Hydrophobic Cations

Small organic cations, such as ethidium, acrydinium, and anthryl/pyrenyl methyl ammonium ions, have served as reporter molecules to study physical/chemical behavior after intercalation into the galleries. Properties of much larger guests, such as metal complexes, differ significantly from these organic guests in several ways, while the intercalation of biological macromolecules, such as proteins and enzymes, provides fascinating details about the physical/chemical nature of the galleries. The protein–inorganic interactions gleaned from these studies are beginning to open new opportunities to understand biomineralization, biocatalysis, and the design of biocompatible materials for artificial organs or implants.

As mentioned previously, intercalation of butyl ammonium ions into α-ZrP results in the opening of the galleries (gallery spacing 18.6 Å) and the resulting solid, BAZrP, is an excellent precursor for the intercalation of hydrophobic cations of appropriate size (Fig. 5). The kinetic barrier for intercalation of large hydrophobic ions is, therefore, lowered, and the binding is controlled by the interactions of a guest with the butyl chains of BA as well as the negatively charged inorganic matrix. The orientation of the bound guest is expected to be aligned with the bilayer packing of the BA, at least at low loadings. At high loadings, the guest–guest interactions become dominant, and these interactions may change the gallery spacings to a significant extent.

Upon protonation, aromatic amines such as 1-naphthlene methylamine HCl (NMAC), 9-anthracene methylamine HCl (AMAC), 4-(1-pyrene) butylamine HCl (PBAC), ethidium bromide (EB), and acridinium HCl (ACR) are expected to have favorable electrostatic interactions with the BAZrP matrix. In addition, these hydrophobic ions are expected to interact favorably with the butyl chains of BA and to sterically fit well within the bilayer structure in the galleries. This is because the length along the short axis of these hydrophobic ions is comparable to the length of the butyl amine spacer. Most of the binding is expected to be in the galleries due to the foregoing favorable criteria, although binding at the edges is also possible to a lesser extent. Binding of these chromophores in the galleries (intercalation) increases their local concentration when compared to that in the bulk phase. This is a simple consequence of the reduction in the total volume accessible to the probes. Increased local concentrations promote bimolecular events such as self-quenching, dimer formation leading to exciton splitting, excimer emission, rapid energy migration among the chromophores bound within the galleries, and accelerated electron transfer reactions. Another important consequence of guests' binding in galleries is that their orientation is no longer isotropic, thereby affecting bimolecular chemical reactions in a predictable manner. The dielectric field of the negatively charged matrix and the head groups of BA provide a charge field unlike solution state conditions, and these electric fields can have a strong influence on electronic transitions, chemical reactivity, and guest orientation. Specific examples of these phenomena are discussed in detail later. First, experimental methods used for the quantitation of binding, distinguishing binding in the galleries vs. binding at the edges, and estimating the average area occupied per guest are presented.

1. Binding Studies

Binding constants, binding stoichiometries, and coopertativity/anticooperativity for binding are determined to quantitate the binding interaction. After guest molecules are equilibrated with BAZrP suspensions, the bound and free guests are separated by centrifugation and by quantitating the concentration of the free guest in the supernatant. Due to the large size of the BAZrP particles (100–1000 nm), they settle easily and carry bound guest molecules with them. Using centrifugation methods, the binding isotherms of AMAC, PBAC, EB, and ACR are obtained. The binding data thus obtained at various guest concentrations, but at a fixed BAZrP concentration, are analyzed to obtain the binding parameters.

The binding isotherm of AMAC indicates, for example, that binding is saturated when the average area occupied per AMAC is 53 $\mathring{A}^2$; this is estimated by considering the available area as 100 m^2/g BAZrP (8). The area occupied per AMAC molecule, in turn, is consistent with a bilayer distribution of AMAC ions in the galleries, with their short axes inclined at 50–60° to the α-ZrP planes. This simple experiment permitted the estimation of the area of cross section of the guest, the length of its long axis, its orientation in the galleries, and its packing behavior as a bilayer (71). These estimates are also consistent with powder XRD data, which indicate a gallery spacing of 17.6 $\mathring{A}$.

Similar binding isotherms for EB, tris(2,2'-bipyridyl)Ru(II) (abbreviated as $Ru(bpy)_3^{2+}$) provide binding constants, binding stochiometry, and the average area occupied per guest ion. These values are consistent with a bilayer packing of EB, while monolayer formation is indicated for $Ru(bpy)_3^{2+}$ These conclusions are consistent with observed gallery spacings (Table 3).

2. Enhanced Excimer Formation

Close packing of chromophores in galleries can drastically change their photophysical properties. The proximity of adjacent guest molecules, within the van der Waals contact distance, is a prerequisite to induce excitonic interactions in the ground state and excimer formation in the excited state (Fig. 7). Intercalation of AMAC (71) and PBAC (72) in the galleries resulted in enhanced ground-state dimer and excimer formation. The ground-state interactions are readily seen in electronic absorption spectra, while excimer formation is evident from fluorescence spectra (71). In the case of AMAC, excimer formation depended on the loading: At low loadings no excimer was observed, whereas in the case of PBAC, excimer emission was observed even at low loadings. Guest–guest interactions are weaker for AMAC when compared to PBAC, and this parallels their relative hydrophobicities. The more hydrophobic PBAC begins to aggregate in the galleries even at low loadings and induces excimer formation. When AMAC/PBAC bind to DNA or to proteins, such guest–guest interactions are not observed. Therefore, this is due to the compartmentalization of the individual guests in these other media (100).

Excimer formation and ground-state complexation requires specific orientations of the participating chromophores and appropriate electronic properties of the participating states. Attainment of parallel geometry between the aromatic ring systems with van der Waals contact in the ground state, for example, is required. If the guest molecules do not form ground-state complexes but exhibit excimer formation, then the excited dimer is produced within the excited-state lifetime, under diffusion-controlled conditions. Under these conditions, the rise time of the excimer emission can be measured in time-resolved fluorescence experiments, and the data are analyzed to estimate the diffusional motion of the guest molecules within the galleries.

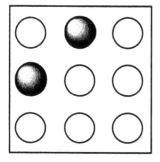 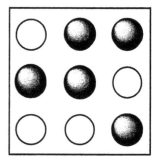

FIGURE 7 Low substitution (left) and the extensive substitution (right) of *n*-butylammonium (BA) ions by guest molecules in the galleries of α-ZrP.

3. Distinguishing Between Binding in the Galleries and that on the Outside

Spectroscopic methods allow one to differentiate between the binding of guest molecules within galleries and that on the outside of the stacks of the α-ZrP plates. Binding of guest molecules in galleries protects them from fluorescence quenchers that reside in the aqueous phase, and enhanced quenching will be observed when the quenchers also bind in the galleries. The binding of the guests in the galleries or on the outside can be investigated by appropriate choice of the quencher molecules. Quenching of fluorescence by iodide, for example, is extensively inhibited after intercalation of AMAC or PBAC in BAZrP. The negatively charged iodide is not expected to enter the galleries, and the inhibition of quenching AMAC fluorescence by iodide is due to intercalation of the guest. Using NMAC intercalated in the galleries, enhanced quenching of NMAC fluorescence by AMAC or PBAC was demonstrated, and this is because the guests as well as the quenchers are bound in the galleries (105). Thus, chromophores that are bound in the galleries are not available for the water-bound iodide anions, but they are readily accessible to quenchers bound within the galleries. Fluorescence quenching studies provide a simple method to distinguish between the binding of guests in the galleries and that on the outside.

4. Excitonic Interactions

As mentioned earlier, the positioning of the guests within contact distance of each other is a prerequisite for strong excitonic interactions. Such intermolecular interactions are also sensitive to the relative orientations of the interacting chromophores, their distance of separation, the symmetries/energies of the interacting electronic states, and the nature of the contact region between them. Face-to-face vs. side-to-side association can be distinguished from the position and/or the intensity of excitonic absorption bands (55).

The foregoing examples illustrate the use of BAZrP as an intermediate in the intercalation of guests. The polarity, gallery dimensions, and other aspects are to be kept in mind in the design of direct intercalation experiments, but intercalation into pristine α-ZrP is possible. Ordered packing of the dye molecules in the galleries of α-ZrP is indicated by excitonic splitting of the absorption spectra of intercalated dyes (106,107). Dye molecules such as thionin bind in the galleries of α-ZrP to form monolayers (18b). Intercalation of large dyes, such as methylene blue and rhodamine derivatives, face a kinetic barrier for direct intercalation, as pointed out earlier, and intercalation of such dyes is accelerated in the presence of alkyl amines (108). The flexible chains of the alkylamines are easier to intercalate than the more rigid guests, such as methylene blue, and the increased gallery spacing due to the intercalation of the amine facilitate dye binding.

Preintercalation of α-ZrP with propylamine, for example, accelerates the binding of rhodamine, (109) and its binding is accompanied by a large blue shift in the emission spectrum of the dye, from 690 nm in the solid state to 675 nm for the intercalated dye. The blue shift in the emission peak is attributed to the aggregation of the dye in the galleries due to the high local concentrations, but a similar scenario is to be expected in the solid state. Clearly, reduced repulsion due to the binding to the matrix and the dielectric properties of the matrix are also to be considered to explain the dye behavior.

Direct intercalation of crystal violet into pristine α-ZrP indicates strong dye–dye interactions, similar to that of rhodamine. At low loadings of the acetate salt, the gallery spacings increase to 11.5 Å, while at high loadings the gallery spacing is nearly 22 Å, indicating the drastic change in the orientation of the crystal violet with loading. At low loadings, the dye is present more parallel to the galleries, while at high loadings it is more likely to be perpendicular to the galleries (110). The strong dispersive interactions between the π-electron systems of adjacent dye molecules results in the formation of crystal violet dimers, and this may be facilitated due to the screening of the electrostatic repulsion between dye molecules by the matrix (111).

Binding of guest ions in the galleries is often accompanied by the transfer of corresponding counterions into the galleries. Close proximity and strong interactions between the guest ions and their counterions can result in interesting processes, such as charge transfer, light-induced electron transfer, and covalent bond formation. Bromide salts of 3-cyanopyridine, 4-cyanopyridine, and isoquinoline, for example, show a new band attributed to the charge transfer between the bromide and the pyridinium chromophores (112).

5. Intercalation of Porphyrins

Porphyrins are biologically important molecules related to chlorophylls, hemes, and phthalocyanins. They can coordinate a number of metal ions, which provides a handle for modifying their catalytic properties. Porphyrins have an attractive chromophore for photochemical studies, and this property is exploited in the photodynamic therapy of cancer (113). Intercalation of porphyrins, therefore, into a number of layered materials has been reported (114). Due to the strong hydrophobicity of most porphyrins, direct intercalation into α-ZrP is not successful. But intercalation of positively charged porphyrins, such as tetraaminopheny-lporphyrin (TAPP), and tetramethyl pyridiniumporphyrin (TMPyP) into α-ZrP has been successful after preintercalation of p-methoxyaniline (PMA) (114). This is yet another example of how the gallery spacings of α-ZrP can be expanded to accommodate guests of various sizes under specific conditions.

Intercalation of TAPP into α-ZrP resulted in expansion of the gallery spacings to 17 Å from 7.6 Å, and this spacing is less than the edge-to-edge distance of the porphyrin ring system. The orientation of the intercalated porphyrin ring system is suggested be at a 45° angle to the phosphate matrix. Alternatively, the porphyrins could form a bilayer with the porphyrin rings parallel to the α-ZrP matrix. When the intercalation reaction was allowed to proceed for several weeks, the gallery spacing increased to 24 Å, which corresponds to the canted orientation of the porphyrins ring systems with respect to the metal phosphate layers (45–60°) (113).

Porphyrin-α-ZrP intercalates are interesting materials for catalytic applications due to the vast array of transition metal ions that can be bound to these ring systems, and the metal ions can assume various oxidation states appropriate for catalytic applications. Intercalation of other biologically significant cofactors, prosthetic groups, and coenzymes into the α-ZrP matrix are yet to be explored in detail, but this activity is certain to expand the catalytic potential of α-ZrP and its derivatives.

B. Probing the Galleries with Inorganic Metal Complexes

Among the many metal complexes that show interesting photophysical properties upon intercalation in the galleries, tris(2,2′-bipyridine)Ru(II) (abbreviated as $Ru(bpy)_3^{2+}$) deserves special attention. This is because of the extensive literature available on this metal complex, as a result of the intense efforts targeted in testing this remarkable metal complex for solar energy harvesting and conversion. Photophysical properties of this metal complex are sensitive to its microenvironment (115,116). and the photophysics of Ru(bpy)-ZrP complexes is significantly different from that of $Ru(bpy)_3^{2+}$ when bound to other organized media, such as layered oxides, (117) zeolites, (118) and smectite-type clay minerals (119).

The metal complex has been intercalated in the galleries by different methods, and the properties of the resulting material depend on the method (120). When the metal complex is present in the reaction mixture during the synthesis of α-ZrP by the HF method, a crystalline solid with a layer spacing of 15.6 Å results. This increased spacing is nearly the same as expected from the known dimensions of the metal complex, and it indicates the formation of a monolayer of the metal complex, with some distortion of the geometry. This distortion was attributed to the restricted environment of the galleries. Consistent with this interpretation, the emission maximum of the intercalated $Ru(bpy)_3^{2+}$ was at 615 nm, while adsorption of the metal complex onto the surface of α-ZrP stacks resulted in an emission maximum of 640–645 nm.

Intercalation of $Ru(bpy)_3^{2+}$ into the galleries of α-ZrP by preintercalation with butylamine resulted in contrasting photophysical results. Binding of the metal complex saturated when the average area occupied per metal complex tallied with the known area of cross section of the metal complex. The gallery spacings expanded to 19.5 Å, (121) and the spacings match well with the known diameter of the metal complex. The electronic absorption spectrum of the intercalated metal complex, furthermore, underwent substantial changes when compared to that observed with aqueous solutions (115). The many changes in its properties include: a red shift of the metal-to-ligand charge transfer (MLCT) absorption band from 452 nm to 481 nm, a blue shift of the emission maximum from 610 nm to 580 nm, and a fivefold increase in the luminescence intensity when compared to the corresponding behavior in aqueous solutions. Photophysical properties such as absorption and emission maxima are independent of metal complex loading, suggesting the intercalation follows the moving boundary model (116). Initial binding of the metal complex results in the formation of pools of the metal complex; as loading increases, these pools expand to accommodate the additional material (see Fig. 8).

The red shift of the absorption spectrum and the concomitant blue shift of the emission spectrum imply a reduced Stoke's shift when the metal complex binds in the galleries. The increase in the luminescence intensity, in addition, suggests the strong role of the solid solvent surrounding the metal complex in controlling the excited state dynamics and its deactivation to the ground state. In solution, the solvent surrounding the excited state rearranges to stabilize the newly created dipole of the MLCT state, but such stabilization is not facile in the galleries due to the solid solvent surrounding the excited state. The smaller Stoke's shift and enhanced emission intensities are likely due to the lack of stabilization of the initially produced MLCT state. This interpretation is supported by the increased emission lifetime from 350 ns to ~1500 ns under air-saturated conditions (115).

The luminescence intensities and lifetimes, however, decrease with the loading of the metal complex, and this indicates self-quenching ($k_q = 7.2 \times 10^9$

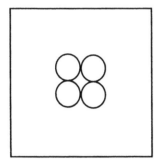 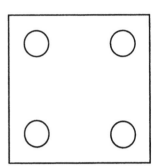

FIGURE 8 Two extreme distributions of guests in the galleries. The formation of islands of guest molecules follow the moving boundary model, where the guest molecules cluster, (left). When the interguest interactions are weak, they may be distributed more uniformly (right).

$M^{-1}s^{-1}$) of emission within the metal complex pools. The quenching rate constant is much larger than the anticipated diffusion-limited rate for quenching in the galleries (10^{-10} cm^2/s) (122). Binding of the metal complex in the galleries was confirmed from luminescence-quenching studies with iodide, and the metal complex was not accessible to the negatively charged quencher. Binding on the outside or at the edges would have exposed the metal complex to anionic quenchers (123,120). Interestingly, addition of potassium chloride produced apparent quenching of emission, but detailed studies indicated the release of the metal complex from the galleries by the electrolyte. Luminescence-quenching experiments with electrolytes, therefore, are to be evaluated carefully before drawing conclusions regarding the location of the bound guest ions.

In related studies, preintercalation of hexylammonium ions into the galleries of α-ZrP did not facilitate the intercalation of Ru(bpy)$_3^{2+}$. XRD/spectral data indicate binding of the metal complex on the outside as well as within the galleries (119b). The absorption spectrum of Ru(bpy)$_3^{2+}$ in either case showed minor differences from that recorded in aqueous solution.

C. Accelerated Energy Transfer in the Galleries

The arrangement of chromophores in the galleries resembles the arrangement of pigment molecules in the natural photosynthetic antenna system, which are suitable for energy/electron transfer reactions. The ordered arrangement of intercalated organic chromophores in the galleries of α-ZrP, as discussed earlier, has been exploited to accelerate energy/electron transfer reactions. Hydrophobic, cationic chromophores that have a high affinity for binding in the galleries of α-ZrP were chosen for the energy transfer studies. For the energy transfer to be efficient, the donor and acceptor chromophores must be placed in proximity, but far enough apart that undesirable excimer or exciplex formation between them is suppressed. Aggregation of the donors or the acceptors, therefore, is undesirable, and they must be evenly distributed (Fig. 8). These criteria are readily met by the NMAC-donor and AMAC-acceptor system; these chromophores also satisfy the electronic and spectroscopic requirements for singlet–singlet energy transfer (105).

In solution-phase conditions, the energy transfer between the singlet excited state of NMAC and the ground state of AMAC is slow and inefficient. Increasing the concentrations in the solution phase to improve the energy transfer is limited by the solubilities of the chromophores. However, intercalation of NMAC (donor) and AMAC or PBAC (acceptor) into α-ZrP results in ordered donor–acceptor assemblies (Fig. 9) that exhibit rapid and efficient energy transfer (71,72). Excitation of NMAC in the presence of micromolar concentrations of AMAC or PBAC results in intense fluorescence from the acceptors, indicating the energy transfer. Quantitative studies of these systems provides clear evidence for donor-to-donor energy migration followed by energy transfer to the acceptor. Direct light absorption by the acceptor was minimized by the appropriate choice of excitation wavelength, and this is one of the first examples of efficient energy transfer among noncovalent donor–acceptor assemblies.

Enhanced local concentrations, reduced distance of separation between the chromophores, and ordered arrangement of the donors and acceptors are factors that are in favor of efficient energy transfer between chromophores bound in the galleries. In the natural photosynthetic system, the excitation migrates from donor to donor until it is trapped by an acceptor. In case of donors bound in the galleries, such donor-to-donor energy migration cannot be ruled out, and the experimental data strongly support such energy migration in α-ZrP galleries loaded with appropriate chromophores. Similar control over the arrangement, distribution, and choice of the chromophores is also achieved in solid crystals, Langmuir–Blodgett films, and thin films (8,124). One advantage of layered materials is that they are mechanically robust and chemically stable and can be opened to accommodate guests of any size, under a variety of conditions.

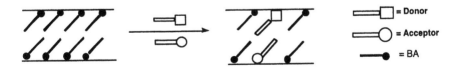

FIGURE 9 Substitution of BA by donor and acceptor ions for energy transfer or electron transfer studies.

Intercalation of donors and acceptors that are much larger than AMAC or PBAC was achieved via the preintercalation of long-chain surfactants such as cetyltrimethylammonium bromide (CTAB) and dodecyldimethylammonium bromide (DDAB), followed by substitution of the surfactants with chromophores (103). CTAB and DDAB intercalate readily into the galleries, with increased layer spacings of 33 Å and 56 Å, respectively. These gallery spacings correspond to a bilayer packing of surfactants if the alkyl chains are oriented at 60° to the metal phosphate. Intercalation of these surfactants facilitated the binding of xanthene dyes into the galleries (see Fig. 10) (103). This route is much more efficient than direct intercalation of these dyes into α-ZrP.

Energy transfer studies with the xanthene dyes after intercalation into the surfactant-laden α-ZrP indicated efficient energy transfer from fluorescein to rhodamine (103). Energy transfer with these systems, however, is not as efficient as the NMAC/AMC systems discussed earlier. Enhanced energy transfer in the fluorescein/rhodamine system is attributed mainly to increased local concentrations of the acceptor in the galleries. Energy transfer between these donor–acceptor systems has been used to measure the intermolecular distances as well as interlayer vs. intralayer energy transfer (125). These studies provide clear evidence for the organization of the donors and acceptors in the galleries, and these chromophores participate in donor-to-donor and donor-to-acceptor energy transfer. Rate accelerations are due primarily to enhanced ordering, increased local concentrations, and even donor-to-donor hopping of the excitation energy, which is unusual non-natural systems at room temperature. Similar systems with improved performance and suitable optical properties are of intense interest for developing artificial light-harvesting systems (103,126).

D. Accelerated Electron Transfer in the Galleries

The efficient, accelerated energy transfer between donor and acceptor systems organized in the galleries of α-ZrP, described earlier, has provided inspiration for the construction of similar assemblies to achieve rapid, light-induced electron transfer reactions. Among the many donor–acceptor systems, $Ru(bpy)_3^{2+}$ has been chosen to function as the electron donor in the excited state due to its

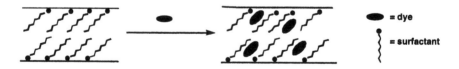

FIGURE 10 Binding of xanthene dyes in surfactant-laden, hydrophobically modified α-ZrP galleries.

attractive photophysical properties, with $Co(bpy)_3^{3+}$ as the electron acceptor (Fig. 11) (127). The success of these systems depends on the improved photophysical properties of $Ru(bpy)_3^{2+}$ upon intercalation in the galleries, increased local concentrations of the donors in the galleries, and enhanced excited state lifetime of $Ru(bpy)_3^{2+}$ intercalated into the galleries.

The donors and acceptor ions, $Ru(bpy)_3^{2+}$ and $Co(bpy)_3^{3+}$, are similar in size, hydrophobicity, shape, and charge density; hence, their intercalation characteristics are expected to be similar. Cointercalation of $Co(bpy)_3^{3+}$ with $Ru(bpy)_3^{2+}$, for example, is expected to distribute the donor and acceptor ions within the galleries in a similar fashion. After cointercalation, the excited-state lifetimes of $Ru(bpy)_3^{2+}$ were monitored as a function of acceptor concentration (127). The rate constants estimated for light-induced electron transfer are in the range of $1-30 \times 10^{11}$ m^2 mol^{-1} s^{-1}, a rate that far exceeds the expected diffusional rate in the galleries. The rate constants are also independent of solvent viscosity, temperature, and loading of the donor, all characteristic of long-range electron transfer (Table 4). Donor-to-donor energy migration followed by rapid electron transfer between the nearest donor–acceptor pairs is a distinct possibility in these systems (103,120,126) (Fig. 11); and this mechanism is reminiscent of the photoinduced electron transfer reactions of the photosynthetic apparatus.

The foregoing metal complexes have served as donor–acceptor pairs in other, related systems, and these also demonstrate some aspects of the preceding rapid electron transfer (128). In a related study, the electron transfer between PBAC as the donor and tris(9,10-phenanthroline)Co(III)$^{3+}$ (abbreviated Co(-phen)$_3^{3+}$) as the electron acceptor was investigated. Even when just micromolar concentrations of the acceptor were intercalated in the galleries of BAZrP, the donor fluorescence was quenched very efficiently. The estimated quenching rate constant ($\sim 3 \times 10^{12}$ dm^3 $mol^{-1}s^{-1}$) far exceeded the expected diffusional rate (69,120).

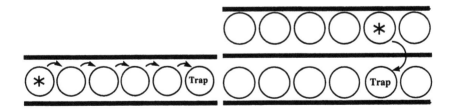

FIGURE 11 Excitation may migrate between donor chromophores within galleries (left) or between galleries (right). The excitation may be trapped by the acceptor via electron or energy transfer reactions with the closest donor.

Table 4 Bimolecular Electron Transfer Rate Constants Measured When the Metal Complexes Were Intercalated in BAZrP (0.008% by Weight)

$[Ru(bpy)_3^{2+}]$	Medium	Temperature (k)	$k_q(1)$ $(m^2mol^{-1}s^{-1})$	$k_q(2)$ $(m^2mol^{-1}s^{-1})$
0.5 μM	Aqueous	298	1.6×10^{12}	4.0×10^{11}
5 μM	Aqueous	298	2.1×10^{12}	6.4×10^{11}
10 μM	Aqueous	298	2.2×10^{12}	7.6×10^{11}
5 μM	EG-W (2 : 1)*	298	3.0×10^{12}	5.9×10^{11}
5 μM	EG-W (2 : 1)*	77	1.8×10^{12}	1.4×10^{11}

The rate constants $k_q(1)$, $k_q(2)$, corresponding to the quenching of the short- and long-lived components, were calculated as indicated in the text.
* EG-W represents a mixture (v/v) of ethylene glycol and water.
Source: Ref. 127.

IV. CATALYTIC PROPERTIES OF α-ZRP AND ITS INTERCALATION COMPOUNDS

The preceding studies of ion exchange and intercalation properties are pertinent to catalytic applications of α-ZrP, since these processes are integral to the mechanistic details of catalysis. α-ZrP has been used to catalyze a number of chemical reactions, including oxidation, dehydrogenation, dehydration, dechlorination, desulfurization, deamination, hydrogenation, hydration, hydroxylation, hydroformylation, alkylation, carbonylation, aldol condensation, H-atom transfer, C—C bond formation, olefin isomerization, and Friedel–Crafts reactions (see Table 5). Often, these reactions are catalyzed by the intercalation of metal oxides, metal ions, and other substances; therefore, intercalation is an important event in most of these catalytic processes. Pillaring of the galleries with large inorganic particles, cyclodextrins, proteins, or catalytic nanoparticles improves the ingress and egress of reactants and products, respectively (Fig. 12). A few specific examples of these mentioned catalytic reactions are provided next.

A. Oxidation

Incorporation of Cu(II) into the galleries of α-ZrP resulted in a catalyst that is active for the oxidation of carbon monoxide in the presence of oxygen, and the catalytic activity was comparable to that of a number of similar catalysts used for the oxidation of CO (129). Porphyrins and phthalocyanins intercalated into α-ZrP were used to oxidize olefins to epoxides by dioxygen, with considerable selectivity. While cyclohexene was oxidized to predominantly the epoxide and smaller amounts of allylic oxidation products, cis-stilbene gave rise to different

Table 5 Examples of the Catalytic Applications of Zirconium Phosphates and Its Derivatives

Substrate	Reagents	Reaction	Catalyst	Ref.
Co NADH		Oxidation	Riboflavin/ZrP	154
Alkylaromatics	Oxygen	Oxydehydrogenation	Ce-ZrP	155
CO		Oxidation	Cu-ZrP	129
Ethylene	H_2O	Hydration	ZrP	156
Acetone	Hydrogen	Reductive aldol condensation	Pd/ZrP	157
Cu(II)	O_2/H_2	Oxidative dehydrogenation	$CuZr(PO_4)_2$	158
Ethylbenzene		Oxidative dehydrogenation	α-ZrP	159
Chlorobenzene	H_2O	Hydrolysis	ZrP-Cu(II)	145
Cyclohexanol		Dehydration	α-ZrP poisoned by Cs^+/quinoline	138
1,3-Cyclohexadiene		Disproportionation	$ZrH(.eta.5\text{-}C_6H_7)(dmpe)_2$	160
Propene	O_2	Oxidation	α-$Zr(HPO_4)_2$ with Mn, Co, Ni, Cu, and Zn	131
Cu(II)	H	Reduction	$ZrCu(PO_4)_2$	161
MeOH		Oxydehydrogentaion	ZrP/CuX zeolite	162
Ethylbenzene	coke	Oxydehydrogenation	ZrP/Zr-TiP	134
Alcohol		Decomposition	ZrP/TiP	163
Cyclohexene, Cu(II)		Oxydehydrogenation or reduction	$ZrCu(PO_4)_4$	164
Cyclohexanol		Dehydration	Ti-ZrP and ZrP	165
Ethylene		Oxidation	Ag/ZrP with K^+, Na^+, Cs^+	166
Cyclopropane, butenes		Isomerization	$Zr(HPO_4)_2$	167
H_2O		Photocatalysis	$Ru(bpy)_3^{2+}$-$Zr(HPO_4)_2$	168
Isopropanol		Dehydration	$Zr(HPO_4)_2$ pillared by Ph, Me_2Ph, and Ph_2	169
Alcohol, butenes		Dehydration isomerization	(-$Zr(HPO_4)_2$)	138
Phenylacetylene	CO/O_2	Carbonylation	Palladium(II)bpy/ZrP	170
Acetic acid		Esterification	Zirconium phosphonates	171
Phenol, phenolic ethers	H_2O_2	Hydroxylation	Microcrystalline ZrP	172
Alkylbenzene		Photo-oxidation	Zirconium phosphate	173
Aniline	HCl/PANI	Polymerization	$CuZr(PO_4)_2$ and phosphonates	174
Anisole	alcohol	Freidal–Crafts alkylation	Amorphous ZrP	175
Alpha- and beta-pinene		Rearrangements	Zr(IV) phosphate polymer	176
Isopropyl alcohol		Decomposition	Alumina/chromia/(-ZrP)	177
Isopropyl alcohol		Dehydration	Fluorinated alumina (-ZrP)	178
Naphthalene	Air/CH_3CN	Photocatalytic oxidation	ZrP and GeP	179
Isopropanol		Decomposition	Ti-ZrP	180
Cyclohexene		Dehydrogenation	Pd/(-ZrP)	134
Isopropanol		Decomposition	Ga(II)-ZrP	181
Benzene		hydrogenation	Ni/Al_2O_3 on ZrP	182
Methylene chloride, but-l-ene		Oxidation, isomerization	GaCr/(-ZrP)	88
Terpene		Rearrangement	ZrP/Zr-organo-substituted phosphonates	142
OH	Fe^{3+}	Photoxidation	$[Agl\text{-}xHxZr_2(PO_4)_3]$	183
Hydroquinone		photoxidation	$[Agl\text{-}xHxZr_2(PO_4)_3]$	184
Isopropyl alcohol		Dehydration/ decomposition	Chromia pillared α-ZrP	185

(Continued)

Table 5 *Continued*

Substrate	Reagents	Reaction	Catalyst	Ref.
1,3-diols		Dehydration/ retroprins reaction	Metal(IV) phosphate	186
Cyclohexanol, methylcyclohexanol		Dehydration	α-ZrP and phosphite	140
MeOH	EtNH₂	Amination	ZrP	187
Thiophene		Thiophene HDS reaction	Ni/Mo/Ni-Mo/alumina/ chromia/α-ZrP	188
Propane		Dehyrogenation	Chromia/ZrP	189
CCl₂F₂		Decomposition	ZrP/AlPO₄	146
Isopropanol, acetic acid, benzaldehyde	Ethanol	Dehydration, esterification, reduction	Zr phenyl phosphonate	190
Aromatic alkanes, cycloalkanes/ alkenes, alcohols	30% H₂O₂	Oxidation	Chromia Zr phenyl phosphonate	132
Unsaturated. alcohols, 1,3-diols		Reverse prins reaction	Zr(IV) phosphates	191
Fructose, inulin		Dehydration	α-ZrP₂O₇	144
m-Phenylenediamine		Synthesis of resorcinol	ZrP (α and γ)	148
Phenyl acetylene	CO/O₂	Carbonylation	Pd(II) Rh(II) ZrP	192
Alkane		Isomerization	Nanocomposite ZrP/ WO₃, MoO₃, and Pt	193
Ethane		Oxydehydrogenation	Cr, Al/Cr, and Ga/Cr ZrP	137
Ethylene benzene			Gamma ZrP/SiO₂	194
Benzaldehyde		Knoevenagel reaction	Zr(O₃POK)₂	195
Tetralin		Hydrogenation, ring opening	Ni/Ni-Mo alumina/α-ZrP	196
NO	Propane	Reduction	Cu/alumina/α-ZrP	197
Propene, 1-hexene		Hydrofomylation	α-ZrP/rhphosphine	152
Cyclohexene, stilbene	Dioxygen	Oxidation	Porphyrin and pthalocyanin/α-ZrP	130
1,n-diols		Cyclodehydration	Zr(IV) phosphate	143
Olefin		Hydrofomylation	Rhodium thiolate/γ-ZrP	198
Isopropanol, toluene		Dehydration,	Silica/mixed oxide/α-ZrP	199

ratios of *cis*- to *trans*-stilbene-oxides.(130) The former reaction proceeds by free radical pathway and is inhibited by radical scavengers, while the latter reaction appears to proceed by radical and alternative mechanisms. Similarly, the catalytic oxidation of propene by oxygen was achieved with Mn-, Co-, Ni-, Cu-, and Zn-impregnated α-ZrP (131). The logarithm of the rate constants for the oxidation correlated linearly with the heats of formation of the corresponding metal oxides, and this indicates the role of the metal oxides in the catalysis. Using hydrogen peroxide as the oxidant, a number of aromatics, alkanes, cycloalkanes, cycloal-kenes, and alcohols were oxidized with chromia-pillared α-ZrP (132). These catalytic activities are higher than those of the Cr-aluminum phosphates and Cr-MCM-41.

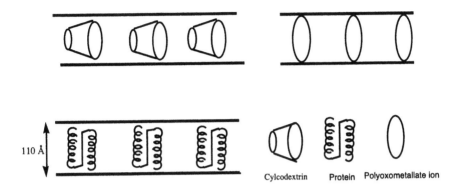

FIGURE 12 α-ZrP galleries may be opened by the intercalation of cyclodextrins, poly-oxometallate ions, or proteins. Such gallery expansion facilitates diffusion of guests, re-agents, or products into or out of the galleries.

Dehydrogenation

Oxydative dehydrogenation of cyclohexene to benzene was catalyzed by $ZrCu(PO_4)_2$, and the type of products formed depended on the sequence of intro-duction of the substrate and oxygen (133). In the absence of oxygen, reduction of Cu(II) was noted; subsequent exposure of the reaction mixture to oxygen resulted in the complete burn-up of the hydrocarbon. If the oxygen was sorbed first and then exposed to cyclohexene, high yields of benzene were noted. Efficient dehydrogenation of cyclohexene to benzene with >80% selectivity at >85% conversion was reported with Pd-exchanged α-ZrP (134). Ethyl benzene was converted to styrene in good yields when α-ZrP was used as a catalyst carrier. The organic polymer initially formed in the galleries, which consisted of oxidative functionalities (coke), served as the catalyst (135), and the activity of the coke was augmented by halogens. The mixed phosphates of Zr-Sn converted ethyl benzene to styrene with >90 selectivity at 50% conversion (136). The catalytic role of surface acidity, and that of the oxidative coke, similar to the earlier exam-ple, are discussed. Chromia-pillared α-ZrP catalyzes the oxidation of propane and ethane, but these conversions have low selectivity for the olefin formation. The selectivity is enhanced at the expense of activity when Ga or Al is partially substituted for Cr (137).

C. Dehydration

Cyclohexanol is dehydrated to cyclohexene by a variety of α-ZrP catalysts. The strong role of Bronsted acid sites in the catalysis was proposed based on the observation that the catalytic sites are strongly inhibited by Cs^+ poisoning. Similar acid catalysis was also reported for the dehydration of isopropanaol and 1-propanol and in the case of acid-catalyzed 1-butene isomerization (138). Dehydration of deuterium-labeled cyclohexanol over amorphous α-ZrP indicates that the reaction proceeds via the carbocation mechanism rather than via a concerted mechanism (139). Similar evidence for carbocation participation was also observed for the dehydration of 4-methylcyclohexanol, but 2-methyl cyclohexanol reacts partly via a concerted mechanism (140). Labeling studies of the dehydration of 1-methylcyclohexanol over metal phosphates indicates that the reaction may proceed via both concerted as well as carbocation pathways (141). Dehydration of nerol and geraniol and the rearrangement of α-pinene are catalyzed by pillared α-ZrP (142). The surface acidity and the surface area of the catalyst correlate with the observed differences in the reaction rates. Dehydration of 1,n-diols on ZrP resulted in the corresponding cyclodehydration products, while dehydration of diethanolamine resulted in the corresponding cyclization product, morpholine (143). The dehydration of fructose and inulin to 5-hydroxymethyl-2-furaldehyde is catalyzed by metal phosphates and proceeds with high selectivity and turnover numbers. The performance of the catalysts correlates with the nature of the surface acidic sites of the catalyst (144).

D. Dechlorination

Dehalogenation of organic pollutants and their conversion to more acceptable, commercially important products is an ongoing activity to safeguard the environment and to achieve greener chemistry. Dehalogenation catalysts are central to this achievement. Conversion of chlorobenzene to phenol was achieved with a catalyst prepared by the ion exchange of Cu(II) into ZrP; the efficiency of the dechlorination reaction depended on the Cu content as well as on the pH at which Cu(II) was introduced. The reaction was suggested to proceed via a radical mechanism to produce HCl, with the loss of HCl from the catalyst surface being the rate-controlling step (145). Another interesting example of dehalogenation is the decomposition of chlorofluorocarbons over the metal phosphate catalysts; both Zr and Al phosphates showed high catalytic activities for these conversions (146).

E. Desulfurization

Pillaring of α-ZrP with chromia and subsequent sulfidation produced catalysts that are highly stable and active for the conversion of thiophene to butane and butenes (147). Such catalysts will be useful in the desulfurization of gasoline to meet the more stringent, aggressive environmental regulations for sulfur emissions.

F. Deamination

α-ZrP was used for the deamination of m-phenylenediamine to resorcinol at elevated temperatures (148). High yields of resorcinol are reported, with ammonia as the byproduct. Ammonia did not poison the acidic sites on α-ZrP, but rather inhibited the active sites on the corresponding gamma phase. In a related study, potassium-exchanged α-ZrR showed high catalytic activity for the desilylation of phenol silyl ethers (149).

G. Hydration

Conversion of ethylene to ethanol was observed over zirconium or aluminum phosphate catalysts impregnated with phosphoric acid, and the activities are slightly better than that of phosphoric acid on silica, the industrial catalyst for ethylene hydration. The active species for this transformation was identified to be liquid phosphoric acid present on the catalyst (150).

H. Hydroxylation

Phenol, anisole, and phenolic ethers are readily hydroxylated by hydrogen peroxide in the presence of microcrystalline ZrP and acetic acid (151). The selectivity for catechol formation was greater than 29% at 47% conversion, and hydroquinone was observed to be the byproduct.

I. Hydroformylation

Intercalation of an amine-substituted phosphine ligand in the galleries of α-ZrP resulted in the formation of a rhodium catalyst after treatment with $Rh(CO)_2$(acetylacetonate). Hydroformylation of 1-hexene and propene indicated moderate activity and reasonable regioselectivity (152).

The foregoing examples are snapshots of a variety of catalytic transformations achieved with α-ZrP and its derivatives. This brief description serves to illustrate the potential of these materials in regio-, chemo-, and stereoselective transformations (153). The properties of the α-ZrP can be improved by the replacement of the OH group by alkyl or aryl functions, and this avenue provides a rational approach to tailor these materials for specific applications.

V. α-ZIRCONIUM PHOSPHONATES

The replacement of the phosphate group of α-ZrP by alkyl or aryl phosphonates results in the corresponding group IV metal phosphonates (α-$Zr(O_3PR)_2 \cdot nH_2O$, where the OH group of the phosphate is replaced by R, abbreviated as α-ZrRP) (see Fig. 2). The major advantage of these materials is that their properties can be controlled in a predictable manner by choosing the type, chemical nature, and extent of substitution of phosphate groups by the phosphonate groups. That is, mixed phosphate-phosphonate phases and mixed phosphonates containing two different R groups of specific compositions are known. The high stability of metal phosphates is carried over to the phosphonates, and these materials can be more stable than the parent phosphates. The orientation of gallery R groups of the alpha phase is the same as that of the OH in phosphates, perpendicular to the metal phosphonate plane. The R groups are packed rather closely, as in a Langmuir–Blodgett film or a self-assembled monolayer, but with much higher mechanical and chemical stabilities. The R group also provides a stable spacer between the galleries to enlarge the spacings; bis-phosphonates may function as pillars to improve the porosity of the materials (Fig. 13). The gallery spacings of α-ZrRPs, therefore, are readily controlled by choosing the appropriate R group.

A. Synthesis and Structure

α-ZrRPs are prepared in a manner similar to that of phosphates by replacing phosphoric acid with the desired phosphonic acid. Highly crystalline phosphonate materials are obtained by using the acid reflux method or the HF method, as described earlier for the synthesis of α-ZrP. α-ZrRPs with alkyl, aryl, halo, carboxy, nitro, amino, hydroxyalkyl, sulfonato, and vinyl derivatives are known,

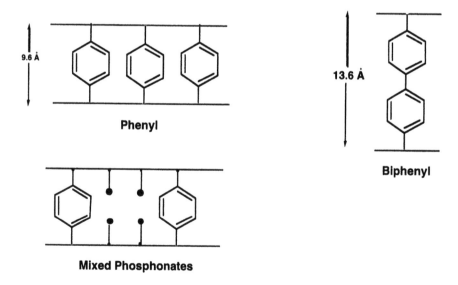

Phenyl

Biphenyl

Mixed Phosphonates

FIGURE 13 Another approach to open the galleries is pillaring of the layers with bisphosphonates or mixed phosphonates.

Table 6 Examples of Zirconium Phosphonates and Some of Their Properties

α-Zr(RPO₃)₂ R =	Type of guest for binding	d (Å)	Ref.
H	Neutral	5.61	206a
OH	Cationic	7.6	2b
CH₃	Hydrophobic	8.9	216
CH₂Cl	Polar	10.1	206a
CH₂OH	Polar	10.1	2a
CH₂CH₂CN	Polar	13.2	206a
(CH₂)₂-NH₂	Anionic	14.6	217
(CH₂)₃-NH₂	Anionic	14.6	210
CH₂-CO₂H	Cationic	11.1	206a
C₆H₅	Neutral	15.0	206a
(CH₂)₅-CO₂H	Cationic	19.0	206a
(CH₂)₃-SO₃H	Cationic	18.8	198b
CH₂(NHCH₂CH₂)₃-NH₂	Anionic/cationic	27.6	205
1, 8-Ocatnediyl		13.5	221
4, 4'-Azobenzenediyl		16	224
[5-[4-[[(6-hydroxyhexyl) sulfonyl]phenyl]azo] phenyl]pentoxide		27	224

(200–222) (see Table 6), and α-ZrRPs have been prepared in a variety of forms, such as layered materials, multilayer films, and mesoporous solids (201,202). Based on physical characteristics such as interlayer distances, density, and the similarity of the chemical properties of α-ZrRPs with α-ZrP, the structures of α-ZrRPs were predicted to be similar to that of α-ZrP (Fig. 2), and this was found to be true (203,204).

Spectroscopic and structural studies on α-ZrRPs indicate that the P—C bond of the R group is perpendicular to the matrix, and in the case of alkyl derivatives the carbon chain is inclined at an angle of ~59° to the metal-phosphonate plane. Within a given layer, the R groups are 5.3 Å apart, alternate above and below the metal plane, and are arranged in a hexagonal pattern in a given layer similar to the phosphate groups of α-ZrP (105). In contrast to α-ZrP, the interlayer interactions can be controlled by placing suitable functionalities at the free ends of the R groups. Interlayer hydrogen bonding, for example, exists between carboxy derivatives of appropriate chain lengths (Fig. 14). The orientation of the carboxyl should be such that it can accept and donate hydrogen-bonding interactions with the layer above (or below).

Amide and ester functions are introduced into the galleries of α-ZrRPs via preparation of the corresponding acid chlorides and subsequent reaction of the

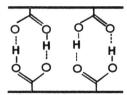

FIGURE 14   Appropriate choice of the R groups can ensure hydrogen bonding between adjacent layers of the phosphonates.

acid chloride with amine or alcohol functionalities via intercalation and subsequent condensation reactions (106). The acid chloride itself was obtained by treatment of the corresponding acid with thionyl chloride after opening the galleries of the acid by intercalation with ammonium ions. Thermal and chemical stabilities of the layered amides were found to be greater than the corresponding noncovalent intercalation compounds, and this fact attests to the improved stabilities of phosphonates (28,207).

Chemical cross-linking of adjacent layers, across the gallery, can be achieved either by linking R groups or by synthesizing the cross-linked phosphonate by starting with a bis-phosphonate in place of mono-phosphonic acid (Fig. 13) (21,206). The interlayer spacings of such cross-linked materials are readily controlled by appropriate choice of bis-phosphonic acids (208,209). By a clever choice of bis-phosphonic acid shape, with a smaller cross section at the center when compared to the two phosphonate termini, cross-linked solids with high crystallinity and high interlayer porosity are obtained (210). Pillared materials with well-defined void spaces and, hence, tailored microcomposites can also be engineered by tailoring the length of the organic pillars and their lateral distances within the α-ZrRP layers. Mixed α-ZrRP materials with engineered pillaring materials are likely to find application as molecular sieves and shape-selective catalysts.

Mixed phosphonates of the type α-Zr(O$_3$PR$_1$)·(O$_3$PR$_2$)·nH$_2$O (abbreviated as α-ZrR$_1$R$_2$P), where the number of R groups can be varied from 0 to 2, have been prepared to produce materials with novel properties (Table 7). Thus, two different functional groups can be arranged in the galleries at specific ratios, and single-phase materials are obtained where the minor component is uniformly distributed within the major component (211). Growth of semiconductor nanoparticles in the galleries of Zr(IV) phosphonates (R = CH$_2$CH$_2$COOH) was achieved by ion exchange with suitable metal ions (Zn, Pb, Cd) followed by exposure to hydrogen sulfide or hydrogen selenide (3). The resulting nanoparticles are of 30–50 Å in size, and they are characterized by diffraction and dehydration studies. In the case of the amine derivative, (R = CH$_2$CH$_2$CH$_2$NH$_2$), the strong Lewis

Table 7 Examples of Mixed Phosphonates, Their Composition, and Their Layer Spacings

R_1	R_2	$R_1:R_2$	d-Spacing (Å)	Ref.
OH	Phenyl	1.15 : 0.85	12.4	218
OH	H	0.66 : 1.34	6.5	2b
OH	$(CH_2)_2$-CO_2H	1 : 1	12.9	216
CH_3	$CH_2)_3$-NH_2	1 : 1	11.3	210
$CH_2C_6H_5$	Me	1 : 1		219
$(CH_2)_3COOH$	OH	1 : 1		217
CH_3	$CH_2C_6H_4NH_2$	1.75 : 0.25		220
CH_3	$C_{16}H_9$	1 : 1		221
OH	H, Ph, $CH_2CH_2CO_2H$	1 : 1	25.5–10.4	209
H	C_6H_5	1 : 1	25.5–10.4	209
H	Phenyl sulfonate		16.1	242
OH	$NO_3NH_3CH_2CH_2PO_3$	0.64 : 1.68	14.6	215
OH	C_6H_5O		16.4	222
$CH_2CH_2COXC_nH_{2n}H$	$R1(X = NH, n = 0$–$18,$ $X = O, n = 2$–$6)$	1 : 1	13.8–53	204
CH_2CH_2COCl	R_1	1 : 1		204
Zn,Cd, Pb(II)	$(CH_2CH_2CO_2)_2$		28.6–28.0	3
$(CH_2)_3NH_3)_2$	Cl_2	0.2 : 1.8	15.3	210
$CH_2)_3NH_3)_2$	CH_3		10.4	210
CH_2CH_2(bipyridinium) $CH_2CH_2PO_3$	X_2 $(X = Cl, Br, I)$			241
$CH_2CH_2NH_3Cl$	R1	1 : 1	14.3	223
$CH_2(N\{CH_2COOH\}_2$	$CH_2N\{CH_2COOHCH_2$ $COO\}$			224

base can be used to coordinate metal ions to produce covalently bound metal complexes for catalytic and photochemical applications (212).

Synthesis of organic–inorganic hybrid materials with high selectivity for the binding of specific ions was achieved by polymerizing phosphonic acids derived from crown ethers. Because crown ethers are well known to show high selectivity for specific metal ions, the resulting crown-metal phosphonates are expected to show high selectivity in ion exchange applications (213). The layer spacings of a crown–Zr(IV) phosphonate hybrid was 15 Å, consistent with the expected value for the incorporation of the crown ether in the galleries. The synthesis of both the α and γ forms of the crown-Zr-phosphonates are reported (213).

A convenient way to produce novel derivatives of α-ZrRP is via direct chemical manipulation of the R groups of α-ZrRP. Treatment of α-ZrRP (R = phenyl) with fuming sulfuric acid results in sulfonation of the aromatic ring system while preserving the inorganic matrix (214). The resulting sulfonic acid derivative displays a large interlayer volume with a layer spacing of 16.1 Å, a feature that is attractive for catalytic applications. The strong acidity of the aromatic sulfonic acid function served as a catalyst for a number of chemical transformations discussed next (215).

B. Catalytic Applications of α-ZrRP Materials

1. Hydrolysis

Introduction of acidic and hydrophobic functions in the galleries of mixed phosphonates yields attractive catalysts for the hydrolysis of ethyl acetate with improved efficiency (225). In another study, α-ZrRP containing sulfophenyl phosphonic acid pendent groups, described earlier, was used to hydrolyze 1,3-ditholanes and 1,3-dithianes to the corresponding carbonyl compounds (226). Catalytic hydrolysis studies often use ethyl acetate as the substrate and monitor the production of acetic acid to evaluate the corresponding catalytic parameters.

2. Esterfication

As in the case of ester hydrolysis, mixed phosphonates consisting of acidic and hydrophobic functions accelerated the esterification of acetic acid when compared to catalysts without the hydrophobic function (169,227).

3. Epoxidation

α-ZrRP thin films containing monolayers of Mn(II) porphyrins anchored via four alkyl phosphonate arms were assembled to form molecular films consist of noninteracting chromophores for catalytic applications. Epoxidation of cyclooctene by iodosylbenzene in the presence of the porphyrin monolayer films was examined, and the α-ZrRP-bound porphyrin showed much better activity when compared to the homogeneous-phase reaction (228). This was attributed to prevention of the formation of mu-oxo porphyrin dimers, which are not catalytically active, and the retention of a favorable conformation of the catalyst at the solid surface.

4. Polymerization

Using a mixed phosphonate α-ZrR$_1$R$_2$P (R$_1$ = CH$_2$CH$_2$CH$_2$COOH, R$_2$ = OH), polymerization of aniline was achieved with a concomitant increase in interlayer spacing. The resulting product was violet in color, and it did not undergo protonation by HCl (172).

5. Production of Hydrogen Peroxide

Efficient catalysts for the production of hydrogen peroxide were constructed by bridging the adjacent layers of the α-ZrRP with viologen spacers(229). Ion exchange of the halide counterion with tetrachloro salts of Pt/Pd followed by reduction of the metal with hydrogen resulted in the formation of metal clusters in the galleries, and these clusters catalyzed the formation of hydrogen peroxide when a stream of hydrogen and oxygen was passed through the samples.

The few examples of catalytic activities of metal phosphonates described above are meant to indicate the potential of these materials for future explorations.

C. Thin Films of α-ZrRPs and Their Applications

Thin films of α-ZrRPs are grown at a variety of surfaces with relative ease due to the dependable chemistry of Zr(IV) phosphonate linkage chemistry and the availability of a variety of phosphonic and bisphosphonic acids. Several examples are discussed later to illustrate these features. Functionalization of silica surfaces, for example, via treatment with an organosilane bearing a phosphonic acid or by treatment with $ZrOCl_2$ can be used for the construction of α-ZrRP layers on silica substrates (230). The thin films are grown layer by layer with controlled molecular thickness (see Fig. 15). This ability to control both thickness and chemical nature of the films provides unprecedented control over the construction of multilayer molecular films.

Such thin-film assembly is not limited to silica surfaces; it has been extended to other surfaces, such as gold and germanium (231). Film construction is initiated by first depositing a bifunctional thiol on gold or amino silane on germanium surfaces such that a spacer consisting of phosphonic acid function is

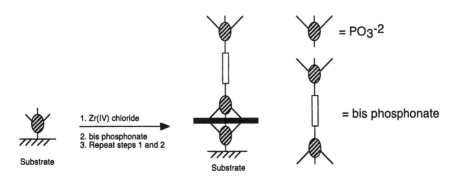

FIGURE 15   Synthesis of molecular multilayers of controlled thickness using zirconium phosphonate chemistry.

attached to the solid. These phospho-derivatives are used for the construction of the next layer using Zr(IV) chemistry. The phosphonate/Zr layers are constructed alternately, until the desired thickness or the desired number of layers is achieved. One advantage with these metal/ZrRP films is that the film growth and thickness and its ordering can be monitored by attenuated total reflectance or plasmon resonance spectral methods (232). Using two different bisphosphonic acids derived from suitable organic spacers, mixed monolayers of the α-ZrRP films are grown on gold surfaces (233). The refractive index of a α-ZrRP monolayer film depends on the relative compositions of the two bisphosphonic acids present in the monolayers, providing a handle on the synthesis of thin films with desired refractive indices. Using this strategy of polymerization of bisphosphonic acids with Zr(IV), smooth thin films containing a viologen moiety as the spacer at gold surfaces for photochemical applications were prepared (234).

Cationic and anionic multilayers can be constructed at gold surfaces in a "mix-and-match" approach by extending the concept of sequential growth of phosphonate films (235), α-ZrRP derivatives are some of the novel layered materials that have a high potential for the rational design of nonlinear optical (NLO) materials (236). A phosphate–phosphonate multiplayer, for example, was prepared from a polar azo dye, 4-{4-[N,N-bis(2-hydroxyethyl) amino]phenylazo}phenylphosphonic acid (Fig. 16). The NLO properties of the material were similar to these of $LiNbO_3$, an efficient NLO material, with excellent thermal stability (150°C for 3 hours). This approach was extended in 2000 to organic chromophores containing electron donor and acceptor functions in the organic spacer. One of the phosphonic acid groups of the bisphosphonate is protected, since the ester and film assembly is carried out in a vectoral manner to produce acentric multilayers (237). The methodology has also been extended to cross-link the functional groups of Langmuir–Blodgett films to produce robust organic/inorganic multilayers (238). Photoactive organic–inorganic multilayer films are constructed using phosphonate/naphthalene-di-imides as the spacers (239). Irradiation of these multilayer thin films results in a pink color characteristic of the naphthalenediimide cation radical.

Using phosphonate derivatives of maleimide vinyl ether monomers, alternating copolymers were synthesized where the phosphonate functions are polymerized using Zr(IV) chemistry. Layer thicknesses of 16–31 Å are achieved, and up to 10 layers of the organic–inorganic multilayers are built, with control over the layer thickness, number of layers, and layer-dependent Zr loading (240). The chemical sensitivities of different layers can be exploited to selectively modify desired layers by preprogrammed layer composition and constitution. Using different metal ions, for example, or using different spacers in each layer, the chemical properties of each layer can be individually controlled. By using two different spacers, such as a tetrahydroxamate and a bisphosphonate, metal coordinated hybrid multilayer materials are prepared such that the layers are assembled using

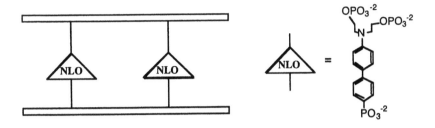

FIGURE 16   Molecular thin films of phosphonates bearing noncentrosymmetric chromophores for NLO applications.

different linkers. The bisphosphonate/Zr layer is acid resistant, while the hydroxamate layer is selectively dissolved under acidic conditions. Such hybrid multilayer materials with different chemical properties of the binders are novel, and this approach will be useful in manipulating the well-ordered multilayer films in a preprogrammed manner (241). Such film assembly is not limited to the examples just discussed but is readily adopted to modify the physicochemical properties of the thin films (242).

D. Photochemical Studies

Photochemical and photophysical studies of layered phosphonates, as in the case of metal phosphates, are helpful in understanding the binding, diffusion, and distribution of guest ions, molecules, and metal complexes in the galleries. One advantage with α-ZrRP materials is that their physical/chemical properties can be modified by changing the R group, and the chemical stability of the covalently linked R group is superior to corresponding intercalated materials. Donor or acceptor chromophores, for example, can be covalently linked to the matrix, avoiding their release from the galleries under high ionic strengths or at elevated temperatures. Alternatively, the R groups of α-ZrRPs can be used to coordinate specific metal ions for photocatalytic applications. Some representative examples are presented here to illustrate the nature of the galleries of the phosphonates.

1. Viologens

One of the attractive examples of photoinduced electron transfer with α-ZrRP materials involves a viologen moiety covalently linked to the matrix (243). Viologen has been studied extensively for solar energy harvesting and conversion via photoinduced charge separation reactions. Incorporation of the viologen moiety in galleries has resulted in the binding of anions in the galleries. Photoexcitation oxidizes halide counterions, followed by reduction of the viologen moiety. The proximity of the counterion and retention of the primary photoproducts in the galleries results in interesting photochemical processes that are similar to those observed with methyl viologen intercalated in clay minerals (244). Another example is the incorporation of viologen in molecular thin films via viologen phosphonates, using Zr(IV) chemistry (245).

2. Uranyl/Europium System

One advantage of phosphonates is that metal ions can be directly coordinated to the functionalities of matrix R groups in order to carry out catalytic or photochemical reactions. This aspect was exploited to anchor photochemically interesting ions in the galleries of α-ZrRP, and the photophysical properties of uranyl and europium ions coordinated to the carboxyl groups, for example, have been studied. This strategy is attractive because the excited states of these ions are localized on the metal, diffusion of the metal ions in the galleries is impeded due to coordination, and interchromophore interactions can be suppressed due to the coordination of the metal ions to the carboxylate functions of α-ZrRP (R = CH_2CH_2COOH) (246). The metal ions bound to adjacent sites, for example, can be as close as 5 Å. This distance is short enough for Forster-type energy transfer to occur, from acceptors to donors, but long enough to prevent contacts between adjacent donors or acceptors. The energy transfer from uranyl excited states to europium ions, when these are coordinated to the carboxyl functions of α-ZrRP (R = CH_2CH_2COOH), was accelerated by a factor of 9. This rate acceleration has been attributed to the increased local concentrations, reduced distances of separation between the ions, and enhanced communication between the metal ions. Because of the reduced diffusion of metal ions in the galleries of phosphonates, the rate of energy transfer should have been reduced, contrary to the observations; therefore, metal coordination to the matrix improves the energy transfer characteristics of the bound ions.

3. Metal Complexes

Another major advantage of α-ZrRP materials is that the local environment of a guest can be modified by changing the nature of R; therefore, the properties of guest molecules can be controlled in a rational manner. Sulfonation of phenyl groups in α-ZrRP (R = phenyl) leads to the formation of strong acid sites in the galleries (ZrPS). These sites can be used for catalysis and for binding guest ions that are otherwise reluctant to bind in the hydrophobic interior of α-ZrRP (R = phenyl). Intercalation of $Ru(bpy)_3^{2+}$, for example, into ZRPS is facile, and if causes significant modification of $Ru(bpy)_3^{2+}$ properties (247). These spectroscopic changes are in contrast to the results obtained with $Ru(bpy)_3^{2+}$ intercalated into BAZrP. The absorption spectra of the metal complex, recorded after intercalation into ZrPS, showed a red shift of the MLCT band to 494 nm, and the luminescence maximum shifted from 604 nm (aqueous medium) to 640 nm. The rate constant for self-quenching of Ru(bpy)$_3^{2+}$ emission was significantly reduced when compared to the corresponding rate constant in solution. The intercalated metal complex is accessible to methyl viologen, and the mechanism of luminescence of the metal complex by methyl viologen was explained by proposing a sphere-of-action model.

VI. BIOLOGICAL APPLICATIONS OF α-ZRP AND α-ZRRP

The catalytic potential of layered materials can be expanded substantially if enzymes are intercalated or covalently immobilized in the galleries. One advantage of this endeavor is that the enzyme–solid composites can be used to carry out complex biochemical transformations under ambient conditions with a high degree of regio-, chemo-, setereo-, and chiral selectivity. The biocatalyst, bound to the solid, can be readily recycled to decrease the cost, and multiple enzymes can be accommodated to carry out a predefined sequence of chemical transformations in a single step. Several strategies have been employed to bind enzymes in solid matrices, (248) but layered materials are unique. This is because the layer spacings can be expanded to accommodate enzymes of any size, and the enzymes bound in the galleries are resistant to microbial degradation. Gallery spacings are quite narrow when compared to the dimensions of typical bacteria, so their access to the bound enzyme is limited. Gallery spacings, at the same time, are comparable to the average pore sizes of other solids, (249) and the spacings are adequate for the diffusion of reagents into the galleries and for the release of products to the outside. Other solids used for enzyme binding include polymer matrices, porous glass, sol-gels, hydrogels, and cellulose (250–252). Large surface areas, high thermal/chemical stability, and relatively easy synthetic access to a variety of α-ZrRP materials with specific chemical characteristics are some advantages of these materials for enzyme binding.

The chemical and physical properties of galleries can be altered as described earlier to accommodate the delicate needs of enzymes, and the local pH can be adjusted to their individual requirements. Chemical characteristics such as hydrophobicity, hydrophilicity, and hydrogen-bonding ability are readily controlled to maximize enzyme stability and activity. Initial results indicate that the enzymes bound in α-ZrP galleries are nearly as active as native enzymes in aqueous media, but higher activity and improved selectivities are also noted.

A. Protein/Enzyme Intercalation

Direct intercalation of proteins into α-ZrP, however, is not successful due to the small gallery spacings and the large kinetic barrier for intercalation of large guest molecules. Forceful intercalation at high temperatures, long times, and extreme pHs can result in enzyme denaturation.

FIGURE 17   Exfoliation of α-ZrP stacks with tetrabutylammonium cations followed by binding of proteins and subsequent reassembly of the α-ZrP plates with proteins trapped in the galleries.

This problem was overcome by adopting the exfoliation route for enzyme binding (see Fig. 17). Exfoliation of α-ZrP stacks with butylammonium ions followed by contact with a variety of proteins/enzymes under controlled pH conditions at room temperature has resulted in facile binding of the biomolecules to α-ZrP with little or no adverse effects on their properties (8). Protein binding in the galleries is confirmed by expanded gallery spacings, as observed in powder XRD studies (253). This method of enzyme binding can be applied to proteins/enzymes of any size, and it does not involve preintercalation of harmful substances or high temperatures. Armed with this benign approach, a number of enzymes/proteins have been intercalated into the galleries of α-ZrP under ambient conditions. The binding is due mostly to electrostatic interactions, with the negative charge field of α-ZrP preferring binding of positively charged enzymes/proteins. However, neutral or even negatively charged biomolecules can be incorporated under suitable conditions (see Table 8).

Cytochrome C (Cyt c), met-myoglobin (Mb), met-hemoglobin (Hb), lysozyme (Lys), glucose oxidase (GO), and α-chymotrypsin (CT) are some of the proteins intercalated in the galleries of α-ZrP. The gallery spacings observed with some of these intercalates are listed in Table 9, which have been correlated very

Table 8   Stochiometries and Binding Constants for Immobilized Protein/α-ZrP Composites

Protein/α-ZrP	Stoichiometry (μM)	K_b/M^{-1}
No protein	—	—
Myoglobin	12	2.0×10^5
Lysozyme	40	1.33×10^6
Hemoglobin	14	5.4×10^6
Chymotrypsin	3	2.5×10^5
Glucose oxidase	1.1	5.6×10^4

Table 9 Physical and Catalytic Properties of Proteins Intercalated into α-ZrRP (R = OH)

Protein	d-Spacing (Å)	Protein size (Å); (Ref.)	K_m (Bound)	V_{max} (Bound)	K_m (Free)	V_{max} (Free)
Lysozyme	47	32 × 32 × 55; (40)	0.5 mM	0.63 μM/s	0.5 mM	0.63 μM/s
Myoglobin	54	30 × 40 × 40; (37)	1.6 mM	0.04 μM/s	1.4 mM	0.08 μM/s
Chymotrypsin	62	40 × 43 × 65; (49)	ND	ND	0.6 mM	ND
Hemoglobin	66	53 × 54 × 65; (57)	0.11 mM	42 nM/s	0.10 mM	34 nM/s
Glucose oxidase	116	43 × 51 × 68; (54)	2.5 mM	0.042 mM/s	0.8 mM	0.037 mM/s

ND = not determined.

The maximum rate of the reaction (V_{max}) and the concentration needed to achieve half the maximum rate (K_m) are also indicated.

Source: Ref. 256.

well with the known diameters of the respective proteins (254). A nonzero intercept of 6 Å was observed when the d-spacings, after subtracting the ZrP plate thickness, were plotted against the protein size, and this intercept has been attributed to the existence of multiple layers of water molecules bound to the intercalated protein. In the case of GO, the d-spacing was double the diameter of GO, which indicated bilayer packing of GO. Binding of proteins to solid surfaces usually results in release of water from the interacting surfaces, but the presence of protein-bound water is attributed to improved stabilities of biomolecules at solid surfaces (255). Binding of the biomolecules at the edges or on the outside would not have increased the gallery spacings, and the XRD data provide a strong evidence for the intercalation of enzymes/proteins. The intercalated enzymes/ proteins have been characterized using a number of spectroscopic techniques as well as biochemical activity tests. All enzymes retained a significant portion of their activities (>90%), while some exhibited improved activities and selectivities.

Secondary and tertiary structures of the proteins after intercalation were examined using circular dichroism (CD) and infrared methods. These spectra are sensitive to protein conformational changes and are used extensively to estimate the alpha helical and beta sheet contents of globular proteins. Even though protein/ ZrP intercalates are suspensions that scatter light below 300 nm, the circular dichroism spectra of the bound proteins could be recorded readily. This is because very short optical path lengths (<0.1 mm) are adequate for these studies, because proteins absorb strongly in the 190- to 300-nm region. CD spectra of the bound proteins are nearly superimposable with those of the corresponding proteins in aqueous buffers, and these data indicate a significant retention of the native conformation of the bound proteins/enzymes.

One of the main features of redox proteins is their ability to undergo rapid electron transfer reactions with appropriate donors and acceptors. For example,

Cyt c bound to α-ZrP retains its redox activity (253). Reduction of the Fe(III) form of Cyt c/α-ZrP by dithionite resulted in the growth of an absorption band at 550 nm, and the reduction of Fe(III) is slower in the galleries than in the aqueous phase. Similarly, addition of ferricyanide to the Fe(II) form of Cyt c/α-ZrP resulted in the loss of the 550-nm band, indicating the oxidation of the Fe(II) to Fe(III) form (253). These observations highlight the accessibility of bound Cyt c and retention of its redox properties when present in α-ZrP galleries.

B. Enhanced Catalytic Activities of Bound Proteins

One of the major applications of enzymes bound on solid surfaces is their utility in catalyzing specific chemical transformations under mild conditions. While Lys, GO, and CT are enzymes, heme proteins such as Hb and Mb do not function as enzymes in natural systems, under ordinary conditions, but they do catalyze a number of chemical reactions in vitro. Peroxidase, monooxygenase, catalase, demethylase, and hydrolase activities are a few examples of the known activities of these heme proteins. These activities are of biological significance in ischemic repurfusion, stroke, and heart attacks, where these proteins are released into the intercellular region and can potentially damage tissue. Intercalation of these inexpensive heme proteins provides high-utility biocatalysts for the aforementioned chemical transformations.

Catalytic activities of biocomposite materials derived from α-ZrP are collected in Table 9; these are comparable to the corresponding activities of the native proteins in aqueous solutions. In some cases they are enhanced; for example, the peroxidase activity of bound Mb is greater when 4-methoxy phenol, and other para-substituted phenols were used as substrates (256). This is one of the rare examples where the bound protein had a higher activity. The catalytic data provided later clearly establish the versatility of α-ZrP in providing a benign medium for the incorporation of delicate proteins/enzymes for biocatalytic applications.

In addition to the improved activities, Mb bound to α-ZrP exhibited improved substrate selectivity (see Table 10). Oxidation of p-methoxy phenol, for example, is accelerated by bound Mb, while the ortho isomer is less reactive, and the selectivity for bound Mb is improved ten fold (256). Similar improved selectivities are observed with other derivatives of phenol, and these improvements do not correlate with the redox potentials of the substrates (256). The improved selectivity is attributed to changes in the active site geometry of the bound protein such that it can accommodate the para and meta isomers better than the ortho isomers (Fig. 18). Such improvements in the selectivities are to be investigated further, and improved selectivities are welcome changes for biocatalytic applications.

C. Enhanced Thermal Stabilities of Intercalated Proteins

In addition to retention of the structure, activity, and improved selectivities, the biocatalyst should also be robust. Thermal stability of the bound enzyme should

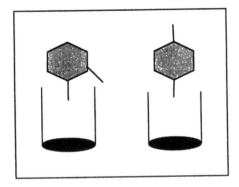

FIGURE 18 Schematic illustration of the proposed basis for the observed selectivity for the oxidation of substituted phenols by myoglobin immobilized in α-ZrP galleries.

be at least comparable to the enzyme in solution, and the operational stability of the bound enzyme should also be adequate for the successful, prolonged use of the biocatalyst over many catalytic cycles. Intercalation of Hb improved the thermal stability of the protein; the denaturation temperature for the intercalated protein was greater than 90°C, while the free protein denatures below 64°C (257). Similar improvement in the thermal stability of HRP/α-ZrP is also noted (258). Stabilization of the native form and destabilization of the denatured form by the α-ZrP matrix is the most likely reason for the improved thermal stability.

Significant improvements in the denaturation temperatures of proteins and enzymes (increases as large 33°C) when intercalated in α-ZrP and α-ZrRP (R = CH_2COOH, CH_2CH_2COOH) are demonstrated (259). The denaturation tempera-

Table 10 Improved Selectivities for the Peroxidase Activity of Mb/α-ZrP with Small Organic Molecules as Substrates

Substrate	Relative rate (bound/free)	Oxidation potential/V[23]
p-Methoxyphenol	2.54	0.406
o-Methoxyphenol	0.24	0.456
Phenol	1.26	1.04
o-Cresol	0.31	0.556
Aniline	0.97	0.70
m-Aminophenol	1.75	—
o-Aminophenol	0.09	0.124

Source: Ref. 256.

tures of met hemoglobin/α-ZrP, glucose oxidase/α-ZrRP (R = OH, CH_2CH_2COOH), and cytochrome c/α-ZrRP (R = OH, CH_2CH_2COOH), for example, were over 95, 55, 65, 100, and 75°C, respectively, which are greater than the corresponding denaturation temperatures of the free proteins (70, 50, and 67°C, respectively). Proteins bound to α-ZrRP (R = OH, or CH_2CH_2COOH) in general indicated greater stabilities, while binding to α-ZrRP (R = CH_2COOH) resulted in diminished stabilities, except in case of Hb. The type and strength of interactions of the surface functions of the solid matrix with the bound protein, therefore, are crucial in determining the bound protein stability. The improved stabilities observed in specific cases are welcome changes for biocatalytic applications. The layered solid may impose a kinetic barrier for the denaturation of the bound protein by constraining the biomolecule in two dimensions; (260) further studies are required to test this hypothesis.

D. High-Temperature Activity of Bound Enzymes/Proteins

Enzymes typically function at normal body temperature, 37°C. Their activities at higher temperatures are desirable for rate acceleration, but free enzymes denature at elevated temperatures. Upon intercalation in α-ZrRP (R = OH), both HRP and Hb showed peroxidase activities at over 85°C, observed for the first time, (258) while the free enzyme/protein deactivated rapidly at these temperatures with no activity. The maximum rate of the reaction (V_{max}) increased 3.6-fold, while the concentration needed to achieve half the maximum rate (K_m) decreased by 20% at these higher temperatures. Such high-temperature activities of enzymes/proteins are unusual, and they indicate the promise of α-ZrRPs for enzyme stabilization in high-temperature applications. This strategy of enzyme stabilization in α-ZrRP may provide alternatives to thermophilic enzymes obtained from thermophiles, and these may supplement thermostable enzymes obtained by protein engineering.

E. Reversible Thermal Denaturation of Proteins

Another interesting observation regarding proteins/enzymes bound in the galleries is that Hb, HRP, and GO show better recovery of activity and structure subsequent to thermal denaturation than the corresponding proteins in aqueous solutions (257,261). Hb contains four subunits; proper folding of each subunit, their reassembly, and binding of the heme at its native site are important for the recovery of Hb structure/activity. Heating Hb/α-ZrRP (R = CH_2COOH) at 95°C for 5 minutes (in nitrogen atmosphere) resulted in protein denaturation. This was indicated by the loss of XRD peak, shifting of amide I and amide II bands in the FTIR spectra, and the loss of 211-, 220-nm bands in the CD spectra. Cooling the denatured samples to room temperature (22°C, for ~24 h) resulted in the recovery of the XRD peak, recovery of the alpha helical structure (CD at 210, 220 nm), and peroxidase activity (>95% recovery). The recovery of activity was poor for free Hb in solution (<20% peroxidase activity) under similar conditions. GO and Mb indicated efficient renaturation at α-ZrRP (R = CH_2CH_2COOH), while Hb refolded more efficiently on α-ZrRP (R = CH_2COOH). The activity recovery of denatured Hb and the rate of recovery are in the order α-$Zr(PO_3CH_2CH_2COOH)_2$ > α-$Zr(PO_3CH_2COOH)_2$ > α-$Zr(HPO_4)_2$ (257). The orientations of hydrogen-bonding functions (CO and/or OH) are distinctly different in these galleries (Fig. 19). The hydrogen-bonding interaction of these solids with bound protein may be important in the mechanism of protein refolding. The surface functions of the support matrix are likely to play a major role in dictating the properties of the bound protein, and these surface functions can be readily controlled with phosphonates. This observation is novel, and it strongly suggests further investigation of protein binding and stability along these lines.

These aspects of protein intercalation and activities in layered materials are essential prerequisites for biocatalytic applications. Most certainly, other layered materials will be tested for their abilities to bind, stabilize, and enhance the catalytic properties of enzymes.

F. Footprinting of Proteins Bound to Solid Surfaces (α-ZrP and α-ZrRP)

Most certainly, the physical and catalytic behavior of the proteins bound to solid surfaces will depend on how these proteins are oriented at the solid and on what specific amino acid residues of the protein interact with the solid. Such detailed information regarding the protein–solid contact regions is currently missing, and it will be of fundamental significance in understanding the behavior of proteins at solid surfaces. This information can be obtained by using reagents that can cleave the peptide backbone of the protein at specific sites. By comparing the

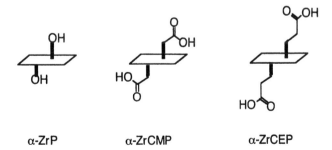

FIGURE 19   Differences in the orientations of carboxyl functions in phosphonate galleries, and the dependence of carbonyl orientation on the number of carbons in the chain.

sites that are cleaved when the protein is in solution with those obtained when the protein is bound to the solid, specific details about these protein–solid contact regions can be obtained.

PentammineaquoCo(III) (CoPA) and tetramminediaquoCo(III)(CoTA) ions cleave proteins such as Hb, or lysozyme, at selected sites, under hydrolytic conditions, (262) and these reagents (artificial metallo-peptidases) have been used to examine the orientation of proteins bound to solid surfaces (263). The amino acid residues at the protein–solid interface, for example, are expected to be protected from the cleavage reagents, and cleavage would occur only at sites that are exposed. Hb/α-ZrP was cleaved by CoPA and CoTA with high efficiencies (40–50% yield) while free Hb was unreactive under similar conditions. CoPA was suggested to cleave Hb/α-ZrP at the helix C-loop-helix D region of the α-subunit, while Lys82-Gly83 is the suspected target site on the β-subunit, for CoTA. Favorable binding of these cationic metal complexes to the negatively charged α-ZrP accelerates the protein cleavage in the galleries, due to the increased local concentrations of the reagents. In contrast to Hb bound to α-ZrP, Hb/α-ZrRP (R = CH_2COOH) was not cleaved, indicating the unavailability of the corresponding cleavage sites when the protein binds to this solid. Accelerated protein cleavage was noted with lysozyme/α-ZrP, but Mb was not cleaved to a significant extent when bound to these solid matrices. Modulation of the protein cleavage, therefore, provides specific information about protein–solid contact regions or the structural changes induced by protein binding at the solid surface.

G. Other Biological Applications of α-ZrP and α-ZrRP

Zirconium phosphates have found their way into several other biological applications, including immunoassays, dialysis, and radiation protection, the last becoming important for the current issues of bioterrorism. Zirconium phosphate, for example, was effective in reducing the levels of cerium-144 in the liver and the femur of rats (264). Among the various zirconium salts tested, ZrP was one of the most effective in reducing the gastrointestinal absorption of cerium-144 and strontium-85. Applications of ZrP are not limited to radiation protection. They are extended to the capture of metabolites such as ammonia in noninvasive dialysis applications. Microencapsulation of urease for the degradation of urea to ammonia and sorption of the resulting ammonia by the ZrP matrix in the gastrointestinal tract was tested in vitro (265). This approach is promising for urea removal from blood via a noninvasive treatment of dialysis patients or to decrease the frequency of dialysis treatments. A microencapsulated zirconium phosphate–urease system was tested for the treatment of chronic uremia for clinical use (266). Zirconium phosphate gel was used in the radioimmunoassay for carcinoembryonic antigen (267). Most certainly, many more biological applications of ZrP are expected in the near future.

VII. OTHER METAL PHOSPHATES AND PHOSPHONATES

Lamellar or linear structures are also formed by various other metal ions, such as zinc, (268) manganese, (269) molybdenum, (270) and vanadium (271,272). Among these other solids, vanadium derivatives attracted significant attention due to their potential as industrial catalysts. The structures of the metal phosphonates are similar to those of the corresponding phosphates, in a manner similar to the zirconium phosphate/phosphonates. Polymerization of phenylphosphonate, for example, with Mo(IV) resulted in a linear double-stranded structure where the phenyl groups are positioned on the outside of the linear structure (273). These other metal derivatives are bound to yield many future investigations regarding the fundamental and applied chemistry of layered materials.

VIII. CONCLUSIONS AND FUTURE WORK

Layered solids provide unique opportunities for the binding, encapsulation, exchange, and activation of specific chemical species for chemical and biochemical applications. The feasibility of these applications is enhanced tremendously by the opportunities to introduce organic functionalities in the galleries of phosphonates in order to modify molecular properties in a rational manner. Introduction of chemical or biological catalysts, in addition to the native catalytic properties of the materials, into the galleries via covalent or noncovalent methods makes these materials versatile. The ability to introduce proteins/enzymes expands the catalytic potential of these materials beyond ordinary expectations. The high thermal stability of a handful of proteins investigated indicates their potential to stabilize biocatalysts for synthetic, high-temperature, and biomedical applications.

ACKNOWLEDGMENTS

The authors thank the National Science Foundation (DMR-9729178, INT-0138401), the University of Connecticut Research Foundation, and the donors of the Petroleum Research Fund (PRF#33821-AC4) for their financial support of this work.

REFERENCES

1a. G. Alberti, U. Costantino. In: M. S. Whittingham, A. J. Jacobson, eds. Intercalation Chemistry. New York: Academic Press, 1982:Chapter 5.

1b. G. Alberti, U. Costantino, F. Marmottini. In: P. A. Williams, M. J. Hudson, eds. Recent Developments in Ion Exchange. New York: Elsevier Applied Science, 1987.

1c. A. Clearfield. Chem. Rev 1988; 88:125.

2a. G. Alberti, U. Costantino. J. Mol. Catal 1984; 27:235.
2b. A. Clearfield. J. Mol. Catal 1984; 27:251.
3.   G. Cao, L. K. Rabenberg, C. M. Nunn, T. E. Mallouk. Chem. Mater 1991; 3:149.
4.   H. Lee, J. Kepley, H. Hong, T. E. Mallouk. J. Amer. Chem. Soc 1988; 110:618.
5a. Z. Li, C. Lai, T. E. Mallouk. Inorg. Chem 1989; 28:178.
5b. D. Rong, Y. I. Kim, T. E. Mallouk. Inorg. Chem 1990; 29:1531.
6.   M. G. Kanatzidis, C. Wu, H. O. Marcy, D. C. DeGroot, C. R. Kannerwurf. Chem. Meter 1990; 2:222.
7.   M. G. Kanatzidis, M. Hubbard, L. M. Tonge, T. J. Marks, H. O. Marcy, C. R. Kannerwurf. Synth. Met 1989; 28:C89.
8.   C. V. Kumar, B. B. Raju. In: V. Ramamurthy, K. Schanze, eds. Mol. Supramol. Photochem.. New York: Marcel Dekker, 2001:505.
9.   C. V. Kumar. Photochem. In: V. Ramamurthy, ed. Organized and Constrained Media. New York: VCH, 1991:783.
10a. A. Clearfield, G. D. Smith. Inorg. Chem 1969; 8:431.
10b. G. Alberti, U. Costantino, S. Allulli, N. Tomasini. J. Inorg. Nucl. Chem 1978; 40: 1113.
10c. G. Alberti, M. Casciola, U. Costantino. J. Colloid Interface Sci 1985; 107:256.
11a. G. Alberti. Acc. Chem. Res 1974; 11:163.
11b. C. F. Lee, M. E. Thompson. Inorg. Chem 1991; 30:4.
11c. R. M. Kim, J. E. Pillion, D. A. Burwell, J. T. Groves, M. E. Thompson. Inorg. Chem 1993; 22:43.
11d. J. Snover, M. E. Thompson. J. Amer. Chem. Soc 1994; 116:765.
11e. G. Alberti, C. Dionigi, S. Murcia-Mascarós, R. Vivani. In: G. Tsoucaris, ed. Crystallography of Supramolecular Compounds. Amsterdam: Kluwer Academic, 1996: 143.
11f. G. Alberti. Acc. Chem. Res 1978; 11:163.
12a. A. Clearfield, Å. Oskarsson, C. Oskarsson. Ion Exchange Membranes 1972; 1:91.
12b. C. B. Amphlett. Inorganic Ion Exchangers. Amsterdam: Elsevier, 1964.
13a. A. Clearfield, L. Kullberg. J. Phys. Chem 1974; 78:1150.
13b. S. E. Horsley, D. V. Nowell. J. Appl. Chem. Biotechnol 1973; 23:215.
13c. G. Alberti, U. Costantino, R. Giulietti. J. Inorg. Nucl. Chem 1980; 42:062.
13d. J. M. Troup, A. Clearfield. Inorg. Chem 1969; 8:431.
13e. C. Trobajo, S. A. Khainakov, A. Espina, J. R. Garcia. Chem. Mater 2000; 12:1787.
14. G. Alberti, E. Torracca. J. Inorg. Nucl. Chem 1968; 30:317.
15. S. K. Shakshooki, L. Szirtes, A. Azrak, M. Ahmed, S. Khalil. J. Radioanal. Nucl. Chem 1989; 132:49.
16. C. V. Kumar, A. Chaudhari. unpublished results. 2001.
17. Y. Inoue, Y. Yamada. Bull. Chem. Soc 1979; 52:3528.
18a. J. Jimenez-Jimenez, P. Maireles-Torres, P. Olivera-Pastor, E. Rodriguez-Castellon, A. Jimenez-Lopez, D. Jones, J. Roziere. Adv. Mater 1998; 10:812.
18b. E. Rodriguez-Castellon, A. Jimenez-Lopez, P. Olivera-Pastor, J. Merida-Robles, F. Perez-Reina, M. Alcantara-Rodriguez, F. Souto-Bachiller, L. Rodriguez-Rodriguez, M. Fortes, J. Ramos-Barrado. Molecular Crystals and Liquid Crystals Science and Technology. Section A: Molecular Crystals and Liquid Crystals 1998; 311: 677.

19. D. Wang, R. Yu, N. Kumada, N. Kinomura. Chem. Mater 2000; 12:956.
20. D. Wang, R. Yu, T. Takei, N. Kumada, N. Kinomura, A. Onda, K. kajiyoshi, K. Yanagisawa. Chem. Lett 2002; 3:398.
21. A. Clearfield, S. D. Smith. J. Colloid Interface Sci, A. Clearfield, S. D. Smith. Inorg. Chem, J. M. Troup, A. Clearfield. Inorg. Chem 1977; 16:3311.
22a. E. Michel, A. Z. Weiss. Naturforsch. B 1965; 20:1307.
22b. A. Clearfield, R. M. Tindwa. J. Inorg. Nucl. Chem 1979; 41:871.
23. J. Albertsson, A. Oskarsson, R. Tellgren, J. O. Thomas. J. Phys. Chem 1977; 81: 1574.
24. X. Mathew, V. U. Nayar. Infrared Phys 1988; 28:189.
25. G. A. Alberti, A. Grassi. Inorg. Chem 1999; 38:4249.
26. K. A. Kraus, H. O. Phillips. J. Am. Chem. Soc, C. B. Amphlett, L. A. McDonald, M. J. Redman. Chem. Ind. (London) A. ed. Clearfield, ed. Inorganic Ion Exchange Materials, CRC Press 1982; 78:1314.
27. I. D. Coussio-, G. B. Bettolo, V. Moscatelli. J. Chromatogr 1963; 11:238.
28. G. Alberti, G. Grassini. J. Chromatog, A. Clearfield, G. D. Smith. J. Inorg. Nucl. Chem 1968; 30:327.
29. A. Clearfield, W. L. Duax. J. Phys. Chem 1969; 73:3424.
30. V. Kotov, I. A. Stenina, A. B. Yaroslavtev. Solid State Ionics. also see Ref. 42 1999; 125:55.
31. G. Alberti, U. Costantino, J. P. Gupta. J. Inorg. Nucl. Chem 1974; 36:2109.
32. N. Mikami, M. Sasaki, T. Yasunaga, K. F. Hayes. J. Phys. Chem 1984; 88:3229.
33. M. Sasaki, N. Mikami, T. Ikeda, K. Hachiya, T. Yasunaga. J. Phys. Chem 1982; 86:5230.
34. A. Clearfield, G. A. Day, A. Ruvarac, S. Milonjic. J. Inorg. Nucl. Chem 1981; 43: 165.
35. A. Clearfield, H. Hagiwara. J. Inorg. Nucl. Chem 1978; 40:907.
36. G. Alberti, R. Bertrami, U. Costantino, J. P. Gupta. J. Inorg. Nucl. Chem 1977; 39:1057.
37. K. G. Varshney, A. H. Pandith. J. Ind. Chem. Soc, C. A. Borgo, Y. Gushikem. J. Colloid Interface Sci 2002; 246:343.
38. G. Alberti. Atti. Accad. Nazl. Lincei, Rend., Classe Sci. Fis., Mat. Nat 1961; 31: 427.
39. M. J. N. Costa, M. A. S. Jeronimo. J. Chromatog 1961; 5:456.
40. A. Clearfield, G. D. Smith. J. Inorg. Nucl. Chem 1968; 30:327.
41. G. Alberti, U. Costantino, M. Pelliccioni. J. Inorg. Nucl. Chem 1973; 35:1327.
42. G. Alberti, U. Costantino, J. P. Gupta. J. Inorg. Nucl. Chem 1974; 36:2103.
43. S. Allulli, A. La Ginestra, M. A. Massucci, M. Pelliccioni, M. Tomassini. Inorg. Nucl. Chem. Lett 1974; 10:337.
44. B. R. Palmer, Fuerstenau M. C.. Proc. Int. Miner. Process. Congr., 10th 1974:1123.
45. G. Alberti, U. Costantino, S. Alluli, M. A. Massucci. J. Inorg. Nucl. Chem 1975; 37:1779.
46. N. Souka, A. S. Abdel-Gawad, R. Shabana, K. Farah. Radiochim. Acta 1975; 22: 180.
47. A. Clearfield, R. A. Hunter. J. Inorg. Nucl. Chem 1976; 38:1085.

Kumar et al.

48. A. Clearfield, J. M. Kalnins. J. Inorg. Nucl. Chem 1976; 38:849.
49. R. Smits, D. L., et al. Massart. Anal. Chem 1976; 48:458.
50. G. Alberti, M. G. Bernasconi, U. Costantino, J. S. Gill. J. Chromatogr 1977; 132: 477.
51. S. Allulli, C. Ferragina, A. La Ginestra, M. A. Massucci, N. Tomasinni. J. Chem. Soc., Dalton Trans 1977; 19:1879.
52. G. Alberti, M. G. Bernasconi, M. Casciola, U. Costantino. Ann. Chim. (Rome) 1978; 68:265.
53. G. Alberti, M. G. Bernasconi, M. Casciola, U. Costantino. J. Chromatogr 1978; 160:109.
54. A. Clearfield, Z. Djuric. J. Inorg. Nucl. Chem 1979; 41:885.
55. U. Constantino. J. Inorg. Nucl. Chem 1979; 41:1041.
56. Y. Hasegawa, S. Kizaki. Chem. Lett 1980; 3:241.
57. M. A. Massucci, A. La Ginestra, C. Ferragina. Gazz. Chim. Ital 1980; 110:73.
58. L. Kullberg, A. Clearfield. J. Phys. Chem 1981; 85:1578.
59. L. Kullberg, A. Clearfield. J. Phys. Chem 1981; 85:1585.
60. Y. Hasegawa, S. Kizaki, H. Amikura. Bull. Chem. Soc. Jpn 1983; 56:734.
61. Y. Hasegawa, G. Yamamine. Bull. Chem. Soc. Jpn 1983; 56:3765.
62. C. Y. Yang, Chen, et al. J. S.. Sep. Sci. Technol 1983; 18:83.
63. C. Ferragina, A. La Ginestra, M. A. Massucci, A. A. G. Tomlinson. J. Phys. Chem 1984; 88:3134.
64. V. A. Borovinskii, E. V. Lyzlova, L. M. Ramazanov. Radiochemistry (Moscow, Russian Federation)(Translation of Radiokhimiya) 2001; 43:84.
65. U. Costantino, L. Szirtes, E. Kuzmann, J. Megyeri, K. Lazar. Solid State Ionics 2001; 141:359.
66. T. Shaheen, M. Zamin, S. Khan. Main Group Metal Chemistry 2001; 24:351.
67. C. H. Kim, H. S. Kim. Synth. Met 1995; 71:2051.
68. A. Clearfield. Comm. Inorg. Chem 1990; 10:89.
69. C. V. Kumar, Z. J. Williams. Spectrum 1995, June:17.
70. R. Tindwa, D. K. Ellis, G. Z. Peng, A. Clearfield. J. Chem. Soc., Fraday Trans, A. Clearfield, R. M. Tindwa. J. Inorg. Nucl. Chem 1979; 41:871.
71. C. V. Kumar, E. H. Asuncion, G. Rosenthal. Microporous Mater 1993; 1:123.
72. C. V. Kumar, E. H. Asuncion, G. Rosenthal. Microporous Mater 1993; 1:299.
73. K. Peeters, R. Carleer, J. Mullens, E. F. Vansant. Microporous Mater 1995; 4:475.
74. M. Casciola, U. Costantino, L. Di Corce, F. Marmottini. J. Inclusion. Phenom 1988; 6:291.
75. G. Alberti, U. Costantino, F. Marmottini, G. Perego. J. Inclusion phenom. Mol. Recognit. Chem 1989; 7:549.
76. M. Kaneno, S. Yamaguchi, H. Nakayama, K. Miyakubo, T. Udea, T. Eguchi, N. Nakamura. Int. J. Inorg. Mater 1999; 1:379.
77. U. Costantino. J. Chem. Soc. Dalton Trans 1979; 2:402.
78. M. Trchova, P. Capkova, P. Matejka, K. Melanova, L. Benes. J. Solid State Chem 1999; 145:1.
79. U. Costantino, R. Vivani, V. Zima, L. benes, K. Melanova. Langmuir 2002; 18: 1211.

80. T. Kijima. Thermochim. Acta 1982; 59:95.
81. Y. Hasegawa, T. Akimoto, D. Kojima. J. Inclusion Phenom. Mol. Recognit. Chem 1994; 20:1.
82. C. Ferragina, M. A. Massucci, G. Mattogno. J. Inclusion Phenom. Mol. Recognit. Chem 1989; 7:529.
83. R. Hoppe, G. Alberti, U. Costantino, C. Dionigi, G. Schulz-Ekloff, R. Vivani. Langmuir 1997; 13:7252.
84. U. Costantino, M. A. Massucci, A. La Ginestra, A. A. Tarola, L. Zampa. J. Inclusion Phenom 1986; 4:147.
85. T. Kijima, S. Ueno. J. Chem. Soc. Dalton Trans 1986; 1:65.
86. K. Yamamoto, Y. Hasegawa, K. Nikki. J. Inclusion Phenom. Mol. Recognition 1998; 31:289.
87. J. W. Johnson. J. Chem. Soc. Chem. Commun, Y. Hasegawa, M. Kamikawaji, K. Sasaki. Bull. Chem. Soc, D. J. MacLachlan, D. M. Bibby. J. Chem. Soc. Dalton Trans, R. C. Yeates, S. M. Kuznicki, L. B. Lloyd, E. M. Eyring. J. Nucl. Inorg. Chem 1981; 43:2355.
88. M. Alcantara-Rodriguez, P. Olivera-Pastor, E. Rodriguez-Castellon, A. Jinenez-Lopez, M. Lenarda, L. Storaro, R. Ganzerla. J. Mater. Chem 1998; 8:1625.
89. K. E. Dungey, J. Coble, 222nd ACS National Meeting, Chicago, II, INOR-375.
90. Y. Yuan, P., et al. Wang. Anal. Bioanal. Chem 2002; 372:712.
91. S. Yamanaka, Y. Horibe, M. Tanaka. J. Inorg. Nucl. Chem 1976; 38:323.
92. T. Kijima, S. Ueno, M. Goto. J. Chem. Soc., Dalton Trans 1982:2499.
93. T. Kijima, Y. Sekikawa, S. Ueno. J. Inorg. Nucl. Chem 1981; 43:849.
94. C. Ferragina, A. La Ginestra, M. A. Massucci, P. Patrono, A. A. G. Tomlinson. J. Phys. Chem 1985; 89:4762.
95. J. Kornyei, L. Szirtes, U. Costantino. J. Radioanal. Nucl. Chem 1985; 89:331.
96. C. Ferragina, M. A. Massucci, P. Patrono, A. A. G. Tomlinson, A. Ginestra. Mater. Res. Bull 1987; 22:29.
97. G. Alberti, U. Costantino, F. Marmottini, G. Perego. Gazz. Chim. Ital 1989; 119: 191.
98. P. Giannoccaro, C. F. Nobile, G. Moro, A. La Ginestra, C. Ferragina, M. A. Massucci, P. Patrono. J. Mol. Catal 1989; 53:349.
99. C. V. Kumar, Z. J. Williams, K. Falguni. Microporous Mater 1996; 7:161.
100. C. V. Kumar, E. H. Asuncion. J. Chem. Soc. Commun, C. V. Kumar, E. H. Asuncion. J. Am. Chem. Soc, C. V. Kumar, L. M. Tolosa. J. Phys. Chem 1993; 97: 13914.
101. P. Capkovi, L. Benes, K. Melanova, H. Schenk. Appl Crystallog 1998; 31:845.
102. D. M. Kaschak, S. A. Johnson, D. E. Hooks, H-N. Kim, M. D. Ward, T. E. Mallouk. J Am Chem Soc 1998; 120:10887.
103. C. V. Kumar, A. Chaudhari. Microporous Mesoporous Mat 2000; 41:307.
104. G. Alberti, C. Dionigi, S. Murcia-Mascarós, R. Vivani. In: G. Tsoucaris, ed. Crystallography of Supramolecular Compounds. Amsterdam: Kluwer Academic, 1996: 143.
105. C. V. Kumar, A. Chaudhari. J. Am. Chem. Soc 1994; 116:403.
106. F. Souto, E. Rodriguez, G. Siegel, A. Jimenezz, L. Rodriguez, P. Olivera, J. Meridaa, F. Perez, M. Alcantara. Mol. Cryst. Liq. Cryst. Sci. Technol., A 1998; 311:817.

107. S. Okuno, G. Matsubayashi. Inorg. Chim. Acta 1996; 245:101.

108. M. Danjo, M. Tsuhako, H. Nakayama, T. Eguchi, N. Nakamura, S. Yamaguchi, H. Nariai, I. Motooka. Bull. Chem. Soc. Jpn 1997; 70:1053.

109. G. G. Aloisi, U. Costantino, F. Elisei, M. Nocchetti, C. Sulli. Mol. Crystals Liquid crystals science Technol 1998; 311:653.

110. H. Hoppe, G. Alberti, U. Costantino, C. Dionigi, G. Schulz-Ekloff, R. Vivani. Langmuir 1997; 13:7252.

111. E. Rabinowitch, L. Epstein. J. Am. Chem. Soc 1941; 63:69.

112. S. Okuno, G. Matsubayashi. Inorg. Chim. Acta 1995; 233:173.

113. E. D. Sternberg, D. Dolphin, C. Bruckner. Tetrahedron, M. G. H. Vicente. Current Medicinal Chem: Anti-Cancer Agents 2001; 1:175.

114. R. M. Kim, J. E. Pillion, D. A. Burwell, J. T. Groves, M. E. Thompson. Inorg. Chem 1993; 32:4509.

115a. D. P. Vleirs, R. A. Schonheydt, F. C. de Schryver. J. Chem. Soc., Faraday Trans 1985; 81:2009.

115b. D. P. Vleirs, D. Collin, R. A. Schoonheydt, F. C. de Schryver. Langmuir 1986; 2:165.

116. C. V. Kumar, Z. J. Williams. J. Phys. Chem 1995; 99:17632.

117. J. Wheeler, J. K. Thomas. J. Phys. Chem, T. Kajiwara, K. Hasimoto, T. Kawai, T. Sakata. J. Phys. Chem, I. Willner, J. M. Yang, C. Laane, J. W. Otvos, M. Calvin. J. Phys. Chem, R. Memming. Surf. Sci, J. W. Perry, A. J. McQuillon, F. C. Anson, A. H. Zewail. J. Phys. Chem 1982; 87:1480.

118a. W. de Wilde, G. Peeters, J. H. Lunsford. J. Phys. Chem 1980; 21:97.

118b. W. H. Quayle, J. H. Lunsford. Inorg. Chem 1982; 21:97.

119a. D. Krenske, S. Abdo, H. Van Damme, M. Cruz, J. J. Fripiat. J. Phys. Chem 1980; 84:2447.

119b. S. Abdo, P. Canesson, M. Cruz, J. J. Fripiat, H. Damme. J. Phys. Chem 1981; 85:797.

119c. H. Nijs, M. Cruz, J. J. Fripiat, H. Damme. J. Chem. Soc. Chem. Commun 1981:1026.

119d. H. Nijs, J. J. Fripiat, H. Van Damme. J. Phys. Chem 1983; 87:1279.

119e. R. A. Schoonheydt, J. Pelgrims, Y. Hereos, J. B. Uytterhoeven. Clay Miner 1978; 13:435.

120a. R. A. Schoonheydt, P. De Pauw, D. Vliers, F. C. de Schryver. J. Phys. Chem 1984; 88:5113.

120b. D. P. Vliers, R. A. shoonheydt, F. C. De Schriyver. J. Chem. Soc., Faraday Trans 1985; 81:2009.

120c. D. P. Vliers, D. Collin, R. A. Schoonheydt, F. C. De Schryver. Langmuir 1986; 2:165.

121. Z. J. Williams. PhD dissertation: University of Connecticut, 1997.

122. J. L. Colon, C.-Y. Yang, A. Clearfield, C. R. Martin. J. Phys. Chem 1990; 94:874.

123. A. Awaluddin, R. N. DeGuzman, C. V. Kumar, S. L. Suib, S. L. Burkett, M. E. Davis. J. Phys. Chem 1995; 99:9886.

124. D. G. Whitten, S. P. Spooner, Y. Hsu, P. L. Penner. React. Polym, S. E. Weber. Chem. Rev, W. F. Mooney, D. G. Whitten. J. Am. Chem. Soc 1986; 108:5712.

125a. R. P. Haugland. In:. Molecular Probes Handbook of Fluorescent Probes and Research Chemicals. Eugene. OR: Molecular Probes, 1992–1994:20.

125b. D. M. Kaschak, T. E. Mallouk. J. Am. Chem. Soc 1996; 118:4222.

126. D. Gust, T. A. Moore, A. L. Moore. Acc. Chem. Res 1993; 26:198.

127. C. V. Kumar, Z. J. Williams, R. S. Turner. J. Phys. Chem. A 1998; 102:5562.

128. C. V. Kumar, J. K. Barton, N. J. Turro. J. Am. Chem. Soc 1985; 107:5518, and Ref. 122.

129. T. J. Kalman, M. Dudukovic, A. Clearfield. Adv. Chem. Ser 1974; 133:654.

130. M. Nino, S. A. Giraldo-, E. A. Mozo. J. Mol. Cat 2001; 175:139.

131. M. Iwamoto, Y. Nomura, S. Kagawa. J. Cat 1981; 69:234.

132. J. Xiao, J. Xu, Z. Gao. Catl. Lett 1999; 57:37.

133. H. C. Cheung, A. Clearfield. J. Catal 1986; 98:335.

134. R. B. Borade, B. Zhang, A. Clearfield. Catal. Lett 1997; 45:233.

135. G. Emig, H. Hofmann. J. Catal 1983; 84:15.

136. G. Bagnasco, P. Ciambelli, M. Turco, A. La Ginestra, P. Patrono. Appl. Catal 1991; 68:69.

137. B. Salsona, J. M. Lopez-Nieto, M. Alcntaran-Rodriguez, E. Rodriguez-Castellon, A. Jimenez-lopez. J. Mol. Cat. A. Chemical 2000; 153:199.

138. A. Clearfield, D. S. Thakur. J. Catal, A. La Ginestra, P. Patrono, M. L. Berardelli, P. Galli, C. Ferragina, M. A. Massucci. J. Catal 1987; 103:346.

139. R. A. Johnstone, J. Y. Liu, D. Whittaker. J. Chem. Soc. Perkin Trans 1998; 6:1287.

140. M. C. Costa, L. F. Hodson, R. A. W. Johnstone. J. Mol. Cat. A: Chem 1999; 142:349.

141. R. A. Johnstone, J. Y. Liu, D. Whittaker. J. Mol. Cat. A: Chem 2001; 174:159.

142. C. M. Conceicao, R. A. W. Johnstone, D. Whittaker. J. Mol. Catal., A: Chem 1998; 129:79.

143. S. M. Patel, U. V. Chudasama, P. A. Ganeshpuri. Green Chem 2001; 3:143.

144. F. Benvenuti, C. Carlini, P. Psatrono, G. Raspolli, G. Sbarana, M. A. Massucci, P. Galli. Appl. Catl. A: General 2000; 193:147.

145. Y. Izumi, Y. Mizutani. Bull. Chem. Soc. Jpn 1979; 52:3065.

146. Y. Takita, M. Ninomiya. Phys. Chem. Chemical Physic 1999; 1:2367.

147. F. J. Perez-Reina, E. Rodriguez-Castellon, A. Jimmenez-lopez. Langmuir 1999; 15:2047.

148. B. Brack, D. W. Gammon, E. van Steen. Micropor. Mesopor. Mater 2000; 41:149.

149. M. Curini, F. Epifano, M. C. Marcotullio, O. Rosati, M. Rossi, A. Tsadjout. Syn. Comm 2000; 30:3181.

150. C. M. Fougret, W. F. Holderich. Appl. Catl. A: General 2001; 207:295.

151. R. C. Wasson, A. Johnstone, W. R. Sanderson. PCT Int. Apl.. Wo (Solvay Interox Ltd., UK).

152. M. Karlsson, C. Andersson, J. Hjortkjaesr. J. Mol. Cat. A: Chemical 2001; 166:337.

153. G. Alberti, R. Vivani, F. Marmottini, P. Zappelli. J. Porous. Mater 1998; 5:205.

154. A. Malinauskas, T. Ruzgas, L. Gorton. Biolectrochemistry Bioenergetics Field Publication 1999; 49:21.

155. G. E. Vrieland, H. H. Beck. U.S 1973:4.

156. M. J. Todd. Brit 1974:4.
157. Y. Watanabe, Y Matsumura, Y. Izumi, Y. Mizutani. Bull. Chem. Soc. Jpn 1974; 47:2922.
158. A. Clearfield, S. P. Pack. J. Catal 1978; 50:431.
159. T. Hattori, H. Hanai, Y. Murakami. J. Catal 1979; 56:294.
160. M. B. Fischer, E. J. James, T. J. McNeese, S. C. Nyburg, B. Posin, W. Wong-Ng, S. S. Wreford. J. Am. Chem. Soc 1980; 102:4941.
161. A. Clearfield, D. S. Thakur, H. Cheung. J. Phys. Chem 1982; 86:500.
162. Z. V. Gryaznova, N. N. Ponomareva, A. R. Nefedova, L. S. Eshchenko, R. N. Dvoskina, Z. I. Yakovenko. React. Kinet. Catal. Lett 1982; 19:393.
163. S. Cheng, Z. Peng, A. Clearfield. Ind. Eng. Chem. Prod. Res. Dev 1984; 23:219.
164. A. Clearfield. J. Catal 1979; 56:296.
165. T. N. Frianeza, A. Clearfield. J. Catal 1984; 85:398.
166. S. Cheng, A. Clearfield. J. Catal 1985; 94:455.
167. K. Segawa, Y. Kurusu, Y. Nakajima, M. Kinoshita. J. Catal 1985; 94:491.
168. D. P. Vliers, R. A. Schoonheydt, F. C. De Schrijver. Mater. Sci. Monogr 1985; 28: 493.
169. B. Z. Wan, R. G. Anthony, G. Z. Peng, A. Clearfield. J. Catal 1986; 101:19.
170. P. Giannoccaro, M. Aresta, S. Doronzo, C. Ferragina. Appl Organometallic Chem 2000; 14:581.
171. K. Segawa, N. Kihara. J. mol. Catal 1992; 74:213.
172. R. C. Wasson, A. Johnstone, W. R. Sanderson. PCT Int. Appl 1992:18.
173. A. Monaci, A. L. Ginestra, P. Patrono. J. Photochem. Photobiol 1994; 83:63.
174. G. L. Rosenthal, J. Caruso, S. G. Stone. Polyhedron 1994; 13:1311.
175. M. Cruz Costa, R. A. W. Johnstone, D. Whittaker. J. Mol. Catal. A: Chem 1995; 103:155.
176. Cruz Conceicao, M. Costa, R. A. W. Johnstone, D. Whittaker. J. Mol. Catal. A: Chem 1996; 104:251.
177. A. Jimenez-Lopez, J. Maza-Rodriguez, E. Rodriguez-Castellon, P. O. Pastor. J. Mol. Catal. A: Chem 1996; 108:175.
178. J. M. Merida-Robles, P. Olivera-Pastor, A. Jimenez-Lopez, E. Rodriguez-Castellon. J. Phys. Chem 1996; 100:14726.
179. A. Monaci, A. L. Ginestra. J. Photochem. Photobiol. A 1996; 93:65.
180. J. Santamaria Gonzalez, M. Martinez Lara, M. Lopaz Granados, J. L. G. Fierro, A Jimenez-Lopez. Appl. Catal. A 1996; 144:365.
181. J. Jimenez-Jimenez, P. Maireles-Torres, P. Oliver-Pastor, E. Rodriguez-Castellon, A. Jimenez-Lopez. Langmuir 1997; 13:2857.
182. J. Merida-Robles, P. Olivera-Pastor, E. Rodriguez-Castellon, A. Jimenez-Lopez. J. Catal 1997; 169:317.
183. H. Miyoshi, H. Kourai, T. Maeda, T. Yoshino. J. Photochem. Photobiol. A 1998; 113:243.
184. H. Miyoshi, H. Kourai, T. Maeda. J Chem Soc, Faraday Trans 1998; 94:283.
185. F. Perez-Riena, P. Olivera-Pastor-, P. Torres, E. Rodriguez-Castellon, A. Jimenez-Lopez. Langmuir 1998; 14:4017.
186. F. A. H. Al-Qallaf, R. A. W. Johnstone, J. Y. Liu, L. Lu, D. Whittaker. J Chem Soc Perkin Trans 2: Phys Organic Chem 1999; 7:1421.

187. C. Dume, J. Kervennal, S. Hub, W. F. Holderich. Appl Catal A: General 1999; 180:421.
188. J. Merida-Robles, E. Rodriguez-Castellon, A. Jimenez-Lopez. J Molec Catal A: Chemical 1999; 145:169.
189. F. Perez-Reina, E. Rodriguez-Castellon, A. Jimenez-Lopez. Langmuir 1999; 15: 8421.
190. J. Xiao, J., et al. Xu. Appl Catal A: General 1999; 181:313.
191. F. A. H. Al-Qallaf, L. F. Hodson, R. A. W. Johnstone, J-Y. Liu, L. Lu, D. Whittaker. J Molec Catal A: General 2000; 152:187.
192. P. Giannoccaro, M. Aresta, S. Doronzo, C. Ferragina. Appl Organometallic Chem 2000; 14:581.
193. V. A. Sadykov, S. N. Pavlova, G. V. Zabolotnaya, R. I. Maximoskaya, D. I. Kochubei, V. V. Kriventsov, G. V. Odegova, N. M. Ostrovskii, O. B. Bel'skaya, V. K. Duplyakin, V. I. Zaikovskii, E. A. Paukshtis, E. B. Burgina, S. V. Tsybulya. Mater Res Innovations 2000; 3:276.
194. G. Alberti, S. Cavalaglio, F. Marmottini, K. Matusek, J. Megyeri, L. Szirtes. Appl Catal A: General 2001; 218:219.
195. D. Fildes, D. Villemin, P-A. Jaffres, V. Caignaert. Green Chem 2001; 3:52.
196. R. Hernandez-Huesca, J. Merida-Robles, P. Maireles-Torres, E. Rodriguez-Castellon, A. Jimenez-Lopez. J Catal 2001; 203:122.
197. R. Hernandez-Huesca, J. Santamaria-Gonzalez, P. Braos-Garcia, P. Maireles-Torres, E. Rodriguez-Castellon, A. Jimenez-Lopez. Appl Catal B: Environmental 2001; 29:1.
198. S. Rojas, S. Murcia-Mascaros, P. Terreros, J. L. G. Fierro. New J Chem 2001; 25: 1430.
199. N. He, Y. Yue, Z. Gao. Microporous Mesoporous Mater 2002; 52:1.
200a. M. B. Dines, P. C. Griffith. Inorg. Chem 1983; 22:567.
200b. P. M. DiGiacomo, M. B. Dines. Polyhedron 1982; 1:61.
200c. M. B. Dines, P. C. Griffith. Polyhedron 1983; 2:607.
200d. G. Alberti, U. Costantino. In: J. L. Atwood, J. E. D. Davies, D. D. MacNicol, eds. Inclusion Compounds, Inorganic and Physical Aspects of Inclusion. Oxford: Oxford University Press, 1991:Chapter 5.
201. M. E. Thompson. Chem. Mater 1994; 6:1168.
202. C. Cao, H.-G. Hong, T. E. Mallouk. Acc. Chem. Res 1992; 25:420.
203. G. Alberti, U. Costantino, C. Dionigi, S. M. Mascarós, R. Vivani. Supramol. Chem 1995; 6:29.
204. D. M. Poojary, H. L. Hu, F. L. Campbell, A. Clearfield. Acta Crystallogr 1980; 42:1062.
205. J. M. Troup, A. Clearfield. Inorg. Chem 1977; 16:3311.
206. D. A. Burwell, M. E. Thompson. Chem. Mater 1991; 3:730.
207. C. Y. Ortiz-Avila, C. Bharadwaj, A. Clearfield. Inorg. Chem 1994; 33:2499.
208a. M. B. Dines, P. M. DiGiacomo. Inorg. Chem 1981; 20:92.
208b. M. B. Dines, P. M. DiGiacomo, K. P. Callahan, P. C. Griffith, R. H. Lane, R. E. Cooksey. In: J. S. Miller, ed. Chemically Modified Surfaces in Catalysis and Electrocatalysis. Washington. DC: ACS Symp. Ser. 192, 1982.

209. J. Alper. Chem. Ind 1986:335.
210. G. Alberti, U. Costantino, F. Marmottini, R. Vivani, P. Zappelli. Angew. Chem. Int. Ed. Engl 1993; 32:1357.
211. G. Alberti, U. Costantino, J. Kornyei, M. L. L. Giovagnotti. React. Polym. Ion Exch. Sorbents, A. Clearfield, C. V. K. Sharma, B. Zhang. Chem. Mater 2001; 13: 3099.
212. G. L. Rosenthal, J. Caruso. Inorg. Chem, W. R. Leenstra, J. C. Amicangelo. Inorg. Chem 1998; 37:5317.
213. B. Zhang, A. Clearfield. J. Am. Chem. Soc 1997; 119:2751.
214. C.-Y. Yang, A. Clearfield. React. Polym. Ion Exch. Sorbents 1987; 5:13.
215. L. Kullberg, A. Clearfield. Solvent Extr. Ion Exch 1990; 8:187.
216. M. B. Dines, P. C. Griffith. J. Phys. Chem 1982; 86:571.
217. L. Maya. J. Inorg. Nucl. Chem 1981; 43:400–401;.
218. G. Alberti, U. Costantino, J. Kornyei, M. L. L. Giovagnotti. Chim. Ind (Milan) 1982; 64:115.
219. G. L. Rosenthal, J. Caruso. J. Solid State Chem 1993; 107:497.
220. J. C. Amicaangelo, W. R. Leenstra, Book of Abstracts, 216th ACS National Meeting, August 1998, Boston.
221. T. C. Castonguay, W. R. Leenstra, In: Abstracts of Papers, 222nd ACS National Meeting August 26, 2001, Chicago.
222. G. Alberti, U. Costantino. Int. Chem 1982:147.
223. M. Casciola, U. Costantino, A. Peraio, T. Rega. Solid State Ionics 1995; 77:229.
224. B. Zhang, D. M. Poojary, A. Clearfield, P. Guangzhi. Chem. Mater 1996; 8:1333.
225. K. Segawa, A. Sugiyama, Y. Kurusu. Stud. Surf. Sci. Catal, C. M. Nam, Y. G. Kim. Korean J. Chem. Eng 1995; 12:586.
226. M. Curini, M. C. Marcotullio, E. Pisani, O. Rosati. Synlett 1997; 7:769.
227. K. Segawa, N. Kihara. Stud. Surf. Sci. Catal, K Segawa, T. Ozawa. J. Mol. Cat. A Chem 1999; 141:249.
228. I. O. Benitz, B. Bujoli. J. Am. Chem. Soc 2002; 124:4363.
229. K. P. Reis, V. K. Joshi, K. Vijay, M. E. Thompson. J. Catal, A. Dokoutchaev, V. V. Krishnan. J. Mol. Str 1998; 470:191.
230. H. G. Hong, D. Sackett, T. E. Mallouk. Chem. Mater 1991; 3:521.
231. B. Frey, D. G. Hanken, R. M. Corn. Langmuir 1993; 9:1815.
232. S. F. Bent, M. L. Schilling, W. L. Wilson, H. E. Katz, A. L. Harris. Chem. Mater 1994; 6:122.
233. D. G. Hanken, R. M. Corn. Anal. Chem, D. G. Hanken, R. R. Naujok. Anal. Chem 1997; 69:240.
234. J. L. Snover, H. Byrd, E. P. Supenova, E. Vicenzi, M. E. Thompson. Chem. Mater 1996; 8:1490.
235. M. Fang, D. M. Kaschak, A. C. Sutorik, T. E. Mallouk. J. Am. Chem. Soc, H.-N. Kim, S. W. Keller. Chem. Mater 1997; 9:1414.
236a. H. E. Katz, G. Scheller, T. M. Putvinski, M. L. Schilling, W. L. Wilson, C. E. D. Chidsey. Science 1991; 254:1485.
236b. H. E. Katz, W. L. Wilson, G. Scheller. J. Amer. Chem. Soc 1994; 116:6636.
236c. T. Coradin, R. Backov, D. J. Jones, J. Roziere, R. Clement. Mol. Cryst. Liq. Cryst. Sci. Technol. A, W. C. Flory, S. M. Mehrens. J. Am. Chem. Soc 2000; 122:7976.

237. G. A. Neff, M. R. Helfrich, M. C. Clifton, C. J. Page. Chem. Mater 2000; 12:2363.
238. M. A. Petruska, G. E. Fanucci, D. R. Talham. Chem. Mater 1998; 10:177.
239. M. A. Rodrigues, D. F. S. Petri, M. J. Politi, S. Brochsztain. Thin Solid Films 2000; 371:109.
240. P. Kohli, G. J. Blanchard. Langmuir, P. Kohli, G. J. Blanchard. Langmuir 2000; 16:695.
241. A. Hatzor, T. Van der Boom-Moav, S. Yochelis, A. Vaskevich, A. Shanzer, I. Rublinstein. Langmuir 2000; 16:4420.
242. J. C. Horne, G. J. Blanchard. J. Am. Chem. Soc, J. C. Horne, Y. Huang, G.-Y. Liu, G. J. Blanchard. J. Am. Chem. Soc 1999; 121:4419.
243a. L. A. Vermeulen, M. E. Thompson. Nature 1992; 358:656.
243b. L. A. Vermeulen, J. L. Snover, L. S. Sapochak, M. E. Thompson. J. Am. Chem. Soc 1993; 115:11767.
244. G. Villemure, C. Detellier, A. G. Szabo. J. Am. Chem. Soc 1986; 108:4658.
245. L. A. Vermeulen, M. E. Thompson. Chem. Mater 1994; 6:77.
246. C. V. Kumar, A. Chaudhary. Microporous Mesoporous Mater 1999; 32:75.
247. J. L. Colón, C.-Y. Yang, A. Clearfield, C. R. Martin. J. Phys. Chem 1988; 92:5777.
248. S. Braun, S. Rappoport, R. Zusman, S. Shtelzer, S. Druckman, D. Avnir, M. Ottolenghi. In: D. Kamely, A. Chakrabarty, S. E., Kornguth, eds. Biotechnology: Bridging Research and Applications. Amsterdam: Kluwer Academic, 1991:205.
249. D. Avnir, D. Levy, R. Reisfeld. J. Phys. Chem 1984; 88:5956.
250. D. Shabat, F. Grynszpan, S. Saphier, A. Turniansky, D. Avnir, E. Keinan. Chem. Mater 1997; 9:2258.
251. J. F. Kennedy, C. A. White. In: A. Weisman, eds. Handbook of Enzyme Biotechnology, G. Gubitz, E. Kunssberg, P. van Zoonen, H. Jansen, C. Gooijer, N. H. Velthorst, R. W. Fei. In:, D. E. Leyden, W. T. Collins, eds. Chemically Modified Surfaces, L. Gorton, G. Marko-Varga, E. Dominguez, J. Emneus. In:, S. Lam, G. Malikin, eds. Analytical Applications of Immobilized Enzyme Reactors. Vol. 2. New York: Blackie Academic & Professional, 1994:1.
252. T. Kunitake, I. Hamachi, A. Fujita. J. Am. Chem. Soc, A. Fujita, H. Senzu, I. Hamachi. Chem. Lett, J. Fang, C. M. Knobler. Langmuir, M. Mrksich, G. B. Sigal, G. M. Whitesides. Langmuir, B. Miksa, S. Slomkowski. Colloid Polym. Sci, K. Yoshinaga, K. Kondo, A. Kondo. Polym. J, F. Tiberg, C. Brink, M. Hellsten. Colloid Polym. Sci, K. Yoishinaga, T. Kito, M. Yamaye. J. Appl. Polym. Sci 1990; 41: 1443.
253. C. V. Kumar, G. M. McLendon. Chem. Mater 1997; 9:863.
254a. T. Takano. J. Mol. Biol 1993; 229:12.
254b. T. Imoto, L. N. Johnson, A. North, D. C. Phillips, J. A. Rupley. In: P. D. Boyer, ed. The Enzymes. New York: Academic Press, 1972:665–808.
254c. H. J. Hecht, H. M. Kalisz, J. Hendle, R. D. Schmid, D. Schomburg. J. Mol. Biol 1993; 229:13.
255. Proteins bound to hydrated surfaces indicated better stabilities. G. Das, K. A. Prabhu. Enzyme Microb. Tech, I. V. Mozhaev, I. V. Berezin, K. Martinek. CRC Crit. Rev. Biochem 1988; 23:235.
256. C. V. Kumar, A. Chaudhari. J. Am. Chem. Soc 2000; 122:830.

257. C. V. Kumar, A. Chaudhari. Chem. Mater 2001; 13:238.
258. C. V. Kumar, A. Chaudhari. Chem. Commun 2002; 2382.
259. C. V. Kumar, A. Chaudhari. Micropor. Mesopor. Mater 2003; 57:181.
260. H.-X. Zhou, K. A. Dill. Biochemistry 2001; 40:11289.
261. C. V. Kumar, A. Chaudhari. Micropor. Mesopor. Mater 2001; 47:407.
262. C. V. Kumar, A. Buranaprapuk, A. Chow, A. Chaudhari. Chem. Commun 2000: 597.
263. C. V. Kumar, A. Chaudhari. Micropor. Mesopor. Mater 2003:In Press.
264. J. Severa. Proc. IRPA Eur. Congr. Radiat. Prot 2nd:327.
265. E. A. Wolfe, T. M. Chang. Int. J. Artificial Organs, C. Kjellstrand, H. Borges. Trans. Am. Soc. Artificial Organs 1981; 27:24.
266. E. A. Wolfe, T. M. S. Chang. Int. J. Artificial Organs 1987; 10:269.
267. M. Koch, T. A. McPherson. Am. J. Clinical Path 1980; 74:465.
268. K. J. Martin, P. J. Squattrito, A. Clearfield. Inorg. Chim. Acta 1989; 155:7.
269. G. Cao, H. Lee, V. M. Lynch, T. E. Mallouk. Inorg. Chem 1988; 27:2781.
270. M. Damodara, D. M. Poojary, Y. Zhang, B. Zhang, A. Clearfield. Chem. Mater 1995; 7:822.
271. W. Guliants, J. B. Benziger, S. Sundaresan, I. E. Wachs, J. M. Jehng. Chem. Mater 1995; 7:1493.
272. G. Huan, A. J. Jacobson, J. W. Johnson, D. P. Goshorn. Chem. Mater 1992; 4:661.
273. M. Ogawa, K. Kuroda. Chem. Rev 1995; 95:399.

8

Layered Double Hydroxides (LDHs)

Paul S. Braterman, Zhi Ping Xu, and Faith Yarberry
University of North Texas
Denton, Texas, U.S.A.

I. INTRODUCTION

A. Organization and Scope

In this chapter we present an overview of what we regard as the most significant areas of recent activity in layered double hydroxide chemistry, using documents known to us through mid-2002, with occasional later examples. While we have attempted to give credit to the initiators in each area, we make no claims to ascribing priority (itself an uncertain matter in a fast-moving field) and will no doubt have overlooked many seminal papers; for this we can only ask our colleagues for their understanding. The division into sections and subsections is, we hope, self-explanatory. Significant results can of course have implications in more than one area, and we have therefore attempted to reduce duplication by cross-referencing.

B. Background and Overview

The layered double hydroxides (LDHs), also known as hydrotalcite-like materials or as anionic (more properly speaking, anion exchanging) clays, are a large group of natural and synthetic materials readily produced when suitable mixtures of metal salts are exposed to base. They consist of layers, containing the hydroxides of two (sometimes more) different kinds of metal cations and possessing an overall positive charge, which is neutralized by the incorporation of exchangeable anions. In general, the materials also contain various amounts of water, hydrogen bonded to the hydroxide layers and/or to the interlayer anions. In the cases that we shall consider, the metal cations are octahedrally coordinated by hydroxide groups. This criterion excludes the calcium aluminum hydroxide family (in which the calcium is 7-coordinated), but includes cases, such as the green rusts discussed later, where the different kinds of cation are derived from the same element in different oxidation states. In most cases, the cations are in the $+2$ and $+3$ oxidation states, but there are some exceptions to this, most notably the lithium aluminum hydroxide derivatives. These materials are of interest as catalysts, as catalyst precursors and supports, as ceramics precursors, as weathering product of rocks, as traps for anionic pollutants, including some kinds of nuclear waste, as antacids and delivery systems for pharmaceuticals, and as additives for polymers. The last of these is the main commercial application, exploiting the ability of these materials to neutralize the large amounts of acid generated, for instance, in the heating or degradation of poly(vinylchloride) (PVC), and to retard flame. Very recently, systematic interest has emerged in the ability of these materials to give rise to self-assembled structures containing anionic surfactants, with implications for the fine-tuning of properties.

Overall, the materials are currently the subject of over 300 scientific publications each year and have been reviewed several times. Given such circumstances, we have not attempted comprehensive coverage, but have concentrated on more recent developments and on aspects that struck us as being of particular interest. Among these are the interrelated issues of physical and spectroscopic properties, control of particle morphology and texture, and catalytic activity. By virtue of their great insolubility, LDHs commonly form as submicron particles under conditions of high supersaturation and, therefore, of kinetic rather than thermodynamic control. In addition, the preparation most commonly involves formation of an intermediate trivalent metal hydroxide, which provides numerous sites for heteronucleation. The properties of the materials, including the degree of order of distribution of the different cations, depend critically on the exact mixing procedure as well as on the details of subsequent aging or other treatment, which can make it difficult to compare results from different groups. A variety of physical methods (using spectroscopy, microscopy, and various kinds of diffraction, among other techniques) can give information about crystallite size and perfection on various length scales, and it seems evident that these variables will in turn affect the performance of the LDHs in all their various applications.

C. Reviews

In addition to a recent multiauthor book (1) and volumes of special journal issues (2,3), the hydrotalcites have been the subject of a number of excellent general reviews (e.g., Refs. 4–7), as well as reviews focusing on catalysis (8–10), the intercalation of carboxylic acids and their anions (11), pillared hydrotalcites (12), physical characterization (13), nanocomposites from LDH interleaved with polymers (14), and layered double hydroxides intercalated with metal coordination compounds and oxometalates (15). An older review by Reichle (16) is still deservedly cited for its summary of preparative methods.

Other more specialized reviews are cited as appropriate in later sections.

D. History

These materials were originally approached from two different directions, namely, mineralogy and descriptive inorganic chemistry, and the identity of natural and synthetic materials was not initially obvious. The hydrotalcite and pyroaurite family of minerals had been known since the mid-19th century and was described by 1910 (17,18). While the earliest authors regarded them as mixed hydroxides, by 1920 they were identified as magnesium aluminum and magnesium iron (hydr)oxide carbonates (19). In 1930, Treadwell and Bernasconi (20) reported that the pH for precipitation of Mg^{2+} in the presence of $Al(OH)_3$ was lower than that required for formation of $Mg(OH)_2$. They attributed this to the formation of an adsorption complex but do not seem to have realized the relationship with the mineral materials. Feitknecht in the late 1930s and 1940s published extensively (see, e.g., Refs. 21–23) on the formation of these materials by addition of alkali to solutions containing M(II) and M(III) ions, but erroneously regarded them as double layer materials, in which (for instance) magnesium-rich and aluminum-rich layers alternated. He described the materials as having a double layered structure (Doppelschichtenstruktur), and our own expression *layered double hydroxide* is perhaps a felicitous misinterpretation of this term. Around 1967, however, several groups (see, e.g., Refs. 24–30) correctly identified the layered structure as one containing both kinds of metal ion and, between them, realized the essential identity between the mineral and laboratory materials. Controversy as to the degree of ordering of the metal ions within the layers dates back to this period (27,31) and is still a topic of interest.

The thermal degradation of hydrotalcite and its conversion into spinel were reported in 1944 (32). The clinical antacid use of Mg-Al mixed hydroxides (recognizable with hindsight as hydrotalcite) dates back at least as far as 1960 (33), while the use of hydrotalcite fillers as flame retardants (still the major bulk use) was patented in 1975 (34). Miyata published his ground-breaking survey of the range of formation and anion exchange in LDH in 1973 (35), and the steady exponential growth of interest in these compounds dates from around that time.

II. SYNTHETIC ASPECTS

Layered double hydroxides are highly insoluble materials, at least at the relatively high pH used in their preparation. Moreover, the order of stability of LDHs containing different anions is well known, and the difference between anions in this regard is large. Thus, in principle, all that appears necessary is to mix together soluble precursor salts, and, if convenient, to replace the initially incorporated anion with the one desired. The reality is much more complicated, as it so often is with seemingly simple inorganic materials. Nonetheless, nearly all preparative methods used are variants on these themes of precipitation by base and selective displacement of anions from precursors. For reasons explained later, the exact form of the product, including such important features as particle size and crystallinity and the base strength of the material, is expected to be extremely sensitive to the precise preparative details.

A. LDHs in Nature

LDH materials can be found in nature as minerals or readily synthesized in the laboratory. In nature, they are formed from the weathering of basalts or precipitation in saline water sources. All LDH minerals found in nature have a structure similar to that of hydrotalcite or its hexagonal analog, manasseite, and the majority adheres to the general formula $[M^{II}_{1-x}M^{III}_x(OH)_{2x}]^{x+}(A^{n-})_{x/n} \cdot mH_2O$, where M^{II} represents a divalent metal, M^{III} a trivalent metal, and A^{n-} an anion. Some of these minerals are listed in Table 1. Unlike clays, however, LDHs are not found in large or commercially useful deposits.

B. Direct Synthesis

Direct synthesis is perhaps the single most widely used method of preparation. This method involves nucleating and growing the metal hydroxide layer by mixing an aqueous solution containing the salts of two metal ions, in the presence of the desired anion, and a base; 50% sodium hydroxide is particularly useful for this purpose, since the common ion effect keeps it relatively free of carbonate (see later). It has been demonstrated that LDH materials form in preference to a mixture of the individual metal hydroxides (68) and that, in the case of aluminum as the trivalent cation, they generally do so through an aluminum hydroxide intermediate. Variations of this method include titration at constant or varied pH and buffered precipitation.

    One inherent limitation of this technique is that it can be used only if the desired interlayer anion is at least as tightly held as the counterion in the metal salts used. For this reason, metal chlorides and nitrates are widely used, while sulfates are to be avoided. A more serious limitation is that the anion to be incorporated in the LDH should not too readily form insoluble salts with the constituent cations. LDH phosphates, for example, cannot be prepared by this method.

    Several LDH materials have been prepared while allowing the pH to vary to study their formation (15,69–70). Whether the divalent metal $(M^{2+}) = Zn^{2+}$,

Table 1 Some Minerals Related to Hydrotalcite

Common name	Chemical composition	System*	Class*	Space* group	a*	c*	Refs.
Hydrotalcite	$Mg_6Al_2(OH)_{16}CO_3 \cdot 4H_2O$	Trigonal	Hexagonal-scalenohedral	R3m-	6.13	46.15	36
Manasseite	$Mg_6Al_2(OH)_{16}CO_3 \cdot 4H_2O$	Hexagonal	Dihexagonal-dipyramidal	P63/mmc	6.12	15.324	37
Pyroaurite	$Mg_6Fe_2(OH)_{16} CO_3 \cdot 4.5H_2O$	Trigonal	Hexagonal-scalenohedral	R3m-	6.19	46.54	30, 38, 39
Sjögrenite	$Mg_6Fe_2(OH)_{16} CO_3 \cdot 4.5H_2O$	Hexagonal	Dihexagonal-dipyramidal	P63/mmc	3.113	15.61	30
Stichtite	$Mg_6Cr_2(OH)_{16} CO_3 \cdot 4H_2O$	Trigonal	Hexagonal-scalenohedral	R3m-	6.18	46.38	37, 40–42
Barbertonite	$Mg_6Cr_2(OH)_{16} CO_3 \cdot 4H_2O$	Hexagonal	Dihexagonal-dipyramidal	P63/mmc	6.17	15.52	37, 40
Takovite	$Ni_6Al_2(OH)_{16}CO_3,OH \cdot 4H_2O$	Trigonal	Hexagonal-scalenohedral	R3m-	3.028	22.45	43
Reevsite	$Ni_6Fe_2(OH)_{16} CO_3 \cdot 4H_2O$	Trigonal	Hexagonal-scalenohedral	R3m-	6.614	45.54	38, 44, 45
Desautelsite	$Mg_6Mn_2(OH)_{16}CO_3 \cdot 4H_2O$	Trigonal	Hexagonal-scalenohedral	R3m, R3m-	6.23	46.78	46
Motukoreaite	$NaMg_{19}Al_{12}(OH)_{54}(CO_3)_{6.5}(SO_4)_4 \cdot 28H_2O$	Trigonal		R-3m	9.172	33.51	47, 48
Wermlandite	$Mg_7AlFe(OH)_{18}Ca(SO_4)_2$ $12H_2O$	Trigonal	Ditrigonal-pyramidal	P3c1-	9.303	22.57	49
Meixnerite	$Mg_6Al_2(OH)_{18} \cdot 4H_2O$	Trigonal	Hexagonal-scalenohedral	R3m-	3.0463	22.93	50
Coalingite	$Mg_{10}Fe_2(OH)_{24}CO_3 \cdot 2H_2O$	Trigonal	Hexagonal-scalenohedral	R3c-	3.12	37.4	51

(Continued)

Table 1 Continued

Common name	Chemical composition	System*	Class*	Space* group	a*	c*	Refs.
Chlormagaluminite	$Mg_{3.55}Fe_{0.27}Na_{0.05}Al_{1.93}Fe_{0.07}Ti_{0.01}(OH)_{12}Cl_{1.48}(0.5CO_3)_{0.24} \cdot 2H_2O$	Hexagonal		P6/mcm, P6cm,P-6c2	5.29	15.46	52
Carrboydite	$(Ni,Cu)_{6.90}Al_{4.48}(OH)_{21.69}(SO_4,CO_3)_{2.78} \cdot 3.67H_2O$	Hexagonal		Unknown	18.28	20.68	53
Honessite	$Ni_6Fe_2(OH)_{16}SO_4 \cdot 4H_2O$	Trigonal	Hexagonal-scalenohedral	R-3m	3.083	26.71	45, 54
Woodwardite	$Cu_4Al_2(OH)_{12}SO_4 \cdot 4H_2O$	Hexagonal		Unknown	6	27.3	55–57
Zincowoodwardite	$Zn_{0.47}Al_{0.38}(OH)_2(SO_4)_{0.18} \cdot 4H_2O$	Trigonal	Rhombohedral	P3-, R-3m	3.0364	8.85	58
Hydrowoodwardite	$Cu_{0.5}Al_{0.5}(OH)_2(SO_4)_{0.25} \cdot 4H_2O$	Trigonal	Hexagonal-scalenohedral	R-3m	3.07	31.9	59
Iowaite	$Mg_4Fe(OH)_{10}Cl \cdot 3H_2O$	Trigonal	Hexagonal-scalenohedral	R3m-	3.1183	24.113	60, 61
Hydrohonessite	$Ni_{5.43}Fe_{2.57}(OH)_{16} \cdot 6.95H_2O(SO_4)_{1.28} \cdot 0.98NiSO_4$	Hexagonal		Unknown	3.09	10.8	45, 62
Mountkeithite	$Mg_{8.1}Ni_{0.9}Fe_{1.3}Cr_{1.0}Al_{0.6}(OH)_{24}CO_3)_{1.1}(SO_4)_{0.4}Mg_{1.8}Ni_{0.2}(SO_4)_{1.9}$	Hexagonal		Unknown	10.698	22.545	63
Woodallite	$Mg_6Cr_2(OH)_{16}Cl_2 \cdot 4H_2O$	Trigonal		R-3m	3.102	24.111	64
Brugnatellite	$Mg_6Fe(OH)_{13}CO_3 \cdot 4H_2O$	Hexagonal	Dihexagonal-dipyramidal	P63/mmc	5.48	16	65, 66

* This information can be found at http://webmineral.com.

Source: Ref. 67.

Ni^{2+}, Co^{2+}, or Mg^{2+} and the trivalent metal (M^{3+}) = Al^{3+} or Cr^{3+}, LDH forms at a pH well below that necessary to form the most soluble hydroxide, a fact that led to the initial recognition of LDH as a distinct phase.

Two primary types of pH curves have been observed. One, which has been observed with Cr(III)-containing LDH, consists of a single plateau at a pH lower than that necessary to precipitate $Cr(OH)_3$ or $M(II)(OH)_2$ (71). This corresponds to direct formation of the LDH from solution.

The more common pH curve exhibits two plateaus (4,68,72,73). The first plateau is associated with the formation of the least soluble metal hydroxide, with the second plateau occurring during LDH formation. This type of titration curve has been observed for almost all materials containing Al as the trivalent metal (Cu(II) is an exception) (74). $Al(OH)_3$ is formed first at a pH of ~4, with LDH resulting from conversion of this initial precipitate to the final product. The materials formed in this way are initially in the form of aggregates of poor crystallinity, and presumably they arise through the adherence of divalent metal and additional hydroxide ions to the $Al(OH)_3$ precipitate, followed by rearrangement, heteronucleation of the LDH, and dissolution of the initial $Al(OH)_3$ precipitate.

One common refinement of this technique is precipitation at a nominally constant pH (75–77), sometimes referred to as *coprecipitation*, suggesting that all the cations precipitate simultaneously in a ratio fixed by the starting solution. This method, as discussed in the literature on layered double hydroxides, involves the steady addition of a solution of cations and a solution of base simultaneously to a flask with vigorous mixing, with the relative rates of addition regulated so that the overall pH is steady. It should, however, be noted that precipitation will be occurring in the mixing region around the metal salt inlet, where the pH will be far from constant, and precipitation of the most insoluble metal hydroxide, at least, will be rapid. The final precipitate will be formed at the nominal reaction pH only if at least one of two conditions is satisfied: the rate of mixing is greater than the rate of conversion of initial precipitate to LDH, or the mixture is left stirring long enough for any LDH formed in the mixing zone to redissolve as part of a process of Ostwald ripening. The pH chosen must evidently be higher than that necessary for LDH formation, but it should also be lower than that required for precipitation of the divalent metal hydroxide itself or its basic salts. In addition, it seems likely that higher pH might lead to incorporation of a certain amount of hydroxide as a counterion, on the surface of the particles or even within the interlayer galleries.

A further refinement is the use of a buffer during precipitation. Such a buffer must not itself act as a source of interlayer anions or as a complexing agent toward the metal cations. The simplest buffer of all is an excess of divalent metal during a titration at varied pH. In this technique, the desired metal–metal ratio is obtained by control of the amount of added base. One advantage of this method is that precipitation will occur at a lower pH than in the absence of excess

M(II) ions. Thus there is less risk of incorporation of an uncertain or unwanted amount of hydroxide anions and less rapid uptake by the reaction mixture of adventitious carbon dioxide.

Such uptake of carbon dioxide, giving rise to carbonate, is a major nuisance in LDH chemistry. Carbonate is among the most strongly held of anions within the LDH lattice, and it is very efficiently incorporated during initial formation and workup. Carbonate can result from the absorption of carbon dioxide from the atmosphere into solution, especially under basic pH conditions. It is therefore desirable to exclude carbon dioxide from the atmosphere within the reaction vessel if another anion is to be intercalated. We recommend that the precipitation be carried out under a stream of carbon dioxide–free air or other gas and that all water used in preparation or workup be freshly deionized or boiled and purged. For all critical experiments and applications, we further recommend collection of the infrared spectra of materials to be used, preferably using a normalized absorbance or log transmission scale, as a way of semiquantitatively monitoring carbonate contamination.

C. Coprecipitation in Nonaqueous Solutions

The majority of precipitation reactions have been carried out in an aqueous solution. Gardner et al. (78) describe the formation of LDH by coprecipitation in various solutions of alcohols to form mixed alkoxide/inorganic anion–intercalated LDH materials. Dispersion of this material into an aqueous solution overnight leads to hydrolysis of the alkoxide anion and the formation of a transparent LDH suspension that, once dried, forms a thin film. This method of preparation therefore can be used to form pillared anionic clays that can be used as precursors for the preparation of transparent LDH films.

D. Anion Exchange Within an LDH Precursor

The interlayer anions in LDH are exchangeable, with the order of preference (6,69,79) being

$$NO_3^- < Br^- < Cl^- < F^- < OH^- < MoO_4^{2-} < SO_4^{2-} < CrO_4^{2-} <$$
$$HAsO_4^{2-} < HPO_4^{2-} < \text{Naphthol Yellow}^{2-} < CO_3^{2-}$$

This order is presumably the consequence of charge, charge density, and hydrogen bonding. A very useful consequence is that a more weakly held anion can be quantitatively replaced by one more strongly held, and this is simply accomplished by stirring the LDH containing the anion to be replaced in a solution containing an excess of the replacement. LDH nitrate and chloride are commonly used as starting materials in this procedure. Anions of an organic and inorganic nature having various sizes, shapes, and charges have successfully been intercalated into LDH using this simple method. We, however, have found, much to our initial surprise, that it is extremely difficult to completely eliminate the characteristic signal of nitrate, even when using the much more strongly held carbonate anion (80a). We speculate that some nitrate monoanions are left isolated within the lattice, and we wonder if the same is true of chloride, which of course lacks such a telltale signal.

There are few important constraints on this technique. It is difficult to use for the most weakly held anions, such as iodide and perchlorate. The minimum area per unit charge of the incorporated anions must not be greater than the area per unit charge of the LDH sheet, otherwise the desired material cannot form. This constraint is more rigorous than it appears, since some anions (such as metal cyanide complexes) have hydrogen-bonding requirements that make their demand for space greater than might have been expected (73,80b,81).

The anion exchange method is widely used in the formation of pillared clays (15,81–84). *Pillaring* refers to the effect that a bulky, highly charged anion imparts on the structure of the LDH. Ideally, the anion forms a perpendicular column between two metal hydroxide layers, increasing the surface area and pore volume associated with the LDH. Polyoxometallates are the most common pillaring agents for anionic clays (85,86). These materials are chosen because they are robust, highly charged species with important catalytic properties (see Section VI.B). Their strength helps impart large gallery heights to the LDH, and the charge helps increase the amount of free space between anions within the interlayer. Preparation of pillared anionic clays can be difficult because of side reactions, the dependence of the nature of the pillar on conditions, and perhaps also the large gallery height itself (6). For this reason, some workers have chosen to introduce the polyoxometallate pillars indirectly, by displacement of a tall organic pillar, such as terephthalate (87,88).

Another important variation of the anion exchange reaction occurs through the treatment of LDH carbonates with acids, especially carboxylic acids (89,90). It may seem surprising that LDH carboxylates can be prepared by displacement of the more strongly held carbonate anion, but the driving force comes from the ease of protonolysis of carbonate to give carbon dioxide and water:

$$Mg_4Al_2(OH)_{12}CO_3 + 2\ HX = 2Mg_2Al(OH)_6X + H_2O + CO_2$$

This method is particularly useful for anions of weak acids HX, since, evidently, the pH must not be allowed to drop to the point where the hydroxide layer itself begins to dissolve. This class of LDH, like the ones containing polyoxometallate complexes, has been studied for their pillaring effect. Long-chain carboxylic acids have been incorporated into these materials, hence making varying gallery heights possible, with the height directly dependent on the length of the alkyl chain (91,92). One application of this technique utilizes the pore size produced and the anion-scavenging abilities of LDH to snare potentially hazardous anions from the environment. Pillared LDHs and their uses are discussed further in Section VI.B.

A further variation of the anion exchange reaction, in which the LDH derivative of an anionic surfactant such as dodecylsulfate is treated with the cationic surfactant salt of the desired anion, is used by Crepaldi et al. (93). The cationic and anionic surfactants interact to form neutral micelles, leaving the LDH free to capture the desired anion. Since the interlayer spacing in the starting materials for this procedure is already quite large, this method offers particular promise for the incorporation of bulky anions.

E. Glycerol-Effected Exchange

This method is still yet another variation of the exchange reaction previously discussed. This reaction, however, introduces glycerol or some other glycol to a solution of LDH carbonate, to promote uptake of carboxylate anions into the interlayer by first expanding the distance between the hydroxide layers. The exact mechanism by which this reaction takes place is incompletely understood. One theory is that the glycerol or glycol penetrates the interlayer, causing it to swell. This has the effect of decreasing the hydrogen-bonding strength that exists between the carbonate anion and hydroxide layers. The reduced bond strength lowers the carbonate's resistance to replacement by a different anion. The second theory is that a negatively charged glycerolate anion reacts with the carbonate of the LDH, which is then exchanged for the pillaring carboxylate species. In reality, it is probably a combination of processes that occur (11).

This reaction has been carried out using two different reaction conditions that vary only in the temperature of the glycerol treatment. The exchange reaction carried out using glycerol at room temperature shows only a slight increase in basal spacing associated with carboxylate anion uptake (94). The same reaction, except at an elevated glycerol temperature, gives material with a much larger increase in basal spacing (95). This difference is most likely associated with a monolayer, compared to bilayer, arrangement of the carboxylate anion.

F. Preparation from Oxides and Hydroxides

There are two preparatory procedures related to LDH formation from oxides and hydroxides. The first is presumably what unintentionally occurs in antacids (96). This reaction involves the hydration of metal oxides and/or metal hydroxides in the presence of an anion. The other form utilizes a memory characteristic of LDH materials. This method is based on calcining LDH materials containing a thermally labile interlayer anion and then rehydrating the resulting amorphous oxides in the presence of the desired anion (97–99).

The hydration of metal oxides and hydroxides has been used to prepare LDHs, such as the Mg/Al LDH in antacids. The reaction presumably proceeds through a dissolution/precipitation process, probably with heteronucleation on one of the original hydroxide phases. LDH formation from the conversion of oxides or hydroxides will result in a surplus hydroxide group for every trivalent metal incorporated into the LDH. This strongly basic interlayer hydroxide can then react with carbon dioxide or some other weak acid.

A related method uses what is sometimes referred to as the *memory effect* of LDH. This technique consists of heating an LDH containing a thermally labile anion. The resulting amorphous oxide is then rehydrated in the presence of the desired replacement anion to form a new LDH. Two factors of importance in this method are the choice of starting material and the temperature of calcination. The starting material must contain a thermally labile anion, and the temperature of calcination must be controlled so as to avoid excessive heating that would result in the formation of spinel, which is resistant to rehydration.

G. Preparation by Sol-Gel Techniques

The term *sol-gel* is derived from the physical properties through which a reaction proceeds (100). This process involves formation of a mobile colloidal suspension (sol) that then gels (gel) due to internal cross-linking. Materials prepared by this technique exhibit good homogeneity, relatively good control of stoichiometry, and high surface and high porosity characteristics. The preparation of LDH is the result of the hydrolysis and polymerization of a solution of metal alkoxides (101); on the atomic level this is another example of preparation from hydroxides. The alkoxides are first dissolved in an organic solvent and refluxed. To this solution water is slowly added, causing cross-linkage to occur. Mg/Al LDH, the most common of the LDHs formed by this technique, has been prepared with Mg/Al ratios close to 6, compared to the more common ratios of 2 and 3 for materials attained by other preparatory techniques.

H. Preparation by Reaction of Metal Oxides/Hydroxides with Salts

Metal ion deposition onto an oxide surface is of importance from colloidal chemistry, surface science, and geochemical points of view. Hydrotalcite-like materials have been shown to form as a result of the impregnation of a metal ion, from a metal salt solution, onto a metal oxide/hydroxide surface (102–107). The first step of the mechanism probably is the formation of a charge on the metal hydroxide surface due to dissociation. Caillerie and Clause (104) observed LDH formation in γ-alumina impregnated with Ni(II) and Co(II) ions, while Mascolo and Marino (103) observed that alumina gel incorporated Mg(II) ions to form $[Mg_{1-x}Al_x(OH)_2]^{+x}[xOH_{(0.81-x)}H_2O]$, with x in the range 0.23–0.33. Nayak et al. found that alumina gel absorbed Li(I) ions from LiOH solution to give the hydrotalcite analog $LiAl_2(OH)_7 \cdot 2H_2O$ (102).

I. So-Called Homogeneous Precipitation

Homogeneous precipitation is a process whereby one of the reagents in formation of a highly insoluble material is slowly generated in situ, thus avoiding massive supersaturation while mixing reagents, reducing the number of nuclei formed, and thereby generating larger, more well-crystallized product particles (108,109). This rationale has been applied to the formation of LDH carbonates by slow hydrolysis of urea, which generates base (ammonia) and hydrogen carbonate in equilibrium with ammonium carbonate (110). The LDH were indeed formed as euhedral platelets (111), but it seems clear from pH titration studies (112) that the true mechanism was slow re-solution of an initial aluminum hydroxide precipitate. It was, however, a homogeneous reaction in that all reagents necessary for LDH precipitation are available, either directly or due to an internal reaction with the starting reaction mixture. We have ourselves applied this method to the precipitation of Co/Al LDH, where it gave rise to hollow aggregates of platelets, which can be ascribed to heteronucleation on initially formed aluminum hydroxide that subsequently redissolves (112). It should in principle be possible to generate Zn/Cr LDH by true homogeneous precipitation, but in our hands complex formation between metal cations and hydroxide precursor has always interfered.

J. Precipitation via Aluminate

All of the foregoing routes to LDHs containing aluminum as the trivalent metal have been shown to proceed via solid $Al(OH)_3$, leading to obvious problems in the preparation of well-controlled product. Recently, research has been performed that utilizes the amphoteric nature of $Al(OH)_3$ to precipitate LDH directly from solution without $Al(OH)_3$ nucleation.

In the aluminate route to LDH, aluminum salt (113,114), aluminum hydroxide (115,116), or aluminate (117–119) is initially treated with the required amount of base (e.g., six hydroxides for each aluminum if the product is to be of type $M(II)_2Al(OH)_6X$), followed by addition of M(II), either in a salt or a carbonate form (113–115,117,119) or in the form of an oxide or hydroxide (116,118,120,121). The pH graph observed for this titration shows endpoints related to the formation of $Al(OH)_3$, aluminate, and LDH. This method of formation yields LDH directly from solution, with no intervening solid phase. The product has an increased degree of order in the metal hydroxide layer and slightly smaller surface area when compared to LDH prepared by other methods, as shown by infrared spectroscopy, powder X-ray diffractometry, scanning electron microscopy, and MAS ^{27}Al and MAS [35]Cl NMR (113,114,122).

K. Preparation from Metals

The formation of LDH from metal starting materials, as compared to preparation from metal salts, involves an additional oxidation step and is rarely used. The most common material formed in this manner is "green rust" (123,124). This occurs as a result of the oxidation of metallic iron to Fe^{II} and Fe^{III} in water and is important in corrosion (125,126). Reactions of this kind could, however, be useful, when carried out in D_2O, for the preparation of layered double deuteroxides for special purposes, such as neutron diffraction studies (127).

L. Preparation by Oxidation

Various metal compositions of LDHs have been prepared by intentionally introducing air or oxygen during the aging phase of an oxidizable metal hydroxide. Zeng et al. used this method of preparation to produce $Co^{II}Co^{III}$-, $Mg^{II}Co^{II}Co^{III}$-, and $Co^{II}Co^{III}Al^{III}$-LDHs (128–136). Delmas et al. prepared $Ni^{II}Co^{III}$-LDH by oxidizing Co^{2+} to Co^{3+} with perchloric acid or peroxide (137,138). Rives et al. also prepared Mg-Mn LDH by oxidizing Mn^{2+} into Mn^{3+} in the solid state (139). Finally, Markov et al. used this method for the preparation of CuCo hydrotalcite nitrate (140). Very interestingly, Hansen et al. synthesized Co^{2+} and Fe^{3+} solid mixture during the coprecipitation at a constant pH of 6.60 by the addition of 0.5M sodium hydrogen carbonate (141).

M. Postpreparative Treatments

LDHs formed under conventional methods result in poorly crystalline materials with disordered metal hydroxide layers. In circumstances where increased uniformity and crystallinity are desired, the LDH is often altered by postpreparative treatments, which consist of aging the precipitate at or often above ambient temperature, for instance, by heating to gentle reflux. The aging presumably occurs through Ostwald ripening (142–144), in which larger and more perfect crystallites grow at the expense of smaller particles in solution by dissolution/reformation processes. In addition to aging, LDH materials are washed thoroughly following formation to remove any excess ions in solution that could continue to react with the material.

Hydrothermal (4,6,89,145) and microwave (146,147) treatments are techniques used to process LDH formed by one of the preparatory methods mentioned previously rather than an individual method of preparation. Although the excitation method differs between these two techniques, the outcome is similar: improved crystallinity of the LDH material. Hydrothermal treatment favors dissolution and recrystallization of the LDH by heating the solution during LDH formation. The chosen temperature for this procedure is controlled by the LDH desired (e.g., $Mg_2Al(OH)_6A^-$ forms $Mg_3Al(OH)_8A^-$ and $Al(OH)_3$ at temperatures above 120°C). The exact nature of microwave interaction with the reactants during LDH formation is unknown. It is believed, however, that this interaction is one of energy transfer inducing rapid heating through resonance and relaxation.

The result of microwave treatment is well-crystalline material in less time than necessary for conventional hydrothermal treatments.

N. LDH Formation Precautions

In addition to the exclusion of carbon dioxide, it is necessary to exclude oxygen from reactions where cations of the metal hydroxide layer are easily oxidized under basic conditions. LDHs of this type include manganese, cobalt (131) and iron in the divalent oxidation state, and vanadium in the trivalent state. Reactions involving these metal ions have formed LDHs with, for instance, Co^{2+}/Co^{3+} and Fe^{2+}/Fe^{3+} metal content, with the latter, known as *green rust*, being discussed earlier in Section II.J.

Incorporation of iron cyanides into LDH material requires a certain degree of precaution as well (84). LDH materials containing magnesium, when exposed to a ferrocyanide solution for an extended period of time, form the cubic material $M^{1+}MgFe(CN)_6$, where $M^{1+} = K^+$ or NH_4^+, by altering the LDH composition (such a possibility should be borne in mind whenever M(II) or M(III) forms extremely insoluble salts with the interlayer anion). In addition, LDH ferricyanides photolyze and therefore must remain unexposed to light.

O. Monitoring and Quality Control

The formation of LDH materials is monitored by several techniques, most of which will be discussed in further detail in Section III. One such technique is the monitoring of the pH. As previously discussed, study of the pH can assist in determination of LDH formation and the mechanism through which the LDH is formed. Use of analytical techniques such as infrared spectroscopy, powder x-ray diffraction, and nuclear magnetic resonance spectroscopy can give information about such matters as the quality of LDH formed, the identity and packing of the interlayer anion, and the internal order within the metal hydroxide layer.

P. Industrial Production

The preparation of LDHs has been the subject of over 50 patents. The claims are specific variants of the general methods previously described, including coprecipitation (148), anion exchange (149), preparation from oxides and hydroxides (150–152), reaction of metal oxides or hydroxides with a salt (152–154), and precipitation via aluminate (115–118,120,155). The principal material described is Mg/Al LDH carbonate; however, other anions, such as those composed of heavy metal complexes (120,149), alcoholates (156), anionic surfactants (148), phosphates (148,157), and halides (115,158a), have also been claimed. LDH can also be produced by controlled addition of $Ca(OH)_2$ and $AlCl_3$ to seawater (158b).

III. STRUCTURE AND CHARACTERIZATION OF LDH

A. Structural Issues

1. Basic Structure

The layered structure of LDH is closely related to that of brucite, $Mg(OH)_2$. In a brucite layer, each Mg^{2+} ion is octahedrally surrounded by six OH^- ions, and the different octahedra share edges to form an infinite two-dimensional layer. The octahedron is slightly flattened, with the distance between OH^- neighbors on the same side of the layer being 0.314 nm (this is the crystallographic a or b spacing, i.e., the length AB in Fig. 1) (4,6,159). However, the distance between OH^- neighbors on opposite sides of the sheet is only 0.270 nm, i.e., the length CD in Figure 1. The bond length of Mg-O is ca. 0.207 nm, and the repeat distance, or layer thickness, is 0.478 nm, as the layers are stacked one on top of another (4,6,159).

Partial replacement of Mg^{2+} ions by Al^{3+} gives the brucite-like layers a positive charge, which in hydrotalcite itself is balanced by carbonate anions, located in the interlayer region (gallery) between the two brucite-like layers (Fig. 1). This gallery also contains water molecules, hydrogen bonded to layer OH and/or to the interlayer anions. The electrostatic interactions and hydrogen bonds between the layers and the contents of the gallery hold the layers together, forming the three-dimensional structure, as shown in Figure 1 (4,6).

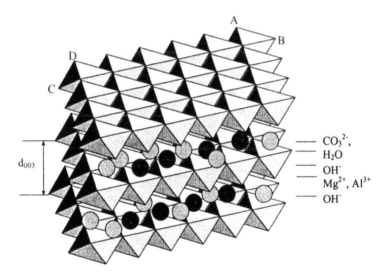

FIGURE 1 Schematic representation of the hydrotalcite three-dimensional structure.

Figure 2 displays a view of the ordered octahedral sheet in the brucite-like layer. Each alternate triangle (shown shaded) contains a central cation. The unit-cell projection in the *001* plane is shown as a diamond with arrows. Normally in LDH, the parameter a or b is about 0.30–0.31 nm, corresponding to the distance between two nearest OH^- ions in the same side layer or between two nearest metal cations (refer to the length AB in Fig. 1). This has been observed repeatedly by powder X-ray diffraction and more directly by atomic force microscopy (AFM) imaging, which shows two-dimensional periodicity with the unit lattice of a = 0.31 ± 0.02 nm, b = 0.31 ± 0.02 nm, and α = 58 ± 3° for the surface of $[Mg_6Al_2(OH)_{16}](CO_3)_{1/2}Cl \cdot 2H_2O$ in contact with an aqueous solution of 0.1M $NaSO_4$ (160).

Unlike the good planarity of brucite layer $(Mg(OH)_2)$, the LDH brucite-like layer is often corrugated due to the cations ordering. Bellotto et al. observed this phenomenon by radial distribution function analyses from XRD and EXAFS data. They found that the cation ordering can proceed when the Mg—O (0.202 nm) and Ga—O (0.195–0.198 nm) bond lengths in Mg_2Ga—CO_3—LDH are not much different, in contrast with the case of Mg_2Al—CO_3—LDH, due to the larger difference between Al—O (0.190 nm) and Mg—O (0.211 nm) bond lengths. The ordering process appears to provoke a corrugation of the layer, which increases the cation–cation distances and thus minimizes repulsion (161).

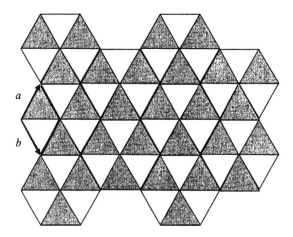

FIGURE 2 View of the ordered octahedral sheet in a brucite-like layer showing the unit-cell projection as a diamond with side length of *a* or *b* in the *001* planed. There is one cation in each shaded triangle while each blank triangle, corresponds to a void tetrahedron.

The brucite-like layers can stack on top of one another with either rhombo-hedral (3R) or hexagonal (2H) sequence (30) (see Sec. III.A.4). For hydrotalcite ($Mg_6Al_2(OH)_{16}CO_3 \cdot 4H_2O$), which has a 3R stacking sequence, the parameters of the unit cell are $a = 0.305$ nm (Fig. 2) and $c = 3d_{003} = 2.281$ nm (Fig. 1), where d_{003} is the interlayer spacing, 0.760 nm (Fig. 1) (30). It is thus inferred that the spacing occupied by the anion (so-called gallery height) is *ca.* 0.280 nm after allowing for a thickness of 0.480 nm for the brucite-like layer. In addition, the average M—O bond length is *ca.* 0.203 nm, and the distance between two nearest OH^- ions in the two opposite side layers (refer to the length *CD* in Fig. 1) is 0.267 nm (30), shorter than a (0.305 nm) and indicative of some contraction along the c-axis.

The isostructural replacement of Mg^{2+} with Al^{3+} brings about excess positive charges into the brucite-like layer. The charge density can thus be estimated from the formula

$$C_d = xe/(a^2 \sin 60°)$$

or

$$C_d \approx 12.0xe/\text{nm}^2$$

for $a = 0.31$ nm. For example, C_d is 4.0 and 3.0e/nm^2 in Mg_2Al—LDH ($x = 0.33$) and Mg_3Al—LDH ($x = 0.25$) respectively.

Outstanding issues regarding LDH compositions and structures involve (a) the range of divalent and trivalent ions possible; (b) the inclusion of various anions; (c) whether the different metal ions are distributed at random (as the previous paragraph implies), or form some kind of regular arrangement, which could lie anywhere between short range statistical correlation and the regularity of a long-range superlattice; and (d) the stacking styles of brucite-like layers in the c-axis. These issues are discussed further later.

2. Within the Hydroxide Layers

There are a number of combinations of divalent and trivalent cations that can form LDHs. Besides Mg^{2+}, the available divalent ions can be Ni^{2+}, Co^{2+}, Zn^{2+}, Fe^{2+}, Mn^{2+}, Cu^{2+} (4,6,16,162), as well as Ti^{2+} (163), Cd^{2+} (35,164), Pd^{2+} (35,163,165) and Ca^{2+} (91,166–168). Similarly, the trivalent ions can be Al^{3+}, Ga^{3+}, Fe^{3+}, Cr^{3+}, Mn^{3+} (4,6,139), Co^{3+} (129–136), V^{3+} (169–171), In^{3+} (172), Y^{3+} (173), La^{3+} (163,165,174), Rh^{3+} (163,165,174–176a,176b), Ru^{3+} (174,176b), Sc^{3+} (177), etc. For these ions, the only requirement is that their radii be not too different from those of Mg^{2+} and Al^{3+}. Table 2 lists the radius data of some divalent and trivalent cations that have been successfully incorporated into the brucite-like layer. To a first approximation, if we assume close stacking of OH^- ions in an octahedron (refer to the length *CD* in Fig. 1), there

Table 2 Ionic Radii of Some Cations with Coordinate Number of 6

M^{2+}	Radius (nm)	M^{3+}	Radius (nm)
Fe	0.061	Al	0.054
Co	0.065	Co	0.055
Ni	0.069	Fe	0.055
Mg	0.072	Mn	0.058
Cu	0.073	Ga	0.062
Zn	0.074	Rh	0.067
Mn	0.083	Ru	0.068
Pd	0.086	Cr	0.069
Ti	0.086	V	0.074
Cd	0.095	In	0.080
Ca	0.100	Y	0.090
		La	0.103
V^{4+}	0.058		
Ti^{4+}	0.061	Li^+	0.076
Sn^{4+}	0.069	Na^+	0.102
Zr^{4+}	0.072		

Source: Ref. 186.

is a hole inside with radius of about 0.070 nm (M–O bond length minus radius of OH^- ion, i.e., $0.203 - 0.267/2 = 0.070$ nm). Therefore, a cation with radius not far from 0.070 nm can be incorporated into the brucite-like layer.

Many common layered double hydroxides consist of such metal ions, as shown in Table 2. Larger ions, such as Mn^{2+}, Pd^{2+}, Cd^{2+}, Ca^{2+}, Y^{3+}, and La^{3+}, can be incorporated into the brucite-like layer by distorting the close-stacking configuration to some extent, i.e., puckering, or combining with other divalent or trivalent cations, as in Mg/Cd/Al and Mg/Al/La systems (163,165,174). An extreme case is Ca_2Al-LDHs (35,164), in which Ca is distorted to be 7-coordinated while Al is still kept 6-coordinated (178).

Although not as common as trivalent cations, some tetravalent cations, such as V^{4+}, Ti^{4+}, Zr^{4+} and Sn^{4+}, have been introduced into the brucite-like layers to replace part of the trivalent cations in LDH compounds (179–182). In a very similar way, monovalent cations, Li or $Li_{1-x}Na_x$, together with Al, can form a similar series of LDH compounds (183–185), with atomic ratio Al/Li or Al/(Li + Na) fixed at 2 and intercalation with various counteranions. However, we focus mainly on $M^{II}M^{III}$-LDH compounds in this review.

In principle, all divalent and trivalent cations (as well as monovalent and tetravalent ones) that meet the radius requirement could form LDHs. For example, V^{3+}, a very air-sensitive species, has been successfully incorporated into the brucite-like layer to form LDHs (169–171). Rives et al. synthesized $MgMn^{2+}Mn^{3+}$-LDH through oxidation of Mn^{2+} to Mn^{3+} in the solid state (139). Xu and Zeng et al. prepared several Co^{3+}-containing LDHs, such as $MgCo^{II}Co^{III}$- and $Co^{II}Co^{III}$-LDHs (129–136). In 1997, Fernandez et al. introduced Y^{3+} into the brucite-like layer (173). However, Cr^{2+} and Ti^{3+} have not yet been incorporated into LDH, being too strongly reducing under basic conditions. Some other potential ions, including Ni^{3+} (0.056nm), Ir^{3+} (0.068nm), and Tl^{3+} (0.089nm) (186), have not yet been reported. In view of the cation size, all the lanthanide trivalent ions (0.086–0.103 nm) could also in principle replace Al^{3+}, at least in part.

Although LDHs ($[M_{1-x}^{II}M_x^{III}(OH)_2]^{x+}(A^{n-})_{x/n} \cdot mH_2O$) are reported to exist for values of x from 0.1 to 0.5 (reviews), many experimental results indicate that it is more usual to obtain pure LDHs for 0.2–0.33. For x values outside this range, hydroxides or other compounds may be formed. Due to the electrostatic repulsion between positive charges, M^{III} ions should ideally not be adjacent in the brucite-like layer, but remain apart. This requires x to be no more than 1/3. This is analogous to Lowenstein's rule for aluminosilicates (187). On the other hand, low values of x will lead to a high density of M^{II} in the layer and fewer anions in the interlayer, reducing the barrier to formation of $M^{II}(OH)_2$. As a consequence, the value of the unit-cell parameter a for MgAl-CO_3-HT is a function of x (5,188). The parameter a linearly decreases with increasing x in the range of 0.2–0.33 (189). It is believed that this effect is a result of the stronger interactions between the central cations and the coordinated OH^- anions as x increases (188). However, there are exceptional examples of LDHs with x values outside of this range. For example, in $Co_{1-x}Al_x$-LDH, x has been reported to be up to 0.67 (190,191).

It is also possible to precipitate multimetal ions to obtain multicomponent LDHs. For example, Morpurgo et al. synthesized several multication LDHs, such as CuZnCoAlCr- and CuZnCoCr-LDHs (192–194). This procedure suggests a general way to synthesize a number of LDHs as the precursors of mixed oxides with versatile functionalities, especially through incorporation of some rare earth metal cations (with radius from 0.085 to 0.100 nm), as has been done by Perez-Ramirez et al. (163,165). However, the overall ratio of divalent to trivalent cations should be kept in the range of 0.2–0.4.

3. Between the Layers

a. Nature of the Anions

As for the anions located in the interlayer gallery, the choice is much more versatile. There is almost no limitation to the nature of anions in the LDHs as long as the anions do not abstract the metal ions from the hydroxide layer and have a sufficient charge density in one cross section, i.e., not much less than

$3.0e/nm^2$. Up to now, many kinds of anions have been reported to be included in the literature:

1. *Common inorganic anions:* halides (X^-), CO_3^{2-}, NO_3^-, OH^-, SO_4^{2-}, $Al(OH)_4^-$ or $H_2AlO_3^-$ (128,195), PO_3^-, PO_4^{3-}, HPO_4^{2-}, $H_2PO_4^-$, $P_2O_7^{2-}$ (164,196–198), AsO_3^- (199), borate and tetraborate (200,201), ClO_4^- (202), TcO_4^-, ReO_4^- (203), MnO_4^- (204,205), CrO_4^{2-}, $Cr_2O_7^{2-}$ (199,206,207), MoO_4^{2-} (208), HVO_4^{2-}, VO_4^{3-} (199,205), silicate anion (205,209), and C_{60} anion (210)

2. *Organic anions:* carboxylates (11,89,91), dicarboxylates (11,91,211,212), benzenecarboxlates (89,91,213–215), alkylsulfates (91,211,212), alkanesulfonates (11,91,211,212), chlorocinnamates (211), *t*-butanoate anion (216), glycolate (217), glycerolate $(CH_2OH—CHOH—CH_2O^-)$ (95), and organic dyes (218)

3. *Polymeric anions:* poly(vinylsulfonate), poly(styrenesulfonate), and poly(acrylate) (219–222), polyaniline (223), ionized poly(vinyl alcohol) (224), poly(ethylene glycol) (225), and even polystyrene oligomer anion (226)

4. *Complex anions:* $CoCl_4^{2-}$, $NiCl_4^{2-}$ (227,228), $IrCl_6^{2-}$ (229), $Fe(CN)_6^{4-}$, $Fe(CN)_6^{3-}$ (230–233), $Mo(CN)_8^{4-}$ and $Mo(CN)_8^{3-}$, $Ru(CN)_6^{4-}$ and $Ru(CN)_6^{3-}$ (234,235), $Co(CN)_6^{3-}$ (236), and $MoO_2(O_2CC(S)Ph_2)_2^{2-}$ (237)

5. *Macrocyclic ligands and their metal complexes:* porphyrin and phthalocyanine derivatives (238,239) and their Cu^{2+} (238,240), Co^{2+} (241,242), Mn^{3+} (243), and Zn^{2+} (244) complexes;

6. *Iso- and heteropolyoxometalates (POMs):* $Mo_7O_{24}^{6-}$ (208,214,245,246), $W_7O_{24}^{6-}$ (245,247) and $H_2W_{12}O_{40}^{6-}$ (245), $V_{10}O_{28}^{6-}$ (214,248,249), Keggin anion α-$(XM_{12}O_{40})^{n-}$ (X = H, Si, P, Ge, etc., M = Mo, V, W, etc.), e.g., $PMo_{12}O_{40}^{3-}$ (250), $PW_{12}O_{40}^{3-}$ (211), $PW_6Mo_6O_{40}^{3-}$ (179), $PV_3W_9O_{40}^{6-}$ (88); α-$(XM_{11}ZO_{40})^{n-}$ (X, M are the same as before, and Z can be the metal ions in the first transition series) (251–253)

7. *Biochemical anions:* (a) various amino acids (254–259); (2) DNA with 500–1000 base pairs (260,261); (3) CMP, AMP, GMP, ATP, ADP, and related species (262,263)

b. Orientation of the Anions

The various anions are very different in their structure, dimensions and charges; however, they are all hosted between the brucite-like layers. This feature is quite generally reflected in the interlayer spacing (usually referred to as d_{003}, although this label is valid only for 3R stacking) or in the gallery height (d_{003} minus the thickness of the brucite-like layer). In general, the anions are oriented to some extent in order to maximize the interaction with their surroundings. For example, planar CO_3^{2-} anions are usually positioned in parallel to the brucite-like layer in order that the three oxygen atoms can interact well with the layer hydroxyl groups by forming hydrogen bonds. This orientation also maximizes the electrostatic interaction between CO_3^{2-} anions and the positively charged layers since the gallery height is minimized, corresponding to the thickness of the planar CO_3^{2-}. However, the CO_3^{2-} changes from flat lying (D_{3h}) to tilted (C_{2v}) in the interlayer when trivalent cation content (x in $Ni_{1-x}Al$-LDH) increases to 0.44, as shown by the change of the IR spectra (264).

Although nitrate has the same shape as carbonate, its lower charge density leads to deviations from the flat-lying structure even at low values of x. From changes in the interlayer spacing, the XRD patterns and characteristic IR bands in $Mg_{1-x}Al_x(OH)_2(NO_3)_x$-LDH ($x = 0.18$–0.34), Xu and Zeng have proposed that the arrangement of NO_3^- changes from a flat-lying model at low x values (0.18–0.26) to an alternating upper-lower gallery surface model at high x values (0.26–0.34) (265).

For tetrahedral anions, such as SO_4^{2-}, MoO_4^{2-}, CrO_4^{2-}, PO_4^{3-}, ClO_4^-, $CoCl_4^{2-}$, and $NiCl_4^{2-}$, there are two possible ideal configurations. The pyramidal configuration (with its C_3 axis perpendicular to the hydroxide layer) has three oxygen (or chlorine) atoms closer to hydroxyl groups in one brucite-like layer and the fourth one pointing toward the opposite hydroxyl plane. The other configuration (with its C_2 axis perpendicular to the hydroxide layer) places two oxygen atoms toward the opposite hydroxyl planes in each of two adjacent brucite-like layers, giving rise to a smaller height.

It has been found from the gallery height that ClO_4^- anions in LDH-ClO_4^- adopt the former configuration, while SO_4^{2-} anions employ the latter one (266). It has also been observed that the relative humidity in the environment affects the interlayer spacing of LDH-SO_4 and LDH-ClO_4. For instance, under the condition of relative humidity higher than 50%, the repeat spacing increases from 0.89 and 0.92 nm to 1.08 and 1.17 nm for LDH-SO_4 and LDH-ClO_4, respectively, as the result of incorporation of hydrogen-bonded water into the gallery.

Octahedral complexes, such as $Fe(CN)_6^{3-}$, adopt the configuration shown in Figure 3, as proposed by Kikkawa and Koizumi (267) from the gallery height. Braterman et al. (232) have used this model, and its implication of hydrogen bonding between ligand cyanide nitrogen and the pendant OH groups of the layer, to explain the perturbation of the vibrational spectrum from that expected for ideal octahedral symmetry.

Some macrocyclic ligand anions and macrocyclic ligand metal complexes have also been intercalated in LDHs. For example, 5,10,15,20-tetra(4-sulfonatephenyl)porphyrin anion is inserted into the interlayer by replacing Cl^- in a MgAl-HT (238). The gallery height (1.76 nm) is very close to the side length of the almost perfect square molecule of this porphyrin (1.80 nm), and thus the

FIGURE 3   Intercalation model of octahedral complexes.

porphyrin anion is intercalated with its molecular plane at or close to perpendicular to the brucite-like layers. An anionic cobalt phthalocyanine complex species has been also inserted into LDHs (268). Numbers of other macrocyclic complex anions have been successfully intercalated into LDHs with the molecular plane perpendicular to the hydroxide layer (15).

For polyoxometalates (POMs), the orientation is dependent on the anion structure, which varies with solution pH. For example, vanadate can have ten forms at different pHs (249):

pH = 1–3   Decavanadate: $V_{10}O_{26}(OH)_2^{4-}$, $V_{10}O_{27}(OH)^{5-}$, $V_{10}O_{28}^{6-}$

pH = 4–6   Metavanadate: $VO_2(OH)^-$, $V_3O_9^{3-}$, $V_4O_{12}^{4-}$

pH = 8–11   Pyrovanadate: $VO_3(OH)^{2-}$, $HV_2O_7^{3-}$, $V_2O_7^{4-}$

pH > 12   Vanadate: VO_4^{3-}

Fortunately, pH can be controlled well and different species could be selectively intercalated. Similar cases include MoO_4^{2-}, WO_4^{2-}, NbO_4^{3-}, TaO_4^{2-}, and their associated polyanions as well as Keggin-type anions α-$(XM_{12}O_{40})^{n-}$ (6,85,269–271). LDHs with POMs and some complexes anions intercalated are sometimes known as pillared LDHs due to the large heights and multiple negative charges carried by these anions, which result in micropores in the interlayers.

The orientations of organic anions are related to the anionic concentration and reaction temperature. For aliphatic monocarboxylates and sulfonates, there are three possible cases: monolayer (strictly speaking, interpenetrating bilayer), simple bilayer (11,272), and partial overlap packing, as shown in Figure 4. The monolayer packing is readily formed when the anion/M^{3+} ratio is close to unity, while the bilayer packing is favored by the presence of excess carboxylate, which is incorporated as the free acid even at high pH (273). Variable partial overlap was noted by Clearfield et al. in the case of alkyl sulfates (274), while our own group has found controlled partial overlap, imposed by the nonlinear geometry of the alkenyl chain, in LDH cis-oleate (Fig. 4(C)); we also find partial overlap with stearate (273). Normally, the carboxylate group is anchored to the brucite-

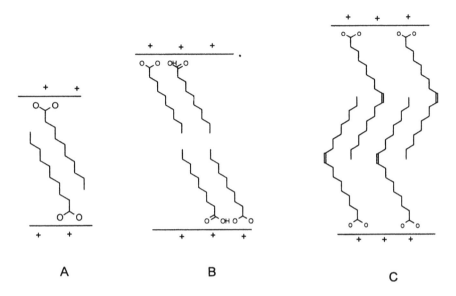

FIGURE 4   Three possible orientations. A: monolayer; B: bilayer of decanoate; and C: partial overlap of oleate in the interlayer.

like layer, while the hydrocarbon chain is slanted at an angle. The ideal angle is 55°, since the two oxygen atoms in carboxylate can then interact equally with the hydroxide groups of the layer. The hydrocarbon chains are fully extended in the energetically favorable all-trans conformation with strong mutual hydrophobic interactions (Fig. 4). The ionized COO^- groups, the H-bonded COOH groups in Figure 4B, and the all-trans conformation have been confirmed by the characteristic vibration peaks in IR spectra (see Sec. III.B.2.a and Ref. 90).

For aliphatic dicarboxylate anions, two COO^- groups are normally anchored to two adjacent brucite-like layers and the hydrocarbon chains are ideally tilted at 55°. However, the real packing always deviates from the ideal one, varying with the preparation conditions. This may be the reason that different researchers reported different gallery heights, ranging from 0.32 to 1.40 nm, for the sebacate anion (35,87,275,276). For aromatic anions, such as benzenesulfonate, and terephthalate, the benzene ring stands erect in the interlayer with the anionic groups toward hydroxyl layers (11,214,277).

In recent years, various biological molecular species, such as amino acids and DNA, have been intercalated into LDHs, in part because of interest in the possible role of minerals in the origin of life (263,278–281). Whilton et al. intercalated aspartic acid and glutamic acid anions into MgAl-LDH, and noted the thermally induced polymerization of aspartate into poly(a,b-aspartate) *in situ* (259). Fudala et al. (257,258) almost quantitatively introduced the deprotonated and the protonated forms of L-tyrosine and L-phenylalanine into the ZnAl-HT interlayer and proposed bilayer spatial packing in the interlayer. Aisawa et al. attempted to intercalate most of the natural amino acids but for some amino acid anions found only a small degree of uptake (255,256). Choy et al. and Kuma et al. investigated the intercalation of various nucleoside monophosphates, DNA with 500–1000 base pairs, and ionized molecular drugs into the LDHs and proposed that LDHs could act as gene reservoir and drug carriers (260–263,282).

c. Grafting

Condensation between the positive hydroxide layers and the negative anions can sometimes lead to an anion-grafting process onto the hydroxide layers. For example, Depege et al. reported that grafting occurs when CuCr-Cl-LDH is exchanged with $V_2O_7^{4-}$ at pH = 10.0 at room temperature (206,283). They noted that the resultant LDH has an interlayer spacing of 0.762 nm, much lower than the expectation (0.888 nm) from the free intercalation of oxoanions at H-bond distances from hydroxide layers. They thus explained this layer contraction by a combined hydrolysis and grafting of $V_2O_7^{4-}$ with the two OH groups of the hydroxide layers:

$$2M\text{-}OH + V_2O_7^{4-} \rightarrow 2\ \dot{M}\text{-}OVO_3^{2-} + H_2O$$

A similar process has been also observed for other MO_4 or M_2O_7 oxoanions in ZnAl-, ZnCr-, and CuCr-LDHs (206,284). Such grafting can easily take place with mild heating treatment, as investigated by Menetrier et al. in great detail (285a).

Grafting has also been observed for phosphates and phosphonates by Costantino et al. and others (196,286). For instance, the interlayer spacing of ZnAl-HPO_4^{2-}-LDH, initially equilibrated at room temperature and 75% relative humidity (1.08 nm), decreases to 0.82 nm after drying over P_2O_5. Heating at temperatures higher than 40°C caused a similar layer contraction. Correspondingly, the IR band specific to the hydrogenphosphates ($1100\ cm^{-1}$) is shifted to $1000\ cm^{-1}$, characteristic of the deprotonated phosphates. Further supportive evidence is the sharp decrease of the a.c. conductivity by an order of magnitude when heated over 40–60°C (196).

In addition, Guimaraes et al. claimed the grafting of ethylene glycolate to the hydroxide layers on the basis of IR observations (217).

d. Water in LDHs

Water molecules normally occupy the sites available in the interlayer that are not occupied by anions (287). Each OH group in brucite-like layers provides one such site. In general the water molecules are held through the formation of hydrogen bonds with the hydroxide layer OH and/or with the interlayer anions. This is reflected in Miyata's formula:

$$n = 1 - Nx/c$$

for the estimation of water content in $[M(II)_{1-x}M(III)_x(OH)_2]A_{x/c}\cdot nH_2O$, where N is the number of sites occupied by each anion and c is the anionic charge. For example, in NiMgAl-CO$_3$-LDH, when $x = 0.29$, the calculated n is 0.565 ($c = 2$ and $N = 3$ for CO$_3^{2-}$), which is in good agreement with 0.530 ± 0.020 obtained from five samples (4,6,37,202). In, addition, the inelastic neutron scattering investigation by Kagunya et al. (288a) shows that the water molecules in the interlayer are not fixed in one position but rotate freely and move about hydroxide oxygen sites. The same group found using Raman spectroscopy that there are three types of structured water: (a) water hydrogen bonded to the interlayer carbonate ion, (b) water hydrogen bonded to the hydrotalcite hydroxyl surface, and (c) interlamellar water (288b).

This water is lost over a temperature range of 120–250°C, depending on the strength of the interactions (128). The gallery water molecules can also cause an expansion in the thickness of the interlayer region, as mentioned previously for LDH-SO$_4^{2-}$ (see Sec. III.A.3.b).

Heating up to 150–200°C can push out the water and leave the framework unchanged. The dehydrated LDHs can also reabsorb water while cooling in humid surroundings (4). In addition, a certain amount of water may be physically absorbed on the surface of small crystallites. This weakly absorbed water can be removed by heating to 100°C.

4. Stacking and Polytype

There are several different ways for brucite-like layers to stack on top of each other (289,290). There are three highly symmetrical ways for the OH groups of one layer to relate to those of an adjacent layer. They may be in register (giving a trigonal prism) or out of register. If out of register, the OH groups of the upper layer can be in either of two positions relating to the lower. The fact that only half of the OH triangles contain a cation (see Fig. 2) leads to further possibilities. It can be shown that for a two-layer repeat there are three polytypes, all hexagonal (2H), while a three-layer repeat gives in all nine polytypes, of which only two (known as 3R$_1$ and 3R$_2$) are rhombohedral (for full discussion see Refs. 289 and 290). LDH minerals are known in both 2H and 3R forms (see Table 1), and synthetic LDHs are usually assigned to 3R$_1$, although in 2002 a 3R$_2$ form of synthetic hydrotalcite was described (291).

We now further consider the arrangement of the intercalated anions. Large organic anions in particular will impose their own packing requirements, as discussed earlier, and it will in general no longer be the case that the OH groups of the adjacent hydroxide layers will be in register for close packing. For this reason, some authors prefer indexing the basal reflections in such systems as *001*, *002*, *003*, etc. Moreover, the coexistence of two anions in the interlayer may give rise to intermediate interstratified phases. Drits et al. found that the coexistence of various amount of SO_4^{2-} and CO_3^{2-} in the MgAl-LDH results in an extra series of diffractions with an interlayer spacing of 1.65 and 1.85 nm. This corresponds to an alternating sequence of (brucite-CO_3^{2-})-(brucite-SO_4^{2-}) (167). Brindley observed a similar interstratified phase containing CO_3^{2-} and SO_4^{2-} in motukoreaite (see Table 1) (47). Grey and Ragozzini prepared LDH-like compounds $[Mg_{0.88}Al_{0.12}(OH)_2]_2(CO_3)_{0.12} \cdot 0.64H_2O$ and found from XRD that they have a brucite–brucite–carbonate repeat sequence, e.g., $[Mg(OH)_2]-[Mg_{0.76}Al_{0.24}(OH)_2]-[(CO_3^{2-})_{0.24} \cdot 0.64H_2O]$ (98). O'Hare et al. (292) observed intermediate phases with coexistence between different dicarboxylates or one dicarboxylate and Cl^-. These intermediates can be isolated as pure crystalline phases. In the case of two different dicarboxylates (50:50), only the interlayer spacing expected for the large anion is observed, implying the presence of both types of anion in the same interlayer. However, in the case of dicarboxylate and Cl^-, the repeated interlayer spacing is the sum of their single LDH interlayer spacings, indicating a brucite–dicarboxylate–brucite–chloride layer sequence. Jones et al. (293,294) made similar observations regarding the coexistence of terephthalate and some inorganic anions.

5. Superlattice Formation

The ideal cation distribution within the brucite-like layer in $M_2^{II}M^{III}$-LDH would obey a rule analogous to Lowenstein's rule for aluminosilicates (187). The rule states that trivalent cations should not be in adjacent metal sites. For LDH with M(II):M(III) = 2:1, this gives a hexagonal superlattice (or supercell) with $a' = 2a \sin 60° = a\sqrt{3}$, where a' is the separation distance of adjacent trivalent cations, as shown in Figure 5A. The X-ray diffraction intensities of the superlattice, estimated by Bookin and Drits for model structures, depend heavily on the scattering of divalent and trivalent cations and on the types of anion in the interlayer (289,290). For example, Mg_2Al-SO_4^{2-}-LDH is expected to show a single but strong *100* superlattice reflection, while Mg_2Fe-SO_4^{2-}-LDH will show a set of superlattice reflection, *100*, *101*, and *102*, etc.

Vucelic et al. reported that the *110* reflection has been observed in Mg_2Al-benzoate-LDH with diffraction distance 0.454 nm, close to the estimated value $d = a' \sin 60° = 2a(\sin 60°)^2 = 1.5a$, e.g., 0.456 nm ($a = 0.304$ nm) (295); this reflection is presumably attributable to the scattering by the anions, since the difference in X-ray scattering power between Mg and Al is small. A complete ordering of cations has been reported for $Ca_4Al_2(OH)_{12}SO_4 \cdot (6H_2O$ (168), but this is not surprising in view of the large size and distorted environment of Ca^{2+} in

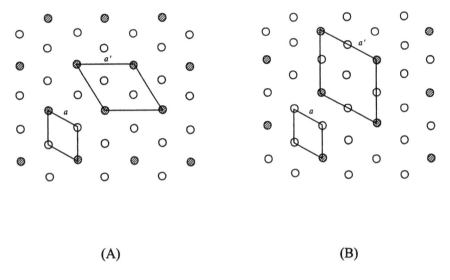

(A)   (B)

FIGURE 5 Hexagonal superlattices resulting from ideal cation ordering within a brucite-like layer in LDH with M(II): M(III) = 2: 1 (A, $a' = 2a \sin 60° = a\sqrt{3}$) and 3:1 (B, $a' = 2a$). Open circles stand for M(II) and shaded circles for M(III).

the hydroxide layers, and it need not imply such regularity in the less distorted LDH considered here. Such cation ordering is also found in several other hydrocalumite-like materials, such as Ca_2Fe-, Ca_2Ga-, and Ca_2Sc-LDHs (177).

Local ordering of cations in the brucite-like layer in Zn_2Cr-LDH intercalated with terephthalate or $V_{10}O_{28}^{6-}$ has been revealed by Evans et al. with EXAFS data in a short range (271). The zinc K-edge EXAFS of both LDHs show there are six O at 0.206–1.207 nm, three Zn at 0.309 nm, three Cr at 0.311 nm, six Zn at 0.536–0.537 nm, as well as three Zn and three Cr at ca. 0.62 nm. Consistent with this, the chromium K-edge EXAFS indicates six O at 0.198 nm and six Zn at 0.311 nm. These observed data are in good agreement with the ideal model in Figure 5A. Similar observations have been made for Ni_2Al-LDHs.

In addition, Solin et al. investigated the cationic and anionic ordering in $Ni_{1-x}Al_x$-CO_3-LDH (x = 0–0.4) and found two cases with superlattice structure (296,297). One is the same case as already discussed for x = 0.33 (i.e., Ni:Al = 2:1) with a $(\sqrt{3} \times \sqrt{3})R30°$ superlattice structure. The diffraction patterns also show a superlattice for x = 0.30 with a $(\sqrt{13} \times \sqrt{13})R13.90°$ superlattice ordering, which arises from diffraction of the guest carbonate anions, thus also providing an explanation for the $(\sqrt{13} \times \sqrt{13})R13.90°$ superlattice reported by Taylor in the $MgFe$-CO_3-LDH system (37).

In a hydrothermally prepared $LiAl_2$-CO_3-LDH, a superlattice with $a' = 0.532$ nm has been observed by XRD, indicating cation ordering in the brucite sheet. This is similar to the model in Figure 5A, with open circles taken to be Al^{III} while shaded ones are Li^I (184).

It seems reasonable to expect a related superlattice for $M_3^{II}M^{III}$-LDH, as shown in Figure 5B, with $a' = 2a$; but to the best of our knowledge this has not yet been observed by XRD. However, this superlattice structure has been seen via atomic force microscopy (AFM) and scanning tunnel microscopy (STM). Cai et al. (110) and Yao et al. (160) have reported that the surface of $Mg_6Al_2(OH)_{16}CO_3$ in contact with Na_2SO_4 solution shows a periodicity with a lattice constant of $a' = 0.62$–0.64 ± 0.03 nm, nearly equivalent to the superlattice constant ($a' = 2a = 0.62$ nm) for Mg_3Al-LDH in Figure 5B.

A superlattice-like arrangement of anions adsorbed on the surface of LDH has also been observed through AFM and STM. Cai et al. studied the adsorption of singly and doubly ionized 5-benzoyl-4-hydroxyl-2-methoxybenzenesulfonic acid ($MBSA^{1-}$ and $MBSA^{2-}$) on Mg_3Al-CO_3-LDH (110). They observed in AFM the periodic lattices with $a = 0.80$ nm and $b = 1.93$ nm for $MBSA^{1-}$-LDH and $a = 0.96$ nm and $b = 1.82$ nm for $MBSA^{2-}$-LDH. These data are in good agreement with those from their proposed models. Similarly, Yao et al. obtained the evidence of the formation of a $Fe(CN)_6^{3-/4-}$ anionic surface layer on Mg_3Al-CO_3Cl-LDH from STM (160,298,299). They observed periodic lattices with $a = 1.40 \pm 0.04$ nm and $b = 1.85 \pm 0.04$ nm and $a = 1.94 \pm 0.04$ nm and $b = 1.90 \pm 0.04$ nm for $Fe(CN)_6^{3-}$ and $Fe(CN)_6^{4-}$, respectively.

B. Characterization Methods

Many characterization techniques have been used to identify and analyze the LDHs. In the following, we briefly introduce the basic principles of the most commonly used techniques in LDH characterization and summarize what kinds of structural and compositional information they can provide.

1. Diffraction Methods

a. X-Ray Diffraction

Powder X-ray diffraction is the main tool for the identification of LDH structure at present, although in general the reflections are not sufficiently defined to provide indexing with any certainty. Figure 6 shows a diffraction pattern obtained for $Zn_2Al(OH)_6(CO_3)_{0.5} \cdot nH_2O$, with diffraction planes indexed in $3R_1$; this is typical of the patterns of LDH compounds with inorganic anions (4,289,290).

For LDH compounds, there usually is a series of strong basal reflections conventionally indexed, assuming a rhombohedral structure, as *003*, *006*, etc. These basal reflections correspond to successive orders of the basal (interlayer)

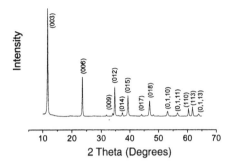

FIGURE 6   XRD of $Zn_2Al(OH)_6(CO_3)_{1/2} \cdot nH_2O$.

spacing c'. The interlayer spacing c' is equal to d_{003}, $2d_{006}$, $3d_{009}$, etc.; thus c' is often estimated by a formula $(d_{003} + 2d_{006} + \cdots + nd_{00(3n)})/n$. The cell parameter c is a multiple of the interlayer spacing c', depending on the layer stacking sequence, $c = 3c'$ for rhombohedral (3R) and $c = 2c'$ for hexagonal (2H) sequences (see earlier). The higher-order reflections ($n > 3$) are more pronounced in organic LDH compounds, for example, in Zn_2Al-stearate-LDH (monolayer packing), as shown in Figure 7, with an interlayer spacing of 3.21 nm. For other examples see Ref. 91.

Among the various nonbasal reflections (*hkl*, either *h* or *k* is not zero), the *110* reflection gives a direct measure of $a(a = 2d_{110})$. It should be noted, however, that the adjacent *113* peak often overlaps the *110* one, especially for those LDHs with a larger interlayer spacing.

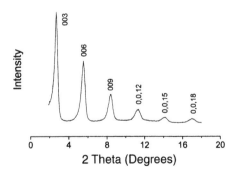

FIGURE 7   XRD of $Zn_2Al(OH)_6(C_{17}H_{35}COO^-) \cdot nH_2O$.

b. Neutron Diffraction

Unlike X-ray diffraction, neutron diffraction has rarely been used to examine the structure of LDHs (300), since suitable instrumentation is much less readily available. However, neutron diffraction has some advantages over XRD in the identification of thermally decomposed mixed oxides from LDHs. This is because different atoms and isotopes have different neutron-scattering length, and thus they can be discriminated from one another; in particular, neutrons can distinguish clearly between Mg and Al, but X-ray cannot (301). Vaccari et al. monitored the decomposition pathway of MgAl-LDH and PdMgAl-LDH from room temperature up to 1050°C by using neutron diffraction and analyzed the phase composition at each temperature stage (302,303). They observed that MgAl-LDH starts to collapse from about 200°C and that the LDH phase is changed into a cubic oxide phase (isostructural to MgO) at 500°C, which is separated into a fine MgO phase and spinel at 800–1000°C by a segregation process. For PdMgAl-LDH, PdO is also separated at 650°C, and large metal Pd crystallites form at 900°C (refer to Sec. V.C).

2. Spectroscopic Methods

a. Infrared and Raman Studies

Vibrational (infrared (IR) as well as Raman) spectroscopy is another useful and powerful tool for the characterization of LDHs, involving the vibrations in the octahedral lattice, the hydroxyl groups, and the interlayer anions. Infrared spectrometers are generally available, and in view of the information available by this technique we strongly recommend that infrared spectra of LDH be collected as a matter of routine.

It is convenient to consider the motions within the lattice $[M_2^{II}M^{III}O_6]$, with the ideal symmetry D_{3d} for the case of $M_2^{II}M^{III}$-LDHs, and the correlation of site group to the factor group gives: $T = 2A_{1g} + 2A_{2g} + 4E_g + A_{1u} + 3A_{2u} + 4E_u$, among which the $3A_{2u} + 4E_u$ modes would be infrared active, but in the range of 250–4000 cm^{-1} only five vibrations ($2A_{2u} + 3E_u$) should be observable. In fact, five infrared peaks in 250–1000 cm^{-1} have been observed for the lattice vibrations in Mg$_2$Al-, Ni$_2$Al-, and Fe$_2$Al-LDH (304). Similar analysis for $M_3^{II}M^{III}$-LDH with lattice $[M_3^{II}M^{III}O_8]$ predicts seven infrared absorption bands, while only five bands are observed, due to the disordered cation distribution in these LDHs. Although this IR vibrational information is not sufficient to fully characterize the brucite-like sheets, some peaks for specific LDH may still be indicative. For example, aged Mg$_2$Al-LDH with well-distributed cations in the brucite-like sheet shows a sharp peak at 445–455 cm^{-1}, while freshly prepared material, and the intrinsic to the less well-ordered Mg$_3$Al-LDH, does not (265,304,305).

Vibrational mode analysis for the hydroxyl groups within a brucite-like sheet identifies two OH stretching modes (A_{1g} and A_{2u}) and two OH librational modes (E_g and E_u) (306a). In the range of 4000–400 cm^{-1}, the A_{2u} (ν_{OH}) at 3400–3500 (OH$\cdots$HOH) and at 3000–3100 cm^{-1} (OH$\cdots$CO$_3^{2-}$) and the E_u (δ_{OH}) at 650–900 cm^{-1} are observed in the IR spectrum. The detailed deconvolution of these broad peaks can also provide information about the local environments (M^{II} or M^{III}) of OH (304), as can the analysis of Raman spectra of the hydroxyl-stretching region (288b). Correspondingly, the A_{1g} (ν_{OH}) at 3400–3600 cm^{-1} and the E_g (δ_{OH}) at 1060 cm^{-1} are also recorded in Raman spectra for Mg$_2$Al-CO$_3$-LDH. In the infrared of hydrotalcite itself, interlayer water coordinated to carbonate gives rise to a pronounced shoulder around 3060 cm^{-1}. In contrast with hydroxide OH vibration, the bending vibration of interlayer water (δ_{H2O}) occurs at 1600–1650 cm^{-1}, giving a characteristic peak rarely overlapped by others. Partial deuteration (306b) confirms these assignments and is also useful in the detailed assignment of the lower-frequency modes, although the bending vibration of D$_2$O could not be observed. The reviewers have independently found extensive deuteration of LDH to be very difficult, and note in this context that zero-point energy considerations will favor preferential uptake of H rather than D into the sites where H bonding to carbonate lowers the stretching frequency.

Monitoring the in situ IR and Raman spectra of LDHs under heating has also provided valuable insight into the LDH decomposition process and the ensuing mixed oxide formation (307,308; see also Sec. V.C).

The power of IR studies is shown most strongly in the characterization of the intercalated anions (4). For common inorganic anions (except for halides), there are specific IR peaks. For example, the peaks at 1350–1380 (ν_3), 850–880 (ν_2), and 670–690 (ν_4) cm^{-1} are indicative of CO$_3^{2-}$. When the symmetry of CO$_3^{2-}$ is lowered from D$_{3d}$ to C$_{2v}$, the ν_1 vibration can be noted at 1050 cm^{-1} together with a shoulder at 1400 cm^{-1} or a double band at 1350–1400 cm^{-1}. This symmetry lowering indicates the orientation change of interlayer carbonates at high charge density, as discussed earlier for Ni$_{1-x}$Al-LDH (264). The case for NO$_3^-$ is very similar, with ν_3 at ~1380, ν_2 at ~830, and ν_3 at 720–750 cm^{-1}. The free SO$_4^{2-}$ and ClO$_4^-$ belong to the point group T_d and have IR active ν_3 (at ~1100 cm^{-1}) and ν_4 (at ~740 cm^{-1}) modes. However, the actual configuration (266) in the interlayer always reduces symmetry to C$_{3v}$ or C$_{2v}$, resulting in the splitting of the ν_3 mode apart from 60 to 140 cm^{-1} and a weak band at 900–950 cm^{-1} (ν_1 mode). On the other hand, the ν_1 mode is very active in Raman spectra (309).

Braterman et al. deposited MgAl·Fe(CN)$_6^{4-}$-LDH particles on a BaF$_2$ disc so that they were oriented with the c-axis perpendicular to the holder surface (232). Under these conditions, the sharper of the two overlapping CN stretching vibration bands was inactivated, as required by the anion orientation model in Figure 3.

IR is even more specific for organic anions. Many characteristic vibration peaks of organic groups, such as CH, CN, COO, CC, SO$_3$, OSO$_3$, and NO$_2$, can be used to identify the existence of one specific organic anion. For example, the IR of Mg_2Al-LDH with dodecylbenzene sulfonate as counterpart anion is shown in Figure 8 and assignments given in Table 3.

The sensitivity of IR vibration modes to the environments of the carboxylate group can be further utilized to confirm the coexistence of neutral COOH groups with anionic COO^- species in the LDH interlayer (Figure 9). This is shown for mono- (A) and bilayer (B) packing of stearate/stearic acid, as shown in Figure 4. In the monolayer-packing LDH, there are two well-defined peaks at 1549 cm^{-1} (COO asymmetric stretching) and 1412 cm^{-1} (COO symmetric stretching). However, in the bilayer-packing LDH, one much stronger peak at 1597 cm^{-1} and other shoulders at 1620 and 1398 cm^{-1} are caused by the H-bonded COOH vibration (90). The acidity of neutral COOH group with pK_a normally ranging from 4 to 6 has been suppressed, with the formation of H bonds between COOH and COO^- in the interlayer (Figure 4) (273). This may be why the neutralization or condensation with the brucite-like hydroxide group does not occur even under

Table 3   Assignments of IR Peaks of Dodecylbenzene Sulfonate in LDH and Sodium Salt

In LDH (cm^{-1})	In Na salt (cm^{-1})	Peak assignment
3010 (br, w)	3010 (br, w)	CH stretching in benzene ring
2958 (sh, m)	2958 (sh, m)	CH_3 asymmetrical stretching
2870 (sd, w)	2870 (sd, w)	CH_3 symmetrical stretching
2926 (sh, vs)	2926 (sh, vs)	CH_2 asymmetrical stretching
2855 (sh, s)	2855 (sh, s)	CH_2 symmetrical stretching
1603 (sh, w)	1603 (sh, w)	
1500 (sh, w)	1500 (sh, w)	CC stretching in benzene ring
1409 (sh, m)	1409 (sh, m)	
1466 (sh, m)	1466 (sh, m)	CH_2 scissoring
1450 (sd, w)	1450 (sd, w)	CH_3 asymmetrical bending
1378 (sh, w)	1378 (sh, w)	CH_3 symmetrical bending
1181 (br, vs)	1191 (br, vs)	S=O asymmetrical stretching
1040 (sh, vs)	1046 (sh, vs)	S=O symmetrical stretching
1132 (sh, s)	1133 (sh, s)	In-plane aromatic CH bending
1012 (sh, s)	1013 (sh, s)	
832 (sh, m)	832 (sh, m)	Out-of-plane aromatic bending
687 (br, s)	693 (br, m)	SO_3 bending
	613 (br, m)	CH_2 rocking
582 (sh, m)	583 (sh m)	

s = strong; m = medium; w = weak; sh = sharp; sd = shoulder; br = broad; v = very

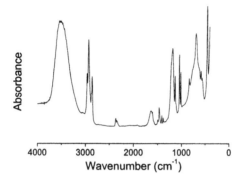

FIGURE 8 FTIR of Mg_2Al-LDH with dodecylbenzene sulfonate.

reflux condition, in contrast with the case of LDH-HPO_4^{2-} (196) and LDH-phenylphosphonate (310).

It should be mentioned that in the expanded region of 1150–1350 cm^{-1} (Fig. 9), there is a progression of bands, corresponding to the CH_2 wagging and indicative of all-trans arrangement of straight hydrocarbon chain (90). This is consistent with the models in Figure 4 proposed from the XRD data.

Recently, other vibrational techniques have been used in the characterization of LDHs and the decomposition process, such as near-infrared spectroscopy (4000–8000 cm^{-1}) (311), Raman spectroscopy, and infrared emission spectroscopy (312,313). These new techniques offer more ways to investigate the structure and properties of LDHs.

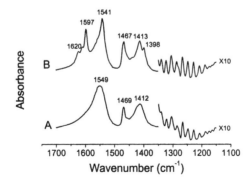

FIGURE 9 IR spectra of Zn_3Al-Stearate-LDH with stearate monolayer packing (A) and bilayer packing (B).

b. Neutron Scattering

Quasi-elastic neutron scattering and inelastic neutron scattering were used by Kagunya et al. to investigate the vibrational modes of LDHs and the properties of water adsorbed on LDHs (306a,314). The water adsorbed in LDHs was found not to be fixed in one location, but to exhibit translational diffusion as well as reorientational motions (314). The technique is particularly useful for vibrations involving the motion of hydrogen atoms, including metal hydroxide stretching modes. These show a broad band of frequencies from 400 to 1000 cm^{-1} (306a), further suggesting that the much more detailed structure observed in the IR spectra are the result of some kind of selection rule.

c. UV/Vis Spectroscopy

The UV/vis spectra of transition metal ions show d–d transitions and charge transfer bands that are highly sensitive to oxidation state and environment (301). There are a number of reports regarding the use of UV/vis spectroscopy in the characterization of LDHs and derived oxides containing transition elements (271,315–321). For example, Roussel et al. used UV/vis spectroscopy as well as EXAFS to examine the formation of ZnCr-Cl-LDH. They studied the species during precipitation at a constant pH = 5 by slow addition of metal chloride and NaOH solutions simultaneously to CO_2–free water. The continuous changes in peak position and intensity of the Cr^{III} species indicated the formation of Cr^{III} monomer, dimer, trimer, and tetramer species at different addition stages (315).

Velu et al. utilized UV/vis diffuse reflectance spectroscopy to investigate the spectral changes with Mn loading in Mn-containing MgAl-LDHs and the calcined mixed oxides (316). The UV/vis diffuse reflectance spectra show that Mn keeps its valence (Mn^{2+}) unchanged at low Mn loading, while it is partially oxidized to Mn^{3+} and/or Mn^{4+} in the high-Mn-loading LDHs. Together with XRD and ESR data, the UV/vis spectra further reveal that calcination of low-Mn-loading LDHs in air brings about the dissolution of Mn^{2+} into the MgO-like phase while the higher Mn-loading LDHs form spinel phases such as Mn$_3$O$_4$ and MnAl$_2$O$_4$ as well (316). They also studied the valence state of cobalt ions in several (Cu,Co,Zn)Al-LDHs (317) and CoSnAl-LDHs (318) by UV/Vis analysis. Prinetto et al. used this technique to examine the interconversion of Ni^{2+} and Ni^{3+} during the calcination and rehydration of NiAl-LDH (319). Rives et al. investigated the UV/Vis diffuse reflectance spectra of (Cu,Ni)Al-LDHs and found them to be similar to the corresponding hexa-aquo complexes, as expected given the locations of the transition metal ions in octahedral sites within the brucite-like layers (320).

d. Electron Paramagnetic Resonance

Electron paramagnetic resonance (EPR), also known as electron spin resonance (ESR), is directly related to the presence of unpaired electrons in atoms, radical or molecules. For most transition metal ions that have unpaired electrons, the corresponding LDHs as well as the oxides can show their own ESR signals. The ESR lineshapes and the hyperfine patterns often provide more information regarding the oxidation states as well as the chemical environments (301). A number of transition metal-containing LDHs have been examined by ESR spectroscopy (316,322–327). Velu et al. studied the effect of Mn substitution in MgAl-LDH on the physicochemical properties with this method (316). Serwicka et al. used ESR to study the thermal decomposition of V-containing LDHs with brucite-like layer doping and interlayer doping (322,323). The ESR results of ZnAl-FeW$_{11}$NiO$_{39}$-LDH indicate that the heteropolyanions are of ordered arrangement in layered clay region (324). No signal was found for NiCl$_4^{2-}$-intercalated LiAl$_2$-LDH, and this was taken to indicate the presence of square planar of NiCl$_4^{2-}$ in the interlayer (325) although in view of the known properties of this anion (328) such a conclusion seems unlikely.

e. Nuclear Magnetic Resonance

Nuclear magnetic resonance (NMR) in principle gives information about the environment of a nucleus that possesses a magnetic dipole. For solid samples, which of course include all LDH derivatives, special techniques must be employed to average out the effects of local anisotropies, and even so it is not possible to obtain the superb resolution typical of solution NMR spectra (301). Despite these limitations, the technique has been extensively applied to LDH derivatives, with ^{1}H (287,329,330), ^{13}C (287), ^{15}N (331), ^{35}Cl (332), ^{77}Se (333), ^{27}Al (195,334–336), ^{51}V (285), ^{71}Ga (330), ^{119}Sn (182), and ^{31}P (337) among the nuclei used. The ^{1}H NMR studies of Zn$_2$Al(OH)$_6$Cl·nH$_2$O by Dupuis et al. reveal the existence of two kinds of protons with very different mobilities (329). The protons in the brucite sheets form the fixed species, while those from the interlayer water and adsorbed water are the mobile species. The ^{1}H and ^{13}C NMR spectra of Mg$_6$Al$_2$(OH)$_{16}$CO$_3$·nH$_2$O indicate that interlayer water possesses rotational freedom around the C$_2$ molecular axis and that the molecular symmetry axes of both carbonate (C$_3$) and water (C$_2$) are oriented parallel to the c-axis of the LDH unit cell (287). ^{1}H NMR can be also used to monitor the decomposition process. For example, ^{1}H NMR spectra reveal that Mg$_{0.714}$Ga$_{0.286}$(OH)$_2$(CO$_3$)$_{0.143}$·mH$_2$O loses the brucite-like layer OH groups on calcination at 550°C (330). Using ^{15}N NMR, Hou et al. showed that in nitrate LDH, interlayer and externally adsorbed nitrate ions exhibit very different dynamic and structural behavior, having the interlayer C$_3$ axis perpendicular to the brucite-like layer (331). The analysis of ^{35}Cl in LiAl$_2$-Cl LDH shows there are three types of Cl$^-$, depending on the temperature and hydration, one on the surface, the second in the interlayer with uniaxial chemical shift anisotropy, and the remaining one also in the interlayer but yielding a structureless NMR peak (332). The second type is dominant and is the only case when fully dried. ^{77}Se data have been used to investigate swelling properties and dynamic behavior in LDH-SeO$_4$, LDH-SeO$_3$ (333).

NMR spectroscopy has often been utilized to identify the coordination of metal cations, where ^{27}Al is extensively examined. Al is most commonly octahedrally coordinated, as expected. However, five- and four-coordinate Al are also detected, giving distinct ^{27}Al NMR spectra (195,334). During the dehydration–rehydration process, Beres et al. observed interconversion of octahedrally and tetrahedrally coordinated aluminum (335). Rocha et al. found that the calcined MgAl-LDH can be reconstructed more or less completely after calcining at 550°C or below and partially after calcining at 1000°C but that all the reconstructed LDHs contain some tetrahedrally coordinated aluminum (336).

If organic anions are intercalated into LDH, the ^{2}H, ^{13}C, and ^{31}P NMR etc. can also be used to study the anion dynamics and state in the interlayer (338,339). The 2D ^{2}H NMR line shapes of MgAl-(AB)-LDH (A = CO$_3^{2-}$ and B = p-C$_6$D$_4$(COO)$_2^{2-}$) indicate that the motion of the organic anions is mainly a rotational motion about the C—COO$^-$ axis at temperatures between -28 to 82°C (338). The ^{31}P NMR data of phenylphosphonic acid in MgAl-LDH revealed the existence of singly and doubly ionized phenylphosphonate in the interlayer and on the surface (339).

f. Extended X-Ray Absorption Fine Structure

The extended X-ray absorption fine structure (EXAFS) is the fine structure in the X-ray absorption coefficient starting somewhat past an absorption edge and typically extending 1 kV in energy range. This structure arises from quantum mechanical constraints on the ability of an ejected electron to propagate, and thus gives information about the neighborhood of the excited atom. The information obtained is in a short range, giving useful information about local ordering even if it does not persist over a large enough range to be detected by diffraction methods (340). This method has been widely used to investigate the local ordering of the cations in the brucite-like layer (315,341–343) and of the anions in the interlayer (271,344a,345). The Zn$_2$Cr- and Cu$_2$Cr-Cl-LDHs have been closely examined by EXAFS (315,341). The EXAFS spectra unambiguously show cation ordering in Cu$_2$Cr-Cl-LDH, but the evidence for similar ordering in Zn$_2$Cr-Cl-LDH is less strong (315,341). As for the formation process of Zn$_2$Cr-Cl-LDH, the EXAFS spectra confirm the formation of the Cr^{III}-oligomeric species at the Cr K-edge and hexaaquozinc(II) complexes at the Zn K-edge before the formation of the LDH phase, in agreement with the results from UV/vis (315). The intermediate oligomer process may be highly relevant to the formation of fixed Zn/Cr atomic ratio (2:1) in Zn:Cr-Cl-LDH. In pyroaurite (MgFe-LDH), EXAFS indicates a high level of local cation ordering and the absence of Fe(III)-Fe(III) neighbors (344b). Local ordering of cations in Co$_2$Al-, Cu$_2$Al-, and Co$_2$Fe$_y$Al$_{1-y}$-brucite-like layers has also been demonstrated by EXAFS (344b,344c). Bigey et al. found grafting of the interlayer CrO_4^{2-} or $Cr_2O_7^{2-}$ onto the brucite-like layer when calcined at 150°C (342,343). In the pillared LDH-polyoxometalate (POM), EXAFS data (Zn, Cr, and Ni K-edge) indicate that little or no change in layer structure occurs after POM intercalation. The data are also best fitted by a model in which the POMs, such as $V_2W_4O_{19}^{4-}$ and $Nb_3W_3O_{19}^{5-}$, are structurally unchanged after intercalation and arranged with their shortest (C_3) axis perpendicular to the hydroxide layer (271,345).

g. Mössbauer Spectroscopy

In Mössbauer spectroscopy, a γ-ray emitted by a nucleus in its excited state is absorbed by a ground-state nucleus in the sample. For ^{57}Fe, the most widely used Mössbauer nucleus, the precise energy of the transition depends on the electron density and the electric field gradient close to the nucleus. There is in general a small energy shift required for resonance, and this is provided by the Doppler shift in the frequency of the γ-ray when source and sample are moved relative to each other (301). The technique has been applied to a range of Fe-containing LDHs and their derived materials (346–352). Mössbauer spectra explicitly show the difference in the Fe^{III} local environment between $Mg_6Fe_2(OH)_{16}CO_3 \cdot 4H_2O$ and $Mg_6Fe_2(OH)_{16}(A) \cdot xH_2O$, where $A = O_2C(CH_2)_nCO_2$ ($n = 0$–14) (346,347). In-situ Mössbauer spectroscopy was used to study the reduction reaction of LDH-derived MgFe- and MgFeAl-oxides in a H_2/N_2 atmosphere. $MgFeAlO_4$ was identified by its characteristic doublet (348,349). Mössbauer spectra also indicate a well-dispersed ferrihydrite phase supported by a $MgFe_2O_4$ matrix in calcined MgFe-LDHs (350–352).

3. Other Physicochemical Methods

a. Adsorption

The adsorption of N_2 gas is often used to evaluate the surface-accessible area and pore size distribution by the Brunauer–Emmett–Teller (BET) method (353). The accessible surface is generally that of the internal pores within the crystallites and the external surface between the crystallites. Correspondingly, the measured pores are those inside of and between the crystallites. Within most common organic and inorganic LDHs, the interlayers are full of the anions as well as water, and thus only the external surface of the crystallites contributes to the accessible surface area. Values normally range from 10 to 200 $m^2\ g^{-1}$ (4,354), and only the intestinal pores between the crystallites compose the pore space. An important exception is provided by pillared LDHs (see Sec. III.A.3.b) that may possess micropores in the interlayers whose sizes vary with the size and charge of the interlayer pillars (355), as estimated by Nijs et al. for the MgAl-Fe(CN)$_6^{3-}$-LDHs and ZnAl-PV$_2$W$_{10}$O$_{40}^{5-}$ (356). They assume that pillared anions are evenly distributed in the interlayer and estimate the size of the interpillar pores. They find values of 0.1–0.6 nm in diameter on average, with a maximum pore volume of 0.38 cm^3 g^{-1}. The lower the hydroxide layer charge density and the greater the charge number carried with pillared anions, the smaller the amount of anions needed to keep charge balance, the bigger the interpillar distance, and thus the larger the interpillar micropores. Additional information about pores between crystallites is discussed in Section IV.B.

When LDH is calcined (see Sec. V.C), the decomposition of anions, the dehydroxylation of brucite-like layers, and the formation of newly born oxide crystallites bring about the formation of new pores or enlarge old pores to macropores. Research by Xu and Zeng has shown the relationship of the specific surface area and pore size distribution to the calcination temperature for Co^{II}Co^{III}-LDHs (129). They found that both the specific surface area and the pore volume between 2 and 50 nm increase with temperature to 200°C and then decrease with temperature to 600°C. The specific surface area is in good agreement with that predicted with a sphere model, where the average radius of the Co$_3$O$_4$ crystallites is estimated from the full-width-half-maximum of several XRD peaks.

The acid/base properties of LDHs and the derived oxides are also widely characterized with the adsorption of probe molecules. LDHs behave as weak bases. For example, Mg$_{1-x}$Al$_x$-, Ni$_{1-x}$Al$_x$-, Co$_{1-x}$Al$_x$-, Zn$_{1-x}$Al$_x$-, and Cu$_{1-x}$-Al$_x$-LDH-CO$_3$ with $x = 0.43–0.48$ show selectivity for the CO$_2$ adsorption. The adsorption behavior curves indicate that NiAl-, MgAl-, and ZnAl-LDH-CO$_3$ irreversibly adsorb CO$_2$ via a possible chemical reaction at room temperature:

$$\text{(on LDH) OH} + \text{CO}_2 \rightarrow \text{(on LDH) O}-\text{CO}_2\text{H}$$

while CuAl- and CoAl-LDH reversibly adsorb CO$_2$ at 25°C (357–359).

When basic LDH is mixed or intercalated with acidic solid component, such as a zeolite, the composite will show a dual functional adsorption for CO$_2$ and NH$_3$. Okada et al. investigated the adsorption behaviors of MgAl-LDH/aluminosilicate composites and found that the adsorptivity of the composites for both of the probe molecules is dependent on the preparation pathways. The composite prepared via sol/precipitation shows a superior adsorption for both acidic and basic gases to those prepared via mechanical mixing and reconstruction methods (360). This is quite similar to the case of the dual acid/base properties of LDH-derived oxides, especially the LDH-polyoxometallate–derived oxides, which have been carefully examined by a similar adsorption method (361).

b. Thermodynamic Studies

The stability and thermodynamic properties of LDH have recently received much attention due to the effect of absorpability and dissolution of LDH-like minerals on the metal migration in the soil (68,104,363–367). It has been found that various divalent metal cations, such as Mg^{2+}, Ni^{2+}, Co^{2+}, Zn^{2+}, and Mn^{2+}, can combine with suspensions containing Al^{3+}, Fe^{3+}, and Cr^{3+} (104,363,364). For example, Ni^{2+} is absorbed by gibbsite to form NiAl-LDH (3,247). Boclair et al. (68,365) have used pH titration to study the stability of LDHs, and thus have shown that their formation occurs either in a two-step process or in a single-step process. In the two-step process, such as MgAl-LDH, $Al(OH)_3$ forms first and is then converted to MgAl-LDH. In the single-step process, such as the formation of ZnCr-LDH, the LDH phase is obtained from the outset, a process assisted by low solubility of the LDH and relatively high solubility of the trivalent metal hydroxide. The formation of LDH rather than individual hydroxides indicates the relative thermodynamic stability of LDH over the pure hydroxides under these conditions.

High-temperature oxide-melt solution calorimetry (366) shows that the enthalpies of formation of some LDHs-CO_3 are only 0 to -10 kJ mol^{-1} different from those calculated by adding the values for mixtures of the corresponding hydroxides and carbonates. Based on this observation, the enthalpy as well as the Gibbs free energy of any LDH can be calculated approximately from the addition of the corresponding component hydroxides and salts, and thus the solubility product of any LDH can be estimated (367).

c. Elemental Analysis

The available methods for metal analysis are numerous, although of course care must be taken to avoid interferences. Atomic absorption (AA) spectrometry is particularly convenient and widely used for this purpose, providing the weight percentage of each metal, and hence the all-important metal atomic ratio in the LDH, which may well be different from that in the initial preparation solution.

In a usual commercial CHN analysis, a sample is combusted to convert carbon, hydrogen, and nitrogen elements into CO_2, N_2, and H_2O, which are then separated on a column and quantified by thermal conductivity changes in helium carrier gas.

Combining the results of metal and CHN analyses gives a nominal chemical formula of the LDH as well as an indication of the degree of any CO_3^{2-} and, for organic ions in particular, the degree of intercalation and of uptake of undissociated acid molecules into the interlayer. For these reasons, we advocate the discipline of analyzing for both metals and nonmetals, much as organic chemists routinely analyze for C, H, and N.

4. Molecular Modeling

Molecular modeling, known as molecular dynamics simulation, has been also widely applied to LDH system in recent years. By choosing a set of suitable force fields, molecular dynamics simulation can improve our understanding of the properties of the target system (332,368–374), such as the anion arrangement in the interlayer (332,368–371), hydration effect (372), and ionic absorption and diffusion (373,374). By minimizing the system energy, the orientation of the anions (e.g., CO_3^{2-}, NO_3^-, SO_4^{2-}, Cl^-, $C_2O_4^{2-}$, benzoate, terephthalate, and amino acids) in the interlayer has been predicted, and the results for the orientation and the gallery height are in good agreement with the experiment data (332,368–371). Through the simulation, Kirkpatrick et al. found the hydration energy of $Mg_2Al(OH)_6$ nH_2O to be minimum at $n \approx 2$, similar to the experimental observation (372). Their simulation also shows that the calculated diffusion coefficient of Cl^- on the outer surface is almost three times that of Cl^- within the interlayer, but is still about an order of magnitude less than that of Cl^- in bulk solution (372–374). There is no doubt that the simulation results help us obtain an understanding of more dynamic properties of LDH, such as the anion absorption, the anion exchange, and the gallery expansion when large anions are intercalated.

C. Summary

In summary, the main structural features of LDH are determined by (a) the nature of the brucite-like layer, i.e., the metal elements and their atomic ratio; (b) the types of anions intercalated, i.e., the inorganic, organic, polyoxometallates, and complexes; and (c) the properties of anions in the interlayer, i.e., the charge number, the size, and the arrangement of anions. The water content in the interlayer, the type of stacking of the brucite-like layer, and the presence or absence of long- and short-range ordering also influence the physicochemical properties of LDHs to some degree. Such features can be identified by various analytical methods, as described earlier. These features, on the other hand, are sensitive to exact preparative details (Sec. II) and in turn affect texture, catalytic performance, and dispersability (Sec. IV, VI, VII). It is therefore of special importance that these subtle materials be characterized as fully as possible.

IV. MORPHOLOGY AND TEXTURE

Morphology and texture, as described by particle size, crystallinity, porosity, and surface area, depend on factors such as preparation and postpreparative treatments as well as the ions incorporated into the layer. These structural details are analyzed by techniques such as N_2 adsorption/desorption, powder X-ray diffraction (PXRD), and optical and electron microscopy.

A. Particle Size and Crystallinity

The primary factor involved in the size and crystallinity of LDH particles formed is the method through which they evolve. Crystallinity is a direct function of the organization within the metal hydroxide layer. Thus reaction time, temperature, concentrations of reactants, postpreparative treatments, and reaction solvent contribute to the ensuing properties of the particles (375,376).

LDH materials formed gradually, by dissolution/reformation processes, possess greater crystallinity and, frequently, increased particle size over those formed rapidly. This can be accomplished through the use of postpreparative treatments like those described in Section II.M (146,333,377) and methods of preparation such as homogeneous precipitation (discussed in Sec. II.I) (109,111). The effect is the evolution of LDH particles through Ostwald ripening, leading to a decrease in the width at half height of the PXRD peaks. For an ideal crystal, this width is directly related to particle size by the Debye–Scherrer formula (378). Additionally, peak broadening arises from strain and defects within each crystallite. Since both of these, as well as small size, increase the solubility of a material, Ostwald ripening generally leads to more perfect as well as larger crystallites. In our experience, however, this does not necessarily lead to larger particles, since immature particles are often aggregates (379). Increased temperature and time both promote Ostwald ripening, but one caveat to using increased temperature is that the LDH composition might be altered, as mentioned in Section II.M.

Additionally, LDH particles formed in a one-step process have been shown to exhibit increased crystallinity. This is presumably due to their direct formation from solution rather than by transformation of a pre-existing solid, such as $Al(OH)_3$, which will offer numerous sites for heteronucleation (101). Thus, LDH formed by the coprecipitation or aluminate method are of higher crystallinity than that formed by precipitation at varying pH (380). MAS NMR studies confirm that the appropriate number of divalent metals surround the trivalent metal, forming a homogeneous metal hydroxide layer, for materials formed via aluminate, compared to a less than ideal ratio in materials prepared by precipitation at varying pH (122). Yet another factor that influences the crystallinity of the material formed is the level of ion saturation in solution. In more concentrated solutions, the LDH is formed with poorer crystallinity (381).

If crystallinity depends only on homogeneous distribution of the metals within the hydroxide layer, then one method of formation that would exhibit an increased degree of crystallinity, compared to that imparted on LDH from coprecipitation, is preparation using LDHs' ''memory effect.'' The moderately calcined product might have been expected to have a greater metal uniformity throughout the mixture that upon rehydration forms a more homogeneous metal hydroxide lattice. However, PXRD data of these materials generally show formation of poorly crystalline material, with, in some cases, separation of divalent metal oxide (320,382,383).

Another reaction condition that can alter the crystallinity and particle size of the LDH is the solvent from which the material is prepared (94). Malherbe et al. (384) showed that Mg/Al LDH carbonate precipitated in a water/organic solvent system formed crystalline LDH particles, using acetone, glycerol, ethylene glycol, propanol, ethanol, or methanol as the organic solvent. Of these solvent systems, acetone and ethylene glycol produced the largest, most crystalline particles. The organic solvent/increased crystallinity relationship varies depending on preparation technique, e.g., from the calcined salt, coprecipitation, or anion exchange.

B. Porosity and Surface Area

Porosity can be used to describe the pore distribution and pores size associated with an LDH. Pore distribution is related to the method of LDH formation (383) and ions associated with the material, whereas pore size is related more to the method of preparation and interconnection of LDH platelets. The porosity of a material is commonly analyzed by N_2 adsorption/desorption and pore size distribution analysis. N_2 adsorption/desorption isotherms are a plot of the volume of N_2 adsorbed versus relative pressures. Pore size distributions are calculated using the Barrett, Joyner, and Halenda method based on the isotherm data (385).

The cavities associated with LDH materials can render the materials either microporous (0–2 nm) or mesoporous (2–10 nm). These cavities can be the result of the interconnection of multiple LDH platelets or can be a characteristic imparted on the LDH structure by the interlayer anion. Frequently the porosity induced by the latter, microporosity, is not easily detected; when it is, the interlayer anion is often a polyoxometallate (POM) (6,356) and metallocyanides (362). Long-chain organic anions, such as carboxylates, should produce this same effect (386), but thus far there has been very little literature regarding the effect of such anions on LDH porosity.

Isotherms associated with N_2 adsorption and desorption are directly correlated with the porosity of an LDH. An adsorption isotherm that rises rapidly at low pressures is a result of the adsorption of multiple layers of N_2, indicating microporosity, whereas a gentle rise in the adsorption isotherm at low pressure represents the formation of a monolayer or mesoporosity. A sharp increase in the adsorption isotherm at relatively high pressures illustrates that condensation is occurring, which is indicative of mesoporosity (320). When the adsorption isotherm is coupled with a desorption isotherm, the result is a hysteresis loop. The structure of this loop can impart information related to the geometry of the pores (regular or irregular) and the distribution of pore dimensions (101,387).

Solvent influence can cause a difference in the shape of the loop observed and hence the pores produced in a material (384). When an ethylene glycol/water solvent system is used in the preparation of LDH carbonate from coprecipitation, the material formed appears to have pores of regular geometry and relatively equal size throughout. A glycerol/water solution on the other hand produces material with mesopores of varying geometry. Analysis of the hysteresis loop that results from the latter solvent combination shows the initial formation of micropores with the later formation of mesopores. This initial formation of micro-porosity may be due either to the formation of much more compact aggregates due to the solvent or to micropores that appear in the grain boundary rather than in the interlayer.

Alteration of the metal cations within the layer can also demonstrate this effect. Mg/Al, Cu/Al, and Fe/Al LDH carbonates, as prepared in water, have varying adsorption/desorption loop structure, which gives information about their pore distribution (387). From the hysteresis loop, Mg/Al LDH carbonates have nonuniform-shaped mesopores, Cu/Al LDHs possess mesopores of a regular geometry and small size distribution, while Fe/Al LDHs have both micro- and mesopores of varying geometry. These variable pore sizes may be due in part to the distortion of the metal hydroxide layer that occurs from iron incorporation due to differences in atomic size of the metals. Scanning electron micrographs have been taken to help explain the results of that observed from N_2 adsorption. In LDHs where the N_2 adsorption/desorption isotherms have suggested initial formation of microporosity followed by mesoporosity, researchers have attributed the formation of microporosity to small aggregates of LDH and the formation of mesoporosity with the connection of multiple aggregates to form cavities (388).

Reichle et al. showed how LDH materials are texturally modified by calcinations (389). The calcined material has an increase in both surface area and pore volume. This is a result of the formation of regularly spaced pores due to the elimination of carbon dioxide and water from the surface and the pores are both micro- and mesoporous in size. Regeneration of the LDH shows the disappearance of the micropores contributing to this increase in surface area and also a broader range of mesopore sizes. Use of organic solvent during the regeneration process produces LDHs with decreased surface areas, smaller pores, and smaller distribution of pore sizes.

C. Scanning Electron Microscopy and Transmission Electron Microscopy

SEM and TEM allow the researcher to observe what the LDH crystallites look like in shape and how LDH crystallites aggregate. LDH materials containing inorganic anions most commonly give materials of a hexagonal platelet shape observable by SEM and TEM (109,376,383). Materials with intercalated organic anions, however, have produced materials of varying shapes and sizes, e.g., ribbon- and barlike substances (259). The LDH carbonate particle has a typical diameter of around 1 micrometer. There seems to be very little correlation between cations used, anions used, or method of preparation and the size of particle obtained.

The growth habit of LDH materials containing inorganic and organic anions has been explained in terms of intermolecular interactions. In LDH materials containing inorganic anions, crystallite growth is along the a and b axes to maximize the exposure of hydroxyl groups to the aqueous phase, since the hydroxides of the metal hydroxide layer are found in the ab plane (390). In materials containing organic interlayer species, factors such as hydrophobicity, critical micelle concentration (CMC), and charge density affect the structure.

Somasundaran and Fuerstenau (391) and later Pavan et al. (392,393) distinguished four stages in the modification of structure by adsorption of organic anions onto LDH, of which the first three are attributed to increasing coverage of the surface of the particles by the anions, while the last occurs after the anion concentration exceeds the CMC. SEM images observed by Pavan, of the adsorption of sodium dodecylsulfate (SDS) onto LDH, tend to coincide with this four-region description, with the first three appearing similar to LDH carbonate and the last showing the formation of a bandlike structure.

Xu and Braterman observed that SEM images of LDH sulfonates (RSO_3^-) varied from platelike sheets to barlike particles and curved sheets (390). These formation differences can be explained using hydrophobicity arguments. In surfactant LDH materials the (ab) plane would exhibit hydrophobic characteristics. The formation of antiparallel structures (or of bilayers in general) by adsorption of surfactants will lead to facile nucleation of new sheets. Therefore, growth along the c axis is a low-energy and kinetically accessible behavior, thus forming barlike particles. The curvature observed in the curved sheets could be brought about by the hydrophobic interaction of internal and external RSO_3^-.

TEM images too look different between LDHs containing inorganic and organic anions. Images have been produced for organic intercalated LDHs that give additional information about the morphology and texture of these species. Images frequently contain dark and light parallel lines, interpreted as the organic and inorganic layers of the LDH (259,394). We interpret the changed morphology in terms of altered surface energetics. The ab face is now hydrophobic and of high energy in water, in contrast to the situation with inorganic ligands. At the same time, the surface can now attract additional surfactant to form a bilayer, which will in turn attract cations. Thus ab growth is now inhibited but nucleation of new layers (c axis growth) facilitated, leading to this reversal of growth habit.

V. REACTIONS

A. General

Most reactions of LDH are to some extent covered elsewhere in this chapter. The most characteristic reaction, exchange of interlayer anion, is considered at sufficient length as a preparative method in Section II.D. Dissolution at low pH is simply the reverse of the precipitation reactions used to form LDH and will be important in environmental applications, giving an advantage to zinc- over magnesium-based systems. Oxidation, either thermal or electrochemical, is also discussed in that section as a preparative route for LDH containing Co^{III} or Mn^{III} Thermal decomposition, on its own or accompanied by redox reaction, is an important route to LDH-based catalysts (Sec. VI.C), while the structural implications of grafting are considered in Section III. The discussion of these topics here is, accordingly, brief.

B. Functionalization and Grafting

In this regard, the LDH can be contrasted with OH-terminated silicas. The latter show an extensive chemistry based on converting this group to a wide range of OR or OX functionalities. Such chemistry is almost absent at this time for LDH, and there are several good reasons for this. Many functionalizing agents condense with OH by eliminating the elements of acid, a process that would evidently destroy the LDH. In addition, functionalization always has to compete with side reactions involving interlayer water. Finally, the layer OH group is considerably more ionic than the corresponding group on silica, with the result that the reaction can often be reversed by hydrolysis. We would not, however, be surprised to see such functionalization of the surfaces or even the interlayers of LDH dispersed or delaminated into organic solvents in the future.

The closest approach to functionalization at present is the grafting reaction, in which a layer OH group is formally replaced by the oxygen atom of an interlayer anion. Grafting by phosphates, vanadates, and glycolate is discussed in Section II.A.3.c. A similar reaction is shown by chromate and dichromate, leading to a stabilized lamellar structure that resists rehydration (395), and by organic phosphonates, sulfonates, and carboxylates. The reaction is shown by aromatic (ortho- or para-substituted) carboxylates, but not, presumably, by ortho, showing the importance of geometric constraints (396).

C. Thermal Decomposition

Thermal decomposition of layered double hydroxides typically involves endothermic processes of dehydration, dehydroxylation, and the removal of anions. These treatment processes are of great importance in the production of oxide and oxide-supported catalysts (see Sec. VI.C). For this reason, because of the wide range of metal and anion combinations of interest in this context and because of various preparative methods available, this remains an active area of research. It is, however, a difficult area because of the problems of observing metastable intermediate phases and the poor crystallinity of the thermolyzed material; for this reason it employs a range of old and new techniques (1,4,13).

For hydrotalcite ($Mg_6Al_2(OH)_{16}CO_3 \cdot 4H_2O$), dehydration takes place at 100–300°C, while dehydroxylation, i.e., the collapse of the hydroxide layers, occurs at 350–500°C, overlapping the decomposition of CO_3^{2-} to CO_2

(25,397–400), as shown by TGA-DTA-MS (357) or DTA-MS alone (307). The case is similar for NiAl- (307,401), ZnAl- (357), CuAl- (357), PbAl- (402), MgFe^{III}- (352,403), MgGa- (404), NiFe^{III}- (405), and ZnGa-LDHs (406). The endothermic peak temperatures of these two events in DTA vary with the metal ion types and ratios (25), the anion types (128,407), the postformation treatments of the LDH (408), and the aging time (409). Moderate calcination of hydrotalcite, for example, at 400–500°C for 2 h, leads to an intermediate phase together with poorly crystallized cubic MgO. This product usually possesses a so-called memory function, i.e., it can be rehydrated to layered double hydroxide, although the reconstructed material is not identical with the parent (202,410,411). Further heating to 1000°C gives cubic crystalline oxides (e.g., MgO) and spinels (i.e., MgAl$_2$O$_4$) (412,413).

The mechanism of LDH thermolysis has been studied by thermal analyses (DTA, TGA, DSC, etc.), powder XRD of the products, preferably in situ at the temperature of their formation, infrared and Raman spectroscopy, gas absorption measurements, Auger and XPS spectroscopies, EXAFS, and MAS ^{27}Al and ^{25}Mg NMR. The heating of hydrotalcite itself is typical. It proceeds successively by loss of adsorbed and interlayer water, of carbon dioxide, and of hydroxyl water and is accompanied by the formation of micropores and the incorporation from an early stage of some of the aluminum ions in tetrahedral sites (389,398,414,415a). (One 2002 study (415b) proposes an alternative scenario, in which loss of hydroxyl occurs at lower temperature than the most extensive loss of carbon dioxide.) Among proposed intermediate phases is one in which carbonate is grafted to the hydroxide layer (416); if so, we suggest this would explain the formation (417–419) of an intermediate metastable phase with reduced interlayer spacing. Lower charge densities lead to more facile loss of carbon dioxide (420a). The kinetics of the dehydration reaction has been studied (420b), as have enthalpies of rehydration of the intermediate phases (420c). The intermediate stages are of considerable complexity and have been investigated by AWAXS and EXAFS, since methods that depend on long-range order only do not give an adequate description of the process (413,414). In the catalytically interesting Cu-Zn-Al LDH carbonate thermolysis products, in situ FTIR during thermolysis provides evidence for a change in carbonate H bonding from interlayer water to layer hydroxyl during dehydration (420d).

For LDHs containing oxidizable divalent cations, such as Co^{II}, Fe^{II}, and Mn^{II}, a spontaneous oxidation reaction occurs during decomposition in air. In this case, dehydroxylation is facilitated and takes place at temperatures below 300°C or even overlaps with the dehydration. For example, in air there are two overlapping endothermic peaks for CoAl-LDH in the range 200–250°C (307,421) and one major unresolved endotherm at around 200°C for Co^{II}Fe^{III}-LDHs (422) in DTA curves. Under nitrogen, this oxidative decomposition takes place only at higher temperatures, and the dehydration and dehydroxylation with decomposition of carbonate become two distinct processes (141,422). An exception is those LDHs containing anions such as NO_3^- as oxidant, where there is not much difference between decomposition in air and in N_2 (129). These oxidative decompositions give rise to the progressive formation of spinel, without the appearance of a crystalline $M^{II}O$ phase (129,422), and with loss of the so-called memory effect. The reaction is thought to be a topotactic process that converts (140), for example, the *ab* plane of carbonate green rust to the *111* plane of magnetite, as shown by TEM/electron diffraction (423).

Oxidation of M^{III} can also occur, for instance during the calcination of LDHs that contain Cr^{III}, V^{III}, or Mn^{III} cations. Thus the calcination of MgCr-, NiCr-, and ZnCr-LDHs at intermediate temperatures (i.e., 350°C) results in the formation of chromate-like (CrO_4^{2-}) species (321,424,425). Similarly, $Mg_{0.71}V_{0.29}(OH)_2(CO_3)_{0.145} \cdot 0.72H_2O$ transforms to an amorphous material with octahedrally coordinated V^{III} at 175°C and thence to tetrahedrally coordinated V^V at higher temperatures (170). Similarly, MgMn(III)-LDH decomposes to Mg_2MnO_4 with minor MgO at temperatures higher than 400°C (139), while Ni_2Mn-LDH changes to $NiMnO_3$ with NiO at 700°C (426).

Transition metal–containing LDHs or their derived oxides are often treated with a reducing gas such as H_2, CO, or CH_4, so as to convert constituents such as Ni^{II}, Cu^{II}, Mn^{III}, Fe^{III}, Co^{II} and Co^{III} to metallic or metallic-like species that can be the active sites as catalysts. A great advantage of this process is the very fine dispersion of the metallic or metallic-like species on the oxide support. The reducibility of the metal cations depends on various factors, such as the cation types and ratios, the formation of spinel, and temperature. One technique widely used here is temperature-programmed reduction (TPR). For example, ZnNiAlCr-LDHs, calcined at 500°C for 3 h with an H_2/N_2 (50:50 vol) gas mixture, can be used as hydrogenation catalysts (427,428) in which metallic Ni is essential for the activity and selectivity. The reduced NiMgAl–mixed oxide shows high catalytic activity for methane reforming with CO_2 to synthesis gas (429). The reduced CuMgAl-oxide can catalytically decompose nitric oxide via active Cu^I species and also reduce the nitric oxide with a coreductant such as propane via Cu^0 sites (430).

Few studies have appeared on the thermal decomposition of LDH organic derivatives, although in our opinion this may give rise to materials and reactions of some interest. Hibino et al. noted similarities in the decomposition behaviors of carbonate and organo-MgAl-LDHs in N_2 and found about 2% carbon left after calcination at 500°C, but they did not report results in air (431). Xu and Zeng (432) investigated in great detail the decomposition of (Co,Mg)Al LDH terephthalate in air and in N_2 by DTA, TGA, and TG-FTIR techniques and noted that: (a) dehydroxylation and decomposition of the organic anion in air occur in a narrow temperature range, showing a vigorous exothermic process and leaving very little carbon; (b) in N_2, dehydroxylation takes place at a much lower temperature than decomposition of organic anion, which shows an exothermic–endothermic–exothermic heat flow; (c) the carboneous residue after calcination in N_2 consists of various carbon nanoparticles. They also observed similar phenomena for CoAl-poly(vinylsulfonate)-LDH decomposition (433). Our recent studies show that the calcination of surfactant-LDHs in air can cause combustion. In addition, AlN is formed in the thermolysis of an LDH–polyacrylonitrile material (434). The decomposition of ZnCr-LDH surfactant sulfate and sulfonates has been also investigated with TG-MS, and results in the intermediates ZnO, ZnS, and sulfate at 500–1000°C and then $ZnCr_2O_4$ and Cr_2O_3 at higher temperatures (435). Thermolysis of MgAl-LDH with tris-(oxalato) ferrate(III) is a more complex process (400).

D. Miscellaneous Reactions

There are a few reported reactions in which ions from the metal hydroxide layer are removed or replaced. Prolonged exposure of MgAl-LDH to potassium or ammonium ferrocyanide leads to the formation of cubic potassium (or ammonium) magnesium ferrocyanide (84), while in a related reaction ferrocyanide-exchanged green rust gave beautiful thin, square crystals (436) of what, with hindsight, was most probably a ferrous ferrocyanide. The reaction in which Mg_2Al-LDHs disproportionate hydrothermally, to give Mg_3Al-LDH and finely divided aluminum hydroxide (see Sec. II.M), also formally belongs in this class.

As-prepared $Co^{II}Co^{III}$-LDHs can incorporate additional Mg^{2+} or Al^{3+} into their lattices (128). LDHs can also show exchange of constituent cations, with MgAl-LDHs being extremely selective for such transition metal cations as Cu^{2+}, Ni^{2+}, Co^{2+}, and Zn^{2+} (437). Reactions of this kind will no doubt receive more attention in the future, as a way of using LDH to remove cationic, as well as anionic, pollutants.

VI. CATALYSIS

As described previously, the cations in LDHs are evenly distributed in the brucite-like layers. Thus, in principle, the catalytic activity of LDHs can be well controlled by varying the cation ratio and incorporating different cations. Catalytically active constituents of LDH include the hydroxide groups and the metal ions themselves, especially if these are redox active. The introduction of catalytically active anions, such as polyoxometalates (POMs), can further modify the properties of LDHs. Thermal decomposition (calcination) of LDH gives mixed basic oxides of high surface area and catalytic activity. Finally, the reduction of LDH can give rise to finely divided catalytically active metal and to the prospect of metal/base bifunctional catalyst.

In order to name various type of catalysts more conveniently, we denote the original LDH(s) (wet or dried by below 150°C) as LDH(s), the calcined

LDH(s) as LDO(s) (layered double oxide), the reconstructed LDO(s) as LDOR(s), and the supported catalyst, such as Pt on LDH or LDO, as Pt/LDH (uncalcined) or Pt/LDO (calcined), respectively. In addition, LDH-POM means the LDH intercalated with polyoxometalate (POM) and LDO-POM the calcined LDH-POM.

In this section, we first consider reactions catalyzed by untreated LDH, classified by reaction type, followed by catalysis using LDH-POM assemblages, catalysts from LDH by calcination and/or reduction, and finally catalysts with an active component loaded on LDH or LDO.

A. LDHs as Catalysts

1. Oxidation

A particularly attractive feature of LDHs as oxidation catalysts is their ability to catalyze reactions using relatively cheap and nonpolluting oxidants, such as peroxides or even oxygen itself. Reasons for this may include the stabilization of oxidation states not normally stable in solution, such as Cu^{III} and Mn^{III}, in the highly basic environments experienced by the metal ions.

a. Hydroxylation

Copper-containing CuMAl-LDHs (M $=$ Co^{2+}, Ni^{2+}, Cu^{2+}, Zn^{2+}, and Fe^{2+}) have been used as catalysts by Liu et al. (438) in the hydroxylation of phenol to diphenols using H_2O_2 as oxygen source. They compared the activities of pristine and calcined LDHs and noted that the pristine LDHs have high activities, the highest of which is found in the CuAl(3:1)-LDH catalyst. Toluene, and o-, m-, and p-xylene undergo similar hydroxylation with H_2O_2 over CuZnAl-LDHs. As might be expected, copper is essential to the catalytic activity (439).

b. Alkyl Oxidation to Carbonyl

The methyl group attached to a benzene ring can be oxidized to aldehyde with H_2O_2 over CuZnAl-LDHs (439), competing with the hydroxylation. Velu et al. (316) prepared MgMnAl-LDHs or the mixed hydroxides, which were used directly as catalysts for the oxidation of toluene to benzaldehyde and benzoic acid, using tert-butyl-hydroperoxide as oxidant. Liu et al. (440) synthesized CoMAl-CO_3-LDHs (M $=$ Cu^{2+}, Ni^{2+}, Mn^{2+}, Cr^{3+}, and Fe^{3+}) and used these LDHs as well as the corresponding oxides as catalysts for the selective oxidation of p-cresol to p-hydroxybenzaldehyde with O_2. They found that the calcined CoCuAl-LDH (3:1:1) has the highest activity and selectivity. Cyclohexene was oxidized mainly to 2-cyclohexene-1-one and 2-cyclohexene-1-ol by oxygen with transition metal–containing LDHs (441). The catalytic activity varied with the type and loading of the transition metal ions in LDHs.

c. *Alcohol Oxidation to Carbonyl*

A number of aliphatic, allylic, and benzylic alcohols are selectively transformed to the corresponding aldehydes or ketones by using Pd^{II}-hydrotalcite mixture as catalyst. The oxidation readily occurs with the atmospheric oxygen as oxidant in high to excellent yields. It is noteworthy that the catalyst is also applicable to the oxidation of unsaturated alcohols, such as geraniol and nerol, without any isomerization of the alkenic part. The Pd^{II}-hydrotalcite mixtures are more active than the Pd^{II} complexes alone for this kind of reaction (442,443). This oxidation can be also efficiently catalyzed by RuCoAl-LDHs, RuMgAl-LDHs (444–446) as well as by RuCuAl-LDH (447) with various suitable cooxidants, including iodosylbenzene, tetra-butyl ammonium periodate, and oxygen.

d. *Ketone Oxidation to Ester*

Ueno et al. (448–452) examined the catalytic activity of many MgAl-LDHs incorporating partial transition metals in the Baeyer–Villiger oxidation of ketones and cyclic ketones to corresponding esters and lactones in the presence of molecular oxygen and aldehydes:

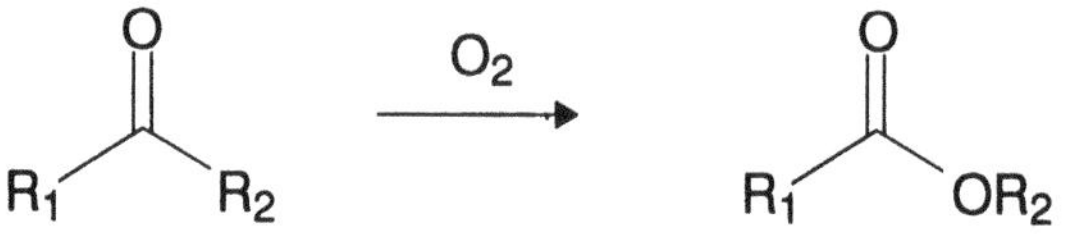

They observed that MgAl-LDHs with Fe^{2+}, Cu^{2+}, and Ni^{2+} partial incorporation are more active and that benzaldehyde is most effective for the oxidation among the various aldehydes tested.

e. *Epoxidation*

Epoxidation of α,β-unsaturated ketones is of great interest in organic synthesis, because the epoxide intermediates can be further converted to various functionalized ketones. MgAl-CO_3-LDH itself can be used to catalyze this oxidation for five-membered, six-membered, and open-chain α,β-unsaturated ketones in excellent yields under mild conditions using H_2O_2 as oxidizing agent. The catalytic activity of LDHs increases as the basicity of their surface increases. For less reactive open-chain ketones, adding a cationic surfactant such as dodecyltrimethylammonium bromide greatly accelerates the reaction (453–456). In particular, Choudary et al. used MgAl-*tert*-butanoate-LDH to catalyze the epoxidation of olefins and α,β-unsaturated ketones at very high rate (457a). Honma et al. reported highly efficient expoxidation of α,β-unsaturated ketones by H_2O_2 over a LDH catalyst made from MgO and Al_2O_3 (457b).

Fraile et al. reported the epoxidation of electron-deficient alkenes by H_2O_2 with LDH as catalyst. The activity of the LDH depends on the basicity of the solid, an LDH with a Mg/Al ratio of 3 having the highest activity. Some chiral alkenes derived from D-glyceraldehyde undergo similar catalytic epoxidation (458).

2. Base-Catalyzed Reactions

a. Isopropanol and 2-Methyl-3-Butyn-2-ol

The acid/base property of various LDHs can be tested through the reactions of some probe organic chemicals. For example, isopropanol undergoes reaction (A) catalyzed by LDH-CO_3 acting as a base (251,459–461):

$$CH_3CH(OH)CH_3 \rightarrow CH_3COCH_3 + H_2 \qquad (A)$$

However, it undergoes reaction (B) to generate propene through acid catalysis by LDH-$SiW_{11}M$ (M = Mn, Fe, Co, Ni, Cu, and Cu divalent ions) with high selectivity (251):

$$CH_3CH(OH)CH_3 \rightarrow CH_3CH{=}CH_2 + H_2O \qquad (B)$$

Another probe organic molecule is 2-methyl-3-butyn-2-ol, which reacts according to the nature of the catalyst (460–462):

$$(CH_3)_2C(OH)\text{-}CCH \rightarrow CH_3COCH_3 + HCCH \qquad \text{Basic catalysis} \qquad (C)$$

$$(CH_3)_2C(OH)\text{-}CCH \rightarrow CH_2{=}C(CH_3){-}CCH \qquad \text{Acidic catalysis} \qquad (D)$$

It was found that over LDH-CO_3 and its activated catalysts the probe molecule follows reaction (C), while over LDH-Cl, LDH-SO_4, and their activated catalysts it decomposes by a mixture of reactions (C) and (D).

b. Aldol Addition Reactions

Arrhenius, Eschenmoser, and colleagues have examined the reactions of the material formed by substitution of the glycolaldehyde phosphate dianion into $Mg_2Al(OH)_6Cl$, in connection with processes possibly related to the origins of life. Incorporation into the LDH promotes self-addition to form four- and six-carbon sugar di- and triphosphates. Of particular interest is the fact that the LDH alters the stereospecificity of the reaction, and that reaction in the presence of formaldehyde leads to pentose-(including ribose)-2,4-disphosphates (279–281).

3. Halide Substitutions

In 1986, Martin and Pinnavania first examined the nucleophilic reaction between alkyl bromides in toluene and the interlayer halides anions in $Zn_2Cr(OH)_6X$ (X = Cl^- or I^-) and found up to 80% conversion at 90°C (463):

$$LDH\text{-}Cl + RBr \rightarrow LDH\text{-}Br + RCl$$

This reaction also occurs between the solid LDH and vapor-phase alkyl halides. The substitution takes place mainly on the external surface of LDH, with rapid moving out of internal chloride and formation of LDH-Br.

Suzuki et al. (464,465) further used LDH-Cl as catalyst for the halide exchange between $C_6H_5CH_2Cl$ and C_4H_9Br in moderate yields:

$$C_6H_5CH_2Cl + C_4H_9Br \rightarrow C_6H_5CH_2Br + C_4H_9Cl$$

They proposed an initial exchange between C_4H_9Br and LDH-Cl, followed by exchange between LDH-Br and $C_6H_5CH_2Cl$. Several other alkyl bromides undergo similar reactions. They reported that LDH-$NiCl_4^{2-}$ and LDH-CN^- behaved similarly to LDH-Cl (227,466).

Hoshino and Shimada found that the catalyzed exchange occurs even between octyl halides and LDH-OAc or LDH-CN. For example, octyl bromide reacted with LDH-CN at 90°C for 40 h to give 72% $C_8H_{17}CN$ (467).

B. Pillared LDHs

Pillared LDHs are special in that there is freely accessible interlayer space resulting from the intercalation of large polyvalent anions, such as polyoxometalates (POMs) and phthalocyanines (Pc) (see Sec. II.D and III.A.3.b). POMs are acidic, introducing acidic properties to the basic LDH or LDO and making LDH-POM an acidic catalyst. The pillared LDHs can be also used as oxidation catalysts with intercalation of redox-active anions, such as Co-phthalocyaninetetrasulfonate (CoPcTs). The usage of pillared LDH is described according to the reaction types in the following.

1. Acid-Catalyzed Reactions

The presence of acidic sites on the POM pillars enables the LDH-POM to act as acid/base bifunctional catalysts or even as strongly acidic catalysts (251–253,468). As mentioned previously, isopropanol is a probe molecule for distinguishing between acidic and basic catalysis. For example, ZnAl-$SiW_{11}MO_{40}$-LDH (M = Mn, Fe, Co, Ni, Cu, and Zn divalent ions) has been used as the catalyst for the decomposition of isopropanol to give propene almost exclusively (98%, reaction (B)) at 90% conversion (251).

Another example is the acid catalyzed esterification of acetic acid and *n*-butanol (251,469,470). High selectivity (>99.8%) with moderate conversion (50–70%) was observed for the esterification in an equimolar mixture at 363 K for 5 h with ZnAl-$GeW_{11}O_{39}^{8-}$-LDH, ZnAl-$GeW_{11}CuO_{39}^{6-}$-LDH, and ZnAl-$GeW_{11}NiO_{39}^{6-}$-LDH catalysts (261). A similar reaction between BuOH and phthalic anhydride was catalyzed by ZnAl-$PW_{12}O_{40}^{3-}$ to give dibutyl phthalate with 100% selectivity in 82% yield (471).

The acidic property of LDH-POM was also demonstrated by the adsorption of basic species, such as NH_3 and pyridine (468,470). An IR investigation of pyridine adsorption showed the presence of both Brönsted and Lewis acid centers in LDH-POM, e.g., $Zn_2Al-XW_{11}O_{39}Co^{n-}$ (X = Ge^{4+}, B^{3+}, and Co^{2+}) (470).

The acid/base activity and their relative strength were found to depend on the preparation route. Pannavaia et al. swelled $MgAl-CO_3$ with glycerol and triethylglycerol, and then carried out the exchange with Keggin anion $H_2W_{12}O_{40}^{6-}$. They found that the catalyst from the glycerolate preswelling was more active than that from the triethylglycerolate route as a catalyst for 2-methyl-3-butyn-2-ol base-catalyzed decomposition (472).

2. Oxidation

a. Disulfide Formation from Thiols

The conversion of thiols to disulfides can be carried out over LDHs intercalated with phthalocyanine complexes or similar macrocyclic complexes at near room temperature and atmospheric pressure with oxygen as oxidant. For example, Iliev et al. (473,474), Liu et al. (475) and Pinnavaia et al. (476) investigated the oxidation of 2-mercaptoethanol, 1-octanethiol, and 1-decanethiol with the effective catalysts of MgAl-LDHs intercalated with Co-phthalocyanine (Pc) and Co-phthalocyaninetetrasulfonate (PcTs) complexes in the interlayer. These LDH catalysts are very stable and can be used repeatedly for long time. Corma et al. (477–479) found that aromatic and aliphatic thiols are stoichiometrically changed to the corresponding disulfides in the presence of ZnAl-LDHs intercalated with the $[MoO_2(O_2CC(S)Ph_2)_2]^{2-}$ anion. In addition, Hirano et al. (480) used the calcined MgAl-HT as the basic catalyst for air oxidation of a variety of aromatic, aliphatic, and alicyclic thiols in hexane to the corresponding disulfides with very good yields under mild and neutral conditions.

In the oxidation of mercaptans (thiols), LDH acts as both base and oxidant catalyst. The following reaction mechanism has been proposed (10):

$$2R\text{-}SH + 2LDH \rightarrow 2R\text{-}S^- + 2LDH\text{-}H^+ \tag{1d}$$

$$2LDH\text{-}Co^{2+}(Pc) + O_2 \rightarrow 2LDH\text{-}Co^{3+}(Pc) + O_2^{2-} \tag{1e}$$

$$2R\text{-}S^- + 2LDH\text{-}Co^{3+}(Pc) \rightarrow 2R\text{-}S^{\cdot} + 2LDH\text{-}Co^{2+}(Pc) \tag{1f}$$

$$2R\text{-}S^{\cdot} \rightarrow R\text{-}SS\text{-}R \tag{1g}$$

$$O_2^{2-} + 2LDH\text{-}H^+ \rightarrow 2LDH + H_2O + (1/2)O_2 \tag{1h}$$

The overall reaction is

$$2R\text{-}SH + (1/2)O_2 \rightarrow R\text{-}SS\text{-}R + H_2O \tag{1i}$$

b. Phenol Oxidation

Pinnavaia et al. also used LDH-M(PcTs)$^{4-}$ (M = metal ions in groups 7–10) as oxidation catalysts in the aerobic oxidation of 2,6-di-*tert*-butylphenol in water at 35°C (482–486), and the catalyst keeps its activity after 3200 turnovers:

This reaction could be useful in removing phenolic pollutants in effluents from paper and pulp mills.

c. Epoxidation

POM-pillared LDHs have also been used for the oxidation of alkenes to epoxides with H_2O_2 alone or more efficiently assisted by bromide. As previously discussed, α,β-unsaturated ketones can be readily epoxidized with the help of normal LDH catalysts. So far, however, the epoxidation of nonactivated olefins requires the use of POM-LDH catalysts. For example, cyclohexene is selectively epoxidized using $Ni_2Al-SiW_{11}O_{40}^{4-}$ as catalyst, with the combination of oxygen and aldehyde as oxidant (487,488). The epoxidation of 2-hexene and β-methylstyrene was also investigated by the same authors.

Sels et al. found that LDH-$V_{10}O_{28}^{6-}$ is a highly active catalyst for the epoxidation of geraniol with tetra-butyl hydroperoxide as an oxidant in anhydrous toluene (489). A series of allylic alcohols of terpene origin undergo the epoxidation under similar conditions. The same authors investigated in detail the catalytic activity of LDH-MoO_4^{2-} and LDH-WO_4^{2-} for the epoxidation of alkenes using H_2O_2 together with Br^- (490). They observed that the molybdate in LDH-MoO_4^{2-} reacts with H_2O_2 to form peroxomolybdate anions, such as $Mo(O_2)_4^{2-}$ and $Mo(O_2)_3O^{2-}$, and more particularly that the latter decays to $Mo(O_2)O_3^{2-}$ and singlet-state $^1O_2$ (491–493). This singlet-state $^1O_2$ takes part in various selective oxygenations, such as Diels–Alder-like cycloaddition to dienes to give their 1,4-endoperoxides (493). Tungstate in LDH also reacted with H_2O_2 to form peroxotangstates that transfer an oxygen atom to olefins, causing epoxidation. In addition, the regioselectivity can be controlled by cointercalating hydrophilic (e.g., inorganic) or hydrophobic (e.g., organic) anions (247,494). For example, LDH-

WO_4 with Cl^- or NO_3^- coexistence is more effective for the epoxidation of allylic alcohols, while $LDH-WO_4$ with p-tosylate favors the epoxidation of hydrophobic olefins, as shown by the following reaction:

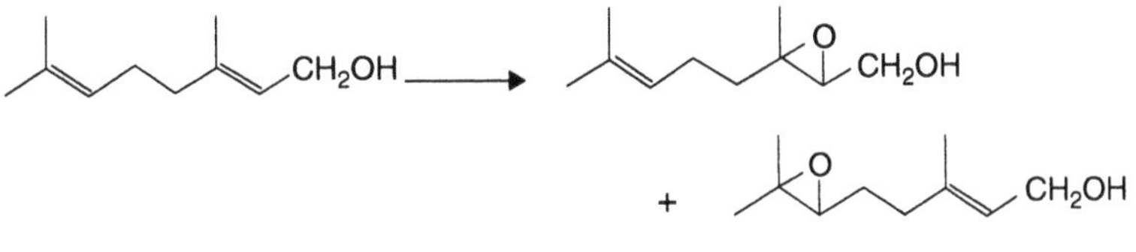

| LDH-WO4 with Cl | 55 : 45 |
| LDH-WO4 with p-Tosylate | 15 : 85 |

In addition, bromide anion has been found to assist epoxidation of olefin to some extent with high stability, high productivity, and high chemo-, regio-, and stereoselectivities (495,496).

d. α-H Oxidation

Cyclohexene was also found to undergo an oxidation reaction to give mainly α-H oxidation products 2-cyclohexene-one and 2-cyclohexene-ol over MAl-$XW_{11}ZO_{39}$-LDH (M = Zn, Mg; X = P, Si; Z = Mn^{2+}, Fe^{2+}, Co^{2+}, Ni^{2+}, Cu^{2+}). The use of P instead of Si increases the catalytic activity, while ZnAl and MgAl show the close activity. The LDHs with Co-, Cu-, and Mn-containing pillars in the LDH show higher activity than those with Ni- and Fe-containing pillars (497–500).

3. Alkylation

Xu et al. studied the alkylation of isobutane by butene with calcined MgAl- and ZnAl-$Si(W_2O_7)_6^{8-}$-LDHs and NiAl-$P(W_2O_7)_6^{7-}$-LDH as catalysts. These catalysts from POM-pillared LDHs show high actitivity and selectivity toward C_8 compounds (501,502).

4. Photocatalysis

MgAl-, ZnAl-, and ZnCr-LDHs pillared with iso- or heteropolyoxometalates $W_7O_{24}^{6-}$, $SiW_{11}O_{39}Mn(H_2O)^{6-}$, $SiW_{11}O_{39}Ni(H_2O)^{6-}$, $SiW_{11}O_{39}^{8-}$, $P_2W_{17}O_{61}Mn(H_2O)^{8-}$, and $NaP_5W_{30}O_{110}^{14-}$ show high photocatalytic activity in the degradation of hexachlorocyclohexane to give CO_2, HCl, and organic acids. The large vacant gallery hosts the guest molecule, and the photogeneration of OH radicals via the catalyst leads to the degradation (503–505).

In addition, $V_{10}O_{28}^{6-}$-pillared ZnAl- and NiAl-LDHs exhibit high activity in the photocatalytic oxidation of isopropanol to acetone (506).

C. Catalyst Formation from LDHs

LDHs are extensively used as precursors of oxide catalysts (LDOs). In general, LDHs can be changed into multimetal oxides by calcination at 400–500°C. As Markov et al. (140) have shown, the dehydroxylation of the brucite-like layers, i.e., the collapse of the layered structures, occurs topotactically at temperatures less than 500°C. This means that the diffusion of metal ions during this stage of calcination can be ignored. Even upon calcination at 600–700°C (this temperature is normally high enough for most catalytic preparation and reactions), the diffusion of cations should not be serious. Therefore, the calcined LDHs usually inherit a good cationic dispersion from the pristine LDHs, a characteristic much sought after in the development of many multimetal oxide catalysts. Moreover, incorporation of a range of different cations into the brucite-like layers makes these materials suitable for fine modifications of chemical composition and hence of catalytic properties.

1. Reduction

a. Hydrogenation

Metallic Ni or Ni compounds are known to effectively catalyze hydrogenation of unsaturated bonds. As described previously, Ni ions can be readily incorporated into the brucite-like layer and well dispersed with various other cations. The calcination of Ni-containing LDHs in a reducing atmosphere, e.g., H_2 or CO, leads to the reduction of Ni ions with the simultaneous formation of LDO. Such prepared hydrogenation catalysts feature fine Ni particles divided by the spinel and/or mixed oxides (LDO). Moreover, the catalytic performance can be optimized by varying the compositions, i.e., the properties of the accompanied spinel and/or mixed oxides (LDO), as shown later.

Rives et al. (427,428) studied the hydrogenation of acetylene to ethylene with multimetallic oxide catalysts prepared by calcining ZnNiAlCr-LDHs at 500°C for 3 h with an H_2/N_2 (50:50 vol) gas mixture. The redox property of Ni is essential for the activity and selectivity of the catalysts and the presence of ZnO decreases the coke formation.

Coq et al. (507–509) employed CoNiMgAl-LDHs as precursors of catalysts for the hydrogenation of acetonitrile and investigated the activity and selectivity of catalysts with different chemical compositions at 80–180°C; they found 98% conversion to ethylamine (507):

$$CH_3CN + H_2 \rightarrow CH_3CH_2NH_2 \tag{1j}$$

The catalysts were prepared by calcining at 120°C and then reducing at 620°C in H_2/N_2 atmosphere, leading to the formation of Ni metallic particles or CoNi bimetallic aggregates that are responsible for the activity and high selectivity.

They extended the catalytic application to the hydrogenation to other nitriles and to α,β-unsaturated aldehydes. These catalysts showed selectivity higher than 90% toward primary amines in hydrogenation of nitriles and 80% selectivity for hydrogenation of the C=C bond of α,β-unsaturated aldehydes (510a). Relatedly, adiponitrile can be selectively hydrogenated to aminocapronitrile by Ni-containing catalysts derived from LDHs (510b).

Similarly, Castiglioni et al. (511) examined the transformation of maleic anhydride to γ-butyrolactone through hydrogenation and dehydration over calcined CuZnM catalysts (M $=$ Cr^{3+}, Al^{3+}, or Ga^{3+}). A prereduced catalyst with a ratio of Cu:Zn:Al $=$ 38:38:24 gave γ-butyrolactone in 96% yield.

b. Hydrogenation of Phenol

Narayanan et al. (512–516) and Chen et al. (517) investigated Pd-containing MgAl-oxide catalysts for the selective conversion of phenol to cyclohexanone at 180°C. The incorporation of Pd into the brucite-like layer of MgAl-CO$_3$-LDHs or impregnation of MgAl-CO$_3$-LDHs with PdCl$_2$ enhances the catalytic performance in selectivity (>90%).

c. Nitro Reduction

A very early patent in the 1970s (518,519) on the application of LDHs concerned the reduction of nitrobenzene into aniline. Bröcker et al. prepared CoMnAl-LDH catalysts from the corresponding LDHs and obtained a 66% yield of aniline at 150°C and 133 atm. Actually, many LDO catalysts are found to promote the reduction of aromatic nitro compounds to anilines (350,520,521). For instance, Kumbhar et al. (350,520) found that calcined MgFe-LDHs are very efficient catalysts for the reduction of 4-nitrotoluene to 4-aminotoluene at 100% selectivity. They noted that these catalysts are also effective for the hydrogenation reduction of various other aromatic nitro compounds.

2. Alkane Dehydrogenation

Dehydrogenation of propane to propene was carried out in the presence of CO$_2$ over CrMgAl-LDO catalysts. The best catalysts had Cr:Mg:Al $=$ 1:1:1 (522). The catalysts were found to have both basic and acidic sites from the adsorption of NH$_3$ and CO$_2$, and this was thought to be responsible for the catalytic activity. Dula et al. examined the influence of interlayer doping using VCl$_3$ versus brucite-layer doping of vanadium species (V$_{10}$O$_{28}^{6-}$) into MgAl-LDHs on catalytic performance in propane dehydrogenation. They concluded that the interlayer-doped catalysts showed higher activity and selectivity than the brucite-layer-doped ones (322).

Fritz et al. used Pt-Sn-LDH-derived LDO catalysts in the dehydrogenation of C$_2$–C$_{10}$ alkanes (523). Li et al. investigated similar Pt/MgAl-LDO and PtSn/

MgAl-LDO (0.37% Pt and Pt-Sn) for the dehydrogenation of butane and observed that n-butane conversion and butene selectivity are both higher than Pt/Al$_2$O$_3$ and PtSn/Al$_2$O$_3$ (524). The reaction kinetics of n-butane dehydrogenation with O$_2$ to butenes (1-, 2-cis-, and 2-$trans$-butene), butadiene, and carbon oxides over calcined V/MgAl-LDOs has been studied by Dejoz et al. Two possible mechanisms were proposed, i.e., a hydrogen abstraction process responsible for the formation of olefins and diolefins and a radical process responsible for the formation of carbon oxides (525).

ZnFeAl-LDH-derived catalysts have been developed for the dehydrogenation of ethylbenzene to styrene in the presence of CO$_2$ as an oxidant, and the highest catalytic activity is reached with an oxide catalyst with Zn:Fe:Al = 6:1:2 (526). Malherbe et al. tested the catalytic activity of MgAl- and ZnAl-LDH with various anions (e.g., CO$_3^{2-}$, Cl$^-$, PO$_4^{3-}$, and V$_2$O$_7^{4-}$) for this reaction. They found that ethylbenzene is dehydrogenated to styrene with high selectivity (98%) but with only 38% conversion at 450°C over a V-containing catalyst (527).

3. Aldol Condensations

Aldol condensations are often catalyzed by basic oxides, of which the LDOs (calcined LDHs) are good examples. The condensations always give byproduct water, which can cause LDOs to be rehydrated (reconstructed) into LDHs. Fortunately, the rehydration does not hurt the catalytic performance, but enhances the activity, because of the high basicity of newly reconstructed LDHs.

Ketones or aldehydes are often catalytically condensed to dimeric or higher-membered β-hydroxyl ketones or aldehydes over base catalysts. The condensation of acetone was extensively investigated by Reichle and other researchers (216,528–532), who examined the activity and selectivity of many calcined LDHs:

Mesityl oxide   Isophorone

The results showed that the catalyst derived from MgAl-CO$_3$-HT is the most active and very selective. Di Cosimo et al. (533,534) have also used MgAl-LDOs to synthesize α,β-unsaturated ketones, and Kelkar et al. (535a) have investigated the optimum conditions for α-isophorone production. In addition, Climent et al. found that activated LDHs show higher activity and selectivity than aluminophosphate or KF-alumina for the aldol condensation between citral and methyl-ethyl ketone (535b).

Another example is the synthesis of 2-ethyl-2-hexenal from the condensation of n-butyraldehyde over MgAl-CO$_3$-LDO catalysts. At lower than 200°C, the reaction occurs at close to 75% conversion with more than 80% selectivity.

More interestingly, the catalytic activities for condensation (base related) and hydrogenation (redox related) can be combined into dual-functionized catalysts. Unnikrishnan et al. (536a) have shown that NiMgAl-LDO selectively gives methyl isobutyl ketone under mild conditions (78% selectivity and 48% conversion at 100°C and 1 atm). Das and Srivastava and Tichit et al. reported a similar reaction over PdMgAl-LDH-derived catalysts (536b,536c). Moreover, the intermediate methyl isobutyl ketone can be further hydrated to methyl isobutyl carbinol (100% selectivity at 18% conversion) over CoMgAl-LDO. This one-step combined condensation-hydrogenation process was further used to synthesize 2-ethyl-hexanol from n-butyraldehyde. Over a Pd/MgAl-LDO catalyst (1 wt% Pd), the aldehyde was transformed to the target compound at 68% conversion with 90% selectivity (537).

Claisen–Schmidt condensation, i.e., the condensation of benzaldehyde and acetone to α,β-unsaturated ketones, has been performed to a high degree at 0°C with rehydrated oxides from LDH precursors by Rao et al. (532).

Tichit et al. (538) and Kantum et al. (539) examined the substitute effect on the aldol condensation of various benezaldehydes with acetone using modified LDHs. As expected for base catalysis, electron-withdrawing substituents on the benzene ring increase the reaction rate. This has been demonstrated by the quantitative reaction of p-chlorobenzaldehyde with excess acetone at 60°C within 1.5 h. Choudary et al. (216) intercalated tetra-butanoate anion into MgAl-LDH in THF, which catalyzes condensation reactions selectively to aldols only at 0°C in quantitative yields in acetone.

There have been many other investigations of the applications of LDO catalysts to Claisen–Schmidt condensation (540). For example, with activated MgAl-CO$_3$-LDH as catalyst, benezaldehyde and 2-hydroxylacetophenone react at 50°C to give the desired compounds (2-hydroxychalcone and flavanone) at 80% conversion and 78% selectivity (541). As another example, the condensation of citral and acetone produces an intermediate that can be finally transformed to chemically useful ionones (542,543). The use of calcined LDH results in 70% selectivity at about 95% conversion. In addition, Dumitriu et al. (544) and Suzuki and Ono (531) have studied the condensation of formaldehyde with acetaldehyde or acetone over LDOs and noted that there are both self- and cross-condensations.

The LDOs as catalysts are also very effective in the Knoevenagel reaction, i.e., cross-aldol condensation of a carbonylic compound with an active hydrogen compound, as shown in following equation (539,545–548)):

(R_1, R_2 = H, alkyl, phenyl, etc.; R_3, R_4 = COR, CN, COOR, SO_2R etc.) For example, 2-furfural and ethyl cyanoacetate react to generate 100% Knoevenagel product using a calcined and rehydrated MgAl-LDH (539).

4. Alkylation

Alkylation of phenol has been studied very extensively in recent years as a route to a range of valuable chemical materials, as shown by the following reaction network:

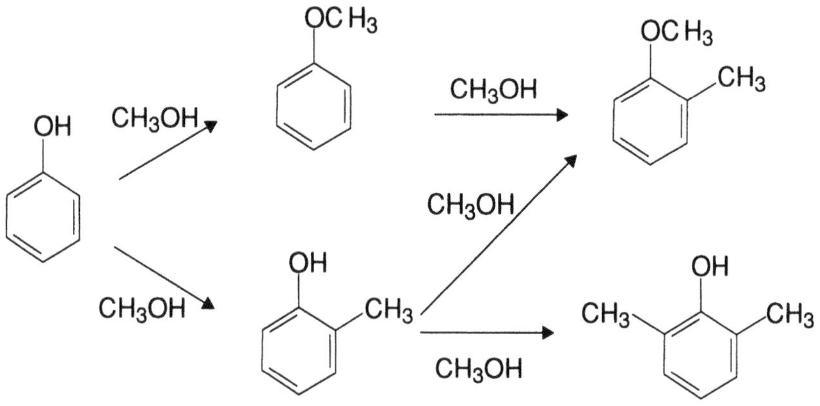

With the MgAl-CO_3-LDOs (Mg/Al = 3–10) as catalysts, Velu et al. noted that the major products are anisole, o-cresol, and 2,6-xylenol. The product distribution, and in particular the ratio of O- to C-methylation, is dependent on temperature, phenol/methanol ratio, contact time, pressure, and catalyst composition (549,550). With the calcined MgFe-, MgCr-, and CuAl-LDHs as catalysts, direct C-alkylation is dominant (461,551,552). Kinetic parameters were obtained from the alkylation of phenol with methanol over various catalysts derived from MAl-LDHs (M = Mg, Mn, Co, Cu, and Zn divalent ions) (553). The alkylation of phenol with various other reagents, such as C_1–C_8 alcohols (554), isobutanol (555a,555b), and dimethyl carbonate $(CH_3O)_2CO$ (556), over various catalysts have also been examined in detail. Calcined MgAl-LDHs (Mg/Al = 3) can be used as an efficient catalyst in the selective O-methylation of phenol and catechol and N-monomethylation of aniline employing $(CH_3O)_2CO$ as methylating reagent in the vapor phase at 275°C. The alkylations of phenol-like compounds, such as *m*-cresol (557a,557b) and catechol (556,558,559) with alcohols, and dimethyl carbonate over LDOs have been further investigated.

Mg$_{2.8}$Al-LDO catalyzes the direct C-alkylation of 2,4-pentanedione with alkyl halides (560). For example, C-monoethylation of 2,4-pentanedione with ethyl iodide occurs at 100% selectivity and 87% conversion. Similarly, Choudhary et al. used MgGa-LDHs that were treated with gaseous HCl as catalysts for benzylation of benzene and toluene with benzyl chloride at high conversion (561a,561b).

The methylation of aniline with methanol over calcined MgAl-LDHs results in an almost single product, i.e., *N*-methylaniline, in up to 68% yield. LDHO catalysts are more active than mixed MgO-Al$_2$O$_3$ (562). Aniline can also be methylated to the same product, with dimethylaniline as a minor product, using dimethyl carbonate over an MgAl-LDO catalyst at 275°C (556). Another example is the LDO-catalyzed alkenylation of formamide with vinyl formate to give *N*-vinylformamide. The intermediate *N*-vinylformamide can then be polymerized into poly(*N*-vinylformamide), which can be converted into polyvinylamine by hydrolysis (563).

5. Acylation

Acylation of compounds with a labile H atom, such as alcohols, NH$_3$, amines, and amides by alkyl carbonate leads to various carbonate esters and carbamates:

$$\underset{R_1}{\overset{O}{\|}}\!\!\!\!\!\!OR_2 \; + \; HXR_3 \longrightarrow \underset{R_1}{\overset{O}{\|}}\!\!\!\!\!\!XR_3 \; + \; HOR_2$$

(X = O, NH, NHCO, etc.). This reaction is catalyzed with solid base catalysts such as LDOs. King et al. and Bastrom used MgAl-LDOs as catalysts to synthesize ethers, amines, aminoethers, and carbamates from suitable precursors, i.e., various groups R$_1$ and R$_3$ (564–570).

Recently, Choudhary et al. treated MgGa-LDH with HCl gas to prepare a catalyst for the acylation of benzene and toluene with benzoyl chloride. They attributed the high activity of this catalyst to the existence of well-dispersed gallium and magnesium chlorides on MgO (561a,561b).

LDOs have been also used as catalysts for *trans*-esterification (571), including the efficient reaction of triglyceride with glycerol to give mono- and diglycerides (572):

$$
\begin{array}{ccc}
\begin{array}{l}\text{CH}_2\text{OCOR}\\ \text{ROCOCH}\\ \text{CH}_2\text{OCOR}\end{array}
& + &
\begin{array}{l}\text{CH}_2\text{OH}\\ \text{HOCH}\\ \text{CH}_2\text{OH}\end{array}
& \longrightarrow &
\end{array}
$$

$$
\begin{array}{l}\text{CH}_2\text{OH}\\ \text{ROCOCH}\\ \text{CH}_2\text{OCOR}\end{array}
$$

$$
+ \;\; \begin{array}{l}\text{CH}_2\text{OCOR}\\ \text{HOCH}\\ \text{CH}_2\text{OCOR}\end{array}
$$

$$
\begin{array}{l}\text{CH}_2\text{OH}\\ \text{ROCOCH}\\ \text{CH}_2\text{OH}\end{array}
$$

$$
+ \;\; \begin{array}{l}\text{CH}_2\text{OCOR}\\ \text{HOCH}\\ \text{CH}_2\text{OH}\end{array}
$$

Block et al. extended their research to the *trans*-esterification of $(CH_3O)_4Si$ or $(C_2H_5O)_4Si$ with alcohols (ROH) to prepare colorless tetra-alkoxysilanes $Si(OR)_4$ in the presence of activated organic and inorganic LDHs. For example, $(CH_3O)_4Si$ and 2-ethylhexanol was heated at 120–200°C for 2 h over lauric acid–treated LDH to give 95% conversion to the colorless product $Si(OCH(C_2H_5)C_5H_{11})_4$ by removing CH_3OH with distillation (573).

6. Cyanoethylation

Cyanoethylation is the addition of compounds with labile H atoms, such as ROH and RSH, to acrylonitrile:

$$
\text{RXH} \;\; + \;\; \overset{}{\diagup}\!\!\diagdown\text{CN} \;\; \longrightarrow \;\; \text{RX}\diagdown\!\!\diagup\!\!\diagdown\text{CN}
$$

Kumbhar et al. found that calcined and rehydrated $MgAl-CO_3-LDH$ is a highly active, reusable and air stable catalyst for the cyanoethylation of primary and secondary alcohols with very high conversion and selectivity at 50°C (574). Choudary et al. reported that MgAl-tetra-butanoate-LDH can be directly used as catalyst in the cyanoethylation of alcohols and thiols with very high conversion and selectivity at room temperature (575).

7. Epoxide Ring Opening

Under the proper conditions, alkylene oxides can react with other chemicals by ring opening over LDH-derived catalysts. For example, the hydration of ethylene oxide over $NiAl-VO_3{}^{-}-LDO$ catalyst at 150°C produces glycol in 95% yield (576,577).

The reaction of ethylene oxide with n-butanol gives 2-butoxyethanol (527,578) in very high yield over MgAl- and CuCr-LDO catalysts. Epoxides, such as propylene oxide and styrene oxide, can also react with CO_2 to form cyclic carbonates with MgAl-oxide catalysts (579). On the industrial scale, ethylene oxide and propylene oxide can be polymerized into water-soluble polyols over an MgAl-CO_3-LDO catalyst (580–583).

8. Miscellaneous

a. Michael Addition

Michael addition, i.e., the 1,4-addition of α,β-unsaturated ketones to a compound with a labile H atom, often leads to the formation of a single compound:

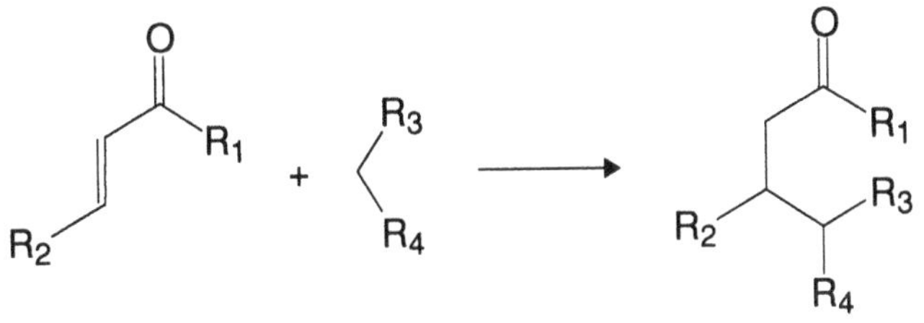

(R_3, R_4 = CN, COR, etc.). Choudary et al. reported that calcined MgAl-CO_3-LDH is a very active catalyst for this addition (584a,584b).

b. Sulfur Oxidation

As an alternative to the hydrodesulfurization reaction (HDS, Sec. VI.D.1.), the removal of refractory sulfur-containing compounds can be accomplished by oxidation with hydrogen peroxide in the presence of a catalyst. Thus Palomeque et al. found that dibenzothiophene can be fully oxidized to sulfone in a nitrile solvent over activated MgAl-LDO at 60°C (585a), where the resulting sulfone can be extracted by the solvent.

c. Methylamine Synthesis

The diversity in catalytic selectivity of substituted LDHs is clearly shown by Carja et al.'s work on methylamine synthesis from methanol and ammonia (585b). It was found that monomethylamine is the favored product over Cu-containing MgAl-LDO, while dimethylamine is the main product over Fe-containing MgAl-LDO. When MgAl-LDO is used as the catalyst, trimethylamine is preferentially synthesized.

d. Meerwein–Ponndorf–Verley Reduction

MgAl-LDO is also an active catalyst for the Meerwein–Ponndorf–Verley reduction of cycloalkanones and substituted cyclohexanones, with the conversion over

95% and selectivity 100%. In particular, the reduction of 4-*tert*-butylcyclohexa-none to 4-*tert*-butylcyclohexanol gives a high stereoselectivity (cis:trans ratio > 12) (585c).

D. LDHs as Catalyst Supports

The LDOs, i.e., calcination derivatives of LDHs, are often used as catalyst supports. As mentioned previously, LDOs are mixed multimetal oxides with the cations well dispersed. Due to the flexibility of compositions (nature and ratios of cations) and the good cationic dispersal, the basic strength of LDOs as supports can be finely tuned. As a consequence, the activity and selectivity of loaded active components can be adjusted or controlled, at least to some degree, to meet the requirement for the specific catalytic reactions.

LDHs are normally moderately calcined, e.g., at 400–500°C into LDOs. The active components, such as Ru, Rh, and Pd, can be loaded by immersing LDOs in the solution containing salts or complexes of the metal of interest. A second calcination then gives rise to the LDO-supported catalysts. LDO-supported catalysts can be also made by simply loading the active components with drying but without calcination.

The use of such prepared catalysts has already been investigated for various reactions, including hydrogenation and hydrodesulfurization (HDS), polymerization of ethylene, synthesis and application of synthesis gas (syngas), and some other reactions, as briefly reviewed next.

1. Hydrogenation and Hydrodesulfurization (HDS)

As described previously, acetone can be transferred to α,β-unsaturated ketone (mesityl oxide) by condensation and then hydrogenated to methyl isobutyl ketone over NiMgAl-LDO catalysts (536a). The same reaction can be accomplished in one step over Pd supported on MgAl-LDO (586,587).

Sulfiding CoMo or NiW oxides supported on LDO are also employed in hydrodesulfurization of thiols (588,589). Xu and Zeng reported that CoAl-MoO_4^{2-}-LDO catalysts are also active for the removal of sulfur from thiophene to give H_2S and butene (1-, 2-*cis*-, and 2-*trans*-butene) as well as butadiene (590a). In addition, Zhao et al. loaded Mo and Co directly onto the dried LDHs and then calcined at 500°C. They report that a catalyst on a hydrotalcite support is superior to catalysts with alumina supports for the selective hydrodesulfuriza-tion of gasoline in catalytic cracking (590b,590c).

2. Polymerization of Ethylene

There are several patents comparing the use of various LDH-supported $TiCl_4$ catalysts in ethylene polymerization (591–598). The use of VCl_3 on an LDH support is also reported as catalyst for this reaction (592).

3. Synthesis Gas (Syngas)

Synthesis gas ($CO + H_2$) can be obtained cheaply from coal, oil, or natural gas and is a key material in some routes to bulk organic compounds, such as methanol, acetic acid, alcohols, and hydrocarbons. LDO catalysts have been intensively explored in both the production and the use of this material.

a. Manufacture

The partial oxidation of methane with oxygen according to the idealized reaction

$$2CH_4 + O_2 = 2CO + 4H_2$$

is catalyzed by many Ni-, Ru-, Rh-, Pd-, Ir-, and Pt-supported active catalysts, as claimed in a series of patents by Amoco (176b,599,600). Basile et al. incorporated noble metal ions, such as Rh, Ru, Ir, Pd, and Pt, together with Ni into LDHs that are then calcined and reduced to the LDO catalysts. As a general observation, the presence of low-valence transition metals and noble metals reduced during the calcination is essential for the oxidation of light alkanes (601–604). Shishido et al. impregnated Ni into MgAl- and CaAl-LDOs and observed the high activity and selectivity for syngas formation through the foregoing reaction. The high activity and selectivity in the CaAl-LDO case is probably due to the stable and highly dispersed Ni metal particles in addition to the basic property of CaAl-LDO as the support (605,606). In a related reaction, partial oxidation of other light paraffins, such as C_3H_8, has been also investigated with LDO-supported catalysts (607,608).

In the preparation of syngas from methane, there is always some total oxidation of methane to CO_2 and H_2O as well as coke formation. Coke formation can be reduced by improving the catalytic selectivity, while CO_2 and H_2O can be reused in reforming reactions:

$$CO_2 + CH_4 = 2CO + 2H_2$$

$$H_2O + CH_4 = CO + 3H_2$$

These reactions can be also catalyzed with LDH-contained or-supported transition metal or noble metal ions (175,609). Bhattacharyya et al. found NiMgAl-oxide catalysts from their pristine LDHs are very active and selective in these two reforming reactions (609). The NiMgAl-oxide catalyst can be also made through intercalation of $Ni(EDTA)^{2-}$ into the interlayer and then calcination, showing high activity as well as stability and reusability (429).

b. Application

Syngas can be converted to common chemicals, such as methanol and hydrocarbons, over LDH-derived catalysts. For example, it converts to methanol over

ZnCr- and CuZnAl-LDOs. LDO catalysts generally give rise to high selectivity for methanol, while surface doping with Cs on mixed oxides increases catalyst stability (162,610–612). Higher alcohols can be also obtained with similar catalysts but at higher temperature and with a lower CO/H_2 ratio (610,611). The doping of alkali, such as Cs, promotes the formation of branched alcohols (613). In addition, Ru supported on LDH-derived oxides, exhibits substantial selectivity toward alcohols, mainly methanol, at moderately low pressures (614).

Hydrocarbon synthesis from syngas (Fischer–Tropsch reactions) can be carried out over the catalysts prepared from Co- and Cu-containing LDHs. The products include methane, higher paraffins, and olefins as well as methanol. The loading of Co and Cu determines the selectivity for each compound. For instance, Co-rich catalysts give more paraffins, while Co-poor ones lead to methanol (615).

4. Steam Reforming of Methanol (SRM)

Catalytic steam reforming of methanol is a well-established route for the production of H_2 for fuel cells:

$$CH_3OH + H_2O \rightarrow 3H_2 + CO_2$$

Cu-containing catalysts, especially CuZn or CuZnAl mixed oxides, have been found to possess high activity and high selectivity (616–622a), and the active sites are thought to be the reduced metallic Cu species (616). As a suitable precursor for the preparation of mixed oxide catalyst, LDH should have its place in this catalytic reforming, and there are actually some research reports (619–622a) regarding this issue. Kearns et al. investigated CuZn-containing catalysts calcined from the their LDH precursors (619,620). Their results show that the presence of an LDH phase has an important effect on the activity and that CuZnCr-catalyst is most active at 250°C, with little deactivation. Velu et al. used NiAl- and CoAl-LDH as the catalyst precursors and obtained very high selectivity of H_2 and CO_2 at 300°C over the CoAl-LDO catalyst (621). More detailed investigation by Segal et al. (622a) has shown that 50–60% methanol conversion and 98–99% selection to H_2 and CO_2 are obtained over CuAl-LDO at 400°C, much more active and selective than NiAl- and CoAl-LDOs. They further noted that the activity of as-prepared CuAl-LDH used directly in situ is higher by 5–10% methanol conversion than that of CuAl-LDO derived via calcination in flowing air at 400°C for 4 h and then reduction in H_2/He stream for 1 h at the same temperature. This difference is probably attributed to the difference in dispersion of Cu species and to the surface area, as well as to the Cu oxidation states. Shishido et al. noted that Ni/MgAl-LDH-derived catalyst has high and stable activity for the CH_4 reforming, presumably due to the formation of highly dispersed and stable Ni metal (622b). Velu et al. have further investigated the oxidative reforming of ethanol over CuNiZnAl-LDOs for hydrogen production (622c).

5. Other Reactions

a. CN⁻ removal

LDH-intercalated phthalocyanines can be used for the detoxification of ground-water polluted with cyanides. Keruk used MgAl-oxide–supported $Co(PcTs)^{4-}$ as catalyst in his patent for conversion of CN^- into environmentally benign CO_2 and N_2 in the presence of O_2 (623):

$$4CN^- + 5O_2 + 2H_2O \rightarrow 4CO_2 + 2N_2 + 4OH^-$$

b. Dehydrocyclization

A platinum catalyst with MgAl-LDH-derived support can catalyze the dehydro-cyclization and isomerization of *n*-hexane and shows higher activity than that of Pt/Al_2O_3 catalysts (624).

c. Wittig Reaction

The Wittig reaction, normally used to synthesize alkenes with unambiguous positioning of the double bond, can also be catalyzed by the LDH-related catalysts. Sychev et al. reported the usage of MgAl- and ZnAl-LDHs and their LDOs for the liquid-phase Wittig reaction and found that the calcined LDHs are more active for this reaction (625a).

d. Heck Reaction

Activated Pd-containing LDH in brucite layers or supports is an efficient catalyst for the Heck reaction, i.e., the reaction between aryl halides and olefins to give C—C coupled products (625b,625c). For example, an LDH with Pd:Mg:Al = 4:100:40 gives 100% conversion, over 80% yield, and over 99% trans product for the reaction of iodobenzene with methyl acrylate at 140°C for 4 h (625b).

E. Summary

The applications of LDHs as catalysts, catalyst precursors, or catalyst supports for a wide range of organic chemical reactions have just been reviewed in detail. Many examples have demonstrated that LDHs can be made in various ways into efficient catalysts for oxidation/reduction, hydrogenation/dehydrogenation, alkylation/acylation, additions and condensations, etc. The catalytic activity and selectivity of these catalysts are determined by chemical compositions (e.g., cations, cation ratios, and anions), activation processes (e.g., drying, calcination, oxidation/reduction, and/or reconstruction), as well as reaction conditions (e.g., temperature, pressure, atmosphere, and solvent). In general, the LDH or LDH-related catalysts with higher activity and selectivity for a specific reaction can be readily obtained due to the versatile choice of cations and anions, the good cationic dispersion, the controllable basic/acidic strength, and the adjustable metal ion oxidation state. It is thus anticipated with confidence that there will be numbers of LDH-related novel catalysts developed for more organic reactions in the future.

VII. OTHER APPLICATIONS

A. Environmental Remediation

1. Removal of Nitrogen and Sulfur Oxides

N_2O is well known as a greenhouse gas and a depletor of the ozone layer (626,627). Its catalytic thermal decomposition to ambient N_2 and O_2 over metal oxides can be dated back to the early 1930s (626–639). Well-dispersed multimetal oxides derived from LDHs (LDOs), especially for those containing small amounts of transition or noble metals, show rather high activity for this process (165,176a,640–657). Kannan et al. (640,641–644) tested the activity of many calcined LDHs (LDOs) and reported that calcined Co–containing LDHs (LDOs) with very small amounts of La^{3+}, Rh^{3+}, and Ru^{3+} (less than 1%) have especially high activities. Zeng et al. (130,133,134,136) also found that the $MgCo^{II}Co^{III}$-LDOs are very active catalysts for N_2O decomposition. Oi et al. (176a,645) examined the dependency of the catalytic activity of ZnAl-LDOs and other oxide catalysts on the loading of Rh, the Zn/Al ratio, and the feed gas compositions (concentration, water content, NO_2 content, etc.). Perez-Ramirez et al. and others (163,164,646,647) further confirmed that calcined Rh-loaded CoAl-LDHs show a very high activity for N_2O and NO_x decomposition and that CoPdLaAl-LDO (3:1:1:1) has comparable activity. Drago et al. (648) and Centi et al. (649) compared the activities among the catalysts derived from different precursors, such as LDHs, loaded ZSM, and supported hydroxides, as a function of several variables in the preparations, catalytic compositions, and feed gas constitutions. In addition, Dandl and Emig (650) studied the kinetics of the decomposition of nitrous oxide over CoLaAl-LDO and calculated the reaction rate constant and activation energy. The most active catalysts can give over 90% decomposition at 400°C and atmospheric pressure (130,133,134,136,640,648).

Nitrogen oxides can also be removed by catalytic reduction (651). For example, Wen et al. added CO into the NO_x mixture as reducing agent and examined the reduction of NO_x over CuMgAl-LDO in the presence of O_2, H_2O, or SO_2. They concluded that the activity-temperature window of catalyst shifts to lower temperature with increasing CuO content (652). Marquez et al. used in situ XPS to study the active copper species derived from CuMgAl-LDOs in the catalytic removal of NO_x by reaction with hydrocarbon. They concluded that Cu(0) is responsible for NO decomposition and Cu(0)/Cu(I) for NO reduction with a hydrocarbon (653). Giroir-Fendler et al. further investigated the effect of support acidity (Pt on hydrotalcite, SiO_2, SiO_2-Al_2O_3, and ZSM-5 zeolite) on the catalytic reduction of NO by C_3H_8 under lean-burn conditions (654a). Christoforou et al. investigated various supporting catalysts, i.e., metals (Rh, Ru, Pd, Co, Cu, Fe, In) on different supports (Al_2O_3, SiO_2, TiO_2, ZrO_2 and calcined hydrotalcite $Mg_3Al(OH)_8 \cdot 2H_2O$), for the decomposition of N_2O to N_2 using various hydrocarbons (CH_4, C_3H_6, C_3H_8) as reducing agent (654b).

Catalysts for NO_x decomposition and reduction are often deactivated in the presence of SO_2, which is formed from sulfur in fossil fuels and usually accompanies NO_x in vehicle exhaust. Vaccari et al. impregnated both Pt and Cu onto MgAl-LDHs and found that the coexistence of both metals enhances the activity for NO_x and considerably promotes the resistance to deactivation after repeated hydrothermal treatment and in the presence of SO_2 (655a,655b). In 2002, Li claimed the development of SO_x-tolerant NO_x trap catalysts from Pt-LDH compounds (656). Corma et al. studied the simultaneous removal of SO_2 and NO_x (657,658) by using CoMgAl-LDOs and CuMgAl-LDOs as catalysts. Their experiments show that the redox property of Cu or Co makes it the active site for the reduction of NO with propane. They further observed that the CuMgAl-LDH-based catalyst can simultaneously remove SO_2 (by an oxidative and/or reductive reaction) and NO (by a reduction and/or decomposition), while the addition of an oxidant such as cerium oxide on CoMgAl-LDH is necessary to oxidize SO_2 to SO_3.

2. Water Purification

LDHs can be used as anion exchangers on account of the accessibility of the interlayer region (199,204). These materials and the derived mixed oxides act as anion scavengers not only for common anions, such as CO_3^{2-}, $C_2O_4^{2-}$, SO_4^{2-}, Cl^- (659a), and SeO_3^{2-} (659b), but also for poisonous oxometalates and oxononmetalates (199,660–663). For example, the removal of chromate ions (CrO_4^{2-} and $Cr_2O_7^{2-}$) in aquatic solution is quite successful, and the sorption capabilities of ZnAl-Cl-LDH for CrO_4^{2-} and $Cr_2O_7^{2-}$ are close to 1 and 0.6 mmol/g, respectively (660–662). In the case of SeO_3^{2-}, the adsorption reaction has been shown to be exothermic (659b). The sorption capabilities of the calcined LDHs have been tested in detail by Kovanda et al. (199) through repeated calcination–rehydration–anion exchange processes. During the processes the removal of arsenate (AsO_4^{3-}) and vanadate (VO_4^{3-}, $V_2O_7^{4-}$, etc) was also successfully carried out. The adsorption of arsenite is also investigated by You et al., with LDHs and their oxides as absorbents (664). In addition, the uptake by calcined LDHs of phosphate anions was examined (158b,665), and Kindaichi found that LDH is an effective material for phosphate removal in the human excrement treatment processes, even in the presence of sulfate and chloride (666).

An ingenious application of LDH properties is the removal of Al^{3+} from the recovery system of a closed kraft pulp mill (667) and seawater (158b). This was realized by adding $MgSO_4$ into the pulp mill, where MgAl-LDH precipitated faster and formed bigger particles than $Al(OH)_3$ on its own, making the solid filtration easier (667). The utilization of the anion exchange properties of LDHs, the memory effect of the calcined LDHs, and the adsorption capability of the positively charged brucite-like surface have been extended to the removal of organic pollutants, such as substituted phenols (668–671), pesticides (672–676), some humic substances (677–679), some surfactants (392,393,680,681a,681b), and acidic organic species (682a,682b) in water purification. Cu- and Zn-containing LDHs have been found to be effective antibacterial disinfectants in this context (683).

3. Nuclear Waste Treatment

Nitrate-exchanged LDHs have been widely studied for nuclear waste treatment by removal of anions formed by ^{235}U fission products, such as $^{131}I^-$, $^{99}TcO_4^-$, $^{99}MoO_4^{2-}$, $H^{99}MoO_4^-$ and neutral $H_2^{99}MoO_4$, $H_2^{132}TeO_4$, $^{103}Ru(NO)(OH)_5^{2-}$, $^{103}Ru(NO)(OH)_4(H_2O)^-$ and neutral $^{103}Ru(NO)(NO_3)(OH)_2(H_2O)_2$, and $^{186}ReO_4^-$ at pH $= 5$ (684–687). Oscarson et al. (684) and Fetter et al. (685) used LDH-nitrates to absorb $^{129}I^-$ and $^{131}I^-$. Carbonate from ambient CO_2 interferes (684), but calcination improves iodide uptake (203,686). The adsorption of ReO_4^- and TcO_4^- with calcined LDHs is about five times less than that of I^-, but I^- and ReO_4^- do not interfere with each other (203,686). Calcination products of a wide range of LDHs ($M_{1-x}^{II}M_x^{III}O_{1+x/2}$ (M^{2+} = Mg, Mn, Fe, Co, Ni, Cu, and/ or Zn; M^{3+} = Al, Fe, Cr, Co, and/or In; 0.1 $x < 0.4$) have been examined for the sorption of $^{129}I^-$ and $^{131}I^-$ in a radioactive waste solution (687), and the sorption behavior of ^{235}U fission products, i.e., ^{99}Mo (anions and neutral species), $^{99}TcO_4^-$, ^{131}I salt, ^{103}Ru (anions and neutral species), and neutral $H_2^{132}TeO_4$ on Mg$_3$Al-LDO has been investigated, showing that the sorbed amount decreases from ^{99}Mo to ^{132}Te in the preceding order (688). Other ^{235}U fission products, including ^{95}Zr, ^{140}Ba, ^{140}La, and ^{141}Ce in cationic and neutral forms, have also been absorbed by Mg$_3$Al-LDO to a significant degree.

B. Polymer-LDH Composites

In recent years, many anionic polymers have been interleaved into LDH to form a new class of LDH-polymer nanohybrids, in which the LDH hydroxide layer and polymer anion layer alternate; this topic has been reviewed by Leroux and Besse (14). LDH compounds have been added to neutral polymer as additives or fillers to improve the properties of polymeric materials, such as thermal stability, flammability, mechanical strength, and hardness. The delamination of LDH into single hydroxide layers offers a route to a new kind of polymer-LDH nanocomposite, analogous to the polymer–aluminosilicate nanocomposites extensively studied since the mid 1990s. We review these three classes next.

1. LDH-polymer Nanohybrids

These nanohybrids are actually LDH intercalated with polymeric anions in the interlayer. The intercalation of polymeric anions can be achieved in several ways.

1. *Coprecipitation.* Polymer-interleaved LDH can be made by precipitating LDH from basic solution in the presence of dissolved polymeric anion. Oriakhi et al. used this method to prepare LDH-poly(acrylate), LDH-poly(vinylsulfonate), and LDH-poly(styrenesulfonate) (219, 220,689a).
2. *In situ polymerization.* The monomer anion, such as acrylate, is first intercalated into the LDH interlayer and then polymerized in situ within the interlayer spacing (223,224).
3. *Reconstruction.* Moderately calcined LDH is immersed in the polymeric anion solution, resulting in LDH with polymeric anions in the interlayer (14).
4. *Exchange.* Bubniak et al. put MgAl-LDH dodecylsulfate into poly(ethylene oxide) (PEO) solution and found that PEO is incorporated in part, resulting in the expansion of the interlayer spacing and well-crystallized material, especially with mild heating (689b).

Physically, the intercalation of polymer components expands the LDH interlayer spacing. It has been suggested that polymeric anions such as poly(vinylsulfonate) in LDH adopt a bilayer packing, with the hydrophobic chains in close contact, while the pendant anion groups ($-SO_3^-$) interact with the hydroxide layers through electrostatic forces (219). However, the charge density of efficiently packed vinylic sulfonate polymer (15–20 e/nm^2) is much greater than the charge density (2.4–5.0 e/nm^2 for $x = 0.2$–0.4) in the hydroxide layer, so there must be many voids in the interlayer if the polymeric anions employ such a packing mode.

The intercalation of polymeric anions further affects the crystallinity, crystallite dimensions, and morphology of the LDHs. More interestingly, Messersmith et al. found that poly(vinyl acetate)/Ca$_2$Al-layered material keeps its layer structure at up to 400°C, while Ca$_2$Al(OH)$_6$OH-LDH without organic polymer decomposes at as low as 125°C (224). Oriakhi et al. noted that the calcination of Mg$_2$Al-poly(styrenesulfonate)-LDH produces a macroporous solid, whereas that of Mg$_2$Al-CO$_3$ resulted in dense aggregates (690).

2. Polymer-LDH Blends

Chlorine-containing polymers such as polyvinylchloride (PVC) undergo degradation under the influence of heat and UV light. This degradation generally gives off gaseous HCl that corrodes molding machines and may catalyze further degra-

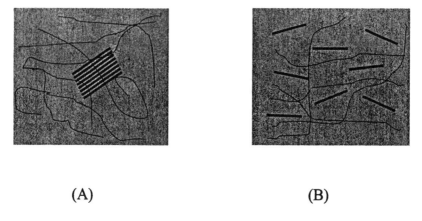

(A)   (B)

FIGURE 10 Mixing between polymers (curved lines) and LDH (bold straight lines) in two extreme cases. (A) LDH crystallites in a polymer matrix (straight parallel lines stand for an LDH crystallites and curved lines polymer chains); (B) The LDH hydroxide sheets delaminated and separated by polymer chains (one straight line stands for a deleminated LDH sheet).

dation, leading to brittleness and discoloration. Layered double hydroxides are environmentally benign neutralizers for this acid.

Fisher et al. (691–693) have shown that MgZnAl-LDHs can achieve close to their theoretical HCl absorbance. For instance, 1 mol $Mg_3ZnAl_2(OH)_{12}(C_{17}H_{3-5}COO)_2 \cdot 4H_2O$ absorbs 13 mol HCl, close to its theoretical possible ability (14 mol HCl) (693). When the anion is itself a weak base, such as carbonate or acetate, its neutralization precedes reaction of HCl with the hydroxide layers.

Miyata (694) has shown that as little as 0.05–1% well-crystallized LDH improves the thermal stability and weather resistance of thermoplastic resins containing halogen and/or acidic components from the manufacturing process.

It is evident that incorporation of intact LDH particles into the polymer matrix, as shown in Figure 10A, will give limited contact area and long diffusion paths, slowing down the removal of destructive species, such as HCl. This is one of the motivations behind current studies of LDH delamination, as discussed next.

3. Polymer-LDH Nanocomposites

This is a newly emerging class of polymer-LDH nanocomposites. Ideally, the completely delaminated LDH hydroxide sheets are uniformly dispersed in the polymer matrix, as shown in Figure 10B. The nanometer-scale interaction between polymer and LDH hydroxide sheets and the tortuosity of diffusion through such a composite may well prove to exert a tremendous influence on such properties as mechanical strength and hardness, thermal stability, flexibility and durability, and impermeability.

Fisher et al. (691,692) claim nanocomposites of this kind in their patents, using LDH at least partially exchanged with anion surfactants and then mixed with monomeric materials, e.g., caprolactam. The claim that delamination of LDH had occurred during the polymerization was based on their XRD results.

By incorporating surfactant anions, LDH was reported by Besse et al. to exfoliate in hexanol on heating (695). In elegant work, Hibino and Jones prepared the glycinate of MgAl-LDH and observed its complete delamination into formamide (696), presumably strongly assisted by glycine-solvent hydrogen bonding. After removal of the solvent, the LDH structure was partially reconstructed in both cases.

We ourselves have exfoliated Mg_3Al-stearate-LDH in hexadecane, which was removed by washing with hexane and drying in the vacuum. The resultant white powder was dispersed into styrene, followed by in situ polymerization. High-resolution TEM images of microtomed sections showed the presence of individual dispersed Mg_3Al-stearate-LDH single layers. O'Leary et al. also found the almost complete delamination of Mg_2Al-dodecylsulfate in polar acrylate monomer solution and then incorporated the delaminated sheets into polymer matrix after polymerization in situ. This polymer/LDH nanocomposite shows a much higher thermal stability than the pure polymer itself (697).

Obviously, the influence of LDH in polymer-LDH nanocomposites is much stronger than that in polymer-LDH blends. As shown in Figure 10B, the contact between polymer and LDH greatly increases when LDH delaminates into single hydroxide sheets. Miyata claimed in his patent that the specific surface area of used LDHs is preferably less than 30 m^2/g (694). However, the completely delaminated hydroxide sheets can provide a contact area of more than 1000 m^2/g. It is not surprising that such extensive interfacial contact between polymeric component and 2D inorganic sheets in nanometer scale offers novel properties different from the conventional composites. In the last decade, nanocompoistes between polymers and cationic clays, such as montmorillonite, have attracted much attention, and we can now expect a similar growth of interest in polymer nanocomposites containing LDH.

C. Reaction Templating

1. Polymerization

Since intercalated anions in LDH are confined to the two-dimensional interlayer gallery and partially oriented, some reactions between them are anticipated to occur with suitable activation. In fact, polymerization could take place in this narrow space. For example, Shichi et al. (698) introduced 4-vinylbenzoate and

4-benzoylbenzoate to the MgAl-LDH interlayer and then employed UV light to initiate radical polymerization to form polyvinylbenzoate. The introduced 4-benzoylbenzoate increases the conversion of 4-vinylbenzoate to polymer but hinders its cyclodimerization. Rhee and Jung intercalated (Z,Z)-muconate anions into LiAl$_2$-LDH and noted the vertical orientation of the anions in the interlayer. When UV light was irradiated on the suspension containing this LDH, the intercalated (Z,Z)-muconate anion underwent polymerization in aqueous media or isomerization to more stable (E,E)-muconate in methanol (699). Depege et al. (209) intercalated silicate anions (HSiO$_3^-$ and SiO$_3^{2-}$) to ZnAl- and ZnCr-LDH via a direct precipitation or anionic exchange method. Hydrothermal treatment caused polymerization of intercalated silicates, grafting to the brucite-like layer, and layer condensation. Geismar et al. in 1991 studied the polymerization of anionic styrene derivatives as well as the redox reactions of interlayer anions with N$_2$H$_4$ or H$_2$O$_2$ (700).

2. Photocycloaddition

Unsaturated organic anions can undergo photochemical cycloaddition within the LDH interlayer, provided that the distance between them is suitable. Tagaki et al. examined the cycloaddition of two kinds of unsaturated carboxylates intercalated in the MgAl-LDH (701–703). They found that (a) the conversion is higher than that of reaction without LDH; (b) the addition is very stereoselective for head-to-head isomers. In contrast, photoaddition without LDH is nonselective, giving both head-to-head and head-to-tail isomers in approximate amounts.

3. Photochemical Support

Robins and Dutta prepared a complicated LiAl$_2$-LDH that contains myristate anions, tetrakis-(4-carboxyphenyl)porphyrinato zinc(II) (ZnTPPC) at about 2 wt% Zn, TiO$_x$ at up to 6 wt% Ti, and SCN$^-$ (244). TiO$_x$ was formed by hydrolyzing the Ti(OBu)$_4$ in the interlayer, showing the semiconductor-like property that photoexcitation onto the bandgap leads to electron–hole formation. SCN$^-$ was used to inhibit the hole–electron recombination, and thus the electron can be transferred to viologens. They found that such intercalated porphyrin promotes the viologen (cationic heptyl viologen or neutral propyl viologen sulfonate) reduction upon excitation in the presence of EDTA as sacrificial electron donor, and they reasoned that the photochemical reactions take place near the edge of the LDH particle according to the product yields.

D. Electrodes, Sensors, and Optical Materials

LDH-modified electrodes have been widely investigated. LDHs can be used in this way not only because of the availability of valence-changeable cations in the brucite-like layers (such as Co, Cr, Ni, Mn, and Fe) (235,704–707), but also because of their ability to host inorganic redox-active anions, such as $Fe(CN)_6^{4-}$, $Mo(CN)_8^{4-}$, and $Ru(CN)_6^{4-}$ (704,708,709), and organic electroactive species, such as *m*-nitrobenzene sulfonate, 2-anthraquinone sulfonate, 2,6-anthraquinone disulfonate, 1,5-anthraquinone disulfonate (710), and ferrocenebutyrate (711) in the interlayer. It has been observed that the improved charge transport in LDH films containing redox active metals in the hydroxide layer allows a much larger fraction of the adsorbed anions to participate in electrochemical reactions (235). In general, the modified electrodes can be prepared and renewed in a simple and reproducible manner (709). In particular, Morigi et al. dispersed $MgAl-SO_4$-LDH particles into polymeric materials (made from PVC and polydimethylsiloxane). The mixed dispersion was then deposited on the electrode surface. The LDH-modified electrode shows a high selectivity for sulfate (712).

The hydration of LDHs is related to environmental humidity. Moneyron et al. (713a,713b) utilized the resultant changes in the ac and dc protonic conductivity of ZnAl-Cl-LDH to make a new type of humidity sensor.

Frequency doubling and other potentially useful optical properties of LDH derivatives have received insufficient attention, probably because of the light-scattering problem, which, we suggest, may be soluble by embedding in a suitable matrix. Thus we can find only one report of nonlinear optical behavior; frequency doubling was found in 4-nitrohippuric acid, intercalated as the neutral molecule in 2:1 $LiAl_2$-LDH chloride (714). This report is particularly interesting because crystals of this acid are themselves centrosymmetric (and thus inactive), and the finding has implications regarding the packing of the molecules between the layers.

E. Separation Medium

Jakupca and Dutta have found that $LiAl_2$-LDH intercalated with myristate is a good candidate for gas chromatographic stationary phase to separate benzene derivatives (715) and alkanes (716). They have observed that this LDH is stable up to 250°C in an anaerobic environment. They attribute the high separation ability and high stability to the membrane-like structure, the hydrophobilicty and suitable porosity after conditioning. LDH could also be coated onto the TLC plates as stationary phase to identify and separate cephalosporins (717). The chromatographic behaviors of inorganic anions as well as cations on such prepared plates have been investigated by Ghoulipour and Husain (718).

The preferential absorption of different anions can be utilized to separate organic isomers in a mixture. For example, O'Hare and Fogg have investigated separation of many organic isomers, such as maleate and fumarate, with $LiAl_2$-Cl-LDH (719), six isomers of pyridinedicarboxylate with $LiAl_2$-Cl-LDH (720),

1,2- and 1,4-benzenedicarboxylates with Ca_2Al-NO_3-LDH (721), and nucleoside monophosphates with various LDHs (722).

F. Clinical and Related Uses

The weak basicity and high capacity of MgAl-LDH makes it a very good neutralizing agent, for example as a pharmaceutical antacid, and it is presumably the active ingredient in magnesium aluminum hydroxide mixtures. Hydrotalcite was found (723–729) to have the highest neutralizing capacity among seven antacids in a hospital formulary. Hydrotalcite tablets maintain a steady optimum pH of 3–5 for around 1.5 hours. The high antipeptic activity of LDH can be attributed both to the adsorption of negatively charged pepsin onto the positively charged surface of LDH and to the persistence of the buffering of the pH at about 4. Furthermore, antacid talcid (hydrotalcite) has been observed to activate the expression of epidermal growth factor (EGF) and its receptors (EGF-R) in normal and ulcerated gastric mucosae, promoting the ulcer healing action (730,731). In addition, the Fe-containing LDHs could be of use in the treatment of iron deficiencies (732).

Recent research has indicated that LDH can be used as drug carrier and controllable drug delivery system (262,282,733–735). O'Hare et al. intercalated various drugs containing carboxylate groups into $LiAl_2$-LDH with quantitative exchange and then tested the release times at pH = 4 and pH = 7 in phosphate buffer solution. Their observation shows that 90% release of some drugs from LDH interlayers takes almost half an hour (282). Choy et al. also observed that LDHs can act as biomolecular reservoirs and gene and drugs carriers (262,734). The plant growth regulator α-naphthaleneacetate can be intercalated in ZnAl-LDH, from which it shows a sustained pH-dependent rate of release (735).

VIII. CONCLUDING REMARKS

The course of LDH chemistry has in some ways typified that of inorganic chemistry as a whole. The earlier work was largely derivative from mineralogy and from the study of interferences in classical gravimetric analysis. The widespread adoption of physical techniques, especially PXRD and, to a lesser extent, FTIR, together with the discovery of simplified preparative methods, made possible the later rapid growth of the subject. In the last 20 years, much excellent research has been driven by the discovery of LDH-related catalysis, rapidly joined by LDO catalysis and bifunctional catalysis, in which LDH or LDO provides an active base that supports an active metal. The blurring of distinctions between organic and inorganic chemistry is shown here by studies of the uptake of large organic anions, while the current emphasis on materials science is most recently leading to interest in morphology and in LDH-containing composite materials.

While it may be too early to speak of a bioinorganic chemistry of LDHs, they are increasingly closely investigated as carriers for pharmaceuticals, and there is a small but steady stream of accounts of the interactions between LDH and amino acids or other biologically significant molecules.

The future (736) is a very difficult thing to predict. Nonetheless, we will hazard some prophesies. We expect:

Increased interest in control of the morphology of LDH by surfactants and by the use of controlled ripening conditions, the preparation of deposited LDH films, and delamination and adhesion in composites with polymers

Continued interest in the use of LDH in decontamination, with emphasis on compositions containing other adsorbants for the simultaneous removal of more than one chemical class of contaminant

A closer study of the dynamics of interaction between LDH and organic or especially biological molecules, with a view to constraining the conformation of the latter by hydrogen bonding, and thus generating or modifying recognition specificity

Growing use of LDH-based materials in "green" catalytic processes, such as the known processes using oxygen or hydrogen peroxide as oxidant, and new processes involving immobilized enzymes or their mimics

The coming of age of LDH electrochemistry, whose applications could include the development of sensors containing highly specific molecules adsorbed on LDH as well as more obvious bulk uses in energy storage and protonic conduction.

We look forward to the next 90 years of LDH research with eager anticipation.

REFERENCES

1. V Rives. Layered Double Hydroxides. Huntington. New York: Nova Science, 2001.
2. F Basile, M Campanati, EM Serwicka, A Vaccari. Appl Clay Sci 2001; 18:1–2.
3. D Tichit, A Vaccari. Appl Clay Sci 1998; 13:311–315.
4. F Cavani, F Trifirò, A Vaccari. Cat Today 1991; 11:173–301.
5. SP Newman, W Jones. Supramolecular Organization and Materials Design. Cambridge. UK: Cambridge University Press, 2002:295–331.
6. F Trifirò, A Vaccari. In: Comprehensive Supramolecular Chemistry. New York: Pergamon Press, 1996:251–291.
7. A Vaccari. Chim Ind 1992; 74:174–181.
8. A Vaccari. Ceramics: Getting into the 2000's, Pt D. Advances in Science and Technology. Vol. 16. Faenza. Italy, 1999:571–584.
9. BF Sels, DE De Vos, PA Jacobs. Catal Rev 2001; 43:443–488.
10. D Tichit, F Fajula. Stud Surf Sci Catal 1999; 125:329–340.
11. S Carlino. Solid State Ionics 1997; 98:73–84.

12. MA Drezdzon. Novel Materials and Heterogeneous Catalysis. ACS Symposium Series, 1990:140–148.
13. V Rives. Mater Chem Phys 2002; 75:19–25.
14. F Leroux, JP Besse. Chem Mater 2001; 13:3507–3515.
15. V Rives, MA Ulibarri. Coord Chem Rev 1999; 181:61–120.
16. WT Reichle. Solid State Ionics 1986; 22:135–141.
17. G Flink. Arkiv Kemi Min Geol 1910; 3:1–166.
18. G Flink. Z Kryst Min 1914; 53:409–420.
19. WF Foshag. Proc US Nat Museum 1920; 58:147–153.
20. WD Treadwell, E Bernasconi. Helv Chim Acta 1930; 13:500–509.
21. W Feitknecht. Z Angew Chem 1936; 49:24.
22. W Feitknecht. Helv Chim Acta 1938; 21:766–784.
23. W Feitknecht, M Gerber. Helv Chim Acta 1942; 25:131–137.
24. MM Mortland, MC Gastuche. Compt Rend 1962; 225:2131–2133.
25. GJ Ross, H Kodama. Am Mineral 1967; 52:1036–1047.
26. L Ingram, HFW Taylor. Mineral Mag 1967; 36:465–479.
27. R Allmann. Am Mineral 1968; 53:1057–1058.
28. MC Van Oosterwyck-Gastuche, G Brown, MM Mortland. Clay Miner 1967; 7: 177–192.
29. G Brown, MC Van Oosterwyck-Gastuche. Clay Miner 1967; 7:193–201.
30. R Allmann. Chimia 1970; 24:99–108.
31. GJ Ross, H Kodama. Am Mineral 1968; 53:1058–1060.
32. S Caillere. Compt Rend 1944; 219:256–258.
33. SM Beekman. J Am Pharm Assoc 1960; 49:191–200.
34. I Soma, H Wakano, H Takahashi, M Yamaguchi. Flame-Resistant Vinyl Chloride Resin Compositions. Jap Patent 50,063,047, May 29, 1975.
35. S Miyata, T Kumura. Chem Lett 1973; 8:843–848.
36. R Allmann, HP Jepsen. Neues Jahrb Mineral Monatsh 1969:544–551.
37. HFW Taylor. Mineral Mag 1973; 39:377–389.
38. Y Song, H-S Moon. Clay Miner 1998; 33:285–296.
39. AE Kapustin, SB Milko. Bulg Chem Commun 1995; 28:147–150.
40. C Frondel. Am Mineral 1941; 26:295–315.
41. LD Ashwal, B Cairncross. Cont Mineral Petrol 1997; 127:75–86.
42. SK Mondal, TK Baidya. Mineral Mag 1996; 60:836–840.
43. DL Bish, GW Brindley. Am Mineral 1977; 62:458–464.
44. SA De Waal, EA Vijoen. Am Mineral 1971; 56:1077–1081.
45. DL Bish, A Livingstone. Mineral Mag 1981; 44:339–343.
46. PJ Dunn, DR Peacor, TD Palmer. Am Mineral 1979; 64:127–130.
47. GW Brindley. Mineral Mag 1979; 43:337–340.
48. KA Rodgers, JE Chisholm, RJ Davis, CS Nelson. Mineral Mag 1977; 41:389–390.
49. PB Moore. Lithos 1971; 4:213–217.
50. S Koritnig, P Suesse. Tschermaks Mineral Petrogr Mitt 1975; 22:79–87.
51. FA Mumpton, HW Jaffe, CS Thompson. Am Mineral 1965; 50:1893–1913.
52. AA Kashaev, GD Feoktistov, SV Petrova. Zap Vses Mineral Obshch 1982; 111: 121–127.

53. EH Nickel, RM Clark. Amer Min 1979; 61:366–372.
54. AV Heyl, C Milton, JM Axelrod. Am Mineral 1959; 44:995–1009.
55. A Livingstone. Mineral Mag 1990; 54:649–653.
56. EH Nickel. Miner Mag 1976; 40:644–647.
57. DL Bish. Bull Mineral 1980; 103:170–175.
58. T Witzke, G Raade. Neues Jahrb Mineral Monatsh 2000; 10:455–465.
59. T Witzke, G Halle. Neues Jahrb Mineral Monatsh 1999; 2:75–86.
60. R Allmann. Am Mineral 1969; 54:296–299.
61. DW Kohls, JL Rodda. Am Mineral 1967; 52:1261–1271.
62. EH Nickel, JE Wildman. Miner Mag 1981; 44:333–337.
63. DR Hudson, M Bussell. Miner Mag 1981; 44:345–350.
64. BA Grguric, IC Madsen, A Pring. Miner Mag 2001; 65:427–435.
65. M Fenoglio. Rev Geol 1938; 19:128.
66. E Artini. Atti Accad Lincei 1910; 18:3–6.
67. D Barthelmy. Mineralogy Database. www.webmineral.com, 2002.
68. JW Boclair, PS Braterman. Chem Mater 1999; 11:298–302.
69. T Yamaoka, M Abe, M Tsuji. Mater Res Bull 1989; 24:1183.
70. P Courthy, C Marcilly. Preparation of Catalysts III. Amsterdam: Elsevier, 1983: 485–517.
71. JW Boclair, PS Braterman, J Jiang, S Lou, F Yarberry. Chem Mater 1999; 11: 303–308.
72. A de Roy, C Forano, K El Malki, JP Besse. Expanded Clays and Other Microporous Solids. New York: Reinhold, 1992:108–169.
73. J Boclair. Thermodynamic and structural studies of layered double hydroxides. PhD dissertation. Denton. TX: University of North Texas, 1998.
74. JW Boclair, PS Braterman. Copper/Aluminum Titration. Unpublished work. Denton. TX: University of North Texas, 1999.
75. FJ Bröcker, W Dethlefsen, K Kaempfer, L Marosi, M Schwarzmann, B Triebskorn, G Zirker. German Patent 2,255,909, 1972.
76. S Miyata, T Kumura, M Shimada. German Patent 2,061,156, 1970.
77. GM Woltermann. US Patent 4,454,244, 1984.
78. E Gardner, KM Huntoon, TJ Pinnavaia. Adv Mater (Weinheim, Ger) 2001; 13: 1263–1266.
79. S Miyata. Clays Clay Miner 1983; 31:305–311.
80a. Braterman group, unpublished observations; close examination of carbonate spectra reported by other groups confirms this observation.
80b. M Richardson, PS Braterman, F Yarberry, ZP Xu. Metallocyanide/carbonate layered double hydroxide competition. Unpublished work. Denton. TX: University of North Texas, 2002.
81. KA Carrado, JE Forman, RE Botto, RE Winans. Chem Mater 1993; 4:472–478.
82. A Bééres, I Pálinkó, I Kiricsi, JB Nagy, Y Kiyozumi, F Mizukami. App Cat A: Gen 1999; 182:237–247.
83. JD Wang, RA Cahill, G Serrette, WL Shes, A Clearfield. Catalytic study of polyoxometalate-pillared layered double hydroxides, American Chemical Society Southwest Regional Meeting, Austin, TX, 1993.

84. JW Boclair, PS Braterman, BD Brister, Z Wang, F Yarberry. J Sol State Chem 2001; 161:249–258.
85. JD Wang, G Serrette, Y Tian, A Clearfield. Appl Clay Sci 1995; 10:103–115.
86. EA Gardner, SK Yun, T Kwon, TJ Pinnavaia. Appl Clay Sci 1998; 13:479–494.
87. MA Drezdon. Pillared Hydrotalcites. US Patent 4,774,212, September 27, 1988.
88. J Wang, Y Tian, RC Wang, A Clearfield. Chem Mater 1992; 4:1276–1282.
89. SP Newman, W Jones. New J Chem 1998; 22:105–115.
90. M Borja, PK Dutta. J Phys Chem 1992; 96:5434–5444.
91. M Meyn, K Beneke, G Lagaly. Inorg Chem 1990; 29:5201–5207.
92. S Carlino, MJ Hudson. J Mater Chem 1995; 5:1433–1442.
93. EL Crepaldi, PC Pavan, JB Valim. Chem Commun 1999:155–156.
94. ED Dimotakis, TJ Pinnavaia. Inorg Chem 1990; 29:2393–2394.
95. HCB Hansen, RM Taylor. Clay Miner 1991; 26:311–327.
96. WJ McLaughlin, JL White, SL Hem. J Colloid Interface Sci 1994; 165:41–52.
97. H Morioka, H Tagaya, M Karasu, J Kadokawa, K Chiba. J Solid State Chem 1995; 117:337–342.
98. IE Grey, R Ragozzini. J Solid State Chem 1991; 94:244–253.
99. G Mascolo. Appl Clay Sci 1995; 10:21–30.
100. HH Kung, EI Ko. Chem Eng J 1996; 64:203–214.
101. T Lopez, P Bosch, E Ramos, R Gomez, O Novaro, D Acosta, F Figueras. Langmuir 1996; 12:189–192.
102. M Nayak, TRN Kutty, V Jayaraman, G Periaswamy. J Mater Chem 1997; 7: 2131–2137.
103. G Mascolo, O Marino. Miner Mag 1980; 43:619–621.
104. JB d'Espinose de la Caillerie, M,O Clause. J Am Chem Soc 1995; 117: 11471–11481.
105. M Rajamathi, PV Kamath. Bull Mater Sci 2000; 23:355–359.
106. JT Sampanthar, HC Zeng. Chem Mater 2001; 13:4722–4730.
107. N Gutmann, B Mueller. J Sol State Chem 1996; 122:214–220.
108. L Gordon, ML Salutzky, HH Willard. Precipitation from Homogeneous Solution. London: Wiley, 1959.
109. M Ogawa, H Kaiho. Langmuir 2002; 18:4240–4242.
110. H Cai, AC Hillier, KR Franklin, CC Nunn, MD Ward. Science 1994; 66:1551–1555.
111. U Costantino, F Marmottini, M Nocchetti, R Vivani. Eur J Inorg Chem 1998: 1439–1446.
112. JW Boclair, PS Braterman, F Yarberry. Morphologies of layered double hydroxides by homogeneous precipitation. Unpublished Work. Denton. TX: University of North Texas, 1999.
113. F Yarberry. Layered double hydroxides via aluminate. PhD dissertation. Denton. TX: University of North Texas, 2002.
114. F Yarberry, PS Braterman, ZP Xu. Improved Technique for the Preparation of Layered Double Hydroxides, Southwest Regional American Chemical Society Meeting, San Antonio, TX, 2001.
115. P Daute, J Foell, I Lange, S Kuepper, P Wedl, JD Klamann. Methods of Preparing Cationic Layer Compounds, Cationic Layer Compounds Prepared Thereby, and Methods of Use Therefor. US Patent 6,362,261, March 26, 2002.

116. RH Goheen, WA Nigro, PJ The. Process for Producing Aluminum Hydroxide of Improved Whiteness. US Patent 4,915,930, April 10, 1990.
117. K Hrnciarova, M Zikmund, C Hybl. Process for Preparing Alum. Czech Patent 277,548, March 17, 1993.
118. C Misra. Adsorbent and Substrate Products and Method of Producing Same. US Patent 4,656,156, April 7, 1987.
119. M Zikmund, K Putyera, K Hrnciarova. Chem Pap 1996; 50:262–270.
120. R Kikuchi, T Fujii. Collection of Heavy Metal–Containing Ions. Jap Patent 10,216,742, August 18, 1998.
121. S Takayama, Y Mikami, N Iyatomi, M Orikasa. Process for Treatment of Sodium Hydroxide Waste Liquor Containing Aluminum. Eur Patent 636,577, February 1, 1995.
122. X Hou, RJ Kirkpatrick. NMR studies of Mg/Al and Zn/Al LDH Prepared by Coprecipitation and Aluminate Methods. Unpublished work. Urbana. IL: University of Illinois at Urbana–Champaign, 2000.
123. G Ona-Nguema, M Abdelmoula, F Jorand. Environ Sci Technol 2002; 36:16–20.
124. HM Cho, H Kawasaki, H Kumazawa. Can J Chem Eng 2000; 78:842–846.
125. P Refait, M Abdelmoula, JMR Géénin. Corros Sci 1998; 40:1547–1560.
126. P Refait, JMR Géénin. Corros Sci 1994; 36:55–65.
127. F Yarberry, PS Braterman, S Chen, J Martin, BD Brister, J Hall, ZP Xu, P Encinias, P Ranguswamy, R VonDreele. Layered double deuteroxides analyzed by high-intensity powder diffractometry. Unpublished work. Los Alamos. NM: Los Alamos National Laboratories, 2001.
128. ZP Xu, HC Zeng. Chem Mater 2001; 13:4555–4563.
129. ZP Xu, HC Zeng. Chem Mater 2000; 12:3459–3465.
130. ZP Xu, HC Zeng. Chem Mater 2000; 12:2597–2603.
131. ZP Xu, HC Zeng. Inter J Inorg Mater 2000; 2:187–196.
132. ZP Xu, HC Zeng. Chem Mater 1999; 11:67–74.
133. ZP Xu, HC Zeng. J Mater Chem 1998; 8:2499–2506.
134. U Chellam, ZP Xu, HC Zeng. Chem Mater 2000; 12:650–658.
135. HC Zeng, ZP Xu, M Qian. Chem Mater 1998; 10:2277–2283.
136. M Qian, HC Zeng. J Mater Chem 1997; 7:493–499.
137. C Faure, C Delmas, P Willmann. J Power Sources 1991; 35:263–277.
138. C Faure, C Delmas, P Willmann. J Power Sources 1991; 36:497–506.
139. JM Fernandez, C Barriga, MA Ulibarri, FM Labajos, V Rives. J Mater Chem 1994; 4:1117–1121.
140. L Markov, K Petrov, A Lyubchova. Solid State Ionics 1990; 39:187–193.
141. HCB Hansen, CB Koch, RM Taylor. J Solid State Chem 1994; 113:46–53.
142. W Ostwald. Z Phys Chem 1897; 22:289.
143. R Boistelle, JP Astier. J Cryst Growth 1988; 90:14–30.
144. HK Henisch. Crystals in Gels and Liesegang Rings: in vitro veritas. New York: Cambridge University Press, 1988.
145. L Hickey, JT Kloprogge, RL Frost. J Mater Sci 2000; 35:4347–4355.
146. S Kannan, RV Jasra. J Mater Chem 2000; 10:2311–2314.
147. KJ Rao, B Vaidhyanathan, M Ganguli, PA Ramakrishnan. Chem Mater 1999; 11: 882–895.

148. A Okada, K Shimizu, S Yamashita. Preparation of Hydrotalcite Compounds of Low Uranium Content as Fillers in Resin Sealants. Eur Patent 989,095, March 29, 2000.

149. Y Mitsuo. Treatment Method of Contaminant to Make it Almost Insoluble. Jap Patent 269,664, February 10, 2001.

150. S Miyata, K Masataka. Method for Inhibiting the Thermal or Ultraviolet Degradation of Thermoplastic Resin and Thermoplastic Resin Composition Having Stability to Thermal or Ultraviolet Degradation. Ger Patent 3,019,632, November 10, 1981.

151. I Norio, I Takeo, H Katuyuki, K Teruhiko, Y Kimiaki. Process for the Preparation of Hydrotalcite. Ger Patent 1,592,126, November 10, 1970.

152. CP Kelkar. Simplified Synthesis of Anion Intercalated Hydrotalcites. US Patent 8,814,291, September 29, 1998.

153. W Jones, P O'Connor, D Stamires. Situ Formed Anionic Clay–Containing Bodies. US Patent 2,002,092,812, July 18, 2002.

154. WE Horn, V Cedro, ES Martin, JM Stinson. Two-Powder Synthesis of Hydrotalcite and Hydrotalcite Like Compounds. US Patent 5,728,364, March 17, 1998.

155. C Misra. Synthetic Hydrotalcite. US Patent 4,904,457, February 27, 1990.

156. A Brasch, K Diblitz, J Schiefler, K Noweck. Process for Producing Hydrotalcites and the Metal Oxides Thereof. US Patent 2,001,001,653, May 24, 2001.

157. JA Kosin, BW Preston, DN Wallace. Modified Synthetic Hydrotalcite. US Patent 4,883,533, November 28, 1989.

158a. WE Horn, V Cedro, ES Martin, JM Stinson. Two-Powder Synthesis of Hydrotalcite and Hydrotalcite-Like Compounds. US Patent 5,728,363, March 17, 1998.

158b. T Kameda, T Yoshioka, M Uchida, A Okuwaki. Phosphorus, Sulfur and Silicon and the Related Elements. 2002; 177:1503–1506.

159. Powder Diffraction File for $Mg(OH)_2$, Card No. 44–1482. Swarthmore. PA: Joint Committee on Powder Diffraction Standards, 1995.

160. K Yao, M Taniguchi, M Nakata, M Takahashi, A Yamagishi. Langmuir 1998; 14: 2410–2414.

161. M Bellotto, B Rebours, O Clause, J Lynch, D Bazin, E Elkaiem. J Phys Chem 1996; 100:8527–8534.

162. C Busetto, GD Piero, G Manara, F Trifiro, A Vaccari. J Catal 1984; 85:260–266.

163. J Perez-Ramirez, J Overeijnder, F Kapteijn, JA Moulijn. Appl Catal B-Environ 1999; 23:59–72.

164. FM Vichi, OL Alves. J Mater Chem 1997; 7:1631–1634.

165. J Perez-Ramirez, F Kapteijn, JA Moulijn. Catal Lett 1999; 60:133–138.

166. W Feitknecht, M Gerber. Helv Chim Acta 1942; 25:106–130.

167. VA Drits, TN Sokolova, GV Sokolova, VI Cherkashin. Clays Clay Miner 1987; 35:401–417.

168. R Allmann. Neuse Jahrb Mineral Monatsh 1977; 3:136–144.

169. P Malet, JA Odriozola, FM Labajos, V Rives, MA Ulibarri. Nucl Instrum Meth Phys Res, B 1995; 97:16–19.

170. FM Labajos, V Rives, P Malet, MA Centeno, MA Ulibarri. Inorg Chem 1996; 35: 1154–1160.

171. V Rives, FM Labajos, MA Ulibarri, P Malet. Inorg Chem 1993; 32:5000–5001.

172. MA Aramendia, V Borau, C Jimenez, JM Marinas, FJ Romero, FJ Urbano. J Mater Chem 1999; 9:2291–2292.

173. JM Fernandez, C Barriga, MA Ulibarri, FM Labajos, V Rives. Chem Mater 1997; 9:312–318.

174. JN Armor, TA Braymer, TS Farris, Y Li, FP Petrocelli, EL Weist, S Kannan, CS Swamy. Appl Catal B-Environ 1996; 7:397–406.

175. F Basile, G Fornasari, E Poluzzi, A Vaccari. Appl Clay Sci 1998; 13:329–345.

176a. J Oi, A Obuchi, A Ogata, GR Bamwenda, R Tanaka, T Hibino, S Kushiyama. Appl Catal B-Environ 1997; 13:197–203.

176b. F Basile, L Basini, G Fornasari, M Gazzano, F Trifiro, A Vaccari. Chem Comm 1996:2435–2436.

177. I Rousselot, C Taviot-Gueho, F Leroux, P Leone, P Palvadeau, JP Besse. J Solid State Chem 2002; 167:137–144.

178. JP Rapin, A Walcarius, G Lefevre, M Francois. Acta Crystallogr, Sect C 1999; 55: 1957–1959.

179. RM Taylor. Clay Miner 1984; 19:591–603.

180. S Velu, V Ramaswamy, A Ramani, BM Chanda, S Sivasanker. Chem Commun 1997:2107–2108.

181. S Velu, DP Sabde, N Shah, S Sivasanker. Chem Mater 1998; 10:3451–3458.

182. S Velu, K Suzuki, M Okazaki, T Osaki, S Tomura, F Ohashi. Chem Mater 1999; 11:2163–2172.

183. JP Thiel, CK Chiang, KR Poeppelmeier. Chem Mater 1993; 5:297–304.

184. CJ Serna, JL Rendon, JE Iglesias. Clays Clay Miner 1982; 30:180–182.

185. ZY Wen, ZX Lin, KG Chen. J Mater Sci Lett 1996; 15:105–106.

186. DR Lide. In: Handbook of Chemistry and Physics, Chapter 12. New York: Chapman & Hall, 1999:14–16.

187. W Lowenstein. Am Miner 1954; 39:92.

188. I Pausch, HH Lohse, K Schuermann, R Allmann. Clays Clay Miner 1986; 34: 507–510.

189. S Miyata. Clays Clay Miner 1980; 28:50–56.

190. G Busca, V Lorenzelli. Mater Chem Phys 1992; 31:221–228.

191. G Busca, V Lorenzelli, VS Escribano. Chem Mater 1992; 4:595–605.

192. S Morpurgo, M LoJacono, P Porta. J Solid State Chem 1996; 122:324–332.

193. S Morpurgo, M Lojacono, P Porta. J Solid State Chem 1995; 119:246–253.

194. P Porta, S Morpurgo. Appl Clay Sci 1995; 10:31–44.

195. AV Lukashin, SV Kalinin, MP Nikiforov, VI Privalov, AA Eleseev, AA Vertegel, YD Tretyakov. Dokl Akad Nauk 1999; 364:77–79.

196. U Costantino, M Casciola, L Massinelli, M Nocchetti, R Vivani. Solid State Ionics 1997; 97:203–212.

197. M Badreddine, A Legrouri, A Barroug, A de Roy, JP Besse. Mater Lett 1999; 38: 391–395.

198. A Ookubo, K Ooi, F Tani, H Hayashi. Langmuir 1994; 10:407–411.

199. F Kovanda, E Kovacsova, D Kolousek. Collect Czech Chem Commun 1999; 64: 1517–1528.

200. JT Lin, SJ Tsai, SF Cheng. J Chin Chem Soc 1999; 46:779–787.

201. LS Li, SJ Ma, XS Liu, Y Yue, JB Hui, R Xu, YM Bao, J Rocha. Chem Mater 1996; 8:204–208.
202. S Miyata. Clays Clay Miner 1975; 23:369–375.
203. MJ Kang, KS Chun, SW Rhee, Y Do. Radiochim Acta 1999; 85:57–63.
204. T Yamagishi, Y Oyanagi, E Narita. Nippon Kagaku Kaishi 1993:329–334.
205. T Sato, T Wakabayashi, M Shimada. Ind Eng Chem Prod Res Dev 1986; 25:89–92.
206. C Depege, C Forano, A de Roy, JP Besse. Mol Cryst Liq Cryst Sci Technol, Sect A 1994; 244:161–6.
207. F Malherbe, L Bigey, C Forano, A de Roy, JP Besse. J Chem Soc, Dalton Trans 1999:3831–3839.
208. J Twu, PK Dutta. Chem Mater 1992; 4:398–401.
209. C Depege, FZ Elmetoui, C Forano, A Deroy, J Dupuis, JP Besse. Chem Mater 1996; 8:952–960.
210. WP Ding, G Gu, W Zhong, WC Zang, YW Du. Chem Phys Lett 1996; 262:259–262.
211. K Chibwe, JB Valim, W Jones. Abstr Pap Am Chem Soc 1989; 198:20–21.
212. HP Boehm, J Steinle, C Vieweger. Angew Chem 1977; 89:259–260.
213. T Sato, A Okuwaki. Solid State Ionics 1991; 45:43–48.
214. MA Drezdon. Inorg Chem 1988; 27:4628–4632.
215. E Kanezaki, K Kinugawa, Y Ishikawa. Chem Phys Lett 1994; 226:325–330.
216. BM Choudary, ML Kantam, B Kavita, CV Reddy, KK Rao, F Figueras. Tetrahedron Lett 1998; 39:3555–3558.
217. JL Guimaraes, R Marangoni, LP Ramos, F Wypych. J Colloid Interface Sci 2000; 227:445–451.
218. IY Park, K Kuroda, C Kato. J Chem Soc, Dalton Trans 1990:3071–3074.
219. CO Oriakhi, IV Farr, MM Lerner. J Mater Chem 1996; 6:103–107.
220. OC Wilson, T Olorunyolemi, A Jaworski, I Borum, D Young, A Siriwat, E Dickens, C Oriakhi, M Lerner. Appl Clay Sci 1999; 15:265–279.
221. M Tanaka, IY Park, K Kuroda, C Kato. Bull Chem Soc Jpn 1989; 62:3442–3445.
222. S Rey, J Merida-Robles, KS Han, L Guerlou-Demourgues, C Delmas, E Duguet. Polym Int 1999; 48:277–282.
223. T Challier, RCT Slade. J Mater Chem 1994; 4:367–371.
224. PB Messersmith, SI Stupp. Chem Mater 1995; 7:454–460.
225. P Aranda, F Leroux, E Ruiz-Hitzky, JP Besse. Report CNRS/CSIC, 2001.
226. GTD Shouldice, PY Choi, BE Koene, LF Nazar, A Rudin. J Polym Sci, Part A: Polym Chem 1995; 33:1409–1417.
227. E Lopez-Salinas, N Tomita, T Matsui, E Suzuki, Y Ono. J Mol Catal 1993; 81: 397–405.
228. E Lopez-Salinas, Y Ono. Microporous Mater 1993; 1:33–42.
229. K Itaya, HC Chang, I Ichida. Inorg Chem 1987; 26:624–626.
230. I Crespo, C Barriga, V Rives, MA Ulibarri. Solid State Ionics 1997; 101:729–735.
231. MJ Holgado, V Rives, MS Sanroman, P Malet. Solid State Ionics 1996; 92:273–283.
232. PS Braterman, CQ Tan, JX Zhao. Mater Res Bull 1994; 29:1217–1221.
233. HCB Hansen, CB Koch. Clays Clay Miner 1994; 42:170–179.
234. JB Qiu, G Villemure. J Electroanal Chem 1995; 395:159–166.
235. JB Qiu, G Villemure. J Electroanal Chem 1997; 428:165–172.

236. E Suzuki, S Idemura, Y Ono. Clays Clay Miner 1989; 37:173–178.
237. A Corma, F Rey, JM Thomas, G Sankar, GN Greaves, A Cervilla, E Llopis, A Ribeira. Chem Commun 1996:1613–1614.
238. IY Park, K Kuroda, C Kato. Chem Lett 1989:2057–2058.
239. S Bonnet, C Forano, A de Roy, JP Besse, P Maillard, M Momenteau. Chem Mater 1996; 8:1962–1968.
240. S Kannan, SV Awate, MS Agashe. Stud Surf Sci Catal 1998; 113:927–935.
241. HC Liu, XY Yang, GP Ran, EZ Min. Wuli Huaxue Xuebao 1999; 15:918–924.
242. ME Perezbernal, R Ruanocasero, TJ Pinnavaia. Catal Lett 1991; 11:55–62.
243. L Barloy, JP Lallier, P Battioni, D Mansuy, Y Piffard, M Tournoux, JB Valim, W Jones. New J Chem 1992; 16:71–80.
244. DS Robins, PK Dutta. Langmuir 1996; 12:402–408.
245. E Gardner, TJ Pinnavaia. Appl Catal, A 1998; 167:65–74.
246. K Putyera, J Jagiello, TJ Bandosz, JA Schwarz. J Chem Soc, Faraday Trans 1996; 92:1243–1247.
247. BF Sels, DE de Vos, PA Jacobs. Stud Surf Sci Catal 1997; 110:1051–1059.
248. J Twu, PK Dutta. J Catal 1990; 124:503–510.
249. J Twu, PK Dutta. J Phys Chem 1989; 93:7863–7868.
250. FAP Cavalcanti, A Schutz, P Bileon. In: Preparation of Catalysts IV. New York: Elsevier, 1987:165–174.
251. L Xu, CW Hu, EB Wang. J Nat Gas Chem 1997; 6:155–168.
252. J Guo, T Sun, JP Shen, D Liu, DZ Jiang. Gaodeng Xuexiao Huaxue Xuebao 1995; 16:346–350.
253. J Guo, QZ Jiao, JP Shen, HJ Lu, D Liu, DZ Jiang, EZ Min. Huaxue Xuebao 1996; 54:357–362.
254. T Ikeda, H Amoh, T Yasunaga. J Am Chem Soc 1984; 106:5772–5775.
255. S Aisawa, S Takahashi, W Ogasawara, Y Umetsu, E Narita. J Sol State Chem 2001; 162:52–62.
256. S Aisawa, S Takahashi, W Ogasawara, Y Umetsu, E Narita. Clay Science 2000; 11:317–328.
257. A Fudala, I Palinko, I Kiricsi. Inorg Chem 1999; 38:4653–4658.
258. A Fudala, I Palinko, I Kiricsi. J Mol Struct 1999; 483:33–37.
259. NT Whilton, PJ Vickers, S Mann. J Mater Chem 1997; 7:1623–1629.
260. JH Choy, JS Park, SY Kwak, YJ Jeong, YS Han. Mol Cryst Liq Cryst Sci Technol, A 2000; 341:425–429.
261. JH Choy, SY Kwak, JS Park, YJ Jeong, J Portier. J Amer Chem Soc 1999; 121: 1399–1400.
262. JH Choy, SY Kwak, JS Park, YJ Jeong. J Mater Chem 2001; 11:1671–1674.
263. K Kuma, W Paplawsky, B Gedulin, G Arrhenius. OLEB 1989; 19:573–602.
264. FM Labajos, V Rives, MA Ulibarri. Spectrosc Lett 1991; 24:499–508.
265. ZP Xu, HC Zeng. J Phys Chem B 2001; 105:1743–1749.
266. GW Brindley, S Kikkawa. Clays Clay Miner 1980; 28:87–91.
267. S Kikkawa, M Koizumi. Mater Res Bull 1982; 17:191–198.
268. H Tagaya, A Ogata, T Kuwahara, S Ogata, M Karasu, J Kadokawa, K Chiba. Microporous Mater 1996; 7:151–158.

269. CW Hu, X Zhang, QL He, EB Wang, SW Wang, QL Guo. Trans Metal Chem 1997; 22:197–199.
270. CW Hu, YY Liu, ZP Wang, EB Wang. Acta Chim Sin 1997; 55:49–55.
271. J Evans, M Pillinger, JJ Zhang. J Chem Soc, Dalton Trans 1996; 14:2963–2974.
272. T Kanoh, T Shichi, K Tagaki. Chem Lett 1999:117–118.
273. ZP Xu, PS Braterman, N Seifollah. Self-Assembly and Multiple Phases in ZnAl-Stearate-Layered Double Hydroxides. 224th ACS National Meeting, Boston, 2002.
274. A Clearfield, M Kieke, J Kwan, JL Colon, RC Wang. J Inclusion Phenom Mol Recognit Chem 1991; 11:361–378.
275. S Carlino, MJ Hudson. J Mater Chem 1994; 4:99–104.
276. K Chibwe, W Jones. J Chem Soc, Chem Comm 1989:926–927.
277. ZP Xu, PS Braterman. J Mater Chem 2003; 13:268–273.
278. JW Boclair, PS Braterman, BD Brister, JP Jiang, SW Lou, ZM Wang, F Yarberry. OLEB 2001; 31:53–69.
279. R Krishnamurthy, G Arrhenius, A Eschenmoser. OLEB 1999; 29:333–354.
280. R Krishnamurthy, S Pitsch, G Arrhenius. OLEB 1999; 29:139–152.
281. V Kolb, S Zhang, Y Xu, G Arrhenius. OLEB 1997; 27:485–503.
282. AI Khan, L Lei, AJ Norquist, D O'Hare. Chem Comm 2001:2342–2343.
283. C Depege, L Bigey, C Forano, A de Roy, JP Besse. J Solid State Chem 1996; 126:314–323.
284. C Forano, A de Roy, C Depege, M Khali, FZ El Metoui, JP Besse. Preprints—Am Chem Soc, Div of Petrol Chem 1995; 40:317–319.
285a. M Menetrier, KS Han, L Guerlou-Demourgues, C Delmas. Inorg Chem 1997; 36:2441–2445.
285b. KS Han, L Guerlou-Demourgues, C Delmas. Solid State Ionics 1996; 84:227–238.
286. Y Bao, L Li. Huaxue Tongbao 2002; 65:194–197.
287. A Vanderpol, BL Mojet, E Vandeven, E Deboer. J Phys Chem 1994; 98:4050–4054.
288a. W Kagunya, PK Dutta, Z Lei. Physica B 1997; 234:910–913.
288b. TE Johnson, W Martens, RL Frost, Z Ding, JT Kloprogge. J Raman Spectros 2002; 33:604–609.
289. AS Bookin, VA Drits. Clays Clay Miner 1993; 41:551–557.
290. AS Bookin, VI Cherkashin, VA Drits. Clays Clay Miner 1993; 41:558–564.
291. SP Newman, W Jones, P O'Conner, EN Stamires. J Mater Chem 2002; 12:153–155.
292. AM Fogg, JS Dunn, D Ohare. Chem Mater 1998; 10:356–360.
293. M Kaneyoshi, W Jones. Chem Phys Lett 1998; 296:183–187.
294. F Kooli, IC Chisem, M Vucelic, W Jones. Chem Mater 1996; 8:1969–1977.
295. M Vucelic, GD Moggridge, W Jones. J Phys Chem 1995; 99:8328–8337.
296. SA Solin, DR Hines, GT Seidler, MMJ Treacy. J Phys Chem Solids 1996; 57:1043–1048.
297. DR Hines, GT Seidler, MMJ Treacy, SA Solin. Solid State Commun 1997; 101:835–839.
298. K Yao, M Taniguchi, M Nakata, A Yamagishi. J Electroanal Chem 1998; 458:249–252.
299. K Yao, M Taniguchi, M Nakata, M Takahashi, A Yamagishi. Langmuir 1998; 14:2890–2895.

300. AV Besserguenev, AM Fogg, RJ Francis, SJ Price, D O'Hare, VP Isupov, BP Tolochko. Chem Mater 1997; 9:241–247.
301. JC Lindon, GE Tranter, JL Holmes. Encyclopedia of Spectroscopy and Sepctrometry. New York: Academic Press, 2000.
302. M Gazzano, W Kagunya, D Matteuzzi, A Vaccari. J Phys Chem B 1997; 101: 4514–4519.
303. F Basile, G Fornasari, M Gazzano, A Vaccari. Appl Clay Sci 2001; 18:51–57.
304. MJ Hernandez-Moreno, MA Ulibarri, JL Rendon, CJ Serna. Phys Chem Miner 1985; 12:34–38.
305. JT Kloprogge, RL Frost. J Solid State Chem 1999; 146:506–515.
306a. W Kagunya, R Baddour-Hadjean, F Kooli, W Jones. Chem Phys 1998; 236: 225–234.
306b. JT Kloprogge, L Hickey, RL Frost. J Mater Sci Lett 2002; 21:603–605.
307. J Perez-Ramirez, G Mul, JA Moulijn. Vib Spectrosc 2001; 27:75–88.
308. J Perez-Ramirez, G Mul, F Kapteijn, JA Moulijn. J Mater Chem 2001; 11:821–830.
309. JT Kloprogge, D Wharton, L Hickey, RL Frost. Am Mineral 2002; 87:623–629.
310. H Nijs, A Clearfiled, EF Vansant. Microporous Mesoporous Mater 1998; 23: 97–108.
311. RL Frost, Z Ding, JT Kloprogge. Can J Anal Sci Spectrosc 2000; 45:96–102.
312. JT Kloprogge, L Hickey, RL Frost. Appl Clay Sci 2001; 18:37–49.
313. JT Kloprogge, RL Frost. Appl Catal, A 1999; 184:61–71.
314. WW Kagunya. J Phys Chem 1996; 100:327–330.
315. H Roussel, V Briois, E Elkaim, A de Roy, JP Besse, JP Jolivet. Chem Mater 2001; 13:329–337.
316. S Velu, N Shah, TM Jyothi, S Sivasanker. Microporous Mesoporous Mater 1999; 33:61–75.
317. S Velu, K Suzuki, S Hashimoto, N Satoh, F Ohashi, S Tomura. J Mater Chem 2001; 11:2049–2060.
318. S Velu, K Suzuki. Stud Surf Sci Catal 2000; 129:451–458.
319. F Prinetto, D Tichit, R Teissier, B Coq. Catal Today 2000; 55:103–116.
320. V Rives, S Kannan. J Mater Chem 2000; 10:489–495.
321. M del Arco, V Rives, R Trujillano, P Malet. J Mater Chem 1996; 6:1419–1428.
322. R Dula, K Wcislo, J Stoch, B Grzybowska, EM Serwicka, F Kooli, K Bahranowski, A Gawel. Appl Catal, A 2002; 230:281–291.
323. K Bahranowski, R Dula, F Kooli, EM Serwicka. Colloids Surf, A 1999; 158: 129–136.
324. X Yu, S Zhang, J Zhang, E Wang, R Zhan, Q Liu. Yingyong Huaxue 1996; 13: 17–20.
325. K Okada, F Matsushita, S Hayashi. Clay Miner 1997; 32:299–305.
326. E Lopez Salinas, Y Ono. Bull Chem Soc Jpn 1992; 65:2465–70.
327. RP Grossor, SL Suib, RS Weber, PF Schubert. Chem Mater 1992; 4:922–928.
328. AF Cotton, G Wilkinson. Advanced Inorganic Chemistry, A Comprehensive Text. 2nd ed.. New York: Interscience, 1966.
329. J Dupuis, JP Battut, Z Fawal, H Hajjimohamad, A De Roy, JP Besse. Solid State Ionics 1990; 42:251–255.

330. MA Aramendia, V Borau, C Jimenez, JM Marinas, FJ Romero, JF Ruiz. J Solid State Chem 1997; 131:78–83.
331. X Hou, RJ Kirkpatrick, P Yu, D Moore, Y Kim. Am Mineral 2000; 85:173–180.
332. X Hou, AG Kalinichev, RJ Kirkpatrick. Chem Mater 2002; 14:2078–2085.
333. X Hou, RJ Kirkpatrick. Chem Mater 2000; 12:1890–1897.
334. M Vucelic, W Jones. NATO ASI Ser, Ser C: Mathematical Phys Sci 1993; 400: 373–378.
335. A Beres, I Palinko, JC Bertrand, JB Nagy, IJ Kiricsi. J Mol Struct 1997; 410–411: 13–16.
336. J Rocha, M del Arco, V Rives, MA Ulibarri. J Mater Chem 1999; 9:2499–2503.
337. G Jun, L Yun-Lun, J Qing-Ze, J Da-Zhen. Chem Res Chin Univ 1998; 14:176–178.
338. S Carlino, MJ Hudson, SW Husain, JA Knowles. Solid State Ionics 1996; 84: 117–129.
339. RS Maxwell, RK Kukkadapu, JE Amonette, H Cho. J Phys Chem B 1999; 103: 5197–5203.
340. DC Koningsberger, R Prins. X-ray Absorption: Principles, Applications, Techniques of EXAFS, SEXAFS, and XANES. New York: Wiley, 1988.
341. H Roussel, V Briois, E Elkaim, A de Roy, JP Besse. J Phys Chem B 2000; 104: 5915–5923.
342. L Bigey, F Malherbe, A de Roy, JP Besse. Mol Cryst Liq Cryst Sci Technol, A 1998; 311:629–634.
343. L Bigey, C Depege, A de Roy, JP Besse. J Phys IV 1997; 7:949–950.
344a. M Vucelic, W Jones, GD Moggridge. Clays Clay Miner 1997; 45:803–813.
344b. F Leroux, EM Moujahid, H Roussel, AM Flank, V Briois, JP Besse. Clays Clay Miner 2002; 50:254–264.
344c. M Intissar, R Segni, C Payen, JP Besse, F Leroux. J Solid State Chem 2002; 167: 508–516.
345. M Doeuff, T Kwon, TJ Pinnavaia. Synth Met 1990; 34:609–615.
346. U Pegelow, M Winterer, BD Mosel, M Schmalz, R Schollhorn, Z Naturforsch. A-Phys Sci 1994; 49:1200–1206.
347. L Raki, DG Rancourt, C Detellier. Chem Mater 1995; 7:221–224.
348. X Ge, M Li, J Shen. J Solid State Chem 2001; 161:38–44.
349. X Ge, Y Wang, JY Shen, HL Zhang. Wuji Huaxue Xuebao 2000; 16:79–83.
350. PS Kumbhar, J Sanchez-Valente, JMM Millet, F Figueras. J Catal 2000; 191: 467–473.
351. J Sanchez-Valente, JMM Millet, F Figueras, L Fournes. Hyperfine Interact 2001; 131:43–50.
352. JY Shen, B Guang, M Tu, Y Chen. Catal Today 1996; 30:77–82.
353. S Loweel, JE Shields. Powder Surface Area and Porosity. 2nd ed.. New York: Chapman and Hall, 1984.
354. A Elm'chaouri, MH Simonot-Grange. Thermochimica Acta 1999; 339:117–123.
355. FAP Cavalcanti, A Schutz, P Biloen. Stud Surf Sci Catal 1987; 31(Prep Catal 4): 165–174.
356. H Nijs, M De Bock, N Maes, EF Vansant. J Pourous Mater 1999; 6:307–321.
357. M Tsuji, G Mao, T Yoshida, Y Tamaura. J Mater Res 1993; 8:1137–1142.

358. G Mao, M Tsuji, Y Tamaura. Clays Clay Miner 1993; 41:731–737.
359. Y Ding, E Alpay. Chem Eng Sci 2000; 55:3461–3474.
360. K Okada, A Kaneda, Y Kameshima, A Yasumori. Mater Res Bull 2002; 37: 209–219.
361. JY Shen, JM Kobe, Y Chen, JA Dumesic. Langmuir 1994; 10:3902–3908.
362. H Nijs, M De Bock, EF Vansant. Microporous Mesoporous Mater 1999; 30: 243–253.
363. KG Scheckel, AC Scheinost, RG Ford, DL Sparks. Geochimica Cosmochimica Acta 2000; 64:2727–2735.
364. AM Scheidedegger, GM Lamble, DL Sparks. J Colloid Interface Sci 1997; 186: 118–128.
365. JW Boclair, PS Braterman, JJ Jiang, S Lou, F Yarberry. Chem Mater 1999; 11: 303–307.
366. RK Allada, A Navrotsky. Science 2002; 296:721–723.
367. N Oreskes, K Shrader-Frechette, K Belitz. Science 1994; 263:641–646.
368. AM Aicken, IS Bell, PV Coveney, W Jones. Adv Mater 1997; 9:496–500.
369. SP Newman, T Di Cristina, PV Coveney, W Jones. Langmuir 2002; 18:2933–2939.
370. AM Fogg, AL Rohl, GM Parkinson, D O'Hare. Chem Mater 1999; 11:1194–1200.
371. SP Newman, SJ Williams, PV Coveney, W Jones. J Phys Chem B 1998; 102: 6710–6719.
372. JW Wang, AG Kalinichev, RJ Kirkpatrick, XQ Hou. Chem Mater 2001; 13: 145–150.
373. AG Kalinichev, RJ Kirkpatrick, RT Cygan. Am Mineral 2000; 85:1046–1052.
374. AG Kalinichev, RJ Kirkpatrick, J Wang, Molecular Dynamics Simulation of Ionic Sorption and Diffusion on Surfaces and Interfaces of Layered Double Hydroxides. Proceedings of 223rd ACS National Meeting, Apr. 7–11, 2002, Orlando, FL.
375. B Zapata, P Bosch, G Fetter, MA Valenzuela, J Navarrete, VH Lara. Int J Inorg Mater 2001; 3:23–29.
376. MA Ulibarri, FM Labajos, V Rives, R Trujillano, W Kagunya, W Jones. Inorg Chem 1994; 33:2592–2599.
377. MA Ulibarri, I Pavlovic, C Barriga, MC Hermosin, J Cornejo. Appl Clay Sci 2001; 18:17–27.
378. HP Klug, LE Alexander. X-Ray Diffraction Procedures for Polycrystalline and Amorphous Materials. New York: Wiley, 1974.
379. F Yarberry, PS Braterman. Ostwald ripening of Mg/Al LDH. Unpublished work. Denton. TX: University of North Texas, 1999.
380. F Yarberry. Layered double hydroxides via aluminate. PhD dissertation. Denton. TX: University of North Texas, 2002.
381. O Clause, M Gazzao, F Trifiro, A Vaccari, L Zatorski. Appl Catal 1991; 73: 217–236.
382. KR Poeppelmeier, SJ Hwu. Inorg Chem 1987; 26:3297–3302.
383. SK Yun, TJ Pinnavaia. Chem Mater 1995; 7:348–354.
384. F Malherbe, C Forano, JP Besse. Microporous Mater 1997; 10:67–84.
385. EP Barrett, LG Joyner, PP Halenda. J Am Chem Soc 1951; 73:373–380.
386. K Putyera, TJ Bandosz, J Jagieo, JA Schwarz. Carbon 1996; 34:1559–1567.

387. M del Arco, V Rives, R Trujilano. Stud Surf Sci Catal 1994; 87:507–515.
388. G Carja, R Nakamur, T Aida, H Niiyama. Microporous Mesoporous Mater 2001; 47:275–284.
389. WT Reichle, SY Kyang, S Everhardt. J Catal 1986; 101:352–359.
390. ZP Xu, PS Braterman. J Mater Chem 2003; 13:268–273.
391. P Somasundaran, DW Fuerstenau. J Phys Chem 1966; 70:90–96.
392. PC Pavan, G de A Gomes, JB Valim. Microporous Mesoporous Mater 1998; 21: 659–665.
393. PC Pavan, EL Crepaldi, G de A Gomes, JB Valim. Colloids Surf, A 1999; 154: 399–410.
394. ZP Xu, PS Braterman. HR-TEM of surfactant LDH. Unpublished work. Denton. TX 76203: University of North Texas.
395. F Malherbe, JP Besse. J Solid State Chem 2000; 155:332–341.
396. V Prevot, C Forano, JP Besse. Appl Clay Sci 2001; 18:3–15.
397. M del Arco, V Rives, R Trujillano. Stud Surf Sci Catal 1994; 87:507–515.
398. KJD MacKenzie, RH Meinhild, BL Sherriff, Z Xu. J Mater Chem 1993; 3: 1263–1269.
399. ML Valchevatraykova, NP Davidova, AH Weiss. J Mater Sci 1993; 28:2157–2162.
400. S Carlino, MJ Hudson. Solid State Ionics 1998; 110:153–161.
401. A Vaccari, M Gazzano. Stud Surf Sci Catal 1995; 91:893–902.
402. ML Valchevatraykova, N Davidova, AH Weiss. J Mater Sci 1995; 30:737–743.
403. JM Fernandez, MA Ulibarri, FM Labajos, V Rives. J Mater Chem 1998; 8: 2507–2514.
404. MA Aramendia, Y Aviles, V Borau, JM Luque, JM Marinas, JR Ruiz, FJ Urbano. J Mater Chem 1999; 9:1603–1607.
405. M Tu, JY Shen, Y Chen. Thermochim Acta 1997; 302:117–124.
406. K Fuda, N Kudo, S Kawai, T Matsunaga. Chem Lett 1993:777–780.
407. O Marino, G Mascolo. Proc Eur Symp Thermal Anal 1981:391–394.
408. S Mohmel, I Kurzawski, D Uecker, D Muller, W Gebner. Cryst Res Technol 2002; 37:359–369.
409. MZ Bin Hussein, YH Taufiq-Yap, R Shahadan, NY Abd Rashid. Chem Pap 2001; 55:273–278.
410. TS Stanimirova, L Vergilov, G Kirov, N Petrova. J Mater Sci 1999; 34:4153–4161.
411. T Sato, H Fujita, T Endo, M Shimada, A Tsunashima. React Solids 1988; 5: 219–228.
412. FM Labajos, V Rives, MA Ulibarri. J Mater Sci 1992; 27:1546–1552.
413. M Bellotto, B Rebours, O Clause, J Lynch, D Bazin, E Elkaim. J Phys Chem 1996; 100:8535–8542.
414. M Bellotto, D Bazin, M Bessiere, O Clause, E Elkaim, J Lynch, B Rebours. Mater Sci Forum 1996; 228–231:347–352.
415a. F Rey, V Fornes, JM Rojo. J Chem Soc, Faraday Trans 1992; 88:2233–2238.
415b. W Yang, Y Kim, PKT Liu, M Sahimi, TT Tsotsis. Chem Engn Sci 2002; 57: 2945–2953.
416. TS Stanimirova, L Vergilov, G Kirov, N Petrova. J Mater Sci 1999; 34:4153–4161.
417. E Kanezaki. Mater Res Bull 1998; 33:773–778.

418. E Kanezaki. Inorg Chem 1998; 37:2588–2590.
419. E Kanezaki. Solid State Ionics 1998; 106:279–284.
420a. T Hibino, Y Yamashita, K Kosuge, A Tsunashima. Clays Clay Miner 1995; 43: 427–432.
420b. SW Rhee, MJ Kang. Korean J Chem Engn 2002; 19:653–657.
420c. N Petrova, T Mizota, T Stanimirova, G Kirov. J Miner Petrol Sci 2002; 97:1–6.
420d. I Melian-Cabrera, ML Granados, JLG Fierro. Phys Chem Chem Phys 2002; 4: 3122–3127.
421. L Chmielarz, P Kustrowski, A Rafalska-Lasocha, D Majda, R Dziembaj. Appl Catal B-Environ 2002; 35:195–210.
422. M del Arco, R Trujillano, V Rives. J Mater Chem 1998; 8:761–767.
423. IR McGill, B McEnaney, DC Smith. Nature 1976; 259:200–201.
424. FM Labajos, V Rives. Inorg Chem 1996; 35(18):5313–5318.
425. K Fuda, K Suda, T Matsunaga. Chem Lett 1993:1479–1482.
426. C Barriga, JM Fernandez, MA Ulibarri, FM Labajos, V Rives. J Solid State Chem 1996; 124:205–213.
427. A Monzon, E Romeo, C Royo, R Trujillano, FM Labajos, V Rives. Appl Catal, A 1999; 185:3–63.
428. V Rives, FM Labajos, R Trujillano, E Romeo, C Royo, A Monzon. Appl Clay Sci 1998; 13:363–379.
429. AL Tsyganok, K Suzuki, S Hamakawa, K Takehira, T Hayakawa. Catal Lett 2001; 77:75–86.
430. IJ Shannon, F Rey, G Sankar, JM Thomas, T Maschmeyer, AM Waller, AE Palomares, A Corma, AJ Dent, GN Greaves. J Chem Soc Faraday Trans 1996; 92: 4331–4336.
431. T Hibino, K Kosuge, A Tsunashima. Clay Clay Miner 1996; 44:151–154.
432. ZP Xu, HC Zeng. J Phys Chem B 2000; 104:10206–10214.
433. ZP Xu, R Xu, HC Zeng. Nano Letters 2001; 1:703–706.
434. Y Sugahara, N Yokoyama, K Kuroda, C Kato. Ceram Int 1988; 14:163–167.
435. EL Crepaldi, PC Pavan, J Tronto, JBA Valim. J Colloid Interface Sci 2002; 248: 429–442.
436. G Arrhenius, B Gedulin, S Mojzsis. In: C. Ponnamperuma, ed. Chemical Evolution: Origin of Life. Virginia: Deepak, 1993:25–50.
437. S Komarneni, N Kozai, R Roy. J Mater Chem 1998; 8:1329–1331.
438. KZ Zhu, CB Liu, XK Ye, Y Wu. Appl Catal A 1998; 168:365–372.
439. K Bahranowski, R Dula, M Gasior, M Labanowska, A Michalik, LA Vartikian, EM Serwicka. Appl Clay Sci 2001; 18:93–101.
440. YM Liu, ST Liu, KZ Zhu, XK Ye, Y Wu. Appl Catal A-Gen 1998; 169:127–135.
441. P Dinka, Z Cvengrosova, M Hronec. Petroleum Coal 1999; 41:57–61.
442. T Nishimura, N Kakiuchi, M Inoue, S Uemura. Chem Commun 2000:1245–1246.
443. N Kakiuchi, Y Maeda, T Nishimura, S Uemura. J Org Chem 2001; 66:6620–6625.
444. T Matsushita, K Ebitani, K Kaneda. Chem Commun 1999:265–266.
445. K Kaneda, T Yamashita, T Matsushita, K Ebitani. J Org Chem 1998; 63:1750–1751.
446. K Kaneda. Ru–Al–Mg Type Synthetic Hydrotalcite, Oxidation Catalyst and Production of Carbonyl Compound Using Same. Jap Patent 2,000,070,723, March 7, 2000.

447. HB Friedrich, F Khan, N Singh, M Van Staden. Synlett 2001:869–871.
448. S Ueno, K Ebitani, A Ookubo, K Kaneda. Appl Surf Sci 1997; 121:366–371.
449. K Kaneda, S Ueno. ACS Symposium Series 1996; 638:300–318.
450. K Kaneda, S Ueno, Heterogeneous Baeyer-Villager Oxidation using Hydrotalcite Catalysts. 211th ACS National Meeting, March 24–28, 1996, New Orleans.
451. K Kaneda, S Ueno, T Imanaka. J Mol Catal A 1995; 102:135–138.
452. K Kaneda, S Ueno, T Imanaka. J Chem Soc, Chem Commun 1994:797–798.
453. K Yamaguchi, K Mori, T Mizugaki, K Ebitani, K Kaneda. J Org Chem 2000; 65: 6897–6903.
454. JM Fraile, JL Garcia, JA Mayoral, F Figueras. Tetrahedron Lett 1996; 37: 5995–5996.
455. K Yamaguchi, K Ebitani, K Kaneda. J Org Chem 1999; 64:2966–2968.
456. S Ueno, K Yamaguchi, K Yoshida, K Ebitani, K Kaneda. Chem Commun 1998: 295–296.
457a. BM Choudary, ML Kantam, B Bharathi, CV Reddy. Synlett 1998:1203–1204.
457b. T Honma, M Nakajo, T Mizugaki, K Ebitani, K Kaneda. Tetrahedron Lett 2002; 43:6229–6232.
458. JM Fraile, JL Garcia, D Marco, JA Mayoral, E Sanchez, A Monzon, E Romeo. Stud Surf Sci Catal 2000; 130B:1673–1678.
459. VRL Constantino, TJ Pinnavaia. Inorg Chem 1995; 34:883–892.
460. VRL Constantino, TJ Pinnavaia. Catal Lett 1994; 23:361–67.
461. S Velu, CS Swamy. Appl Catal A 1997; 162:81–91.
462. TJ Pinnavaia, M Chibwe, VRL Constantino, SK Yun. Appl Clay Sci 1995; 10: 117–129.
463. KJ Martin, TJ Pinnavaia. J Am Chem Soc 1986; 108:541–542.
464. E Suzuki, M Okamoto, Y Ono. Chem Lett 1989:1485–1486.
465. E Suzuki, M Okamoto, Y Ono. J Mol Catal 1990; 61:283–294.
466. E Suzuki, A Inoue, Y Ono. Chem Lett 1998:1291–1292.
467. M Hoshino, H Shimada. J Adv Sci 1995; 7:51.
468. J Guo, DZ Jiang, EZ Min. Huaxue Yanjiu Yu Yingyong 1998; 10:610–616.
469. CW Hu, QL He, T Tang, EB Wang. Wuli Huaxue Xuebao 1995; 11:193–195.
470. CW Hu, YY Liu, ZP Wang, EB Wang. Huaxue Xuebao 1997; 55:49–55.
471. JB Hui, QF Liu, YX Ma, HZ Liu, LS Li, RR Xu. Guocheng Gongcheng Xuebao 2001; 1:152–156.
472. SK Yun, VRL Constantino, TJ Pinnavaia. Microporous Mater 1995; 4:21–29.
473. VI Iliev, AI Ileva, L Bilyarska. J Mol Catal A 1997; 126:99–108.
474. VI Iliev, AI Ileva, LD Dimitrov. Appl Catal A 1995; 126:33–340.
475. HC Liu, XY Yang, GP Ran. Acta Phys-Chim Sin 1999; 15:918–924.
476. ME Perez-Bernal, R Ruano-Casero, TJ Pinnavaia. Catal Lett 1991; 11:55–62.
477. A Corma, V Fornes, F Rey, A Cervilla, E Llopis, A Ribera. J Catal 1995; 152: 237–242.
478. A Cervilla, E Llopis, A Ribera, A Corma, V Fornes, F Rey. J Chem Soc, Dalton Trans 1994:2953–2957.
479. A Cervilla, A Corma, V Fornes, E Llopis, P Palanca, F Rey, A Ribera. J Am Chem Soc 1994; 116:1595–1596.

480. M Hirano, H Monobe, S Yakabe, T Morimoto. J Chem Res Synop 1999:374–375.
481. TJ Pinnavaia, EM Perez-Bernal, R Ruarno-Casero, M Chibwe. Polyaryl-Metallic Complex Intercalated Layered Double Hydroxides. US patent 5,302,709, April 12, 1994.
482. TJ Pinnavaia, EM Perez-Bernal, R Ruarno-Casero, M Chibwe. Salcomine-Metallic Complex Intercalated Layered Double Hydroxides. US patent 5,453,526, September 26, 1995.
483. TJ Pinnavaia, EM Perez-Bernal, R Ruarno-Casero, M Chibwe. Polyaryl-Metallic Complex Intercalated Layered Double Hydroxides. US patent 5,459,259, October 17, 1995.
484. TJ Pinnavaia, EM Perez-Bernal, R Ruarno-Casero, M Chibwe. Polyaryl-Metallic Complex Intercalated Layered Double Hydroxides. US patent 5,463,042, October 31, 1995.
485. M Chibwe, TJ Pinnavaia. J Chem Soc, Chem Comm 1993:278–279.
486. M Chibwe, ME Perez-Bernal, R Ruano-Casero, TJ Pinnavaia. Layered Double Hydroxide Supported Cobalt(II) Phthalocyanines as Possible Environmental Remediation Oxidation Catalysts. Adelaide. Australia: Proceedings of the International Clay Conference, 1995:87–91.
487. Y Watanabe, K Yamamoto, T Tatsumi. J Mol Catal, A 1999; 145:281–289.
488. T Tatsumi, K Yamamoto, H Tajima, H Tominaga. Chem Lett 1992:815–817.
489. AL Villa, DE De Vos, F Verpoort, BF Sels, PA Jacobs. J Catal 2001; 198:223–231.
490. BF Sels, DE De Vos, PA Jacobs. Stud Surf Sci Catal 2000; 129:845–850.
491. F van Laar, DE De Vos, D Vanoppen, B Sels, PA Jacobs, A Del Guerzo, F Pierard, A Kirsch-De Mesmaeker. J Chem Soc, Chem Comm 1998:267–268.
492. DE De Vos, J Wahlen, BF Sels, PA Jacobs. Synlett 2002:367–380.
493. F van Laar, DE De Vos, F Pierard, A Kirsch-De Mesmaeker, L Fiermans, PA Jacobs. J Catal 2001; 197:139–150.
494. BF Sels, DE De Vos, PA Jacobs. Tetrahedron Lett 1996; 47:8557–8558.
495. BF Sels, DE De Vos, M Buntinx, F Pierard, A Kirsch-De Mesmaeker, PA Jacobs. Nature 1999; 400:855–857.
496. BF Sels, DE De Vos, PA Jacobs. J Am Chem Soc 2001; 123:8350–8359.
497. J Guo, QZ Jiao, G Xiong, HJ Lu, DZ Jiang, EZ Min. Chinese Chem Lett 1996; 7: 531–534.
498. J Guo, QZ Jiao, JP Shen, DZ Jiang, GH Yang, EZ Min. Catal Lett 1996; 40:43–45.
499. J Guo, N Wang, QZ Jiao, DZ Jiang, GH Yang, EZ Min. Chem Res Chinese U 1995; 11:256–258.
500. J Guo, QZ Jiao, SH Duan, DZ Jiang, EZ Min. Gaodeng Xuexiao Huaxue Xuebao 1998; 19:129–131.
501. Z Xu, Y Wu, H He, DZ Jiang. Stud Surf Sci Catal 1994; 90:279–84.
502. Z Xu, H He, DZ Jiang, Y Wu. Wuli Huaxue Xuebao 1994; 10:6–8.
503. Y Guo, D Li, C Hu, Y Wang, E Wang. Int J Inorg Mater 2001; 3:347–355.
504. Y Guo, D Li, C Hu, Y Wang, E Wang, Y Zhou, S Feng. Appl Catal B 2001; 30: 337–349.
505. YH Guo, DF Li, CW Hu, YH Wang, EB Wang, B Yue. Gaodeng Xuexiao Huaxue Xuebao 2001; 22:1453–1455.

506. T Kwon, GA Tsigdinos, TJ Pinnavaia. J Am Chem Soc 1988; 110:3653–3654.
507. B Coq, D Tichit, S Ribet. J Catal 2000; 189:117–128.
508. FM Cabello, D Tichit, B Coq, A Vaccari, NT Dung. J Catal 1997; 167:142–152.
509. F Medina, R Dutartre, D Tichit, B Coq, NT Dung, P Salagre, JE Sueiras. J Mol Catal A-Chem 1997; 119:201–212.
510a. D Tichit, B Coq, S Ribet, R Durand, F Medina. Stud Surf Sci Catal 2000; 130A: 503–508.
510b. D Tichit, R Durand, A Rolland, B Coq, J Lopez, P Marion. J Catal 2002; 211: 511–520.
511. GL Castiglioni, M Ferrari, A Guercio, A Vaccari, R Lancia, C Fumagalli. Catal Today 1996; 27:181–186.
512. S Narayanan, K Krishna. Catal Today 1999; 49:57–63.
513. S Narayanan, K Krishna. Appl Catal, A 1998; 174:221–229.
514. S Narayanan, K Krishna. Stud Surf Sci Catal 1998; 113:359–363.
515. S Narayanan, K Krishna. Chem Commun 1997:1991–1992.
516. S Narayanan, K Krishna. Appl Catal A 1996; 147:L253–L258.
517. YZ Chen, CW Liaw, LI Lee. Appl Catal A 1999; 177:1–8.
518. FJ Bröcker, L Kainer. Catalysts and Catalysts Carriers Having Very Finely Divided Active Components and Their Preparation. Ger Patent 2,024,282, December 25, 1973.
519. FJ Bröcker, L Kainer. UK Patent 1,342,020, 1971.
520. PS Kumbhar, J Sanchez-Valente, F Figueras. Tetrahedron Lett 1998; 39: 2573–2574.
521. SM Auer, JD Grunwaldt, RA Koppel, A Baiker. J Mol Catal A 1999; 139:305–313.
522. R Shangguan, X Ge, H Zou, K Li, W Zhang, J Shen. Cuihua Xuebao 1999; 20: 515–520.
523. PM Fritz, H Boelt. Catalytic Dehydrogenation of C2–10 Alkanes. Ger Patent 19,858,747, June 21, 2000.
524. Z Li, M Tu, J Shen, J Jia, Z Xu, L Lin. Cuihua Xuebao 1998; 19:1–2.
525. A Dejoz, JML Nieto, F Melo, I Vazquez. Ind Eng Chem Res 1997; 36:2588–2596.
526. N Mimura, I Takahara, M Saito, Y Sasaki, K Murata. Catal Lett 2002; 78:125–128.
527. F Malherbe, C Forano, B Sharma, MP Atkins, JP Besse. Appl Clay Sci 1998; 13: 381–399.
528. WT Reichle. J Catal 1985; 94:547–557.
529. WT Reichle. J Catal 1980; 63:295–306.
530. WT Reichle. Chemtech 1981; 11:698–702.
531. E Suzuki, Y Ono. Bull Chem Soc Jpn 1988; 61:1008–1010.
532. KK Rao, M Gravelle, SV Jaime, F Fugueras. J Catal 1998; 173:115–121.
533. JI Di Cosimo, VK Diez, CR Apesteguia. Appl Clay Sci 1998; 13:433–449.
534. JI Di Cosimo, VK Diez, M Xu, E Iglesias, CR Apesteguia. J Catal 1998; 178: 499–510.
535a. CP Kelkar, AA Schutz. Appl Clay Sci 1998; 13:417–432.
535b. MJ Climent, A Corma, S Iborra, A Velty. Green Chem 2002; 4:474–480.
536a. R Unnikrishnan, S Narayanan. J Mol Catal A 1999; 144:173–179.
536b. NN Das, SC Srivastava. Bull Mater Sci 2002; 25:283–289.

536c. D Tichit, B Coq, S Cerneaux, R Durand. Catal Today 2002; 75:197–202.

537. BJ Arena, JS Holmgren. Direct Conversion of Butyraldehyde to 2-Ethylhexanol-1. US patent 5,258,558, November 2, 1993.

538. D Tichit, M Lhouty, A Guide, B Chiche, F Figueras, A Auroux, D Bartalini, E Garrone. J Catal 1995; 151:50–59.

539. ML Kantam, BM Choudary, CV Reddy, KK Rao, F Figueras. Chem Comm 1998: 1033–1034.

540. MJ Climent, A Corma, S Iborra, J Primo. J Catal 1995; 151:60–66.

541. J Lopez, JS Valente, JM Clacens, F Figueras. J Catal 2002; 208:30–37.

542. C Noda, GP Alt, RM Werneck, CA Henriques, JLF Monteiro. Braz J Chem Eng 1998; 15:120–125.

543. JC Roelofs, AJ van Dillen, KP de Jong. Catal Today 2000; 60:297–303.

544. E Dumitriu, V Hulea, C Chelaru, C Catrinescu, D Tichit, R Durand. Appl Catal A 1999; 178:145–157.

545. MJ Clíment, A Corma, R Guil-Lopez, S Iborra, J Primo. Catal Lett 1999; 59:33–38.

546. A Corma, S Iborra, J Primo, F Rey. Appl Catal A 1994; 114:215–25.

547. V Corma, V Fornes, RM Martin-Aranda, F Rey. J Catal 1992; 134:58–65.

548. A Ramani, BM Chanda, S Velu, S Sivasanker. Green Chemistry 1999; 1:163–165.

549. S Velu, CS Swamy. Appl Catal A 1994; 119:241–252.

550. S Velu, CS Swamy. Appl Catal A 1996; 145:225–230.

551. S Velu, CS Swamy. Appl Catal A 1996; 145:141–153.

552. S Velu, CS Swamy. Catal Lett 1996; 40:265–272.

553. S Velu, CS Swamy. React Kinet Catal Lett 1997; 62:339–346.

554. S Velu, CS Swamy. Res Chem Intermed 2000; 26:295–302.

555a. AH Padmasri, VD Kumari, PK Rao. Stud Surf Sci Catal 1998; 113:563–571.

555b. AH Padmasri, A Venugopal, VD Kumari, KSR Rao, PK Rao. J Mol Catal A: Chem 2002; 188:255–265.

556. TM Jyothi, T Raja, MB Talawar, K Sreekumar, S Sugunan, BS Rao. Synth Comm 2000; 30:3929–3934.

557a. S Velu, S Sivasanker. Res Chem Intermed 1998; 24:657–666.

557b. M Bolognini, F Cavani, D Scagliarini, C Flego, C Perego, M Saba. Catal Today 2002; 75:103–111.

558. TM Jyothi, T Raja, MB Talawar, BS Rao. Appl Catal, A 2001; 211:41–46.

559. MB Talawar, TM Jyothi, T Raja, BS Rao, PD Sawant. Green Chem 2000; 2: 266–268.

560. C Cativiela, F Figueras, JI Garcia, JA Mayoral, MM Zurbano. Synth Commun 1995; 25:1745–1750.

561a. VR Choudhary, SK Jana, AB Mandale. Catal Lett 2001; 74:95–98.

561b. VR Choudhary, SK Jana, VS Narkhede. Appl Catal A: General 2002; 235:207–215.

562. J Santhanalakshmi, T Raja. Appl Catal A 1996; 147:69–80.

563. T Ruhl, J Henkelmann, M Heider, B Fiechter. Preparation of N-alkenylcarboxamides. US patent 5,654,478, M Heider, T Ruhl, J Henkelmann. Preparation of N-alkenyl Carboxamides 5,710,331, January 20, 1998.

564. SW King. Process for the Preparation of Hydroxyl-Containing Compounds. US Patent 5,191,123, April 1, 1992.

565. SW King. Process for the Preparation of Amino Ethers. US Patent 5,214,142, March 25, 1992.
566. SW King. Catalytic Processes for the Preparation of *N,N,N*-Trisubstituted Nitrogen-Containing Compounds. US Patent 5,245,032, March 25, 1992.
567. SW King. Catalytic Processes for the Preparation Triethylenediamine. US Patent 5,280,120, January 18, 1994.
568. SW King, KD Olson. Process for the Preparation of Ethers. US Patent 5,210,322, March 25, 1992.
569. SW King, KD Olson. Process for the Preparation of Cyclic Ethers. US Patent 5,247,103, September 21, 1993.
570. VC Bastrom. Process for the Preparation of Tertiary Aminocarbonates and Aminoethers. US patent 5,563,288, October 8, 1996.
571. Y Watanabe, T Tatsumi. Micropor Mesopor Mater 1998; 22:399–407.
572. A Corma, S Iborra, S Miquel, J Primo. J Catal 1998; 173:315–32.
573. C Block, W Breuer, H Endres, J Hachgenei. Ger Patent 4,040,679, June 25, 1996.
574. PS Kumbhar, J Sanchez-Valente, F Figueras. Chem Commun 1998:1091–1092.
575. BM Choudary, ML Kantam, B Kavita. Green Chem 1999; 1:289–292.
576. JH Robson, BC Ream, H Soo. Monoalkylene Glycol Production Using Mixed Metal Framework Compositions. US Patent 4,967,018, October 30, 1990.
577. JH Robson, BC Ream, H Soo. Mixed Metal Framework Compositions for Monoalkylene Glycol Production. US Patent 5,064,804, November 12, 1991.
578. F Malherbe, JP Besse, SR Wade, WJ Smith. Catal Lett 2000; 67:197–202.
579. K Yamaguchi, K Ebitani, T Yoshida, H Yoshida, K Kaneda. J Am Chem Soc 1999; 121:4526–4527.
580. S Kohjiya, T Sato, T Nakayama, SM Yamashita. Chem, Rapid Commun 1981; 2: 231–233.
581. DE Laycock. Synthesis of Stereoregular Poly(Propylene Oxide). US Patent 4,962,281, October 9, 1990.
582. DE Laycock. Catalytic Process for the Preparation of Polyols. US Patent 4,962,237, October 9, 1990.
583. DE Laycock, RJ Collacott, DA Skelton, MF Tchir. J Catal 1991; 130:354–358.
584a. BM Choudary, ML Kantam, CRV Reddy, KK Rao, F Figueras. J Mol Catal, A 1999; 146:279–284.
584b. BM Choudary, M Lakshmi Kantam, V Neeraja, RK Koteswara, F Figueras, L Delmotte. Green Chem 2000; 3:257–260.
585a. J Palomeque, JM Clacens, F Figueras. J Catal 2002; 211:103–108.
585b. G Carja, R Nakamura, H Niiyama. Appl Catal A: General 2002; 236:91–102.
585c. MA Aramendia, V Borau, C Jimenez, JM Marinas, JR Ruiz, FJ Urbano. J Chem Soc, Perkin Trans 2 2002:1122–1125.
586. N Das, D Tichit, R Durand, P Graffin, B Coq. Catal Lett 2001; 71:181–185.
587. YZ Chen, CM Hwang, CW Liaw. Appl Catal A 1998; 169:207–214.
588. RK Sharma, ED Olson. Coal Sci Technol 1991; 18:377–384.
589. S Chakka. Selective Hydrodesulfurization of Cracked Naphtha Using Hydrotalcite-Supported Catalysts. US Patent 5,851,382, December 22, 1998.
590a. ZP Xu, HC Zeng. Proceedings of Chemical & Process Engineering Conference 2000, Singapore, in conjunction with Regional Symposium on Chemical Engineering. Singapore, Dec. 11–13, 2000.

590b. R Zhao, C Yin, H Zhao, C Liu. Preprints—American Chemical Society, Division of Petroleum Chemistry 2002; 47:309–311.

590c. C Yin, R Zhao, C Liu. Preprints—American Chemical Society, Division of Petroleum Chemistry 2002; 47:63–65.

591. K Tsubaki, H Morinaga, K Maeda, S Yamamoto, S Kamiyama. Ger Patent 2,621,591, November 25, 1976.

592. N Okada, H Morinaga, H Ktaguchi. Jap Patent 49,035,486, April 2, 1974.

593. N Kashiwa. Jap Patent 48,066,179, 1973.

594. T Takahashi. Jap Patent 48,026,275, April 6, 1973.

595. H Muller-Tamm, H Frielingsdorp, G Schweier, L Reuter. Ger Patent 2,163,851, June 28, 1973.

596. A Bhattacharyya, WD Chang, MS Kleefisch, CA Udovich. Method for Preparing Synthesis Gas Using Nickel Catalysts. US Patent 5,399,537, March 21, 1995.

597. WD Chang, CA Udovich, A Bhattacharyya, MS Kleefisch. Catalyst Prepared from Nickel-Containing Hydrotalcite-Like Precursor Compound. US Patent 5,767,040, June 16, 1998.

598. WD Chang, CA Udovich, A Bhattacharyya, MS Kleefisch. Method for Preparing and Using Nickel Catalysts. US Patent 5,939,353, August 17, 1999.

599. WD Chang, CA Udovich, A Bhattacharyya, MS Kleefisch. Method for Preparing Synthesis Gas Using Nickel Catalysts. WO Patent 9,915,459, March 21, 1995.

600. A Bhattacharyya, MP Kaminsky. Simplified Preparation of Hydrotalcite-Type Clays. Eur Patent 536,879, April 14, 1993.

601. F Basile, L Basini, M D'Amore, G Fornasari, A Guarinoni, D Matteuzzi, G Del Piero, F Trifiro, A Vaccari. J Catal 1998; 173:247–256.

602. F Basile, L Basini, G Fornasari, M Gazzano, F Trifiro, A Vaccari. Stud Surf Sci Catal 1998; 118:31–40.

603. F Basile, G Fornasari, F Trifiro, A Vaccari. Stud Surf Sci Catal 2000; 130A: 449–454.

604. F Basile, G Fornasari, F Trifiro, A Vaccari. Catal Today 2001; 64:21–30.

605. H Morioka, Y Shimizu, M Sukenobu, K Ito, E Tanabe, T Shishido, K Takehira. Appl Catal A 2001; 215:11–19.

606. T Shishido, M Sukenobu, H Morioka, M Kondo, Y Wang, K Takaki, K Takehira. Appl Catal A 2002; 223:35–42.

607. K Schulze, W Makowski, R Chyzy, R Dziembaj, G Geismar. Technologia Chemiczna na Przelomie Wiekow 2000:103–106.

608. AM Gaffney, R Song, R Oswald, D Corbin. Cobalt-Based Catalysts and Process for Producing Synthesis Gas. WO Patent 0,136,323, May 25, 2001.

609. A Bhattacharyya, VW Chang, DJ Schumacher. Appl Clay Sci 1998; 13:317–328.

610. JG Nunan, RG Herman, K Klier. J Catal 1989; 116:222–229.

611. JG Nunan, PB Himelfarb, RG Herman, K Klier, CF Bogdan, GW Simmons. Inorg Chem 1989; 28:3868–3874.

612. F Trifiro, A Vaccari, O Clause. Catal Today 1994; 21:185–96.

613. TH Vanderspurt, RJ Koveal. Isoalcohol Synthesis. US Patent 5,703,133, December 30, 1997.

614. M Rameswaran, EG Rightor, ED Dimotakis, TJ Pinnavaia. Syngas Conversion over an Acidic Pillared Clay and a Basic Layered Double Hydroxide. Proc Int Congr Catal. Vol. 2. Ottawa. Ontario, 1998:783–790.

615. G Fornasari, S Gusi, F Trifiro, A Vaccari. Ind Eng Chem Res 1987; 26:1500–1505.
616. GC Shen, S Fujita, S Matsumoto, N Takezawa. J Mol Catal A 1997; 124:123–136.
617. BA Peppley, JC Amphlett, LM Kerns, RF Mann. Appl Catal A 1999; 179:21–29.
618. JP Breen, JRH Ross. Catal Today 1999; 51:521–533.
619. LM Kearns, JC Amphlett, E Halliop, HM Jensen, RF Mann, BA Peppley. Prepr of Symposia—ACS, Div of Feul Chemistry 1999; 44:905–908.
620. LM Kearns, JC Amphlett, RF Mann, BA Peppley. Correlation of Morphology and Activity of Cu/ZnO/Al$_2$O$_3$ Catalyst Used in Steam Reforming of Methanol. Book of Abstracts, 215th ACS National Meeting. Dallas, March 29–April 2, 1998.
621. S Velu, K Suzuki, T Osaki. Catal Lett 2000; 69:43–50.
622a. SR Segal, KB Anderson, KA Carrado, CL Marshall. Appl Catal A 2002; 231: 215–226.
622b. T Shishido, P Wang, T Kosaka, K Takehira. Chem Lett 2002:752–753.
622c. S Velu, N Satoh, CS Gopinath, K Suzuki. Catal Lett 2002; 82:145–152.
623. PR Kurek. Oxidation and Hydrolysis of Cyanides Using Metal Chelates on Supports of Metal Oxide Solid Solutions. US Patent 5,476,596, December 19, 1995.
624. X Yang, J Yuan. Ranliao Huaxue Xuebao 1998; 26:61–64.
625a. M Sychev, R Prihod'ko, K Erdmann, A Mangel, RA van Santen. Appl Clay Sci 2001; 18:103–110.
625b. TH Bennur, A Ramani, R Bal, BM Chanda, S Sivasanker. Catal Comm 2002; 3: 493–496.
625c. BM Choudary, ML Kantam, NM Reddy, NM Gupta. Catal Lett 2002; 82:79–83.
626. Y Saito, Y Yoneda, S Makishima. Acres Congr Int Catalyse 1961:1937–1950.
627. EM Serwicka. Polish J Chem 2001; 75:307–328.
628. RJ Meyer, E Pietsch. In: Gmelins Handbuch der Anorganischen Chemie. Berlin: Verlag Chimie, 1936:558–597.
629. F Kapteijn, J Rodriguez-Mirasol, JA Moulijn. Appl Catal 1996; 9:25–64.
630. A Cimino, F Pepe. J Catal 1972; 25:362–377.
631. A Cimino, M Schiavello. J Catal 1971; 20:202–216.
632. A Cimino, V Indovina. J Catal 1970; 17:54–70.
633. A Cimino, V Indovina, F Pepe, M Schiavello. J Catal 1969; 14:49–54.
634. ERS Winter. J Catal 1969; 15:144–152.
635. M Schiavello, M Valigi, F Pepe. Trans Faraday Soc 1975; 71:1642–1648.
636. F Pepe, M Schiavello, G Ferraris. Z Physik Chem Neue Folge 1975; 96:297–310.
637. C Angeletti, F Pepe, P Porta. J Chem Soc Faraday Trans 1 1978; 74:1595–1603.
638. R Sundararajan, V Srinivasan. Appl Catal 1991; 73:165–171.
639. J Christopher, CS Swamy. J Mol Catal 1990; 62:69–78.
640. JN Armor, TA Braymer, TS Farris, Y Li, FP Petrocelli, EL Weist, S Kannan, CS Swamy. Appl Catal B-Environ 1996; 7:397–406.
641. CS Swamy, S Kannan, Y Li, JN Armor, TA Braymer. Method for Decomposing N$_2$O Utilizing Catalysts Comprising Calcined Anionic Clay Minerals, US Patent, 5,407,652, March 1, 1995.
642. S Kannan, CS Swamy. Catal Today 1999; 53:725–737.
643. S Kannan. Appl Clay Sci 1998; 13:347–362.
644. S Kannan, CS Swamy. Stud Surf Sci Catal 1995; 91:903–914.

645. J Oi, A Obuchi, GR Bamwenda, A Ogata, H Yagita, S Kushiyama, K Mizuno. Appl Catal B 1997; 12:277–286.
646. J Perez-Ramirez, JM Garcia-Cortes, F Kapteijn, MJ Illan-Gomez, A Ribera, CSM de Lecea, JA Moulijn. Appl Catal B 2000; 25:191–203.
647. F Kapteijn, G Marban, J Rodriguez-Mirasol, JA Moulijn. J Catal 1997; 167: 256–265.
648. RS Drago, K Jurczyk, N Kob. Appl Catal B 1997; 13:69–79.
649. G Centi, A Galli, B Montanari, S Perathoner, A Vaccari. Catal Today 1997; 35: 113–120.
650. H Dandl, G Emig. Appl Catal A 1998; 168:261–268.
651. Y Liu, X Yang, Z Zhang, Y Wu. Cuihua Xuebao 1999; 20:450–454.
652. B Wen, MY He, JQ Song, BN Zong, XT Shu. Shiyou Xuebao, Shiyou Jiagong 2000; 16:72–78.
653. F Marquez, AE Palomares, F Rey, A Corma. J Mater Chem 2001; 11:1675–1680.
654a. A Giroir-Fendler, P Denton, A Boreave, H Praliaud, M Primet. Top Catal 2001; 16/17:237–241.
654b. SC Christoforou, EA Efthimiadis, IA Vasalos. Catal Let 2002; 79:137–147.
655a. G Centi, G Fornasari, C Gobbi, M Livi, F Trifiro, A Vaccari. Catal Today 2002; 73:287–296.
655b. G Fornasari, F Trifiro, A Vaccari, F Prinetto, G Ghiotti, G Centi. Catal Today 2002; 75:421–429.
656. Y Li. PCT Int. Appl 2002:49.
657. AE Palomares, JM Lopez-Nieto, FJ Lazaro, A Lopez, A Corma. Appl Catal B 1999; 20:257–266.
658. A Corma, AE Palomares, F Rey, F Marquez. J Catal 1997; 170:140–149.
659a. AJ Perrotta, FS Williams, L Stonehouse. Light Met 1997:37–48.
659b. J Das, D Das, GP Dash, KM Parida. J Colloid Interf Sci 2002; 251:26–32.
660. B Houri, A Legrouri, A Barroug, C Forano, JP Besse. J Chim Phys Phys—Chim Biol 1999; 96:455–463.
661. B Houri, A Legrouri, A Barroug, C Forano, JP Besse. Collect Czech Chem C 1998; 63:732–740.
662. SW Rhee, MJ Kang, H Kim, CH Moon. Environ Technol 1997; 18:231–236.
663. M Lehmann, Al Zouboulis, KA Matis. Chemosphere 1999; 39:881–892.
664. YW You, HT Zhao, GF Vance. Environ Technol 2001; 22:1447–1457.
665. HS Shin, MJ Kim, SY Nam, HC Moon. Water Sci Technol 1996; 34:161–168.
666. T Kindaichi, K Nishimura, T Kitao, K Kuzawa, Y Kiso. Haikibutsu Gakkai Ronbunshi 2002; 13:99–105.
667. P Ulmgren. Nord Pulp Pap Res J 1987; 2:4–9.
668. MC Hermosin, I Pavlovic, MA Ulibarri, J Cornejo. Water Res 1996; 30:171–177.
669. MC Hermosin, I Pavlovic, MA Ulibarri, J Cornejo. J Environ Sci Heal A 1993; 28:1875–1888.
670. MC Hermosin, I Pavlovic, MA Ulibarri, J Cornejo. Fresenius Environ Bull 1995; 4:41–46.
671. MA Ulibarri, I Pavlovic, MC Hermosin, J Cornejo. Appl Clay Sci 1995; 10: 131–145.

672. R Celis, WC Koskinen, AM Cecchi, GA Bresnahan, MJ Carrisoza, M Ulibarri, I Pavlovic, MC Hermosin. J Environ Sci Heal B 1999; 34:929–941.

673. MV Villa, MJ Sanchez-Martin, M Sanchez-Camazano. J Environ Sci Heal B 1999; 34:509–525.

674. M Lakraimi, A Legrouri, A Barroug, A de Roy, JP Besse. J Chim Phys PCB 1999; 96:470–478.

675. HT Zhao, GF Vance. J Inclusion Phenom Mol 1998; 31:305–317.

676. J Inacio, G Taviot-Gueho, C Forano, JP Besse. Appl Clay Sci 2001; 18:255–264.

677. G Onkal-Engin, R Wibulswas, DA White. Environ Technol 2000; 21:167–175.

678. Y Seida, Y Nakano. Water Res 2000; 34:1487–1494.

679. S Amin, GG Jayson. Water Res 1996; 30:299–306.

680. K Esumi, S Yamamoto. Colloid Surf A 1998; 137:385–388.

681a. I Pavlovic, MA Ulibarri, MC Hermosin, J Cornejo. Fresen Environ Bull 1997; 6: 266–271.

681b. PC Pavan, EL Crepaldi, JB Valim. J Colloid Interface Sci 2000; 229:346–352.

682a. J Bhatt, HM Mody, HC Bajaj. Clay Res 1996; 15:28–32.

682b. Y You, H Zhao, GF Vance. Appl Clay Sci 2002; 21:217–226.

683. S Sunayama, T Sato, A Kawamoto, A Ohkubo, T Suzuki. Biocontrol Sci 2002; 7: 75–81.

684. DW Oscarson, HG Miller, RL Watson. An Evaluation of Potential Additive to a Clay-Based Buffer Material for the Immobilization of $^{129}I^-$, at Energy Can Ltd, 1986.

685. G Fetter, MT Olguin, P Bosch, VH Lara, S Bulbulian. J Radioanal Nucl Ch 1999; 241:595–599.

686. SD Balsley, PV Brady, JL Krumhansl, HL Anderson. J Soil Contam 1998; 7: 125–141.

687. J Serrano, F Granados, V Bertin, S Bulbulian. Sep Sci Technol 2002; 37:329–341.

688. E Narita, Y Umezu, S Takahashi. Removal Method for Radioactive Iodine. Jap Patent 11,084,084, March 26, 1999.

689a. OC Wilson Jr, T Olorunyolemi, A Jaworski, L Borum, D Young, E Dickens, C Oriakhi. Ceram Trans 2000; 110:93–102.

689b. GA Bubniak, WH Schreiner, N Mattoso, F Wypych. Langmuir 2002; 18: 5967–5970.

690. CO Oriakhi, IV Farr, MM Lerner. Clays Clay Miner 1997; 45:194–202.

691. HR Fisher, LH Gielgens. Nanocomposite Material. US Patent 6,372,837, April 2, 2002.

692. HR Fisher, LH Gielgens. Nanocomposite Material. US Patent 6,365,661, April 2, 2002.

693. L van der Ven, MLM van Gemert, LF Batenburg, JJ Keern, LH Gielgens, TPM Koster, HR Fisher. Appl Clay Sci 2000; 17:25–34.

694. S Miyata, M Kuroda. Method for Inhibiting the Thermal or Ultraviolet Degradation of Thermoplastic Resin and Thermoplastic Resin Compositions Having Stability to Thermal or Ultraviolet Degradation. US Patent 4,299,759, November 10, 1981.

695. F Leroux, M Adachi-Pagano, M Intissar, S Chauviere, C Forano, JP Besse. J Mater Chem 2001; 11:105–112.

696. T Hibino, W Jones. J Mater Chem 2001; 11:1321–1323.
697. S O'Leary, D O'Hare, G Seeley. Chem Comm 2002:1506–1507.
698. T Shichi, S Yamashita, K Takagi. Supramol Sci 1998; 5:303–308.
699. SW Rhee, D-Y Jung. Bull Korean Chem Soc 2002; 23:35–40.
700. G Geismar, J Lewandowski, E De Boer. Chem Ztg 1991; 115:335–339.
701. K Takagi, Y Sawaki. J Synth Org Chem Jpn 1996; 54:280–288.
702. T Shichi, K Takagi, Y Sawaki. Chem Lett 1996:781–782.
703. K Takagi, T Shichi, H Usami, Y Sawaki. J Am Chem Soc 1993; 115:4339–4344.
704. B Ballarin, M Gazzano, R Seeber, D Tonelli, A Vaccari. J Electroanal Chem 1998; 445:27–37.
705. JX He, K Kobayashi, M Takahashi, G Villemure, A Yamagishi. Thin Solid Films 2001; 397:255–265.
706. JX He, K Kobayashi, YM Chen, Y-A Gillesvillemure. Electrochem Comm 2001; 3:473–477.
707. E Scavetta, M Berrettoni, R Seeber, D Tonelli. Electrochim Acta 2001; 46: 2681–2692.
708. B Keita, A Belhouari, L Nadjo. J Electroanal Chem 1993; 355:235–251.
709. J. Labuda, M Hudakova. Electroanal 1997; 9:239–242.
710. C Mousty, S Therias, C Forano, JP Besse. J Electroanal Chem 1994; 374:63–69.
711. P Wang, G Zhu. Electrochem Comm 2002; 4:36–40.
712. M Morigi, E Scavetta, M Berrettoni, M Giorgetti, D Tonelli. Anal Chim Acta 2001; 439:265–272.
713a. JE Moneyron, A Deroy, JP Besse. Solid State Ionics 1991; 46:175–181.
713b. JE Moneyron, A Deroy, JP Besse. Sensor Actuat B 1991; 4:189–194.
714. S Cooper, PK Dutta. J Phys Chem 1990; 94:114–118.
715. M Jakupca, PK Dutta. Chem Mater 1995; 7:989–994.
716. M Jakupca, PK Dutta. Chem Ind 1997; 69:595–606.
717. SZ Qureshi, RMAQ Jamhour, N Rahman. J Planar Chromatogr—Mod TLC 1996; 9:466–469.
718. V Ghoulipour, SW Husain. J Planar Chromatogr—Mod TLC 1999; 12:378–382.
719. DM O'Hare, AM Fogg. Separation Process. US Patent 6,391,823, February 21, 2002.
720. L Lei, F Millange, RI Walton, D O'Hare. J Mater Chem 2000; 10:1881–1886.
721. F Millange, RI Walton, L Lei, D O'Hare. Chem Mater 2000; 12:1990–1994.
722. B Lotsch, F Millange, RI Walton, D O'Hare. Solid State Sci 2001; 3:883–886.
723. MS Lin, P Sun, HY Yu. J Formos Med Assoc 1998; 97:704–710.
724. G Simoneau. Eur J Drug Metab Ph 1996; 21:351–357.
725. HC Dollinger, M Liszkay. Eur J Gastroen Hepat (Suppl. 3) 1993; 35:S133–S137.
726. CL Peterson, DL Perry, H Masood, HH Lin, JL White, SL Hem, C Fritsch, F Haeusler. Pharmaceut Res 1993; 10:998–1007.
727. M Dreyer, D Marwinski, N Wolf, HG Dammann. Arzneim Forsch 1991; 41: 738–741.
728. Z Kokot. Pharmazie 1988; 43:249–251.
729. D Mendelsohn, L Mendelsohn. South African Med J 1975; 49:1011–1014.
730. AS Tarnawski, M Tomikawa, M Ohta, IJ Sarfeh. J Physiol (Paris) 2000; 94:93–98.

731. AS Tarnawski, R Pai, R Itani, FA Wyle. Digestion 1999; 60:449–455.
732. Medications for treatment of iron deficiency, Kyowa Chem. Ind. Co., Jpn. Kokai Tokkyo Koho JP 6,006,619, 1985.
733. G Fardella, V Ambrogi, L Perioli, G Grandolini. Intercalation of Indomethacin and Tolmetin in Hydrotalcite-Like Compounds and in vitro Release. Proceedings of the International Symposium on Controlled Release of Bioactive Materials. 1998: 774–775.
734. JH Choy, SY Kwak, YJ Jeong, JS Park. Angewandte Chemie, International Edition 2000; 39:4042–4045.
735. MZ bin Hussein, Z Zainal, AH Yahaya, DWV Foo. J Controlled Release 2002; 82:417–427.
736. Berra, Yogi, attrib.

9

Layered Manganese Oxides: Synthesis, Properties, and Applications

Jia Liu, Jason P. Durand, Laura Espinal, Luis-Javier Garces, Sinue Gomez, Young-Chan Son, Josanlet Villegas, and Steven L. Suib
University of Connecticut
Storrs, Connecticut, U.S.A.

I. INTRODUCTION

This chapter discusses the major areas in the field of layered manganese oxide materials, including synthesis, properties, and applications. The section on synthesis and properties is divided into three parts based on three different types of layered manganese oxide materials: (1) feitknechtite and pyrochroite, which have the CdI_2 type of structure; (2) birnessite and buserite type of materials (OL-1), in which inorganic/organic cations and water molecules are introduced into the interlayer region; and (3) pillared layered manganese-based materials (POL) and manganese-based mesoporous materials (MOMS)/manganese-based porous mixed oxides (MANPOs). In the latter materials, organic pillaring or surfactant molecules are introduced into layered manganese oxide precursors with small interlayer distances to achieve layered structures with large spacings and porosity. In each subsection, conventional and novel synthesis methods are discussed, including precipitation routes (i.e., oxidation method), sol-gel routes, ion exchange, framework doping, and intercalation/pillaring. The characterization and properties of these materials are also included in this section. The second section concerns five major areas of layered manganese oxide material applications, including sorption, catalytic oxidation reactions, degradation of organic compounds, battery materials, and mesoporous solids and manganese-based porous mixed oxides. These applications are not all-inclusive with regard to recent developments, but they are of considerable interest as shown by the large number of publications in these areas. A final section covers conclusions and possible new directions.

II. LAYERED MANGANESE OXIDES

A. Feitknechtite and Pyrochroite

1. Feitknechtite

The natural mineral feitknechtite is found in Franklin, New Jersey, USA. In 1945, Feitknecht and Marti, who referred to this mineral as hydrohausmannite, first studied feitknechtite systematically. Mixed phases of $MnOOH$ and Mn_3O_4 were prepared by oxidation of an aqueous suspension of $Mn(OH)_2$ [1]. In 1962, Feitknecht, Brunner, and Oswald suggested that the material referred to as hydrohausmannite was a mixture of Mn_3O_4 and β-$MnOOH$, which was thereafter named feitknechtite, on the basis of electron microscopy data. [2] Feitknechtite (β-$MnOOH$) comprises 62.47% Mn, 1.15% H, and 36.39% O, where 89.76% of the product is represented by Mn_2O_3 and the molecular weight is 87.94 g/mol. The mineral is made up of $MnO_3(OH)_3$ units consisting of edge-shared MnO_6 octahedra and has the CdI_2 layered structure. Meldau et al. [3] found that the crystal structure of feitknechtite is hexagonal, with $a_o = 3.32$, $c_o = 4.71$ and a calculated d-spacing of 3.249 Å and $Z = 1$ in the space group D33d-P3 m1. Feitknechtite is isotypic with pyrochroite.

Feitknechtite, in fact, has been prepared from the precursor material pyrochroite. The oxidation of pyrochroite, prepared both with and without Mg^{2+} dopants, can be done in air or water. According to Luo et al. [4], feitknechtite can be prepared using a Mg^{2+} dopant by the following method (SF-1): Two solutions are prepared separately and then mixed. Solution 1 is prepared by dissolving 20 g of NaOH in 80 mL of distilled deionized water (DDW). A second solution of 7.5 g of $MnCl_2 \cdot 4H_2O$ and 1.5 g of $MgCl_2 \cdot 6H_2O$ in 50 mL of DDW is slowly mixed with solution 1 and stirred for 20 min. The white slurry formed is known as pyrochroite, which is then filtered and washed in order to decrease the pH to ~10. The filtrate is allowed to oxidize in air at room temperature and a dark brown fine powder is produced. The overall formula of feitknechtite via the SF-1 preparation is $Na_{0.3}Mg_{0.3}O_{0.45} MnO_{1.08}OH$, and the average manganese oxidation state is 3.15.

Feitknechtite using Mg^{2+} dopants has also been synthesized without oxidizing pyrochroite in air by Luo et al. [4]. This synthetic pathway is referred to as SF-2. A solution of 1.5 g of $KMnO_4$ in 80 mL of DDW is added to the basic slurry of pyrochroite, and a black precipitate of feitknechtite with some amorphous MnO_x is formed (SF-2). The overall formula of feitknechtite via the SF-2 preparation is $K_{0.03}Na_{0.14}Mg_{0.21}O_{0.31}MnO_{1.04}OH$, and the average manganese oxidation state is 3.08.

The synthesis of feitknechtite without Mg^{2+} dopants utilizes a different procedure [4], referred to here as SF-3. Magnesium-free pyrochroite is prepared as follows: A solution of 7.5 g of $MnCl_2 \cdot 4H_2O$ in 50 mL of DDW is added to a solution of 20 g of NaOH in 80 mL of DDW. A 10% hydrogen peroxide solution (200 mL) is added dropwise to the pyrochroite slurry, producing a black precipitate. The overall formula of the feitknechtite via this SF-3 preparation is $K_{0.16}MnOOH$. The synthesis of feitknechtite without Mg^{2+} dopants is also carried out by the reduction of large d-spacing layered manganese oxides (SF-4) [4]. The reduction of MnO_4^- with alcohols occurs in the presence of organic ammonium salts [4a]. The overall formula for feitknechtite via the SF-4 method is MnO(OH).

2. Pyrochroite

Pyrochroite $(Mn(OH)_2)$ comprises 61.76% Mn, 2.27% H, and 35.97% O, which is 79.75% MnO, and it has a molecular weight of 88.95 g. This mineral is the product formed kinetically after hausmannite formation. Aminoff [5] found that the crystal structure of pyrochroite is ditrigonal scalenohedral. The hexagonal elementary parallelopiped has the dimensions $c = 4.68 \times 10^{-8}$ and $a = 3.34 \times 10^{-8}$.

Pyrochroite has been prepared by making alkali solutions of Mn^{2+} [4a]. Two methods to prepare pyrochroite, with and without Mg^{2+} dopants, were mentioned earlier, where it was synthesized as the precursor to feitknechtite. Yang and Wang [6] have used yet another approach. All three of these approaches have in common the oxidation of pyrochroite. Giovanoli et al. [7] proposed a method for the oxidation of pyrochroite in oxygen (designated as the OPO method) to form birnessite. The reaction pathway is shown in Figure 1. Many researchers in the 1980s adopted this method for the preparation of birnessite; however, a side reaction of this method yields hausmannite.

To avoid the side reaction of hausmannite formation in the preparation of birnessite using pyrochroite, an alternative method developed by Luo et al. [8] oxidizes pyrochroite with permanganate instead of oxygen (designated as the OPP method). They prepared a mixture of solutions as follows: A solution of 13.5 g of $MnSO_4 \cdot H_2O$ (or 19.8 g of $Mn(Ac)_2 \cdot 4H_2O$) and 3.95 g of $MgSO_4 \cdot 7H_2O$ (or 3.2 g of $Mg(Ac)_2 \cdot 4H_2O$) in 160 mL DDW is added slowly to a solution of 40 g NaOH in 180 mL DDW with vigorous agitation, to form pink gels of $Mn(OH)_2$. A solution of 5.1 g of $KMnO_4$ in 160 mL DDW is then added slowly into the gel with vigorous agitation to produce a black suspension of MnO_x, the volume of which is about 530 mL. The suspension is aged at room temperature and atmospheric pressure. Feitknechtite (β-MnO(OH)) forms along with haus-

OPO method

Mn(OH)$_2$ → β-MnOOH → birnessite

⇓   Topotactic transformation
⇓
⇓   Recrystallization (side reaction)
⇓
Hausmannite

FIGURE 1   Reaction pathway of the OPO method for the oxidation of pyrochroite in oxygen.

mannite. The initial main product observed in XRD is feitknechtite. The induction period is approximately 32 h, after which birnessite begins to form. A total of 75 days is needed for feitknechtite to completely disappear. The induction period is shortened by two-thirds upon increasing the temperature from 27° to 40°C. At 65°C the induction period is shortened to less than 2 h, and above 85°C there is no visible induction period, with the crystallization and phase transformation of birnessite complete after 1 day. The induction period and crystallization rate were also monitored at room temperature at MnO_4^-/Mn^{2+} = 0.4. When the concentration of NaOH decreases from 1.6 mol/L (typical synthesis) to 0.74 mol/L, the induction period increases from 30 h to 60 h. The crystallization of birnessite became very slow, and the formation and phase transformation of feitknechtite were slowed considerably. When the NaOH concentration increases to 3.6 mol/L, the induction period is lowered to 8 h and the crystallization rate increases to 3.6 h^{-1} [8].

Recently, a third method has been studied by Yang et al. [6]. They modified the conventional OPO method and developed an oxidation-deprotonation reaction (designated as the ODPR method). This work is important because the prevention of hausmannite formation was achieved. Pyrochroite is prepared in the absence of Mn^{3+} using the ODPR method. Both NaOH and Mn^{2+} are purged with N$_2$ gas before they are mixed in order to exclude oxygen from each solution. According to Yang and Wang [6], the formation of hausmannite is favored by two factors: the presence of Mn^{3+} ions before NaOH mixing and temperature. A temperature higher than 283 K favors hausmannite formation, while a temperature below 283 K inhibits the formation of Mn^{3+} ions during the mixing process. Randomly stacked birnessite is more favored during the fast oxidation of pyrochroite. Thus, pyrochroite oxidation is carried out ·under an O$_2$ flow rate of 5 L min^{-1} for the first 30 min. Thereafter, the flow rate is cut to 1 L min^{-1} and continued for about 5 h.

B. Birnessite and Buserite

Birnessite ($Na_4Mn_{14}O_{27}\cdot7H_2O$, JCPDS 32-1128) and buserite ($Na_4Mn_{14}O_{27}\cdot7H_2O$, JCPDS 23-1046) are the two most important layered manganese oxide materials, existing as the primary manganese oxide phase in deep-sea nodules [9–13]. These two layered materials both consist of manganese oxide sheets with exchangeable hydrated cations between these sheets. Each of the manganese oxide layers is made up of edge-shared MnO_6 octahedral units. Birnessite has only a monolayer of hydrates with an interlayer distance of 7 Å, while buserite has double layers of hydrates with an interlayer distance of 10 Å (Fig. 2) [14–16]. The manganese in these layered manganese oxide materials is mixed valent, with Mn^{4+}, Mn^{3+}, and Mn^{2+} [16,17]. Buserite is the initial form of birnessite before dehydration. Natural birnessite is poorly crystalline and impure in composition. Detailed studies on the crystal structures and applications of natural birnessite have therefore been limited. To obtain birnessite and buserite with good crystallinity and pure composition, syntheses of birnessite have been intensively examined, including precipitation, sol-gel, and hydrothermal routes, as well as ion exchange, isomorphous substitution for framework doping, and high-temperature solid-state reactions.

1. Precipitation Routes

In the precipitation routes, birnessite is usually prepared via the oxidation of aqueous Mn^{2+} (oxidation method). In the work of Giovanoli et al. [7] and Golden

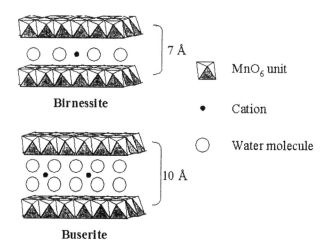

FIGURE 2 Structural models of birnessite and buserite.

et al. [14,17], oxygen was used to oxidize $Mn(OH)_2$ in NaOH or KOH solution (OPO method). Oxidation syntheses using oxygen are complete in a few hours. When H_2O_2 or $S_2O_8^{2-}$ is used as the oxidant, the synthesis is complete in a half hour [18,19]. However, reproducibility of these oxidation syntheses is not satisfactory due to a lack of understanding of the factors that influence the oxidation of Mn^{2+}. Furthermore, these oxidizing agents are too expensive for large-scale preparation of birnessite. In the recent work of Cai et al. [20], air was used instead of oxygen to synthesize birnessite materials. Preparative parameters such as the rate and extent of oxidation, the concentration of OH^-, and the aging process were investigated. Birnessite prepared by this air oxidation method was more thermally stable than those prepared using oxidants such as H_2O_2 and $S_2O_8^{2-}$.

Kinetics and mechanistic studies of the oxidation method using H_2O_2 have been studied by Luo et al. [21] A model was proposed by Luo et al. [4] of the topotactic conversion occuring during oxidation of pyrochroite to feitknechtite to birnessite (Fig. 3). When H_2O_2 is added, $Mn(OH)_2$ is first oxidized to β-MnOOH, with further oxidation to MnO_6 units. The oxidation state of Mn in pyrochroite is $2+$, and the average Mn(II)—O length is 2.2 Å. The interplanar distance between $Mn(OH)_6$ sheets is 4.8 Å. When H_2O_2 is added, $Mn(OH)_2$ is oxidized to β-MnOOH and the basal d-spacing shrinks to 4.7, Å, with a Mn(I-II)—O length of about 2.0 Å. The oxidation process of pyrochroite to feitknechtite is accompanied by the deformation of the original pyrochroite structure due to shrinkage of Mn—O bonds and partial removal of hydroxyl groups. When MnOOH is further oxidized, the $MnO_3(OH)_3$ units change to MnO_6 units gradually. The MnO_6 sheets become negatively charged due to the existence of Mn^{3+}, which makes possible the introduction of hydrated cations such as Na^+ and/or K^+ to compensate the interlayer charges. Birnessite is thus formed with an interplanar spacing of 7.2 Å. Another precipitation pathway was developed by Shen et al. and Suib et al. [22,23] in which birnessite is formed from the oxidation of $Mn(OH)_2$ in NaOH solution by $Mg(MnO_4)_2$ or $KMnO_4/MgCl_2$ (redox method). The redox reaction between Mn^{2+} and MnO_4^- in aqueous media is almost stoichiometric. This feature makes control of the process and final products relatively simple.

A detailed investigation has been performed by Luo and Suib on redox reactions, various synthetic parameters and their effects on the crystallinity, and properties of resultant birnessite and buserite materials [24]. The use of redox methods to synthesize birnessite consists of three stages: an induction period, a fast crystallization period, and a slow crystallization period. The lengths of each of these periods strongly depend on temperature. The increments of temperature can shorten the time of crystallization. However, high temperature can decrease the ion exchange capability of the resulting birnessite. Increments of basicity have effects that are similar to but less important than those of increased temperature.

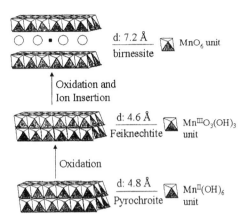

d: 7.2 Å
birnessite MnO_6 unit

↑ Oxidation and
Ion Insertion

d: 4.6 Å
Feiknechtite $Mn^{III}O_3(OH)_3$ unit

↑ Oxidation

d: 4.8 Å
Pyrochroite $Mn^{II}(OH)_6$ unit

FIGURE 3 Topotactic conversion occuring during oxidation synthesis of pyrochroite to feiknechtite to birnessite. (Adapted from Ref. 4.)

Birnessite prepared at higher temperature and/or higher basicities has better layer ordering and/or larger crystal size. Mg^{2+} addition results in oxidation of low-valent manganese oxide. Due to the larger ionic strength of Mg^{2+} compared to Na^+, Mg^{2+}-added birnessite can retain interlayer hydrates at a relatively high temperature. The different diameters of Mg^{2+} and Na^+ cause more disordered products. In acid solutions, Mg^{2+}-free birnessite forms a stable H-birnessite, while the Mg^{2+}-added material dissolves, which is due to the weak Mg—O bonds that are easily broken by H^+. The ratio of MnO_4^-/Mn^{2+} has an important effect on the induction period, the crystallization rate, the phase transformation process, the purity of the final product, and the average oxidation state of manganese in birnessite that ranges from 3.40 to 3.99. The crystallization rate also can be affected by anions used in the synthesis, from an investigation on acetate, chloride, nitrate, and sulfate anions. Acetate accelerates crystallization the most, which is probably because it is the most basic of these anions. This redox synthesis method is very slow as compared to oxidation synthesis, and usually requires days and even months for the reaction and aging process to be complete.

Birnessite has also been synthesized from amorphous manganese oxides prepared from the reduction of MnO_4^- by alcohols in basic media (reduction method) [21,25,26]. The reaction process starts with the formation of an amorphous manganese oxide, followed by the crystallization of birnessite from amorphous manganese oxide. Increasing the aging temperature can accelerate the crystallization rate. This reduction reaction process is similar to the redox reaction process.

2. Sol-Gel Routes

The first report of manganese oxide gels was in 1915 by Witzemann [27]. However, further studies on the sol-gel syntheses of manganese oxides have been carried out only since 1990 [28–30]. Sol-gel synthesis provides homogeneous reaction systems that allow for mixing of reactants on the molecular level as well as control over the identity, shape, morphology, and particle size of the resulting phase. The sol-gel synthesis of birnessite-type manganese oxides was first reported by Bach et al. [28]. In their work, fumaric acid was used as a reducing agent to reduce $NaMnO_4$ or $KMnO_4$. A monolithic gel was formed and birnessite-type materials formed upon drying and calcination.

Various sugars and other organic polyalcohols have been used as reducing agents to prepare birnessite materials by Ching and coworkers [29]. The sol-gel synthesis is initiated with an aqueous $NaMnO_4$ or $KMnO_4$ and glucose solution under vigorous stirring for less than 1 min. The resulting mixture is then allowed to stand without further agitation. The exothermic reaction produces a reddish sol that becomes a brown gel in a few seconds. Gelation is due to the cross-linking of the manganese oxide sites by partially oxidized organic fragments. This gel is then dried overnight at 110°C and calcined at 400°C for 2 h. For potassium systems, the redox reactions between permanganates and sugar are strongly dependent on the ratio of sugar/permanganates and the concentration of solution, as illustrated in Figure 4 [29]. High ratios of glucose/$KMnO_4$ can result in K-birnessite, while lower ratios produce a flocculent gel, yielding amorphous manganese oxide or a cryptomelane phase after calcination. The overall concentration of reactants in solution also determines the crystalline structure of final products. Low amounts of glucose can cause the formation of more flocculant gels because there is not enough organic cross-linking agent to form a monolith, which results in cryptomelane and Mn_2O_3. For sodium systems, the formation of gels is less dependent on the glucose/permanganate ratio. No cryptomelane-type product is obtained because the K^+ cation is believed to be the ideal dimension to fit in the 2×2 tunnels, and Na^+ cations are too small to support the 2×2 tunnels. Two different Na-birnessite materials are produced, one with a typical 7-Å interlayer spacing and the other with a 5.5-Å interlayer spacing. The 5.5-Å Na-birnessite is dehydrated from the 7-Å phase and converts to 7-Å Na-birnessite after prolonged exposure to water.

K- and Na-birnessite prepared using these sol-gel methods have unique properties, such as particulate morphologies, not found in materials synthesized by the conventional precipitation routes. Buserite with a 10-Å interlayer spacing cannot be formed using these sol-gel methods. Unlike their conventional counterparts, these birnessite materials have poor ion exchange capability and cannot undergo 100% exchange. These unusual ion exchange properties may be due to the smaller amounts of hydrates existing in the interlayers, which can decrease

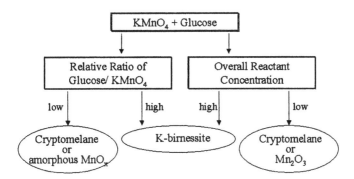

FIGURE 4 Scheme of the sol-gel reactions between $KMnO_4$ and glucose under different reaction conditions. (Adapted from Ref. 29.)

the mobility and accessibility of the cations between the layers. This drawback, however, has been circumvented using another sol-gel pathway, in which tetraalkylammonium cations are introduced as structure directors to form larger layer spacings.

In some recent work of Brock et al. [31], nanosize layered manganese oxide materials were prepared as colloidal solutions via reduction of tetraalkylammonium (TAA = methyl, ethyl, propyl, and butyl) permanganate salts in aqueous solutions with 2-butanol and ethanol. TAA-permanganates are made from tetraalkylammonium bromide salts and potassium permanganate. TAA-manganese oxide aqueous sols are produced by adding the TAA-permanganate salt to a stirred mixture of water and 2-butanol or ethanol with the result of dark red-brown sols. These sols are aged at 25–85°C for several hours to days and form gels. Gelation time is dependent on the cation, the amount of alcohol, the aging temperature, and the concentration of manganese. These gels can undergo self-assembly, upon drying, to produce nanosize layered manganese oxide structures with tetraalkylammonium cations in the interlayers (TAA-OL-1). A schematic illustration of the self-assembly process is shown in Figure 5 [31a]. The colloidal sols are very stable for up to months at room temperature. The key to the stabilization of the colloids is the incorporation of TAA cations and the exclusion of hard metal cations such as alkali metals and alkaline earths. The use of a secondary alcohol neutral aqueous matrix, instead of the salt aqueous solutions with high basicity used in conventional syntheses, provides a more homogeneous environment that prevents small TAA-manganese oxide particles from coagulating or condensing. The average oxidation states of manganese ranges from 3.49 to 3.73 for the layered manganese oxide materials prepared by this method.

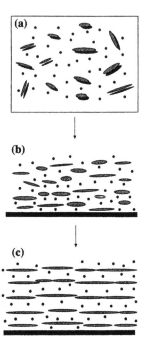

FIGURE 5 Schematic illustration of the self-assembly process that takes place on glass slides to produce a layered manganese oxide structure with tetraalkylammonium cations intercalated between the layers: (a) unassociated particles free-floating in solution; (b) initial structure consisting of small regions of crystallinity formed from evaporation of colloids onto glass slides; (c) well-ordered structure produced from heating the colloid in solution or annealing the thin-film sample described in (b). The disks represent the negatively charged disklike manganese oxide particles, and the filled black circles are the tetraalkylammonium cations. (Adapted from Ref. 31a).

TAA-manganese oxide layered materials show very fast and quantitative ion exchange ability [32]. The ion exchange of lamellar films of these manganese oxides with Cd^{2+} is complete in a few seconds, which is not observed with any other layered manganese oxide material. Fast ion exchange reactions have been performed with alkali metal cations, alkaline earth cations, and transition metal cations [31b]. The keys to the outstanding ion exchange ability of these TAA-manganese oxide materials are: (a) the variety of pore types, (b) their small particle size, and (c) the low charge density coupled with large hydration sphere of TAA$^+$ cations, which causes them to be easily displaced by cations of higher charge density.

3. Interlayer/Framework Substitution

In the conventional precipitation and sol-gel routes, the resultant birnessite or buserite materials are usually Na type or K type, in which the cations are located in the interlayers, and MnO_6 octahedral units form the CdI_2-type structure in the layers. The structure and properties of birnessite and buserite can be modified to load other cations, including alkali metals, alkaline earths, and transition metals, via two pathways. One pathway is interlayer substitution via ion exchange method at room temperature [14,22]. In this process, Na- or K-birnessite are stirred in a solution of dopants. Divalent metal cations, such as Mg^{2+}, Ni^{2+}, Co^{2+}, have been successfully ion exchanged with Na-birnessite materials and used to prepare todorokite-type materials with tunnel structures [22].

Ion exchange methods also have been used to prepare birnessite-type materials with organic cations intercalated into the interlayer regions [33]. K-birnessite materials were used to produce H-OL-1 by ion exchanging with H$^+$ three times in a dilute HNO_3 acid solution. TAA hydroxides and diamines are then introduced. The structural models for TAA-and diammonium ion–intercalated layered materials are illustrated in Figures 6 and 7. The particles sizes of TAA-OL-1 products are on the nanometer scale, while DA-OL-1 materials are of relatively large particle size. This intercalation process is different from the redox reaction of TAA and secondary alcohol described earlier. The relatively large particles of H-birnessite precursor were changed into nanometer-sized particles with a large interlayer spacing by ion exchange methods. The schematic illustration is shown in Figure 8. In the early stages of gel formation, the H-OL-1 retains the large particle size of about 20–50 microns. Further intercalation of TAA ions leads to larger distances between the manganese oxide layers. The number of stacks of layers decreases simultaneously because the interactions between the inorganic layers and the TAA ions are weaker than those between inorganic layers and hard cations (such as K$^+$ and Na$^+$) or DA ions. The layers of manganese oxides are relatively easy to split, with the result of smaller particle sizes. The preparation methods used here are similar to the pillaring processes discussed in detail in the following subsection. The other pathway for modifying birnessite materials is framework substitution by adding metal dopants into the initial solution [20,34]. Compared with interlayer substitution, framework substitution is less understood due to difficulties in finding suitable isomorphous dopants in the MnO_6 octahedral units. Mg^{2+} is probably the perfect cation to fit in Mn positions in these octahedral units, and it forms very stable buserite materials [22,24]. Other dopants that are

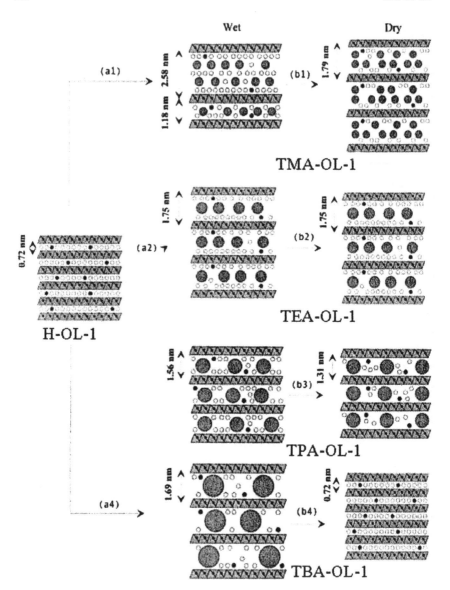

FIGURE 6 Structural models of TAA-birnessite, TAA-OL-1. TMA ion intercalated into (a1) H-OL-1 and (b1) dried TMA-OL-1; TEA ion intercalated into (a2) H-OL-1 and (b2) dried TEA-OL-1; TPA ion intercalated into (a3) H-OL-1 and (b3) dried TPA-OL-1; TBA ion intercalated into (a4) H-OL-1 and (b4) dried TBA-OL-1. The small white circles represent water molecules, small black circles are potassium ions, and large black circles are TAA cations. The layered black rectangular circles are the manganese oxide layers. (Adapted from Ref. 32.)

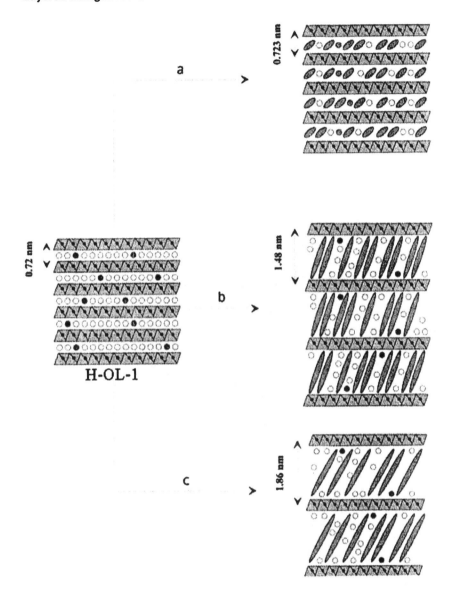

FIGURE 7 Structural models of diammonium ion–intercalated birnessite DA-OL-1. (a) Ethylenediamide intercalated into H-OL-1; 1,6-DHA intercalated into (b) H-OL-1, (c) 1,10-DOA-OL-1. XRD patterns of wet and dried states of DA-OL-1 were the same. The small white circles represent water molecules, small black circles are potassium ions, and large black circles are TAA cations. The layered black rectangular circles are manganese oxide layers. (Adapted from Ref. 32.)

488   Liu et al.

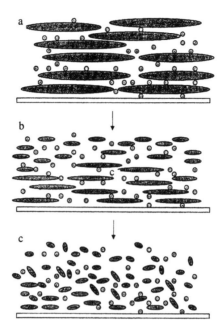

FIGURE 8 Schematic illustration of the ion exchange process that takes place in the TAA-OL-1 gels at different times. (a) At the beginning of formation of the gel, the H-OL-1 retains its normal particle sizes of about 20–50 μm. (b) Upon further intercalation of TAA ions, the stacks of layers are smaller and the particles become smaller. (c) Colloidal suspensions of nanometer-sized particles of manganese oxides form after 3 days of stirring. The disks represent negatively charged disklike manganese oxide particles, and the filled black circles are the tetraalkylammonium cations. (Adapted from Ref. 32.)

stable in the framework of birnessite are Fe^{3+}, Co^{2+}, and Ni^{2+}, which were investigated in 2002 by Cai et al. [20].

C. Pillared Layered Manganese-Based Materials and Manganese-Based Mesoporous Materials

1. Pillared Layered Manganese-Based Materials

Pillared layered materials have attracted wide interest since 1983 due to their molecular sieving and catalytic properties [35–37]. In an effort to broaden the diversity of pillared lamellar materials and to enhance their physical, chemical, and catalytic properties, various bulky species have been used as pillaring species. Robust metal oxides, such as Al, Ga, Zr, and Si oxides, and their mixtures have been used as guest molecules to increase thermal stability. Transition metal oxides have been introduced to probe potential catalytic applications with respect to redox organic reactions due to their variable valence states [38]. The interest in synthesizing pillaring materials has traditionally focused on clays as the host layers [37]. A variety of nonclay layered materials, such as phosphates [35,39], titanates [36,40], and manganese oxides [38,41–45], have also been intensely investigated for intercalating large species (see also other chapters in this book).

   Layered manganese oxide materials are well known for their excellent cation exchange and molecular adsorptive properties. However, only a few reported studies have been carried out on the pillaring reactions of layered manganese oxides or manganese-related oxides [38,40–44]. Pillared layered manganese oxide systems consist of edge-sharing MnO_6 octahedral sheets, such as birnessite, which are held apart by cluster molecules (pillars) that are located in the interlayer space [38,41–45]. These pillars not only prevent the collapse of the layered structure but also allow the formation of large spaces between the intercalated species, leading to stable microporous structures with high surface areas. The unpillared layered materials, unlike clays, are not readily swelled in water or intercalated because of the high charge density on the framework. The high charge density causes a strong attractive force between the layers, making the intercalation of bulky guest ions difficult. Therefore, layered manganese oxides are not accessible to pillaring by conventional methods.

   To pillar layered manganese oxide materials (OL-1), it is necessary initially to modify the charge density of the system by incorporating large organic molecules, which causes expansion of the interlayer space [35]. Upon this treatment, the layered compounds are ready to intercalate large pillaring species and the layered structure is preserved. The synthesis of pillared layered manganese oxides (POL) materials is based on a two-step intercalation method, as shown in Figure 9. First, the starting layered manganese oxide material (birnessite or buserite), prepared by the precipitation methods mentioned earlier in this chapter, is treated with acid at room temperature for a period of 1–3 days [38,41,42,44]. Protonated samples called H-OL-1 are produced after this treatment. The H-OL-1 is then reacted with alkylamine at room temperature. Hexylamine [38,41] and octylamine [42] have been used as preintercalating agents. The resultant material, alkylammonium ion–pillared OL-1 (AAOL), is the direct precursor for the pillaring process that is the second step of the synthesis.

   The pillaring process occurs via cation exchange, where the expanded solid is reacted with an aqueous solution of the pillaring agent. Typically, AAOL is stirred in excess of the pillaring molecules for 1–4 days at 50–65°C, washed, and dried in air. The final microporous materials are produced by calcination for several hours at 200–400°C. Calcination removes the water molecules and preswelling alkylammonium ions and yields the POL product. The pillaring conditions and calcination temperatures depend on the intercalating species. Various

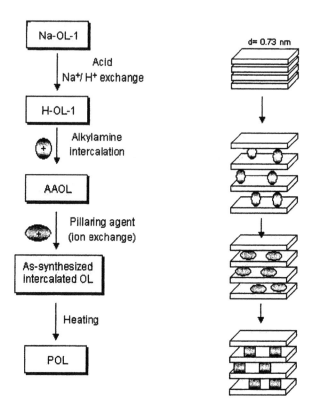

FIGURE 9 Schematic representation of the pillaring process of OL-1.

cation oligomers have been used to pillar OL-1, including polyoxo cations of aluminum (known widely as Keggin ions $[Al_{13}O_4(OH)_{24}(H_2O)_{12}]^{7+}$) [41], tetraethylorthosilicate (TEOS) [42], transition metal oxides (i.e., chromium oxide) [42], and organic monomers, which lead to polymerization between the layers [43]. Investigations have been done on the intercalating process, the structures of the pillared materials produced, and their thermal properties.

Pillaring processes have been investigated using X-ray diffraction (XRD) and X-ray absorption spectroscopy (XAS) [38,41,42]. In such studies the basal spacings of the materials pillared with silica, Keggin ions, or chromium oxide are compared with the interlayer distance (7.3 Å) of the initial birnessite. Keggin ion–pillared manganese oxide has an interlayer space of 14 Å, almost twice that of the starting birnessite, which corresponds closely to the sum of the basal thickness of birnessite and the diameter of the Keggin ions (sum = 13.66 Å).

This confirms the presence of Keggin ions between the manganese oxide layers [41].

Besides the two-step intercalation method commonly used to prepare POLs, additional techniques have been reported to improve the properties of the resultant materials [38,42]. An intercalation/solvothermal reaction has been used to prepare silica-pillared layered manganese oxide with high surface area and high thermal stability [42]. In this method the silica-pillared manganese oxide materials, which are obtained from the cationic exchange reaction, are soaked in tetraethylorthosilicate (TEOS) liquid in Teflon-lined stainless steel vessels and autoclaved at 140°C under autogeneous pressure for 48 h. The solvothermal treatment results in an increasing amount of silica in the interlayer. The Si/Mn molar ratio increases from 0.63 to 0.84. Silica-POL obtained via this method shows higher porosity and thermal stability than that synthesized by the conventional two-step method. Details of the formation reaction of this material can be found in the original reference [42]. The silica-pillared layered manganese oxide materials show XRD patterns that are consistent with incorporation of TEOS molecules into the layer space. The basal spacing of the layer material increases to 24.3 Å [42,44]. The patterns also show weak reflections in high-order diffraction peaks, suggesting that the stacked structure of the pillared manganese oxide materials is more disordered.

An intercalation process under reflux conditions for 4 days has been performed by Ma et al. to prepare chromium oxide–pillared manganese oxides [38]. In this reaction chromium hydroxyl acetate clusters, $Cr_3(OAc)_7(OH)_2$, are used as the pillaring species. Cr^{3+} self-polymerizes under reflux to form polynuclear clusters between the manganese oxide layers. Intercalation of this trinuclear chromium complex constitutes a challenge because redox reactions may occur between manganese oxide and the transition metal oxide during the process, which could destroy the layer structure.

The preparation of the CrO_x POL is based on the reaction of hexylammonium ion–pillared OL-1 (HAOL) with $Cr_3(OAc)_7(OH)_2$ solution under reflux conditions. This Cr trimer has a size of ~5 Å × 6.5 Å, which is smaller than the interlayer space of HAOL at 12.9 Å [7] and thus is expected to enter the gallery region. The cation exchange occurs rapidly (~20 min), which is indicated by the disappearance of the dark green color of the Cr trimer solution. Porous CrO_x-pillared manganese oxide is obtained after heating at 200°C in N_2. There is no significant redox reaction observed between MnO_x and Cr^{3+} pillaring species during the intercalation process. The fast intercalation of the Cr^{3+} species into the starting birnessite and the robustness of the Cr^{3+} clusters in the interlayer space may act to prevent redox reactions. [38]

XRD and XAS studies of chromium oxide–pillared manganese oxide reveal the presence of Cr species in the interlayer, as shown in the Figure 10 [38]. These materials exhibit poor crystallinity due to irregularity and disorder of the layers.

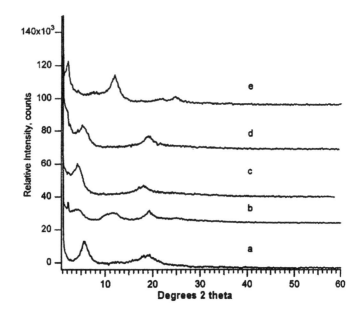

FIGURE 10 XRD patterns of CrO_x-pillared OL-1 materials prepared with different Cr trimer solution concentrations and different initial Cr/Mn molar ratios: (a) [Cr trimer] = 0.008 M, Cr/Mn = 0.17; (b) [Cr trimer] = 0.008 M, Cr/Mn = 0.44; (c) [Cr trimer] = 0.017 M, Cr/Mn = 0.44; (d) [Cr trimer] = 0.008 M, Cr/Mn = 0.88; (e) [Cr trimer] = 0.004 M, Cr/Mn = 0.18. The interlayer distances vary with these parameters. (Adapted from Ref. 38).

The structure of the pillared layers is constructed of Cr—O—Mn bonds formed via corner-shared linkages of CrO_6 and MnO_6 octahedra. The intercalated Cr species have a structure similar to Cr_2O_3.

Recently a new synthetic route has been developed to prepare novel layered manganese oxide nanocomposites with a stacked structure of higher order than the silica-POL obtained by conventional intercalation reactions [45]. This new material has been synthesized via a delamination/reassembling process using propylamine-containing silica as the pillaring agent. The morphology of these pillared materials is similar to that of pure birnessite, consisting of platelike particles corresponding to the layered structures. The success of pillaring layered manganese oxide can also be confirmed by surface area studies and thermogravimetric analysis (TGA). Unlike birnessite and buserite, pillared materials exhibit high surface area and good porosity. Silica-pillared manganese oxide materials exhibit surface areas in the range of 150–260 m^2/g, depending on the calcination temperature [42]. The thermal stability of these materials is also enhanced after the formation of micropores via the pillaring process. Pillared manganese oxide materials are stable until 600°C. At higher temperatures, the pillars collapse and the porous structure is destroyed.

2. Manganese-Based Mesoporous Materials and Manganese-Based Porous Mixed Oxides

In the early 1990s, Mobil researchers discovered a group of manganese materials that possessed interesting properties. These mesoporous materials were denoted M41S [46]. Since the discovery of M41S materials, the research involved in synthesizing mesoporous structures has grown dramatically. M41S materials contain uniform channels in the mesopore size range (15–100 Å). The first three structures to be reported were the following phases: hexagonal (MCM-41), cubic (MCM-48), and lamellar (MCM-50). Despite the regularity and long-range order of M41S materials, their wall structure is essentially amorphous [47]. MCM-41 has been the most intensively studied phase in the M41S family. MCM-41 possesses a high surface area, a narrow pore size distribution, and a well-defined pore structure. MCM-41 is thermally, hydrothermally, chemically, and mechanically a stable structure.

M41S materials are potential candidates for numerous applications, including catalytic reactions involving large molecules, chromatography, and molecular host materials [47]. In the past few years, the preparation of transition metal oxide mesoporous materials has gained attention. Antonelli et al. [48] have developed a new family of electroactive alkali metal–doped mesoporous Nb, Ta, and Ti oxides. Various synthetic routes have been developed to substitute transition metal oxides into the M41S framework. For example, iron has been substituted into MCM-48 [49,50] and manganese into MCM-48 [51] and MCM-41 [52].

The focus of this chapter is Mn-based mesoporous materials rather than Mn-substituted mesoporous silica materials. Although the catalytic applications thus far are limited for Mn-based mesoporous structures, they show great potential for certain catalytic systems.

a. Manganese Oxide Mesoporous Solids (MOMS)

A synthetic route has been developed by Tian and coworkers for the synthesis of MOMS by combining known syntheses of layered and microporous manganese oxides with procedures for the synthesis of M41S mesoporous materials [53,54]. Manganese oxide mesoporous solids (MOMS) have been synthesized as both hexagonal and cubic phases, via the incorporation of surfactant micelles and partial oxidation of $Mn(OH)_2$ followed by oxidation of Mn^{2+} to Mn^{3+} and Mn^{4+} and removal of surfactant by calcination [53]. To synthesize MOMS, long-chain organic ammonium surfactants, such as cetyltrimethylammonium bromide (CTAB), are used as micellar templates in water, and $MnCl_2$ is used to prepare the $Mn(OH)_2$ precursor.

The precursors for the synthesis are prepared in two different steps. The first step consists of making both a $Mn(OH)_2$-layered phase and an aqueous surfactant solution. The former is prepared by mixing $MnCl_2$ and NaOH in air. The micellar solution is prepared by diluting the surfactant in water in proportions dependent on the mesophase product of interest. CTAB is used in concentrations of 25 wt% and 10 wt% for the hexagonal and cubic phases, respectively [53]. The second step consists of mixing the $Mn(OH)_2$ crystallites with the surfactant solution in air, leading to the formation of a manganese mesophase. During this step, the crystalline $Mn(OH)_2$ phase is mildly oxidized in air and a mixed-valent manganese oxide shell is formed. Once the manganese mesophase is formed, the surfactant templates are removed by calcination at 600°C. The calcination step also causes further oxidation of manganese species, which leads to the final manganese mesoporous structure.

The high degree of ordering in MOMS materials is in direct relation to the M41S mesoporous materials and OMS systems [53]. As mentioned previously, MOMS materials have been synthesized in hexagonal and cubic structures. A typical XRD pattern for the MOMS hexagonal phase is shown in Figure 11. The walls of the MOMS are composed of microcrystallites containing dense phases of Mn_2O_3 and Mn_2O_4, with MnO_6 octahedra as the primary building blocks [53,54]. The hexagonal array of MOMS has an open porous structure with thick walls of approximately 1.7 nm [54].

A high-resolution transmission electron microscopy (HR-TEM) micrograph for hexagonal MOMS is shown in Figure 12. A hexagonal lattice morphology is

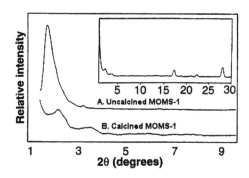

FIGURE 11 (A) XRD pattern of the uncalcined hexagonal MOMS-1. Note the strong (100) ($d = 4.7$ nm) and weak (110) ($d = 2.7$ nm) peaks. (B) XRD pattern of calcined hexagonal MOMS-1. The positions of the (100) and (110) peaks are shifted to lower d-spacings, and the peak intensities are decreased. (Inset) The XRD patterns of calcined samples show two additional broad peaks ($d = 0.50$ and 0.30 nm), which are Mn_2O_3 (gamma phase) and Mn_3O_4 (hausmannite) microcrystallite phases. (Adapted from Ref. 53.)

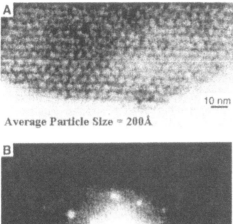

FIGURE 12 (A) Lattice morphology of the calcined hexagonal MOMS (CTAB concentration = 28%) shown by HRTEM. Crystallites withan average particle size of 200 Å are observed, with no evidenceof any other phase. (B) Convergent beam electron diffraction (CBED) pattern of the calcined hexagonal MOMS showing hexagonal symmetry. Ordering of MOMS in two dimensions shows nine orders of observed reflections. (Adapted from Ref. 53.)

observed. The surface area is 170 m^2/g and 47 m^2/g for the hexagonal and cubic phases, respectively, with an average pore size of 3 nm in both cases [53]. The MOMS hexagonal phase and synthetic todorokite have similar average oxidation state values (~3.5). However, MOMS has a higher conductivity at 8.13×10^{-6} Ω^{-1} cm^{-1} compared to $2 \times 10^{-7} \Omega^{-1}$ cm^{-1} for todorokite, which suggests that the structural features in the MOMS material play an important role in the electrical properties [53]. Both the hexagonal and the cubic MOMS possess thermal stability up to 1000°C, which is unusually high for manganese oxide systems [53].

Luo et al. first reported the transformation of a typical layered manganese oxide to a mesoporous structure [55]. The precursor for the transformation of layered birnessite to a mesoporous manganese oxide is synthesized K-birnessite, which is ion exchanged in HNO$_3$ to obtain H-birnessite. Details of this preparation and the ion exchange step can be found elsewhere in this chapter. The H-birnessite is then slurried in water, and a solution of 20–40 wt% of organic ammonium hydroxide salt is added, which introduces ammonium ions into the birnessite layers [55]. This process leads to the formation of mesoporous manganese oxide. The structural changes starting from the H-birnessite to the manganese oxide mesostructure are clearly recognizable by XRD. To date, no further work has been published with regard to characterization or catalytic applications for this kind of material. However, these materials are cetain to have interesting properties with regard to oxidative catalysis.

b. Manganese-Based Porous Mixed Oxides (MANPOs)

Before the appearance of MANPO materials, no other routes had been developed for the synthesis of large-surface-area Mn-based mesoporous materials. Their synthesis is based on the hydrolysis of a trinuclear pyridine manganese complex, $[Mn_3O(CH_3COO)_6(pyr)_3]ClO_4$, in the presence of metal ions such as La, Fe, Al, Ni, Co, Ce, and Sr along with combinations of these ions [56–58]. These mixed oxide materials are prepared by the dropwise addition of an aqueous solution containing the metal ion in its acetic or nitrate form into an acetonic solution containing the water-sensitive trinuclear manganese complex [55,56]. The manganese complex is readily soluble in warm acetone but hydrolyzes instantly in water, yielding a brown precipitate [57]. The precipitate is then filtered, washed, dried, and heated to 300–500°C in air. The reader is directed to the literature for detailed preparation procedures [56,57].

BET surface areas of MANPO materials vary from 180 to 900 m^2/g [56], which are significantly higher than those reported for MOMS (47–170 m^2/g) [53]. The surface areas are strongly dependent on the nature of the heterocation used in the synthesis [56,57]. The tortuosity factors for MANPO materials were measured using the corrugated pore structure model (CPSM-nitrogen). The values are in the range 3–5.7 Å, which are typical of conventional porous catalysts and indicate that they posses a complex pore structure, in contrast to the highly ordered pore structure of MCM-41 solids [59].

III. APPLICATIONS OF LAYERED MANGANESE OXIDE MATERIALS

A. Ion Removal from Water

Layered manganese oxide materials have porous structures ranging from ultramicropore to mesopore dimensions and cation exchange and molecular adsorptive properties. Therefore, they can be used as ion sieves, molecular sieves, and catalysts, similar to zeolites [61]. Spinel-type ion sieves have shown good adsorptive properties for lithium from seawater and other dilute solutions. Repeated adsorption–desorption using an ion sieve granulated system of birnessite-type materials with polyvinyl chloride binder has shown that this is a chemically stable system for the recovery of lithium from seawater [61]. Another method to recover lithium using layered manganese oxide materials is by electrochemistry using a redox-type spinel ion sieve electrode (Pt/λ-MnO$_2$). This system has been used to remove lithium from geothermal water [61].

Porous layered birnessite-type materials have been used as ion exchange materials for the cleaning of aqueous radioactive wastes from nuclear power plants and research centers [62]. Trace amounts of cesium (^{137}Cs), strontium (^{89}Sr), and cobalt (^{57}Co) are removed from aqueous solution with synthetic sodium birnessite. Synthetic sodium and potassium birnessite are very effective for the removal of (^{57}Co) from liquid solutions in a wide pH range, with the optimum range from 7 to 10. The mechanism proposed is the exchange of Co(II) for Mn(II) ions, with further oxidation of Co(II) to Co(III). For cesium and strontium, sodium or potassium ions at sites located on the crystal water sheet are involved in the ion exchange mechanism. Birnessite-type materials are not selective for the removal of trace amounts of cesium from liquid solutions, according to selectivity coefficients determined by the batch method using radioactive tracer ions [62]. The performance for removal of strontium is slightly better than that of cesium [62].

Birnessite has been used to study the sorption behavior of uranium, which is also a contaminant from the nuclear industry. The sorption behavior of uranium has been investigated as a function of uranium solution pH using the batch distribution factor (K_d) and the percentage sorption selectivity factor [63]. Several metal-exchanged birnessites were used to ion exchange aqueous uranium solutions. Layered manganese oxide materials show different sorption properties for the uptake of uranium as a function of the metal ions exchanged into the birnessite. Cs-birnessite treated in an autoclave has been found to have a high distribution factor (100% sorption) after 2 h of equilibrium at pH 6. The maximum sorption of uranium is 100% for K-birnessite at pH 2. The sorption mechanism of uranium is a function of its aqueous chemistry as well as of the structural features of the microporous materials at different pH values. Uranium uptake may take place via ion exchange when pH $\sim$3, with uranium present as the uranyl cation. At higher pH, sorption probably occurs on the surface of birnessite, followed by precipitation [63].

The retention of heavy metal ions over birnessite (MnO$_2$) surfaces and on TiO$_2$ (the rutile phase) has been studied in an effort to understand the reaction mechanism [64]. Various mechanisms have been proposed for the sorption of cationic heavy metals on oxide surfaces such as birnessite. The formation of surface precipitates prior to bulk precipitates has been observed using transmission electron microscopy [64]. At pH 5 and Al(III) concentration of 400 uM, a surface precipitate is observed on birnessite but not on TiO$_2$. It is important to consider surface precipitation reactions before modeling the sorption mechanisms of hydrolyzable metal ions [64].

The transformation of iron has been studied in the $FeCl_2$-NH_4OH system at different Mn/Fe ratios with birnessite as a catalyst at pH = 4–6 [65]. Birnessite materials promote precipitation of iron oxide. Oxidation of iron oxide by birnessite has been confirmed by the presence of Mn(II) in the solution [65]. Birnessite materials increase the crystallization processes of hydrolytic products of iron. The amount of precipitate formed in $FeCl_2$-NH_4OH is influenced by Mn/Fe molar ratios and pH values. The main products formed during this catalytic reaction were Fe-oxyhydroxides. These iron products play an important role in soil formation as well as dynamics and the fate of nutrients and environmental pollutants [65].

B. Oxidative Catalysis

Pillared layered manganese oxides have potential applications as catalysts because of their high surface area, good porosities, and high thermal stabilities. Although the major motive for the preparation of these materials has been to create new porous, highly active catalysts, very few catalytic studies have been done. [41,38,66]. Pillared layered manganese oxides containing CrO_x and Al-Keggin ions as intercalated species have shown catalytic activity in oxidative dehydrogenation reactions of cyclohexane and ethane, respectively.

Oxidative dehydrogenation of cyclohexane over CrO_x-pillared manganese oxides has been studied in a fixed-bed U-type quartz reactor by Ma et al. [38]. The main products of this reaction are benzene and cyclohexane. Typically, only cyclohexane is detected when OMS materials are used as catalysts, while benzene is the main product when the oxidation reaction is carried out over chromia-pillared clays [38]. The activation of chromia pillars occurs at 450°C and involves the formation of reversible structural O_2. This reversible oxygen produces reversible oxygen vacancies that selectively oxidize the reactants rather than completely oxidizing them to CO_2. At this high temperature (450°C) in the presence of O_2, a large production of benzene is observed.

Catalytic oxy-dehydrogenation of ethane over Keggin-ion pillared layered manganese oxide has been reported by Wong et al. [41,66]. The reaction is carried out on a fixed-bed flow-through system with a diluted catalyst. Keggin-ion pillared catalysts show better performance than Na-buserite, which may be due to the pillaring. However, the selectivity toward ethane is comparable in both systems.

In addition to the pillared manganese oxide materials, birnessite itself also has been widely used as a catalyst. The catalytic activity of birnessite for liquid-phase decomposition of acetone, methanol, and 2-propanol has been reported [67]. Sodium-, potassium-, bismuth-, lead-, and zinc-exchanged birnessite materials have shown superior catalytic activities compared to pyrolusite in the decomposition of these organic substrates. For the oxidation of methyl alcohol using metal-exchanged birnessite materials, the conversion ranged from 30% to 53%; the products are carbon dioxide, water, and HCOOH. For the oxidation of acetone, the conversion ranged from 41% to 54%; the products are carbon dioxide and water. For the oxidation of 2-propanol, the conversions range from 38% to 57%, with carbon dioxide, water, and acetone as the products. When pyrolusite is used as the catalyst for the foregoing three oxidation reactions, the conversions decrease to 8% for methyl alcohol, 5% for acetone, and 12% for 2-propanol. The conversions of oxidation reactions are effected by the metal ions exchanged into birnessite as: Bi(III) > Zn(II) > Pb(II) > K(I) > Na(I). A change in the redox behavior due to the catalytic activity is observed in the cyclic voltammograms. An estimation of manganese valency shows a decrease of Mn(IV), with an increase in Mn(III) and Mn(II) content in the spent birnessite compared with that of the fresh sample before the oxidation reaction. This suggests that it is the Mn(IV) oxidation state that contributes to the catalytic activity for the oxidation of organic substrates [67].

Partial oxidation of 1-butene over Cu-OL-1 or synthetic birnessite has been carried out by Krishnan and Suib [68]. A yield of 11.8% and a selectivity of 26% are reported with the product 1,3-butadiene. A 2% oxygen feed is optimum, giving a conversion of about 50%. According to temperature-programmed reduction (TPR) data, the Cu-OL-1 system provides oxygen at lower temperatures. This could explain the higher conversions achieved as well as the change in phase from Cu-OL-1 to MnO, as compared to less active systems, where the transformation is to Mn_3O_4 [68].

The oxidation of CO to CO_2 has been carried out with Cu-OL-1, with an increase in conversion from 42% to 90%, depending on the MnO_4^-/Mn^{2+} ratio [69]. The surface oxygen species are thought to play a critical role, as is a synergistic effect that is attributed to manganese and copper [69]. The activity of Cu-OL-1 is low at relatively low temperature but increases with increasing temperature. Cu-OMS-1 prepared from Cu-OL-1 displays a higher conversion. TPR data show more surface oxygen species for Cu-OMS-1 than for Cu-OL-1, suggesting that surface active oxygen is the key for catalytic activity of the manganese oxide materials [69].

Peroxides (H_2O_2) have been used as oxygen-generating agents to increase soil aeration to help plant growth and to decompose organic substances in wastewater. Magnesium peroxide (MgO_2) also decomposes to release oxygen. Catalytic decomposition of aqueous hydrogen peroxide and solid magnesium peroxide using birnessite has been reported by Elprince and Mohamed [70]. A mechanism involving H_2O_2 (aq.) is proposed based on measurements of the activation energy, the rate constant (k), and the pH value. Reaction between an ion and a neutral molecule is proposed, which is indicated by a linear relationship between the rate constant (k) and the ionic strength. Further verification of this mechanism was obtained using birnessite with different factional coverages of Co, which indicated that the heterogeneous reaction is catalyzed by Mn^{z+1}/Mn^z active centers. In the case of MgO_2, the results suggest a shrinking-core model for fixed-sized particles, with a reaction rate controlled by diffusion through the $Mg(OH)_2(s)$ product layer rather than by reaction at the core surface [70].

Supported MnO_x on high-surface-area SiO_2 has been used for the oxidative dehydrogenation of ethylbenzene to styrene [71]. The catalytic activity of MnO_x is due to an ability to form oxides, such as MnO_2, MnO_3, Mn_3O_4, and MnO, that are the source of oxygen for the oxidation. Manganese can be either a reducing agent or an oxidizing agent, due to its various oxidation states. The active component in this catalytic reaction has been identified as the MnO_2 phase. Analyses of spent catalysts provide information regarding how manganese is reduced from Mn^{4+} to Mn^{2+}. Oxygen is provided to the ethylbenzene reactant and leads to a less active Mn_3O_4 phase [71].

Lvov et al. and Espinal [72] have successfully used stable protein/nanoparticle films for electrocatalytic applications using a layer-by-layer assembly technique. Higher turnover numbers are obtained for these films as compared to protein/surfactant films. Figure 13 shows a conceptual model of the multilayer film assembly of protein and nanoparticles developed by Lvov [72a]. The films have a three-dimensional architecture, with growth perpendicular to the solid support, resulting in a porous structure. Myoglobin (Mb) has been assembled with MnO_2 nanoparticles in mutilayered films that can be used for such applications as electrochemical biosensors and electrocatalysis. Optimum performance for the catalytic oxidation of styrene was obtained using MnO_2 nanoparticle films without Mb. They also are found to be more effective under O_2 and electrolytic conditions.

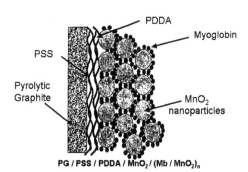

PG / PSS / PDDA / MnO₂ / (Mb / MnO₂)ₙ

FIGURE 13   Conceptual model of the assembly of a Mb/MnO_2 film (PG = pyrolytic graphite; PSS = poly(styrenesulfonate); PDDA = poly(dimethyldiallylammonium) cation; Mb = myoglobin). (Adapted from Ref. 72a.)

C. Degradation of Organic Compounds

The abiotic transformation of organic chemicals in soil can be used for environmental remediation [73a]. The transformation of organic pollutants in soil, such as 2,4-dichlorophenoxyacetic acid (2,4-D), as well as of solvents such as ethyl ether has been carried out using birnessite [73a]. The reaction proceeds via the 2,4-D-assisted dissolution of birnessite to produce Mn^{2+}, with simultaneous hydrocarbon oxidation. The rates of degradation are determined using a calorimetric method [73a]. GC analysis shows that oxygen is involved in the reaction and that CO_2 is a major reaction product. Solid-state degradation can occur at significant rates in the presence of birnessite. Heat-evolution data plotted for this reaction suggest first-order kinetics. Thin-layer chromatograph (TLC) and high-performance liquid chromatography (HPLC) analyses indicate that 2,4-D is transformed by a light-grinding procedure to 2,4 dichlorophenol [73b]. Extractable manganese formed during 2,4-D breakdown indicates that the concentration increases with incubation time, which suggests that Mn(IV) is reduced to Mn(II) in the oxide mineral. Manganese is involved in the removal of chlorine from 2,4-D. A positive correlation is found between the number of moles of CO_2 evolved and the number of moles of Mn produced during incubation. This result suggests that the oxygen for oxidation of 2,4-D is coming from the oxide mineral [73b].

Atrazine (2-chloro-4-ethylamino-6-isopropylamino-s-triazine) is a common postemergent herbicide used for weed control on agricultural lands, and it is frequently detected in surface and ground waters in the United States. Mechanochemical degradation of atrazine adsorbed on birnesite, pyrolusite, and cryptomellane have been studied by Shin et al. [74]. Birnessite follows first-order kinetics for this degradation. Oxygen-independent atrazine dealkylation has been reported to occur on the surface of birnessite, with mainly mono- and dialkyl-atrazine as products [75]. Conversion of the alkyl chains to the alkenes is reported to occur in a nonredox process independent of O_2. A dealkylation mechanism is proposed where protons transfer to Mn(IV)-stabilized oxo and amido bonds. In the presence of O_2, secondary reactions involving olefin oxidation are observed. The oxidation of 2,4-dichlorophenol and 2,4,5-trichlorophenol in the presence of birnessite and pyrolusite also has been studied [76]. Birnessite materials are very effective as abiotic catalysts in the oxidation reaction [76]. Kinetics and the effect of pH have been studied, and a mechanism involving oxide surfaces has been proposed [76].

Birnessite has been used for the sequestration of phenolic compounds in natural sorbents via oxidative coupling with manganese oxide [77]. The rate of polymer formation and the molecular size of the coupling products are controlled by birnessite concentration. A low concentration of birnessite favors the oxidative coupling of organic compounds [77]. The oxidative coupling of phenolic compounds catalyzed with birnessite, with manganese oxide functioning as the electron acceptor, also has been reported [79]. Birnessite materials catalyze the transformation of 2,6-dimethoxyphenol. A comparison is made for enzyme-catalyzed reactions and birnessite-catalyzed reactions. Manganese oxide functions as an electron acceptor, whereas in the enzyme-catalyzed reactions, O_2 functions as an electron acceptor [78] The herbicide propanil and its metabolite DCA (3,4-dichloroaniline) are oxidized to humic monomers. Transformation of DCA is increased five times when it is incubated with laccase and birnessite together, as compared to laccase alone [79]. Various aspects of the oxidative coupling of chlorinated phenols using birnessite and oxidoreductases have been discussed [80], and dehalogenation patterns have been found.

The decomposition of pinacyanol chloride dye using octahedral layered mixed-valent manganese oxides has been published [81]. Catalytic reduction reactions using birnessite have been tried for removing pentachlorophenol (PCP) from soil and water (detoxyfication) [82]. Transformation and dechlorination of PCP incubated with peroxidase, laccase, or birnessite is decreased in the presence of humic monomers as cosubstrate. The dehalogenation number for birnessite is 3.3, compared with 3.5 for peroxidase and 1.5 for laccase [82].

Birnessite and tyrosinase catalyze the transformation of cathecol to oligomers, polycondensates, and fragments [83]. Products of the tyrosinase–catechol system have been compared with products of the birnessite–catechol system. Products of birnessite catalysis contain polycondensates and fragments including aliphatics. FTIR analysis indicates that an organic coating on the birnessite is formed by the reaction products. Cathecol–melanin is green in the presence of birnessite and brown in the presence of the tyrosinase. Birnessite concentration and pH play important roles in the amount of CO_2 released, and they are related to ring cleavage of catechol. In the case of the birnessite, various polycondensates of lower molecular weights are formed. The formation of phenolic–enzyme complexes has been investigated with Mn(IV) oxide birnessite as the catalyst [84]. Birnessite favors the formation of phenolic–lysozyme complexes in the hydroquinone (diphenol)–lysozyme (enzyme) system. The role of birnessite in the formation of humic–enzyme complexes in soil ecosystems requires further study. Birnessite has also been examined for its catalytic activity in performing the Maillard reaction between glucose and glycine [85]. The Maillard reaction was first observed in 1912 and described as reducing sugars and amino acids or proteins reacting together. In foods, the Maillard reaction is responsible for the changes in the flavor, color, and nutrition and the formation of mutagenic compounds. Results show the importance of birnessite catalysis in the Maillard reaction in the natural abiotic formation of humic substances.

The oxidative polymerization of diphenols such as hydroquinone, catechol, and resorcinol to form humic acids using Mn(IV) birnessite as a possible remediation technique has been studied [86]. The yields of humic acid production are: hydroquinoline > cathecol > resorcinol [86]. The polycondensation of pyrogallol and glycine catalyzed with birnessite to form N-polymers resembling natural humic acids has been reported by Wang and Huang [87]. Birnessite-type materials are useful for the deamination and decarboxylation of glycine. Catalytic deamination and decarboxylation of acids may be a new pathway of C turnover and N transformation, with oxygen playing an important role in the catalysis [87].

Birnessite-type materials have also been used for the degradation of organic compounds to form phenolic polymers, with further oxidation to produce humic substances. Birnessite promotes the abiotic formation of nitrogenous polymers in hydroquinoline systems more quickly near neutral conditions than under acidic conditions [88]. Pyrogallol ring cleavage has been studied using Mn(IV), Fe(III), Al, and Si oxides. The activities are high, especially in the case of Fe and Mn oxides [89]. Aliphatic fragments are detected as well as CO_2. Carbon dioxide production is possibly due to the adsorption and polarization of oxygen molecules at the surface of high oxidation state manganese species such as Mn(III) and Mn(IV) [89].

D. Applications as Battery Materials

Birnessites, including microporous manganese oxides, are useful as battery materials because of their relatively high surface areas and transition metal oxide content [90,91]. Battery performance measured as battery discharge activity is improved by ion exchanging Na^+ and K^+-birnessites with Bi^{3+} or Pb^{2+}. Retention capacity and cyclability also are improved by the ion exchange of sodium birnessite with Bi^{3+}, Pb^{2+}, or Li^+ [91]. Thermal stabilities and lithiation capacity of Zn and lithium birnessite in nonaqueous solutions have been investigated as well. The Zn- and Li-birnessite materials can be lithiated up to Li/Mn molar ratios of 0.45 and 1.08, respectively, by reaction in LiI-acetonitrile solution. The greater thermal stability found in the lithiated Zn-birnessite is attributed to a "pillar structure" formed by the Zn species in the interlayer, but this effect is not observed in the case of lithiated Li-birnessite [92]. According to the authors, Zn^{2+} and $Zn(OH)^+$ were exchanged into the layers as a "pillar," but XRD data do not show a zinc oxide phase.

E. Applications of Manganese Oxide Mesoporous Solids and Manganese Based Porous Mixed Oxides

The catalytic oxidation of stable alkanes to more valuable products such as alcohols and ketones is known to occur with microporous manganese oxide molecular sieves (OMS). MOMS materials have Lewis acid sites as well as Brönsted acid sites coexisting on the active surface, although the Lewis acid sites dominate, as is typical for manganese oxides [53].

Hexagonal MOMS materials have been tested for the catalytic oxidation of *n*-hexane and cyclohexane in aqueous solution. This reaction was chosen because previous work performed using OMS materials for alkane catalysis with hydrogen

Table 1 Yields for the Oxidation of Cyclohexane and Hexane over Hexagonal MOMS

	Yield (%)	
Product	n-Hexane feed	Cylohexane feed
Cyclohexanol	N.P.	2.6
Cyclohexanone	N.P.	5.5
1-Hexanol	0.16	N.P.
2-Hexanol	1.64	N.P.
3-Hexanol	1.19	N.P.
2- and 3-Hexanol	3.01	N.P.

Source: Ref. 53.
N.P. indicates that no product was observed.

peroxide was promising. The results suggest that the hexagonal MOMS phase is active for the oxidation of both alkanes. The catalytic results obtained by Tian et al. are shown in Table 1, with conversions of approximately 10% and 8% for cyclohexane and n-hexane, respectively [53]. The hexagonal MOMS activity is proposed to be due to a redox mechanism that is a direct result of their mixed valency and unusually high thermal stability [53,54].

MANPO materials have been tested for environmentally important catalytic reactions, such as lean de-NO_x ($NO/CH_4/O_2$) and CH_4/O_2 combustion [58,60]. The overall catalytic performance was found to be superior to other catalysts. For example, Stahtopoulos et al. found MANPOS catalysts containing Ce, Sr, and La to be very selective (98% to N_2) as well as active materials under conditions of 0.67% CH_4/0.2% NO/5% O_2 for the lean de-NO_x reaction in the low-temperature range of 200–300°C [60]. The Al-containing MANPO materials achieve the second highest selectivity to N_2 (73%) [59]. MANPO materials perform impressively in the presence of 4% H_2O in the feed stream, reaching 98% selectivity for N_2. In addition to high activity, MANPO materials show excellent stability for a 24-h period in a temperature range of 200–500°C. [60] The excellent performance in water is a remarkable result in terms of practical applications. A Pt/SiO_2 catalyst shows a maximum reaction rate at 270°C about five times higher than that observed for the MANPO catalysts, but it loses activity drastically when the oxygen content is increased. This loss of activity is not observed with the MANPO catalyst.

IV. CONCLUSIONS

The synthesis, properties, and applications of different layered manganese oxide materials have been discussed. The synthesis of layered manganese oxide materials such as feitknechtite and birnessite has been thoroughly studied. Properties of these materials can be controlled by using different synthesis routes and by adjusting the preparation parameters. Pillared manganese oxide layered materials and manganese-based mesoporous materials are relatively new, and they have stimulated interest in studying their synthesis, properties, and applications only recently. There are still many unknown factors concerning their preparation in order to achieve designed properties. Research on the applications of layered manganese oxide materials focuses mainly on birnessite-type materials due to their economic cost and unique properties, such as the mixed valence of manganese, their porosity, and conductivity. Promising applications of birnessite-type materials include sorption, selective oxidation catalysis, and battery applications. Detailed mechanistic and kinetic investigations need to be done to better design the materials. In situ characterization studies may help to elucidate the mechanisms of these applications. Pillared manganese oxide layered materials and manganese-based mesoporous materials need to be better characterized and tested for advanced applications, such as chiral organic syntheses, room-temperature homogeneous catalysis reactions, oxidation, photocatalysis, acid–base catalysis, and electromagnetic applications.

V. ACKNOWLEDGEMENTS

The authors acknowledge support in this area of research from the Chemical Sciences, Geosciences, and Biosciences Division, Office of Basic Energy Sciences, Office of Science, U.S. Department of Energy.

REFERENCES

1. VM Feitknecht, W Marti. Helv Chem Acta 1945; 28:129–148, 148–156.
2. VM Feitknecht, P Brunner, HR Oswald. Zeit Anorg Allgem Chem 1962; 316: 154–160.
3. R Meldau, H Newesely, H Strunz. Naturwissenschaften 1973; 60:387.
4a. J Luo. Preparative parameters, mechanisms, and kinetics in the syntheses of layered and tunnel manganese oxides. PhD dissertation. Storrs. CT: University of Connecticut, 1999:5–6, 31–32, 56–61.
4b. J Luo, SR Segal, JY Wang, ZR Tian, SL Suib. Mat Res Soc Symp Proc. Vol. 431, 1996:3–8.
5. G Aminoff. Geol For Forh 1919; 41:407–433.
6. DS Yang, MK Wang. Clays Clay Miner 2002; 50:63–69.
7. R Giovanoli, E Stahli, WM Feitknecht. Helv Chim Acta 1970; 53:209–220.
8. J Luo, A Huang, SH Park, SL Suib, C O'Young. Chem Mater 1998; 10:1561–1568.
9. P Jones, AA Milne. Mineral Mag 1956; 31:283–288.
10a. RG Burns, VM Burns. Philos Trans R Soc, London (A) 1977; 286:283–301.
10b. RG Burns, VM Burns, HW Stockman. Am Mineral 1983; 68:972–980.

11. O Bricker. Am Mineral 1965; 50:1296–1354.
12. R Giovanoli, P Burki. Chimia 1975; 29:266–269.
13a. S Turner, PR Buseck. Science 1979; 203:456–458.
13b. S Turner, PR Buseck. Science 1981; 212:1024–1027.
13c. MD Siegel, S Turner. Science 1983; 219:172–174.
14. DC Golden, JB Dixon, CC Chen. Clays Clay Miner 1986; 34:511–520.
15. JE Post, DR Veblen. Am Miner 1990; 75:477.
16a. VA Drits, E Silvester, AI Gorshkov, A Manceau. Am Mineral 1997; 82:946–961.
16b. E Silvester, A Manceau, VA Drits. Am Mineral 1997; 82:962–978.
17a. DC Golden, CC Chen, JB Dixon. Science 1986; 231:717–719.
17b. DC Golden, CC Chen, JB Dixon. Clays Clay Miner 1987; 35:271–280.
18. P Strobel, JC Charenton. Rev Chim Miner 1986; 23:125–137.
19a. Q Feng, K Yanagisawa, NJ Yamasaki. J Ceram Soc Jpn 1996; 104:897–899.
19b. Q Feng, K Yanagisawa, NJ Yamasaki. Chem Commun 1996; 14:1607–1608.
20. J Cai, J Liu, SL Suib. Chem Mater 2002; 14:2071–2077.
21. J Luo, Q Zhang, SL Suib. Inorg Chem 2000; 39:741–747.
22a. YF Shen, RP Zerger, RN DeGuzman, SL Suib, L McCurdy, DI Potter, CL O'Young. Science 1993; 260:511–515.
22b. YF Shen, SL Suib, CL O'Young. J Am Chem Soc 1994; 116:11020–11029.
23. SL Suib. Stud Surf Sci Catal 1996; 102:47–74.
24a. J Luo, SL Suib. J Phy Chem B 1997; 101:10403–10413.
24b. J Luo, A Huang, SH Park, SL Suib, CL O'Young. Chem Mater 1998; 10:1561–1568.
25. J Luo, SL Suib. Chem Commun 1997; 11:1031–1032.
26. Y Ma, J Luo, SL Suib. Chem Mater 1999; 11:1972–1979.
27. EJ Witzemann. J Am Chem Soc 1915; 37:1079–1091.
28. S Bach, M Henry, N Baffier, J Livage. J Solid State Chem 1990; 88:325–333.
29a. S Ching, JA Landrigan, ML Jorgensen. Chem Mater 1995; 7:1604–1606.
29b. S Ching, DJ Petrovay, ML Jorgensen, SL Suib. Inorg Chem 1997; 36:883–890.
30. NG Duan, SL Suib, CL O'Young. Chem Commun 1995; 13:1367–1368.
31a. SL Brock, M Sanabria, SL Suib, V Urban, P Thiyagarajan, DI Potter. J Phy Chem B 1999; 103:7416–7428.
31b. SL Brock, M Sanabria, J Nair, SL Suib, T. Ressler. J Phy Chem B 2001; 105: 5404–5410.
32. O Giraldo, SL Brock, WS Willis, M Marquez, SL Suib, S Ching. J Am Chem Soc 2000; 122:9330–9331.
33. Q Gao, O Giraldo, W Tong, SL Suib. Chem Mater 2001; 13:778–786.
34. RN DeGuzman, YF Shen, EJ Neth, SL Suib, CL O'Young, S Levine, JM Newsaw. Chem Mater 1994; 6:815–821.
35. A Clearfield, BD Roberts. Inorg Chem 1988; 27:3237–3240.
36. S Cheng, T Wang. Inorg Chem 1989; 28:1283–1289.
37. TJ Pinnavaia. Science 1983; 220:365–371.
38a. Y Ma, SL Suib, T Ressler, J Wong, M Lovallo, M Tsapatsis. Chem Mater 1999; 11:3545–3554.
38b. Y Ma. Studies of layered and pillared manganese oxide materials. PhD dissertation. Storrs. CT: University of Connecticut, 1999.

39. P Olivera-Pastor, P Maireles Torres, E Rodriguez-Castellon, A Jimenez-Lopez, T Cassagneau, DJ Jones, J Roziere. Chem Mater 1996; 8:1758–1769.
40. S Yamanaka, K Kunii, ZL Xu. Chem Mater 1998; 10:1931–1936.
41. ST Wong, S Cheng. Inorg Chem 1992; 31:1165–1172.
42. ZH Liu, K Ooi, H Kanoh, W Tang, X Yang, T Tomida. Chem Mater 2001; 13: 473–478.
43. B Ammundsen, E Wortham, DJ Jones, J Roziere. J Mol Cryst Liq Cryst 1998; 311: 735–740.
44. ZH Liu, K Ooi, H Kanoh, W Tang, T Tomida. Chem Lett 2000; 4:390–391.
45. ZH Liu, X Yang, Y Makita, K Ooi. Chem Lett 2002; 7:680–681.
46. CT Kresge, ME Leonowicz, WJ Roth, JC Vartuli, JS Beck. Nature 1992; 359:710.
47. P Selvam, SK Bhatia, CG Sonwane. Ind Eng Chem Res 2001; 40:3237–3261.
48. M Vettraino, M Trudeau, DM Antonelli. Inorg Chem 2001; 40:2088–2095.
49. W Zhang, TJ Pinnavaia. Catal Lett 2001; 28:261–265.
50. M Stockenhuber, RW Joyner, JM Dixon, MJ Hudson, G Grubert. Microporous Mesoporous Mater 2001; 44–45:367–375.
51. J Xu, Z Luan, M Hartmann, L Kevan. Chem Mater 1999; 11:2928–2936.
52. M Yonemitsu, Y Tanaka, M Iwamoto. J Catal 1998; 178:207–213.
53. ZR Tian, W Tong, JY Wang, NG Duan, VV Krishnan, SL Suib. Science 1997; 276: 926–930.
54. ZR Tian. From microporous to mesoporous manganese oxide crystalline phases: syntheses, characterization, and applications. PhD dissertation. Storrs. CT: University of Connecticut, 1998.
55. J Luo, SL Suib. Chem Commun 1997; 11:1031–1032.
56. VN Stathopoulos, DE Petrakis, M Hudson, P Falaras, SG Neofytides, PJ Pomonis. Stud Surf Sci Catal 2000; 128:593–602.
57. AD Zarlaha, PG Koutsoukos, C Skordilis, PJ Pomonis. J. Colloid Interface Sci 1998; 202:301–312.
58. VN Stathopoulos, VC Belessi, CN Costa, SG Neofytides, P Falaras, PJ Pomonis. Stud Sur Sci Catal 2000; 130:1529–1534.
59. CE Salmas, VN Stathopoulos, PJ Pomonis, GP Androutsopoulos. Langmuir 2002; 18:423–432.
60. VN Stathopoulos, CN Costa, PJ Pomonis, AM Efstathiou. Topics Catalysis 2001; 16:231–235.
61. Q Feng, H Kanoh, K Ooi. J Mater Chem 1999; 9:319–333.
62. A Dyer, M Pillinger, R Harjula, S Amin. J Mater Chem 2000; 10:1867–1874.
63. L Al-Attar, A Dyer. J Mater Chem 2002; 12:1381–1386.
64. M Fendorf, R Gronsky. J Colloid Interface Sci 1992; 148:295–298.
65. G Krishnamurti, P Huang. Can J Soil Sci 1987; 67:533–543.
66a. ST Wong, S Cheng. J Chin Chem Soc 1993; 40:509–516.
66b. ST Wong, S Cheng. J Therm Anal 1993; 40:1181–1192.
67. V Jha, R. Uma, R. Renuka. J Sci Ind Res 2000; 59:829–832.
68. VV Krishnan, SL Suib. J Catal 1999; 184:305–315.
69. YF Shen, SL Suib. J Catal 1996; 161:115–122.
70. AM Elprince, WH Mohamed. Soil Sci Soc Am J 1992; 56:1784–1788.

71. R Craciun, N Dulamita. Ind Eng Chem Res 1999; 38:1357–1363.
72a. Y Lvov, B Munge, O Giraldo, I Ichinose, SL Suib, JF Rusling. Langmuir 2000; 16: 8850–8857.
72b. L. Espinal. Protein/MnO_2 nanoparticles multilayer films for electrocatalytic and sensor applications. M.S. thesis. Storrs. CT: University of Connecticut, 2001.
73a. MA Cheney, G Sposito, AE McGrath, RS Criddle. Colloids Surf A 1996; 107: 131–140.
73b. A Nasser, G Sposito, MA Cheney. Colloids Surf A 2000; 163:117–123.
74. JY Shin, CM Buzgo, MA Cheney. Colloids Surf A 2000; 172:113–123.
75. D Wang, JY Shin, MA Cheney, G Sposito, TG Spiro. Environ Sci Technol 1999; 33:3160–3165.
76. M Pizzigallo, P Ruggiero. Fresenius Environ Bull 1992; 1:428–433.
77. th II, H Selig, CY Payne, WJ Weber. Preprints of American Chemical Society, Division of Environmental Chemistry. Vol. 40, 2000:147–150.
78. S Pal, JM Bollag, PM Huang. Soil Biol Biochem 1994; 26: 813–820.
79. TD Kwon, JE Kim. Han'guk Nonghwa Hakhoechi 1998; 41:384–389.
80. J Dec, JM Bollag. Environ Sci Technol 1994; 28:484–490.
81. SR Segal, SL Suib, L Foland. Chem Mater 1997; 9:2526–2532.
82. JW Park, JE Kim. Han'guk Nonghwa Hakhoechi 1999; 42(4):330–335.
83. A Naidja, PM Huang, JM Bollang. Soil Sci Soc Am J 1998; 62:188–195.
84. H Shindo, T Oshita, N Matsudomi, J Usui, TB Goh. Soil Sci Plant Nutr (Tokyo) 1996; 42:141–146.
85a. A Jokic, AI Frenkel, PM Huang. Can J Soil Sci 2001; 81:277–283.
85b. A Jokie, AI Frenkel, MA Vairavamurthy, PM Huang. Geophysical Res Lett 2001; 28:3899–3902.
86. H Shindo, PM Huang. Sci Total Environ 1992; 103:117–118.
87. MC Wang, PM Huang. Sci Total Environ 1987; 62:435–442.
88. H Shindo, PM Huang. Soil Sci Soc. Am. J 1984; 48:927–934.
89. MC Wang, PM Huang. Soil Sci 2000; 165:934–942.
90. M Nitta. Appl. Catal 1984; 9:151–176.
91. R Armstrong, PG Bruce. Nature 1996; 381:499–500.
92. L Liu, Q Feng, K Yanagisawa, G Bignall, T Hashida. J of Materials Sience 2002; 37:1315–1320.

10

Layered Metal Chalcogenides

Christopher O. Oriakhi

Hewlett-Packard Corporation
Corvallis, Oregon, U.S.A.

Michael M. Lerner

Oregon State University
Corvallis, Oregon, U.S.A.

I. INTRODUCTION

For more than 30 years, there has been a sustained academic interest and technical development effort directed at layered metal chalcogenides (LMCs) and their intercalation compounds. This in part is due to the intriguing structural and physical properties of these two-dimensional (2D) inorganic materials [1]. In the general case, these structures consist of infinite metal chalcogenide layers; within each layer the atoms are bound by strong covalent interactions, but the layers themselves interact only by weaker van der Waals forces [2]. For the most part, the metals involved are transition metals, although SnS_2 is a well-known example of a main-group LMC.

Many of the chemical and physical properties of these materials derive from this anistropic layered structure. Intercalation chemistry, the insertion of ions and/or molecules between individual LMC layers, is possible due to these weak interlayer bonding interactions. Intercalation can be accomplished by electrochemical or chemical methods, as will be discussed further on in this chapter. The highly anisotropic nature of the layered structure also leads to characteristic and often technologically useful mechanical, electrical, magnetic, and optical properties. On the other hand, since a broad range of metals and oxidation states is included in this structural class, a broad range of properties is also observed.

Depending on the metal and the specific structure, LMCs may be semiconductors, semimetallic, or metallic [3,4]. The metallic chalcogenides may exhibit interesting electronic effects, such as charge density waves and superconductivity [5].

Early technological interest was in part focused on the importance of the layered metal disulfides, MS_2, as the catalytic agent in hydroprocessing or hydro-treating reactions of organic sulfur- and nitrogen-containing feedstocks. For the past several decades, MoS_2 and WS_2 in particular have been used as a hydrodesul-furization and denitrogenation catalysts in petroleum processing [6–8]. Other important applications have also been developed or proposed for these materials. For example, MoS_2 is currently an important material in high-temperature solid-state lubrication [9,10]. Other applications have not been realized but have in-volved considerable research and technical interest. For example, LMCs such as TiS_2, MoS_2, and $NiPS_3$ have also been explored as cathodic materials in lithium-anode batteries [11–13]. Many of the MS_2 compounds are sufficiently conductive to allow for potential applications as conductive additives or use as other elec-troactive components. In other cases, novel properties suggest potential applica-tion in new devices or other new roles for solid materials. For example, intercala-tion compounds derived from TaS_2, NbS_2, and $MnPS_3$ demonstrate interesting superconductivity and second-order-harmonic nonlinear optical properties [14,15]. Current and potential applications will be further discussed later in this chapter.

A. Scope

There have recently been intense research activities on the syntheses, structure/property relationships, and applications of a number of LMCs. This chapter fo-cuses on the structure, characteristic chemistry, and technologically significant properties and applications of transition metal LMCs. There is sufficient literature to require some selective reporting of topics. In this chapter, particular emphasis will be placed on intercalation reactions of the dichalcogenides (MX_2) ($X = S$ or Se), metal phosphorus trichalcogenides (MPX_3), and the applications of LMCs in catalysis. A brief overview of layered chalcogenides (MoS_2 and WS_2) with novel fullerene-like and nanotube structure will also be presented. However, a discussion of many of the well-known ternary chalcogenides, such as Chevrel phases A_xMoS_2 and AMo_3Se_3 ($A = Li$ or Na), and of framework structures such as Nb_3S_4 and $Tl_xV_6S_8$ will not be included. Extensive reviews of these subjects can be found elsewhere.

B. Structural Classifications

LMCs are generally classified by their composition. The binary compounds in-clude metal dichalcogenides (MX_2) and metal trichalcogenides (MX_3). Ternary compounds include AM_2S_5 or $(AS)_n(MS_2)_2$, otherwise known as misfit layer compounds, and metal phosphorus trichalcogenides (MPX_3). All of these structure types contain layers with no net charge; i.e., these are the neutral layer structures. In contrast, ternary structures such as A_xMS_2 and AMS_2 contain negatively charged layers. Table 1 illustrates these classes of compounds and provides examples of each.

II. SYNTHESIS OF METAL CHALCOGENIDES

Some main-group and transition metals will react with stoichiometric amounts of sulfur, selenium, and tellurium to form metal chalcogenides at elevated temperatures. While the reaction is conceptually simple, there can be disadvantages to such direct synthetic methods. Solid–solid reactions with slow diffusion rates may produce inhomogeneous products. The physical, and sometimes chemical, properties of metal chalcogenides can be greatly influenced by the method of preparation and depend strongly on product purity and composition as well as on the structural and electronic effects that arise due to nonstoichiometry. Desired products may not be stable at the elevated temperatures required to activate the elemental solids to complete reactions. Alternately, the direct high-temperature reactions of elements often yield highly crystalline materials, i.e., products with low surface areas. For applications such as catalysis, however, it is often desirable to obtain homogeneous, single-phase, and high-surface-area powders. With these target characteristics, a low-temperature method can be highly desirable.

Table 1 Classes of Layered Metal Chalcogenides, with Examples

Layered host lattice	Representative examples	Layer charge
Metal dichalcogenides (MX_2)	M = Sn, Cr, Hf, Ta, Ti, Zr, Nb, Mo, W, or V X = S, Se, or Te	Neutral
Metal trichalcogenides (MX_3)	M = Nb or Zr X = S, Se, or Te	Neutral
Metal phosphorous trichalcogenides (MPX_3)	M = Cd, Fe, Mg, Ca, Mn, Ni, V, Sn, Pb, or Zn X = S or Se	Neutral
Misfit layered compounds ($(RX)_m(MX_2)_n$)	R = Rare Earths, Pb, or Sn M = Ta, Nb, V, Ti, or Cr X = S or Se	Neutral
Ternary transition metal sulfides (AMX_2)	A = Li, Na, K, Rb, or Cs M = Cr, Ti, V, Zr, Nb, or Ta X = S or Se	Negative

Several novel low-temperature preparation routes have been developed to prepare the various types of LMCs. The following discussion contains some important examples for common structure types.

A. Syntheses of Layered Metal Dichalcogenides

1. High-Temperature Syntheses

Most metal dichalcogenides can be prepared by direct combination of the elements at an elevated temperature ($>450°C$) in an evacuated silica tube. For example, microcrystalline MoS_2 is obtained by direct combination of the stoichiometric quantities of the elements at $1100°C$ for an extended period [16]. High temperature and long reaction times are required to achieve favorable kinetic and thermodynamic conditions and to obtain complete reaction. Generally, highly crystalline, multiphase, and nonstoichiometric materials with low surface are obtained, with major ramifications for the physical and chemical properties of the materials.

2. Vapor-Phase Reactions

The vapor-phase method has been used to synthesize a number of transition metal dichalcogenides by reaction of a suitable metal chloride with hydrogen sulfide gas at elevated temperature [17]. The strategic advantage of this method lies in using precursors that do not require the conversion of a reactant solid phase to product. The typical reaction proceeds according to

$$MX_4 + 2H_2S \rightarrow MS_2 + 4HX$$

For example, Thompson and coworkers described the preparation of TiS_2 by reaction of $TiCl_4$ with H_2S at $450°C$ [17]. The gas-phase reaction is represented by the equation

$$TiCl_4 + 2H_2S \rightarrow TiS_2 + 4HCl$$

This approach often yields highly crystalline products with low surface area. SnS_2 has also been prepared by a vapor-phase reaction [18].

3. Chemical Vapor Transport

Chemical vapor-phase transport is one of the earliest methods developed for the synthesis of new materials [19–26]. The method can also be used for the growth of high-quality single crystals or for the purification of solid-state compounds. The method has been used to synthesize a wide range of LMCs.

Chemical vapor transport consists of two basic steps. The crystalline or semicrystalline powder of the MX_2 compound is first prepared by heating a stoichiometric ratio of the elements in an evacuated silica tube in the first step. In the second step, the MX_2 powder, together with a gaseous transporting agent such as iodine, is placed on the hot side of an evacuated silica ampoule while maintaining a temperature gradient of 25–50°C along the length of the tube. A large single crystal of the MX_2 will often grow slowly on the cold end of the tube over a period of a few days. The MX_2 compound reacts with iodine gas, for example, to form a gaseous intermediate product, MI_2, that eventually decomposes to reform crystals of MX_2 at the cold end of the tube. The following equilibria are believed to be responsible for the transport and formation of the MX_2 single crystals [27]:

$$2MX_2 \text{ (powder)} + 2I_2 \rightleftharpoons 2MI_2 + 2X_2$$

$$2MI_2 + 2X_2 \rightleftharpoons 2MX_2 \text{ (crystal)} + 2I_2$$

Using the vapor transport method with bromine or iodine as the transporter, Brixner [28] and Nitsche [29–30] have prepared a wide range of transitional metal dichalcogenides, including: $MoSe_2$, $MoTe_2$, NbS_2, $NbSe_2$, $NbTe_2$, TaS_2, $TaSe_2$, $TaTe_2$, TiS_2, $TiSe_2$, $TiTe_2$, HfS_2, $HfSe_2$, VS_2, VSe_2, WS_2, WSe_2, WTe_2, ZrS_2, and $ZrSe_2$. Nitsche provides a summary of the factors that favor the successful growth of metal chalcogenide crystals by this method. Several researchers [31] have extended the vapor transport method to prepare a range of solid solutions in systems such as TaS_xSe_{2-x}, TiS_xSe_{2-x}, $TiSe_xTe_{2-x}$, SnS_xSe_{2-x}, $TiSe_xTe_{2-x}$, HfS_xSe_{2-x}, and ZrS_xSe_{2-x} for $0 \leq x \leq 2$.

4. Low-Temperature Syntheses

Chianelli and Dines developed a solution method for preparing Group 4, 5, and 6 transition metal dichalcogenides [32], where a low-temperature metathesis occurs between a transition metal halide (MX_n) and an alkali metal sulfide (A_2S) in an aprotic polar organic solvent such as tetrahydrofuran (THF) or ethyl acetate. The method is described by the following general equation:

$$MX_n + (n/2)A_2S \rightarrow MS_{n/2} + nAX$$

For example, in THF at ambient temperature:

$$TiCl_4 + 2Li_2S \rightarrow TiS_2 + 4LiCl$$

M is generally a Group 4, 5, or 6 transition metal, and A is an alkali metal ion such as Li^+, Na^+ or the ammonium ion, NH_4^+. The products of this reaction are mainly amorphous and may be converted into a crystalline product by thermal treatment in an evacuated tube at 400–600°C. Using this method, Li_2S and $TiCl_4$ can be reacted in refluxing THF at only 65°C to prepare single crystals of TiS_2 as large as 100 μm in diameter. The disulfides HfS_2, MoS_2, NbS_2, TaS_2, TiS_2, VS_2, and ZrS_2 have also been prepared using this method.

Another low-temperature route to amorphous metal dichalcogenides involves the reaction of a transition metal chloride and fluoride with organic sulfur compounds such as di-*tert*-butyldisulfide, di-*tert*-butylsulfide, and *tert*-butylmer-captan [33–35]. Amorphous products (for M = Ta, Nb, or Mo) can be converted to the crystalline form by further heating in evacuated silica tubes or in the presence of sulfur. Final product stoichiometries may be controlled by varying the partial pressure and temperature of sulfur in this latter step.

The tin dichalcogenides SnX_2 (X = S or Se) are the only main-group dichalcogenides that show a layered structure. An amorphous material may be prepared from tin tetrachloride and sodium sulfide in a similar manner to that described earlier; however, in this case the reaction may occur in aqueous solution [36]:

$$SnCl_4 \text{ (1)} + 2Na_2S \text{ (aq)} \rightarrow SnS_2 \text{ (s)} + 4NaCl \text{ (aq)}$$

The amorphous powder can subsequently be converted into a highly crystalline form by the application of the iodine vapor transport reaction method [29,30].

5. Other Synthetic Methods

Modern applications of layered MS_2 in chemical sensors and other micro- and molecular electronic devices require that these materials be available as thin films or as nanocolloidal particulates. A range of methods has therefore been developed for preparing thin films and nanoparticles of MS_2 compounds. Thin-film preparations include pulse laser evaporation [37,38], metal organic vapor deposition (MOCVD) [39,40], and radio frequency sputtering [41]. The chemical methods for exfoliating or delaminating MS_2 and other metal chalcogenides will be discussed under intercalation of layered metal chalcogenides and nanocomposite preparation.

B. Syntheses of Inorganic Fullerene-Like Materials and Nanotubes from Layered MS₂

Graphite is the carbon allotrope with a layered structure. It has been known since the 1980s that, under special conditions, carbon can also be prepared in the form of closed polyhedra known as *fullerenes* (C_{60}, C_{70}, etc.) [42]. If the polyhedra comprise multiple shells (such as the multiple skins in an onion), these are known as *nested-fullerene nanoparticles*. Both single-shell and nested-shell structures are also known to grow as elongated allotropes in the form of tubes or rods with nanoscale diameter, which are called *single-walled* or *multiwalled nanotubes* [42–44]. Carbon-based fullerenes display unique structural morphologies, stereo-chemistry, and a range of interesting chemical and physical properties [43]. Because of the many similarities in structure and physical properties, structural analogs were soon sought, and found, for the layered MS_2 compounds. At present, the known fullerene-like structures as well as nanotubes are based on MoS_2 [45–48] and WS_2 [47,49,50]. In addition, related structures have been prepared from other layered inorganic compounds, including V_2O_5 [51] and BN [52]. The layers comprising these materials form a range of folded cages, rings (onions), and hollow structures. Such materials are now commonly referred to as *inorganic fullerene-like* (IF) *structures.*

Layered dichalcogenides with highly folded and disordered structure have been recognized for several decades. The so-called rag and tubular structures in selected MS_2 compounds were first reported in 1979, and they gained wider interest following the discovery of fullerene and carbon nanotubes [53]. Fundamentally, the composition of the IF materials are similar to those of the bulk layered MS_2.

Several methods have been developed to synthesize inorganic fullerenes and nanotubes based on layered MS_2 [45–50]. They include arc discharge, laser ablation techniques, electron beam irradiation of MS_2 crystals, chemical transport reaction, and precursor synthesis approach [50,54–58]. The chemical vapor transport method, which was discussed earlier, has also been used to synthesize both MoS_2 and WS_2 nanotubes.

The IF structures and nanotubes for WS_2 were first observed during hydrodesulfurization using thin films of tungsten [50]. Tenne and coworkers [59–62] developed a precursor synthetic method for obtaining both IF-MS_2 (M = Mo, W) and nanotubes. The method is based on a gas-phase reaction between partially reduced metal oxides and hydrogen sulfide. The synthesis consists of three steps. First is the formation of the oxide MO_3. For M = W, this step involves the preparation of WO_3 by heating a tungsten filament in the presence of water vapor. The second step is the partial reduction of the oxide to an amorphous suboxide MO_{3-x}. For W, the reduction of WO_3 produces WO_{3-x} in the form of needle-like particles. The final step is the sulfidization of the suboxide to obtain the metal sulfide. To obtain the IF and nanotube forms of WS_2, the WO_{3-x} is reacted with H_2S under mild reducing conditions. The following equations summarize the steps in this example:

$$W \text{ (s)} + 3H_2O \text{ (g)} \rightarrow WO_3 \text{ (s)} + 3H_2 \text{ (g)} \tag{1}$$

$$WO_3 \text{ (s)} + H_2 \text{ (g)} \rightarrow WO_{3-x} \text{ (s)} + xH_2O \text{ (g)} + (1-x)H_2 \text{ (g)} \tag{2}$$

$$WO_{3-x} \text{ (s)} + (1-x)H_2 \text{ (g)} + 2H_2S \text{ (g)} \rightarrow WS_2 \text{ (s)} + (3-x)H_2O \text{ (g)} \tag{3}$$

For IF-MoS_2 and nanotubes the steps are similar, and a schematic representation of the experimental setup for the synthesis and growth mechanism has been proposed [62]. The nanotubes produced are multiwalled; Figure 1 provides a transmission electron micrograph that shows the structural features of these products. Related methods have been used to prepare several layered metal dichalco-

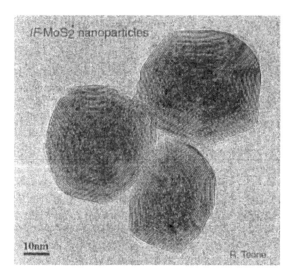

FIGURE 1   Transmission electron micrograph for multiwalled IF-MoS$_2$. (From Ref. 62, reprinted with permission from Elsevier.)

genides with IF structures. Recent examples include the synthesis of IF-SnS$_2$ and IF-VS$_2$ [63].

IF-MS$_2$ can also be synthesized from thiosalts, such as ammonium thiomolybdate, (NH$_4$)$_2$MoS$_4$, and ammonium thiotungstate, (NH$_4$)$_2$WS$_4$ [64,65]. This is a single-precursor approach whereby the salts are thermally decomposed at elevated temperature to yield amorphous MoS$_3$ and WS$_3$, respectively. The MoS$_3$ and WS$_3$ thereby obtained are heated in a hydrogen atmosphere between 1200 and 1300°C to obtain the MoS$_2$ or WS$_2$ nanotubes [64]. The reaction sequence is indicated by the two following steps:

$$(NH_4)_2MS_4 \text{ (s)} \rightarrow MS_3 \text{ (s)} + H_2S \text{ (g)} + 2NH_3 \text{ (g)} \qquad (4)$$

$$MS_3 \text{ (s)} + H_2 \text{ (g)} \rightarrow MS_2 \text{ (s)} + H_2S \text{ (g)} \qquad (5)$$

A one-step method involving the direct heating of thiosalts in a hydrogen atmosphere has also yielded MS$_2$ nanotubes, as follows:

$$(NH_4)_2MS_4(s) + H_2 \text{ (g)} \rightarrow MS_2 \text{ (s)} + 2H_2S \text{ (g)} + 2NH_3 \text{ (g)} \qquad (6)$$

where M = Mo or W. An advantage to this direct approach is that it avoids the formation of solid-state side products. In a variation of this method, the voids of a nanoporous aluminum oxide membrane were permeated with solutions of (NH$_4$)$_2$MoS$_4$ and (NH$_4$)$_2$Mo$_3$S$_{13}$ [65]. When these samples were subjected to thermal decomposition, a templated growth of MoS_2 was observed. Nanotubes were obtained by the selective dissolution of the membrane template in a basic solution.

Laser ablation of the layered structure under inert atmosphere has been used to prepare IF-MoS_2 and WS_2 with metal encapsulated and hollow onion-like structures at 450, 650, 850, and 1050°C [66]. The nanoparticles of WS_2 grown at 450 and 650°C contain three concentric WS_2 outer layers with metals encapsulated within. Laser ablation at higher temperature (1050°C) yields WS_2 nanoparticles with hollow closed-cage polyhedral nanotubes; these have diameters ranging from 10 to 15 nm and contain 4–8 concentric WS_2 shells. Similar results have been obtained for MoS_2 at 1050°C.

C. Syntheses of Metal Phosphorus Trichalcogenides (MPX_3)

Metal phosphorus trichalcogenide compounds form a class of layered materials of general formula MPX_3, where M = Mg, Ca, Zn, V, Mn, Cd, Ni, Sn, Fe, Pb, or Co and X = S or Se. Klingen and coworkers have published an extensive investigation regarding the synthesis and characterization of these compounds [67,68]. There are at least four general methods for their preparation. These include (a) the direct combination of elements, (b) vapor sublimation, (c) chemical vapor transport using a halogen as a transport agent, and (d) room-temperature precipitation from aqueous solution using metathesis reactions between cationic M^{2+} and the $(P_2S_6)^{4-}$ anion [69–71]. The synthetic routes based on the direct combination of the elements, vapor sublimation, and chemical vapor transport are similar to those discussed earlier for the metal dichalcogenides.

An amorphous MPS_3 phase can be prepared by dissolving soluble $A_2P_2S_6$ (A = Li, Na) in water and subsequently adding the appropriate divalent metal ions in order to precipitate the MPS_3. The amorphous material can then be crystallized into the MPS_3 structure by drying at elevated temperature [70,71]:

$$2M^{2+} \text{ (aq)} + P_2S_6^{4-} \text{ (aq)} \rightarrow 2MPS_3 \cdot xH_2O \text{ (s)}$$

For these reactions, the soluble $A_2P_2S_6$ (A = Li, Na) precursors are prepared at elevated temperature by the solid-state reaction of Li_2S or Na_2S with phosphorus and sulfur in a silica tube.

D. Syntheses of Layered Ternary Transition Metal Chalcogenides

Ternary transition metal chalcogenides can be represented by the general formula AMX_2, where A is typically an alkali and alkaline earth metal, M is a transition metal, and X is a chalcogen such as S, Se, or Te. These have highly anisotropic structures and are closely related to the metal dichalcogenides. There are four primary methods for preparing these ternary chalcogenides. These include: (a) the direct combination of elemental and binary metal chalcogenide precursors at high temperature, (b) the high-temperature sulfurization of oxide precursors such as alkali metal carbonates, transition metal oxides, and (c) the intercalation and ion exchange reactions of metal dichalcogenides.

Most ternary transition metal chalcogenides can be prepared either by the direct combination of the elements or by the combination of an element with a binary metal chalcogenide at temperature ranging from 600 to 1200°C. The reactant mixture is placed in a quartz tube and heated in an appropriate temperature gradient. In some cases a small amount of transporting agent is added to the reactants in order to assist crystal growth. Typical reaction times are several days to reach completion. Layered ternary niobium tellurides $NbMTe_2$ (M = Fe or Co) have been prepared by the high-temperature solid-state reaction of stoichiometric quantities of the elements by Li and coworkers [72]. Tellurides such as $TaNiTe_2$ and $NbNiTe_2$ have been synthesized by a similar method. The structure and properties of these materials are described in Ref. 72.

The ternary phases AMX_2 (A = Li, Na, or K; M = Cr, Ti, or V; X = S or Se) have been synthesized by the reaction of mixtures of the alkali metal carbonate and the transition metal oxide in a hydrogen sulfide gas atmosphere at elevated temperature [73]. The reactions are carried out in graphite crucibles and it is usually found to be advantageous to use an excess of the metal carbonate. Structural characterization by X-ray and neutron diffraction confirmed their layered structure. These ternary phases may serve as a synthetic intermediate to new metastable layered metal dichalcogenides by oxidative removal (i.e., deintercalation) of the alkali metal. For example, this method has been used to prepare VS_2 and $CrSe_2$, which are difficult to synthesize by conventional high-temperature combination of the elements [74,75].

Intercalation and ion exchange reactions can provide a viable method to prepare many ternary chalcogenides that cannot be obtained otherwise. Ternary phases can be obtained by exposing layered MX_2 to appropriate metal vapor [76], alkali metal in liquid ammonia solution, or organometallic reductant such as *n*-butyl lithium or sodium naphthalide [77–80]. One advantage of the intercalation approach is that the ternary phases can be obtained at low temperatures. Further details of this synthetic method will be described within the section on intercalation reactions.

E. Misfit Layered Compounds

Misfit layered compounds are ternary chalcogenides of the general formula $(AX)MX_2$ (A = Pb, Bi, Sn, or rare earth; M = Cr, Nb, Ta, or V; X = S or Se). This class of compounds has a unique structure composed of alternate stacking of regular (AX) slabs with hemioctahedral coordination of A and (MX_2) layers derived from edge-sharing trigonal prisms or octahedra.

Generally the misfit chalcogenides are synthesized by direct combination of the elements or the binary chalcogenides at elevated temperatures [81–83]. It can be helpful to use excess chalcogen for these preparations. The ternary sulfide phase may alternately be obtained by sulfurization of the corresponding ternary oxide precursor. Common sulfiding agents include CS_2, H_2S, and elemental sulfur. The use of a transport agent such as chlorine or iodine can enhance the crystal growth process.

III. STRUCTURES OF LAYERED METAL CHALCOGENIDES

The majority of metal chalcogenide phases belonging to the structural class MX_2, MPX_3, AMX_2, and AMX_3 crystallize in a layered structure. In each layer, the metal atoms are arranged hexagonally in sheets sandwiched between two hexagonal sheets of chalcogen such that it supports a quasi-2D electron band. The relative orientation of metal cations and chalcogenides can result in two different coordination types for the metal cations, either octahedral or trigonal prismatic geometries. Here, we will describe the structure of MX_2 and MPX_3 as examples.

A. Structures for Layered MX_2

Layered transition metal dichalcogenide (MX_2) consists of hexagonally arranged atomic X-M-X sheets constructed from MX_6 octahedra or trigonal prisms interconnected by edge sharing. Figure 2 illustrates 2H-MoS_2, which has trigonal prismatic coordination about Mo and a hexagonal unit cell. Atoms within the layers are held by strong covalent bonds, whereas the individual layers are held together by weak van der Waals forces. This structural anisotropy makes it possible to insert ionic or molecular species between the MX_2 layers. The metal ion coordination geometry is found to depend on the relative ionicity of the M-X bond.

There are, in addition, numerous polymorphs or polytypes of many MX_2 structures due to the possibility of different layer-stacking orientations. These include the hexagonal polytype with two layers per unit cell (2H), rhombohedral polytype with three layers per unit cell (3R), and the trigonal polytype (1T) with one layer per unit cell. Figure 3 illustrates the difference in layer stacking between the 2H and 3R polytypes of MoS_2.

B. Structures for Layered MPX_3

The common structure of MPX_3 compounds for first-row transition metals and X = S is related to that of cadmium chloride $(CdCl_2)$, with the metal ions and P—P pairs occupying the cadmium positions and the X (sulfur or selenium) ions

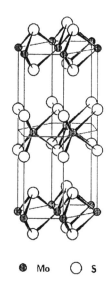

● Mo   ○ S

FIGURE 2 Structure of 2H-MoS$_2$. (From Ref. 148.)

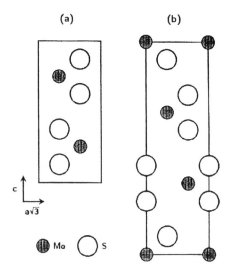

FIGURE 3 Stacking of MoS$_2$ layers in the 2H (a) and 3R (b) polytypes of MoS$_2$. (From Ref. 148.)

occupying the chloride positions. Each MPX_3 layer is made up of M^{2+} and $P_2S_6^{4-}$ ions, with the MS_6 octahedra containing three S ligands from other MS_6 octahedra and three S ligands from P_2S_6 groups. No strong bonds exist between the MPX_3 layers, leading again to an anisotropic, layered structure. An illustration of the $MnPS_3$ structure is provided in Figure 4.

IV. INTERCALATION REACTIONS

Layered transition metal dichalcogenides are an important class of host structure for intercalation reactions. The intercalation of guest cations and neutral molecules between the MS_2 layers can result in a significant modification of the compositional, electrical, magnetic, optical, and structural properties of the host compound. The intercalate guest species can include metal cations, organic molecules, organometallic compounds, and even polymers. The intercalation of alkali metals is achieved by the chemical or electrochemical reduction of the MS_2 host. One typical chemical method employs the lithiating agent n-butyl lithium, which acts as a strong reducing agent and generates intercalation compounds containing the Li^+ intercalate, as in

$$C_4H_9Li \text{ (solv)} + MoS_2 \text{ (s)} \rightarrow LiMoS_2 \text{ (s)} + (1/2)C_8H_{18} \text{ (solv)}$$

More complex intercalate guests can be introduced as well, either by ion exchange or by delamination of the compounds and subsequent reaggregation with new intercalate guests. Some examples of such compounds will be discussed further in the application section.

As suggested by the foregoing example, there are several methods to obtain intercalation compounds. These include the direct insertion of ions, and some-

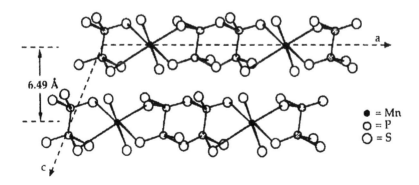

6.49 Å

● = Mn
○ = P
○ = S

FIGURE 4 $MnPS_3$ structure. (From Ref. 144.)

times accompanying neutrals, the application of ion exchange reactions, and a delamination and flocculation method. In these syntheses, solvent molecules may also be cointercalated along with the cation.

The MPX_3 structures also undergo a wide range of intercalation reactions. Table 2 lists some common intercalate species for these hosts.

V. POLYMER NANOCOMPOSITES OF LAYERED METAL CHALCOGENIDES

The energetic terms that favor intercalation chemistry can also be considered in the generation of nanocomposites, where the intercalating species are polymeric.

Table 2   Some Intercalates for MPX_3 Host Structures

Host lattice	Guest species	Basal Spacing (Å)	Host interlayer expansion (Å)	Ref.
$MnPS_3$		6.5		144
	K^+	9.3	2.8	86
	$Ru(bpy)_3Cl_2$	15.2	8.7	140
	PEO	15.4	8.9	86
	Styrylpyridinium cation	12.4	5.9	15, 136
	Pyridine	12.4	5.9	139
	Tetrathiafulvalene	12.9	6.4	138
$CdPS_3$		6.6		15
	K^+	9.4	2.8	144
	PEO	15.6	9.0	86
	Styrylpyridinium cation	12.6	6.0	15, 136
	Tris(bipyridyl) cation	14.6	8.0	137
	Methylviologen cation	9.7	3.1	137
$FePS_3$		6.4		138
	Tetrathiafulvalene	12.6	6.2	138
	$[Co(\eta^5 - C_5H_5)_2]$	11.7	5.4	141
	$[Cr(\eta^5 - C_5H_5)_2]$	12.4	6.0	141
	$C_4H_9NH_2$	10.4	4.0	142
$NiPS_3$		6.3		144
	2,2-bipyridyl	9.8	3.5	143
	$[Co(\eta^5 - C_5H_5)_2]$	11.7	5.4	141
	$[Cr(\eta^5 - C_5H_5)_2]$	12.3	6.0	141

Positively charged polymers, called *ionomers*, will intercalate to form favorable electrostatic interactions with the negatively charged inorganic surfaces. In other cases, the polymeric species can act to solvate smaller intercalate ions. In general, the introduction of nonpolar and uncharged polymers to form nanocomposites is more difficult, since these can neither form ionic interactions nor act effectively as solvents to cationic species. The adsorption or intercalation of polymers is often entropically favorable because the polymer can displace smaller molecules from the interlayer galleries.

The slow kinetics of polymer diffusion usually means that different synthetic strategies are employed in preparing nanocomposites. The direct intercalation reaction method is often not suitable for large intercalate guests, since the incorporation into the galleries that open between the chalcogenide layers can be slow. Novel synthetic strategies can involve either intercalation followed by in situ polymerization of the guest or delamination of the inorganic component in a polymer-containing solution. The latter method, called the *exfoliation-adsorption* method, has the advantage of using a preformed polymer. In either case, the nanocomposites result from the growth of one component (organic or inorganic) around the other, rather than the topotactic intercalation of a polymeric guest.

The first method for the incorporation of polymers within layered chalcogenides requires the intercalation of the monomeric precursor and subsequent in situ polymerization within the host. This in situ method is favorable when the polymeric guest is not tractable, such as with insoluble polymers. One important class of such polymers is that with pi-conjugated backbone structures, such as poly(aniline) and poly(pyrrole). These polymers are typically insoluble or require solvents that cannot be used for intercalation reactions with layered chalcogenides. The precursor monomers, however, are often easy to incorporate into the host structures, which can then be polymerized [85].

For some elastomeric polymers with low glass transition temperatures, T_g, it may be possible to form nanocomposites by direct reaction, similar to when smaller ions or molecules are used. For example, poly(ethylene oxide), $T_g = -65°C$, is an elastomer that can be directly intercalated between host MPS_3 sheets by simply heating pelletized mixtures of the polymer and host material above the glass transition temperature [86]. This method does not require solvents, and it is amenable to important technological processing methods, such as extrusion, which can produce useful shapes and lower cost for products. Although the mechanism for such reactions may be complex, facile segmental motion at the interaction temperatures appears to be required for direct syntheses of this type.

Figure 5 illustrates schematically the exfoliation-adsorption method. Layered host sheets are delaminated into single-sheet particles and then dispersed as a colloidal suspension. In the 1980s, the layered metal disulfides MoS_2 and WS_2 were observed to form such single-sheet colloidal suspensions when their lithiated forms were rapidly hydrolyzed [87–89]. More recently, a layered sheet surface-

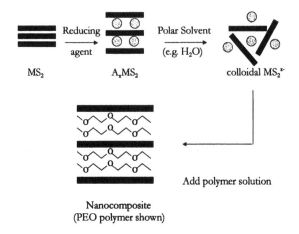

Nanocomposite
(PEO polymer shown)

FIGURE 5 Schematic representation of the exfoliation/adsorption method.

to-charge ratio of $40-120$ $\text{Å}^2/e^-$ has been proposed as appropriate for stabilizing colloidal dispersions for inorganic hosts [90]. Following this pretreatment, the interaction of entire sheet surfaces with soluble species becomes possible:

$$\text{Li}_x\text{MoS}_2 + x\text{H}_2\text{O} \rightarrow \text{MoS}_2 \text{ (coll)} + (x/2)\text{H}_2 + x\text{LiOH (aq)} \qquad (7)$$

$$\text{Polymer (aq)} + \text{MoS}_2 \text{ (coll)} \rightarrow \text{Polymer/MoS}_2 \text{ (coll)} \qquad (8)$$

$$\text{Polymer/MoS}_2 \text{ (coll)} \rightarrow \text{Poly/MoS}_2 \text{ (ppt)} \qquad (9)$$

The nanocomposite structure can be obtained by the reaggregation of the nanosheets with incorporation of the polymeric species. This reaggregation can be facilitated and sometimes controlled by solvent type, pH, or concentration changes.

Nanocomposite galleries typically require a separation of $3-10$ Å between encasing chalcogenide surfaces, and the structure can be very well ordered along the stacking direction. In such cases, the polymeric guests can form monolayers or bilayer structures. These are often referred to as *intercalated nanocomposites structures*. Poly(ethylenimine) forms 4.0- to 4.5-Å galleries in several layered chalcogenides, as does poly(ethylene oxide) when combined in a limiting stiochiometry. Simple steric considerations suggest that these must be polymer monolayers between inorganic layers. When poly(ethylene oxide) is combined in excess, however, the interlayer expansion can increase to 9 Å, indicating the presence of a polymer bilayer [91].

A list of proposed applications and nanocomposites obtained is provided in Table 3.

VI. PROPERTIES AND APPLICATIONS OF LAYERED METAL CHALCOGENIDES

The novel physical properties and applications of metal chalcogenides arise primarily because of their layered structures.

A. Tribological Applications

Layered metal chalcogenides such as MoS_2, NbS_2, and WS_2 (platelets of the 2H polytype) are used both as solid lubricants and as additives in liquid lubricants in several technological applications [55,92]. The layers of MX_2 are held together by weak (van der Waals) forces, and this makes it easy for them to slide past each other to reduce friction and wear. Because of their good oxidation resistance at extreme operating conditions and their excellent friction and wear resistance, these LMCs are used in applications such as lubricating gears, cams, ball bearings, threaded connections, and gimbals in moving mechanical assemblies [92]. However, rapid oxidation in air over 400°C and in moisture is detrimental to perfor-

Table 3 Potential Applications for Several Nanocomposites with LMC Hosts Prepared by the Exfoliation/Adsorption Method

Polymer intercalate	LMC host	Potential applications
Polyaniline	MoS_2	Rechargeable-battery electrodes, chemical sensors, magnetocalorific effect
Polystyrene	MoS_2	Catalysts, barrier applications
Poly(ethylene oxide)	MoS_2, $MoSe_2$ or TiS_2	Solid electrolytes, electrodes
	$NbSe_2$	Superconductors, solar energy cells
	$MnPS_3$ or $CdPS_3$	Electrochromic devices
Poly(ethylene glycol)	$NbSe_2$	Solid electrolytes, electrodes, superconductors, solar energy cells
Poly(vinyl pyrrolidinone)	$NbSe_2$	Solid electrolytes, electrodes, superconductors, solar energy cells
Poly(ethylenimine)	$MnPS_3$ or $CdPS_3$	Solid electrolytes, electrodes
	MoS_2, $MnPS_3$ or $CdPS_3$	Solar energy cell

mance. The larger dimensions of the MS_2 layered platelets can also hinder their inclusion in the pores of metal parts, and the accumulation and protrusion of particles can lead to higher surface reactivities and degraded performance of the lubricated parts [62,93].

The application of IF-MoS_2 and WS_2 powders as solid lubricants has also been demonstrated. These nanoparticles, with microstructures resembling those of carbon fullerenes or nanotubes, are reported to have superior tribological properties within a defined loading range (PV $\sim$ 150 Nm/s) as compared with that of typical platy metal dichalcogenides [94]. Unlike the 2H-WS_2 and 2H-MoS_2 platelets, the IF nanoparticles have a pseudo-spherical or -cylindrical shape as well as an inert sulfur–terminated surface [95]. The rounded microstructures and absence of dangling bonds lead to superior performance when compared with platy metal dichalcogenides. As solid lubricants, the IF-MS_2 nanoparticles act like nano-ball bearings, which may exfoliate or mechanically deform into a rugby-shape ball in the presence of an applied mechanical stress [94]. Apart from their application as solid lubricants, IF-MS_2 nanoparticles may be used to modify lubrication fluids, oils, and greases.

B. Insertion Materials for Positive Electrodes

Many layered metal chalcogenides may be used as a cathodic (positive) insertion electrode in rechargeable alkali metal batteries because of their ability to reversibly intercalate alkali metal cations [11,12,96]. Any useful electrode material must fulfill most of the following requirements:

> Allow intercalation/deintercalation of a large amount of alkali metal cation for maximum energy density (a wide range of reversibility for x in A_xMS_2)
> Good solid-state electronic and Li^+ ionic conductivities for higher rate performance
> Highly reversible intercalation/deintercalation process to sustain the specific charge during charge/discharge cycles
> A high free energy of reaction to provide significant cell potential
> Stability and insolubility in the electrolyte
> Limited intercalation or cointercalation of the electrolyte
> Limited toxicity and safety issues
> Low materials cost

Several transition metal dichalcogenides, such as MX_2 (M = Ta, Ti, V, Mo, or Cr; X = S or Se) meet some or several of these requirements. Prototype cells that operate at ambient or elevated temperatures have been tested with these materials [11,12]. At present, one of the more promising cells is based on $LiTiS_2$, which provides relatively high energy density and reversibility [12]. A lithium battery based on the Li/Li$_x$TiS$_2$ intercalation reaction consists of a lithium metal anode and a TiS$_2$ cathode, both of which are immersed in a lithium ion–containing electrolyte solution. The electrode reactions during cell discharge are as follows:

Anode reaction: $Li \rightarrow Li^+ + e^-$
Cathode reaction: $Li^+ + e^- + TiS_2 \rightarrow LiTiS_2$

The reaction is highly reversible, and cycling involves intercalation and deintercalation of Li$^+$ ions in and out of the TiS$_2$ interlayer space to give Li$_x$TiS$_2$ ($0 \leq x \leq 1$). The electronic conductivity of TiS$_2$ increases upon reductive intercalation of the layers. Cells can operate at room temperature with an open-circuit voltage of about 2.0 V, an energy density up to approximately 480 Wh/kg, and cycle life of 350–400 cycles. A major drawback for using TiS$_2$ as a cathode material lies in the relatively low open-circuit voltage of 2 V. Table 4 gives the theoretical energy densities of several metal dichalcogenide cathodes.

Commercial cells currently employ LiCoO$_2$, with an open-circuit potential of approximately 4 V, and the major research focus has been on the use of nickel or manganese oxides.

Layered metal dichalcogenides have been used as electrode materials in solid-state battery systems and electrochemical photocells for converting solar energy into chemical energy. In solar cells based on WS$_2$, energy conversion efficiency as high as 17% has been achieved [97].

Recent studies have explored changes in the electrochemical properties of metal dichalcogenide cells that arise from the intercalation of polymer electrolytes between layered dichalcogenide sheets [98]. Results for the incorporation of solid polymer electrolyte poly(ethylene oxide), PEO, for example, into TiS$_2$ and MoS$_2$, have indicated an enhancement of ionic conductivities. In the case of the MoS$_2$ nanocomposites Li$_x$MoS$_2$(PEO)$_{0.5}$ and Li$_x$MoS$_2$(PEO)$_{1.0}$, the diffusion coeffi-

Table 4   Electrochemical Parameters for Several Metal Dichalcogenide Cathodes

Cathode material	ΔG for intercalation (kJ/mol)	Initial cell voltage (v)	Final cell voltage (v)	Theoretical energy density (Wh/kg)	Current density (mA/cm^2)	%utilization in 1st discharge
TiS$_2$	206	2.5	1.9	481	10	100
TiSe$_2$	186	2.1	1.7	243	0.5	92
NbSe$_2$	200	2.3	1.7	216	0.5	85
TaS$_2$	211	2.9	1.5	233	1	60
VS$_2$	222	2.5	2.1	505	0.2	42
VSe$_2$	191	2.0	2.0	246	2	97

cient, D, increases with temperature but decreases with increasing lithium ion content (x value). However, lithium ion diffusion rates are often significantly higher for the nanocomposites than for the pristine MoS_2 [98].

Transition metal trichalcogenides such as TiS_3 and $NbSe_3$ with a one-dimensional host structure and the layered MPX_3 (M = Ni or Fe; X = S or Se) compounds have also demonstrated interesting electrochemical properties for battery application. Some details of the electrochemical performance on these and other MS_2 materials can be found in the literature [13].

C. Electronic and Optical Properties

The electronic properties of layered metal dichalcogenides cover a broad range. Depending on the metal and the chalcogen involved, MX_2 compounds exhibit electrical properties ranging from metallic to semimetallic to semiconducting behavior [99–101]. For example, isoelectronic MoS_2, $MoSe_2$, WS_2, and WSe_2 are semiconductors with bandgaps between 1.0 and 1.5 eV [100], whereas MX_2 (M = Ta or Nb; X = S, Se, or Te) are metallic. Studies on the electronic structure of TiS_2 indicate it is a narrow-gap semiconductor with a semiconductor–semimetallic phase transition [100]. $TiSe_2$ is a semimetal.

When electron donors or Lewis bases such as alkali metals or organic compounds are intercalated into the interlayer space of these layered compounds, the electronic and optical properties of the host are significantly modified. For example, charge transfer due to intercalation can raise the Fermi level, E_F, thus increasing the population of mobile electrons by several orders of magnitude. Semiconducting layered MX_2 may be used as an active element in such solid-state devices as Schottky and p-n photovoltaic and photoelectrochemical solar cells [102]. In these applications, the diselenides are more efficient in solar cell fabrication than the disulfides. Advantage can also be taken of the high absorption constant of most layered semiconducting MX_2 by using them as absorber materials in solar energy conversion [103]. The high mobility of the carriers and optical absorption characteristic of these materials make them ideal candidates for low-dimensional conductors and in optoelectronic applications. Other technology applications of layered MX_2 include photocatalysis, thermoelectrics, and optical storage materials [102].

D. Superconductivity

The group V transition metal dichalcogenides MX_2 (M = Ta or Nb; X = S, Se, or Te) are metallic and exhibit a charge density wave phenomenon. They are superconductors, and their transition temperature (T_c) depends on the polymorph studied. For example, both the superconducting metals TaS_2 and NbS_2 are polymorphic. The superconducting property and the critical transition temperature of a given MX_2 can be altered by inserting an organic molecule or inorganic cation into the interlayer space. The T_c for pristine TaS_2 is approximately 0.8 K. Upon intercalation, the T_c can be as high as 5.3 K, depending on the nature of the intercalated species [27,104]. By intercalation chemistry one can also induce semiconductor-to-metal and metal-to-superconductor transition. The polytype $2H$-WS_2 is a semiconductor, but reductive intercalation of potassium ions between layers results in a semiconductor-to-metal transition [105]. Similarly, the nonsuperconducting $2H$-MoS_2 becomes superconducting with a T_c ranging from 3.7 to 6.3 K following intercalation with alkali metal cations [106]. Table 5 shows the superconducting T_c values for some intercalated MX_2 compounds. Metallic $NbSe_2$ has the highest superconducting temperature of all layered metal dichalcogenides, with a T_c of 7.2 K. A metallic polymer-$NbSe_2$ nanocomposite has been prepared from exfoliated $NbSe_2$ and water-soluble polymers such as poly(vinyl pyrrolidone) (PVP), poly(ethylene oxide) (PEO), and poly(ethylene glycol) (PEG). The nanocomposites exhibit electronic conductivities ranging from 140 to 250 S/cm and a T_c of 6.5–7.1 K [14]. Some potential applications of these materials include their use in electromagnetic shielding and in magnetic resonance imaging (MRI).

Table 5 T_c Transition Temperatures for Some Intercalated MX_2 Compounds

Host	Intercalate	T_c onset	Ref.
2H-MoS$_2$	Li	3.7	27
	Na	4.1	27
	K	6.1	27
	Rb	6.3	27
	Cs	6.3	27
2H-NbS$_2$		6.1	144
	Pyridine	4.1	144
2H-NbSe$_2$		7.2	144
	PVP	7.1	14
	PEO	6.5	14
	PEG	7.0	14
SnSe$_2$		n/a	144
	Co(Cp)$_2$	8.1	144
2H-TaS$_2$		0.8	145
	NH$_3$	4.2	145
	Pyridine	3.5	146
	Aniline	3.1	147
	KOH	5.3	147
	LiOH	4.5	147
	Hydrazine	4.7	147
	Stearamide	3.0	147

E. Scanning Probe Microscopy

Inorganic nanotubes derived from WS_2 may find application as tips in scanning probe microscopy. The feasibility of using such tips in inspecting microelectronic circuitry has now been demonstrated. The tips produced from WS_2 nanotubes can outperform the microfabricated Si tip counterparts with respect to resilience and surface passivity [107]. Additional advantages of WS_2 nanotubes include tunable electrical conductivities and strong light absorption in the visible spectrum. These properties may provide additional application opportunities in areas such as nanoelectronics and photocatalysis.

F. Catalytic Applications of Layered Metal Disulfides

Layered metal disulfides (MS_2) are used as heterogeneous catalysts in several petroleum refining and industrial chemical processes. The catalytic properties of these materials are largely attributed to their anisotropic structures. Some of the major reactions catalyzed by layered MS_2 include catalytic hydrodesulfurization (HDS), hydrodenitrogenation (HDN), and hydrodemetallation (HDM). Heteroatoms, such as N and S, or metals are thereby removed from petroleum feedstocks. Additional catalytic reactions include hydrogenation of hydrocarbons and ketones and dealkylation and ring opening of aromatics [6,7,108–110]. Layered MS_2 compounds have also been used to catalyze numerous additional organic reactions; examples include Fischer–Tropsch and alcohol syntheses, the isomerization of parafins, amination, the dehydrogenation of alcohol, the hydrogenation of olefins, coal liquefaction, and syntheses of thiols and thiophene [108–110].

The need for efficient hydrodesulfurization processes first became evident in the latter half of the 20th century, for several reasons. First, there was new environmental legislation regulating the level of sulfur in gasoline in an effort to control pollution. Second, the bimetallic hydrotreatment catalysts in use were extremely sensitive to the sulfur content in the feedstock and were rapidly poisoned or deactivated. And finally, lower sulfur content in feedstocks is helpful in odor and corrosion control. Efforts to achieve these objectives led to the discoveries that some MS_2 compounds, particularly MoS_2, ReS_2, TaS_2, and WS_2, were efficient HDS catalysts [108]. HDS catalysts have been widely investigated, and several reviews are available (see references in the next section). The following discussion will focus mainly on the catalytic hydrodesulfurization reaction.

1. Catalytic Hydrodesulfurization (HDS)

The hydrodesulfurization process is the removal of sulfur from organic sulfur compounds, such as thiols, sulfides, thiophene, and substituted thiophenes from petroleum feedstocks, to avoid or minimize sulfur poisoning of metallic or bimetallic reforming catalysts and to avoid introduction of S into the environment.

The most widely used catalysts in the petroleum industry are MoS_2 and WS_2. Cobalt (Co) or nickel (Ni) metal are usually added to the sulfide as a promoter. The Co-or Ni-containing MS_2 is dispersed uniformly on high-surface-area carriers, such as alumina (Al_2O_3). The Co-or Ni-promoted, alumina-supported MoS_2 system may be represented as $Co(Ni)$-MoS_2/Al_2O_3. Generally the activity and selectivity of the promoted catalyst is higher for hydrodesulfurization and hydrogenation reactions as compared to the nonpromoted phase. It is generally accepted that the improved activity is related to the presence of the $Co(Ni)$—Mo—S structures made up of nanoclusters of MoS_2 or WS_2, with promoter atoms (Co/Ni) located at the particle edges [110]. Thus, the active phase of the MoS_2 catalyst is designated as CoMoS or NiMoS, depending on the promoting metal used. The Co or Ni promoter atom located at the MS_2 particle edges interacts with the host by inducing charge transfer to Mo and W in MoS_2 and WS_2. However, a detailed understanding of the origin for this promotion effect remains to be fully determined. Available analytical tools have proven to be inadequate in determining this effect, and only a limited set of data on the promoted catalysts is available. For example, the promoted system does not form a crystallographic phase amenable to X-ray studies, and only indirect structural information is obtained from spectroscopic studies [111,112]. However, with recent developments in density functional theory calculations and the deployment of surface probes such as AFM and STM, further progress can be expected in understanding the details of the promotion effect of Co or Ni on MS_2 catalysts [113,114]. The commercial success of layered MS_2 catalysts is largely due to their excellent resistance to sulfur poisons and carbon deposition, because other metallic or oxide catalysts are often rapidly deactivated in the presence of sulfur compounds.

Many HDS investigations have used thiophene and its derivatives as model reactants because of the difficulties associated with the desulfurization of these compounds in petroleum feedstocks. The reaction sequence includes HDS followed by hydrogenation reactions. Figure 6 illustrates the HDS of thiophene, benzothiophene, and dibenzothiophene in the presence of layered MS_2 catalyst. The general mechanism for HDS reaction is widely accepted to involve hydrogenolysis of the C—S bond. Detailed mechanisms for HDS catalysis are available in the literature [110,113].

Several studies on the hydrogenation and isomerization of dienes employing the MoS_2/Al_2O_3 catalyst system have been reported [115]. Their catalytic activity and the product distribution are greatly influenced by the sulfur content of the ($\bar{1}10$) edge plane of the MoS_2 slabs, where all the Mo ions are identical. Both hydrogenation and isomerization reactions are sensitive to the coordination chemistry of Mo ions [115]. There are several factors that can affect the catalytic activities of layered metal disulfides. These include the electronic structure of MS_2, the nature and number of adsorption and catalytic sites on the MS_2 particles, the crystal structure of MS_2, the size and morphology of MS_2, the support material,

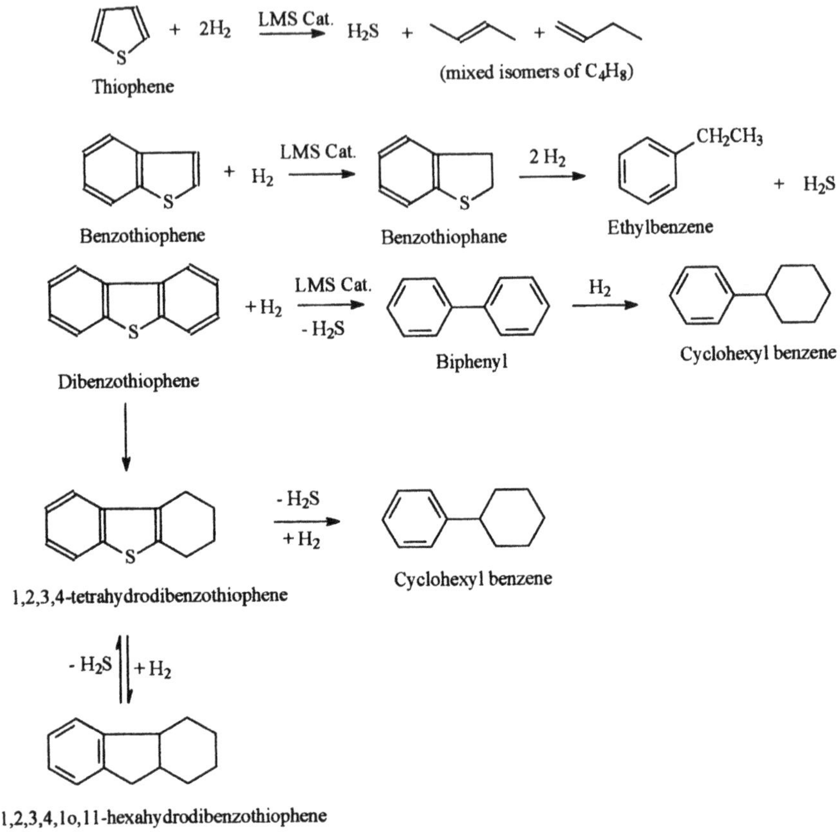

FIGURE 6 Hydrodesulfurization (HDS) mechanisms for thiophene, benzothiophene, and dibenzothiophene in the presence of layered MS_2 catalyst.

and the presence of sulfur vacancies. These factors can influence both the activities and selectivity of MS_2 catalysts. Chianelli and his coworkers have published an extensive review on the role of these factors [6].

2. Catalytic Hydrodenitrogenation

Hydrodenitrogenation (HDN) is a hydrotreating process used to reduce or remove organic nitrogen-containing compounds from petroleum feedstocks, heavy oil feedstocks, coal-derived distillates, and oil shale [116–119]. The removal of nitrogen during refining is particularly important in order to reduce ultimate NO_x emissions and to prolong the lifetime (i.e., to avoid poisoning) of acidic catalysts.

Molybdenum- and tungsten-based hydrotreating catalysts supported on Al_2O_3 are widely used in the industry for HDS and HDN processes. Cobalt and nickel are the promoters of choice for HDS and HDN catalyst system based on MS_2 (M = Mo or W), and their presence greatly increases both HDS and HDN activity. For HDN, Ni is the promoter of choice, whereas Co is preferred for HDS. Other additives, such as fluorine, chlorine, boron, and phosphorus, have been incorporated into both promoted and unpromoted MoS_2-type catalysts with beneficial effect on HDN and HDS processes [116].

The mechanism for HDN has been investigated extensively using various aliphatic and aromatic nitrogen-containing model compounds. Aliphatic amines are very reactive and are not a major component of coal or petroleum feedstocks. However, they are formed as intermediates during HDN of aromatic and alicyclic organic N-compounds. Generally, the mechanism of HDN follows a Hoffmann degradation reaction [116,120]. For aromatic N-compounds such as aniline, pyridine, pyrrole, indole, and carbazole, the initial step involves the saturation of the heterocyclic ring. This is followed by quanternization of the leaving nitrogen and deamination by elimination or nucleophilic substitution. All HDN reactions involve hydrogenolysis or C—N bond cleavage and may proceed via elimination or nucleophilic substitution reaction, depending on the nature of the catalyst and the structure of the organo-nitrogen compounds involved [116].

Studies have shown that the HDN of. 1,2,3,4-tetrahydroquinoline and 1,2,3,4-tetrahydroisoquinoline catalyzed by sulfided $NiMo/Al_2O_3$ occur via a nucleophilic substitution mechanism [121]. On the other hand, HDN of aliphatic amines with the same catalyst—$NiMo/Al_2O_3$—occurs by β-elimination [117]. The nature of the base and the amine structure dictate whether the elimination will proceed via a monomolecular (E1) or a bimolecular (E2) mechanism. Similarly, for HDN reactions that occur via nucleophilic substitution, these same factors determine if the reaction will follow a monomolecular (S_N1) or a bimolecular (S_N2) mechanism.

3. Fischer–Tropsch Process/Methanation

The Fischer–Tropsch process was developed by F. Fischer and H. Tropsch in 1921 to produce clean alternative fuel from coal, natural gas, and low-grade refinery products for use in automobile and diesel engines. The process entails the synthesis of hydrocarbons and other aliphatic compounds, such as alcohols, from a mixture of hydrogen and carbon monoxide (synthesis gas, or *syngas*). The following equation illustrates the chemical reactions involved in the process:

$$n\mathrm{CO(g)} + (2n + 1)\mathrm{H_2\ (g)} \rightarrow \mathrm{C}_n\mathrm{H}_{2n+2} + n\mathrm{H_2O\ (g)} \tag{10}$$

For the methanation reaction, $n = 1$:

$$\mathrm{CO\ (g)} + 3\mathrm{H_2\ (g)} \rightarrow \mathrm{CH_4\ (g)} + \mathrm{H_2O\ (g)} \qquad \text{(methanation)} \tag{11}$$

and the reaction products are methane and steam. By adjusting the reactant feed, different types of hydrocarbons can be produced as well:

$$4CO \text{ (g)} + 9H_2 \text{ (g)} \rightarrow C_4H_{10} \text{ (g)} + 4H_2O \text{ (g)} \tag{12}$$

$$6CO \text{ (g)} + 13H_2 \text{ (g)} \rightarrow C_6H_{14} \text{ (g)} + 6H_2O \text{ (g)} \tag{13}$$

These reactions require the presence of a catalyst to proceed at useful rates. Catalysts such as noble metals, Co, Ni, as well as the metal disulfides MoS_2 and WS_2, have been demonstrated to be highly selective and active for CO conversion in these Fischer–Tropsch and methanation reactions [122–125]. The reaction products are strongly influenced by the nature of the catalyst structure and number of active sites. Miremadi and Morrison have reported the formation of a new $MoWS_4$ structure by flocculating exfoliated single-layer suspension of MoS_2 and WS_2 [125]. This material, believed to consist of alternating layers of MoS_2 and WS_2, was reported to exhibit an extremely high catalytic activity for the methanation reaction [125].

4. Pillared MS_2 Structures

Pillaring of aluminosilicate clays and layered double hydroxides results in microporous materials that can be applied as reaction catalysts or as size-selective molecular sieves [126–129]. Pillaring is an intercalation reaction whereby highly charged large ionic species (known as pillaring agents), which are typically inorganic cations, are inserted between the sheets of a layered host lattice. The rigid pillars have the effect of opening the galleries to provide internal pore volume accessible by smaller molecules. These pillars afford structural integrity to the expanded galleries and can result in size- and shape-selective access to the internal pore volume. These can clearly be very useful and desirable properties for the rational control of catalytic activity [126]. The pillaring process can be accomplished by well-established procedures such as ion exchange or the exfoliation–adsorption method. Some host materials, such as clays, can swell in water or other polar solvents significantly, and these highly expanded structures can make the pillaring process relatively facile via ion exchange of small for pillaring ion. In some cases where the host lattice can be delaminated into single layers, the exfoliation–adsorption method can be used to produce the pillared structure by reconstituting the sheets in the presence of the pillaring ion.

There is a growing interest in pillaring layered metal dichalcogenides MS_2 (M = Mo, W, or Ta) as a way to improve and induce new properties and applications. In 1986, Nazar and Jacobson reported the first example of a pillared MS_2 material [128]. That study involved the intercalation a large iron–sulfur cluster, $[Fe_6S_8(P(C_2H_5)_3)_6]^{2+}$, between TaS_2 sheets using the exfoliation–adsorption method. These large divalent cationic species are not incorporated directly into the host; rather, the product is obtained by delamination of the starting host into a colloidal suspension and subsequent reaction with a solution containing the cationic species. The same strategy has also been used to intercalate pillaring agents such as the polyoxocation $Al_{13}O_4(OH)_{24}(H_2O)_{12}^{7+}$ or the cobaltocenium cation, Cp_2Co^+, into this host [129]. Pillared MoS_2 and WS_2 are of particular interest because both are well-known commercial catalysts for HDS, HDN hydrogenation, methanation, and water–gas shift reactions [113,116,124]. More recently, pillared layered sulfides derived from MoS_2 and WS_2 hosts and Co clusters have been prepared by the addition of a soluble form of the ionic clusters to a colloidal dispersion of MoS_2 [130–135]. These products can be used to model the active sites in commercial hydrotreatment catalysts. Other polyoxocations, such as $Al_{13}O_4(OH)_{24}(H_2O)_{12}^{7+}$, as well as nanoparticle CdS quantum dots, have also been intercalated into MoS_2 and WS_2. These novel pillared materials show an increase in surface area (as measured by BET methods), increased thermal stabilities, and high electronic conductivities. This combination of properties suggests that such microporous, pillared materials may find additional applications in photocatalysis and microelectronics [131,135].

REFERENCES

1. RH Friend, AD Yoffe. Adv Phys 1987; 18:1–94.
2. MS Whittingham, AJ Jacobson, eds. Intercalation Chemistry. New York: Acadademic Press, 1982.
3. C Schlenker, J Dumas, M Greenblatt, S van Smaalen, eds.. Physics and Chemistry of Low-Dimensional Inorganic Conductors. NATO ASI Series B: Physics. Vol. 354. New York: Plenum Press, 1996.
4. DR Allan, AA Kelsey, SJ Clark, RJ Angel, GJ Ackland. Phys Rev B 1998; 57: 5106–5110.
5. JA Wilson, FJ Di Salvo, S Mahajan. Adv Phys 2001; 50:1171–1248.
6. RR Chianelli, M Daage, MJ Ledoux. Adv Catal 1994; 40:177–232.
7. A Sobczynski, A Yildiz, AJ Bard, A Champion, MA Fox, T Mallouk, SE Weber, JM White. J Phys Chem 1988; 92:2311–2315.
8. O Weisser, S Landa. Sulfide Catalysts: Their Properties and Applications. Oxford: Pergamon Press, 1973.
9. IL Singer. Solid lubrication processes. In: IL Singer, HM Pollock, eds. Fundamentals of Friction: Macroscopic Processes. Dordrecht: Kluwer, 1992:237.
10. JS Zabinski, MS Donley, VJ Dyhouse. Thin Solid Films 1992; 214:156–163.
11. MS Whittingham. Sol St Ionics 2000; 134:169–178.
12. MS Whittingham. Prog Sol State Chem 1978; 12:41–99.
13. M Winter, JO Besenhard, ME Spahr, P Novak. Adv Mater 1998; 10:725–763.
14. H Tsai, JL Schindler, CR Kannewurf, MG Kanatzidis. Chem Mater 1997; 9: 875–878.
15. I Lagadic, PG Lacroix, R Clement. Chem Mater 1997; 9:2004–2012.
16. B Elvers, S Hawkins, G Schulz, eds. Ullmann's Encyclopedia of Industrial Chemistry VCH. Vol. A16. New York, 1990.

17. AH Thompson, FR Gamble, CR Symon. Mat Res Bull 1975; 10:915.
18. K Kourtakis, J Dicarlo, R Kershaw, K Dwight, A Wold. J Sol St Chem 1988; 76: 186.
19. LE Conroy, KC Park. Inorg Chem 1968; 7:459.
20. B Palosz. Phys Status Sol A 1983; 80:11.
21. G Said, PA Lee. Phys Status Sol A 1973; 15:99.
22. DL Greenway, R Nitsche. J Phys Chem Solids 1965; 26:1445.
23. CR Whitehouse, AA Balchin. J Crys Growth 1979; 47:203.
24. J George, CK Kumari. Sol St Commun 1984; 49:103.
25. Y Ishizawa, Y Fugiki. J Phys Soc Jpn 1973; 35:1259.
26. B Palosz, W Palosz, S Gierlotka. Bull Mineral 1986; 109:143.
27. AA Balchin. In: F Levy, ed. Crystallography and Crystal Chemistry of Materials with Layered Structures. Dordrecht: Reidel, 1976:1–50.
28. LH Brixner. J Inorg Nucl Chem 1962; 23:257.
29. R Nitsche. J Phys Chem Sol 1960; 17:163.
30. R Nitsche, HU Bolsterli, M Lichtensteiger. J Phys Chem Sol 1961; 21:199.
31. BE Brown, DJ Beernsten. Acta Cryst 1965; 18:31.
32. RR Chianelli, MB Dines. J Inorg Chem 1978; 17:2758–2762.
33. DM Schleich, MJ Martin. J Sol St Chem 1986; 64:359.
34. A Bensalem, DM Schleich. Mat Res Bull 1988; 23:857–868.
35. A Bensalem, DM Schleich. Mat Res Bull 1990; 25:349–356.
36. RL Bedard, LD Vail, ST Wilson, EM Flanigen. US Pat No 4880761, (14 Nov. 1989).
37. MS Donley, NT McDevitt, TW Haas, PT Murray, JT Grant. Thin Solid Films 1989; 168:335–344.
38. MS Donley, PT Murray, SA Barber, TW Haas. Surf Coat Technol 1988; 36:329.
39. WK Hoffman. J Mater Sci 1988; 23:3981–3986.
40. J Cheon, JE Gozum, GS Girolami. Chem Mater 1997; 9:1847–1853.
41. R Bichsel, F Levy. J Phys D Appl Phys 1986; 19:1809.
42. HW Kroto, AW Allaf, SP Balm. Chem Rev 1991; 91:1213–1235.
43. D Ugarte. Nature 1992; 359:707–709.
44. S Iijima. Nature 1991; 354:56–58.
45. Y Feldman, E Wasserman, D Srolovitz, R Tenne. Science 1995; 267:222–225.
46. M Remskar, Z Skraba, F Cleton, R Sanjines, F Levy. Appl Phys Lett 1996; 69: 351–353.
47. CM Zelenski, PK Dorhout. J Am Chem Soc 1998; 120:734–743.
48. M Remskar, Z Skraba, C Ballif, R Sanjines, F Levy. Surf Sci 1999; 435:637–641.
49. M Remskar, Z Skraba, M Regula, R Sanjines, F Levy. Adv Mater 1999; 10: 246–249.
50. R Tenne, L Margulis, M Genut, G Hodes. Nature 1992; 360:444–446.
51. PM Ajayan, O Stephen, P Redlich, C Colliex. Nature 1995; 375:564–567.
52. O Stephen, PM Ajayan, C Colliex, P Redlich, JM Lambert, P Bernier, P Lefin. Science 1994; 266:1683–1685.
53. S Harris, RR Chianelli. J Catal 1986; 98:17–31.
54. R Tenne, M Homyonfer, Y Feldman. Chem Mater 1998; 10:3225–3238.

55. R Tenne, AK Zettl. Topics Appl Phys 2001; 80:81–112.
56. M Chhowalla, GA Amaratunga. Nature 2000; 409:164–167.
57. M Jose-Yacaman, H Lorez, P Santiago, DH Galvin, IL Garzon, A Reyes. Appl Phys Lett 1996; 69:1065–1067.
58. PA Parrila, AC Dillon, KM Jones, G Riker, DL Schulz, DS Ginley, MJ Heben. Nature 1999; 397:114–114.
59. A Rothschild, J Sloan, R Tenne. J Am Chem Soc 2000; 122:5169–5179.
60. Y Feldman, V Lyakhovitskaya, R Tenne. J Am Chem Soc 1998; 120:4176–4183.
61. A Zak, Y Feldman, V Alperovich, R Rosentsveig, R Tenne. J Am Chem Soc 2000; 122:11108–11116.
62. R Tenne. Coll Surf A 2002; 208:83–92.
63. Y Feldman, GL Frey, M Homyonfer, V Lyakhovitskaya, L Margulis, H Cohen, G Hodes, JL Hutchison. J Am Chem Soc 1996; 118:5362–5367.
64. M Nath, A Govindaraj, CNR Rao. Adv Mater 2001; 13:283–286.
65. CM Zelenski, PK Dorhout. J Am Chem Soc 1998; 120:734–742.
66. R Sen, A Govindaraj, K Suenaga, S Suzuki, H Kataura, S Iijima, Y Achiba. Chem Phys Lett 2001; 340:242–248.
67. W Klingen, R Ott, H Hahn. Z Anorg Allg Chem 1973; 396:271–278.
68. BE Taylor, J Steger, A Wold. J Sol St Chem 1974; 7:461–467.
69. W Klingen, R Ott, H Hahn. Z Anorg Allg Chem 1973; 396:271–278.
70. BE Taylor, J Steger, A Wold. J Sol St Chem 1973; 7:461–467.
71. R Clement, O Garnier, J Jegoudez. Inorg Chem 1986; 25:1404–1409.
72. E Prouzet, G Ouvrard, R Brec, P Seguineau. Sol St Ionics 1988; 31:79–90.
73. J Li, ME Badding, FJ DiSalvo. Inorg Chem 1992; 31:1050–1054.
74. W Bronger. In: F Levy, ed. Crystallography and Crystal Chemistry of Materials with Layered Structures. Dordrecht: Reidel, 1976:93–125.
75. R Schollhorn, A Lerf. Z Naturforsch 1974; 29B:804.
76. DW Murphy, C Cros, FJ DiSalvo, JV Waszczak. Inorg Chem 1976; 16:3027.
77. WS Onuchi, W Jaegermann, C Pettenkofer, BA Parkinson. J Am Chem Soc 1989; 5:439.
78. MB Dines. Mater Res Bull 1975; 10:287–292.
79. W Murphy, PA Christian, FJ DiSalvo, JV Waszczak. Inorg Chem 1976; 15:17–21.
80. W Rudorff. Chimia 1965; 19:489.
81. W Murphy, PA Christian. Science 1979; 205:651.
82. Y Goto, M Gotoh, K Kawagachi, Y Oosawa, M Onoda. Mater Res Bull 1990; 25:307–314.
83. K Suzuki, T Enoki. Mater Sci Forum 1992; 91:369–374.
84. K Suzuki, T Enoki, K Imaeda. Sol St Comm 1991; 78:73–77.
85. C Oriakhi. J Chem Edu 2000; 77:1138–1146.
86. CO Oriakhi, MM Lerner. Chem Mater 1996; 8:2016–2022.
87. MA Gee, RF Frindt, SR Morrison. Mat Res Bull 1986; 21:543–549.
88. P Joensen, RF Frindt, SR Morrison. Mat Res Bull 1986; 21:457–461.
89. WMR Divigalpitiya, RF Frindt, SR Morrison. Science 1989; 246:369–371.
90. A Jacobson. Mater Sci Forum 1994; 1:152.
91. JP Lemmon, MM Lerner. Chem Mater 1994; 6:207–210.

92. SR Cohen, Y Feldman, H Cohen, R Tenne. Appl Surf Sci 1999; 145:603–607.
93. M Chhowalla, GAJ Amaratunga. Nature 2000; 407:164–167.
94. L Rapoport, Y Feldman, M Homyonfer, H Cohen, J Sloan, JL Hutchison, R Tenne. Wear 1999; 225:975–982.
95. R Tenne. Prog Inorg Chem 2001; 50:269–315.
96. I Samaras, SI Saikh, C Julien, M Balkanski. Mat Sci Eng 1989; B3:209–214.
97. A Aruchamy, ed. Physics and Chemistry of Materials with Low-Dimensional Structure. Dordrecht: Kluwer, 1992.
98. E Ruiz-Hitzky, P Aranda, B Casal, JC Galvan. Adv Mater 1995; 7:180–184.
99. G Betz, H Tributsch. Prog Sol St Chem 1985; 16:195–290.
100. A Klein, S Tiefenbacher, V Eyert, C Pettenkofer, W Jaegermann. Phys Rev B 2002; 64:205411–205414.
101. RA Gordon, D Yang, ED Crozier, TD Jiang, RF Frindt. Phys Rev B 2002; 65: 125407/1–125407/9.
102. A Aruchamy, ed. Photoelectrochemistry and Photovoltaics of Layered Semiconductors. Dordrecht: Kluwer, 1992.
103. D Canfield, KK Kam, G Kline, BA Parkinson. Sol Energy Mater 1981; 4:301.
104. A Zak, Y Feldman, V Lyakhovitskaya, G Leitus, R Popovitz-Biro, E Wachtel, H Cohen, S Reich, R Tenne. J Am Chem Soc 2002; 124:4747–4758.
105. FS Ohuchi, W Jaegermann, C Pettenkofer, BA Parkinson. J Am Chem Soc 1989; 195:439.
106. RB Somoano, V Hadek, A Rembaum. J Chem Phys 1973; 58:697.
107. R Tenne, In: K Karlin, ed. Progress in Inorganic Chemistry. Vol. Vol 50. New York: Wiley, 2001:269–315.
108. H Topsoe, BS Clausen, N Topsoe, E Pedersen. Ind Eng Chem Fund 1986; 25: 25–36.
109. PT Asudevan, JL Fierro. Catal Rev 1996; 38:161–188.
110. AN Startsev. Catal Rev Sci Eng 1996; 37:353–423.
111. A Travert, H Nakamura, RA Santen, S Cristol, J Paul, E Payen. J Am Chem Soc 2002; 124:7084–7095.
112. A Startsev. Catal Rev 1995; 37:353–423.
113. PT Vasudevan. Catal Rev 1996; 38:161–188.
114. JV Lauritsen, S Helveg, E Laegsgaard, I Stensgaard, BS Clausen. J Catal 2001; 197:1–5.
115. S Kasztelan, A Wambeke, L Jalowieki, J Grimblot, JP Bonnelle. J Catal 1990; 124: 12–21.
116. R Prins. Adv Catal 2001; 46:399–464.
117. JL Portefaix, M Cattenot, M Guerriche, J Thivolle-Cazat, M Breysse. Catal Today 1991; 10:473.
118. M Cattenot, JL Portefaix, J Alfonso, M Breysse, G Perot. J Catal 1998; 173:366.
119. TC Ho, AJ Jacobson, RR Chianelli, CRF Lund. J Catal 1992; 138:351.
120. N Nelson, RB Levy. J Catal 1979; 58:485.
121. L Vivier, V Dominguez, S Kaztelan, G Perot. Catal Today 1991; 10:156.
122. J Klose, M Baerns. J Catal 1984; 85:105.
123. RE Hayes, WJ Thomas, KE Hayes. J Catal 1985; 92:312.

124. BK Miremadi, SR Morrison. J Catal 1987; 103:334–345.
125. BK Miremadi, SR Morrison. J Appl Phys 1990; 67:1515–1520.
126. RW McCabe, In: DW Bruce, D O'Hare, eds. Inorganic Materials. New York: Wiley, 1992:295–351.
127. V Rives, ed. Layered Double Hydroxides: Present and Future. New York: Nova Science, 2001.
128. LF Nazar, AJ Jacobson. J Chem Soc Chem Commun 1986:570–571.
129. LF Nazar, AJ Jacobson. J Mater Chem 1994; 4:1419–1425.
130. MD Curtis, In: EI Stiefel, K Matsumoto, eds. Transition Metal Sulfur Chemistry: Biological and Industrial Significance. Washington. DC: American Chemical Society, 1996:154–175.
131. J Heising, F Bonhomme, MG Kanatzidis. J Sol St Chem 1998; 139:22–26.
132. E Prouzet, J Heising, MG Kanatzidis. Chem Mater 2003; 15:412–418.
133. J Brenner, CL Marshall, L Ellis, N Tomczyk, J Heising, M Kanatzidis. Chem Mater 1998; 10:1244–1257.
134. KE Dungey, MD Curtis, JE Penner-Hahn. Chem Mater 1998; 10:2152–2161.
135. J Lee, W Lee, T Yoon, G Park, J Choy. J Chem Mater 2002; 12:614–618.
136. L Lomas, P Lacroix, JP Audiere, R Rene. J Mater Chem 1993; 3:499–503.
137. R Jakubiak, AH Francis. J Phys Chem 1996; 100:362–367.
138. A Leaustic, JP Audiere, PG Lacroix, R Rene, L Lomas, A Michalowicz. Chem Mater 1995; 7:1103–1111.
139. PA Joy, S Vasudevan. J Am Chem Soc 1992; 114:7792–7801.
140. R Clement. J Am Chem Soc 1981; 103:6698–6699.
141. R Clement, MLH Green. J Chem Soc Dalton 1979:1566–1568.
142. PA Joy, S Vasudevan. Chem. Mater 1993; 5:1182–1191.
143. PG Hill, PJS Foot, D Budd, R Davis. Mat Sci Forum 1993; 122:185–194.
144. D O'Hare, In: DW Bruce, D O'Hare, eds. Inorganic Materials. New York: Wiley, 1992:165–235.
145. FR Gamble, BG Silbernagel. J Chem Phys 1975; 63:2544–2552.
146. FR Gamble, JH Osiecki, FJ DiSalvo. J Chem Phys 1971; 55:3525–3530.
147. FE Gamble, JH Osiecki, M Cais, R Pisharody, FJ DiSalvo, TH Geballe. Science 1971; 172:493–497.
148. A Wold, K Dwight. Solid-State Chemistry: Synthesis, Structure, and Properties of Selected Oxides and Sulfides. New York: Chapman & Hall, 1993.

11

Alkali Silicates and Crystalline Silicic Acids

Wilhelm Schwieger
Universität Erlangen-Nürnberg
Erlangen, Germany

Gerhard Lagaly
Universität Kiel
Kiel, Germany

I. INTRODUCTION

Silicates occur in a great number of minerals and rocks as well as in numerous technical products. Materials containing silicates constitute the basis of many different branches of industry. Apart from their principal applications in the construction, glass, and ceramic industries, siliceous materials have achieved an outstanding significance as raw materials and additives in adsorbents, catalyst supports, and ion exchangers. Precipitated silicic acid, fumed silica, silica gels, molecular sieves, mesoporous materials, and clay minerals are currently the most common classes within this group.

Aluminum-free layered silicate hydrates broaden the group of silicates and provide interesting alternatives to the different forms of silica. They are prepared from common raw materials, mostly by hydrothermal processes. The special properties of this group of substances, including resistance to acids, intercalation, and ion exchange, have encouraged a large amount of research concerning their synthesis, characterization, and applications.

These special compounds include natural and synthetic silicates, in particular the minerals kanemite [1,2], grumantite [3,4], revdite [5,6], makatite [7], magadiite [8–10], and kenyaite [8,11] (Table 1). A synthetic crystalline layered silicate hydrate, the so-called octosilicate, also belongs to this family. It is sometimes named *ilerite* to honor Ralph Iler, who synthesized this material for the first time [12]. So far, this silicate has not been found in nature.

In this chapter we present a summary of the actual knowledge on these aluminum-free layered silicate hydrates, classify these types of silicates on the basis of structural aspects (Sec. II), and describe their preparation (Section III). The different types of reactions and possible applications are discussed in sections IV–VI.

II. STRUCTURE AND COMPOSITION

A. Classification of Silica Materials

The number of technically interesting SiO_2 materials has been significantly complemented and enriched over the last three decades. A large number of synthetic, crystalline high-silica compounds with framework or layer structure have been described, including microporous silicalite (an example of a zeolite), crystalline silicic acids, and their salts, the layered metal silicate hydrates.

The high-silica compounds, written as $[SiO_2]_n$ or simply SiO_2 (Fig. 1), are divided into amorphous and crystalline substances, indicating materials with,

TABLE 1 Naturally Occurring Layered Sodium Silicate Hydrates

Mineral	Idealized composition	Discovered at:	Ref.
Kanemite	$NaHSi_2O_5 \cdot 3H_2O$	Tschad (Tschad)	1
		Borgia (Kenya)	235
			236
Makatite	$Na_2Si_4O_{10} \cdot 5H_2O$	Magadi (Kenya)	98
Magadiite	$Na_2Si_{14}O_{29} \cdot 11H_2O$	Magadi (Kenya)	9
		Trinity County (U.S.A.)	237
		Alkali Lake, Oregon	35
		Malheur County, Oregon	238
		Tschad (Tschad)	10
		Tschad (Nigeria)	239
		Borgia (Kenya)	236
Kenyaite	$Na_2Si_{22}O_{45} \cdot 10H_2O$	Magadi (Kenya)	9
		Tschad (Nigeria)	239
Silhydrite	$3\ SiO_2 \cdot H_2O$	Trinity County (U.S.A.)	240

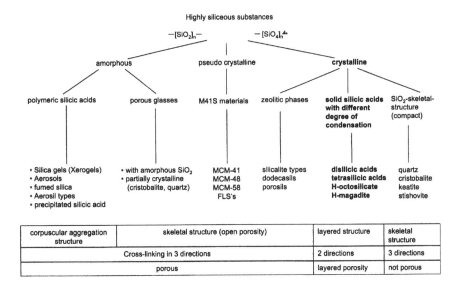

FIGURE 1 Classification of the high-silica compounds.

respectively, short- and long-range order. The amorphous SiO_2 substances[*] comprise the polymeric silicic acids such as silica gel, precipitated silicic acids, aerosols, xerogels, fumed silica, and porous glasses. The crystalline substances include the microporous materials such as silicalite (a zeolite-like product), crystalline silicic acids, and the compact SiO_2 modifications (the high- and low-temperature phase of quartz, cristobalite, and tridymite as well as moganite, stishovite, coesite, keatite, and melanophlogite). Also now applicable are the siliceous MSX mesoporous compounds, which were discovered by Mobil coworkers in the early 1990s [13–15]. These types of SiO_2 compounds are not truly crystalline materials. They are characterized by a regular, cubic, or hexagonal arrangement of tubelike rods, but the walls of the tubes are amorphous. Because of the presence of a regular network of mesopores, they are considered molecular sieve materials. On the other hand, they have properties similar to those of amorphous silica.

Another classification can be made by considering the structure-determining step in manufacturing. These include precipitation (amorphous silica), melting (porous glass), and hydrothermal crystallization (layered materials, zeolites). The

---

[*] Quartz glass (vitreous silica) can also be regarded as a SiO_2 compound. Due to its different applications it will not be discussed here.

manufacturing process* affects particularly the structure formation [corpuscular, layered structure, and open (porous) or closed (nonporous) three-dimensional structures]. It has a crucial influence on the degree of cross-linking and porosity and thus defines the properties of the substance. Sodium silicate solutions are used to produce nearly every compound. The exclusion or expulsion of the alkali ions or additives (e.g., borate in porous glass) in the manufacturing process occurs during the condensation step of silica gels, precipitated silicic acids, and silicalites or, for porous glass and crystalline silicic acids, in a consecutive step by elution (leaching).

The formation of a layered structure requires, in addition to hydrothermal (solvothermal) conditions, a specific ratio between the cationic components and the siliceous fraction in the reaction mixture to achieve the two-dimensional extended framework.

B. Analytical aspects

A general description of the hydrates (M-SHs) in oxide notation is

$$M_{2/n}O \cdot xSiO_2 \cdot yH_2O$$

M stands for cations such as protons, alkali, or alkaline earths, n stands for their valency, x indicates numbers between 2 and 40, and y indicates numbers between 1 and 20. The metal ions are exchangeable; $xSiO_2$ comprises the tetrahedral $[SiO_4]$ units of the framework, and nH_2O comprises all structural hydroxyl groups and the interlayer water molecules. This formulation refers to layered silicate hydrates of different structures, cationic forms, and degrees of hydration. It does not differentiate the M-SHs from other silicate types, such as nesosilicates, inosilicates, and phyllosilicates. Liebau [16] proposed a classification criterion (Table 2) based on the O/Si ratio in the framework. The layered metal silicate hydrates with values between 2.25 and 2.1 are positioned between the traditional phyllosilicates (O/Si ratio 2.5) and the tectosilicates (O/Si ratio 2.0).

The dimension of cross-linking (e.g., 1-, 2-, or 3-dimensional), the extension (as applies to, e.g., the inosilicates), and the number of cross-linked elements (e.g., 2 for double layers or double chains) are identified by the anion complex notation as proposed by Liebau [16]. The symbol $2/\infty$ of the anion complex of the metal silicate hydrates describes the two-dimensional cross-linking of the silicon oxygen tetrahedra to tetrahedral sheets. Following an older model, one, two, three, and five tetrahedral sheets are connected to form the bulk layer of kanemite or makatite, ilerite, magadiite, and kenyaite.

---

* The production of aerosols by flame hydrolysis illustrates the case of precipitation under extreme conditions.

TABLE 2 Classification of the Layered Silicates According to the System of Liebau

O/T ratio	Oxygen bridging atoms	Anion complex	Silicate structure	Dimension of cross-linking
4.0	0	$[SiO_4]^{4-}$	Nesosilicates	0
3.5	1	$[Si_2O_7]^{6-}$	Double tetrasilicates	0
3.0	2	$[Si_3O_9]^{6-}$	3-ring silicates	0
3.0	2	$^1_\infty[SiO_3]^{2-}$	Chain silicates	1
2.75	2	$2^1_\infty[Si_4O_{11}]^{6-}$	Double chain silicates	1
2.5	3	$^1_\infty[Si_4O_8]^{4-}$	Layered silicate	2
2.5	3	$^2_\infty[Si_4O_{10}]^{4-}$	**Kanemite**	2
2.5	3	$^2_\infty[Si_4O_{10}]^{4-}$	**Makatite**	2
2.25	3.5	$2^2_\infty[Si_8O_{18}]^{4-}$	**Octosilicate (ilerite)**	2
2.17	3.67	$3^2_\infty[Si_{12}O_{26}]^{4-}$	**Magadiite**	2
2.1	3.8	$4^2_\infty[Si_{20}O_{42}]^{4-}$	**Kenyaite**	2
2.0	4	$^3_\infty[SiO_2]$	Framework silicate	3

Source: Ref. 16.
Bold face: the metal silicate hydrates, M-SHs

A classification of silicates with predominantly two-dimensional cross-linking of $[SiO_4]$ tetrahedra leads to the trichotomy presented in Figures 2 and 3. Layered silicate hydrates, disilicates, and clay minerals differ in the repetitive characteristic binding element within the layers. The silicon oxygen tetrahedra within a sheet are connected by three bridging oxygen atoms. In the case of clay minerals, the fourth valency is saturated by foreign metal ions (e.g., Al^{3+}, Mg^{2+}) within the octahedral sheet. This leads to the formation of two-layer (e.g., kaolinite) and three-layer clay minerals (e.g., montmorillonite, vermiculite, micas). The fourth valency in the disilicates (e.g., natrosilite) is compensated by cations, mainly sodium or potassium ions. In the case of the layered silicate hydrates, the fourth valency is not completely balanced by metal cations. Here, the $\equiv Si$—O^-M^+ groups, $\equiv Si$—O—$Si\equiv$ bridges, and $\equiv Si$—OH groups, the protons of which are not replaced by cations, are structure- and property-determining. The presence of $\equiv Si$—O^-M^+ and $\equiv Si$—OH groups at the surface of the silicate layers seems to be independent of the number of $\equiv Si$—O—$Si\equiv$ bridges within the silicate layer. This is one of the characteristic features of the layered metal silicate hydrates.*

---

* Materials of this type with layered structures have various names in the literature (cf. Tables 2, 3, 4). We use the term *metal silicate hydrate*, abbreviated as M-SH.

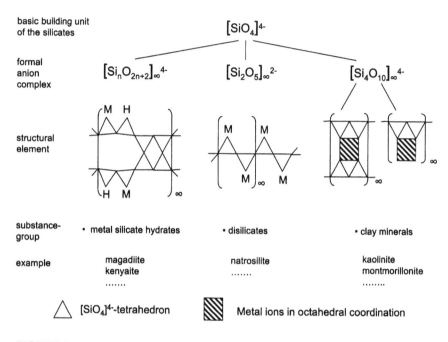

basic building unit
of the silicates

$$[SiO_4]^{4-}$$

formal
anion
complex

$$[Si_nO_{2n+2}]_\infty^{4-} \qquad [Si_2O_5]_\infty^{2-} \qquad [Si_4O_{10}]_\infty^{4-}$$

structural
element

substance-
group

• metal silicate hydrates   • disilicates   • clay minerals

example

magadiite   natrosilite   kaolinite
kenyaite   ....   montmorillonite
....

△ $[SiO_4]^{4-}$-tetrahedron   ▨ Metal ions in octahedral coordination

FIGURE 2   Arrangements of the $[SiO_4]$ tetrahedra as structural building units of silicatic layered materials.

Layered metal silicate hydrates synthesized with different cationic compounds are listed in Tables 3 and 4. The sodium silicate hydrates (Table 3) can be subdivided into five groups according to the SiO_2/Na_2O molar ratio:

≈ 2 is kanemite.
≈ 4 is makatite.
$\approx 8–10$ is octosilicate.
$\approx 10–20$ is magadiite.
≥ 20 is kenyaite.

This formal classification is confirmed in many cases by the basal spacing, a characteristic lattice distance (see later) (Fig. 4, Table 5). However, manufacturer-dependent differences in the positions and intensities of this spacing are often observed. Therefore, a compound is often identified based on its reaction behavior rather than on the basal spacing alone.

The following fundamental reactions are related to the structural properties of M-SH compounds [17,18]:

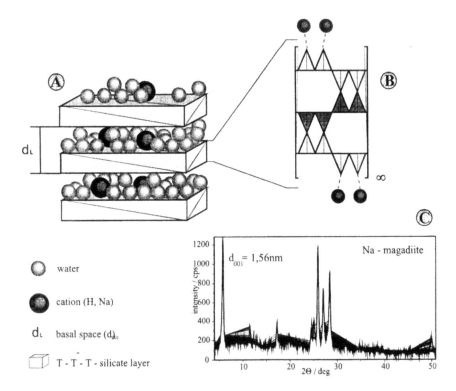

FIGURE 3 Schematic representation of the building principle of the metal silicate hydrate magadiite. (A) Schematic representation of a layered material with the bulk (silicate) layer and the interlamellar layer (cation-containing water layers); (B) structural arrangement of the silicon oxygen tetrahedra in the bulk layer, based on a structural model of Schwieger (26); (C) typical X-ray diffraction pattern of Na-magadiite with the characteristic basal spacing of 1.56 nm.

Exchange of sodium ions for inorganic or organic cations
One-dimensional intracrystalline swelling
Transformation of the metal silicate hydrates into crystalline silicic acids
Intercalation and adsorption of neutral polar organic compounds
Transformation into other structures

These reactions proceed in the interlayer space of the silicates and generally involve changes of the basal spacing (Sec. IV–VI will cover this in great detail).

TABLE 3 Synthetic Sodium Silicate Hydrates with Varying SiO_2/Na_2O ratios (Selection of First Reports)

Structural type	Name	Molar ratio of oxides			Ref.
		Na_2O	SiO_2	H_2O	
Kanemite $SiO_2/Na_2O = 4$	Kanemite	1	4	4	18
Makatite $SiO_2/Na_2O = 4$	Tetrasilicate	1	4.33	3.66	29
		1	4	x	30
	Makatite	1	4	x	23
Unnamed $SiO_2/Na_2O = 8$	Octosilicate	1	8	x	12
		1	12.5	23	32
	PS 2	1	8–10	9–11	33
		1	7.2	8.1	74
	Ilerite	1	8	9	38
	RUB 18	1	8	9	24
Magadiite $SiO_2/Na_2O = 14$–16		1	13.1	x	30
		1	19.5	22.5	32
	PS 1	1	9–16	9	33
	magadiite	1	14	9	42
	magadiite	1	13.8	10	38
	SiO_2-Y	1	21–29	12–14	36
	Na-SKS-1	1	14	x	[DE-OS 34 00 130]
	SiO_2-Y				39
Kenyaite $SiO_2/Na_2O \geq 20$	Kenyaite	1	19	10	38
	Kenyaite	1	20	x	11
	Na-SKS-1	1	21	x	[DE-OS 34 00 130]
	PZ 1		—		[DE-OS 32 11 433]
	—	1	14.3–20	9.3–15	[DD-WP 22 34 26]

C. Structure and Structural Models

Similar to clay minerals, the structure of M-SH is described by the superposition of (bulk) silicate layers* composed of [SiO$_4$] and [SiO$_3$OH] tetrahedra. The interlayer spaces between these layers contain the cations to balance the charges of

---

* Please note the difference between *sheet* and *layer*. (In German, both units have the same term: *Schicht*.) However, *layer* can have two meanings: the unit composed of the sheets (the "bulk layer"), and or the unit composed of the bulk layer and the interlayer space. This last meaning is often used for interstratified structures.

Table 4 Synthetic Metal Silicate Hydrates, Pure and Mixed Cationic Forms with Varying SiO_2/Na_2O Ratios (Selection of First Reports)

Cation (M)	Name	Molar ratios of oxides			Ref.
		$M_{2/n}O$	SiO_2	H_2O	
Single cationic systems					
Li^+	L 1	1	1–6.7	0.3–3.8	37
	L 2	1	0.33–1.1	0.3–1.1	37
	Li-MAG[a]	1	20	x	18
K^+	SiO_2-X		—		31
	SiO_2-X_1		—		74
	SiO_2-X_2	1	22–35	8–9	74
	SiO_2-X_3	1	35–48	8–9	74
	K-1	1	3.4–3.7	20.4–46.5	74
	K-2	1	20.4	7	74
	K-3	1	29–37	8–16	74
	K-4	1	6,7	4–4.5	74
		1	8	x	99
	SiO_2-X				39
	—	1	20	x	40
	K-form	1	23.8	9.8	38
Cs^+, Ca^{2+}, Sr^{2+}, Ba^{2+}, Ag^+		1	20	x	40
Multicationic systems					
Li^+/Na^+	Silinaite	1	2	2	140
Na^+/Cs^+/HMT[b]	RUB-18	1	8	8.4	20
Na/NR_4^+ [c]	SH-P20		—		96
	SH-P30		—		96
	SH-P40	1	19.8	5	96

[a]Lithium form of the magadiite (MAG), directly synthesized.
[b]HTM: hexamethylentetramine, which was added but not incorporated in the silicate structure (also Cs^+ has not been found in the silicate).
[c]NR_4^+: poly-{dimethyl diallylammonium} (in the form of the chloride). The organic cation was incorporated in the as-synthesized silicate.

the layers and water molecules. The silicate layer itself may consist of one sheet of $[SiO_3OH]$ tetrahedra or may be composed of two and more condensed tetrahedral sheets. The layers are held together by electrostatic forces, hydrogen bonds, and van der Waals interactions (Fig. 3). The characteristic features of such layered silicates are the basal spacing d_L and the interlayer distance Δd. The basal spacing d_L (d_{001} or ld_{001} in cases where several integral basal reflections are observed) represents the smallest repeating unit, perpendicular to the cleavage plane in the

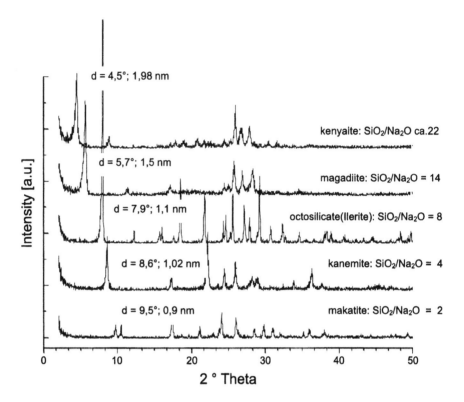

FIGURE 4   X-ray diffraction pattern of synthetic sodium silicate hydrates with increasing SiO$_2$/Na$_2$O ratios.

c-direction comprising the thickness of one (bulk) silicate layer plus the interlayer space.

Figure 4 shows the X-ray powder diffractograms of typical Na-SH materials together with a schematic illustration of the changes in the thickness of the bulk silicate layer. All reactions that characterize the M-SH (e.g., dehydration, ion exchange, intercalation) proceed in the interlayer space. The basal spacing d_L, therefore, varies, but the principle structure of the bulk layer remains unchanged.

Due to the small size of crystallites and the small number of sharp diffraction lines in the X-ray diffraction pattern, only a few crystal structure determinations have been published. The number of distinct reflections decreases from kanemite to kenyaite, i.e., with increasing thickness of the bulk layer and increasing SiO$_2$/Na$_2$O ratio (Fig. 4). Only the crystal structures of kanemite [2], makatite [19], RUB-18, an "ilerite-type" silicate [20], the lithium sodium silicate silinaite [21], and the boron-containing mineral searlesite [22] thus far have been solved.

Table 5 Structural Aspects of the Metal Silicate Hydrates (n = natural, s = synthetic): Strongest X-Ray Powder Reflections and Unit-Cell Parameters

Name	Composition	X-ray peak position in nm, intensity in %			Lattice constant				Ref.
					a (nm)	b (nm)	c (nm)	$\beta(°)$	
Kanemite, n	$NaH[Si_2O_4(OH)_2]\cdot 2H_2O$	1.037 / 100	0.4014 / 100	0.3435 / 90	0.7282	2.0507 Orthorhombic	0.4956		1
Kanemite, n	**$NaHSi_2O_5\cdot 3H_2O$ (Z = 4)**				**4.946(3)**	**20.502(15)**	**7.275(3)**		**2**
Kanemite, s	**$NaHSi_2O_5\cdot 3H_2O$ (Z = 4)**				**4.946(1)**	**Orthorhombic: Pbcn 20.510(1)**	**7.277(1)**		**24**
Makatite, n	$NaSi_2O_3(OH)_3\cdot H_2O$	0.509 / 100	0.904 / 53	0.2996 / 57	1.6840	1.0256 Orthorhombic	1.9146		238
Makatite, s	**$Na_2H_2Si_4O_{10}\cdot 4H_2O$**	**1.550 / 100**	**0.3419 / 90**	**0.2996 / 80**	**0.7388**	**1.8094 Orthorhombic**	**0.9523**	**90.64**	**19**
RUB-18, s	**$Na[Si_4O_8(OH)]\cdot 4H_2O$ (Z = 8)**				**7.3276(1)**	**Monoclinic: space group P2₁/c / Space group: I4₁/amd (choice 2)**	**44.3191(6)**		**20**
Octosilicate, s	$Na_2O\cdot 8SiO_2\cdot 9H_2O$	1.110 / 320	0.480 / 100	0.3062 / 100	0.7345	1.274 Monoclinic	1.125		25
SiO₂-Y, s	$Na_2O\cdot 21\text{-}29SiO_2\cdot 12\text{-}14H_2O$	1.550	0.344 / 90	0.315 / 90	1.550	1.529 Tetragonal	0.660		39
					1.126	Orthorhombic	1.279		39
Magadiite, n	$NaSi_7O_{21}(OH)_3\cdot 3H_2O$	1.541 / 100	0.3435 / 80	0.3146 / 50	1.262	Tetragonal	1.5573		9
Magadiite, n	$NaSi_{17}O_{13}(OH)_3\cdot 3H_2O$	1.577 / 100	0.343 / 80	0.314 / 50	0.722	1.570 Monoclinic	0.691	95.1	237
Magadiite, n Air dried	$NaSi_7O_{13}(OH)_3$	1.556 / 100	0.3434 / 75	0.3145 / 60	0.725	0.725 Monoclinic	1.569	96.8	
Vacuum dried		1.350 / 100	0.3581 / 40	0.3314 / 80	0.730	0.730 Monoclinic	1.373	100.5	71, 72
Kenyaite (n)	$NaSi_{11}O_{20.5}(OH)_4\cdot 3H_2O$	1.968 / 100	0.3428 / 85	0.3198 / 55	1.281	Tetragonal	1.9875		9

Boldface: based on a complete structural solution.

A single crystal of kanemite suitable for refinement was found in nature [2]. Makatite single crystals were synthesized in the presence of triethanolamine [19,23]. The other M-SHs could not be synthesized with crystals large enough for a single crystal refinement. Based on a methodical approach using an ab initio crystal structure solution, Vortmann et al. [20] solved the structure of RUB-18 from X-ray powder diffraction data.

The unit-cell dimensions of silicates based on known structures are listed in Table 5, along with several reported indexing attempts. Kanemite from Aris phonolite, Namibia, with the ideal composition $NaHSi_2O_5 \cdot 3H_2O$, is orthorhombic (space group Pbcn) with unit-cell parameters $a = 4.946(3)$ Å, $b = 20.502(15)$ Å, $c = 7.275(3)$ Å, and $Z = 4$ [2]. The structure (see Fig. 5) consists of alternating corrugated (010) sheets of $[Si_2O_4OH]_n{}^{n-}$ units and hydrated sodium ions. The silicate sheets contain six-membered rings of $[HOSiO_3—SiO_4]^-$ units. The $\equiv Si—O^-$ groups of the layers alternately point up and down, and the sodium ions are octahedrally coordinated by water molecules.

The crystal structure of a synthetic product identical to kanemite was solved by direct methods from integrated intensities of powder diffraction data and subsequently refined by the Rietveld technique [24]. This synthetic kanemite also crystallizes in the orthorhombic space group Pbcn with cell parameters $a = 4.946(1)$ Å, $b = 20.510(1)$ Å, and $c = 7.277(1)$ Å.

Makatite is described as a single-layer silicate [19]. The repeat unit of the layer silicate chain consists of four $[SiO_4]$ tetrahedra, and the silicate layers are considered vierer single layers. The interlayer spaces contain chains of $[Na(OH_2)_6]$ octahedra that connect the silicate layers via electrostatic forces and hydrogen bonds (Fig. 6). The structure is monoclinic with space group $P2_1/c$ and parameters of $a = 7.3881(5)$ Å, $b = 18.094(3)$ Å, $c = 9.5234(5)$ Å, $= 90.64(1)$, and $Z = 4$.

An ab initio structure solution based on the X-ray powder data of RUB-18, a synthetic silicate identical with the octosilicate, was reported by Vortmann et al. [20] (see Fig. 7). The unit cell was indexed as tetragonal, with the space group $I4_1/amd$ (choice 2), $a_0 = 7.3276(1)$ Å, and $c_0 = 44.3191(6)$ Å. The cell content is $Na[Si_4O_8(OH)]$ with $Z = 8$. The topology of RUB-18 is unique. The basic unit is a $[5^4]$ cage that consists of four five-membered rings. This cage contains eight tetrahedra that are connected to four neighboring cages by eight oxygen bridges. The remaining silicon atoms carry the hydroxyl groups. The negative charges are compensated by sodium ions. This is the first report of silicate layers composed of cages, although, as pointed out by Vortmann at al. [20], similar units are found in the structure of several zeolites. The charge-balancing sodium ions reside in pockets formed by two silicate layers. They are octahedrally coordinated by oxygen atoms of the layer and adsorbed water molecules (Fig. 7).

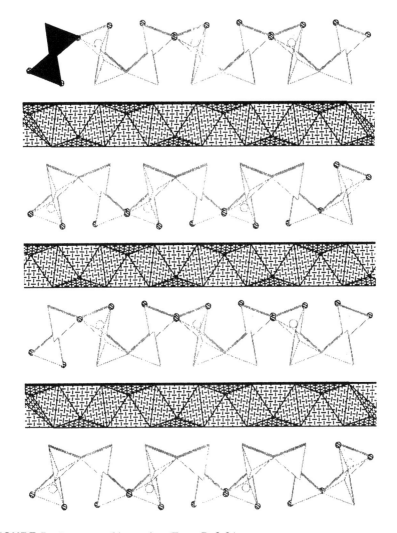

FIGURE 5   Structure of kanemite. (From Ref. 2.)

The ab initio calculations, NMR, and analytical data for RUB-18 indicate a unit-cell chemical composition of $Na_8[Si_{32}O_{64}(OH)_8] \cdot 32H_2O$. This composition is consistent with the results of Borbely et al. [25]. The former model of octosilicate [26], which was established considering the structure of makatite and similarities in the family of the sodium silicate hydrates, therefore has been discarded. Based on the same approach, Gies et al. [27] and Borowski et al. [28]

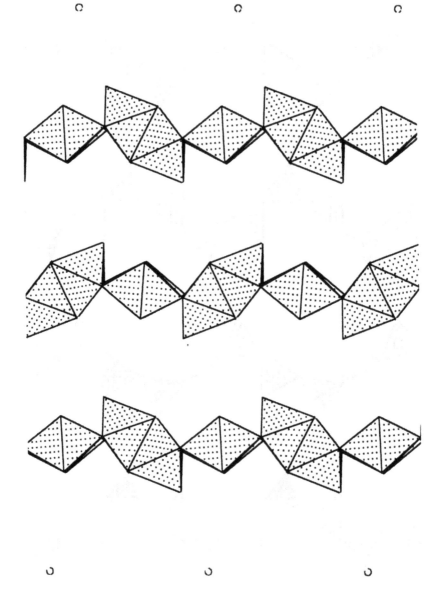

FIGURE 6   Structure of makatite. (From Refs. 19 and 23.)

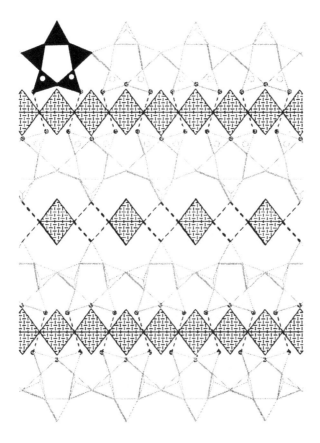

FIGURE 7 Structure of RUB-18 (analogous to octosilicate). (From Ref. 20.)

have derived the structure of H-RUB-18 (tetragonal, space group $I4_1/amd$ (choice 2), $a_0 = 7.383(2)$ Å, $c_0 = 29.759(3)$ Å).

Thus far, no other structure determinations of other metal silicate hydrates have been solved, although the lattice dimensions have been reported [1,12,19,23,28,29–40]. A comparison of the positions and intensity ratios of the strongest X-ray reflections confirms the classification of M-SHs into five groups according to the molar ratio SiO_2/Na_2O (Table 5). Often the authors arrive at different conclusions, even for products of the same type (e.g., SiO_2-Y [39] or magadiite [34]). In some cases indexing is not consistent with the structure determination, e.g., for octosilicate [20,25]. Selected results of these indexing attempts are summarized in Table 5 and compared with the results of the structure determination. No indexing attempts were made for other silicates [29,30].

Crystallographic data for potassium silicates has been reported only for K-SH, SiO_2-X_2, and SiO_2-X_3 synthesized by Micjuk and coworkers [36,41].

In addition to structural solutions, two other lines of investigations concerning structural aspects of the sodium silicate hydrates are:

1. Analysis of the arrangement of intercalated molecules
2. Analysis of the arrangement and linkage of the [SiO_4] tetrahedra in the silicate layers

The first group includes studies of intercalation, ion exchange, swelling, and grafting reactions [11,17,42,43]. Here the structure of the silicate layer is considered unknown and is recorded merely in terms of a specified thickness. The detailed structure plays a secondary role to reaction chemistry.

In a second group [19,26,44,45,46], spectroscopic investigations of the silicate layer and its surface are described. The structure of the surface and the flexibility and stability of the silicate layers are consequences of the type of bonds between the [SiO_4] tetrahedra of the layer and determine the reactivity of different M-SH compounds.

These investigations, therefore, comprise (a) evaluation of the bonding type inside the silicate layer and (b) characterization of the surface $\equiv$Si—OH groups. The results of NMR measurements [25,26,46–51] have been combined with IR studies [47,52–54], thermoanalytic data [25,55–59], as well as analytical data.

Schwieger et al. [26] and Brandt et al. [56] carried out ^{29}Si-MAS NMR experiments of makatite, octosilicate, magadiite, and kenyaite. Based on the XRD data, analytical data, Q^3/Q^4 ratios* from NMR, and the knowledge of the structure of makatite [19,23] with vierer single layers, they proposed the first structural models of Na-SH by combining makatite layers. Figure 8 shows the resulting arrangement of the [SiO_4] tetrahedra in the (001) plane.

According to these data, ilerite is composed of two, magadiite of three, and kenyaite of five makatite-type four-membered single layers. Differences arise, especially for kenyaite, between the measured (1.97 nm) and calculated basal spacing (2.76 nm). This discrepancy is eliminated if the [SiO_4] tetrahedra are combined in different ways [61]. The reaction behavior and the stability effects can be explained on the basis of these models. Nevertheless, the model of octosilicate has to be discounted due to the similarity with RUB-18. The presence of the [5^4] cages as building units of RUB-18 might explain why, to date, this silicate has not been discovered in nature. Garcés et al. [47] proposed five-ring units as structural elements of octosilicate and natural magadiite and discussed similarities

---

* The Q^n group nomenclature was introduced by Lipmaa et al. [60], where n denotes the number of bridging oxygen atoms of the [SiO_4] tetrahedra under study. For instance, Q^3 and Q^4 are assigned to OSi(OSi$\equiv$)$_3$ and Si (OSi$\equiv$)$_4$ groups, respectively.

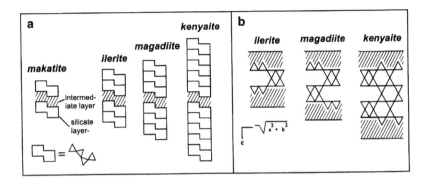

FIGURE 8 Earlier structural models of the sodium silicate hydrates. (From Ref. 26.)

to zeolitic structures, especially zeolites of the mordenite and pentasil groups. In contrast, the known structures of all naturally occurring silicates contain structural elements that are at least similar to the pyroxenic layer of most disilicates [24].

The ^{29}Si-MAS NMR spectra show the existence of different [SiO$_4$] tetrahedral environments [26], which cause shifts and pronounced splitting of the Q^4 group signal of magadiite and kenyaite (Table 6). Kenyaite shows several signals with larger shifts. This supports the assumption that the structural environment around the [SiO$_4$] tetrahedra is more differentiated than in magadiite. This indicates a multilayered arrangement of the tetrahedra, which contradicts Pinnavia's view [46], who proposed a silicate double layer in magadiite. It should be mentioned that the ^{29}Si MAS NMR spectrum of octosilicate at 130 K shows four Q^4 signals at -113.0, -113.7, -114.1, and -114.9 ppm instead of the singlet that is observed at room temperature. So far, low-temperature measurements have not been reported for the other members of this silicate family [62].

Huang and coworkers [52–54] carried out structural investigations of different layered silicate hydrates by vibrational spectroscopy (IR, Raman) considering the known structures of silinaite, kanemite, and makatite. The predictions of the vibrational modes made on this basis were not in agreement in all cases with the observed spectra. Agreement was excellent for silinaite and supported the proposed structural model. The comparison was not fully consistent for makatite. The studies of octosilicate, magadiite, and kenyaite confirmed (a) the multilayer model, (b) the assumption of five- and six-membered rings in the structure, and (c) that the structure of ilerite was not related closely to magadiite and kenyaite as previously suggested. We conclude that the structure of octosilicate might be exceptional and not a member of the (homologous) family of sodium silicate hydrates in spite of the analytical composition and similarity of the reactions.

Table 6 Chemical Shifts (Related to TMS) in the ^{29}Si MAS NMR Spectra of Sodium Silicate Hydrates (Selection)

Silicate	Chemical shifts, Δ_{si} (ppm) Q^3 units	Q^4 units	characteristic Si—O—Si bond angle for the Q^4 unit/deg.	Intensity ratio $Q^4:Q^3$	Ref.
Makatite				no Q^4	
Kanemite	−99.5(*); −100.5	−111.7	n.v.	no Q^4	47
Octosilicate	−97(*); −99(*), −101.9	−112.2	150	0.9:1	26
Octosilicate	−97(*); −99(*), −101.9	−112.2	150	1:1	25
Ilerite (RT) (130 K)	−100.2	−111.4	n.v.	1:1	142
		−113.0; −113.7; −114.1; −114.9	No information		
Octosilicate	−100	110.9	n.v.	1:1	47
Magadiite	−97.7(*), −99.4(*), −100.0	−112.2; −114.8; −117.5	150; 152; 117.5	2.07:1	26
Magadiite (I)	−99.1	−109.5; −111.2; −113.6	n.v.	1.0:1	47
Magadiite (II)	−99.1	−109.5; −111.2; −113.6	n.v.	2.5:1	
Magadiite (Trinity County)	−99.7	−109.5; −111.2; −113.6	n.v.	1.2:1	
Magadiite	99.1	−109.6; −111.1; −113.7	n.v.	2.5:1	59
Magadiite[a] (Na-66R)	−99.2	−110.0; −111.0, −113.5	n.v.	2.3:1	77
Magadiite	−99.0	−110.7; −113.2	n.v.	n.v.	188
Magadiite	−99.0	−110.1; −111.1; −113.7	n.v.	2.86:1	79–81
Kenyaite	97.9(*);−99.8(*),−101.1	−109.7; −112.2(*); −114.1; −117(*); −120.5	146; 150; 152; 160; 170	4.07:1	26
SH-P20[b]	100	−109(*); −112.2; −114.; −119	n.v.	3:1	96,111

[a]Selected from many data measured on different samples.
[b]Taken from published figure.
Note: *lines assigned to small shoulders, n.v. = no values.

From DTA, IR, and H^1 NMR spectra, Rojo and coworkers [48] concluded that magadiite contains two types of $\equiv$Si—OH groups. The entire silanol group density is 3.5 SiOH per nm^2 at an average distance of 0.25 nm. Siloxane bridges are formed between the layers, and only one type of silanol group, with an average distance of 0.5 nm, remains after heating to 500°C. These $\equiv$Si—OH groups are directed into the pockets of the opposite silicate layer (Fig. 9).

Only a few papers deal with the status and dynamics of the hydrogen bonds, for instance, in octosilicate [62] and RUB-18 [28,63,64]. Internal hydrogen bonds are formed between neighboring $\equiv$Si—OH and $\equiv$Si—O$^-$ groups. This strong interaction arises from the short O—O distance of about 0.23 nm and causes a downfield chemical shift in the proton NMR spectra of about 16.3 ppm (relative to TMS). However, this bonding behavior is temperature dependent. At higher temperature the protons became more mobile and the silicate can be considered a proton conductor. A low-temperature network with immobile protons therefore transforms into a high-temperature network with delocalized protons [28].

The hydration of the interlayer cations in magadiite has been studied by adsorption isotherms, thermal analysis, X-ray diffraction, and spectroscopic measurements [57–59]. The water adsorption isotherms increase in two steps, corre-

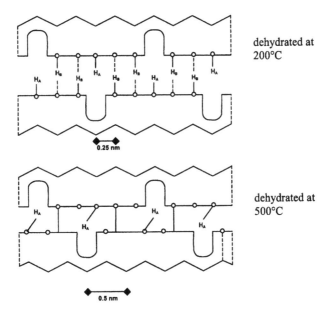

dehydrated at 200°C

dehydrated at 500°C

FIGURE 9 Schematic representation of the OH group arrangements in a H$^+$ magadiite. (From Ref. 44.)

sponding to the increase of the basal spacing. At higher relative pressure, two types of water are found: water molecules linked to the cations (Na^+, K^+) and water molecules interacting with the surface silanol and $\equiv Si—O^-$ groups. Interlamellar Mg^{2+} and Ca^{2+} ions interact more strongly with the surface groups and remain directly coordinated to two $\equiv Si—O^-$ groups, even at high water contents. Considering the analytical and thermoanalytical data, the vibrational and NMR spectroscopic results, and the X-ray diffraction studies, the chemical composition of this M-SH is represented by

$$(Na \cdot 2H_2O)_a H_a \cdot [a(Si_2O_5) \cdot b(Si_2O_4)] \cdot (cH_2O)$$

where a is an integer equal to or greater than 1, b is an integer equal to or greater than 0, and $c = 2$ in the fully hydrated state. This formula expresses the different silicon environments (Q^3 and Q^4 groups), the state of the interlayer space with $\equiv SiOH$ and $\equiv SiO^- Na^+$ groups, the hydration of the interlayer cations, and the intercalated water molecules. Borbely and coworkers [25] first described octosilicate and its H^+ form in this way with $a = 1$, $b = 1$, $c = 2$ and $a = 1$, $b = 1$, $c = 1$, respectively.

According to TG and NMR investigations, the following parameters express the ideal composition:

Kanemite	$NaHSi_2O_5 \cdot xH_2O$	$a = 1, b = 0$ (no Q^4 units)
Makatite	$Na_2H_2Si_4O_{10} \cdot xH_2O$	$a = 2, b = 0$ (no Q^4 units)
Octosilicate	$Na_2H_2Si_8O_{18} \cdot xH_2O$	$a = 1, b = 1$ (Q^3 and Q^4 units)
Magadiite	$Na_2H_2Si_{14}O_{30} \cdot xH_2O$	$a = 2, b = 5$ (Q^3 and Q^4 units)
Kenyaite	$Na_2H_2Si_{20}O_{42} \cdot xH_2O$	$a = 1, b = 4$ (Q^3 and Q^4 units)
	$Na_2H_2Si_{22}O_{46} \cdot xH_2O$	$a = 2, b = 9$ (Q^3 and Q^4 units)

The c-values depend on the hydration state of the silicates.

Despite the success in solving structures of poorly crystalline materials by direct methods and the numerous spectroscopic results, the structure of magadiite- and kenyaite-type silicates still remains unsolved. The information obtained by analytical, thermoanalytical, and spectroscopic methods leads to a very close picture of different structural aspects and contributes to the understanding of the reaction behavior of the silicate hydrates. Nevertheless, a picture of the structural relationships between members of the sodium silicate hydrate family remains elusive.

D. Morphology

All synthetic M-SHs are obtained as fine white powders. Electron microscopic images (Fig. 10) show the typical "*cauliflower*" morphology of magadiite and

makatite   ilerite   magadiite   kenyaite

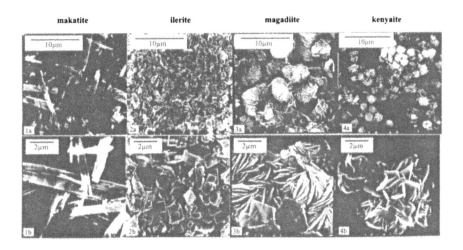

FIGURE 10   Scanning electron micrographs of makatite, octosilicate (illerite), magadiite, and kenyaite, showing the typical morphology of the sodium silicate hydrates.

the more open aggregation of plates in the case of kenyaite. Platelike structures of small crystallites that are intergrown to larger aggregates are detected for both materials. Makatite and octosilicates show a different morphology. Makatite exhibits more fibrous forms similar to some zeolites (e.g., erionite), whereas octosilicate forms nearly perfect platelike particles.

III.   THE SYNTHETIC METAL SILICATE HYDRATES

A.   History

In spite of the great diversity of the synthetic alkali silicates, natural products were unknown until 1967, when Eugster [9] discovered two sodium silicate hydrates, later named magadiite and kenyaite, after the place where they were found, in the deposits of Lake Magadii in Kenya. The first synthetic products with SiO_2/Na_2O ratios of 9.4 and 13.1 were reported by McCulloch [30]. He found such silicates as precipitates of water glass solutions. Later, Iler [12] obtained a similar crystalline hydrated sodium silicate with a SiO_2/Na_2O ratio of 8 and a H_2O/Na_2O ratio of 9. According to the chemical composition, the silicate was named octosilicate [12,26]. This is somewhat misleading due to the fact that this name does not reflect any structural aspects of the material. Earlier, the name "ilerite" was introduced [12,26]. Iler [12] was one of the first scientists who synthesized and investigated such types of sodium silicate hydrates.

A comparison between naturally occurring silicates and the typical synthetic counterparts, based mostly on the similarities in the X-ray diffraction patterns, is shown in Tables 1 and 3. Kanemite was synthesized hydrothermally for the first time by Beneke and Lagaly [18]. Kalt and Wey [66] obtained a silicate similar to kanemite. They prepared $KHSi_2O_5$ by an annealing process and converted it into the proton form, which was subsequently transformed into $NaHSi_2O_5 \cdot xH_2O$ by NaOH.

Makatite was synthesized by Annehed et al. [19] and Beneke and Lagaly [67]. Products similar to makatite, according to the classification in Section II, were synthesized by Baker et al. [29] and McCulloch [30]. Several authors considered synthetic products as identical to magadiite [17,68,69,71–73]; patents DE-OS 34 00 130; DD-WP 223426), whereas the silicates from Ilin et al. [32,33], Turitina and Ilin [74], Mičjuk et al. [36], McCulloch [30], and Kitahara et al. [39] can be considered similar to magadiite. Assuming that the phase PS2 is a kenyaite-type silicate, Ilin et al. [32,33] were the first who synthesized kenyaite. A broad overview of the synthesis was provided by Lagaly et al. [11,67].

A layered sodium silicate containing aluminum ions, PZ 1, was considered by Baacke and Kleinschmidt (DE-OS 3211433) as resembling magadiite. The H^+ form, PZ 2, was obtained by acid treatment of PZ 1, but this ion exchange was irreversible. Synthesis of kenyaite with SiO_2/Na_2O molar ratios equal to or greater than 20 was reported by Schwieger et al. [38,76], Beneke and Lagaly [11], and Rieck [75] (DE-OS 3400130).

A classification of potassium and lithium silicate hydrates analogous to the sodium silicates is currently not possible (Table 4). Beneke et al. [40] described the synthesis of alkaline (Li^+, K^+, Cs^+), earth alkaline (Ca^{2+}, Sr^{2+}, Ba^{2+}), and silver salts of the silicic acid $H_2Si_{20}O_{41}$. The composition of the potassium salt was specified as $K_{1.8}H_{0.2}Si_{20}O_{41} \cdot xH_2O$.

B. General aspects

Metal silicate hydrates (M-SHs) (Tables 3 and 4) are synthesized in the ternary system M_2O—SiO_2—H_2O, mostly by hydrothermal crystallization. Scholzen et al. [77] described the reason for the large diversity of products using magadiite as an example. The diversity results from differences in synthesis conditions, such as composition, SiO_2 source, temperature, reaction time, and additives. Since the metal cations are exchangeable, the SiO_2/M_2O ratio is strongly influenced by the washing conditions during the process of separatins the solids from the mother liquor after synthesis. Exchangeable cations are easily leached out, i.e., replaced by protons. Unfortunately, these conditions are often not specified in the literature. The water content varies with the drying conditions that are also often not specified, or at least not with the required accuracy.

The synthesis fields of sodium, lithium, and potassium silicate hydrates according to Ilin, Turitina, and Schwieger [33,37,76] are illustrated in Figure 11. The crystallization fields are located in the low-alkali region (rich in SiO_2) of the ternary systems. Most authors use distinct compositions of reaction mixtures corresponding to the synthesis fields specified by Ilin et al. [33]. The limiting values of the typical composition of reaction mixtures for SiO_2-X ($SiO_2 \leq 12\%$ w/w, $K_2O \leq 8\%$ w/w, $H_2O \geq 80\%$ w/w) [31] lie close to the fields of the phases called K3 and K4. Data of several authors, e.g., Mičjuk et al. [36,41], cannot be entered in Figure 11 because some of the experimental conditions are not precisely given.

The different M-SH syntheses are shown in Figure 12, independent of the position in the synthesis fields or cations used. The most applicable route is hydrothermal synthesis (route I) (preparation pathway I). A silicate hydrate suspension is obtained from an aqueous alkaline dispersion of SiO_2 when hydrothermal conditions between 80 and 300°C are applied. Closed glass ampoules, steel cylinders, or stirred autoclaves are generally used as pressure vessels. In the past, crystallization was accomplished exclusively in a discontinuous manner. The further processing stages in Figure 12 are needed to obtain pure products and can be arranged according to solid–liquid finishing process. Filtration and centrifugation procedures were recommended (DD-WP 238 536, DD-WP 234878) for the separation of the layered silicate hydrates. Elution is used to eliminate excessive alkalinity as well as soluble byproducts in order to attain higher product purity.

Fine powders are obtained after drying at ~100°C. In this form, the powder can be used for many applications. A granulate of silicate hydrates for special applications (e.g., additives for detergents) can be produced by spray-drying or fluidized-bed drying of the entire synthesis suspension (DD-WP 234 878). Depending on the application (e.g., as phosphate substitute in the builder system of washing powders), spray-drying or direct use of synthesis slurries is recommend (75,78; DD-WP 234 878).

In addition to the hydrothermal routes, organic solvents can be used to synthesize kanemite [18,78–84]; EP 627383) or magadiite [85]. Lagaly et al. [17] proposed several synthesis routes for kanemite. Most procedures include dispersing, drying, calcining, and redispersing the material in water (preparation pathway II of Fig. 12). The technical production of Na-SKS-6, structurally almost identical to δ-$Na_2Si_2O_5$, follows these steps. The Na-SKS-6 acts as the anhydrous precursor for kanemite formation (75,78; EP 627383). After redispersing Na-SKS-6 in water, kanemite is formed.

Recently a new synthesis route (preparation pathway III in Fig. 12) for magadiite and octosilicate was proposed by Crone et al. [86] involving a solid-state transformation process. Solid sodium metasilicate, Na_2SiO_3, and SiO_2 in the form of a xerogel were intensively ground and reacted at different tempera-

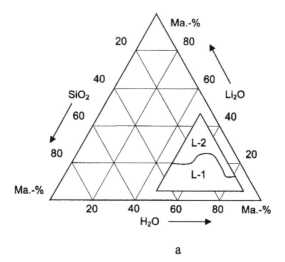

a

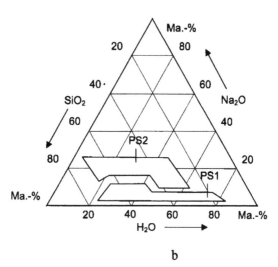

b

FIGURE 11 Typical crystallization fields of the metal silicate hydrates in the triangular coordinates: (a) Li_2O—SiO_2—H_2O; (b) Na_2O—SiO_2—H_2O; (c) K_2O—SiO_2—H_2O; (d) Na_2O—SiO_2—H_2O. (a–c: from Refs. 32 and 33; d: from Ref. 76.)

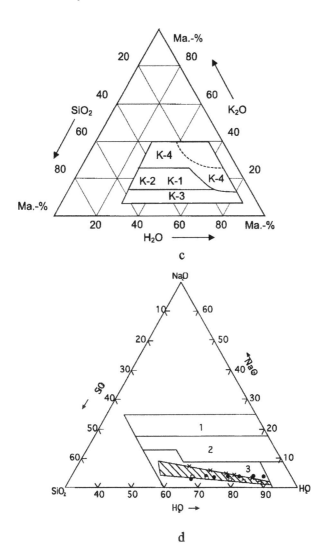

FIGURE 11　Continued.

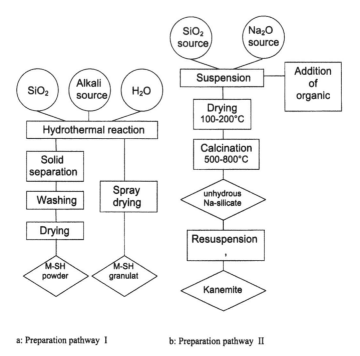

a: Preparation pathway I   b: Preparation pathway II

FIGURE 12   Preparation pathways. Route I: hydrothermal crystallization; route II: combination of precipitation and thermal treatment; route III: solid-state transformation.

tures. The preferred temperatures were 200°C or lower for magadiite formation, 400–500°C or power for the octosilicate, and, as already known from many other publications, 600–800°C for kanemite formation.

C. The Crystallization Process

In analogy to zeolites, synthesis of the layered metal silicate hydrates can proceed by the following consecutive crystallization stages: aging, incubation, nucleation and mass growth, a period of stability, and recrystallization.

Figure 13 shows schematically the transformation of an amorphous material into crystalline phases I, II, and III (e.g., amourphous gel → magadiite → kenyaite → quartz). The aging period[*] (A) is followed by an incubation period (B), which precedes the nucleation and growth step for the first crystalline phase in this

---

[*] In the broadest sense the aging period can comprise just mixing and heating of the raw materials.

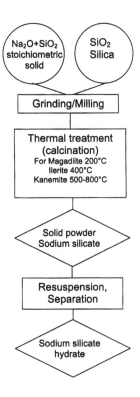

c: Preparation pathway III

FIGURE 12 Continued.

sequence by complex interactions between several unknown processes. A period of stability (D) is followed by the recrystallization process (E).

If this phase transformation is a sequential process where even crystalline phases are transformed to other crystalline phases, characteristic changes observable in the stability range (D) of phase I must correspond to the incubation period (B′) for the second crystalline phase (II), where the nucleation of the subsequent product takes place. As a consequence, the recrystallization process (period E) of phase I and the growing period (C′) of phase II are of the same time scale. This sequential recrystallization process finally leads to thermodynamically more stable products such as cristobalite and quartz.

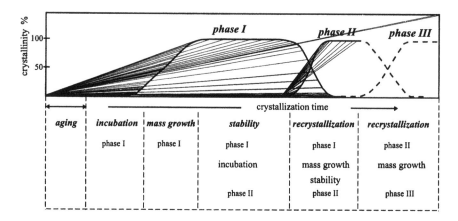

FIGURE 13   Schematic representation of hydrothermal crystallization.

The different stages of such a sequential recrystallization process are influenced by a large variety of parameters and conditions. As a result, the reported synthesis fields and existence regions of the individual layered silicate hydrates constitute an approach that is valid only for the specified conditions. Changing one of the parameters (e.g., temperature, pressure, reactivity of raw materials) can result in different equilibrium diagrams or products with different X-ray diffraction patterns (Sec. II) [76,77].

All of the layered silicate hydrates listed in Tables 3 and 4 represent metastable phases during the process of quartz formation. The stability depends on the reaction conditions and the composition when the products are separated from the mother liquor. After isolation and drying they are stable and can be stored under ambient conditions for long periods of time without any structural changes.

Numerous authors have described different pathways of quartz formation where at least one of the intermediate products is a layered silicate [11,31–33,36,37,39,41,69,74,76,87–89]. Starting from amorphous SiO_2 and depending on the conditions, different sequences of various silicate hydrate phases were observed, e.g., amorphous $SiO_2 \rightarrow$ magadiite $\rightarrow$ kenyaite $\rightarrow$ quartz (containing a sequence of two consecutive layered structures) or, without the layered compounds, e.g., amorphous $SiO_2 \rightarrow$ cristobalite $\rightarrow$ quartz. Other sequences are reported in the cited literature and summarized partially in Table 7. The crystallization sequences are correlated with the increasing density of the crystalline products (magadiite 2.2–2.33 g/cm^3 [36]; cristobalite 2.33 g/cm^3 [74]; quartz 2.65 g/cm^3 [90]) and the degree of condensation [26]. The solubility of the SiO_2 products also decreases in this order [91].

Table 7 Product Sequences During Hydrothermal Quartz Formation with Metal Silicate Hydrates as Intermediate Phases

Starting material	Product sequence	Refs.
amorphous SiO_2 →	cristobalite → quartz	89
amorphous SiO_2 →	keatite → quartz	88, 241
amorphous SiO_2 →	quartz	88, 242
amorphous SiO_2 →	cristobalite → keatite → quartz	88
amorphous SiO_2 →	disordered cristobalite → keatite → cristobalite → quartz	243
amorphous SiO_2 →	disordered cristobalite → quartz	244
amorphous SiO_2 →	SiO_2-X → cristobalite → quartz	31
amorphous SiO_2 →	SiO_2-X → keatite → quartz	245
amorphous SiO_2 →	Opal-like SiO_2 → SiO_2-X or	36
amorphous SiO_2 →	SiO_2-Y → disordered cristobalite → quartz	74
amorphous SiO_2 →	PS1 → cristobalite → quartz	
SiO_2 solution[b] →	magadiite → kenyaite → quartz	
SiO_2 solution[b] →	ilerite	
amorphous SiO_2 →	magadiite → ilerite	68
	magadiite → cristobalite → quartz	
amorphous SiO_2 →	magadiite → cristobalite → quartz	
	magadiite/quartz → kenyaite	111
amorphous SiO_2[a] →	SH-P20 (or SH-P30; SH-P40) → quartz	79, 80 226
SKS-2 (magadiite)[a] →	FLS → KLS2 → FLS → quartz	
magadiite[a] →	HLS → quartz	

[a]Transformations in the presence of organic solvents or cations resulting in organic intercalated layered materials.
[b]Due to the low crystallization temperature, the transformation stops at the stage of a layered silicate.

Both the structure of the layered silicate hydrates and the product sequences are influenced primarily by the composition of the reaction mixture and the synthesis temperature [70–74]. The presence of neutral salts [73,76] and impurities of the raw materials, especially aluminum ions, must also be taken into consideration [25] when the varying phase compositions or impurities shift the synthesis fields or induce changes in the X-ray diffraction patterns of the crystallized products.

Thus far, most studies are related to syntheses in the ternary system Na_2O—SiO_2—H_2O [11,12,17,18,23,29,30,32,33,36,37,41,42,62,66,69,70–

72,74,76,87,92–94]. Several papers dealing with the synthetic aspects of the layered sodium silicate hydrates are listed in Table 4.

Information is also available for the systems K_2O—SiO_2—H_2O [31,36,37,40,74,76,87,95]; DE-OS 27 42 912; DD-WP 220 584) and Li_2O—SiO_2—$_2O$ [37,76]. In contrast, other alkali sources or mixtures of alkali and/or alkali and earth-alkali have not been systematically investigated [37,40,76]. Therefore, the field of layered silicate hydrate research is still very open and may yet provide surprising results.

Direct synthesis of alkaline earth metal and silver–exchanged layered silicates of silicic acids $H_2Si_{20}O_{41} \cdot H_2O$ was reported by Beneke et al. [40]. The synthesis of transition metal layered silicate hydrates has not yet been achieved; they are available, however, from alkaline metal layered silicate hydrates via ion exchange [76].

D. Special Aspects

The general point of view of M-SH formation (usually the first step in the above-mentioned product sequences), i.e., transformation of an amorphous matrix into the crystalline layered silicate hydrate, will be discussed based on literature reports and personal experience with the ternary system Na_2O—SiO_2—H_2O [38,69,70–73,76,96]. Most of the principles, with appropriate adjustments, apply to other systems. Several trends were found that are analogous to zeolite formation, which is carried out under the same conditions and is influenced essentially by the same parameters but usually in the presence of aluminum ions.

Table 8 shows a compilation of the most important parameters influencing M-SH synthesis and their effects on (a) the rate of formation and (b) the formation of impurities. All crystalline substances that do not belong to the desired M-SH are designated foreign phases. These phases include other M-SHs as well as three-dimensionally cross-linked SiO_2 products such as cristobalite and quartz. Some of the parameters will be discussed in the following subsections.

1. Composition of the Reaction Mixture

The most important parameters for the crystallization of a certain product are the molar composition of the reaction mixtures and the type of raw materials employed. The overall composition of the synthesis mixtures is usually given in molar ratios $Na_2O:xSiO_2:yH_2O$. Because they define the position in the synthesis field, these molar ratios basically determine the synthesis conditions. With increasing SiO_2 and decreasing Na_2O, the crystallization yields layered silicates richer in SiO_2, or, in other words, with more highly condensed structures (higher Q^4/Q^3 ratio).

Typical molar ratios for the syntheses of kanemite, octosilicate (ilerite), magadiite, and kenyaite are given in Table 9. The Na_2O and water content deter-

Table 8 Parameters Controlling the Metal Silicate Hydrate Crystallization

Parameter	Influence on the formation rate	Effect on side-product formation
Position in the synthesis field	Accelerative or retarding	Low in the center, promotion on the brink
Reactivity of starting materials	Some effect[a]	Some effect[a]
Impurities in the starting materials	Retarding	Strongly promoting
Increasing synthesis temperature	Strongly accelerative	Strongly promoting
Seed addition	Strongly accelerative	Retarding
Increase of shear stresses	Strongly accelerative	Not investigated
Ageing time	Some effect[a]	Not investigated
Addition of neutral salts	Retarding	Promoting
Addition of organic compounds	Some effect[a]	Not investigated
Synthesis time		See product sequences

[a]Inconsistent or incomplete literature data.

mine the alkalinity of the reaction medium (and thus the dissolution rate of the raw materials) and the nucleation and crystal growth processes. The rate of M-SH formation normally increases with increasing alkalinity (see Fig. 14), as long as the synthesis field is not abandoned. However, the yield of layered silicate hydrate falls with increasing alkalinity because the solubility of sodium silicate rises. When the composition of the reaction mixture is outside the synthesis field, formation of byproducts is favored and impurities are obtained.

Depending on the type of raw materials, feed proportions, and crystallization times, significant differences in the morphology of kenyaite products have been observed [11]. In addition to the oxide ratios, the solid/solution ratio at a constant SiO_2/Na_2O ratio (different alkalinity) plays an important role in the direction of the crystallization. Sakamoto et al. [97] found that magadiite was preferred with increasing solid content, whereas kenyaite was formed at lower solid contents.

2. Raw Materials

The type of raw materials has an influence on M-SH synthesis. Starting materials with different reactivity (amorphous/crystalline, more or less aggregated, different solubility) are common. Table 10 contains an assortment of the raw materials that have been used for the syntheses of different Na-SH compounds. For magadi-

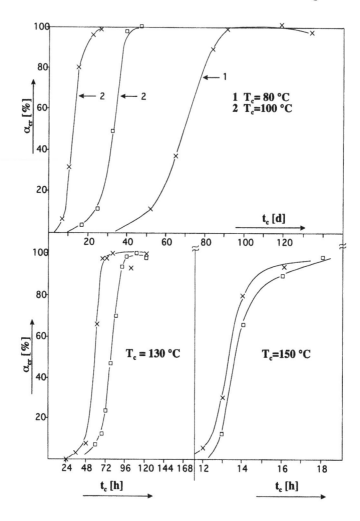

FIGURE 14 Kinetics of magadiite formation at different temperatures and for molar compositions $5SiO_2/1Na_2O/75H_2O$ ($\times$) and $9SiO_2/1Na_2O/75H_2O$ ($\square$) (From Schwieger, unpublished). α_{cr} is crystallization degree (crystalline fraction of the solid product); T_c is crystallization temperature; t_c is crystallization time in hours (h) or days (d).

Table 9 Summary of Typical Synthesis Conditions for the Formation of Selected Sodium Silicate Hydrates

Silicate	Molar composition of the reaction mixture molar ratios	Crystallization time (days)	Crystallization temperature (°C)	Refs.
Octosilicate (ilerite)	$Na_2O:4SiO_2:30H_2O$	21–28	100	26, 142
Magadiite	$Na_2O:9SiO_2:75H_2O$	5	130	26
Magadiite	$Na_2O:9SiO_2:75H_2O$	28	100	17
Magadiite[b]	$NaOH + Na_2CO_3:3–5SiO_2:$ $100–150H_2O$	2–4	150–170	101
Kenyaite[a]	$Na_2O:32SiO_2:70H_2O$	3 months	125	11
Kenyaite	$Na_2O:4SiO_2:35H_2O$	5–7	150	26
Kenyaite[b]	$NaOH +$ $Na_2CO_3:3–20SiO_2:150–2$ $00H_2O$	ca. 3–4	100–150	101

[a]Synthesis run 37 L (Ref. 11).
[b]$NaOH:Na_2CO_3$ ratio = 1:2.

ite and kenyaite, the alkaline constituents are generally hydroxides or sodium salts, especially carbonates [17,18,26,32,33,42,69,98–101]. However, basic silicates (water glass), which act as both alkali and SiO_2 source (e.g., Refs. [11,12,29,30]) are used. The SiO_2/Na_2O ratio required is adjusted with various acids or low-alkali SiO_2 products (e.g., Refs. 11,12; DE-OS 34 00 132; DD-WP 22 34 26). Na-SH compounds are formed from crystalline compounds by recrystallization; crystalline materials can, therefore, be employed as raw materials [18,70–72,76]. Silica freshly precipitated from aqueous sodium silicate solutions with sulfuric acid is a good raw material for magadiite and kenyaite crystallization [102]. Muraishi et al. [103–105] and Kosuge et al. [85] have reported the use of silica gel in the presence of sodium or potassium hydroxide solutions to prepare SiO_2-X_2 or kenyaite-type silicates.

An overview of the diversity of raw materials that can be used for the production of layered silicate hydrates is presented by Beneke and Lagaly [18] for kanemite, which is obtained from numerous amorphous and crystalline raw materials with appropriate reaction processing. Thus, kanemite, or the kanemite precursor δ-$Na_2Si_2O_5$ [106], is obtained not only by hydrothermal reaction but also by annealing of raw materials at around 700°C followed by hydration of the sodium silicate ([18,66,75,78,107]; EP94–106794). Bergk et al. [70–72,108] reported synthesis of magadiite from low-cost raw materials such as solid water glass, technical silicas, and recycled SiO_2 products. They observed decreasing

Table 10 Raw Materials for Crystallizing Metal Silicate Hydrates

Na$_2$O source	SiO$_2$ source	Adjustment of the molar ratios by addition of:	Manufactured Na-SH (example)	Refs.
NaOH	Silica sol (colloidal silica)		Octosilicate	32, 33, 38, 74
Na$_2$CO$_3$	Silica sol (collodial silica)		Magadiite, kenyaite	38; DE-OS 34 00 130, DD-WP 22 34 26
NaOH	Silica gel		Kanemite	42, 74
Na$_2$CO$_3$	Precipitated silica-silicic acids aerosol		Octosilicate (ilerite)	11; DE-OS 34 00 130; DD-WP 22 34 26
			Magadiite	40, 87
			Kenyaite	17
Sodium water glass (solution) Solid sodium water glass ·		Low-alkali-content SiO$_2$ products • silica sol • silica gel • precipitated silica • SiO$_2$ waste products from the silicon, ferrosilicon, or phosphate-production	Octosilicate (ilerite) Magadiite Kenyaite	12, 40, 70, 71 DD-WP 22 34 26; Bergk et al., 1987; DD-WP 221 722, DD-WP 221 724, DE-OS 34 00 132
Sodium water glass (solution) Solid sodium water glass		Diverse organic or inorganic acids: H$_2$SO$_3$, H$_2$SO$_4$, HCl, acetic acid, EDTA, fatty acids	Ilerite, magadiite, kenyaite	72, 73; DD-WP 234 878, DD-WP 220 585, DD-WP 221 722, DD-WP 221 723, DD-WP 221 724, DD-WP 234 878, DD-WP 235 062, DE-OS 34 00 132
NaOH	H$_2$Si$_2$O$_5$-111		Kanemite	66
NaOH	Magadiite	Na silicate solution, water, CH$_3$OH	Kenyaite	DD-WP 220 586
NaOH	Ilerite		Kanemite	18
NaOH	Magadiite		Kanemite	18
NaOH	α,β Na$_2$Si$_2$O$_5$		Kanemite	18

formation rates and varying phase compositions of the reaction products, depending on the impurity levels of Al_2O_3 and Fe_2O_3. Increasing concentrations of the zeolite ZSM-5 is observed when the Al_2O_3 content in the raw materials is too high. Magadiite and kenyaite also appear as undesired byproducts at the borderlines of the ZSM or mordenite zeolite synthesis fields (109; DE-OS 30 48 819; EP. 42 225). Reaction mixtures in which Al_2O_3 or other foreign elements play a role should, strictly speaking, be embraced in the corresponding quaternary systems because they no longer represent pure layered silicate hydrate systems.

3. Crystallization Temperature

The synthesis temperature can be considered the second most important parameter of synthesis. Because M-SH syntheses take place almost exclusively under hydrothermal conditions at autogenous pressure, the pressure is linked strictly to the partial pressure of the solvents, mostly water, taking part in the reaction. The rate of formation of layered silicate hydrates increases with increasing temperature, as shown in Figure 14 for formation of magadiite at four crystallization temperatures (80, 100, 130, and 150°C) and SiO_2/Na_2O ratios of 9 and 5. The kinetic constants of magadiite formation evaluated by the approach of Schwieger et al. [69] are listed in Table 11. The crystallization rate of the low-alkali silicates increases more strongly with the crystallization temperature than for high-alkali silicates. This is indicated by the higher activation energy of seed formation and the longer crystal growth periods. Higher temperatures also accelerate the subsequent recrystallization processes, leading to more dense phases. The formation of kenyaite and quartz as foreign phases is, therefore, favored by higher temperatures [8,74,76].

Metal silicate hydrates in hydrothermal synthesis are not formed above certain temperatures. According to our experience, this temperature lies at about 120°C for octosilicate, 200°C for magadiite, and 240°C for kenyaite if the hydrothermal crystallization is carried out discontinuously (batchwise). With the appropriate choice of other reaction conditions these limits may be shifted.

In accordance with (DE-OS 34 00 130), increasing the synthesis pressure above the autogenous pressure by the use of an inert gas has nearly no effect on the synthesis of kenyaite. On the other hand, a strong acceleration of quartz formation and the appearance of different potassium silicate hydrates under enhanced pressure has been described [6,41,87].

4. Seeding Effects

The influence of crystal nuclei addition on M-SH synthesis is analogous to the effect on zeolite formation. A decrease of the crystallization time was detected in all cases reported (30,107; DE-OS 34 00 130; DD-WP 221 723; DD-WP 221 724; DE-OS 34 00 132;) and yields purer products (DE-OS 34 00 130; DE-OS 34 00 132).

Table 11 Kinetics of Magadiite Crystallization

a. Kinetic data of the incubation period (Schwieger, unpublished). Incubation period t_0 of magadiite crystallization at two molar ratios and four crystallization temperatures and the resulting activation energy of nucleation (estimation of t_0: time to reach a crystallization degree $\alpha_{Kr} = 0.1$ or 1%.

Temperature	Composition of the reaction mixture (molar ratios)			
	$9SiO_2/1Na_2O/75H_2O$		$5SiO_2/1Na_2O/75H_2O$	
T_K (°C)	t_0 for $\alpha_{Kr} = 0.1\%$ (h)	$\alpha_{Kr} = 1\%$ (h)	$\alpha_{Kr} = 0.1\%$ (h)	$\alpha_{Kr} = 1\%$ (h)
80	—	—	60.0	648.0
100	204.0	312	24.0	84.0
130	20.4	40.8	21.6	36.5
150	10.0	11.7	8.1	10.0
E_K (kcal/mol)	19.2	20.1	8.8	16.9

b. Kinetic data of the growth period (Schwieger, unpublished). Crystallization rates of magadiite at two different molar ratios and four crystallization temperatures and the resulting activation energy of crystallization. Crystallization rate (%/h) derived from the crystallization curve between transformation rates of 40% and 60% and the corresponding times; k_{Kr}; crystallization constant assuming a formal first-order kinetic model of crystallization

Temperature	Composition of the reaction mixture (molar ratios)			
	$9SiO_2/1Na_2O/75H_2O$		$5SiO_2/1Na_2O/75H_2O$	
T_K (°C)	%/h	k_{Kr} (h^{-1})	%/h	k_{Kr} (h^{-1})
80	—	—	0.119	0.004
100	0.32	0.008	0.505	0.021
130	4.63	0.106	8.330	0.179
150	66.70	1.620	60.600	1.080
E_w (kcal/mol)	32.7	32.5	26.6	23.2

Source: Ref. 69.

5. Isomorphic Substitution

The isomorphic substitution of framework atoms, especially of Si by different framework-building atoms, is a well-known method in zeolite chemistry to modify the properties of the framework. Only a few cases have been reported for M-SHs. In order to change the properties of layered silicate materials systematically, Schwieger et al. [110–112] studied the substitution of silicon by aluminum and/

or boron. As proven by [11]B, ^{27}Al, and ^{29}Si solid-state NMR, addition of boron and/or aluminum compounds to the reaction mixtures for hydrothermal processes yields boron- and aluminum-containing magadiite- or kenyaite-type materials, without the need of posttreatment [110]. However, the effect of the substitution on the properties has not been elucidated as yet. An isomorphic substitution also takes place if the raw materials contain M_2O_3 compounds as impurities. Three effects are seen [25,111,113]: (a) Isomorphic substitution is possible; (b) additives generally decrease the nucleation and crystallization rate; and (c) high-silica zeolites form if the added amount is too large. In Figure 15 the X-ray diffraction patterns of some novel boron-containing layered materials are shown in comparison to the well-known diffraction patterns of magadiite and kenyaite. The identity of the materials is still unknown. Addition of different amounts of boron yields more or less well crystallized materials, but the boron content in the structures is relatively low.

Pál-Borbély et al. [114] proposed a new method for the synthesis of high-SiO_2 ferrierite (Si/Al = 15–18) by recrystallization of Al^{3+}-containing magadiite varieties. In this case the Al^3-containing magadiite acts as a precursor for the

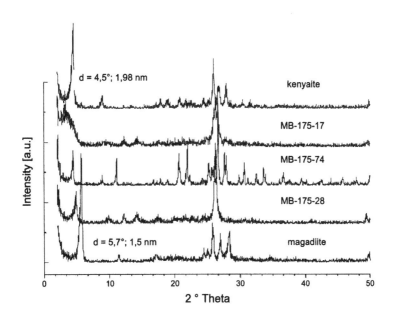

FIGURE 15 Isomorphous substitution of silicon in magadiite by boron during synthesis: X-ray diffraction pattern of the boron-containing products compared with the patterns of magadiite and kenyaite (Schwieger, unpublished) (MB-175–17: boron-containing magadiite synthesized at 175°C, run 17).

crystallization of a microporous material. The synthesis and structural characterization of a novel layered alumophosphate (named AlPO-ntu) with a structure mimicking that of the naturally occurring silicate mineral kanemite is described by Chen et al. [115]. This is the first report in which, similar to the field of zeolitic materials, both framework elements in a layered aluminum phosphate are in tetrahedral surrounding. This new compound was synthesized hydrothermally, using n-alkylamines as structure directing agents. From the results of thermal and elemental analysis, the chemical formula is $AlPO_3(OH)_2[NH_2(CH_2)_xCH_3]$ (x = 3, 5, and 7) for butyl-, hexyl-, and octylamine as structure-directing compounds. Similar to kanemite with a single layer structure, the aluminophosphate layers are able to reorganize and condense to form porous materials when the interlayer alkylammonium ions are exchanged by surfactant cations with longer chains.

6. Addition of Inorganic Salts

The sequence of consecutive products and also the morphology of the synthesis products can be altered by addition of inorganic salts to the synthesis mixture [36,73]. Addition of LiCl, KCl, RbCl, and $MgCl_2$ retards nuclei formation. However, it accelerates recrystallization to cristobalite, kenyaite, and quartz [76]. According to (DE-OS 31 23 000), the formation of magadiite from silica gel and sodium hydroxide solution is accelerated by copper sulfate and ammonia. Bergk et al. [73] observed that the sulfate anion shortens the incubation period of the magadiite crystallization process much more than nitrate or chloride anions.

     Fletcher et al. [100] and Kwon et al. [101,116] synthesized highly crystalline Na^+ magadiite and Na^+ kenyaite when they reduced the alkalinity by replacing NaOH with sodium salts, e.g., Na_2CO_3.

7. Addition of Organic Compounds

The influence of organic additives on M-SH formation has not been systematically investigated. Organic additives modify the solubility of the substances in the reaction, can be bearers of additional alkalinity, or act as structure-directing compounds, analogous to the formation of high-silica zeolites. Only occasional attempts have been described focusing on different aims. M-SHs synthesized with an admixture of polyfunctional alcohols such as glycols and water retain about 1% of the organic compounds (DD-WP 22 34 26). The organic compounds are not included in the structure of the layered silicates as occurs for many structure-directing compounds in zeolite chemistry. Triethanolamine was tested in M-SH syntheses with the aim of producing large crystals suitable for X-ray structure analysis. An effect was detected only for makatite. Annehed and coworkers [23] synthesized a makatite sample from which a thin platy crystal (0.090 × 0.052 × 0.006 mm^3) was selected for the crystal structure determination.

Vortmann et al. [20] also introduced an organic compound into the reacting system to synthesize well-ordered, powdered RUB-18 suitable for structure determination by direct methods as described earlier. According to the first report in 1997, RUB-18 was crystallized in the multicationic system $SiO_2/Na_2O/Cs_2O/$ hexamethylentetramine/triethanolamine/H_2O in Teflon-lined autoclaves at 95°C for 21–28 days. Compared to the standard $Na_2O/SiO_2/H_2O$, system from which octosilicate can be synthesized, the applied synthesis mixture seems to be very complex. No information was given about the function of the different cationic or organic components of the reaction mixture in respect to the crystallization behavior. Organic material was not intercalated or incorporated in the structure of RUB-18.

In the presence of organic compounds, metal silicate hydrates often appear as byproducts in the manufacturing process of high-silica zeolites. Addition of ethylenediamine in the synthesis of ZSM-5 promoted formation of magadiite (DE-OS 30 48 819), whereas addition of alcohols yielded kenyaite (EP 42 225). The properties of the M-SH byproducts were not specified. Kosuge et al. [85] reported that the addition of butanol was very effective for magadiite formation. The crystallization time was strongly reduced at lower temperature and alkalinity.

A group of three new layered silicates, designated SH-P20, SH-P30, and SH-P40, with organic compounds in the interlayer space were synthesized in the quaternary system $Na_2O/SiO_2/H_2O$/poly(diallyl dimethylammonium) salt [111,113,117]. The polycations were intercalated into the silicate during the crystallization process without any decomposition. According to the X-ray diffraction patterns, the novel silicates do not correspond to any known layered silicate. From the structural point of view, many similarities with the alkali silicate hydrates were observed. The new silicates show a high structural stability against a wide range of different energetic influences. Due to this relatively high stability, high-resolution electron microscopy could be applied to study the layer structure of these silicates as representatives of M-SHs. Thus, Schwieger et al. [117] demonstrated for the first time with a direct method that single silicate sheets are combined to form the bulk layer of the silicate.

Pastore et al. [118] prepared cetyltrimethylammonium- and tetradecyltrimethylammonium-intercalated magadiites by direct synthesis, starting from sodium metasilicate with a molar ratio $Na_2O/SiO_2 = 1$ and nitric acid. Total substitution of Na^+ by cetyltrimethylammonium or tetradecyltrimethylammonium cations was not achieved. This indicates that the formation of magadiite is greatly dependent on the presence of sodium cations in the reaction mixture.

8. Recrystallization

As described earlier, recrystallization of thermodynamically unstable intermediates to more stable products is one of the characteristic features of M-SH compounds. Product sequences with two consecutive layer silicates are often reported

[17,69–72,76,77]. In addition to these more or less common recrystallization effects, Kooli et al. [79–84] reported a precise method to recrystallize magadiite or kanemite resulting in new types of layer structures, named KLS and FLS. The transformation takes place in the presence of alkali cations, tetramethylammonium cations (hydroxides), water and 1,4-dioxane. Dioxane was found to be essential to formation of these unique structures (see also Sec. VI.A).

E. Conclusions

Even though the bibliography related to the formation of M-SHs is not very extensive, the comparison with the extensively investigated zeolite crystallization is complementary. The large number of similarities was pointed out in the discussion about synthetic parameters. The diversity of raw materials and product sequences, as well as the absolute necessity of a liquid phase for the formation of M-SH, points toward a solution–crystallization mechanism as discussed for zeolite formation [119–121]. The synthesis of metal silicate hydrates and zeolites often differ solely by the absence or presence of aluminum-containing compounds and, therefore, in the dissolution behavior of the reacting components [11,39,41,109].

IV. SURFACE PROPERTIES AND REACTIONS

A. Surface Acidity

The silanol groups on the surface of silica are only weakly acidic, although they are more acidic than the orthosilicic acid H_4SiO_4, which has a pK_a value of ~ 10. A simple but reliable method for measuring the acidity of surface silanol groups is their reaction with certain dye molecules called Hammett indicators. The acid strength of a solid surface can be defined as the proton-donating ability of surface groups, expressed by the Hammett and Deyrup function:

$$H_0 = -\log \frac{a_{H^+} f_B}{f_{BH^+}} \equiv pK_{ind} \tag{eq. 1}$$

where a_{H^+} is the proton activity of the surface and f_B, f_{BH^+} are the activity coefficients of the adsorbed indicator molecule [122,123]. The color change of an adsorbed Hammett indicator can be used to bracket the surface H_0 values. When the color of the acid form of the indicator is observed, H_0 is equal or lower than the pK_a value of the indicator. Using a series of indicators, the surface H_0 value is bracketed between two pK_a values, one lower and one higher than H_0. The brackets depend on the series of Hammett indicators that are available or can be applied [124]. In rare cases, an acidic color can also be produced by processes other than simple proton transfer [123].

Crystalline silicic acids are distinctly more acidic than all types of silica ($4.9 < H_0 < 6.8$), with a range of H_0 values between -5 and $+3.3$. Most acidic is the acid prepared from magadiite, $H_2Si_{14}O_{29} \cdot xH_2O$, with a surface acidity of $-5 < H_0 < -3$. The acids from kenyaite, the "Iler silicate," and makatite are slightly weaker, $-3 < H_0 < +1.5$. The different modifications of $H_2Si_2O_5$ (from α-, β-$Na_2Si_2O_5$, kanemite, $KHSi_2O_5$, $K_2Si_2O_5$, $Li_2Si_2O_5$, silinaite) are distinctly less acidic ($2.3 < H_0 < +3.3$). This acidity corresponds to aqueous sulfuric acid concentrations of 0.03–0.003% (w/w), whereas H_0 values of $+1.5$, -3, and -5 correspond to equivalents of 0.2%, 48%, and 71%, respectively [123,124].

When the hydration water of $H_2Si_{14}O_{29} \cdot xH_2O$ and $H_2Si_{20}O_{41} \cdot xH_2O$ is desorbed above 200°C, the acidity decreases to $2.3 < H_0 < 3.3$. Transformation into cristobalite increases H_0 to the values of silica ($4.9 < H_0 < 6.8$). Note that the dye molecules cannot penetrate between the layers themselves; therefore, all of these values correspond to the acidity of the external silanol groups.

The acidity of hydrated silicic acids is independent of the method of preparation and the mineral acid used. Reaction with aqueous 2M HCl at 90°C (which increases the acidity of amorphous silica gel from $4.9 < H_0 < 6.8$ to $2.3 < H_0 < 3.3$) has no influence. However, this reaction increases the acidity of the $H_2Si_2O_5$ acids to $-3 < H_0 < 1.5$ [124]. Phosphate adsorption can increase the surface acidity of oxides, e.g., TiO_2 [125], including $H_2Si_2O_5$, which increases to $-3 < H_0 < +1.5$, although it has no effect on the acidity of the more strongly acidic silicic acids [124].

The increased acidity of the silica forms in comparison to the aqueous solution of orthosilicic acid has been explained by formation of $d\pi$–$p\pi$ bonds when the orthosilicic acid molecules condense to amorphous silica. The still higher acidity of the crystalline silicic acids is probably caused by the formation of a regular array of hydrogen bonds on the surface [124].

B. Ion Exchange by Inorganic Ions

The first experiments on the exchange of interlayer sodium ions by other inorganic cations (Li^+, Na^+, Mg^{2+}, Ni^{2+}, Cu^{2+}) and hexadecyltrimethylammonium ions were reported by Iler [12] for the 8:1 type of polysilicate ($8SiO_2/Na_2O$). The exchange of the interlayer cations is less pronounced and does not comprise the diversity of reactions reported for 2:1 clay minerals, in particular, smectites. Typical is the exchange by protons and transformation into crystalline silicic acids, which will be discussed in Section IV.C.

Wolf and Schwieger [65] described the exchange of sodium ions of the Iler silicate by potassium ions and protons as well as the exchange of magadiite sodium ions by potassium ions, lithium ions, and protons. The diagrams in Figure 16 clearly show the selectivity for sodium versus lithium and potassium ions; i.e., the equivalent fraction of the lithium and potassium ions on the silicate is distinctly lower than in the equilibrium solution. The equilibrium constants for the exchange reactions of Na^+ by K^+ or Li^+ are, therefore, very small (on the order of 0.1). Adsorption studies with Cd^{2+}, Cu^{2+}, and Zn^{2+} nitrate solutions on Na^+ magadiite showed preference in the order of $Cd^{2+} > Cu^{2+} > Zn^{2+}$ [126].

The kinetics of exchanging the sodium ions of magadiite by calcium ions was studied by Bergk et al. [92]. The exchange in aqueous solutions was complete after 2 hours at pH > 9.5 (adjusted with ammonia solution) and 20°C.

The pH value during exchange reactions is an important factor because protons compete with the cations at pH ≤ 7. Magadiite, for instance, transforms into the acid even at pH ≈ 7 (Fig. 16). Since the silanol groups are only weakly acidic, the cation exchange decreases at pH < 7 (for Zn^{2+}, see Ref. 127). Sodium ions in the polysilicates cannot be exchanged by Fe^{3+} or Al^{3+} because these cations exist only in acidic aqueous solutions, where the silicates are transformed into crystalline silicic acids. However, the adsorption of La^{3+}, Eu^{3+}, and Ce^{3+} can be studied because these cationic aquo complexes are not acidic. Magadiite shows a preference for La^{3+} and Ce^{3+} (Fig. 16) [128,129]. The amount exchanged (2.34 meq/g La^{3+}, 2.11 meq/g Ce^{3+}) in fact exceeds the cation exchange

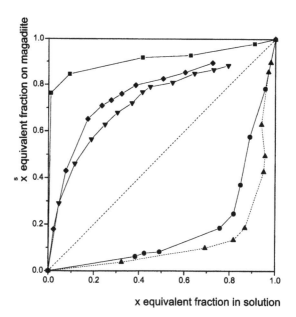

FIGURE 16 Exchange of sodium ions of magadiite by H^+ (■), Li^+ (●), K^+ (▲) at 20°C and La^{3+} (◆), Ce^{3+} (▼) at 25°C. (From Refs. 65 and 128.)

capacity (1.90 meq/g for the composition $1.09Na_2O \cdot 14SiO_2 \cdot 8.14H_2O$) because some cations are bound as divalent cations LaX^{2+} or CeX^{2+}. The basal spacings of air-dried Ce^{3+} and Na^+ magadiites are identical ($d_L = 1.56$ nm), although they differ after dehydration at 400°C (Ce^{3+} $d_L = 1.39$ nm, Na^+ $d_L = 1.29$ nm). It was also noted that the interlayer cerium ions are not completely displaced by protons [128].

An example of ion exchange with complex cations is the adsorption of hexammine cobalt(III) cations $[Co(NH_3)_6]^{3+}$ in kanemite [130]. The amount exchanged increases in the presence of hexadecyltrimethylammonium ions but still remains below the cation exchange capacity. Ion exchange reactions of $[Pt(NH_3)_4]^{2+}$ ions with the crystalline silicic acid from Iler's octosilicate, after calcination, provides a novel method for preparing silica-supported Pt nanoparticles of 2 to 5-nm diameter [131].

In search of novel photofunctional systems, Ogawa and Takizawa [132] tested the reaction of magadiite with tris-(2,2'-bipyridine) ruthenium(II) cations:

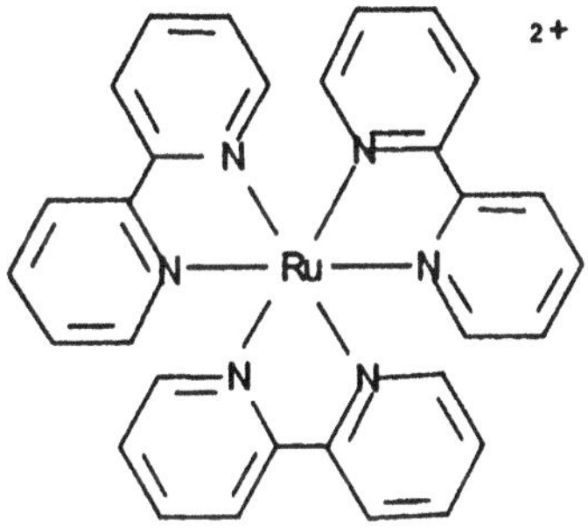

These cations were found to displace the sodium ions of magadiite only in the presence of a crown ether. The reaction increases the basal spacing of the dispersed magadiite from 1.5 nm to 1.8 nm, indicating a monolayer arrangement of $Ru(bpy)_3^{2+}$. The exchange is not fully quantitative; only 0.66 $Ru(bpyr)_3^{2+}$ cations are bound per $Si_{14}O_{29}$, with the remainder of the sodium ions replaced by protons. The authors arrived at the conclusion that the reaction is a two-step process where the crown ether is first intercalated into the interlayer space of magadiite and complexes the sodium ions, which in turn are replaced by the Ru(II) cations.

C. Exchange Reactions with Protons

The Na^+/H^+ exchange of a layered alkali silicate was first studied by Schwarz and Menner [133], who prepared the crystalline silicic acids $H_2Si_2O_5$-I and $H_2Si_2O_5$-II by proton exchange of α-$Na_2Si_2O_5$. The formation of crystalline silicic acids from alkali silicates was not studied until decades later [12,34,66,134–138].

Protons are preferentially exchanged for the interlayer alkali ions of all types of alkali polysilicates. The high selectivity is evident from Figure 16. The thermodynamic equilibrium constants for the Na^+/H^+ exchange in the 8:1 polysilicate ($K_{th} = 81$) and magadiite ($K_{th} = 140$) are much higher than for the Na^+/K^+ or Na^+/Li^+ exchange ($K_{th} \sim 0.1$) [65]. Titration curves reveal that replacement of sodium ions by protons starts in alkaline medium [11,95,139].

The Na^+/H^+ exchange of alkali silicates changes the hydration/dehydration behavior. The basal spacing of the fully hydrated kenyaite (sodium form) and the crystalline silicic acid is 1.97–1.99 nm. Air-drying reduces only the spacing of the H^+ form to 1.76–1.80 nm; after dehydration the spacing remains 1.86 nm or less [11].

When the crystalline silicic acid prepared from natural and synthetic kenyaite is neutralized with potassium hydroxide, the potassium form has a basal spacing of 2.18 nm in the wet state, which decreases to 1.99 nm upon air-drying [11]. The crystalline silicic acid derived from the potassium form retains the higher hydration capacity of the parent material and has basal spacings of 2.13–2.18 nm that decrease to 1.96 nm after air-drying. Clearly, the interlamellar hydration processes of the sodium and potassium forms of kenyaite and their crystalline silicic acids are different. Transformation of the potassium silicate into the sodium silicate, and vice versa, proceeds through two different acid forms [11]:

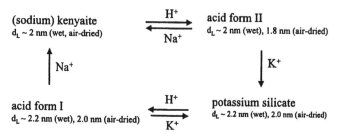

It is likely that the potassium ions between the kenyaite-type layers cause some structural changes that are difficult to detect in the X-ray powder diagrams but clearly differentiated by the hydration and intercalation behavior.

Comparable behavior was found for a potassium silicate prepared either from SiO_2 dispersions in KOH or from potassium water glass solutions [95]. This silicate is related to magadiite but differs by the reversible transition between hydration states with basal spacings of 1.51 nm and 1.73 nm. (Sodium magadiite does not expand above 1.56 nm.) The air-dried acid with a basal spacing of 1.34 nm rehydrates to ~1.5 nm, in contrast to the H^+ form of magadiite, which remains at 1.32 nm. The structural difference between these crystalline silicic acids is also evident in enhanced intercalation behavior; the acid form of the potassium silicate intercalates, for instance, nitriles and ketones, which are not intercalated by the crystalline silicic acid from magadiite.

The exchange of alkali metal ions in silicates consisting of single layers ("Einfachschichten," Ref. 16), such as $Na_2Si_2O_5$, kanemite, and silinaite, usually leads to the formation of a mixture of phyllodisilicic acids. Depending on the experimental conditions, kanemite, for instance, transforms into $H_2Si_2O_5$-III as the main compound, and $H_2Si_2O_5$-II and $H_2Si_2O_5$-IV are formed to a lesser extent. Pure $H_2Si_2O_5$-III is obtained only under special conditions [18]. The reaction of mineral acids with silinaite yields the corresponding acid, but $H_2Si_2O_5$-I (from α-$Na_2Si_2O_5$) is also formed [140]. One of the reasons for the transformation of monophyllosilicates into a mixture of different modifications of $H_2Si_2O_5$ is the flexibility of the single layers. The reaction of silinaite with concentrated acids causes a certain delamination of the silicic acid into the individual layers, or packets of them, so that the X-ray powder diagram is poorly resolved. During washing and air-drying, the layers and lamellae reaggregate, and sharper and more intense reflections are observed.

The exchange of interlayer protons by Na^+ or K^+ ions starts at pH $\sim$ 5 and then increases very steeply [65,124]. The H^+/Na^+ exchange of magadiite and the acid of the "Iler silicate" proceeds at pH $\sim$ 7 and is complete at pH = 9–10, whereas $H_2Si_{20}O_{41} \cdot xH_2O$ is fully exchanged only at pH $\sim$ 12 (Fig. 17) [124].

When coarse particles (fraction 63–200 μm) of the crystalline silicic acid $H_2Si_2O_5$-I (from α-$Na_2Si_2O_5$) are titrated with NaOH, the exchange proceeds in a sequence of small steps [124]. This is indicative of the cooperativity of the exchange reaction. When a small amount of NaOH is added to the crystalline silicic acid, a certain number of protons at an edge site are displaced by sodium ions. The hydrogen-bonding system is strongly disturbed, therefore, further sodium ions can penetrate into this region. However, depending on the acidity of the silanol groups, the layer–layer interaction, the hydrogen bonds, and the elasticity of the layers, the reaction slows down and a new nucleation step is required to intercalate a further amount of sodium ions.

D. Exchange with Organic Cations

Similar to exchange with inorganic ions, protons compete with the organic cations at pH $\leq$ 7. At high pH values (pH $\geq$ 10), large amounts of the base are needed to adjust the pH. The Na^+ or K^+ ions compete with the organic cations and reduce the amount of exchanged organic cations. In comparison to smectites, quantitative exchange of inorganic by organic cations often requires higher concentrations of the organic cations, up to initial concentrations of 0.1 eq/L. These concentrations can be near or even above the solubility limit of long-chain organic salts in water; therefore, ethanol is added to obtain a clear solution. However,

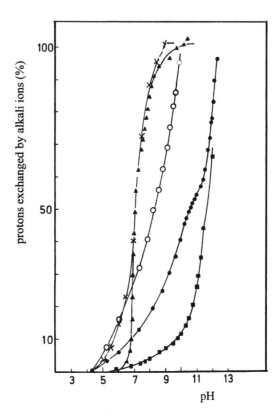

FIGURE 17 Exchange of protons of crystalline silicic acids by titration with NaOH or KOH. $\times$ $H_2Si_{14}O_{29} \cdot 5.4H_2O$ (from magadiite) with NaOH, $\blacktriangle$ $H_2Si_8O_{17} \cdot 1.1H_2O$ (from octosilicate) with NaOH, $10H_2Si_{20}O_{41} \cdot 2.75H_2O$ (from $K_2Si_{20}O_{41} \cdot xH_2O$) with NaOH; $\bullet$ $H_2Si_{20}O_{41} \cdot 2.75H_2O$ (from $K_2Si_{20}O_{41} \cdot xH_2O$) with KOH, $\blacksquare$ $H_2Si_2O_5$-V (from $K_2Si_2O_5$) with KOH.

alcohol contents that are too high can reduce the degree of exchange [42]. As a rule, the ethanol content should be 20% (v/v) or less.

Exchange reactions with cationic surfactants, i.e., long-chain organic cations, are an easy way to change the surface character of the silicate from hydrophilic to hydrophobic. Hydrophobization is an important process for tailoring adsorbents, for instance, to increase the adsorption of toxic compounds such as benzene from aqueous solutions [141].

$$\text{C}_5\text{H}_5\overset{+}{\text{N}} - (\text{CH}_2)_{n-1} - \text{CH}_3 \qquad \text{(sch. 11.3)}$$

Alkylpyridinium ions are quantitatively exchanged [17,42]. In equilibrium with the supernatant (pH = 7.5–8, 10% ethanol, surfactant concentration ≤ 0.1 mol/L), the basal spacing of derivatives increases linearly with the number of alkyl-chain carbon atoms from $d_L \sim 3$ nm ($n = 8$) to $d_L \sim 4.4$ nm ($n = 18$) (Fig. 18). The mean increase $\Delta d_L/\Delta n = 0.150$ nm/carbon atom indicates bilayers of alkylpyridinium ions with the alkyl chains tilted in an angle of 35° to the layer. The pyridinium ring is assumed to lie flat on the surface. Water molecules intercalated between the surfactant cations are essential for the stability of the interlamellar structure.

When a well-crystallized magadiite is partially loaded* with hexadecylpyridinium ions (cetylpyridinium ions), the X-ray powder diagrams show basal spacings of 3.88–4.02 nm, corresponding to the fully exchanged magadiite, and a spacing of 1.55 nm of unreacted magadiite [142]. Thus, the samples are not uniform, and packets of fully exchanged layers alternate with unreacted layers (Fig. 19). It may also be that the organic cations have penetrated into the magadiite crystals to only a certain degree so that core-shell particles are formed. When these samples are dispersed in water (1g/100 mL) and stirred for 4 h at 60°C, a

---

* 27%–131% related to the theoretical cation exchange capacity, 0.170 meq/g for $0.9Na_2O \cdot 13.9SiO_2 \cdot 9.3H_2O$.

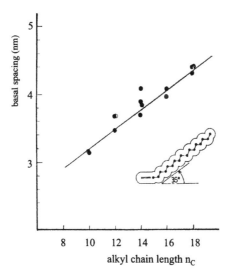

FIGURE 18 Basal spacings d_L of the alkylpyridinium magadiites (in equilibrium with ~0.1 M alkylpyridinium chloride solutions). Calculated values for bilayers of alkylpyridinium ions with the alkyl chains tilted 35° to the magadiite layer. n_C = number of carbon atoms in the alkyl chain. (From Ref. 42.)

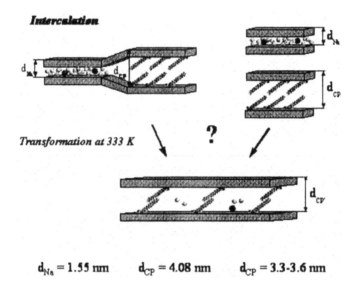

Intercalation

Transformation at 333 K

$d_{Na} = 1.55$ nm $d_{CP} = 4.08$ nm $d_{CP} = 3.3-3.6$ nm

FIGURE 19 Rearrangement of alkylpyridinium ions inside the interlayer spaces of magadiite. Above: nonuniform distribution after ion exchange; below: uniform distribution after equilibration in water. (From Ref. 142).

homogeneous distribution of the surfactant cations is achieved, and the X-ray powder diagrams show (001) and (002) reflections with basal spacings varying between 3.3 nm and 3.6 nm. Unreacted magadiite ($d_L = 1.55$ nm) was observed only at a loading of less than 72%. No surfactant was detected in the aqueous solution after rearrangement. These experiments clearly illustrate that the surfactant cations are mobile and move within and between the interlayer spaces.

Intercalation of dodecylpyridinium ions into magadiite and subsequent displacement by sodium ions with aqueous NaOH solution yields fine platelike particles. This type of magadiite has been used to prepare transparent silica–magadiite monoliths [143].

Like alkylpyridinium derivatives, dialkyldimethylammonium derivatives show basal spacings that increase linearly with alkyl-chain length from ~4 nm ($n = 14$) to ~5 nm ($n = 18$). (Fig. 20). The slope $\Delta d/\Delta n = 0.180$ nm indicates that the alkylammonium ions form bilayers with the chains tilted 45° to the layer [42]. The tetrahedral orientation of ammonium C–N bonds causes the alkyl chains in all-trans conformations to form a "V" (Fig. 21a), and space filling in mono- or bimolecular films subsequently would be poor. Transformation of two trans bonds into two gauche bonds (Fig. 21b) brings the alkyl chains in parallel orientation (Fig. 21c). This configuration is adopted by dialkyldimethylammonium ions

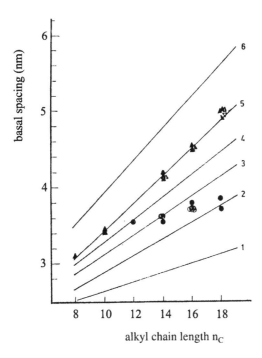

basal spacing (nm)

alkyl chain length n_C

FIGURE 20 Basal spacing d_L of the dialkyl dimethylammonium (▲) and alkyltrimethyl ammonium (●) magadiites in equilibrium with ~0.1 M alkylammonium chloride solutions. Calculated values for monolayers with tilting angles $\alpha = 35°$ (1), $55°$ (2), $90°$ (3) and for bilayers with tilting angles of $35°$ (4), $55°$ (5), $90°$ (6). (From Ref. 42.)

in stable monomolecular films. Nevertheless, a less regular interlamellar arrangement was also discussed [17]. The observed basal spacings were very similar to the high-temperature phases of alkylammonium-alkylamine magadiite or beidellite (see later). Thus, the interlamellar arrangement of the dialkyldimethylammonium magadiite could also correspond to a gauche block structure.

The alkyltrimethylammonium derivatives (Fig. 20) behaved quite differently because the basal spacing increases only minimally with alkyl-chain length (d_L ~3.55 nm for $n = 12$, ~3.8 nm for $n = 18$). The spacings for $n \geq 14$ correspond to monolayers of perpendicular ($n = 14$) or tilted ($n > 14$) alkyl chains. The reason is that, unlike the organic cations mentioned before, the degree of exchange strongly decreases with alkyl-chain length. Whereas about 83% of the sodium ions were displaced by dodecyltrimethylammonium ions, only 30% were displaced by hexadecyltrimethylammonium ions [17]. The reason for this is not quite clear. The trimethylammonium groups may be not anchored on the

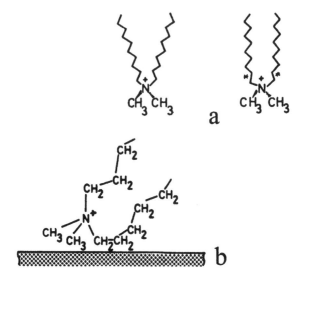

FIGURE 21 Conformation of dialkyldimethylammonium cations. (a) Arrangement with all C—C bonds in trans conformation, (b) with the first C—C bond in gauche conformation so that the surfactants can be densely packed in mono- or bilayers (c).

surface oxygen atoms as tightly as in clay minerals, where the exchange is quantitative.

The alkylammonium derivatives of magadiite also show different interlamellar structures. Due to pH = 7.5–8 and initial concentrations of ~0.1 mol/L, the exchange of sodium ions is accompanied by the intercalation of alkylamines because of the following equilibrium:

$$R\text{-}NH_3^+ + H_2O \rightleftharpoons R\text{-}NH_2 + H_3O^+$$

In aqueous equilibrium solutions, interlamellar alkylammonium ions and alkylamine molecules aggregate as bimolecular films with the all-trans alkyl chains perpendicular to the layers (Fig. 22) [42]. The basal spacings decrease when ethanol (10–15%, v/v) is added to the surfactant solution. Ethanol addition reduces the amount of intercalated alkylamine. Alkylammonium ions with 8–14 carbon atoms form bilayers and are tilted to the layer at angles of $\leq 55°$. Furthermore, the basal spacings correspond to those of gauche block structures. The longer chain cations form monolayers.

Gauche-block structures are formed as a result of alkyl-chain conformational changes when trans bonds rotate into gauche bonds [144–146]. The stepwise decrease of the basal spacings of alkylammonium-alkanol montmorillonites with increasing temperature indicates cooperative formation of kinks (gtg conformations) in the alkyl chains. At temperatures of 70–90°C, and depending on the chain length, these kink-block structures can collapse to gauche blocks. In gauche blocks, longer parts of the chains are tilted to the layers because of a relatively high concentration of isolated gauche bonds or gtg conformations.

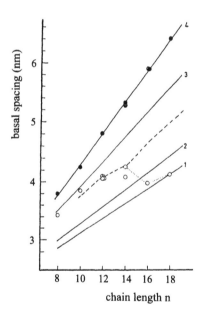

FIGURE 22 Basal spacings d_L of the n-alkylammonium magadiites in equilibrium with ~0.1 M aqueous solutions of alkylammonium chloride (●) and with ~0.1 M aqueous solutions of alkylammonium chloride containing 10–15% (v/v) ethanol (○). Calculated values: 1 monolayer of alkylammonium ions, $\alpha = 90°$; 2, 3, 4 bilayers of alkylammonium ions, $\alpha = 35°, 55°, 90°$, resp. (From Ref. 42.)

The typical basal spacing decrease at the kink-block/gauche-block transition is also observed for alkylammonium-alkylamine magadiite, and it is comparable with the spacings of tetradecylammonium-tetradecylamine beidellite (Fig. 23). Also shown in Figure 22 are the basal spacings of the ditetradecyldimethylammonium derivatives of magadiite and beidellite to illustrate that these large cations, very likely, form gauche-block structures as well.

Unlike the derivatives of montmorillonite and vermiculite, the organic derivatives of the polysilicates, even the air-dried samples, retain high amounts of water in the interlayer space. The air-dried organic derivatives of magadiite contain $\sim$5–8 moles of free water per 14SiO_2, i.e., not bound in the form of silanol groups. The silanol and SiO^- groups projecting into the interlayer space may provide preferential adsorption sites for water. This water essentially contributes to the stability of the interlamellar structures [147].

Amine transfer has been used to prepare a broad variety of organo-magadiite derivatives starting, for example, from dodecylammonium magadiite:

$$(C_nH_{2n+1}NH_3^+)_2Si_{14}O_{29} \cdot zH_2O + 2B \rightarrow (BH^+)_2Si_{14}O_{29} \cdot z'H_2O + 2C_nH_{2n+1}NH_2 + (z - z')H_2O$$

Here B represents diaminodecane, piperidine, pyridine, piperazine, pyrazine, cyclohexylamine, aniline, and benzylamine (see Table 12) [17].

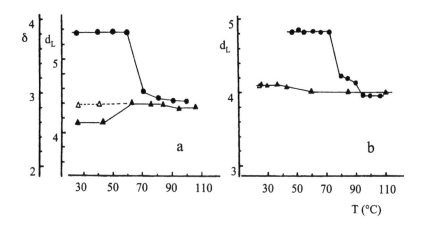

FIGURE 23 Kink- and gauche-block structures: basal spacing d_L and interlamellar film thickness δ as a function of temperature. Tetradecylammonium-tetradecylamine ($\bullet$) and ditetradecyldimethylammonium ($\blacktriangle$,$\triangle$) derivatives of magadiite (a) and beidellite (b). δ = d_L: − layer thickness, 1.54 nm for magadiite, 1.0 nm for beidellite. $\triangle$ metastable spacings after cooling down (From Ref. 17.)

Table 12 Magadiite Amine Complexes Prepared from Dodecylammonium Magadiite by Amine Exchange

Amine	Basal spacing (nm)	Amine	Basal spacing (nm)
n-Hexylamine	2.98	Piperidine	2.30
Daminododecane	2.42	Pyridine	1.61
Cyclohexylamine	2.42	Piperazine	1.90
Aniline	1.52	Pyrazine	1.82
Benzylamine	2.42		

Source: Ref. 17.

The basal spacings of the quarternary alkyltrimethylammonium and dialkyl-dimethylammonium derivatives of kenyaite are about 0.5–0.7 nm higher than those of magadiite derivatives, indicating nearly identical orientation [11]. The difference between the basal spacings of pure kenyaite and magadiite dried at 120°C is 0.62 nm (1.77 nm–1.15 nm), for example. In contrast, the arrangement of the primary alkylammonium ions is different because the basal spacings were lying between the fully expanded phase (Fig. 22) and the structure with tilted chains. Unlike magadiite, kenyaite also reacts with phenyltrimethyl- and phenyl-triethylammonium ions as well as benzylammonium cations (see Table 13).

The alkyltrimethylammonium derivatives of kanemite have been of recent interest because they can be transformed into mesoporous silicates (see Sec. V.C). Kanenite easily reacts with different types of long chain cations: alkylammonium ions, alkyldimethyl- and alkyltrimethylammonium ions, alkylbenzyldimethylam-

Table 13 Organic Derivatives of Kenyaite: Basal Spacings in Equilibrium with 0.1 M Aqueous Alkylammonium Chloride Solutions

Alkylammonium ion	Basal spacing (nm)	Alkylammonium ion	Basal spacing (nm)
Decylammonium	4.06	Decyltrimethylammonium	3.92
Dodecylammonium	4.44	Dodecyltrimethylammonium	4.25
Didecyldimethylammonium	3.92	Phenyltrimethylammonium	2.46
Didodecyldimethylammonium	4.36	Benzyltrimethylammonium	2.72
		Benzyltriethylammonium	2.95

Source: Ref. 11.

monium ions, alkylpyridinium ions, etc. In most cases the basal spacing varies between the theoretical values for bilayers and monolayers with the alkyl chains perpendicular to the layers (for details, see Ref. 18). The degree of exchange at pH = 8–9 was below the cation exchange capacity. Because of the high interlayer cation density, the sodium ions not exchanged by organic cations are replaced by protons:

$$NaHSi_2O_5 \cdot xH_2O + nR_4N^+Cl + (1 - n)H_3O^+ \rightleftharpoons (R_4N^+)_n H_{2-n} Si_2O_5$$
$$\cdot x'H_2O + zH_2O \text{ with } z = x + (1 - n) - x'$$

As in other cases of more highly charged layer compounds, e.g., for the niobate $K_4Nb_6O_{17}$ and the nickel arsenate $KNiAsO_4$, the cation exchange proceeds to an extent that a stable interlamellar structure can be formed [148].

A further reason for the nonquantitative exchange of organic cations on single-layer phyllosilicates such as kanemite and $KHSi_2O_5$ is the tendency for intralamellar condensation of silanol groups [149]. The high degree of folding brings two silanol groups of the same layer into positions that siloxane bridges can form [24,27]. Interlamellar condensation of silanol groups is inhibited by the large organic cations between the layers.

By swelling organic kanemites in alkanols and alkylamines, intralamellar bimolecular films can be prepared with different combinations of alkanol or alkylamine molecules and organic cations; this has been observed for the corresponding montmorillonite derivatives [18].

A good example of a photochemical reaction of an intercalated guest species is the trans–cis isomerization of the azo dye shown here in magadiite:

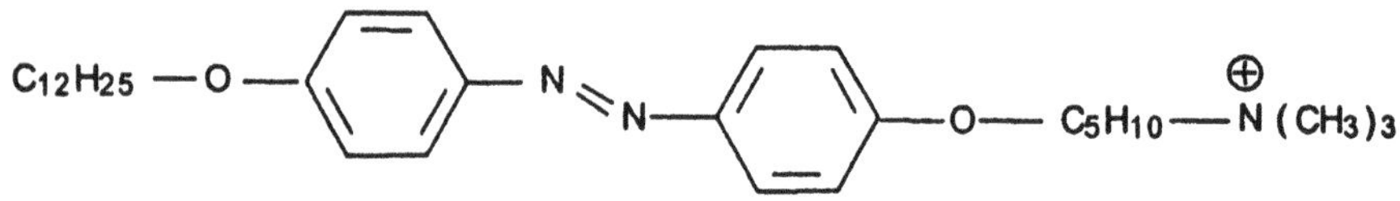

azo dye I

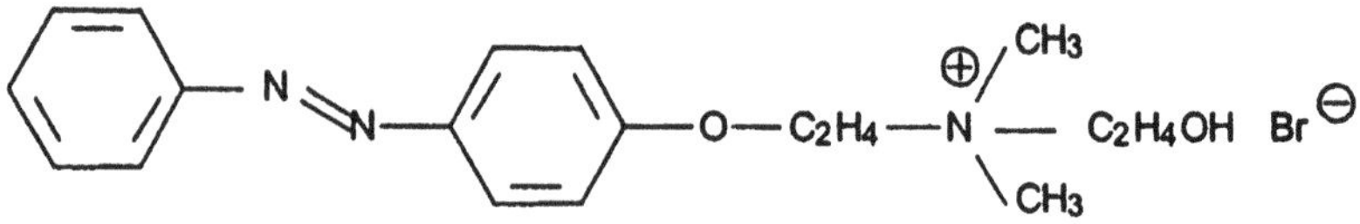

azo dye II

Magadiite reacts with this dye cation in aqueous medium in a conventional exchange reaction, and 70% of the sodium ions are replaced by the dye cations, the rest by protons [150]. After washing with acetone and air-drying, the product is yellow. The large basal spacing of 4.21 nm and the Vis spectrum indicate formation of J-type aggregates* [150]. When irradiated with UV light, the trans form changes into the cis isomer, and the cis isomer changes into the trans isomer under visible light irradiation. The accompanying change of the basal spacing was observed with a similar but shorter azo dye cation, $A_{20}C_2N^+$. The basal spacing of the yellow intercalation product increased from 2.69 nm to 2.75 nm after UV radiation and changed back to ~2.70 nm after irradiation with visible light [151]. The cations in the trans form very likely form densely packed aggregates in the interlayer space. Upon UV irradiation, half of the cations isomerize into the cis form and coexist with the trans form. Such basal spacing changes were not observed with the longer-chain cation $C_{12}A_{20}C_5N^+$ because conformational changes of the alkyl chains disguise the changes due to the trans–cis isomerization.

E. Layer Solvates

Layer solvates of the polysilicates are formed when interlayer water molecules are replaced by polar organic molecules. This can occur by direct exchange of the water molecules or by intercalation of the organic molecules into the dried silicates.

Dried magadiite (basal spacing $d_L = 1.37$ nm) does not expand with organic liquids, but the interlamellar water molecules of air-dried magadiite (basal spacing 1.56 nm) can be replaced by ethylene glycol (basal spacing 1.78 nm; Ref. 34). Much more reactive is the magadiite-type potassium silicate, which easily forms intercalates with a variety of compounds, such as methanol (basal spacing 1.60 nm), ethylene glycol (1.76 nm), glycerol (1.72 nm), trimethylamine-N-oxide (2.13 nm), ephedrine (2.42 nm), imidazole (2.14 nm), and piperazine (1.96 nm) [95]. Kenyaite also forms similar layer solvates [11].

F. Intercalation Reactions of Crystalline Silicic Acids

A characteristic property of the crystalline silicic acids is their ability to intercalate polar organic molecules. Various types of guests are intercalated: short-chain fatty acid amides, urea and derivatives, S- and N-oxides, amines, aromatic bases,

---

* Head-to-tail orientation of the azo dye dipole moment causes a red shift of the absorption bands.

and alcohols. Table 14 shows the basal spacings of the intercalated crystalline silicic acids from magadiite, the magadiite-type potassium silicate, and kenyaite.

The silicic acids differ largely in their various levels of ability toward intercalating certain guest molecules [11,42,95,139,152]. Most reactive are the acids prepared from the potassium silicates ($K_2Si_{14}O_{29} \cdot xH_2O$ and $K_2Si_{20}O_{41} \cdot xH_2O$), followed by H-kenyaite and H-magadiite. Distinctly less reactive are the different modifications of the phyllodisilicic acids $H_2Si_2O_5$.

All acids intercalate alkylamines, and the interlayer expansion can be very large for long-chain amines (Table 15). Other types of amines and aromatic bases and the other groups of guest compounds are only intercalated by certain acids. The interlamellar adsorption of urea derivatives, alcohols, nitriles, and ketones is restricted to the most reactive acids (Table 16). The reactivity of guest molecules follows this order: alkylamines > N- and S-oxides > acid amides > various bases > urea derivatives > alcohols.

The layers of phyllodisilicic acids $H_2Si_2O_5$ and of dehydrated forms of other acids are held together by hydrogen bonds and van der Waals interactions. To separate the layers, the guest molecules have to disrupt hydrogen bonds. This requires guest molecules with high dipole moments: acid amides 3.6–3.8 Debye, dimethyl sulfoxide 4 Debye, N,N'-dimethyl urea 4.8 Debye. Molecules with di-

Table 14 Basal Spacings (nm) of Various Intercalated Crystalline Silicic Acids from (A) magadiite $Na_2Si_{14}O_{29} \cdot xH_2O$, (B) magadiite-type potassium silicate $K_2Si_{14}O_{29} \cdot xH_2O$, (C) kenyaite $Na_2Si_{20}O_{41} \cdot xH_2O$. Starting materials: air-dried acids.

	A	B	C
Formamide	1.32	1.58	2.29
N-Methyl formamide	1.59	1.70	2.26
Dimethyl formamide	1.67	1.68	2.26
Acetamide	1.40	1.58	
Urea	1.55	1.37	~2.2
N-Methyl urea	1.55	1.57	2.13
N, N-Dimethyl urea	1.48	1.61	2.18
N,N'-Dimethyl urea	1.62	1.67	2.24
Dimethyl sulfoxide	1.58	1.66	2.31
Trimethylamine-N-oxide	1.59	1.58	2.21
Ethylene glycol	1.54	1.71	2.30
Glycerol	—[a]	1.65	2.26
Dispersed in water	1.32	~1.5	1.86
Air-dried	1.31	1.34	1.80
Dried at 200°C	1.12	1.16	1.77

[a]Basal spacing <1.32 nm; pure glycerol dehydrates the crystalline silicic acid.

Table 15 Intercalation of Amines and Aromatic Bases into the Crystalline Silicic Acid $H_2Si_2O_5$-I (A) and $H_2Si_{14}O_{29} \cdot xH_2O$ (from Magadiite, Air-Dried) (B)

Guest molecule	pK_B	Basal spacing (nm) A	Basal spacing (nm) B
Hexylamine	3.4	2.48	2.96
Decylamine	3.4	3.46	4.02
Benzylamine	4.67	—[a]	2.39
Pyridine	8.75	1.24	1.62
Quinoline	9.10	—	1.73
Imidazle	7.05	—	1.44
Pyrazole		—	1.50
Dispersed in water		0.77	1.32
Air-dried		0.77	1.32
Dried at 200°C		0.77	1.32

[a]No intercalation.

Table 16 Basal Spacings of the Crystalline Silicic Acid from $K_2Si_{14}O_{29}\cdot xH_2O$ after Intercalation of Alcohols, Ketones, and Nitriles

	Guest molecule	Basal spacing (nm)
Alcohols	Methanol	1.65
	Butanol	1.59
	Octanol	1.56
	Ethylene glycol	1.71
	Glycerol	1.65
	Hexanediol-1,6	1.53
Ketones	Acetone	1.57
Nitriles	Acetonitrile	1.59
	Butyronitrile	1.62
Dispersed in water		~1.5
Air-dried		1.34
Dried at 200°C		1.16

pole moments less than 3.5 Debye are usually not intercalated. The second condition is that the guest molecule possess a sufficiently strong acceptor site for hydrogen bonds, such as $C=O$ and $\equiv N$. Using IR, Rojo and Ruiz-Hitzky [153] detected the formation of hydrogen bonds between the $C=O$ groups of N-substituted amides and silanol groups. For N-methylformamide they observed formation of hydrogen bonds between at least some of the NH groups and internal surface oxygen atoms. Nitriles, in spite of dipole moments of 4 Debye, are not directly intercalated, because they are only weak acceptors of hydrogen bonds [154].

The loss of van der Waals energy that occurs when the layers are separated is often compensated by the van der Waals interaction between guest molecules and between guest molecules and the layers. A close packing of the guest molecules onto the surface oxygen atoms, however, may require arrangements that are not optimal for hydrogen bond formation. This interplay between hydrogen bond formation and van der Waals interaction determines the arrangement and orientation of the guest molecules in the interlayer space. A direct consequence is that homologous guest molecules, for instance, the acid amides, assume different orientations in the interlayer space (see Ref. 139, Fig. 14) and the layer separation does not continuously increase with the molecular volume of the intercalated molecules [43,139].

The driving force for intercalating many amines is the acid–base reaction. The most reactive acids intercalate bases with pK_B values of less than 9.4 [139]. A few exceptions reveal that steric factors are also of influence: pyridine (pK_B = 8.75) is intercalated, but the methyl pyridines (pK_B = 8–8.3) and dimethyl pyridines (pK_B = 7.0–7.9) are not. The influence of steric factors, such as close packing of the guest molecules onto the surface, become more important for the less reactive acids, such as $H_2Si_2O_5$ modifications.

Alkylamines form bilayers between silicate layers when the alkyl chains contain more than five carbon atoms (for details of the interlamellar structures, see Refs. 43,139, and 152). The van der Waals interaction between parallel alkyl chains is decisive for the stability of these intercalation compounds with large layer separation and a high degree of order.

In the most reactive silicic acids the layers of the air-dried samples are separated by water molecules. The intercalation of guest molecules is then a displacement of interlamellar water molecules and does not require the high energy needed for layer separation of the anhydrous acids. Therefore, compounds with dipole moments below 4 Debye are also intercalated, for instance, ethylene glycol, with 2.28 Debye, or alcohols (1.7 Debye). Nevertheless, the interplay between formation of hydrogen bonds and arrangements with optimal van der Waals energy also determines the intercalation reaction.

As an example of alcohol intercalation, Figure 24 shows the basal spacings of the crystalline silicic acids from $K_2Si_{20}O_{41} \cdot xH_2O$ dispersed in an excess of the alcohols. Short-chain alcohols $C_nH_{2n+1}OH$ with $n < 5$ lie flat in monolayers.

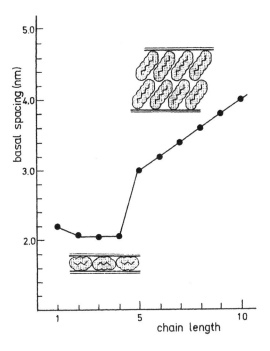

FIGURE 24   Intercalation of alcohols into $H_2Si_{20}O_{41} \cdot 3.6H_2O$: basal spacing and interlamellar arrangement of the alcohol molecules (From Ref. 159.)

Longer-chain alcohols form bilayers, with the chain axis tilted to the layers ($\alpha = 52°$).

An entropy effect may also contribute to the enhanced reactivity of the hydrated acids. The displacement of water molecules, i.e., the transfer of water molecules from the constrained interlayer space into the dispersion medium, compensates or even overcompensates for the loss of entropy of the intercalated guest molecules. In contrast, the penetration of guest molecules between the layers of dehydrated acids is accompanied by a loss of entropy.

Differences in the reactivity of both modifications of the acids $H_2Si_{14}O_{29} \cdot xH_2O$ and $H_2Si_{20}O_{41} \cdot xH_2O$ are also very likely related to differences in the hydrogen-bonding system. It appears that the hydrogen bonding system becomes weaker, which may be a consequence of a smaller number of hydrogen bonds and/or smaller bonding energies, in the following order:

$$H_2Si_{14}O_{29} \cdot xH_2O \text{ from magadiite} > H_2Si_{20}O_{41} \cdot xH_2O \text{ from kenyaite} >$$
$$H_2Si_{14}O_{29} \cdot xH_2O \text{ from } K_2Si_{14}O_{29} \cdot xH_2O \text{ and } K_2Si_{20}O_{41} \cdot xH_2O.$$

Thus, air-dried H-magadiite does not intercalate a further layer of water, and H-kenyaite expands from 1.8 nm to 1.86 nm only when dispersed in water. In contrast, the acids from the potassium silicates intercalate a second layer of water. The parent potassium silicates also show enhanced water intercalation as: $K_2Si_{14}O_{29} \cdot xH_2O$, $d_L = 1.51$ nm (air-dried), 1.73 nm (in water); $K_2Si_{20}O_{41} \cdot xH_2O$, $d_L = 1.94$ nm (air dried), 2.18 nm (in water). The potassium ions influence the structure of the layers and the hydrogen bonds between the silanol groups and the interlamellar water in such a way that the intercalation of a further layer of water is promoted. Transformation into the acid forms retains the enhanced hydration behavior.

Gude and Sheppard [155] described the first natural silicic acid with the composition $3SiO_2 \cdot H_2O$. This mineral, termed *silhydrite*, was found in the Trinity County magadiite deposit in California. Field evidence indicates that silhydrite may have formed by leaching magadiite by near-surface water. Na_2CO_3 transforms silhydrite into magadiite. However, the basal spacing of silhydrite, $d_L = 1.45$ nm, is distinctly different from the H^+ form of magadiite. A carbon content of 0.4% supports the assumption that silhydrite is a natural intercalation product of $H_2Si_{14}O_{29} \cdot xH_2O$. The small amounts of organic material increase the basal spacing from 1.32 nm (typical of $H_2Si_{14}O_{29} \cdot xH_2O$) to 1.45 nm and also impede the lattice expansion when silhydrite is reacted with several guest compounds [156].

Hydrocarbons are not intercalated into crystalline silicic acids. However, high adsorption enthalpies [157] and high values of the dispersive component of the surface free energy [158] in comparison with silica gels are indicative of a partial penetration of the alkyl chains into the interlayer space at the edge regions.

G. Adsorption from Binary Liquid Mixtures

Intercalation reactions are usually studied by dispersing the host material in an excess of the liquid or melted intercalant. When hydrated acids are dispersed in alcohols, the alcohol displaces the hydration water. For instance, $H_2Si_{20}O_{41} \cdot 5.2H_2O$ loses 4.14 moles H_2O (in methanol), 4.21 moles H_2O (in ethanol), and 4.18 moles H_2O (in butanol). (Note that 1 mole of water in $H_2Si_{20}O_{41}$ xH_2O is bound in the form of silanol groups.) In a similar way, 0.73–0.82 moles of water are displaced from $H_2Si_{14}O_{29} \cdot 1.80H_2O$ ($= H_4Si_{14}O_{30} \cdot 0.80H_2O$) [159].

When the guest compound is dissolved in a solvent or mixed with a second liquid, both compounds are often adsorbed and intercalated. The adsorption isotherms then assume the typical shape of composite isotherms. Typical plots contain the surface excess of component 1 or 2, $n_1^{\sigma(n)}$ or $n_2^{\sigma(n)}$, versus the molar fraction of component 1 or 2, x_1 or x_2, in the equilibrium mixture. The surface excess, which is not identical with the amount adsorbed, is derived directly from the concentration of 1 (or 2) before and after the adsorption: $n_1^{\sigma(n)}/m = n^\circ \Delta x_1/$

$m = f(x_1)$. The quantity $n°$ is the total amount of components 1 and 2 ($n° = n_1 + n_2$, in moles) offered per gram of adsorbent, and Δx_1 indicates the change of the molar fraction of component 1 by adsorption. The specific reduced surface excess $n_1{}^{\sigma(n)}/m = f(x_1)$ represents the composite isotherm from which the true adsorption isotherms has to be derived. The adsorption from binary liquid mixtures and the analysis of the composite isotherms have been discussed in great detail by Dékány and coworkers [160–170].

The composite isotherms representing adsorption from methanol or ethanol (component 1) and water (component 2) on the acid from $K_2Si_{20}O_{41} \cdot xH_2O$ (dried in vacuum) shows a long linear section, and at $x_1 > 0.1$, only methanol or ethanol is adsorbed (Fig. 25a). The basal spacing is the same as that for the acid in pure alcohols. Quantitative analysis of the isotherms reveals not only that the interlayer space shows a strong preference for alcohol molecules, but considerable amounts of the alcohols are adsorbed on the external surface, corresponding to about four layers of methanol and three layers of ethanol [159].

The composite isotherm for propanol (1)/water (2) is S-shaped with an azeotropic point at $x_1 = 0.35$ (Fig. 25b), which indicates that both compounds, propanol and water, are adsorbed. As in the case of methanol and ethanol, propanol was enriched at the external surface in amounts corresponding to two layers. The molar ratio of propanol/water depends on the conditions of parent potassium silicate preparation. For $0.1 < x_1 < 0.8$, i.e., for the linear section of the isotherm, it varies between 1.4/4.1 and 1.25/2.6 for acids prepared from two different samples of $K_2Si_{20}O_{41} \cdot xH_2O$ [159]. This is again indicative of the influence of preparation conditions on the structure (fine structure of the layers, hydrogen-bonding system) of alkali silicates and their corresponding crystalline silicic acids.

The specific reduced surface excess isotherms for methanol, ethanol, and propanol on $H_2Si_{14}O_{29} \cdot xH_2O$ (from magadiite, dried in vacuum) are similar to that of $H_2Si_{20}O_{41} \cdot xH_2O$, but they show a stepwise increase at alcohol fractions $x_1 < 0.2$ (Fig. 26) [159]. The step arises because the alcohol molecules cannot penetrate the interlayer space (basal spacing $d_L = 1.12$ nm) at very low concentrations. The alcohol molecules penetrate between the layers only at molar fractions of 0.1 or greater, and the basal spacing increases to 1.52 nm.

The high preference of the crystalline silicic acids for methanol and ethanol from water is clearly expressed when the alcohol molar fraction in the adsorption layer, $x_1{}^s$, is plotted against this fraction, x_1, in the equilibrium solution (Fig. 27). Such a high preference for methanol and ethanol from aqueous solutions was also found for silicalite and ZSM-5 zeolites [171–174]. The parent compound $K_2Si_{20}O_{41} \cdot xH_2O$ also takes up high amounts of ethanol from dilute ethanol/water mixtures [108].

From mixtures of methanol or ethanol with butanol, the shorter-chain alcohols are preferentially adsorbed by $H_2Si_{20}O_{41} \cdot xH_2O$. Butanol was preferentially adsorbed from propanol–butanol mixtures.

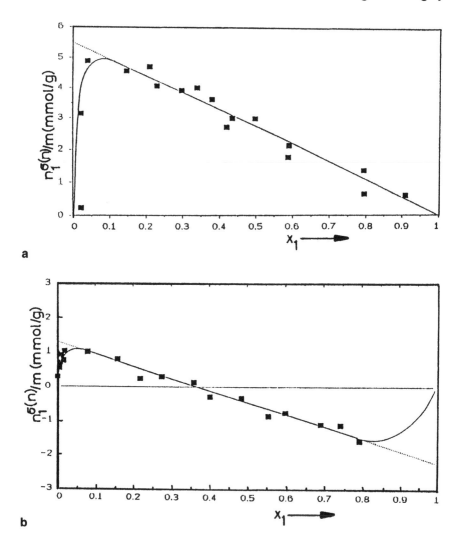

FIGURE 25 Competitive adsorption of alcohols and water on crystalline silicic acids. (a) Methanol (index 1)/water (2) on $H_2Si_{20}O_{41} \cdot 3.6\ H_2O$; (b) propanol (1)/water (2) on $H_2Si_{20}O_{41} \cdot 5.2H_2O$. $n_1^{\sigma(n)}/m$ specific reduced surface excess of alcohol, x_1 molar fraction of alcohol in the equilibrium solution. (From Ref. 159.)

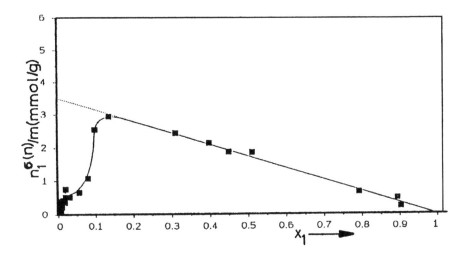

FIGURE 26 Composite adsorption isotherm of methanol (index 1)/water (2) on $H_2Si_{14}O_{29} \cdot 1.8H_2O$. $n_1^{\sigma(n)}/m$ specific reduced surface excess of methanol. x_1 molar fraction of methanol in the equilibrium solution. (From Ref. 159.)

The competition between two types of structures (monolayer for butanol, paraffin-type bilayers for longer-chain alcohols) is also expressed by the composite isotherms for butanol–pentanol and butanol–hexanol mixtures. At $x_1 < 0.2$ (1 = butanol, 2 = pentanol), the bilayers of pentanol molecules determine the interlayer distance and pentanol is preferentially adsorbed (d_L = 3.0 nm) (Figs. 28a, 29). Higher amounts of butanol disturb this arrangement (d_L = 2.05 nm), and the adsorption of butanol became predominant. Bilayers of hexanol molecules (d_L = 3.2 nm) were more resistant against incorporation of butanol molecules, and the basal spacing collapsed only at high molar fractions of butanol (x_1 = 0.8) (Figs. 28b, 29).

The adsorption of alkylamines from alcoholic solutions (ethanol, butanol, etc. to decanol) is also competitive, and both molecules are cointercalated. As long as the alkylamines form monolayers of flat-lying molecules, i.e., in water, ethanol, and butanol, the isotherms show a plateau. This plateau, however, does not give the true amount of amine adsorbed because water or alcohol molecules are also adsorbed. With longer-chain alcohols, the composite isotherms as a function of the molar fraction of amine increase to a second plateau that corresponds to the formation of paraffin-type bilayers (Fig. 30) [175].

Kwon et al. [176] have studied the intercalation of octylamine into $H_2Si_{14}O_{29} \cdot xH_2O$ (from magadiite) from n-hexane, hexadecane, mesitylene, and 2,2,4-trimethyl pentene solvents. These systems form gels, and the competitive

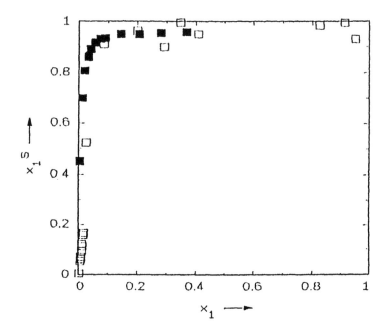

FIGURE 27 Preferential adsorption of ethanol from water on $H_2Si_{20}O_{41} \cdot 3.6\ H_2O$ (□) and silicalite (■) (172). x_1^s molar fraction of ethanol in the adsorption layer and equilibrium solution. (From Refs. 172 and 159.)

adsorption of the solvents increases the basal spacing distinctly; e.g., $d_L = 3.61$ nm in hexane, $d_L = 4.20$ nm in hexadecane at an initial octylamine concentration of 1.66 mol/L (d_L in pure octylamine: 3.40 nm).

H. Intercalation of Anionic Surfactants and Delamination

A unique property of the highly condensed silicic acids is their considerable adsorption of anionic surfactants [177]. Whereas $H_2Si_{14}O_{29} \cdot 1.8H_2O$ (from magadiite) adsorbs about 0.04 mmol/g sodium dodecylsulfate (SDS, plateau value at SDS concentrations >30 mmol/L), the adsorption isotherm for $H_2Si_{20}O_{41} \cdot xH_2O$ (from $K_2Si_{20}O_{41} \cdot xH_2O$) increases to a plateau value of 0.475 mmol/g $H_2Si_{20}O_{41} \cdot 4H_2O$. The SDS adsorption is independent of pH as long as the pH is less than 7. At the beginning of the transformation of acid into an alkali silicate at pH ~7.5, the adsorption decreases sharply from 0.475 mmol/ g to 0.05 mmol/g.

Whereas H-magadiite dispersed in SDS solutions retains a basal spacing of 1.40 nm, X-ray powder diagrams of $H_2Si_{20}O_{41} \cdot 4H_2O$ do not show basal

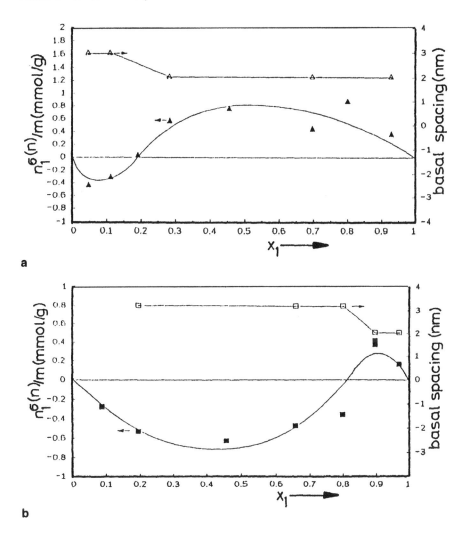

FIGURE 28 Competitive adsorption of n-butanol (1)/n-pentanol (2) (a) and n-butanol (1)/n-hexanol (2) (b) on $H_2Si_{20}O_{41} \cdot 3.6H_2O$. $n_1^{\sigma(u)}/m$ specific reduced surface excess of butanol, x_1 molar fraction of butanol in the equilibrium solution. (From Ref. 159.)

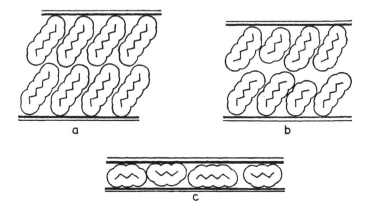

FIGURE 29 Interlamellar arrangement of hexanol in the presence of increasing amounts (a → c) of butanol: transition from bilayers with tilted alcohol molecules to monolayers of flat-lying alcohol molecules. (From Ref. 159.)

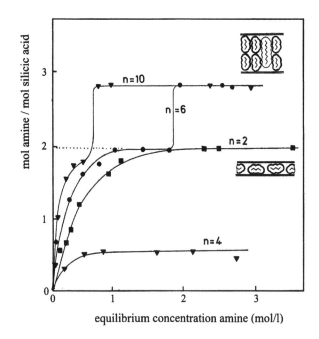

FIGURE 30 Butylamine adsorption from alcoholic solutions on $H_2Si_{20}O_{41} \cdot 4.96H_2O$. Alcohols $C_nH_{2n+1}OH$, x_1 equilibrium concentration of butyl amine (From Ref. 175.)

reflections at equilibrium concentrations greater than 5 mmol/L. When the acid is separated from the SDS solution and dried, the (001)-reflection is again observed, with a basal spacing of $\sim$2 nm. Washing with water reconstitutes the (001)-reflection at $d_L \sim$2.2 nm with an intensity as high as in the original material. In addition, the ^{29}Si-MAS-NMR spectra do not show differences between the original and the reaggregated silicic acid. Thus, the silicic acid $H_2Si_{20}O_{41} \cdot 4H_2O$ is delaminated (or exfoliated) in SDS solution. The particles disperse into smaller packets of silicic acid layers or single layers. However, not all layers can completely exfoliate, because many of them are intergrown, as in the parent silicate. Along with delamination into single layers, therefore, there are also thin aggregates that have a fanlike morphology (Fig. 31). The process is not very fast. The particle size of $H_2Si_{20}O_{41} \cdot 4H_2O$ in 0.05 SDS solution (molar) reaches a minimum value after 50 hours. Delamination also occurs with longer-chain n-alkylsulfates and -sulfonates (four or more carbon atoms in the alkyl chain). Decyl-5-sulfate delaminates the particles but not decyl-(1,10)-disulfate. Long-chain fatty acid anions break up the particles to only a modest extent [177].

In contrast to the interaction with cationic surfactants, anionic surfactants cannot interact by ionic forces with the layer. Instead, it is probable that the terminal methyl group interacts by van der Waals forces with the layer. This group fits well on the layer surface oxygen atoms, and the dense packing strongly increases the van der Waals interaction. This interaction also accounts for the preferential adsorption of methanol and ethanol from aqueous solutions (Sec. IV.E). The structural model developed by Brandt et al. [55] contains channels between the surface [SiO$_4$] tetrahedra that represent appropriate anchor sites for methyl groups. When the terminal methyl groups are anchored at the surface, the electrostatic repulsion between the sulfate or sulfonate groups then drives apart the layers. There are estimates [177] that the charge density is about 0.09 C/m^2,

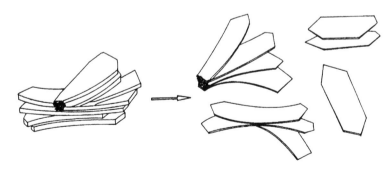

FIGURE 31   Delamination of $H_2Si_{20}O_{41} \cdot xH_2O$ in the presence of anionic surfactants. Surfactants and counterions of the layer charges are not shown. (From Ref. 177).

which is comparable to smectites $(0.07-0.17 \ C/m^2)$. The system may be compared with betaine smectites. In this case the trimethylammonium groups of long-chain betaines $(CH_3)_3N^+-(CH_2)_n-COO^-$ are anchored at the negative surface charges, and the repulsion between the carboxylate groups causes delamination [178].

I. Adsorption of Polymers and Nanocomposites

The interaction of alkali silicates and crystalline silicic acids with macromolecules has been reported for only a few polymer systems. Dörfler et al. [179] studied the adsorption of dodecyloctaethylene oxide on sodium magadiite (amount adsorbed at saturation 109 mg/g) and calcium magadiite (111 mg/g) and the corresponding silicic acid (96 mg/g). The decisive role of hydration of the ethylene oxide groups was indicated by an adsorption that increased with temperature and salt addition.

In contrast to the adsorption of polyethylene oxides on silica [180,181], the adsorption on $H_2Si_{40}O_{21} \cdot 5H_2O$ is independent of the molecular weight, with a saturation value of 130 mg/g (Fig. 32) [182]. The basal spacing of 2.26 nm indicates that the polyethylene oxide chains require an interlayer separation of $2.26-1.77 = 0.49$ nm, which corresponds to the thickness of the chain of ~0.46 nm. The lack of correlation between molecular weight and interlayer separation clearly indicates that the polymer is adsorbed as a single monolayer only.

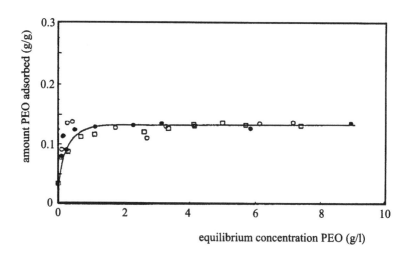

FIGURE 32 Adsorption of poly(ethylene oxide) on $H_2Si_{20}O_{41} \cdot 4.96H_2O$. Molecular weight of the poly(ethylene oxide): ○ 5000, ● 15,000, □ 35,000.

This model is also in agreement with specific surface area values, as is demonstrated in the following example. An amount of 130 mg polyethylene oxide corresponds to $130/44 = 2.95$ mmol ethylene oxide. The maximum area occupied by 1 mole ethylene oxide $(0.16 \text{ nm}^2$ per unit) is $0.16 \cdot 10^{-18} \cdot 6.02 \cdot 10^{23} = 0.96 \cdot 10^5 \text{ m}^2$. One mole ethylene oxide in the interlayer space covers an area of $2 \cdot 0.96 \cdot 10^5 \text{ m}^2$. Thus, 2.95 mmol ethylene oxide are in contact with an area of 566 m^2/g, which corresponds to the estimated total surface area of 530 m^2/g (Sec. V.A).

Wang et al. [183] used octadecylammonium-octadecylamine derivatives of magadiite with a basal spacing of 3.82 nm (paraffin-type arrangement of the alkyl chains with an inclination angle of ~65°) as a precursor for nanocomposites. The sodium ions were completely replaced by the ammonium ions, and a small fraction of neutral amine was essential for forming the paraffin-type structure. The composition in the interlayer space was $(C_{18}H_{37}NH_3^+)_2(C_{18}H_{37}NH_2)_{0.48}$ per $Si_{14}O_{29}$ unit. The polymer matrix was an epoxide resin that was cured with an amine. The absence of (00l)-reflections provided strong evidence of exfoliation during the thermoset curing process. A certain benefit of magadiite exfoliation in polymer reinforcement was illustrated by an increased tensile strength compared to conventional composites prepared from Na^+ magadiite or octadecylammonium magadiite.

The use of secondary, tertiary, or quaternary alkylammonium ions $(C_{18}H_{37}NH_{3-m}(CH_3)_m^+$ with $m = 1, 2, 3$) caused improvement in the elastomeric properties of epoxide silicate nanocomposites formed via in situ polymerization [184]. The problem with primary alkylamine and alkylammonium ions is that the NH_2 and NH_3^+ groups behave similarly to the curing amine and also react with the epoxide; this leads to dangling chain conformations and interrupts the epoxy matrix cross-linking process. In all cases, the tilted alkylammonium ions reorient to perpendicular in order to accommodate the resin. The intercalation of epoxide and curing agent into a magadiite containing secondary ammonium groups increases the basal spacing within 1 hour at 75°C to 6.27 nm. This phase appears to be crucial for achieving the exfoliated state. With increasing aging time, the galleries continue to expand and finally lose their parallel orientation.

Derivatives with ternary and quaternary ammonium ions behave quite differently. Magadiite containing ternary alkylammonium ions forms a new type of nanocomposite with a sufficiently regular layer stacking and d-values of about 8 nm. The catalytic polymerization rate therefore appears to be quite uniform in all interlayer spaces, leading to an exfoliated state with high regular periodicity. In contrast, quaternary alkylammonium ions were noncatalytic, and the interlayer spacing was determined mainly by the initial loading of resin and curing agents. The sample did not exfoliate, and it retained a spacing of 4.10 nm, characteristic of intercalation.

Concerning mechanical properties, the improvement in tensile properties provided by exfoliated smectites is superior to that of delaminated magadiite. The most significant improvement over corresponding epoxy-smectite nanocomposites is the optical transparency of the magadiite composites [143,183].

J. Grafting Reactions

The term *grafting* indicates that covalent bonds are formed between an organic molecule and the inorganic host. Typical is the reaction of silanol groups with silylating agents such as chlorosilanes (alkyl)$_{4-n}$SiCl$_n$ or disilazanes such as hexamethyl disilazane [(CH$_3$)$_3$Si]$_2$ NH. Due to the presence of silanol groups in the interlayer space of crystalline silicic acids, grafting reactions are much more common for these compounds and also for the more highly condensed alkali silicates (magadiite, kenyaite) than for 2:1 clay minerals, which possess silanol groups only at the edges. In most cases, the interlayer space must be preswelled by intercalating appropriate compounds so that the interlayer silanol groups become available to the silylating agent.

As an example, crystalline silicic acids from magadiite and the potassium silicate K$_2$Si$_{20}$O$_{41}$ · xH$_2$O were reacted with several chlorosilanes and disilazanes after intercalation of N-methyl formamide, N,N'-dimethyl formamide, and dimethyl sulfoxide in dioxane [185]. When reacted with hexamethyl disilazane, 100 mmol trimethylsilyl groups were grafted per mol SiO$_2$ of H-magadiite; i.e., 35% of the silanol groups of magadiite (calculated for H$_4$Si$_{14}$O$_{30}$ · xH$_2$O) reacted with the silylating agent. A similar value (112 mol/mol SiO$_2$) was obtained for H$_2$Si$_{20}$O$_{41}$ · xH$_2$O, corresponding to 56% of the total amount of silanol groups (4 mol/20 mol SiO$_2$). More bulkier silylating agents reduced the degree of grafting. Only 37% of the silanol groups of H$_2$Si$_{20}$O$_{41}$ · xH$_2$O reacted with triethylsilyl groups, and just 5% with triphenylsilyl groups. Interlamellar grafting of trimethylsilyl groups increased the basal spacing of H$_2$Si$_{14}$O$_{29}$ · xH$_2$O from 1.12 nm to 1.94 nm (Δd_L = 0.82 nm) and of H$_2$Si$_{20}$O$_{41}$ · xH$_2$O from 1.77 nm to 2.45 nm (Δd_L = 0.68 nm). In contrast to the pure silicic acids, the organic derivatives were hydrophobic and could easily be dispersed in organic solvents, including such apolar liquids as aliphatic and aromatic hydrocarbons.

Higher degrees of grafting can be achieved by starting from the octylamine intercalate of H$_2$Si$_{20}$O$_{41}$ · xH$_2$O from K$_2$Si$_{20}$O$_{41}$ · xH$_2$O [186]. The reaction of surface silanol groups with alkyl trimethoxysilanes (alkyl = methyl, ethyl, propyl) is nearly quantitative, as indicated by a marked reduction in Q^3 signal intensity. The ^{29}Si-MAS-NMR spectra reveal that ethyl and propyl trimethoxy derivatives contain silicon atoms with two silanol groups and one siloxane bridge, silicon atoms with one silanol group and two siloxane bridges, and those with three siloxane bridges. Methyl trimethoxy derivatives with the smaller alkyl groups show preferential condensation to siloxane bridges, probably because silicon atoms with two free silanol groups were absent. Dimethyl dimethoxysilane also reacts with all silanol groups, and each molecule forms two silanol bridges, either with vicinal silanol groups at the surface or with one surface group and a second silane molecule. In contrast, only a few surface silanol groups reacted with trimethyl methoxysilane. It is likely that the trimethyl group retards or even impedes the diffusion of the molecule into the interlayer space. Calcination of these silylated samples at 450°C yielded porous materials with micropore volumes of 97–125 μL/g and mesopore volumes of 87–143 μL/g (for the tetraethoxy derivatives).

The interlamellar silanol groups of silicic acids can be esterified with alcohols. When H-magadiite is refluxed in methanol for 48 hours, the basal spacing increases from 1.15 nm to 1.35 nm, and changes in ^{13}C-HD/MAS-NMR, ^{29}Si CP/MAS-NMR, and ^{1}H-MAS-NMR spectra proved formation of the ester [187]. Longer-chain alcohols (butanol and longer) were introduced after preceding intercalation of N-methyl formamide. The basal spacing of the derivatives with butanol and longer alcohols (up to $C_{16}H_{33}OH$) remained at 1.4 nm, indicating a flat arrangement between the layers. The amount of alkoxy groups, therefore, decreased from ~0.9 mol/$Si_{14}O_{29}$ (for ethanol to pentanol) to 0.2 mol/$Si_{14}O_{29}$ for tetradecanol and hexadecanol. Esterification also makes the crystalline silicic acid hydrophobic, and the derivatives were dispersed in toluene and cast on glass substrates to form transparent films.

The interlamellar silanol groups of alkali silicates can also be reacted with chlorosilanes when the interlayer space is preswelled. Yanagisawa et al. [188–190] used dodecyltrimethylammonium ions to make the interlamellar silanol groups of magadiite and kenyaite available to the silylating agent. During the chlorosilane reaction, most of the dodecyltrimethylammonium ions were displaced. The basal spacing of the precursor material (d_L = 2.94 nm for magadiite, d_L = 3.56 nm for kenyaite) decreased to 1.86 nm (magadiite) and 2.48 nm (kenyaite). Elemental analysis revealed that 2.25 trimethylsilyl groups per $Si_{14}O_{29}$ unit of magadiite and 2.05 groups per $Si_{20}O_{41}$ unit of kenyaite reacted [188]. In the idealized formula, two silanol groups are present in addition to two Si—O$^-$ groups in magadiite and kenyaite. The displacement of alkylammonium ions can provide additional centers for silylation:

$$\equiv SiO^- R_4N^+ + ClSiR_3 \rightarrow \equiv SiOSiR_3 + R_4N^+Cl^-$$

The bulkier diphenyl methyl chlorosilane reacted with 0.79 silanol groups per $Si_{14}O_{29}$ unit of kenyaite [189]. During reaction with an allyl dimethyl chlorosilane, the allyl groups were eliminated, and $SiOSi(CH_3)_2$ OH groups were found in the interlayer space [190].

Magadiite modified with octyl dimethyl chlorosilane intercalates long-chain alcohols with basal spacings that increase linearly with chain length (n = 6–12)

[191]. This behavior illustrates the possible application of organically modified metal silicate hydrates as adsorbents.

Silylation of dodecyltrimethylammonium magadiite with (2-perfluorohexyl-ethyl)dimethyl chlorosilane yields a material with increased thermal stability and film-forming ability [192].

Due to the high flexibility of single layers of kanemite, which consist of six-membered rings of $[SiO_4]$ tetrahedra, the reaction with alkyl trichlorosilanes and alkyl methyl dichlorosilanes causes substantial changes of the layers themselves, and new five- and six-membered rings can be formed [193].

V. POROSITY

A. Gas Adsorption

The specific surface area of alkali silicates such as magadiite and kenyaite is small, ~20 m^2/g, but it increases strongly when the silicate is transformed into the crystalline silicic acid. Typical for H-magadiite are specific BET surface areas of 40–80 m^2/g [179,182,194]. The acids from $K_2Si_{20}O_{41} \cdot xH_2O$ have an even higher specific surface area, at about 130–200 m^2/g. This increase in specific surface area is, at first glance, surprising because the morphology of the particles does not change [69,116,194–196]. However, the crystalline silicic acids, unlike their parent compounds, are microporous. The specific BET surface area is not the actual area but rather an apparent value that includes a contribution due to micropore filling.

The micropores of layered compounds are plate-shaped, not cylindrical. Therefore, when considering the adsorption in such micropores, a different view may be more appropriate [197]. The silicate layers of clay minerals and metal silicate hydrates show not a dense packing of surface atoms but, rather, holes (ditrigonal holes in the surface of clay minerals), so the arriving gas molecules encounter sites of different adsorption energies. The "derivative isotherm summation" method of analyzing gas adsorption isotherms (nitrogen and argon) at very low relative pressure was introduced by Villiéras et al. [198]. This provides information about the energetic and geometrical heterogeneity of lamellar minerals. The method, using certain assumptions, also allows the estimation of particle sizes and aspect ratios. In the case of magadiite, this method has revealed the influence of exchangeable cations on the stacking of elementary layers. The high-energy sites observed for magadiite and kenyaite (not for kanemite) were related to six-membered rings of silica tetrahedra similar to those in clay minerals [197].

Kruse et al. [195] have provided a detailed study concerning the nitrogen gas adsorption and porosity of $H_2Si_{20}O_{41} \cdot 5.4H_2O$ from different samples of $K_2Si_{20}O_{41} \cdot xH_2O$. Typical are type II isotherms with hysteresis loops open to very low relative pressures (Fig. 33). A certain amount of nitrogen remains adsorbed at

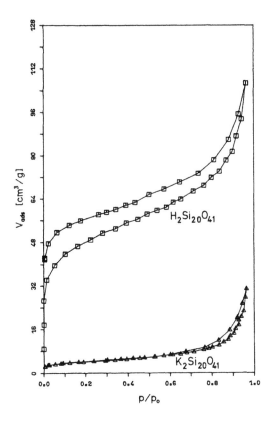

FIGURE 33 N_2 adsorption isotherm (at 77 K) of $H_2Si_{20}O_{41} \cdot xH_2O$, dried in vacuum at 75°C ($\triangle$) and $H_2Si_{20}O_{41} \cdot x H_2O$, dried in vacuum at 60°C ($\square$). $\Delta p/\Delta t = 33$ Pa/min (see text).

the end of the adsorption–desorption cycle at 77 K, and it is not desorbed by evacuation. Instead it is desorbed when the sample is heated to room temperature. It is likely that this nitrogen was trapped in ultramicropores (pores with diameters of 0.7 nm or less; supermicropores: 0.7–2 nm). The ultramicropore volume of $H_2Si_{20}O_{41} \cdot 5.4H_2O$ decreased from 62 µL/g (silicic acid dried at 60°C) to 13 µL/g after calcination at 400°C. The supermicropore volume (27 µL/g) was derived from the *t*-plot only for samples calcined at 400°C. The *t*-plots of samples under 400°C were curved and difficult to evaluate. A final interesting observation was an increased adsorption when the pressure gradient during the adsorption branch was reduced, e.g., from $\Delta p/\Delta t = 33$ Pa/min to 6.7 Pa/min.

The following pore-filling mechanism was proposed by these workers. The ultramicropores (the adsorption sites of highest energy) are localized between the silicate layers. When the water is desorbed from the interlayer spaces, some water molecules remain in the structure and prevent the layers from collapsing completely. The ultramicropore volume (i.e., the volume of the nitrogen molecules at the highest-energy sites), therefore, decreases only slightly up to a calcination temperature of 150°C (basal spacing d_L = 1.77 nm). The steeper decrease of this volume at higher calcination temperatures reflects the desorption of the remaining interlamellar water molecules and, above 250°C, the formation of siloxane bridges between the layers (basal spacing 1.68–1.62 nm). At this temperature, the ability to intercalate guest molecules is lost. However, a small number of highest-energy sites still remain between the collapsed layers up to 400°C.

The ultramicropores are widened to supermicropores at the crystal edges (Fig. 34). The time-dependent adsorption is related to this pore structure. When the pressure increases during adsorption, the nitrogen molecules penetrate deeper into the interlayer region. These ultramicropores are filled irreversibly; the nitrogen molecules trapped between the layers are not desorbed at 77 K. Penetration of the nitrogen molecules deeper into the ultramicropores is accompanied by filling of the supermicropores. A unique feature of this adsorbent is that during this process the supermicropores at the crystal edges are widened. This process is slow and reversible.

The external surface area of crystalline silicic acids can be determined after blocking the pores with dodecylamine or transformation into the potassium salt. These products do not contain micropores (*t*-plot analysis), and they show a specific BET surface area of 15 m²/g (after preadsorption of dodecylamine) and

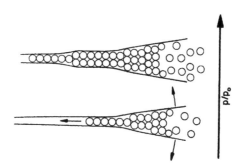

FIGURE 34 Pore-filling and pore dimension changes typical of crystalline silicic acids. The nitrogen molecules penetrate between the layers to some extent. These ultramicropores between the layers open at the crystal edges and are widened with increasing pressure. (From Ref. 195.)

24 m^2/g (potassium silicate). The original potassium silicate has a specific surface area of 15 m^2/g.

The total surface area (including the internal surfaces) of $H_2Si_{20}O_{41} \cdot xH_2O$ can be estimated from an assumed density of the silicate layer (about $2.25 \cdot 10^6$ g/m^3) and the layer thickness (~ 1.68 nm at the beginning of interlamellar siloxane formation) [159,175]: $S = 2/(1.68 \cdot 10^{-9} \cdot 2.25 \cdot 10^6) = 529$ m^2/g ($H_2Si_{20}O_{41}$ $\cdot H_2O$). For the hydrated form $H_2Si_{20}O_{41} \cdot 5.4H_2O$, the specific surface area is $529 \cdot 1237/1316 = 497$ m^2/g. In a similar way one obtains for $H_2Si_{14}O_{29} \cdot xH_2O$ (with a layer thickness of 1.12 nm) $S = 2/(1.12 \cdot 10^{-9} \cdot 2.25 \cdot 10^6) = 794$ $m^2/$ g ($H_2Si_{14}O_{29} \cdot H_2O$) and $S = 794 \cdot 876.8/894.8 = 778$ m^2/g for $H_2Si_{14}O_{29} \cdot 2H_2O$.

B. Pillared Compounds

In 1992 Dailey and Pinnavaia [194] initiated studies about pillaring the silicate layers of H^+-magadiite. The acid was reacted with octylamine to form an octyl-ammonium–octylamine magadiite gel with a basal spacing of 3.4 nm. When this compound was reacted with tetraethyl orthosilicate (TEOS), octylamine was displaced and the basal spacing of the air-dried sample decreased to 2.3–2.8 nm, depending on the TEOS/magadiite ratio. During the replacement of octylamine, the silanol groups of the host layers reacted with the $SiOC_2H_5$ groups of TEOS to form siloxane bonds and ethanol. Further hydrolysis of intercalated TEOS occurred during air-drying. Calcination at 360°C caused oxidation of the octylam-monium cations and remaining octylamine molecules, and hydrated silica pillars formed between the layers. The total surface area varied between 525 and 705 m^2/g, and the surface area related to micropores was between 480 and 670 $m^2/$ g (from t-plots). The differences of 35–45 m^2/g corresponds very well to the specific area of the parent H^+-magadiite (45 m^2/g).

An interesting conclusion of their CP ^{29}Si-MAS-NMR studies was that the connectivity of the $[SiO_4]$ tetrahedra in the siloxane pillars closely mimicked the connectivity of the magadiite layers. Also, acid- or base-catalyzed hydrolysis of TEOS and polycondensation influenced the physical properties of the silica-pil-lared magadiites and kenyaites [116,199,200]. The apparent specific BET surface area of silica-pillared kenyaite varied between 533 and 606 m^2/g (micropore surface: 427–509 m^2/g), depending on the method of hydrolysis. The basal spac-ing of the uncalcined TEOS-octylammonium kenyaites varied between 4.0 and 4.3 nm, and after calcination at 538°C between 2.95 and 4 nm [199,200]. It was confirmed that the intercalated octylamine itself acts as a base catalyst for the hydrolysis of interlamellar TEOS. Scanning electron micrographs show that the original plate structure and morphology of the H^+-kenyaite are maintained when the hydrolysis is carried out in water or 0.05% NH_3 solution. Decomposition into smaller particles was observed during hydrolysis in 0.1% NH_3 solution [116,201].

In another report, silica pillars with Si^{4+} substituted by Al^{3+}, Ti^{4+}, and Zr^{4+} were prepared between the silicate layers of octosilicate by intercalation of octylamine into the H^+ form before mixtures of TEOS and aluminum-tri-sec-butoxide (or tetraethyl orthotitanate and zirconium butoxide) were added [202,203].

Fudala et al. [204] have synthesized MCM-41 phases in the presence of magadiite and Al^{3+}-doped magadiite. A composite material consisting of magadiite plates and MCM-41 tubes is formed. It is likely that some of the tubes are intercalated. Nitrogen adsorption isotherms are typical of mesoporous MCM phases. An advantage of this composite material is the distinctly higher mechanical and thermal stability compared with the MCM phases. Catalytic test measurements also were reported (1-butene isomerization, toluene alkylation).

C. Mesoporous Materials from Kanemite

Inagaki et al. [205,206] have prepared a new type of highly ordered mesoporous material from kanemite in the following way: Kanemite was dispersed in an aqueous solution of hexadecyltrimethylammonium ions (HDTMA) for 3 hours at 70°C. After cooling to room temperature the pH was adjusted to 8.5. The addition of HCl seems to be a necessary step in the synthesis because it may promote condensation of silanol groups [207]. The product was washed, air-dried, and calcined at 700°C and 1000°C. Transmission electron micrographs (TEM) showed a highly ordered mesoporous material with hexagonal arrays of uniform channels with diameters of 3 nm. The X-ray powder diagram revealed several reflections in the low-angle region due to the regular arrangement of the silica tubes; the silica material itself was amorphous. The apparent specific BET surface area was very high, at ~ 1100 m^2/g. The nitrogen adsorption isotherms are typical of MCM-41 phases [51,208–213]. The pore diameter was estimated from the lattice constants of the hexagonal structure (XRD), gas adsorption, and TEM. The as-synthesized material had pore walls 0.4 nm thick, corresponding to slightly less folded single [SiO$_4$] tetrahedral layers, as in the parent kanemite. The apparent pore diameter was 4.2 nm (XRD and TEM). Calcination increased the pore wall thickness to 1.6 nm (from nitrogen adsorption; 1.2 nm by TEM, 0.8 nm by XRD) and reduced the pore diameter to 2.7 nm (from nitrogen adsorption, 3.1 nm by TEM, 3.4 nm by XRD) [209].

The pore diameter can be adjusted by the use of alkylammonium ions of different chain lengths. For instance, alkyltrimethylammonium ions with 8, 10, 12, 14, and 16 carbon atoms in the alkyl chain yielded materials with pore diameters of 1.7 nm, 2.1 nm, 2.3 nm, and 2.8 nm [204]. However, hexadecyltrimethylammonium ions appear to be optimal for the formation of the most ordered hexagonal mesoporous materials [207].

The thickening of the pore walls during calcination is a unique feature of the kanemite-derived materials. The thickness corresponds to a double layer of folded tetrahedral sheets and is too large to be explained by a higher degree of folding of the single layers. Inagaki et al. [209] assumed that dissolved silicate species remaining in the uncalcined material are deposited at the walls during calcination. One has to consider that the TEMs usually show selected areas of the highly ordered material that often are embedded in amorphous material.

Kanemite is not the only source of silicate for the preparation of such porous materials. Other single layer silicates, such as silinaite Na_2SiO_3, δ-$Na_2Si_2O_5$, and even water glass solutions can be used as starting materials [51], as can leached saponite [215]. Kanemites with Al^{3+}- and Ga^{3+}-for-Si^{4+} substitutions were also transformed into mesoporous materials [216]. A Sn^{4+}-incorporated material was prepared from water glass, $SnCl_4$, and NaOH as starting materials [211].

Inagaki et al. [206,208] proposed that the alkylammonium kanemite transforms into the mesoporous material by a folding mechanism. The flexible silicate layers bend in such a way that the alkylammonium ions form micellar-like cylindrical aggregates (Fig. 35). Due to the undulated shape, the layers condense to a regular mesoporous structure when calcination burns out the organic materials. Because of the proposed mechanism these porous materials have been named folded sheet mesoporous materials (FSM-16), where 16 indicates the alkyl chain length. The model is questionable because it requires the alkylammonium ions to cover the silicate layers in patches. Another mechanism can be invoked to explain the formation mechanism (see Fig. 36) [51,207,217]. Exchange reactions with large organic cations can largely disintegrate the structure of single-layer silicates such as kanemite, δ-$Na_2Si_2O_5$, and silinaite [51]. The colloidal fragments and dissolved species then reaggregate under the influence of the alkylammonium ions. On the basis of time-resolved X-ray powder diffraction studies, O'Brien et al. [207] observed an intermediate phase due to the ordering of fragmented anionic silicate layers and surfactant cations. Similar studies by O'Hare et al. [218] revealed that the hexagonal mesophase (which by calcination gives the FSM material, Fig. 34) forms before complete disintegration of the kanemite. Therefore, there is a continuous transport of fragments from the lamellar alkylammonium kanemite to the mesophase.

The aggregation of the alkylammonium ions with caps of fragmented silicate layers is comparable to the aggregation of surfactants, which is directed by the differences in the size of the hydrophobic tail and the polar head group [219,220]. An instructive example of the importance of geometrical conditions and electrostatic factors was reported for the formation of the lamellar precursor phases of MCM-41 alumosilicates [217]. In a similar way, the dimension of the alkyl chain and the size of the layer fragments may control the aggregation. Thus, other types of aggregates and porous materials are conceivable. By carefully adjusting the acid concentration and the pH of the dispersion of the alkylammo-

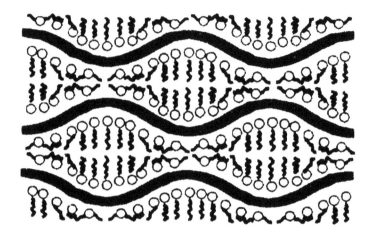

FIGURE 35 The folding mechanism proposed by Inagaki et al. to explain the formation of ordered mesoporous materials. (From Ref. 208.)

nium kanemite, Kimura et al. [221] prepared an ordered mesoporous material (called KSW-2) with square channels with a periodic distance of ~3.3 nm. Considering this mechanism and the presence of colloid-sized, disintegrated silicate layers, the thickening of mesopore walls via condensation with amorphous silicates is more readily explained.

Water adsorption isotherms and calorimetric measurements have revealed that the surface of freshly calcined FSM materials is hydrophobic because of the small number of silanol groups. The material becomes more hydrophilic once

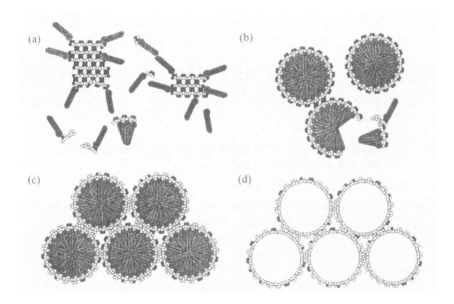

FIGURE 36 Formation of mesoporous materials from single-layer silicates and other silicate sources. (a) Partial disintegration of the silicate structure by alkylammonium ions that carry fragments of the silicate layer; (b, c) aggregation of alkylammonium ions + fragments of the silicate layers; (d) formation of meso- and micropores during calcination. In the structure model of the silicate only the oxygen atoms between two silicon atoms (○) and of the SiO$^-$ and SiOH groups (·) are shown. (From Ref. 51).

water molecules are condensed in the pores, because siloxane bridges hydrolyze into silanol groups [212,213]. However, the water uptake at $p/p_o > 0.5$ is reduced. As in the case of MCM-41 materials [222,223], hydrolysis of siloxane bonds by adsorbed water under the influence of capillary forces promotes the collapse of many pores and reduces the water uptake.

A molecular sieving effect of mesoporous FSM silicas with different pore diameters was reported by Hata et al. [224]. Taxol, an anticancer substance, was not adsorbed in the channels with pore sizes less than 1.6 nm. Taxol contains C=O and OH groups and was adsorbed only from dichloromethane and toluene solution. It is not adsorbed from methanol or acetone due to the low degree of hydrophilicity of the porous materials. Reaction with trimethyl chlorosilane impeded adsorption because the surfaces are too hydrophobic. By a special adsorption–desorption procedure, Taxol could be enriched from yew needle extracts using certain FSMs.

The surface of FSM materials can be modified by silylation with chlorosilanes and ethoxysilanes [225]. It is advantageous to add pyridine in order to trap the hydrogen chloride generated during the silylation reaction. The pore diameter of FSM-18 was reduced from 3.6 nm to 2.2–2.9 nm, depending on the silylating agent. The number of grafted groups varies between 1.1 and 2.4 groups/nm^2. The strongly reduced water vapor adsorption was due mainly to the hydrophobization of the pore walls.

VI. TRANSFORMATION INTO NEW SILICATES, ZEOLITES, AND SILICA MODIFICATIONS

A. New Types of Layer Silicates

Kooli et al. [79–84] reported the transformation of magadiite and H-magadiite into new types of layer silicates, called KLS and FLS. Magadiite was heated in mixtures of tetramethylammonium hydroxide, dioxane, and water. This mixture of solvents was used to transform silica into a layer silicate, called HLS, with unusual helical morphology [226,227]. The presence of dioxane was not critical for the preparation of KLS, but it accelerated the reaction. The reaction started with ion exchange of sodium by TMA$^+$ ions ($d_L = 1.8$ nm). After 2 hours at 150°C, X-ray reflections of KLS 1 appeared; after 3 hours the reflections of TMA$^+$ magadiite were absent, indicating an almost complete transformation into KLS 1. The splitting of the magadiite layers was indicated by a strong increase of the Q^3/Q^4 ratio from 0.3 to 3.9. Particles of KLS 1 showed the platy morphology of the magadiite used. Long-chain alkylammonium ions could be exchanged for the interlayer sodium ions. The composition $H_{0.45}Na_{0.13}SiO_{2.4}[(CH_3)_4N]_{0.22} \cdot 0.47H_2O$ reported is similar to HLS.

Other phases were obtained by slightly changing the experimental conditions: KLS 2 is, probably, a layer silicate $H_{0.4}Na_{0.17}SiO_{2.4}[(CH_3)_4N]_{0.22} \cdot 0.57H_2O$ with $Q^3/Q^4 \sim 3.9$. It is formed by a dissolution and recrystallization process. The FLS phases (e.g., $H_{0.065}SiO_{21}[(CH_3)_4N]_{0.135} \cdot 0.08H_2O$ with a high content of Q^4 atoms $Q^3/Q^4 \sim 0.3$) seem to consist of three-dimensional networks. In contrast to magadiite and KLS 2, the calcined products possess pronounced microporosity. Since many other three-letter combinations are possible, the opportunities for new synthetic products abound.

It is often difficult to decide whether the described transformations are really topotactic processes or if they proceed by dissolution and recrystallization. Mixtures of water with organic solvents are useful in synthesizing new silicates from different silica sources. An example is the formation of the zeolite $Na_8Si_{12}O_{28} \cdot 4H_2O$ from silica, NaOH, ethylene glycol, and water [228].

B. Recrystallization into Zeolites

Tetrapropyl (TPA$^+$) and tetrabutylammonium ions (TBA$^+$) are well-known structure-directing compounds in zeolite synthesis. It is, therefore, not very surprising that alkali polysilicates in the presence of these ions are transformed into zeolites. Magadiite with isomorphous Al^{3+}/Si^{4+} substitution and the corresponding crystalline silicic acids have been recrystallized in TPA^+OH^- and TBA^+OH^- solutions at 135°C to form ZSM-5 and ZSM-11 zeolites [229]. Ferrierite was obtained in the presence of piperidine from Al^{3+}-containing magadiite and kanemite [114,230].

Kanemite can be transformed at 70°C into silicalite-1 (ZSM-5) in dispersions with TPA^+ ions as the structure-directing species [231]. An interesting alternative was discovered by Shimizu et al. [232]. Kanemite was transformed into silicalite-1 (pure silica ZSM-5) and silicalite-2 (pure silica ZSM-11) starting from gels prepared by dispersing kanemite into aqueous solutions of TPA^+Br^- and TBA^+Br^-. The dried gels were shaped into disks and transformed into silicalite by heating to 130°C. As intercalation of the organic cations progresses, the concentration of Q^4 silicon atoms increases due to the high flexibility of the layers. The formation of these siloxane bonds may initiate the transformation into silicalite [232].

Synthesis of zeolites from polysilicates has been examined for two reasons. First, the particle sizes and shapes can be very different from those formed by conventional synthesis. For instance, the solid-state transformation of kanemite gels yields distinctly smaller silicalite particles that are more cubic in shape than by using hydrothermal synthesis. A second reason is the search of zeolite systems that can be shaped without needing binding additives. A possibility would be the transformation of raw materials into a zeolite after being shaped, as illustrated by Shimizu et al. [232].

C. New Types of Silica

Thermal condensation of crystalline silicic acids can create a new family of silica modifications. Because the layers of the silicic acids carry silanol groups, interlamellar condensation yields three-dimensional networks of $[SiO_4]$ tetrahedra. The required dehydration temperatures are over 500°C [99]. The oxides contain small amounts of water, related to some uncondensed silanol groups. They are obtained from different silicic acids and are identified by their X-ray powder diffractograms.

It is characteristic of the dehydration of the crystalline silicic acids that the condensation between layers depends strongly on the first steps of dehydration. As a consequence, the water content of the oxides can differ when the silicic acids are heated in different ways, for instance, directly to 600°C or stepwise from 150°C to 300°C to 600°C. It can also depend on the total water content of the starting material. As examples, the different oxides $SiO_2 \cdot 0.06H_2O$ and $SiO_2 \cdot 0.11H_2O$ are obtained from $H_2Si_{14}O_{29} \cdot 5.4H_2O$; $SiO_2 \cdot 0.01 H_2O$, $SiO_2 \cdot 0.02H_2O$, and $SiO_2 \cdot 0.04H_2O$ from $H_2Si_{20}O_{41} \cdot 2.8H_2O$ [99].

Silica-X was described as a new modification of silica [31,233]. It is characterized by a strong reflection at $d = 1.80$ nm, which differs in half-width and intensity between different samples. Heating to 300°C decreases the d-value to 1.73 nm. It was shown that silica-X is the partially dehydrated form of the silicic acid $H_2Si_{20}O_{41} \cdot xH_2O$ (derived from the potassium silicate $K_2Si_{20}O_{41} \cdot xH_2O^*$) [234].

Another example of the importance of initial condensation steps occurs for silicic acids with intercalated organic molecules, which, upon calcination, can yield different oxides. The oxide $SiO_2 \cdot 0.08H_2O$ was obtained by heating $H_2Si_{14}O_{29} \cdot xH_2O$ intercalated with N,N'-diethyl urea to 200°C [99]. The X-ray powder pattern was different from that of the dehydrated pure silicic acid. Also, transformation into cristobalite occurred at considerably lower temperature (800°C). The calcination of intercalated silicic acids often yields more highly ordered oxides. Decomposed organic material between the layers may facilitate gliding of the silicic acid layers into positions favorable for interlamellar condensation.

The presence of organic compounds in the starting material can also change the morphology of the silica particles and can even influence the shape and size of the cristobalite particles. Cristobalite formed by calcination of $H_2Si_{14}O_{29} \cdot xH_2O$ intercalated with N,N'-diethyl urea shows a morphology very different from the usual particle shapes. Scanning electron micrographs show disintegration of the silicate layers into rods a few microns in length that are intergrown and form net-shaped structures [99].

REFERENCES

1. Z Johan, GF Maglione. Bull Soc Franc Mineralog Cristallogr 1972; 95:371–382.
2. LAJ Garvie, B Devouard, TL Groy, F Camara, PR Buseck. Am Min 1999; 84: 1170–1175.
3. AP Khomyakov, MF Korobitsyn, TA Kurova, GE Cherepivskaya. Zap Vses Mineral Ova 1987; 116:244–248.
4. J Puziewicz. Am Min 1988; 73:440.
5. AP Khomyakov, GE Cherepivskaya, TA Kurova, VP Vlasyuk. Zap Vses Mineral Ova 1980; 104:317–324.
6. M Fleischer. Am. Min 1982; 67:1076.
7. RL Hay. Contr Miner Petrol 1968; 17:255–260.
8. HP Eugster. Contr Miner Petrol 1969; 22:1–31.
9. HP Eugster. Science 1967; 157:1177–1180.
10. G Maglione. Bull Serv Carte Geol 1970; 23:177.
11. K Beneke, G Lagaly. Am Min 1983; 68:818–826.

---

* Formulated as $K_2Si_8O_{17} \cdot xH_2O$ in Beneke and Lagaly [234]

12. RK Iler. J Colloid Sci 1964; 19:648–657.
13. CT Kresge, ME Leonowicz, WJ Roth, JC Vartuli, JS Beck. Nature 1992; 359: 710–712.
14. JS Beck, JC Vartuli, WJ Roth, ME Leonowicz, CT Kresge, KD Schmitt, CTW Chu, DH Olson, EW Sheppard, SB McCullen, JB Higgins, JL Schlenker. J Am Chem Soc 1992; 114:10834–10843.
15. JC Vartuli, CT Kresge, ME Leonowicz, AS Chu, SB McCullen, ID Johnson, EW Sheppard. Chem Mater 1994; 6:2070–2077.
16. F Liebau. Structural chemistry of silicates. Structures, bonding and classification. Berlin: Springer Verlag, 1985.
17. G. Lagaly, K Beneke, A Weiss. Am Min 1975; 60:642–649.
18. K Beneke, G Lagaly. Am Min 1977; 62:763–771.
19. H Annehed, L Fälth. Internat Conf on Zeolites, Napoli 1980. Recent progress reports and discussion. R Sersale, C Collella, R Aiello, eds, 1980:5–10.
20. S Vortmann, J Ruis, S Siegmann, H Gies. J Phys Chem B 1997; 1101:1292–1297.
21. JD Grice. Can Min 1991; 29:363–367.
22. S Ghose, C Wang. Am Min 1976; 61:123–129.
23. H Annehed, L Fälth, FJ Lincoln. Z Kristallogr 1982; 159:203–210.
24. S Vortmann, J Rius, B Marler, H Gies. Eur J Mineral 1999; 11:125–134.
25. G Borbely, KH Beyer, HG Karge, W Schwieger, A Brandt, KH Bergk. Clays Clay Min 1991; 39:490–497.
26. W Schwieger, D Heidemann, KH Bergk. Rev Chem Miner 1985; 22:639–650.
27. H Gies, B Marler, S Vortmann, U Oberhagemann, P Bayat, K Krink, J Rius, I Wolf, C Fyfe. Microporous Mesoporous Mater 1998; 21:183–197.
28. M Borowski, B Marler, H Gies. Z Krystallogr 2002; 217:1–9.
29. CL Baker, LR Jue, JH Wills. J Am Chem Soc 1950; 72:5369–5382.
30. L McCulloch. J Am Chem Soc 1952; 74:2453–2456.
31. A Heydemann. Beitr Min Petrogr 1964; 10:242–259.
32. VG Ilin, LF Kiricenko, VJ Ivanov, V.J. Kondracuk, ZZ Vysocku. Dokl Akad Nauk SSSR 1967; 174:880.
33. VG Ilin, YE Neimark, NV Turitina. Adv Chem Ser (Am Chem Soc, Washington, DC) 1973; 121:235–240.
34. GW Brindley. Am Min 1969; 54:1583–1591.
35. TP Rooney, F Jones, JT Ncal. Am Min 1969; 54:1034–1043.
36. BM Micjuk, LL Gorogoékaja, AL Rastrenenka. Geochimica 1976:803–814.
37. NV Turitina, VG Ilin, MS Kurilenko. Adsorption und Adsorbenzien 1977; 5:42.
38. W Schwieger. PhD dissertation, Universität Halle—Wittenberg, Springer Verlag, 1979.
39. S Kitahara, T Matuie, H Muraishi. Proceed. 1st Int Symp Hydrotherm React 1982. S Shiggaguki, G Bakea, F Kai, eds. Tokyo. Japan, 1983.
40. K Beneke, HH Kruse, G Lagaly. Z anorg allg Chem 1984; 518:65–76.
41. BM Micjuk. Geochimija 1974:1641–1646.
42. G Lagaly, K Beneke, A Weiss. In: JM Serratosa, ed. Proceedings of the International Clay Conference 1972. Madrid: Division de Ciencias C. S. I. C., 1973:663–673.
43. G Lagaly, K Beneke, A Weiss. Am Min 1975; 60:650–658.

44. JM Rojo, E Ruiz-Hitzky, J Sanz, JM Serratosa. Rev Chim Mineral 1983; 20: 807–816.
45. EKH Wittich, J Voitländer, G Lagaly. Z Naturforsch 1975; 30a: 1330–1331.
46. TJ Pinnavaia, ID Johnson, M Lipsicas. Solid State Chem 1986; 63:118–121.
47. JM Garcés, SC Rocke, CE Crowder, DL Hasha. Clays Clay Min 1988; 36:409–418.
48. JM Rojo, E Ruiz-Hitzky, JM Serratosa. Z anorg Allg Chem 1986; 540/541:227–233.
49. GG Almond, RK Harris, KR Franklin. J Mater Chem 1997; 7:681–687.
50. Y Komori, M Miyoshi, S Hayashi, Y Sugahara, K Kuroda. Clays Clay Min 2000; 48:632–637.
51. P Thiesen, K Beneke, G Lagaly. J Mater Chem 2000; 10:1177–1184.
52. Y Huang, Z Jiang, W Schwieger. Microporous Mesoporous Mater 1998; 26: 215–219.
53. Y Huang, Z Jiang, W Schwieger. Can J Chem 1999; 77:495–501.
54. Y Huang, Z Jiang, W Schwieger. Chem Mater 1999; 11:1210–1217.
55. A Brandt, W Schwieger, KH Bergk. Rev Chim Min 1987; 24:564–571.
56. A Brandt, W Schwieger, KH Bergk. Cryst Res Techn 1988; 23:1201–1203.
57. C Eypert-Blaison, E Sauzeat, M Pelletier, LJ Michot, F Villieras, B Humbert. Chem Mater 2001; 13:1480–1486.
58. C Eypert-Blaison, B Humbert, LJ Michot, M Pelletier, E Sauzeat, F Villieras. Chem Mater 2001; 13:4439–4446.
59. C Eypert-Blaison, LJ Michot, B Humbert, M Pelletier, F Villieras. J Phys Chem B 2002; 106:730–742.
60. E Lipmaa, M Mägi, A Samson, GH Engelhard, AR Grimm. J Am Chem Soc 1980; 102:1880.
61. A Brandt, W Schwieger, KH Bergk, P Grabner, M Porsch. Cryst Res Techn 1989; 24:47–54.
62. U Brenn, H Ernst, D Freude, R Hermann, R Jähnig, HG Karge, J Kärger, T König, B Mäadler, UT Pingel, D Prochnow, W Schwieger. Microporous Mesoporous Mater 2000; 40:43–52.
63. H Gies, M Borowski, B Asmussen. Proceedings of the ILL Millenium Symposium, 2001.
64. I Wolf, H Gies, CA Fyfe. J Phys Chem B 1999; 103:5933–5938.
65. F Wolf, W Schwieger. Z anorg allg Chem 1979; 457:224–228.
66. A Kalt, R Wey. Bull Groupe franç Argiles 1968; XX:205–214.
67. K Beneke, G Lagaly. GIT Fachz Lab 1984; 28:516–527.
68. W Schwieger, KH Bergk, S Franze. Z Chem 1987; 27:268–271.
69. W Schwieger, W Heyer, KH Bergk. Z Anorg Allg Chem 1988; 559:191–200.
70. KH Bergk, W Schwieger, M Porsch. Chem Techn 1987; 39:459–466.
71. KH Bergk, W Schwieger, M Porsch. Chem Techn 1987; 39:508–514.
72. KH Bergk, D Kaufmann, M Porsch, W. Schwieger. Seifen Öle Fette Wachse 1987; 113:555–561.
73. KH Bergk, P Grabner, W Schwieger. Z Anorg Allg Chemie 1991; 600:139–144.
74. NV Turitina, VG Ilin. Geochimija 1974:611.
75. HP Rieck. Nachr Chem Technol Lab 1996; 44:699–704.
76. W Schwieger, W Heyer, F Wolf, KH Bergk. Z Anorg Allg Chem 1987; 548: 204–216.

77. G Scholzen, K Beneke, G Lagaly. Z Anorg Allg Chem 1991; 597:183–196.
78. J Himmrich, W Gohla. SOFW J 1994; 120:784, 787–792.
79. F Kooli, Y Kiyozumi, F Mizukami, Y Akiyama. J Mater Chem 2001; 11: 1946–1950.
80. F Kooli, Y Kiyozumi, F Mizukami. Chem Phys Chem 2001:549–551.
81. F Kooli, Y Kiyozumi, F Mizukami. New J Chem 2001; 25:1613–1620.
82. F Kooli, Y Kiyozumi, V Rives, F Mizukami. Langmuir 2002; 18:4103–4110.
83. F Kooli, Y Kiyozumi, F Mizukami. Mater Chem Phys 2002; 7:134–140.
84. F Kooli, Y Kiyozumi, F Mizukami. Mater Chem Phys 2003; 77:134–140.
85. K Kosuge, A Tsunashima, R Otsuka. Natl Res Inst Pollut Resour 1991; 10: 1398–1401.
86. LA Crone, KR Franklin, P Graham. J Mater Chem 1995; 5:7–11.
87. BM Micjuk, LL Gorogoékaja, AL Rastrenenka. Dokl. Akad. Nauk SSSR 1973; 209:926–928.
88. P Bettermann, F Liebau. Contrib Mineralog Petrogr 1975; 53:25–36.
89. AS Campbell, WS Fyfe. Am Min 1960; 45:464–468.
90. OW Flörke. Fortschr Mineralog 1967; 44:181.
91. OP Bricker. Am Min 1969; 54:1026–1033.
92. KH Bergk, G Nietzold, W Schwieger. Z Chem 1988; 28:78.
93. LA Beljakova, VG Ilin, TF Peresunko. Dokl Akad Nauk SSSR 1974; 219:610–613.
94. LA Beljakova, VG Ilin. Teoret i Iksper Chim 1975:337–445.
95. K Beneke, G Lagaly. Am Min 1989; 74:224–229.
96. W Schwieger, KH Bergk, D Heidemann, G Lagaly, K Beneke. Z Kristall 1991; 197:1–12.
97. T Sakamoto, S Kobayashi, H Koshimizu. Daigaku 1996; 22:47–55.
98. RA Sheppard, AJ Gude, RL Hay. Am Min 1970; 55:358–366.
99. G Lagaly, K Beneke, H Kammermeier. Z Naturforsch 1979; 34b:666–674.
100. RA Fletcher, DM Bibby. Clays Clay Min 1987; 35:318–320.
101. OY Kwon, Jeong Soon-Yong, Suh Jeong-Kwon, Lee Jung-Min. Bull Korean Chem Soc 1995; 16:37–741.
102. P Chu, GW Kirker, S Krishnamurthy, JC Vartuli. US Patent 5236681, 1993.
103. H Muraishi. Bull Chem Soc Japan 1992; 65:761–770.
104. H Muraishi. Bull Chem Soc Japan 1995; 68:3027–3033.
105. H Muraishi. Fac Int Stud Cult 1996; 36:22–34.
106. V Kahlenberg, G Dörsam, M Wendschuh-Josties, R Fischer. Solid State Chem 1999; 146:380–386.
107. A DeLucas, L Rodriguez, J Lobato, P Sanchez. Ind Eng Chem Res 2001; 40: 2580–2584.
108. KH Bergk, W Schwieger, A Schäfer. Z Chem 1989; 29:151–152.
109. W Schwieger, K-H Bergk, D Freude, M Hunger, H Pfeifer. ACS Symposium Series 1989; 398:274–289.
110. W Schwieger, K Pohl, U Brenn, CA Fyfe, H Grondey, G Fu, GT Kokotailo. Studies Surface Sci and Catal 1995; 94:47–54.
111. W Schwieger, D Freude, P Werner, D Heidemann. In ML Occelli, H Robson, eds. Synthesis of Microporous Materials. Expanded Clays and Other Microporous Solids. Vol. II. New York: Van Nostrand/Reinhold, 1992:229–244.

112. W Schwieger, KH Bergk, KP Wendlandt, W Reschelilowski. Z Chem 1985; 25: 228.
113. W Schwieger, E Brunner. Colloid Polym Sci 1992; 270:935–938.
114. G Pál-Borbély, HK Beyer, Y Kiyozumi, F. Mizukami. Microporous Mesoporous Mater 1998; 22:57–68.
115. S Chen, J-N Tzeng, B-Y Hsu. Chem Mater 1997; 9:1788–1796.
116. OY Kwon, SW Choi. Bull Korean Chem Soc 1999; 20:69–75.
117. W Schwieger, P Werner, KH Bergk. Colloid Polym Sci 1991; 269:1071–1073.
118. HO Pastore, M Munsignatti, AJS Mascarenhas. Clays Clay Min 2000; 48:224–229.
119. F Janowski, KH Bergk. Z Chem 1982; 22:277–288.
120. RM Barrer. Hydrothermal Chemistry of Zeolites. London: Academic Press, 1982.
121. DW Breck. Zeolite Molecular Sieves. New York: Wiley, 1974.
122. HA Benesi. J Phys Chem 1957; 61:970–973.
123. HA Benesi, BHC Winquist. In H Pines, PB Weisz, eds. Advanced Catalysis. Vol. 27. New York: Academic Press, 1978:97–182.
124. HJ Werner, K Beneke, G Lagaly. Z anorg allg Chem 1980; 470:118–130.
125. H Cornejo, J Steinle, HP Boehm. Z Naturforsch 1979; 33b:1238–1241.
126. SY Jeong, JM Lee. Bull Korean Chem Soc 1998; 19:218–222.
127. DC Apperley, MJ Hudson, MTJ Keene, JA Knowles. J Mater Chem 1995; 5: 577–582.
128. KH Bergk, C Schütz, W Schwieger. Silikattechnik 1990; 41:241–243.
129. N Mizukami, M Tsujimura, K Kuroda, M Ogawa. Clays Clay Min 2002; 50: 799–806.
130. MTJ Keene, JA Knowles, MJ Hudson. J Mater Chem 1996; 6:1567–1573.
131. W Schwieger, O Gravenhorst, T Selvam, F Roessner, R Schlögl, D Su, GTP Mabande. Colloid Polymer Sci 2003; 281:584–588.
132. M Ogawa, Y Takizawa. J Phys Chem B 1999; 103:5005–5009.
133. R Schwarz, E Menner. Ber Dtsch Chem Ges 1924; 57:1477–1481.
134. R Schwarz, HW Hennicke. Z anorg allg Chem 1956; 283:346–350.
135. A Pabst. Am Min 43:970–980.
136. F Wodtcke, F Liebau. Z anorg allg Chem 1965; 335:178–188.
137. F Liebau. Z Kristallogr 120:427–449.
138. MT Le Bihan, A Kalt, R Wey. Bull Soc fr Minéral Cristallogr 1971; 94:15–23.
139. G Lagaly. Adv Colloid Interf Sci 1979; 11:105–148.
140. K Beneke, P Thiesen, G Lagaly. Inorg Chem 1995; 34:900–907.
141. CS Kim, DM Yates, PJ Heaney. Clays Clay Min 1997; 45:881–885.
142. U Brenn, W Schwieger, K Wuttig. Colloid Polym Sci 1999; 277:394–399.
143. K Kikuta, K Ohta, K Takagi. Chem Mater 2002; 14:3123–3127.
144. G Lagaly, S Fitz, A Weiss. Clays Clay Min 1975; 23:45–54.
145. G Lagaly. Angew Chem Int Ed Engl 1976; 15:575–586.
146. G Lagaly, A Weiss, E Stuke. Biochim Biophys Acta 1977; 470:331–341.
147. JM Rojo, E Ruiz-Hitzky. J Sanz Inorg Chem 1988; 27:2785–2790.
148. K Beneke, G Lagaly. Clay Min 1982; 17:175–183.
149. T Kimura, D Itoh, N Okazaki, M Kaneda, Y Sakamoto, O Terasaki, Y Sugahara, K Kuroda. Langmuir 2000a; 16:7624–7628.

150. M Ogawa, M Yamamoto, K Kuroda. Clay Min 2001; 36:263–266.
151. M Ogawa, T Ishii, N Miyamoto, K Kuroda. Adv Mater 2001; 13:1107–1109.
152. G Lagaly, K Beneke, P Dietz, A Weiss, Angew Chem 1974; 86:893–894.
153. JM Rojo, E Ruiz-Hitzky. J Chem Phys 1984; 81:625–628.
154. D Siöberg, WAP Luck. Spectroscopy Lett 1977; 10:613–618.
155. AJ Gude, RA Sheppard. Am Min 1972; 57:1053–1065.
156. K Beneke, G Lagaly. Clay Min 1977; 12:363–365.
157. G Lagaly, HM Riekert, HH Kruse. In R Setton, ed. Chemical Reactions in Organic and Inorganic Constrained Systems. Dordrecht: Reidel, 1986:361–379.
158. H Hadjar, H Balard, E Papirer. Colloids Surfaces 1995; 99:45–51.
159. J Döring, G Lagaly, K Beneke, I Dékány. Colloids Surfaces A 1993; 71:219–231.
160. I Dékány, F Szánto, LG Nagy, GH Foti. J Colloid Interf Sci 1975; 50:265–271.
161. I Dékány, F Szanto, LG Nagy, G Schay. J Colloid Interf Sci 1983; 93:151–161.
162. I Dékány, F Szánto, LG Nagy. J Colloid Interf Sci 1985; 103:321–331.
163. I Dékány, F Szánto, LG Nagy. J Colloid Interf Sci 1986; 109:376–384.
164. I Dékány. Pure Appl Chem 1992; 64:1499–1509.
165. T Marosi, I Dékány, G Lagaly. Colloid Polym Sci 1992; 270:1027–1034.
166. I Regdon, Z Kiraly, I Dékány, G Lagaly. Colloid Polym Sci 1994; 272:1129–1135.
167. T Marosi, I Dékány, G Lagaly. Colloid Polym Sci 1994; 272:1136–1142.
168. I Dékány. In A Dabrowski, VA Tertykh, eds. Adsorption on New and Modified Inorganic Sorbents. Amsterdam: Elsevier, 1996:879–897.
169. I Regdon, Z Kiraly, I Dékány, G Lagaly. Progr Colloid Polym Sci 1998; 109: 214–220.
170. I Regdon, I Dékány, G Lagaly. Colloid Polm Sci 1998; 276:511–517.
171. FA Farhadpour, A Bono. J Colloid Interf Sci 1988; 124:209–227.
172. FA Farhadpour, A Bono, U Tuzun. European Brewery Convention. Britain, 1983: 203–217.
173. WD Einicke, U Messow, R Schöllner. J Colloid Interf Sci 1987; 122:280–282.
174. WD Einicke, W Reschetilowski, M Heuchel, M v Szombathely, P Bräuer, R Schöllner, W Schwieger, KH Bergk. J Chem Soc Farad Trans 1991; 87:1279–1282.
175. J Döring, G Lagaly. Clay Min 1993; 28:39–48.
176. OY Kwon, SY Jeong, JK Suh, BH Ryu, JM Lee. J Colloid Interf Sci 1996; 177: 677–680.
177. J Döring, K Beneke, G Lagaly. Colloid Polym Sci 1992; 270:609–616.
178. CU Schmidt, G Lagaly. Clay Min 1999; 34:447–458.
179. HD Dörfler, KH Bergk, K Müller, E Müller. Tenside Detergents 1984; 21:226–234.
180. E Killmann, R Eckart. Makromol Chem 1971; 144:45–61.
181. J Rubio, JA Kitchener. J Colloid Interf Sci 1976; 57:132–142.
182. D Döring. Adsorption an kristallinen Kieselsäuren. PhD dissertation, Kiel University, Kiel, Germany, 1991.
183. Z Wang, T Lan, TJ Pinnavaia. Chem Mater 1996; 8:2200–2204.
184. Z Wang, TJ Pinnavaia. Chem Mater 1998; 10:1820–1826.
185. E Ruiz-Hitzky, JM Rojo, G Lagaly. Colloid Polym Sci 1985; 263:1025–1030.
186. P Thiesen, K Beneke, G Lagaly. J Mater Chem 2002; 12:1–7.
187. Y Mitamura, Y Komori, S Hayashi, Y Sugahara, K Kuroda. Chem Mater 2001; 13:3747–3753.

188. T Yanagisawa, K Kuroda, C Kato. React Solids 1988; 5:167–175.
189. T Yanagisawa, K Kuroda, C Kato. Bull Chem Soc Jpn 1988; 61:3743–3745.
190. T Yanagisawa, M Harayama, K Kuroda, C Kato. Solid State Ionics 1990; 42:15–19.
191. M Ogawa, S Okutomo, K Kuroda. J Am Chem Soc 1998a; 120:7361,7362.
192. M Ogawa, M Miyoshi, K Kuroda. Chem Mater 1998; 10:3787–3789.
193. A Shimojima, D Mochizuki, K Kuroda. Chem Mater 2001; 13:3603–3609.
194. JS Dailey, TJ Pinnavaia. Chem Mater 1992; 4:855–863.
195. HH Kruse, K Beneke, G Lagaly. Colloid Polym Sci 1989; 267:844–852.
196. HH Kruse. PhD dissertation, Kiel University, Kiel 1987.
197. C Eypert-Blaison, F. Villiéras, L. J. Michot, M. Pelletier, B. Humbert, J. Ghanbaja, J. Yvon. Clay Min 2002; 37:531–542.
198. F Villiéras, JM Cases, M Francois, LJ Michot, F Thomas. Langmuir 1992; 8:1789–1795.
199. SY Jeong, OY Kwon, JK Suh, H Jin, JM Lee. J Colloid Interf Sci 1995; 175:253–255.
200. SY Jeong, JK Suh, H Jin, JM Lee, OY Kwon. J Colloid Interf Sci 1996; 180:269–275.
201. OY Kwon, HS Shin, SW Choi. Chem Mater 2000; 12:1273–1278.
202. K Kosuge, A Tsunashima. J Chem Soc, Chem Commun 1995:2427–2428.
203. K Kosuge, PS Singh. Chem Mater 2000; 12:421–427.
204. A Fudala, Z Konya, Y Kiyozumi, SI Niwa, M Toba, F Mizukami, PB Lentz, J Nagy, I Kiricsi. Microporous Mesoporous Mater 2000; 35–36:631–641.
205. S Inagaki, Y Fukushima, A Okada, T Kurauchi, K Kuroda, C Kato, R. von Ballmoos et al, eds.. Proceedings 9th International Zeolite Conference, Montreal, 1992: Butterworth-Heinemann, 1993:305–311.
206. S Inagaki, Y Fukushima, K Kuroda. J Chem Soc Chem Commun 1993b; 8:680–682.
207. S O'Brien, RJ Francis, A Fogg, D O'Hare, N Okazaki, K Kuroda. Chem Mater 1999; 11:1822–1832.
208. S Inagaki, A Koiwai, N Suzuki, Y Fukushima, K Kuroda. Bull Chem Soc Jpn 1996; 69:1449–1457.
209. S Inagaki, Y Sakamoto, Y Fukushima, O Terasaki. Chem Mater 1996; 8:2089–2095.
210. P Branton, K Kaneko, F. Setoyama. Langmuir 1996; 12:599–600.
211. Y Kitayama, H Asano, T Kodama, J Abe. J Porous Mater 1998; 5:139–146.
212. A Matsumoto, T Sasaki, N Nishimiya, K Tsutsumi. Langmuir 2001; 17:47–51.
213. A Matsumoto, T Sasaki, N Nishimiya, K Tsutsumi. Colloids Surfaces 2002; 203:185–193.
214. M Katoh, K Sakamoto, M Kamiyamane, T Tomida. Phys Chem Chem Phys 2000; 2:4471–4475.
215. T Linssen, P Cool, M Baroudi, K Cassiers, EF Vansant, O Lebedev, JV Landuyt. J Phys Chem B 2002; 106:4470–4476.
216. Q Kan, V Fornés, F Rey, A Corma. J Mater Chem 2000; 10:993–1000.
217. G Fu, CA Fyfe, W Schwieger, GT Kokotailo. Angew Chem Int Ed Engl 1995; 34:1499–1502.
218. D O'Hare, JSO Evans, A Fogg, S O'Brien. Polyhedron 2000; 19:297–305.
219. JN Israelachvili, DJ Mitchell, BW Ninham. J Chem Soc Farad Trans 2 1976; 72:1525–1568.

220. H Hoffmann, G Ebert. Angew Chem 1988; 100:933–944.
221. T Kimura, T Kamata, M Fuziwara, Y Takano, M Kaneda, Y Sakamoto, O Terasaki, Y Sugahara, K Kuroda. Angew Chem Int Ed 2000; 39:3855–3859.
222. XS Zhao, F Audsley, GQ Lu. J Phys Chem B 1998; 102:4143–4146.
223. MML Ribeiro Carrot, AJ Estevao Candeias, PJM Carrot, KK Unger. Langmuir 1999; 15:8895–8901.
224. H Hata, S Saeki, F Kimura, Y Sugahara, K Kuroda. Chem Mater 1999; 11: 1110–1119.
225. T Kimura, S Saeki, Y Sugahara, K Kuroda. Langmuir 1999; 15:2794–2798.
226. Y Akiyama, F Mizukami, Y Kiyozumi, K Maeda, H Izutsu, K Sakaguchi. Angew Chem Intern Ed 1999; 38:1420–1422.
227. T Ikeda, Y Akiyama, F Izumi, Y Kiyozumi, F Mizukami, F Kodaira. Chem Mater 2001; 13:1286.
228. A Matijasic, AR Lewis, C Marichal, L Delmotte, JM Chézeau, J Patarin. Phys Chem Chem Phys 2000; 2:2807–2813.
229. G Pál-Borbély, HK Beyer, Y Kiyozumi, F Mizukami. Microporous Mesoporous Mater 1997; 11:45–51.
230. G Pál-Borbély, Á Szegedi, HK Beyer. Microporous Mesoporous Mater 2000; 35–36:573–584.
231. M Salou, Y Kiyozumi, F Mizukami, F Kooli. J Mater Chem 2000; 10:2587–2591.
232. S Shimizu, Y Kiyozumi, K Maeda, F Mizukami, G Pál-Borbély, RM Mihályi, HK Beyer. Adv Mater 1996; 8:759–762.
233. R Greenwood. Am Min 1967; 52:1662–1668.
234. K Beneke, G Lagaly. Z Naturforsch 1979; 34b:648–649.
235. JJ Tiercelin, RW Renaut, G Delbrias. Palaeocology of Africa 1981; 13:105–120.
236. G Perinet, JJ Tiercelin, CE Barton. Bull Mineralogie 1982; 105:633–639.
237. JL McAtee, R House, HP Eugster. Am Min 1968; 53:2061–2069.
238. AJ Sheppard, AJ Gude. J Res US Geol Surv 1971; 2:625–630.
239. G Maglione, M Serrant. CR Acad Sci Paris 1973; 277:1721.
240. AJ Gude, RA Sheppard. Am Min 1972; 57:1053–1056.
241. PP Keat. Science 1954; 120:328.
242. T Kameyama, S Naka. Am Ceram Soc 1974; 57:499–503.
243. B Siffert, R Wey. Silicates Ind 1967; 32:415.
244. CT Li. Z Krist 1973; 138:216.
245. E Robarick. PhD dissertation, Bochum University Bochum, Butterworth-Heinemann, 1974.

Index